V. K. Antonov

Chemistry
of Proteolysis

With 134 Figures

Springer-Verlag Berlin Heidelberg GmbH

Prof. Dr. Vladimir K. Antonov †

Shemyakin Institute of Bioorganic Chemistry
Academy of Sciences of the USSR
ul. Miklukho-Maklaya 16/10
117871, Moscow V-437,
Russia

ISBN 978-3-662-00981-9

Library of Congress Cataloging-in-Publication Data: Antonov, V. K. [Khimiia proteoliza. English] Chemistry of proteolysis / V. K. Antonov. p. cm. Translation of: Khimiia proteoliza. Includes bibliographical references and indexes.
ISBN 978-3-662-00981-9   ISBN 978-3-662-00979-6 (eBook)
DOI 10.1007/978-3-662-00979-6
1. Proteolytic enzymes. I.
Title. QP609.P78A5813 1992 547.7'58--dc20

© Springer-Verlag Berlin Heidelberg 1993
Originally published by Springer-Verlag Berlin Heidelberg New York in 1993
Softcover reprint of the hardcover 1st edition 1993

Production Editor: Renate Münzenmayer
Typesetting: Macmillan India Ltd., India
31/3145-543210 – Printed on acid-free paper

Preface to the English Edition

A first edition of this book appeared in Russian language in 1983. Since those days the interest in proteolytic enzymes has acquired a new impulse due to advances in genetic engineering techniques, which facilitated structural and functional studies of the enzymes. Much more data on a crucial role of proteases in biological processes in norm and pathology are available now.

Information that appeared in the past 8 years prompted a renewal of the book and furnishing a supplement especially made for the English edition. The book retains the presentation of the previous edition and yet enlarges the scope to cover the methods for site-directed mutagenesis of amide hydrolases, catalytically active antibodies, etc. I have tried to preserve the book as a means of reference for the reader. A list of enzymes has been expanded and the bibliography up to 1990 enlarged by half as much.

I am grateful to the translators T. E. Chernichko, N. I. Loboda and my co-worker Dr. S. L. Alexandrov for their assistance in the preparation of this edition and I feel happy that my book is now available for enzymologists abroad.

Moscow, Summer 1992   Vladimir K. Antonov

Preface to the Russian Edition

Information on proteolytic enzymes, i. e., enzymes catalyzing hydrolysis of amide bonds is of utmost interest to those who try to understand how biological catalysts function, what mechanisms underlie the enormous acceleration of the enzymatic processes, and how a cleavable compound is chosen among multiple similar substances. This is quite understandable, first because, proteolytic enzymes have been most intensively studied and, secondly, because numerous data are available on the mechanism of hydrolysis of amides and other carboxylic acid derivatives in model systems. Proteolytic enzymes contain no complex cofactors or coenzymes, therefore we observe "pure" catalysis by a protein molecule.

Proteolysis still challenges the investigators not only as a good illustration of enzymatic catalysis. Its biological significance has been underestimated for a long time, but it only nowadays became clear that proteolysis plays a key role in the regulation of the majority of physiological processes of the organism. This aspect seems to prevail in current publications.

The book pays minimum attention to biological functions of proteolytic enzymes, but provides a systematic description of chemical events occurring at enzyme-catalyzed amide hydrolysis. This implies the knowledge of the structure of reactants, major regularities of homogeneous catalysis as well as the most important data on specificity, efficiency and regulation of the enzymatic catalysis. Bearing in mind these problems somehow or other the author has presented a wealth of information, so that enzymologists could find the references they need. The literature available is overwhelming, therefore its choice is to a large extent of personal taste. I do hope the reader will find the most essential publications on the aspects outlined here, among which are the most recent reviews. I could not exclude information illustrating my viewpoint on the enzymatic hydrolysis of carboxylic acid derivatives. Apparently, in line with the ideas of other scientists it has been evolved in the course of the experiments performed in the laboratory of enzyme chemistry of the Shemyakin Institute of Bioorganic Chemistry, USSR Academy of Sciences. The studies have been carried out together with L. M. Ginodman, L. V. Kozlov, L. D. Rumsh, S. L. Alexandrov, L. Ya. Bessmertnaya, E. D. Dyachenko, B. L. Dyakov, A. G. Gurova, Yu. V. Kapitannikov, Yu. I. Krylova, A. G. Mikhailova, T. V. Rotanova, T. I. Vorotyntzeva, A. M. Zhelkovsky, A. A. Zinchenko, and others. This edition is a result of our joint effort. I am grateful to M. Ya. Karpeisky, G. I. Yakovlev, E. M. Popov, G. I. Likhtenstein, D. S. Chernavsky, Yu. I. Khurgin for fruitful discussions indispensable for my concept about enzymatic catalysis. I am particularly indebted to L. M. Ginodman, E. M. Popov, K. Martinek and L. I. Krishtalik for their critical comments on individual chapters. I wish to thank T. N. Barshevskaya, N. A. Gavrilova, A. A. Mishin and L. L. Zavada for their assistance in the preparation of the manuscript.

I welcome comments and criticism from readers.

Vladimir K. Antonov

Contents

Contents   XIII

Introduction

Proteolysis – enzymatic hydrolysis of amide bonds in proteins and peptides – is one of the most significant chemical processes universal in nature. With more or less intensity it occurs in every living thing on earth. Evolution has faced a sophisticated problem in creating the mechanism of rapid and selective cleavage of proteins that fits two conditions: first, a constant renewal of the cell protein material without large energy expenditure for constructing primary "building blocks" – amino acids, second, a high chemical stability of the amide bond. When life originated, the proteolytic enzymes – miraculous biological catalysts accelerating protein hydrolysis by a factor ranging from a trillion to dozens of trillions – appeared to solve the problem.

For a long time the functions of proteolytic enzymes were referred to exclusively for their role in destruction (catabolic) processes. In fact, they are of primary significance. The proteins have considerably differing half-lifetime values in the cell, for example it is 11 min for rat liver ornithin decarboxylase, while for RNA polymerase from the same source it is 1.3 h, and for myosin from rabbit skeletal muscles about 30 days; hemoglobin decay, if any, takes much time [1, 2]. How are these processes regulated in the organism, how are they synchronized? To find the answer to these questions is the major aim of biology.

However, proteolytic enzymes are involved not only in destruction processes but also perform regulatory functions, which attract more and more the attention of specialists [3–6]. Regulatory functions of proteases are characterized by unidirective and irreversible action. This is a special regulation type differing from other types of biological control of the organism. One can hardly enumerate all the regulatory processes mediated by proteolytic enzymes. Among them, the so-called reaction of limited proteolysis is comprehensively studied. Limited proteolysis is the most significant stage of protein biosynthesis. Apparently, post-translation cleavage of precursors of biologically active proteins ("processing") is an obligatory step in protein synthesis. Proteolysis reactions are of great importance for the formation of polypeptide hormones. Proteases often operate as triggers of some cascade processes. The blood coagulation system is a good example here. The enzyme function not only as a signal for triggering but also as a signal amplifier. They are also involved in the coupling of some physiological processes.

For a long time the question regarding the function of proteolytic enzymes has been challenging scientists. First, an idea about the chemical nature of biological catalysis was formulated by Berzelius as early as the 1830's [7]. However, only at the beginning of the twentieth century emerged a qualitative approach to these problems. The studies performed by E. Fischer helped forward enzyme and protein chemistry. About one hundred years later after the Berzelius investigation, enzymes were obtained in crystal state: these were proteolytic enzymes.

During the time preceding the first X-ray investigations of enzymes, the ideas and methods of physics and chemistry deeply penetrated into this area. In my opinion, the review by M. Bender (1960) *Mechanisms of catalysis of nucleophilic reactions of carboxylic acids derivatives* was a milestone here [8]. For the first time it was clearly indicated that the mechanism of enzyme catalysis can be comprehended on the basis of general chemical regularities of homogeneous catalysis if the peculiarity of the protein's macromolecular structure is taken into account. This conclusion was completely confirmed by the development of chemical enzymology and physical organic chemistry.

The versatility of proteolytic enzymes is really striking. Interestingly, for catalysis of the same reaction – amide bond hydrolysis – nature created diverse biocatalysts. Thorough analysis shows, however, that all the studied proteases operate via the chemical mechanisms of two rather similar types: nucleophilic (covalent) catalysis performed by the groups of the enzyme molecule itself, and general base catalysis of cleavage of the amide group under the action of an enzyme-bound water molecule. The difference here is nonessential: in the first case a nucleophile is a protein group, while in the second a water molecule. Moreover, the second type of catalysis occurs also in case of nucleophilic catalysis at later stages (deacylation). Thus the chemical mechanism of proteolysis seems to be fairly similar for all or the majority of proteolytic enzymes. The same mechanism is apparently used by other enzymes splitting the derivatives of carboxylic acids (esterases, amidases, etc.). Are there any mechanisms of amide hydrolysis differing in principle from those known nowadays? It seems impossible, though no definite answer has been found yet.

Why are proteases so versatile? First, it might follow from the difference in the enzyme specificity which, on the one hand, is determined by the structure of a substrate bound to the protein molecule and, on the other hand, by the ability of the protein molecule to utilize the energy of the substrate binding to accelerate its chemical transformations. W.P. Jencks [9] has formulated this property of enzymes as the "Circe" effect – ability of "attracting" and transforming a substrate. Second, the variety of proteases arises from different conditions of their functioning (pH, temperature, composition of a milieu) and their different location in cell organelles (cytoplasm, membranes or biological fluids).

Proteolytic enzymes are, in many respects, the touchstone of chemical enzymology. First of all, the enzymes of this class serve to verify numerous methodical approaches and theoretical concepts. What is more, the regularities discovered for these enzymes might be of general significance, i.e., true for many other types of biocatalysts. This makes reasonable a careful consideration of recent data on the structure, properties and mode of action of these enzymes, the more so as a systematic review on the chemistry of proteases has not appeared yet. Here, when necessary, other enzymes splitting amide bonds are described along with proteases. In this book the enzymes catalyzing hydrolysis of amides are termed "amide hydrolases".

The material is arranged here to follow, as closely as possible, the successive steps of the catalytic event. At first, reactants – substrates and enzymes, their structure and properties – are considered. The enzymes are divided into four groups differing in the structure of the catalytic apparatus. The classification is based mainly on the relation of enzymes to inhibitors specific for a certain group. Naturally, for many amide hydrolases their assignment to one or the other group is not final, unless data on the active site structure are available. Besides, the list of enzymes is not complete, since virtually each new issue of a biochemical journal reports isolation of new amide hydrolases.

The mode of enzyme action cannot be described without data on the mechanism of hydrolysis of derivatives of carboxylic acids in the presence of nonbiological catalysts and without basic laws, which govern homogeneous catalysis. A separate chapter deals with these problems.

Experimental results concerning catalytic properties of amide hydrolases (kinetics, specificity, pH dependence, etc.) are brought together in the following two chapters. Along with the X-ray data they underlie the understanding of the catalytic event. It seems reasonable first to analyze the structure of the enzyme-substrate complexes and only then the chemical cleavage of the bond. Nowadays, the first step of the analysis is based on relatively few X-ray data on the enzyme-inhibitor complexes or very slowly cleaved substrates (quasi-substrates). Extrapolation of these data to the true enzyme-substrate complexes requires certain precaution, if not, any conclusion on the reaction mechanism is just a hypothesis.

The available data show that an elementary step of chemical changes of the amide group on the enzyme occurs practically spontaneously, i.e., with very low activation barriers, at least with the most adequate substrates. This inspires an idea that all the events leading to effective catalysis with amide hydrolases take place at the stages preceding the formation of the transition state. The reader will realize that the author tries here to escape analysis of the "structure" of transition states. I am convinced that it is less informative, especially in enzymatic catalysis, as compared to the structure investigation of the ground state of the reaction system based on definite experimental facts. This has led the author and his colleagues to formulating the concept of enzymatic catalysis presented in the final part of the book.

Thus, my aim is to try, as much as possible, to consider chemical aspects of enzymatic hydrolysis of the amide bond from the single point of view of elucidating general and unique features that allow the positioning of amide hydrolases in an individual and very important group of enzymes and, finally, to try again to answer the question why enzymes have unique efficiency and specificity.

Chapter 1   Substrates

1.1   General Characteristics

The compound, whose chemical transformation the enzyme accelerates, is called the substrate. This book considers the processes of enzymatic hydrolysis of amide bonds of various substrates.

A great majority of the substrates of proteases are proteins or peptides. Here an amide bond, known also as a peptide bond, forms the structure, which embraces certain amino acid residues:

$$\ldots \text{NHCH}(R_1)\text{CONHCH}(R_2)\text{CO} \ldots$$

The hydrolyzable amide bond is in the middle of the chain, its cleavage is catalyzed by endopeptidases; the amide bond located at the N- or C-end of the polypeptide chain is a target of exopeptidases. The properties of the amide group only slightly depend on its position in the polypeptide chain. There are many enzymes[1] catalyzing hydrolysis of the amide bonds other than peptide bonds [10]. Some types of such bonds are given below:

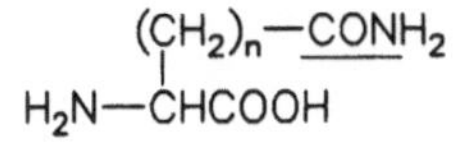

$$\begin{array}{l}(\text{CH}_2)_n\text{—CONH}_2 \\ \text{H}_2\text{N—CHCOOH}\end{array}$$

asparagine (n = 1)
glutamine (n = 2)

R—CONH$_2$

amides of carboxylic
acids (R = Alk or Ar)

R—CONHCH(R$_1$)COOH

formyl (R = H) and acyl
(R = Alk) amino acids

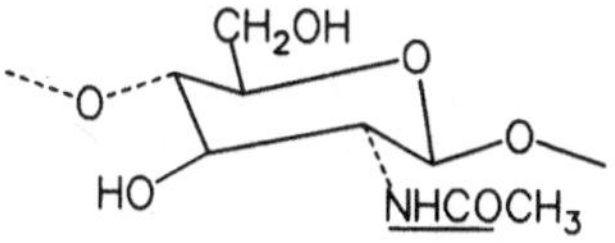

N–acetylglucosamine

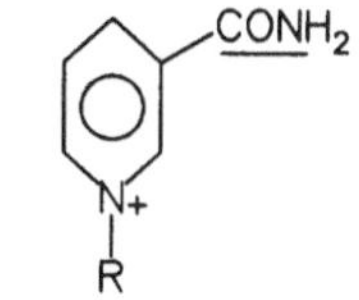

nicotinamide and
its derivatives

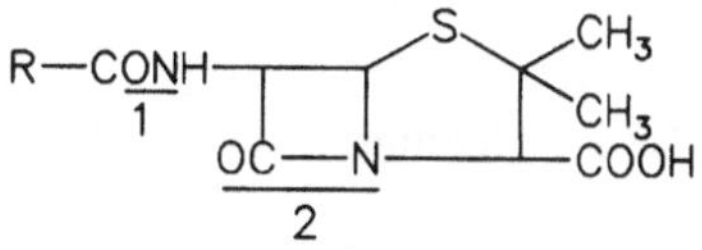

penicillins

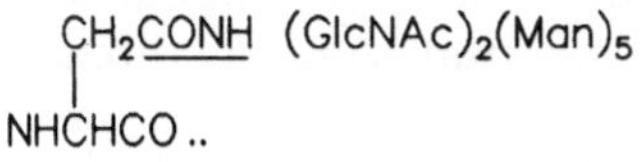

CH$_2$CONH (GlcNAc)$_2$(Man)$_5$
NHCHCO..

asparaginylsugars

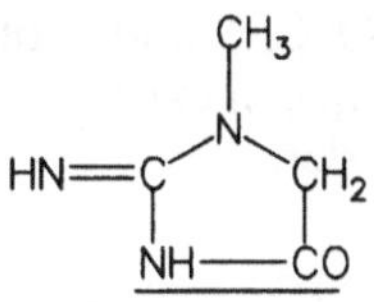

creatine

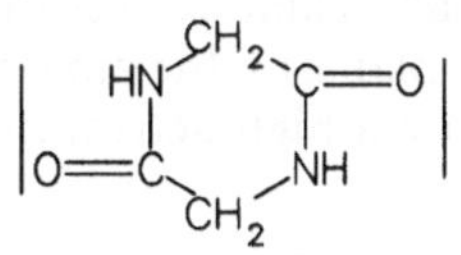

2,5–dioxopiperazines

---

[1] See Table 12 (EC 3.5.1 and 3.5.2)

Thus the cleavable amide bond can enter diverse structures (both linear and cyclic) of naturally occurring compounds. Besides, many proteolytic enzymes catalyze hydrolysis of amide and other bonds in carboxylic acid derivatives, which are not their natural substrates. For example, such proteases as α-chymotrypsin, papain, etc., effectively catalyze hydrolysis of amides, anilides and esters of N-acylamino acids.

As to peptide-protein substrates of proteolytic enzymes, their structural peculiarities should be analyzed step by step. It is reasonable, to consider first the cleavable amide (peptide) group, then an amino acid residue by itself and within a short oligopeptide, and at last the whole protein molecule.

1.2 Amide Group

At first sight, the amide group is a combination of aldehyde (ketone) and amine groups within the single structure. However, this definition is wrong, since the properties of the amide group are not a sum of properties of the groups comprising it (see reviews [11–13]). This is mainly due to the interaction of the unshared pair of nitrogen electrons with π-electrons of the double carbon-oxygen bond. That is why the amide group can be considered as a structure resulting from the superposition of two resonance forms: neutral (I) and zwitterion (II), their portions being 0.6 and 0.4, respectively [14].

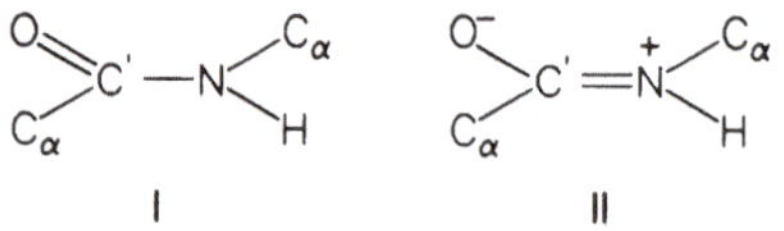

This explanation fits well many properties of the amide group, however, it is not complete.

1.2.1 Geometry

The concept of the interaction of the unshared electron pair of N-atom with π-electrons of double bond C=O in amides entails a number of consequences: (1) the C–N bond in amides should be shorter than in amines, and bond C=O – longer than in ketones; (2) atoms O, C, N and H of the amide group should lie within one plane; (3) the amide group can exist in cis- and trans-conformations separated by a high activation barrier:

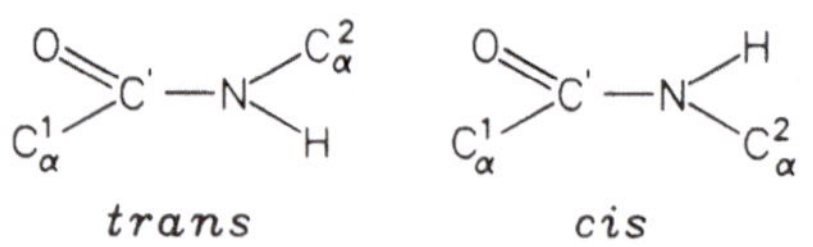

Let us see how these consequences are realized in real molecules. However, first one should say a few words about the nomenclature used to describe the amide group geometry.

The amide group structure can be characterized by five bond lengths: $r_{C_\alpha-C'}$, r_{C-O}, r_{C-N}, r_{N-C_α}, and r_{N-H} (of which, bond lengths C–O and C–N are the most important) and by six valence angles: $\angle C_\alpha CO$, $\angle C_\alpha C'N$, $\angle OC'N$, $\angle C'NC_\alpha$, $\angle C_\alpha NH$, and $\angle C'NH$. According to the recommendations of the Commission on Biochemical Nomenclature [15] the valence angles are designated by the letter τ with corresponding indices, i.e., $\tau_{C_\alpha C'O}$, etc.

In a planar amide group the first three angles are related as follows:

$$\tau_{C_\alpha C'O} + \tau_{C_\alpha C'N} + \tau_{OC'N} = 360° \ .$$

The same is true for the other three angles. Besides, the values of dihedral (torsion) angles also apply:

$$\omega_1 = \omega_{C_1 C'NC_2}, \qquad \omega_3 = \omega_{OC'NC_2} \ ,$$

$$\omega_2 = \omega_{OC'NH}, \qquad \omega_4 = \omega_{C_1 C'NH} \ .$$

For *cis*-amides $\omega_1 = \omega_2 = 0°$ and $\omega_3 = \omega_4 = 180°$, and for *trans*-amides $\omega_1 = \omega_2 = 180°$ and $\omega_3 = \omega_4 = 0°$. Torsion angles in the amide group are related by ratio $(\omega_1 + \omega_2) - (\omega_3 + \omega_4) = 0$. Usually only one angle ω_1, designated as ω, is employed. Some authors use torsion angles denoted θ_C and θ_N [16] or χ_C and χ_N [17] and outlined as follows:

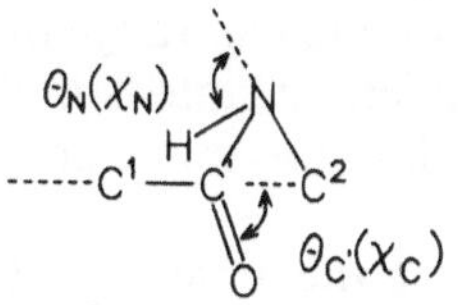

These angles characterize deviations of the amide group from the planar structure and connected with angles ω by the following dependences:

$$\chi_{C'}(\chi_C) = \omega_1 - \omega_3 + \pi = -\omega_2 + \omega_4 + \pi \ ,$$

$$\theta_N(\chi_N) = \omega_2 - \omega_3 + \pi = -\omega_1 + \omega_4 + \pi \ .$$

Finally, the angle between bisectors θ_C and θ_N is sometimes [17] depicted as τ, while

$$\omega_1 = \tau + 0,5(\theta_{C'} - \theta_N) \ ,$$

$$\omega_2 = \tau - 0,5(\theta_{C'} - \theta_N), \text{ etc} \ .$$

Interatomic distances in amides can be obtained either from the X-ray, including also neutrongraphic, analysis [12, 18, 19] or from microwave spectroscopy [20, 21]. The latter provides geometric parameters of the molecule in an isolated (gaseous) state.

Table 1 shows that, as expected from the resonance model of the amide group, the C–N bond in amides is much shorter than in amines. However, for

Table 1. Interatomic distances and angles in amides, methylamine, and acetone

Substance and its state	r_{C-N}, Å	$r_{C=O}$, Å	$\angle NC'O$, degree	$\angle NC'H(C)$, degree	$\angle OC'H(C)$, degree	Reference
$HOCNH_2$						
Gas	1.376	1.19	123.8	113.24	122.96	[20]
"	1.352	1.219	124.7	112.7	122.6	[21]
Crystal	1.300	1.255	121.5	–	–	[22]
CH_3CONH_2						
Gas	1.36	1.21	125	–	122	[23]
Crystal	1.38	1.28	122	109	129	[24]
CH_3-NH_2	1.47	–	–	–	–	[25]
$(CH_3)_2C=O$	–	1.21	–	–	–	[26]

individual amide molecules (in gaseous phase) the length of bond $C=O$ does not differ from that in acetone. All this suggests that the concept of the superposition of resonance structures (I) and (II) is insufficient to describe the amide group.

The condensed state of the amide molecule entails the lengthening of bond $C=O$. Numerous X-ray investigations of the derivatives of amino acids, peptides [18] and proteins [27] lead to mean values of bond lengths and valence angles of the amide group (Fig. 1).

It is noteworthy that these parameters vary depending on the conformation of the amino acid residue [28]. So the length of the $C'-N$ bond can range from 1.343 to 1.371 Å. Thus, the geometry of amides in a condensed state is in line with the above concept; however, it should be refined to describe amide groups in the absence of intermolecular interactions.

Is the second consequence of the concept representing the amide bond as a resonance hybrid of the neutral and ion forms realized, i.e., is the amide bond planar?

The study of microwave spectra of formamide has led to the conclusion [20] that the nitrogen atom of this compound is exposed outside the plane formed by atoms C', H_1, and H_2, torsion angles $HCNH_2$ and $OCNH_1$ being

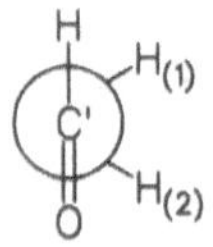

$12° \pm 5°$ and $7° \pm 5°$, respectively. These data, however, are not confirmed by other researchers [21, 29].

The problem of deviation of the amide group structure from the planar system has been studied theoretically and experimentally [16, 17, 19, 28, 30–42]. These investigations reveal a noticeable deviation from the planar structure in many cyclic [17, 30, 34, 43] and linear [32, 33, 35, 40] amides and peptides, atoms C' and N can be outside the plane of $C_\alpha^1 ON$- and $C'C_\alpha^2 H$-atoms, respectively (Table 2).

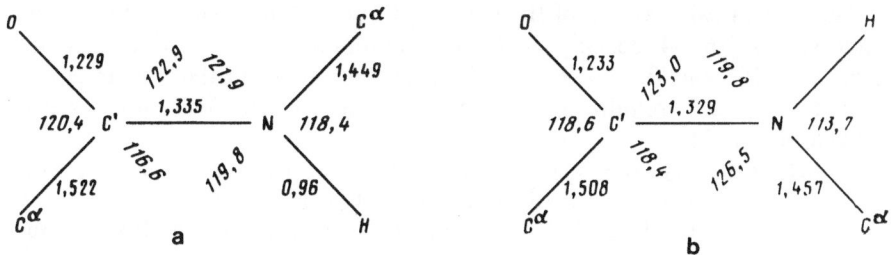

Fig. 1a, b. Bond lengths and valence angles in *trans* (**a**) and *cis*- (**b**) peptide groups [18]

Table 2. Diversions from a planar structure of an amide group

Substance	$\Delta\omega$, degree	θ_N, degree	θ_C, degree	Reference
Glu-Leu	−11.4	+24.1	+0.2	[33]
Gly-Gly	+3.6	+8.2	+3.3	[33]
Gly-Ala (HCl)	−10.7	+17.2	+4.5	[33]
Ala-Ala	+0.8	−21.1	−2.7	[33]
Ac-*D*-Ala-*D*-Ala	−12.0	+19.0	+3.0	[39]
AcPhe-Tyr	−17.7	+11.4	+6.2	[35]

High resolution X-ray analysis also shows the presence of a number of nonplanar amide bonds in some proteins [27, 44]. The bonds at the nitrogen atom (angle θ_N) are most deformed as compared to those at the carbonyl carbon atom (angle $\theta_{C'}$). The pyramidal position of the nitrogen atom in amides makes possible the existence of two states, i.e., rotation of configuration:

The barrier to such a rotation is about 1.1 kcal/mol in formamide [20]. Similarly, the pyramidal position of the carbonyl carbon can result in two conformations: *syn*-($\theta_N \cdot \theta_{C'} < 0$) and *anti*-($\theta_N \cdot \theta_{C'} > 0$) [40].

The direction and value of pyramidalization depend on the electron properties of the neighboring σ-bond [40]. There is a linear correlation (R = 0.893) between lengths of the C′–N bond and the extent of pyramidalization of the nitrogen atom (θ_N) [42]:

anti   *syn*

Theoretical calculations of the energy of out-of-plane deformation of the amide group [16, 34, 35, 38] show that minimum values of the energy are within $\omega = 10°$ and $\theta_N = -20°$ (Fig. 2a). Deviations from the planar amide bond to $\omega = 20°$ require energy expenditure, which does not exceed 2–3 kcal/mol [35].

Figure 2b displays that extraplanar deformation of C'-atom corresponding to this atom exposure out of the planes of C_α-, N- and O-atoms by 0.1 Å consumes energy of about 5 kcal/mol. However, further deformation is connected with a large energy loss.

Thus the amide group can much differ from the planar structure due to changes of the dihedral angle (ω) and the exposure of atoms C' and N out of the plane, i.e., changes of angles θ_N and $\theta_{C'}$. These changes do not expend much energy. Apparently, the energy diagram of the amide group is rather a wide "valley" in the vicinity of θ_N and $\theta_{C'}$ values equal to 0 and $\omega = 0°$ and $180°$, so that pronounced deviations from these values virtually do not change the energy of this system. However, large changes in the ω values, particularly the transition from the trans- ($\omega = 180°$) to the cis-form ($\omega = 0°$), require large energy since it is accompanied with a loss of the resonance stabilization energy in the transition state.

Theoretical [14, 50] and thermochemical [51] determination of the resonance energy in the amide bond gives about 21–23 kcal/mol. According to

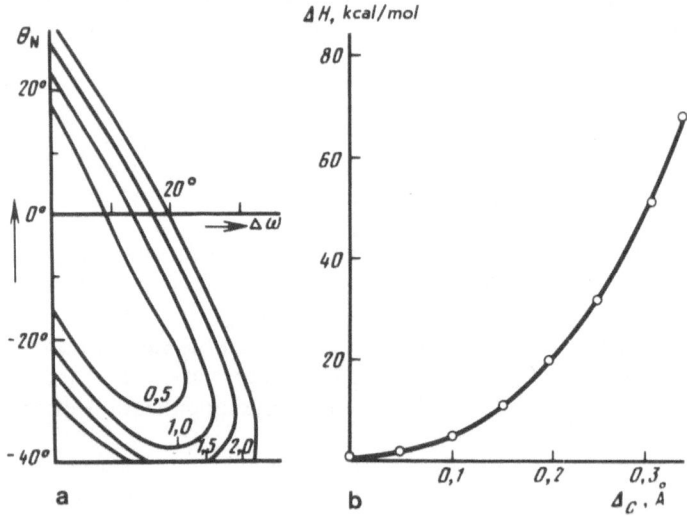

Fig. 2a, b. Out-of-plane deformation of amide group. **a** Isoenergetic contours of out-of-plane deformation of the amide group in acetamide depending on angles θ_N and $\Delta\omega$ ($\Delta\omega = 180° - \omega$) (calculated by INDO); **b** energy change at synchronous hybridization of C- and N-atoms (calculated by CNDO-2, program GEOMO)

Table 3. Activation barriers of *cis-trans*-transition in some amides

Substance	ΔG^+, kcal/mol	ΔH^+, kcal/mol	ΔS^+, e.u.	Reference
HOCNH$_2$	17.7	18.5	+2.7	[45]
HOCN(CH$_3$)$_2$	19.5	18.5	−5.1	[46]
CH$_3$CON(CH$_3$)$_2$	17.65	10.6	−23.5	[47]
CF$_3$CON(CH$_3$)$_2$	17.7	9.3	−28.0	[47]
(CH$_3$)$_3$CCON(CH$_3$)$_2$	10.5	13.5	+10.0	[48]
HOCN(CH$_3$)CH$_2$C$_6$H$_5$	21.6	19.5	−6.5	[49]

Table 3 this value is close to that of the free-energy change for the *cis-trans* transition [50–52].

Generally speaking, numerous data concerning the barriers of the *cis-trans* transitions in amides are rather contradictory [12]. Still, for a certain series of amides it is possible to connect the values of the free energy of rotation (ΔG^+) with such parameters of substituents as sterical constants (E_s or V), inductive (δ_I and σ^*) and resonance (σ_R) constants via the linear free-energy equations [48, 53]. The sterical parameters of substituents contribute much to alteration of the rotation barrier [53]. The barrier to rotation about bond C–N increases with an increase in polarity of the solvent [54]. Quantum-chemical methods [55] have been used to calculate the curve of the energy change upon the *cis-trans* transition in formamide (Fig. 3). The calculations take into account the nonplanar structure of formamide [20] in such a way that the change in angle ω characterizes the transition from the nonplanar *cis-* to the nonplanar *trans*-bond.

Though these calculations show practically the same stability of *cis-* and *trans*-forms, experimental and theoretical data on N-methylacetamide and other peptides reveal the essential difference in energies (3–5 kcal/mol) of both forms [36, 56]. It occurs due to repulsive nonvalence interactions in *cis*-amides [36, 57]. The *cis*-amide bond is predominant in lactams with no more than 11–12 atoms in the cycle. It is observed in linear amides when a hydrogen atom is a substituent at the carbonyl C-atom or a nitrogen atom is of the tertiary character [50, 58, 59].

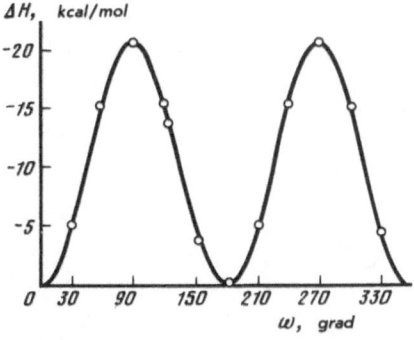

Fig. 3. Potential curve of inner rotation around bond C–N in formamide calculated ab initio [55]

1.2.2 Electron Structure

The atoms, which compose the amide group, have hybrid electron atomic orbitals forming σ- and π-bonds. Moreover, heteroatoms N and O have one or two unshared electron pairs, respectively. The amide group's peculiarity is the ability of the unshared electron pair of the nitrogen atom to interact with valence p-orbital of the carbon atom, thus conferring a partial double bond character to the C–N bond (Fig. 4). This is depicted by resonance formulas (I) and (II) [14].

The calculations by semi-empiric and nonempiric methods of quantum chemistry [12, 38, 55, 60–63] make it possible to evaluate the charge values on each atom and the bond order in the amide group. Though the obtained values depend on the calculation technique, the general pattern of the distribution of electron density is rather similar (Table 4). Force constants and bond orders are found from vibrational spectra of amides [61, 66]. They accord well with the resonance structure of the amide bond and with experimentally found dipole moments and spectral characteristics; however, there are some peculiarities that should be taken into account.

The orders of the C=O bond of the isolated amide molecule and of the amide molecule in acetone are close, the π-orders of bonds C–N and C–O differ much. However, the calculation of π-orders of the bonds from IR spectra of N-methylacetamide crystals and from the values of interatomic distances in the crystal entails practically the same values for bonds C–N and C–O [61]. Thus, in an isolated amide molecule both the C=O distance and the bond order do not correlate with the simple resonance model.

Apparently, in an isolated amide group the state of hybridization of O-atom is intermediate between sp^2 and sp^3, therefore the orbitals of n-electrons of O-atom and π-electrons of the C=O group are nonorthogonal and interact with each other. Such an interaction is distorted in crystals or solutions due to the formation of the bonds with neighboring solvent molecules

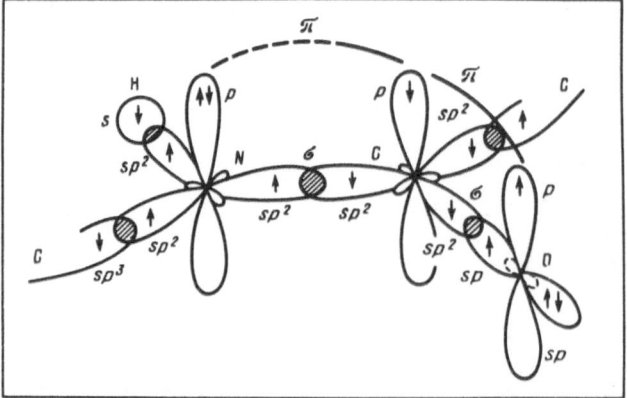

Fig. 4. Electron configurations of atoms of the peptide group

Table 4. Distribution of charges, bond orders, and force constants for amides

Substance	Charges			π-bond ordes		Force constants, mdin/Å		Reference
	O	C'	N	C=O	C–N	K_{CO}	K_{CN}	
Formamide	−0.403	+0.389	−0.608	–	–	–	–	[60]
	–	–	–	1.1	0.6	12.08	8.14	[64]
	−0.473	+0.247	−0.063	–	–	–	–	[65]
	−0.377	+0.258	−0.758	–	–	–	–	[12]
	−0.62	+0.53	−0.88	–	–	–	–	[63]
Formamide (crystal)	–	–	–	0.85	0.85	10.18	9.22	[61]
Acetamide	−0.63	+0.72	−0.91	1.0	0.6	11.45	8.14	[63, 64]
N-Methyl-acetamide	−0.64	+0.73	−0.85	0.98	0.65	11.32	8.27	[63, 64]
Acetone	−0.316	+0.132	–	1.0	–	11.45	–	[61]
Methylamine	–	–	−0.688	–	–	–	4.96	[61]

[61]. This accords well with values of the $n-\pi^*$ transitions in absorption spectra of amides shifted to 200–220 nm, whereas for ketones this transition lies at 270–300 nm. So the orbitals of unshared electron pairs of the oxygen atom in amides have, evidently, the partial bonding character.

The independence of the C=O distance on the amide group hybridization [19] can be also explained by the charge transfer from the nitrogen atom to the carbonyl carbon but not to the oxygen atom.

One way or another, the electron structure of the amide bond is highly labile, but at the same time provides its stability.

1.2.3   Properties of the Amide Group

The structure of the amide group described above specifies many of its physicochemical properties [64]. So partial negative charges on oxygen and nitrogen atoms (Table 4) induce an ability in amides for attaching a proton, thus behaving as a weak base:

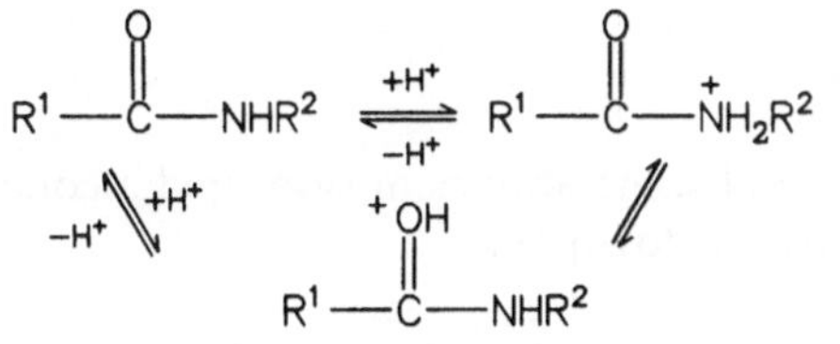

The amides are protonated mainly by the oxygen atom [65] that stems, first, from the charge value on O- and N-atoms and, secondly, from the lower loss of the resonance stabilization energy upon oxygen protonation rather than the proton attachment to the only free pair of N-atom electrons.

In protonated amides the barrier to rotation around bond C–N remains the same which is evident from splitting proton signals at N-atom in NMR spectra at low temperature [67]. Thus, the resonance is maintained in the protonated amide:

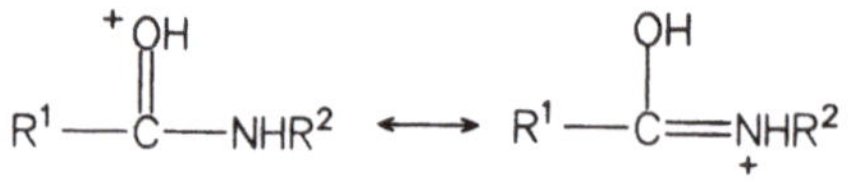

The barrier to rotation around bond C–N somewhat increases in the protonated amide. So the values determined by one and the same method for liquid dimethylformamide and the same amide in 100% sulphuric acid are ΔG^+ 17.84, ΔH^+ 8.9 kcal/mol; $\Delta S^+ - 30$ e.u., and ΔG^+ 18.55. ΔH^+ 12 kcal/mol and $\Delta S^+ - 22$ e.u. [68]. Apparently, protonation hinders the extraplanar deformation of the amide group to a certain extent [34].

Amides are very weak bases. Values pK_a of some amides are presented below [64]:

$$R^1CONHR^2 + H^+ \xrightleftharpoons{K_a} R^1C(OH^+)NHR^2.$$

Substance	$-pK_a$	Substance	$-pK_a$
Acetamide	0.35	N-Butyrylglycine	1.73
Propionamide	0.57	N-Acetylglycine	2.1
N-Methylacetamide	1.0	p-Nitrobenzamide	2.22
N,N-Dimethylbenzamide	1.2	Glycinamide	3.1
N-Methylbenzamide	1.5	Butyrolactam	0.4
Benzamide	1.54	Valerolactam	− 0.6

Substituents in the amide group affect the basicity due to inductive, resonance, and sterical effects. So in the raw of alkyl substituted acetamides K_a and the Taft constant σ^* with $\rho^* = - 1.2$ correlate well [69]. For aromatic amides substituted in a ring the correlation of basicity with the Hammet σ-constants is observed [70].

Amides can behave as weak acids as well:

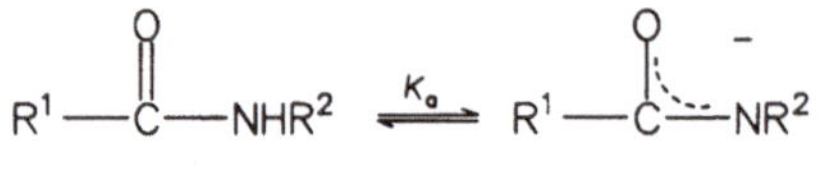

Values of pK_a for acidic dissociation of some amides in isopropyl alcohol containing sodium isopropylate are given below [71]:

Substance	pK_a	Substance	pK_a
Formamide	17.2	N-Methylbenzamide	> 19.0
Acetamide	17.6	p-Nitrobenzamide	19.85
Benzamide	> 19.0	Benzanilide	16.5

The ability to form a hydrogen bond is a typical property of an amide group [72]. In this case it can serve as a donor and an acceptor of such a bond. According to data on protonation of amides and IR spectra an oxygen atom is an acceptor [64]. The ability of amides to form an H-bond correlates with that of protonation, i.e., with their basicity. However, enthalpy of the hydrogen bond is by six or seven times lower than the protonation enthalpy [73].

The studies of hydrogen bonds in crystals [74] provide the following values for hydrogen bond lengths in amides and peptides (mean values for 46 compounds are given):

$$N\text{---}H \ldots X$$

$X=$	COO^-	$\overset{\displaystyle \mid}{\underset{\displaystyle OH}{-C=O}}$	$C=O$	H_2O	N (imidazole)	Cl^-
r_{N-X}	2.97 ± 0.1	2.99 ± 0.11	2.97 ± 0.11	2.90	3.0	3.26

The distances r_{N-X} have a slight dependence on the acceptor structure. However, there exists a strong dependence of this distance on the nature of the donor of the hydrogen bond as follows from experimental data [75–78] and calculations [62, 63, 79–81]. Enthalpy makes the major contribution to free energy of the hydrogen bond (Table 5).

According to theoretical and experimental data the amide-amide hydrogen bond is stronger by 3 kcal/mol than the amide-water and amide-aldehyde bonds [77–79].

The direction is a substantial characteristic of the hydrogen bond. It is directed to a lone pair of electrons of the atom-acceptor [85]. The optimum is almost linear bonds (angle N–H ... O > 150°). The estimation of dihedral angles N–H ... O=C entails the value of θ as about 60° [79].

Table 5. Thermodynamic parameters of hydrogen bonds between amides and phenol

Acceptor of the H-bond	$-\Delta G_0$, kcal/mol	$-\Delta H_0$, kcal/mol	ΔS_0, e.u.	Reference
N,N-Dimethylformamide	2.49	6.1	12.7	[82]
N-Methylacetamide	2.87	4.0	–	[75]
N,N-Dimethylacetamide	2.94	6.4	11.7	[82]
N,N-Diphenylacetamide	2.35	5.2	9.7	[76]
N,N-Dimethylchloroacetamide	2.18	4.7	8.5	[82]
N,N-Dimethyltrichloroacetamide	2.08	3.8	5.5	[82]
N-Benzoylpiperidine	2.21	4.2	6.9	[76]
δ-Valerolactam[a]	2.90	5.3	8.2	[83]
ε-Caprolactam[b]	2.92	5.2	7.8	[83]

[a] In CCl_4 at 25 °C.
[b] The same values are obtained for a self-association [84].

Since the formation of H-bond can stabilize the charge separation in the amide group, the barrier of the inner rotation around the C–N bond can increase. Really, free energy of the activation of *cis-trans*-transition is shown to increase by 1.2 kcal/mol in dimethylformamide upon the formation of the H-bond [86]. However, the studies of the correlation dependences of free energies of the activation of *cis-trans*-transition yield an opposite conclusion regarding a decrease of the activation energy, when a hydrogen bond is formed [53].

An oxygen atom of the carbonyl group of amides can accept two H-bonds ("three-center" bond). This mainly occurs if an atom-donor of a proton carries a formal positive charge [87].

The ability of amides to serve as donors and acceptors of hydrogen bonds predetermines their ability for hydration in aqueous solutions [88] as well as association in nonpolar solvents. Free energy of association of two molecules of N-methylacetamide in chloroform is −3.5 kcal/mol [89].

Hydrogen bonding with amides is only one, though very important, example of the ability of the amide group to form the complex. Another example is the amide-metal complexes, whose investigation is significant for understanding the structure and function of numerous metal-containing enzymes (see review [91]).

Amides form complexes with various metals [92]. As a rule, a shift to the region of low frequencies of the amide I band is observed. This gives evidence in favor of involvement of an amide carbonyl group in the complex formation. Transition metals group can use the maximum number of coordination bonds when the complexes with amides are formed. So, nickel(II) and chromium(III) form octahedral complexes of the type:

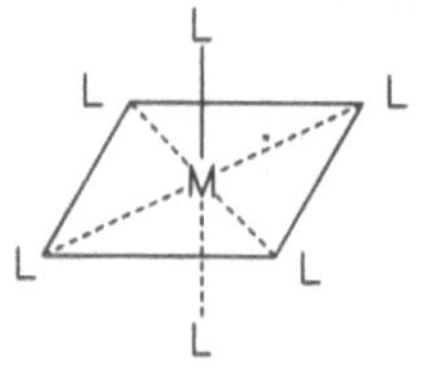

where L is amide or lactam [93]. Addition of some univalent ions yields complexes with amides nitrogen atom. A silver ion forms a complex with dimethylacetamide yielding a complex of type [94]:

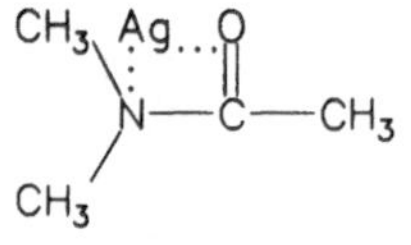

Interestingly, the barrier of the inner rotation around the C–N bond in metal complexes decreases [94, 95]. Thus, free energy of the activation of *cis-trans*-transition falls from 21 kcal/mol (for free amide) to 19 kcal/mol in the silver complex [94]. Quantum-chemical studies of lithium complexes of amides show that metal can form a bridge complex of type [95]:

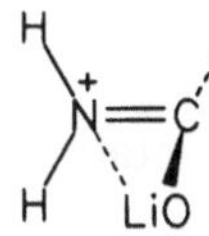

Just the formation of these complexes explains a decrease of the barrier of the inner rotation, as their structure resembles the structure of the transition state at *cis-trans*-isomerization.

The formation of amide-metal complexes changes the geometry of the amide group. So $r_{C=O}$ is decreased by ≈ 0.02 Å and certain, though slight, deviations from a planar configuration of the amide group [92, 96] are observed.

Another type of complexes capable of forming amides is π-complexes. Thorough research into NMR spectra of dimethylformamide in various aromatic solvents [97] shows that the amide forms with them complexes of the structure:

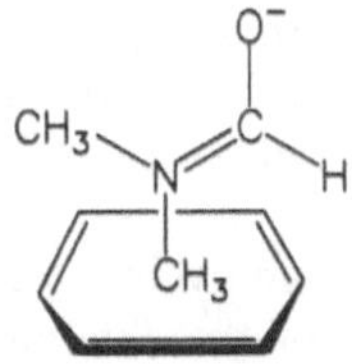

Finally, the concept of the amide group structure as a resonance hybrid of a neutral and an ionic form yields several additional consequences. First, the amide group must possess diamagnetic anisotropy. The polypeptides free of aromatic groups were shown to have distinct diamagnetic anisotropy. This phenomenon was theoretically grounded, the idea of electron mobility in the amide group underlies this concept, the constant of molar diamagnetic anisotropy of the amide group is equal to -5.36×10^{-6} cm·g·el. unit [99]. Secondly, a partially double character of the C–N bond in amides should result in a possible transition of the electron effects of substituents via the amide group. Initial data [100, 101] on such a transition via the amide bond are rather contradictory. But thorough consideration of this matter [102] testifies to the existence of this transition, where an inductive effect of the substituent is transferred weaker from the carbonyl part of the amide bond, as compared to its amine part.

It is noteworthy that further investigations did not confirm [103] amide-imidol tautomerism in linear amides [104].

1.3 Unusual (Nonpeptide) Amide Bonds and Other Bonds of Carboxylic Acid Derivatives

This part discusses two types of bonds cleaved with amide hydrolases – β-lactam and ester.

1.3.1 β-Lactams

Antibiotics of the penicillin(III) and cephalosporin (IV) group contain the β-lactam ring cleaved with β-lactamases:

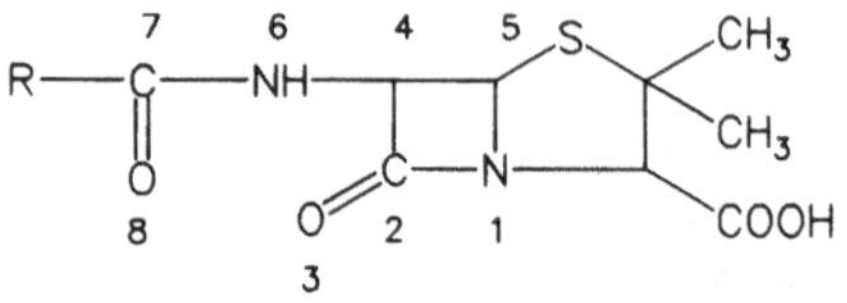

III 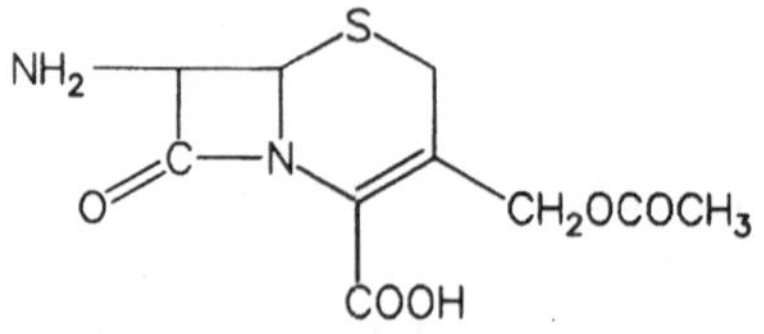   IV

The amide group constituting β-lactams has unusual properties, in particular its hydrolysis in the absence of enzymes stipulates unstability of penicillins in aqueous solutions. This is caused by low resonance in the amide group of β-lactams [105, 106]. The nitrogen atom in penicillins is shown to be located outside the plane of the substituents ($\Delta_N = 0.4$ Å) [108]. A similar situation prevails in cephallosporins [105].

X-ray studies of penicillins [107, 109] show that the β-lactam cycle is very strained. Table 6 presents interatomic distances and angles of the β-lactam cycle in penicillins.

Thus the β-lactam cycle is an almost planar rectangle. It is typical that bond C=O in the lactam cycle is essentially longer than that in common amides. Besides, the β-lactam cycle and thiazoline ring are located at an angle of ≈120° to each other. All this excludes a possibility of interaction of the unshared electron pair of the nitrogen atom with π-electrons of the double bond:

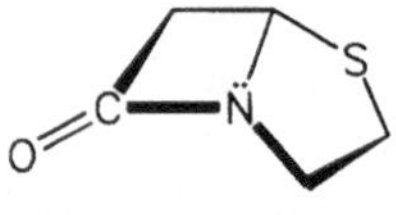

The frequency of the amide I band in penicillins is shifted to 1770 cm^{-1}, while this band has a value 1730 cm^{-1} in monocyclic β-lactams and in common amides – 1665 cm^{-1} [110].

Table 6. Interatomic distances and angles in sodium salt of benzylpenicillin [107]

Bond	r, Å	Angle	τ, degree	Bond	r, Å	Angle	τ, degree
N_1–C_2	1.39	$C_5N_1C_2$	95	C_2–C_4	1.49	$C_2C_4C_5$	86
C_2–O_3	1.29	$N_1C_2C_4$	92	C_5–N_1	1.38	N_1CO_3	134
						$C_4C_5N_1$	87

1.3.2   Ester Bond

Ester bonds cleaved with some proteolytic enzymes are the bonds in acylamino acid esters, in some usually activated esters of carboxylic acids and depsipeptides, made up of alternating amino- and hydroxyacid residues:

$$\ldots HNCH(R^1)COOCH(R^2)CONHCH(R^3)CO \ldots$$

Both in the amide group and esters the interaction of lone pairs of the ester oxygen atom with the π-bond of a carbonyl group, which can be designated as resonance structures (V) and (VI), seems possible. However, due to higher electronegativity of atom O as compared to N the contribution of structure VI is 15% [14].

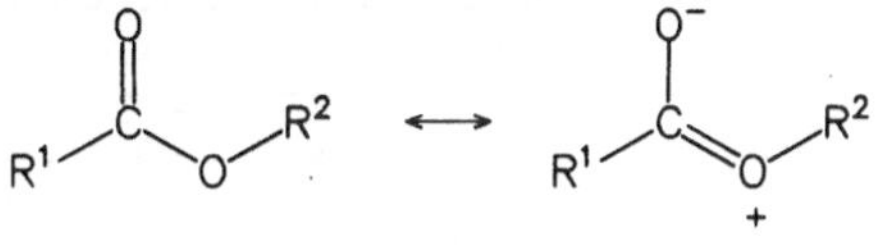

So esters have two forms:

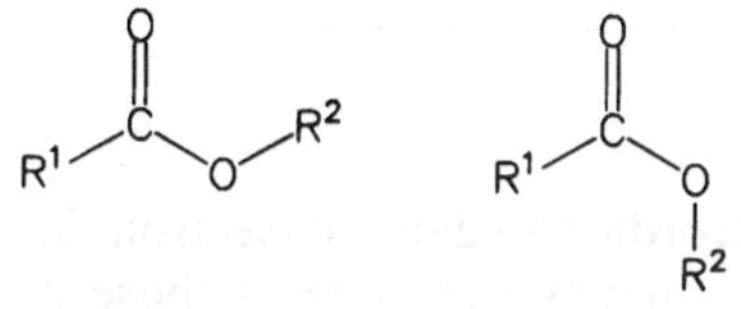

s–trans (Z)   s–cis (E)

The energy of resonance stabilization in the ester group is similar to that for amides (≈ 20 kcal/mol) [14] or even a little bit higher [111]. But a change of free energy of activation of s-cis-s-trans-transition at esters is lower than at amides and is equal to 7–10 kcal/mol [19, 112, 113].

This contradiction may arise [111] because the ester bond in a transition state of cis-trans-isomerization maintains partially the energy of resonance stabilization due to the interaction of one of two lone electron pairs of alcohol oxygen atom with π-electrons of the double bond (Fig. 5).

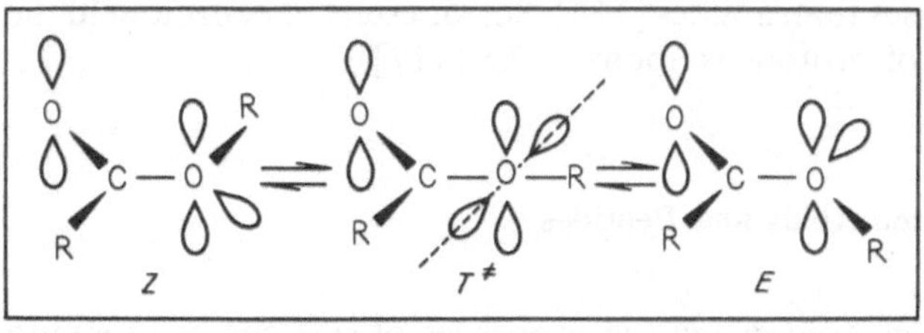

Fig. 5. Trans (Z)-cis (E) isomerization of esters. Transition state designated as $T^{\neq}$ [111]

Table 7. Geometric parameters of an ester group

Substance	r, Å			τ, degree		Reference
	C=O	C–OR	O–R	O=C–O	C–O–C	
Methylformiate	1.22	1.37	1.47	123	112	[114]
Methylacetate	1.22	1.36	1.46	124	113	[114]
Diethylterephtalate	1.28	1.32	1.47	125	117	[115]

Table 8. Charges on atoms, bond orders and force constants for esters (calculated by CNDO/2 method)

Substance	Charges			π-Bond orders		Force constants, mdin/Å	
	C′	$O_{carbonyl}$	$O_{alcohol}$	C–O	C=O	C–O	C=O
$HCOOCH_3$	0.381	−0.290	−0.199	0.053	0.875	6.74	11.83
				0.30[a]	1.09[a]		
CH_3COOCH_3	0.395	−0.355	−0.224	0.016	0.777	7.00	11.51
				0.35[a]	1.02[a]		
$CH_3CONHCH_2COOCH_3$	0.372	−0.334	−0.217	0.028	0.777	–	–

[a] Calculated from force constants [61].

The bond lengths in the ester group, according to data on electron diffraction in a gaseous phase [114], are approximately the same as those in the amide group and they change at the transition from individual molecules to crystals like in the amide group (Table 7). Force constants in group COOR decrease as compared to the amide group, therefore π-orders for bonds C–OR calculated from these data are lower than for the C–N bond in amides (Table 8).

All this testifies to less resonance stabilization of the ester bond than of the amide one.

It is noteworthy that the ester group is not planar in a gaseous phase. According to data on electron diffraction in methylformiate, methylacetate and other esters the angle between the planes R^1–C=O and C–O–R^2 is about 25° [114]. Such a distortion in a planar configuration is not observed in crystals [115].

Esters are weaker bases than amides [116]. So, for esters of benzoic acid the pK_a value for addition of protons is about −7.5 [117].

1.4 Derivatives of Amino Acids and Peptides

Besides the state of the cleaved group the properties of side chains of amino acid residues and their conformation are important characteristics of proteases substrates. The structure and characteristics of side chains of native amino acids vary within a wide range. From the viewpoint of an enzymatic catalysis a charge of the side chain, its size, hydrophobicity, conformation of both the backbone of the polypeptide and side chains, and conformational mobility are the most significant characteristics. These will be considered in the example of acylamino acids and simple peptides.

1.4.1 Ionization and Hydrophobicity

The side chains of amino acid residues according to their ionization state in neutral aqueous solutions can be divided into three groups: neutral, weakly acidic and basic (Table 9). The side chains of aspartic and glutamic acid residues have the negative charge, the side chains of lysine and arginine residues are almost protonated and bear the positive charge, while the side

Table 9. pK_a Values and hydrophobicity of amino acids side chains

Amino acid	R	pK_a	ΔG, kcal/mol; alcohol-water [118]	π [119]	ΔG, kcal/mol; vacuum-water [120]
Glycine	H		0	0	+2.39
Alanine	CH_3		+0.73	+0.50	+1.94
Valine	$CH(CH_3)_2$		+1.69	+1.53	+1.99
Leucine	$CH_2CH(CH_3)_2$		+2.42	+1.98	+2.28
Isoleucine	$CH(CH_3)C_2H_5$		+2.97	+2.04	+2.15
Serine	CH_2OH		+0.04	−1.03	−5.05
Threonine	$CH(CH_3)OH$		+0.44	−0.77	−4.88
Asparagine	CH_2CONH_2		−0.01	−1.68	−10.95
Glutamine	$(CH_2)_2CONH_2$		−0.1	−1.22	−10.20
Aspartic acid	CH_2COOH	3.65	+0.54	−0.72	−9.68
Glutamic acid	$(CH_2)_2COOH$	4.25	+0.55	+0.29	−9.38
Histidine	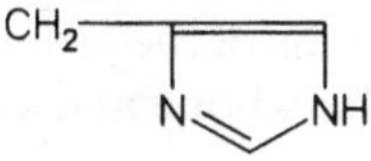	6.0	+1.4	−	−10.27
Lysine	$(CH_2)_4NH_2$	10.5	+1.5	+1.45	−9.52
Arginine	$(CH_2)_3NHC (:NH)NH_2$	12.5	+0.73	−	−19.92
Cysteine	CH_2SH	8.2	+0.65	−	−1.24
Tyrosine	$CH_2C_6H_4-OH$-p	10.0	+2.87	−	−6.11
Phenylalanine	$CH_2C_6H_5$		+2.65	+2.69	−0.76
Tryptophan	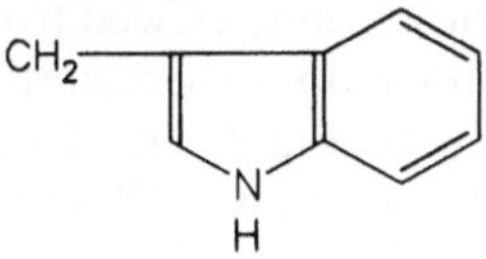		+3.0	−	−5.88
Cystine	$-CH_2-S-S-CH_2$		+1.3	−	−1.48
Methionine	$(CH_2)_2SCH_3$		+2.6	−	−
Proline	$-(CH_2)_3-$		−	+1.2	−

chains of histidine and cysteine, though to a lesser extent, have only a partial positive and negative charge, respectively. The phenol oxygen of tyrosine at pH 7 is protonated, being electrically neutral. The indole group of tryptophan (pK about -2) is considered as neutral.

A change of free energy in the side chain of amino acid at its transition from nonpolar or little polar solvent into water is a quantitative evaluation of hydrophobicity of the amino acid residue. There exists a set of scales of hydrophobicity, depending on whether ethanol [118], n-octanol [119] (π coefficient), heptane [121], etc. are used as a solvent. The scales based on the retention index upon high performance liquid chromatography of amino acids [122] as well as an index equal to a change of free energy upon transition of the side chain from vacuum into water are also applied [120]. These indices often characterize differently side chains of amino acids in respect to their hydrophobicity. For example, at alcohol-water transition lysine, arginine and histidine residues are characterized as weakly hydrophobic, which is stipulated by the presence of methylene and methine groups, and their contribution exceeds that of the charge. The change of free energy at transition from vacuum into water shows these residues to be most hydrophobic (see Table 9).

The additivity of hydrophobicity indices appears not to be always taken into account. As noticed [123], the polar side chains are more hydrophobic in proteins and oligopeptides than in free amino acids due to the flanking peptide bonds. The formation of the α-helical conformation induces a slight decrease in hydrophobicity of most of the amino acid side chains.

1.4.2 Conformation

Each amino acid residue of the polypeptide chain (except for proline) has two bonds, $N-C_\alpha$ and $C_\alpha-C'$, around which free rotation is possible. Besides, the side chains of amino acids have mainly several ordinary bonds admitting free rotation. Due to interactions of groups valently unbound within the amino acid residue as well as of groups of neighboring residues in the chain ("nearest" interactions [124]), not all feasible reciprocal positions of the groups appear to be resolved. Let us consider conformations of the backbone and side chains of amino acid residues (see reviews [11, 13]).

Nomenclature. According to the recommendations of the International Commission on Nomenclature [15] the conformation of the backbone is represented with two dihedral angles, ϕ and ψ, characterizing mutual orientation of substituents at atoms N and C_α, C_α and C'. The chain is viewed from the N- to C-terminal; the position, when the N-H bond is in *cis*-orientation to bond $C_\alpha-C'$, is taken for a zero value of the angle ϕ (Fig. 6). A change of angle ϕ occurs due to rotation of bond $C_\alpha-C'$ round bond $N-C_\alpha$. Clockwise rotation entails the value of the angles from $0°$ to $+180°$, while counterclockwise rotation yields the values from $0°$ to $-180°$. For the zero value of angle ψ one takes the position when the $C_\alpha-N$ bond is in *cis*-orientation as compared to bond $C'=O$. The values of angles are the same here as for angle ϕ. Besides the

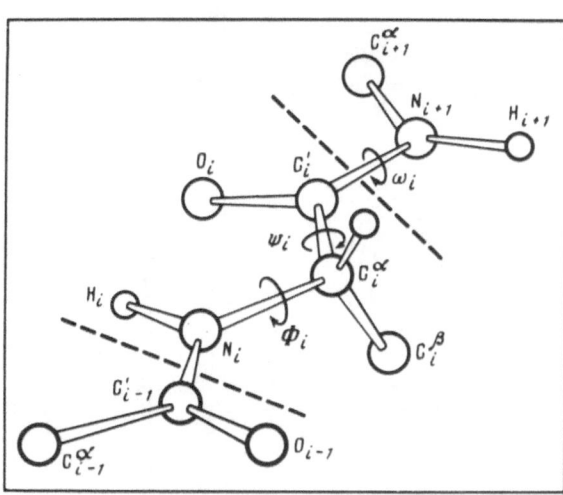

Fig. 6. Designation of dihedral angles in the peptide chain according to recommendations of the Commission on Biochemical Nomenclature [15]

angles ϕ and ψ, dihedral angle ω characterizing the amide group geometry determines the conformation of the backbone.

The conformational state of the side chains is described by χ_1, χ_2, etc. starting with the C_α-atom. For the zero value one takes the conformation, when bond C_β–C_γ is in *cis*-orientation as regards the C_α–N bond:

The Backbone. Numerous X-ray studies on peptides [11, 13, 18] and proteins [125–127] as well as studies on peptides in solutions performed mainly by NMR spectroscopy methods [128–131] show that the conformation of the backbone almost always coincides with the regions of the minimum energy calculated by the methods of atom-atom potentials or by quantum-chemical methods [28, 132–138].

The values of angles ϕ and ψ allowed by the energy are represented, as a rule, by the Ramachandran diagrams [11] (Fig. 7a–c). For methyl amides of N-acetylalanine and N-acetylvaline, the allowed regions are the three ones designated as R, B and L. The conformational freedom is much higher for a glycine residue. The region B corresponds to the β-structure of the polypeptide chain, while regions R and L correspond to the right and left helical structures, respectively. The transition from one region to another requires the overcoming of an essential activation barrier. Regions R and L are divided with a barrier of about 12 kcal/mol; however, the barrier between the R and B regions is equal only to 1–3 kcal/mol.

It should be stressed that isoenergetic contours marked on the map are those of equal potential energies. These values do not allow evaluation of each conformation, since the entropy contribution of each can be different. The

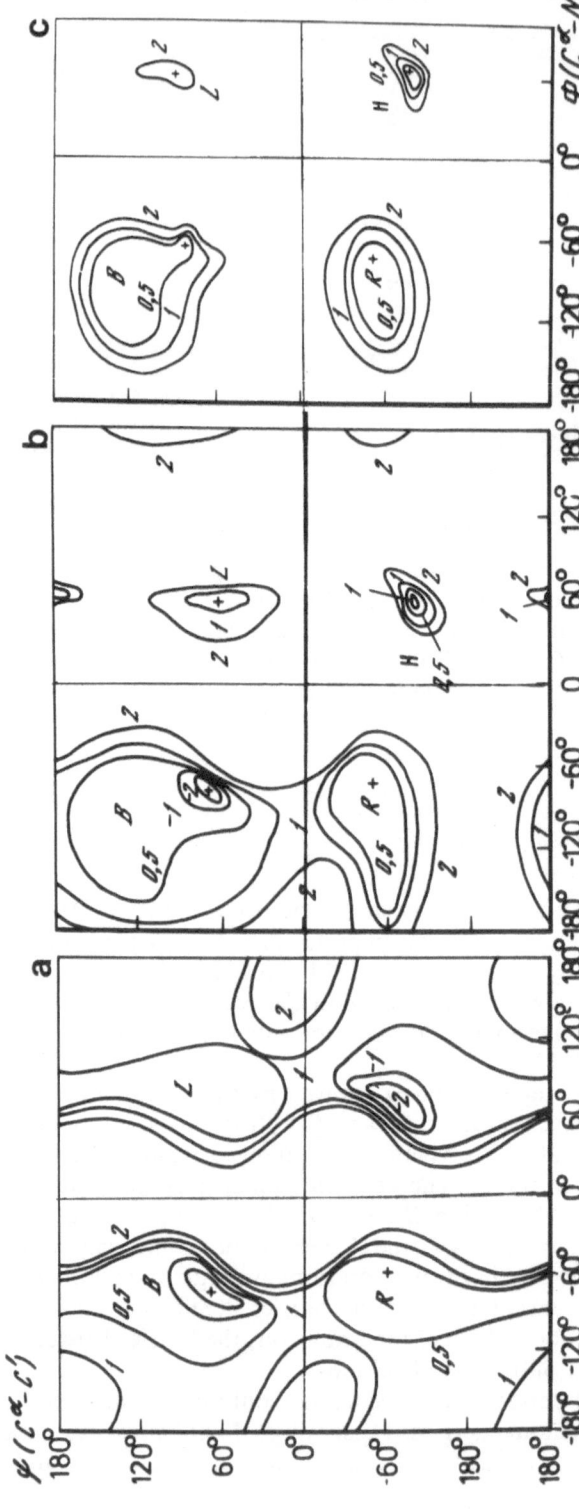

Fig. 7a–c. Conformation maps ϕ and ψ for methylamides of N-acetylglycine (**a**), N-acetyl-L-alanine (**b**) and N-acetyl-L-valine (**c**) [132]

minimum area on the conformational map is a qualitative evaluation of entropy of the given conformation. The attempts at the quantitative evaluation of the entropy for a few peptide conformations are considered in [136, 139, 140]. According to these data (e.g. [136]) for methyl amide of N-acetyl-trialanine the conformation with parameters of the α-helix ($\phi - 60°$, $\psi - 60°$) is less advantageous by free energy 4.7 kcal/mol than the conformation with parameters of the β-structure ($\phi - 80°$, $\psi + 80°$). This corresponds to the concentration ratio of the folded conformation to the α-helical equal to 2.5×10^3.

Quantum-chemical studies [114] lead to the conclusion that the conformation of small peptides is defined mainly by the three factors: repulsion of nonbonding orbitals, the formation of hydrogen bonds, and torsion barriers around bonds C–C and C–N. Besides, π–π-interactions can show a stabilizing effect, for example, between the aromatic ring and the amide bond [142, 143], as well as H–π-interactions between bond N–H of one residue and group C=O of the other [144].

When studying the peptide conformational states, the question regarding conformation influence on the environment is a key problem. Coincidence of the experimentally found conformations with those calculated by means of the method of atom-atom potentials gives the evidence that this method provides good modeling of the peptide state in the aqueous medium. As shown [145, 146], hydration does not change much the conformation and the fluctuation dynamics of dipeptides of the $AcAlaNHCH_3$ type. At the same time, taking

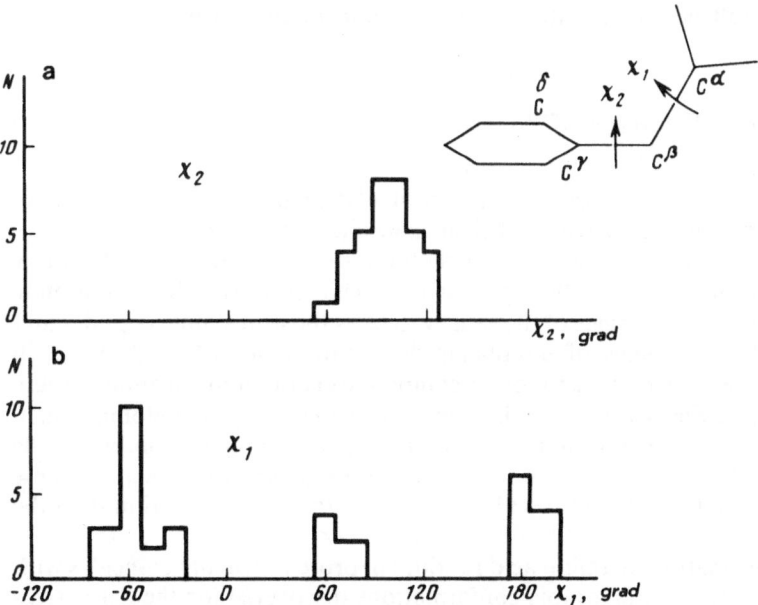

Fig. 8a, b. Histograms of angles χ_2 of the phenylalanine side chain (**a**), and χ_1 of the tyrosine side chain (**b**) [18]

Table 10. Side chain conformations of derivative N-acetylphenyl-alanine [147]

Substance	Solvent	Quota of conformation with χ_1		
		$+180°$	$+60°$	$-60°$
AcPheOEt	$C_6H_5CH_3$	0.34	0.33	0.33
"	$(CH_3)_2SO$	0.58	0.31	0.11
"	D_2O	0.52	0.32	0.16
AcPheNH$_2$	D_2O	0.58	0.31	0.11
AcPheAlaOMe	D_2O	0.66	0.16	0.18

into account hydration calculations [135], the minimum energy of methyl-amide of N-acetylalanine was found to shift from $\phi = -84°$ and $\psi = 79°$ to $\phi = -142°$ and $\psi = 150°$. However, the difference in energies of these conformations is rather slight.

Side Chains. A change of the side-chains conformation does not demand the significant energetic expenditure as a rule. The difference in conformations does not exceed 3 kcal/mol. Conformations with $\chi_1 = -60°$, $+60°$ and $+180°$ are preferably realized in peptide crystals. Figure 8 shows that the three values of χ_1 run approximately with the same frequency, as for χ_2 its values from 60° to 120°.

NMR spectra of solutions of various derivatives of N-acetylphenylalanyl [147] show that conformation $\chi_1 = 180°$ is predominant, while the solvent itself slightly affects the distribution of the conformation (Table 10).

1.5 Proteins as Substrates of Proteases

Proteins are intrinsic substrates of most proteolytic enzymes. Therefore, it is of use to consider the peculiarities of their macromolecular structure.

Amide bonds in the protein molecule do not differ from the bonds in low molecular compounds by their parameters. This concerns the interatomic distances and valence angles of the same values as those in simple peptides and amides [27]. The values of nonplanar deformation of the amide bond in proteins do not exceed those for other compounds containing an amide group. As an example, Fig. 9 presents a histogram of angles ω in myoglobin, which shows that the majority of amide bonds have *trans*-configuration, while the variations of the ω angle are about 15°. A *cis*-amide group appears in some proteins [148–150] and peptides [18], but this is rather an exception than the rule.

The conformation of amino acid residues in proteins usually coincides with the energetically most favorable conformations discovered for the derivatives of amino acids and peptides [124, 151]. This is stipulated by the predominating

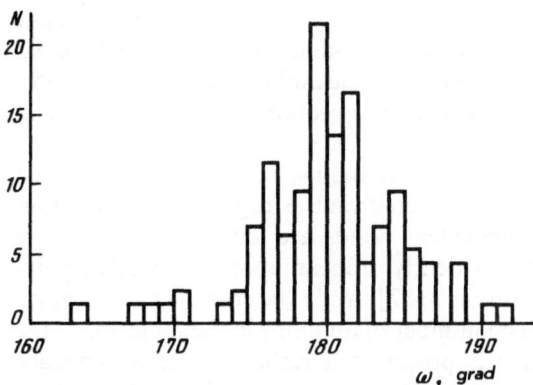

Fig. 9. Histogram of angles ω in myoglobin [27]

role of "neighboring" interactions in the conformational stabilization of the backbone. Most of the residues have angles ϕ and ψ, which coincide with the energetically favorable conformations of methylamides of N-acetylamino acids [124].

The values of angles ϕ and ψ of amino acid residues establish the type of the secondary structure of the whole protein molecule.

In accordance with the dominating secondary structure type the proteins are divided into four groups (Fig. 10) [152]: I, totally α-helical; II, mostly containing the β-structure; III, the proteins, where the α- and β-structures are separated along the polypeptide chain ($\alpha + \beta$) and IV, the proteins with the mixed α- and β-structures (α/β).

The secondary structure is predetermined by the nature, sequence and conformational state of amino acid residues. Some attempts are made to predict the secondary and tertiary structures on the basis of statistic analysis of

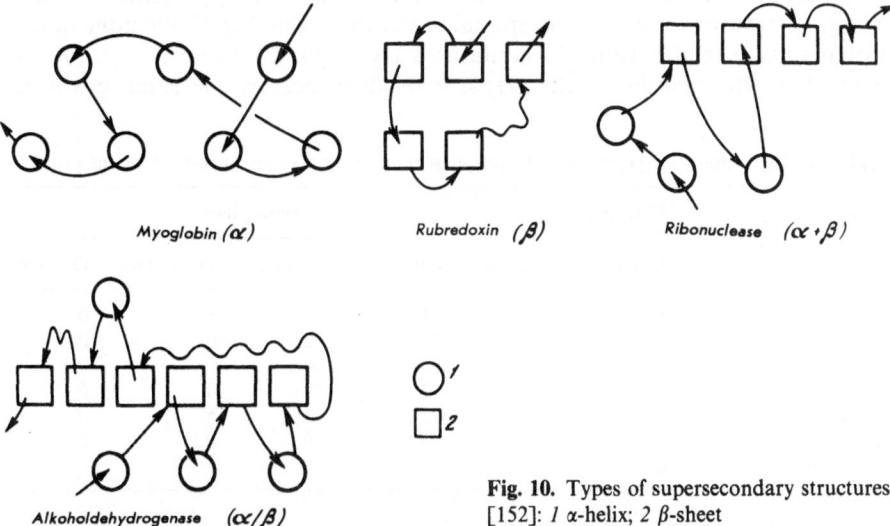

Myoglobin (α) Rubredoxin (β) Ribonuclease $(\alpha + \beta)$

Alkoholdehydrogenase (α/β)

Fig. 10. Types of supersecondary structures [152]: *1* α-helix; *2* β-sheet

amino acid sequences [153] and even of the amino acid composition [154]. It is noteworthy that many functionally close proteins have a similar spatial structure, though they have little in common in the amino acid sequence [155]. And vice versa, identical sequences with various conformations are found in some functionally different proteins [156].

It is typical of many proteins to have domains, i.e., spatially separated compact parts of the protein globule connected via the fragment of the random polypeptide chain. Numerous noncovalent bonds are located between these domains [13, 157]. The domains, evidently, appear upon the protein poly-peptide chain folding at its biosynthesis [157], while the so-called "nucleation centers" play the key role [158]. The number of domains in the molecule of the protein depends on its size. There is proposed a rather simple dependence [159] connecting the number of domains (D), amino acid residues (N), and the number of residues in a certain, randomly chosen structural unit of the molecule (u):

$$D = 2^L - 1, \quad \text{where } L = \log_2(N/u) + 1 \; .$$

Finally, many proteins consist of the subunits which are not covalently bound. These subunits are arranged in the protein quaternary structure and seem to be functionally important.

The most essential peculiarities of proteins as substrates of proteolytic enzymes are as follows: (1) inaccessibility of many side chains of amino acid residues since they are buried in the protein globule, or owing to some other steric hindrances; (2) limited possibility of conformational transitions, par-ticularly of the protein backbone. The accessibility of amino acid residues to an external reagent or a solvent depends on the position of this residue in the protein globule. The majority of polar and charged side chains are located on the globule surface. However, there are charged residues masked inside the globule or within the subunit contacts. These residues, as a rule, form ionic pairs or yield hydrogen bonds with neutral polar groups [160–162].

Hydrophobic residues, for example of tyrosine and tryptophan, are un-evenly distributed in the protein spatial structure (Table 11). While none of the proteins listed in the Table 11 contains a tryptophan residue that is totally exposed in the medium, many tyrosine residues occupy the same position.

Table 11. Distribution of tyrosine and tryptophan residues in crystals of some proteins [163]

Protein	Tyrosine			Tryptophan		
	Inside	On surface	Outside	Inside	On surface	Outside
Carboxypeptidase	1	13	4	2	6	0
α-Chymotrypsin	0	2	2	2	6	0
Elastase	1		10	4	3	0
Ferricytochrome C	2	2	0	1	0	0
Myoglobin	1	0	2	1	5	0
Ribonuclease A	2	1	3	0	0	0
Subtilisin	0	10		0	3	

Most of the residues of aromatic amino acids can be regarded as "surface" residues, i.e., those which are partially embedded in the protein globule or adjoined with their aromatic ring to other groups.

The position of the residue in the protein globule is assigned to its possible hydration. There was found a distinct linear correlation (Fig. 11) between the location of amino acid side chains in the protein and the level of their hydration [120].

Amino acid residues can be masked by carbohydrate chains in glyco-proteins or lipids, if the protein is firmly associated with the membrane [164].

Though the proteins are compact, thermodynamically stable entities, their conformational changes and a certain intramolecular mobility appears possible even in the absence of the outside effect action [165, 166]. Even side chains embedded in the globule in many cases are capable of free rotation [167].

All this concerns the native protein structure. However, many proteins are cleaved with proteolytic enzymes in a denatured state, i.e., in the state where their spatial organization is significantly changed under the outside influence. These can be, for example, pH media, temperature, etc. Digestive enzymes, which function in the stomach, attack the proteins, which are subjected to acid denaturation.

A set of publications are devoted to protein denaturation (see reviews [168, 169]). Here we do not intend to consider this problem, but just mention that denaturation leads to substantial "demasking" of most amino acid residues, so that they are accessible to the action of outside reagents including enzymes. The cleavable bond, even under such extreme conditions as those of the acid medium in the stomach (pH about 1), changes as do the low molecular substances. As to the conformation of the backbone and side chains of amino acids in the denatured protein, it remains, obviously, within the energetically allowed regions, certainly with regard to a changed ionization of the charged groups. The consequence of denaturation is an increase of a number of amino acid residues accessible to external reagents as well as the elimination of

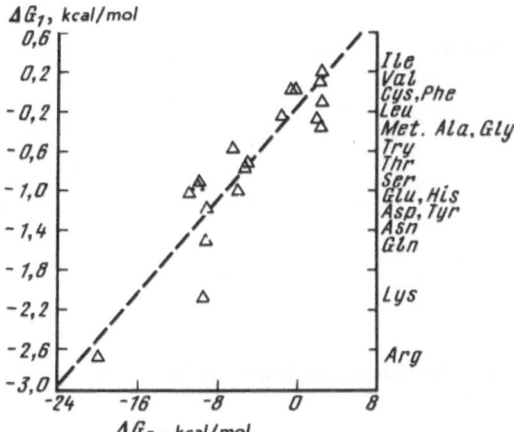

Fig. 11. Correlation of free-energy changes at transfer of the side chain of amino acid residues (12 proteins) from the internal part of the protein globule to its surface (ΔG_1) with the free-energy change at transfer of a residue from vacuum to water (ΔG_2) at pH 7 [120]

limitations of the conformational mobility of the backbone caused by "distant" interactions in the native structure. These factors provide a cleavage of amide bonds catalyzed by enzymes.

1.6 Water

According to the definition, hydrolytic reactions include water as one of the reagents. Therefore, the structure of this "substrate" of amide hydrolases should be briefly characterized. Many studies deal with the structure and properties of water (see, for example, [170, 171]). The molecule structure of water is presented in Fig. 12.

Water has a considerable dipole moment ($\mu = 1.85$ D) due to different electronegativity of oxygen and hydrogen atoms and a nonlinear structure. Therefore, water molecules in a condensed state intensely interact with each other as well as with other polar molecules and form a three-dimensional net of hydrogen bonds:

Many of the estimated and experimental characteristics of water are presented in [172, 173]. The following values of constants of ionization in the reaction should be noted:

$$H_2O + H^+ = H_3O^+ ,$$

where $pK_a = -1.74$, and in the reaction:

$$H_2O = OH^- + H^+ ,$$

where $pK_a \doteq 15.74$.

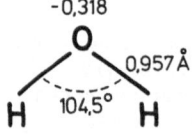

Fig. 12. Bond length, dihedral angle and charge on the oxygen atom in water molecule

1.7   Conclusion

Belying the apparent uniformity of amide hydrolase substrates, thorough analysis reveals their amazing variety. The difference in the spatial structure of proteins and in the accessibility of these or other amino acid residues to external reagents as well as the diversity of their conformational mobility predetermine the variety of the proteolytic enzymes as regards these proteins. This also depends on a set of properties of side chains and a limited conformational state of the backbone leading in many cases to strictly specific enzyme-substrate interactions. The cleavable amide group is a unique structure with two rather contradictory properties, such as stability and lability. The stability of the amide bond is characterized by very low rates of its disruption in the absence of the catalytic influence on this reaction. Lability of both the geometry and the electron structure of the amide group strictly determines its sensitivity to an external effect and its ability for diverse changes under the catalyst's action, in particular the enzymes, upon hydrolytic cleavage.

Chapter 2   Enzymes

2.1   Classification

Hydrolysis of amide bonds in various compounds including mainly proteins and peptides described in the previous chapter is catalyzed by numerous enzymes referred by a modern classification to the class of hydrolases (class 3). The enzymes, which catalyze hydrolysis of peptide bonds, form a subclass of peptide hydrolases (3.4). Since we consider not only the enzymes catalyzing hydrolysis of amide bonds in proteins and peptides, but also many of the enzymes that cleave other types of amide bonds, it seems reasonable to introduce the term "amide hydrolases" as common for all of them. Amide hydrolases are composed of two groups: proteases (3.4) splitting the amide bonds in proteins, and peptides and amidases (some of the enzymes of subclass 3.5) which cleave the amide bonds other than the peptide bonds.

Two important criteria underlie the classification of these subclasses: specificity and structure of the enzyme catalytic apparatus [10]. According to their specificity amide hydrolases can be divided as shown in Fig. 13 (cf. [174]). While peptidases are classified by a position of the cleavable bond in the peptide chain (carboxypeptidase – by the active site as well), proteases (endopeptidases) are identified according to the active site structure.

This seems obscure, since two absolutely different features are confused here and, besides, more and more data on the enzymes with both exo- and endopeptidase activities appear now (see below).

The advances in the chemistry of amide hydrolases make it possible to assign them to one of four structural types: serine, cysteine, aspartic or metal-containing amide hydrolases [175]. Certainly, for a set of enzymes such an assignment is tentative as it is based on indirect data concerning the action of some or other inhibitors on an enzyme, rather than on the knowledge of the structure and the mechanism of catalysis. Moreover, one cannot exclude that among the unclassified amide hydrolases there can be found enzymes differing in the structure and the mode of action from those mentioned above. However, the advantage of the classification is evident: it provides an appropriate basis for considering the enzyme structure and the mechanism of action from the chemical point of view.

Table 12 gives an allotment of known amide hydrolases by the types of their catalytic site structures. Undoubtedly it will be improved and refined in due course.

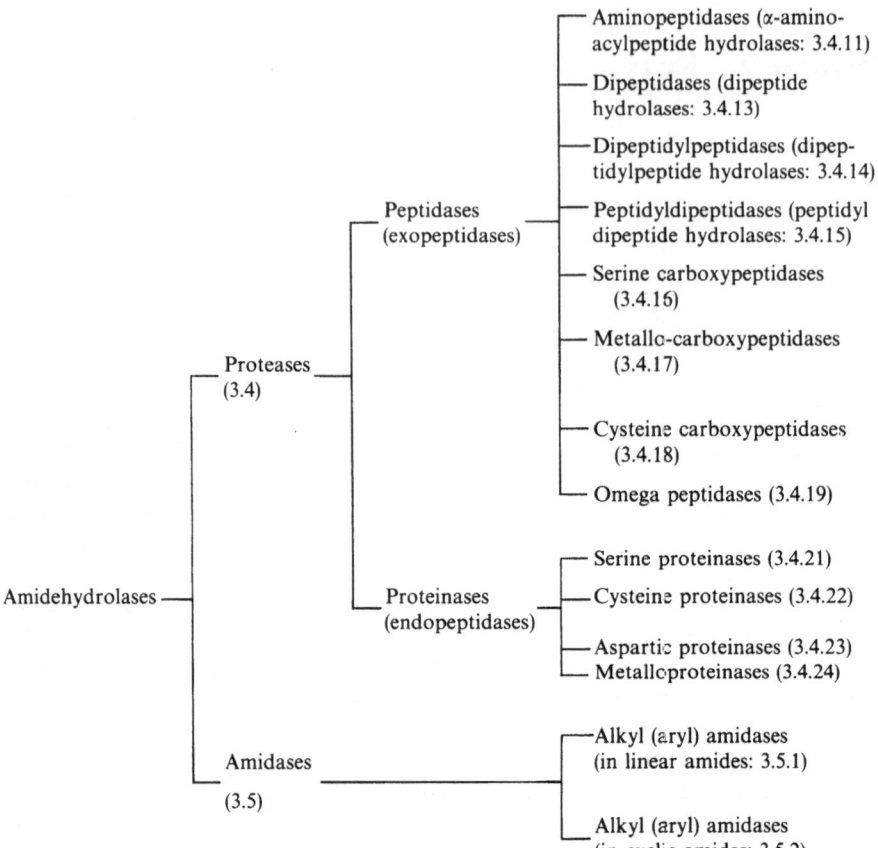

Fig. 13. Classification of amide hydrolases. EC numbers given in *parentheses* [10]

2.2 Distribution

Amide hydrolases compose one of the largest groups. Table 12 lists nearly 600 enzymes of this type, described quite well. It is evident that their number is much higher. Amide hydrolases occur in all (without exception) living organisms – from viruses to humans.

Microorganisms are especially rich in these enzymes. Some kinds of bacteria, e.g., *Bacillus subtilis*, form up to 1 g of proteolytic enzymes per 1 l of nutrient medium [176]. About 25 various amide hydrolases have been isolated from such a classical object of molecular biology as *Escherichia coli* and characterized.

In the last few years artificial, often unusual, sources of amide hydrolases emerge due to the development of the DNA recombinant technique [177], which allows obtaining the enzymes in microbial, yeast and some animal cells,

Table 12. Amide hydrolases

N	EC Number	Name of enzyme	Source	M_r, kDa	Structural data[a]	pH_{opt}	Comments	References[b]
Serine amide hydrolases								
1	3.4.14.2	Dipeptidyl peptidase II	Bovine pituitary	130		5–5.5		[213]
2		The same	Porcine ovary	110	(2 × 54), gl.	6		[214]
3		"	Xenopus laevis	98	gl.	8		[215]
4		"	Sacharomyces cerevisiae yscV	40	1ch.	7–7.5	Membrane	[216]
5		"	Flavobacterium meningosepticum	160	(2 × 75)	7.4–7.8		[217]
6	3.4.14.4.	Dipeptidyl peptidase III	Mammalian lens	80	1ch.	8–9	SH-dependent	[218]
7		The same	Rat brain	80	1ch.	8–9		[219, 220]
8	3.4.14.5	Dipeptidyl peptidase IV	Human fibroblasts	400	2ch. (125, 135)	7.8–8	Membrane	[221]
9		The same	Pig kidney	280	(2 × 130)	7.7		[222–224]
10		"	Lamb kidney	230	(2 × 115)	7.8	"	[225]
11		"	Different organs of rat	190–266	(2 × 90–130)	7.5–8	"	[226, 227, 228]
12		"	Pig liver	360	2 subunits	7.8–8	Soluble	[229, 230]
13	3.4.14.8	Tripeptidyl peptidase I	Hog ovaries	55	Associate	4.5–5		[231]
14		The same II	Liver, erythrocytes	135		7.5		[232–234]
15	3.4.16.1	Serine carboxypeptidase	Rat liver A, S	100	3ch. (20, 55, 25)	5.5	Cathepsin A	[235–237]
16		The same	The same A, L	400	"			"
17		"	Malted barley I	60	2ch. (34, 26) gl., s.	5.5	SH-dependent	[238, 239]
18		"	The same II	60	"	4.8–5.7		[238, 240]
19		"	" III	48	1ch., gl.	4–6		[241]
20		"	Wheat brain II	55	2ch. (37, 26) gl., s.	4–5	SH-dependent	[242]
21		"	The same	126	2ch. (60, 63)	4.4	SH-dependent	[243]
22		"	French beans	120		6	"Phaseolin"	[244]
23		"	Citrus plants C	93	1ch., gl., s.	5.5		[245–247, 248]
24		"	The same	96	1ch., gl., s.	5.5		[247]

Table 12. Continued

N	EC Number	Name of enzyme	Source	M_r, kDa	Structural data[a]	pH_{opt}	Comments	References[b]
25		„	*Sacharomyces cerevisiae*	59	1ch., gl., s.	4.3	Carboxypeptidase Y, SH-dependent	[249, 250, 251]
26		„	*Penicillium janthitelum* S-1	48	1ch.	4.3		[252]
			S-2	128	(2 × 65)	3.9		
27	3.4.16.2	Proline carboxypeptidase	Human lung	115	2 × (66, 45)	5–5.5	Angiotensinase C	[236, 253, 254]
28		The same	Swine kidney cortex	210	(8 × 25)	5–5.5		[255]
29	3.4.16.4	D-Alanyl-D-alanine carboxypeptidase	*Streptomyces* R-39	53	1ch., s.	7		[256, 257–260]
30		The same	*Streptomyces* R-61	37	1ch., s.	7		[256, 258, 260]
31		„	*Bacillus* sp.	50	1ch.	7		[261–263]
32	3.4.19.1	Acylamino-acid-releasing enzyme	Human erythrocytes	300	1ch.	7.5–8	SH-dependent	[264]
33		„	Rat tissue	320	(5 × 75)	7.2–7.6	„	[265, 266]
34	3.4.19.2	Peptidyl-glycinamidase	Pig kidney	442	1ch.	7.5–8		[267]
35		„	Skin of *Bufo marinus*	100	(2 × 48), gl.			[268]
36	3.4.21.1	Chymotrypsin A and B	Human pancreas	25.8	2ch., s.	8	Three forms IA, IB, II	[269, 270]
37		The same	Bovine „	25	3ch., s.	7–9		[271, 272–274]
38		„	Pig „	25.8	3ch., s.	8–9		[275]
39		„	Sheep „	24	3ch., s.	8–8.5	Activated by Ca	[276]
40		„	Fin whale „	17	1ch.	8–9	Two forms	[277, 278]
41		„	Chicken „	26.3		8–9		[279]
42		„	Rat „ D	25	s.			[280]
43		„	Turtle „	26	s.	7–8		[281, 282]
44		„	Prawn hepathopancreas I	27	1ch.	7		[283]
45		„	„ II	26	1ch.	7		[283]
			Insects:					

No.	EC	Enzyme	Source	Mol. wt.	Chains	pH	Remarks	Refs.
46	,,		Hornet *Vespra orientalis*	13.6	1ch., s.	8–9		[284, 285]
47	,,		,, *Vespra crabra*	14.5	1ch., s.	8.2		[284, 285, 286]
48	,,		Butterfly *Pieris brassica*	32		8–9		[284]
49	,,		Bee *Apis mellifera*	19.5	1ch.	8–9		,,
50	3.4.21.2	Chymotrypsin C	Mammalian pancreas	24	2ch., s.	7–9	Part of subunit II of procarboxypeptidase A	[287–289]
51	3.4.21.3	Proteinase A	Sea anemone	25	1ch.	8		[290]
52	3.4.21.4	Trypsin	Human pancreas	23.8	1ch., s.	7–8	Two forms I, II	[291, 293]
53		The same	Bovine ,,	23.4	1ch., s.	7–8	Ca-dependent	[292, 294, 295]
54			Pig ,,	25.5	1ch., s.	7–8		[296, 297]
55			Sheep ,,	26.2	1ch., s.	7–8	Ca-dependent	[298]
56			Rat ,,	30	1ch., s.	7–8	Cationic and anionic	[299]
57			Turkey ,,	18.6	1ch., s.	7–8		[300, 301, 302]
58			Dogfish ,,	21	1ch.	7–8		[279]
59			Cod ,,	24.2	1ch., s.	7–8	3 forms	[303, 304]
			Invertebrates:					
60			Sea urchin	29	1ch.	8	pI 5	[305]
61			Ascidian	27	1ch.	8.5–9	"spermosin"	[306]
62			Starfish	23.6	1ch.	8–8.5	2 forms A, B	[307]
63			Crayfish *Astacus Pluviatilis*	25	1ch., s.	8		[308]
			Insects:					
64			Hornet *Vespra orientalis*	17	1ch., s.	7–8		[309, 310]
65			Butterfly *Pieris brassica*	32	1ch.	7–8		[284, 311]
66			Bee *Apis mellifera*	20	1ch.	7–8		[284]
67			Grasshopper *Locusta migratoria*	17	1ch.	7–8		,,
68			Tobacco vorm *Manduca sexta*	24	1ch.	7–8		,,
69			Silk moth *Bomrix mori*	28	1ch., gl.	7–8	"Cocoonase"	[312]

Table 12. Continued

N	EC Number	Name of enzyme	Source	M_r, kDa	Structural data[a]	pH_{opt}	Comments	References[b]
70	3.4.21.5	Thrombin	Human plasma	40	2ch., gl., s.	7–8		[313, 314–316]
71		"	Bovine "	39	2ch. (32, 6) gl., s.	7–8	3 forms	[313, 317–319]
72	3.4.21.6	"	Pig "	40	2ch., gl., s.	7–8		[317]
73		Coagulation factor Xa	Human "	59	2ch. (42, 17) gl., s.	7–8	Stuart factor	[320–322, 323]
74	3.4.21.7	"	Bovine "	56	2ch. (39, 17) gl., s.			[322]
75		Plasmin	Human "	76	2ch. (26, 49) s.	7–8		[324, 325–327]
76	3.4.21.9	"	Bovine and other plasma	ca. 80	2ch., s.	7–8		[326, 328]
77		Enteropeptidase	Human duodenum	ca. 300	3ch., gl.	8–9	Enterokinase	[329]
78	3.4.21.10	"	Bovine "	150	2ch. (115, 35) gl.	8.4		[330, 331]
79		"	Pig "	190	2ch., gl.			[332]
80		Acrosin	Human sperm, etc.	44	2ch., gl., s.	8–8.2	Ca-activated	[333–335]
81		"	Boar "	41	2ch. (37, 4.2) gl., s.	8.5–9		[336]
82	3.4.21.12	"	Ascidian "	32–34	gl.	8.5		[307]
83		α-Lytic proteinase	Lysobacterium enzymogenus	19.7	1ch., s	8		[337, 338]
84	3.4.21.–3.4.21.14	Lytic proteinase	Bacillus subtilis	28	1ch.	7.8–8.5		[339]
		Microbial serine proteinases						
85		Subtilisin Carlsberg	" licheniformis	26.3	1ch., s.	10–11	Subtiopeptidase A	[340–343]
86		Subtilisin E	" subtilis	27.7	1ch., s.	8–10		[344, 345]
87		Subtilisin BPN'	" amiloliquefaciens	27.5	1ch., s.	8.5–9.5	Subtilopeptidase B, nagarse	[342, 343, 346–348]
88		Subtilisin DY	" subtilis DY	27.5	1ch., s.	7–9		[349]
89		Proteinase	" subtilis W	28.8	(2 × 14)	7–9		[350]
90		Intracellular proteinase	" subtilis	30	1ch., s. dimer	7–9	Ca-dependent	[351, 352]
91		Protease I	Escherichia coli	21	1ch.	7–9	Periplasma	[353, 354]

No.	Name	Source	MW	Form	pH	Remarks	References
92	,, II		58	1ch, gl.	7–9	Cytoplasma	[355, 356]
93	,, IV		34	1ch, s.	7.5	Signal, outer membrane	[357, 358, 359]
94	,, V		(67)	tetramer associated with SDS	7–7.2	Transmembrane	[360]
95	,, VI		43	in SDS, s.	7.8	Membrane	[359]
96	,, VII		180	(2×70)	6	Product of *ompT* gene	[361]
97	,, So		140		6.5	Cytoplasma	[362, 363]
98	,, Re		82	1ch.	8	Periplasma and cytoplasma	[362]
99	,, Mi		110			Periplasma	,,
100	,, Fa		110			Cytoplasma	,,
101	Alkaline proteinases	Bacillus mesentericus	24	1ch, s.	7–9	Mesentericopeptidase	[364]
102	,, B	Aspergillus oryzae	17.7	1ch, gl.	10.5	Aspergillopeptidase B	[365, 366]
103	,, C	The same	19.6	1ch, gl.	10.5	,,	[365]
104	,,	Aspergillus flavus sojae	18	1ch.	7–10	C	[365, 367]
105	,,	Streptomyces griseus	25.7	1ch.	7–10		[365, 366]
106	Proteinases of "Pronase" complex A	The same	18	1ch, s.	8–10		[368, 369, 370]
107	B	,,	20	1ch, s.	8–10		[368, 369, 371]
108	C		16–18	1ch.	8–10		[368, 369]
109	D		27	1ch.	8–10		[372]
110	E		27	1ch.	8–10		,,
111	Extracellular proteinase A	Streptomyces rutgersensis	20	1ch, s.	9–10		[373]
112	The same D	The same	28	1ch.	8–10	Na-dependent 2 forms	[374]
113	Proteinase SSPB	Streptomyces spheroides	28		10–11	Fibrinolytic	[375]
114	,, K	Arthrobacter sp.	23	1ch.	6–9		[376, 377]
115	,, K	Tritirachium album	29	1ch, s.	8		[378, 379, 380]
116	Thermomicolin	Fungi Malbranchea pulchella	33	1ch, s.	8.5	Ca-dependent	[387]

Table 12. Continued

N	EC Number	Name of enzyme	Source	M_r, kDa	Structural data[a]	pH_{opt}	Comments	References[b]
117		Thermostable proteinase	*Thermomonospora*	14.5		9	t_{opt} 80°C	[382]
118		The same	*Thermus caldophilus*	75	gl.	7.2–7.8	At denaturation 7 forms M_r, 31–62	[383]
119		Aqualysin I	*Thermus aquaticus*	28.5	1ch., s.	10	Elastolytic	[384, 385, 386]
120		Bacillopeptidase F	*Bacillus subtilis*	33.5	1ch., gl., s.	7.5–8	2 forms, autolyzed	[387]
121		Proteinase	*Flavobacterium arborences*	19		8–10.5		[388]
122		Cuticle-degrading proteinase Pr-1	*Metarhizium anisopliae*	25	1ch.		Chymoelastase	[389]
123		,, Pr-2	The same	28.5	1ch.		Trypsin-like	,,
124		Proteinase	*Trichophyton rubrum*	90	(2 × 44)	8	Ca-activated	[390]
125		Proteinase of archea-bacteria	*Desulfurococcus* sp.	52	at gelfiltration M_r 13		Still active at 125 C!	[391]
126		,, cyanobacteria	*Anabena variabilis*	52			Ca-dependent	[392]
127		Proteinase	*Lactococcus lactis*	135	1ch., s.		In cell envelope	[393]
128		Thermitase	*Thermoactinomyces vulgaris*	35	1ch., s.	8–9	Inhibited by Hg	[394, 395, 396]
129	3.4.21.16	Proteinase	*Alternaria fenussima*	24	1ch.	8–10		[397]
130	3.4.21.18	Proteinase α	Insect *Tenebrio molitor*	24	1ch.	7–8		[398]
131		,, β	,, The same	60		7–8		,,
132		,,	Larva *Hypoderma lineatum*	27	1ch., s.	7–8.5	Hypodermin F	[399]
133		,,	,,	23	1ch., s.	8–10	Hypodermin D	[400]
134	3.4.21.19	Staphylococcal serine proteinase b	*Staphyllococcus aureus* V8	29	1ch., s.	7.7	Glu-specific	[401, 402, 403, 404]
135		The same a	,, 8325 N	12	1ch.	4 and 7.8	Product of "b" degradation?	[401, 405]
136		Glu, Asp-specific proteinase	*Streptomyces griseus*	20	1ch.	8.8		[406]
137		The same	,, *thermovulgaris*	25	1ch., s.	8.5		[407, 408]

No.	EC No.	Enzyme	Source	M.W.	Subunits/chains	pH	Notes	Ref.
138		Proteinase I	*Acremonas chrysogenum*	28	1ch., s.	10–11.5	Inhibited by Hg	[409]
139		"	Sunflower	25	1ch.	7.8	"	[410]
140	3.4.21.20	Cathepsin G	Human granulocyte	28	1ch., s.	7–8		[411, 412]
141		"	Leukocytes of animals	28	1ch.	7.5		[413, 414]
142		"	Pig neutrophils	28–29	1ch.	8–8.5		[415]
143	3.4.21.21	Coagulation factor VIIa	Bovine blood plasma	53	2ch. (24, 29) gl., s.			[416, 417, 418]
144	3.4.21.22	The same IXa	Human " "	44	2ch. (27, 17) gl., s.	7.5	Christmas factor	[323, 419, 420]
145	3.4.21.–	" Va	"	170	2ch. (94, 74) gl., s.	8	Subunits are linked by Ca	[421, 422, 423]
146	3.4.21.–	" VIIIa	"	160	2ch. (90, 69) gl., s.	8	von Willebrand factor	[424, 425, 426–428]
147	3.4.21.–	Protein C	Human and bovine blood plasma	62	3ch. (40, 21, 2), gl., s.	7	Ca and lipid-dependent	[429, 430, 431]
148	3.4.21.–	Factor C	Horseshoe crab hemocytes	123	3ch. (80, 34, 8) s.			[432, 433]
149	3.4.21.23	Snake proteinase	*Vipera russelli*	124	(2 × 60)	7–9	Ca-dependent	[434]
150	3.4.21.24	Neutral endopeptidase	Sheep red cells	340	(6 × 60)	8	Allosteric	[435]
151	3.4.21.25	Cucumisin	Melon *Cucumis melo*	50	1ch., gl.	9–11		[436]
152	3.4.21.26	Prolyl endopeptidase	Human plasma I	130	2 subunits, gl.	6		[437]
153		The same	The same II	62		6–7		"
154		"	Rat tissues, etc.	68–74		6–7.5		[438–440]
155		"	Bovine brain, etc.	74–77	1ch.	7	Inhibited by PCMB	[441–445]
156		"	Carrot	75		7.3		[446]
157		"	*Basidiomycetes*	76		6.8		[447]
158		"	*Flavobacterium meningosepticum*	74	1ch.	7		[448]
159	3.4.21.27	Coagulation factor XIa	Human blood plasma	130	2ch. (55, 55) gl., s.	7.5	Antihemophilic factor	[449]
160	3.4.21.28	Snake proteinase	*Agkistrodon rhodostroma*	34	1ch., gl.	8.5	"Ankord"	[450]
161	3.4.21.–	The same	" *contorix*	37	1ch., s.	8		[451, 452]
162	3.4.21.29	"	*Botrops atrox* I	29	1ch., gl., s.	7.4	"Batrachobin"	[453, 454]
163		"	II	31.4		7.4		[453]
164	3.4.21.30	"	*Crotalus adamanteus*	33.2	1ch., gl., s.	8	"Crotolase"	[455, 456]
165		"	" *atrox* II	31	1ch., gl.	7	Fibrinogenase 2	[457–459]
166		"	" III	24	(2 × 13)	7		"

Table 12. Continued

N	EC Number	Name of enzyme	Source	M_r, kDa	Structural data[a]	pH_{opt}	Comments	References[b]
167	3.4.21.–	,,	,, *horidus*	29.5	1ch., gl.	8.5	"Flavoscobin"	[460]
168	3.4.21.–	,,	*Trimersures flavoviridis*	23.5	1ch., s.	7.5		[461, 462]
169	3.4.21.31	Plasminogen activator	Mammalian kidney, urine, etc.	31 and 55	1 or 2 ch. s.	7.5	Urokinase, several forms	[463, 464, 465, 466, 467]
170		The same	Mammalian tissues	67	1ch., s.		Tissue activator	[463, 468]
171			Human fibroblasts	52	2ch. (39, 19)	7.5		[469]
172	3.4.21.32	Collagenolytic proteinase	Crab *Uca pugilator*	25.5	1ch., s.	8		[470, 471, 472]
173	3.4.21.33	The same	Fungi *Entomophthora coronata*	23	1ch.	7.2		[473, 474]
174		,,	Crab *Paralithodes camtschatic*	27	1ch.	7.9	3 Isoforms	[475]
175			*Parasilus asotus*	29.5	1ch.	7.5		[476]
176	3.4.21.34	Plasma kallikrein	Human blood plasma	87	3ch. (34, 28, 22), gl., s.	7.6	2 Forms	[477–482, 483]
177		,,	Bovine ,,	95	2ch.	8.8		[484]
178		,,	Pig ,,	88	1ch.	8		[485]
179	3.4.21.35	Tissue kallikrein	Human pancreas and urine	35	1ch., gl., s.	7		[486]
180		,,	Dog pancreas	27	1ch., gl., s.	7		[487]
181		,,	Rat ,,	30	1ch., gl., s.			[488, 489]
182		,,	Mouse submaxillary gland	29	1ch., gl., s.			[490, 491]
183	3.4.21.–	Seminin	Human seminal plasma	26	1ch., gl., s.	7–8	Prostate-specific antigen	[492, 493]
184	3.4.21.36	Pancreatic elastase	Mammalian pancreas	26–28	1ch., gl., s.	8–10		[494, 495, 496–498, 499]

No.	EC	Name	Source	MW	Structure	pI	Other names	References
185	3.4.21.-	Elastase II and III	The same	25	1ch., gl., s.	8–10		[299, 499–501]
186	3.4.21.-	Protease E	"	31	1ch., s.	8		[502–504, 505, 506]
187	3.4.21.37	Leukocyte elastase	Human leukocytes	30	1ch., s.	8–10		[507, 508, 509, 510]
188	3.4.21.-	Medullasin	Human bone marrow cells	32	1ch., s.	8.5		[511, 512, 513]
189	3.4.21.-	Elastolytic proteinase	Schistosoma mansoni	30	1ch., s.	9		[514, 515]
190	3.4.21.38	Coagulation factor XIIa	Human and other blood plasma	77	2ch. (48, 28) gl., s.	7.4	Ca-dependent Hageman factor	[417, 516, 517, 518, 519]
191	3.4.21.39	Chymase I	Human mast cells	30				[520]
192	"	"	Rat, mouse mast cells	26	1ch., s.	8.6–9	Intestinal tissue	[521–523, 524, 525]
193	"	" II	Mouse atypical mast cells	25	1ch., s.	8.6–9	Intestinal mucosa	[521, 522, 526–528]
194	3.4.21.-	Tryptase	Human lung mast cells	144	(2 × 37, 2 × 35)			[527, 529]
195	"	"	Other mammalian tissue	110–120	(4 × 30–36) gl., s.			[530, 531, 532]
196	3.4.21.-	Granzyme A	Human, mouse cytotoxic T-limphocytes	60	(2 × 35), s.	8–8.5	HuTSP, MTSP-I, BLT-esterase, H-factor, HPL	[533–537, 538–541]
197	"	", B(G, H)	The same	29	1ch., s.	8		"
198	"	", D, E, F	"	35–50	1ch., s.	8		"
199	3.4.21.-	Memsin	Walker sarcoma cells	32.5	1ch.			[542]
200	3.4.21.-	Adipsin	Adipocyte	28	1ch., s.			[543, 544]
201	3.4.21.40	Proteinase A	Mouse submandibular gland	30	1ch., s.	9.3	Esteropeptidase	[545–549, 550]
202	"	", B	The same	30.5	1ch., s.	8		"
203	"	", D	"	28	1ch., s.	8		"
204	3.4.21.-	Tonin	"	30	1ch., s.	6–7		[550, 551, 552]
205	3.4.21.-	Protease RSP-V	"	25	2ch. (19.5; 6)	ca. 10		[548]

Table 12. Continued

N	EC Number	Name of enzyme	Source	M_r, kDa	Structural data[a]	pH_{opt}	Comments	References[b]
206	3.4.21.41.	Complement subcomponent C1r	Human blood plasma	170	2ch. (56, 27) dimer, gl., s.	7		[553–555, 556, 557]
207	3.4.21.42	Complement subcomponent, ,, C1s	,,	83	2ch. (56, 27) s.	7.6		[558, 559, 560, 561]
208	3.4.21.43	Classical-complement-pathway convertase C3/C5	,,	280	2ch. (75, 35) gl., s.	7	Factors C4b; C2b, Mg-dependnet	[550, 555, 559, 562, 563, 564]
209	3.4.21.45	Complement factor I	,,	88	2ch. (50, 38) gl., s.	7	C3b-inactivator	[555, 565]
210	3.4.21.46	Complement factor D	,,	24	1ch., s.	7	Activated C3b+Mg	[553, 566–570]
211	3.4.21.47	Alternative-complement-pathway convertase C3/C5	,,	223	2ch. (63, 160) s	7	Factors C3b and Bb, Mg-dependent	[571, 572, 573, 574]
212	3.4.21.48	Proteinase B	Sacharomyces cerevisiae	32	1ch., gl., s.	6.5–7	Inhibited by PCMB	[575, 576–578]
213	3.4.21.–	The same	Candida albicans	30	1ch., gl.	6	The same, protease yscB	[579]
214	3.4.21.–	Proteinase	Human insulinoma cells	66	gl., s.		Similar to yscF	[580]
215	3.4.21.–	Proteinase yscF	Sacharomyces cerevisiae	130	1ch., gl., s.	5.5	Product of KEX-2 gene in [10c]-3.4.22.23	[581–583, 584]
216	3.4.21.49	Collagenolytic proteinase	Hypoderma lineatum	25	1ch., s.			[585, 586]
217	3.4.21.50	Proteinase I	Achromobacter lyticus	27	1ch., s.	8–9	Lys-specific	[587, 588, 589]
218	3.4.21.51	The same	Pseudomonas aeruginosa	30	1ch.	8–9		[590]
219	3.4.21.52	Neutral endopeptidase	Leukocyte membrane	75		7.4	,,	[591]
220	3.4.21.52	Cathepsin R	Rat liver ribosome	25–30	1ch.	7	,,	[592–594]
221	3.4.21.–	Cortical granule protease	Sea urchin eggs	47 (22?)	1ch.	8	Membrane Ca-activated	[595, 596]
222	3.4.21.–	Hatching enzyme	Xenopus leavis	62	gl.	7.7	Compare 3.4.24.12	[597]
223	3.4.21.–	The same	Paracentrotus lividus	52	gl.	8	,,	[598]

No.	EC number	Name	Source	M_r	Subunit	pH	Remarks	References
224	3.4.21.-	Tryase	Rat liver	32.5	1ch.	7.5	Has ATP-ase activity	[599]
225	3.4.21.53	Protease La	Escherichia coli	360	(4 × 87), s.	7–8		[600–602, 603, 604]
226	3.4.21.-	,, Ti	The same	340	(80, 12 × 23) s.	7–8	Also	[605, 606, 607]
227	3.4.21.-	,, Do	,,	520		6–8.5	"Ingensin", forms "proteosomes", inhibited by PCMB, compare 346	[362, 608]
228	3.4.21.-	Multicatalytic proteinase	Animal reticulocytes, bacteria	700–900	(ca. 20 × 25–30), s.	7–9		[609–613, 614–616]
229	3.4.21.-	ATP-dependent-proteinase	Rat, etc, mitochondria	ca. 600		8–9	Inhibited by vanadate	[617–619]
230	3.4.21.-	Ubikitin	In all eukariotes	8.5	1ch., s.	8	Component of ATP-dependent system of proteolysis	[620]
231	3.4.21.54	γ-Renin	Mouse submaxillary gland	26	In reduct. conditions M_r 7.4 and 4.4, s.	8		[621]
232	3.4.21.55	Gabonase	Bitis gabonica	30.6	1ch., gl., s.			[622]
233	3.4.21.56	Euphorbain d1	Euphorbia lathyris latex	117	(4? × 30)	6–8		[623, 624]
234	3.4.21.-	,, d2	The same	65 (43)	(2? × 30)	7		[623]
235	3.4.21.57	Plant Leu-proteinase	Spinach leaves					[625]
236	3.4.21.58	Prohormone serine proteinase	Pituitary, liver	ca. 170	2ch. (2 × 38, 50), s.	8.5	IRCM-I (Hepsin)	[626, 627, 628]
237	3.4.21.-	Processing protease	Rat blood plasma	38	1ch.	7	Cleaves pro-ANF	[629]
238		The same	,, small intestine	33	1ch.	7.5	Cleaves iso-citratde-hydrogenase, etc.	[630]
239		,,	The same	55		7.5	Cleaves prosomatostatin	[631]
240		,,	Neuroblastoma	28	1ch.	8.5	Cleaves dinorphin	[632]
241		,,	Human cerebrospinal fluid	40		6–8	,,	[633]
242		,,	Human erythrocytes	57	1ch.	7.2	Cleaves factor 3C	[634]
243		,,	Golden-brown alga	46	1ch.	6	Activates galactosil-transferase	[635]
244	3.4.21.-	Chromatin proteinase I	Rat liver	18	1ch.	10		[636, 637]
245		The same II	The same	27	1ch.	8		,,

Table 12. Continued

N	EC Number	Name of enzyme	Source	M_r, kDa	Structural data[a]	pH_{opt}	Comments	References[b]
246	3.4.21.–	Endopeptidase of nerve growth factor (γNGF)	Mouse submaxillary gland	26	1ch., s.	6	γ-Subunit of NGF	[638, 639]
247	3.4.21.–	Epidermal growth factor binding protein	The same	29	1ch., gl., s.	8		[640, 641]
248	3.4.21.–	Elastase-like proteinase	Platelet	26	1ch.	7	Latent form M_r 28	[642]
249	3.4.21.–	Chemotactic proteinase	Human skin	30	1ch.	7.8	Similar to Chimase 1	[643]
250	3.4.21.–	Proteinase	Sheep lung-lymph	70–75		7.5		[644]
251	3.4.21.–	Signal peptidase	Escherichia coli	37	1ch., s.	8	Nonsensitive to PMSF	[645, 646]
252	3.4.21.–	The same	Eukariotes	21	1ch., s.		Complex of 5 proteins	[647]
253	3.4.21.–	Prolipoprotein signal peptidase	Escherichia coli	18	1ch., s.			[648, 649]
254	3.4.21.–	Rec A proteinase	The same	40	1ch., s.		ATP-dependent	[650, 651, 652]
255	3.4.21.–	Viral proteinases	Adenoviruses type 2	19	1ch., s.	7		[653, 654]
256	3.4.21.–	The same	Sobemoviruses	22	1ch., s.			[655, 656]
257		"	Luteoviruses	22	1ch., s.			[656]
258		"	Alfaviruses	26	1ch., s.		Core protein	[657]
259	3.5.1.1	Asparaginase	Guinea pig	135	(4 × 35)	8		[658]
260		"	Bacteria	120–140	(4 × 32–35) s.	7–8	Escherichia coli – 2 forms	[659, 660, 661–663]
261		"	Yeast	400		8.5	Dimer (?) 2 Forms	[664, 665]
262	3.5.1.11	Pencillinamidase	Escherichia coli	70	(3–4 × 20)	7–9		[666–668]
263	3.5.2.6	β-Lactamase I	Bacteria	28–32	1ch., s.	7–8		[669–671, 672–675]
264	3.5.2.–	Cephalosporinase amp C	Escherichia coli K-12	39.6	1ch., s.			[676]

Cysteine amide hydrolases

N	EC Number	Name of enzyme	Source	M_r, kDa	Structural data[a]	pH_{opt}	Comments	References[b]
265	3.4.11.5	Proline iminopeptidase	Bacillus megaterium	58	1ch.	7		[677]

No.	EC	Enzyme	Source	Mol. wt.	Subunits	pH	Remarks	Ref.
266		The same	,, coagulans	40	1ch.	7.3		[678]
267		,,	Apricot seeds	220	(4 × 55)	7.5–8		[679]
268	3.4.11.6	Arginine aminopeptidase	Human muscle	72	1ch.	6.5	Aminopeptidase B, activated by Cl, 2 forms	[680]
269		The same	Human erythrocytes	95	(4 × 25)	7	Also, active as dimer	[681]
270		,,	Pig liver	58	1ch.	7.5	Also	[682]
271	3.4.11.–	,, I	Streptococcos mitis	62	1ch.	8.6	Inhibited by Cl	[683]
272		,, II	The same	360		7.6		
273	3.4.11.–	Bleomycin hydrolase	Rabbit lung	250	(5 × 50), s.	7.4		[684]
274	3.4.11.–	Aminopeptidase	Bovine kidney	56	(4 × 14)	7		[685]
275	3.4.11.–	,, I	Guinea pig small intestine	68		7.6		[686]
276		,, II	,,	72		7.6		
277	3.4.13.9	Proline dipeptidase	Human liver	54	Dimer (?) gl., s.	8	Prolidase	[687]
278		,,	Pig kidney	ca. 150		8	Mn-dependent	[688]
279		,,	Bovine small intestine	116	(2 × 56)	7.2		[689]
280		,,	Guinea-pig brain	132	2ch. (64, 68)	8		[690]
281	3.4.13.10	β-Aspartyldipeptidase	Rat, etc. liver	120		7.5		[691, 692]
282		,,	Escherichia coli	120		7.4–8	Activated by NaCl	[693]
283	3.4.13.–	Lysosome dipeptidase II	Rat liver	120		4.5		[694]
284	3.4.14.1	Dipeptidyl peptidase I	Bovine spleen, etc.	210	(8 × 25) gl., s.	6	Cathepsin C	[695, 696]
285	3.4.14.–	Dipeptidyl peptidase V	Rat brain	64 in SDS		7	Activated by Mn	[697]
286	3.4.16.3	Tyrosine carboxypeptidase	Pig thyroid gland			4.1	In [10]-serine enzyme	[698]
287	3.4.17.8	Muramoyl-pentapeptide carboxypeptidase	Escherichia coli	65	(2 × 32)	7–8	3 Forms A, B, C	[699]
288	3.4.18.1	Lysosome carboxypeptidase B	Bovine spleen, etc.	52	(2 × 25), gl.	5	Cathepsin D2	[700–702]
289	3.4.19.3	5-Oxoprolyl-peptidase	Bovine, etc. liver, brain	24–33	1ch.	6–8	Pyroglutamyl aminopeptidase 1	[703, 704]
290		,,	Pseudomonas fluorenses, etc.	70–80	(3 × 24)	7–8		[705–709]
291	3.4.22.1	Cathepsin B (B1)	Human liver	24	2ch., (22.5; 5), gl., s.	6	Has exopeptidase activity	[709, 710–712]
292		,,	Mammalian lysosomes	25–29	2ch. (23–25, 5), gl., s.	6–7	Also	[713–719, 720–723]

Table 12. Continued

N	EC Number	Name of enzyme	Source	M_r, kDa	Structural data[a]	pH_{opt}	Comments	References[b]
293	"		Chicken liver	30	2ch. (25, 5)			[724]
294	3.4.22.–	Proteinase	Mouse tumor	33	1ch.	5.5–7	Latent form M_r 40	[716, 719]
295	3.4.22.2	Papain	Melon tree Carica papaya	23.4	1ch., s.	6		[725, 726]
296	3.4.22.3	Ficin	Latex Ficus glabrata, etc.	23.8	1ch., gl., s.	6.5–9.5		[727, 728, 729]
297	3.4.22.4	Bromelain	Pineapple fruit	18	1ch.	8.3		[730]
298		"	Pineapple stem	35	1ch., gl., s.	5–6		[730, 731, 732]
299	3.4.22.–	Calotropin	Latex Calotropis gigantea	24	1ch.	7.8–8	Several forms	[733]
300	3.4.22.–	Ananain	The same	25	1ch., s.	6.8		[734]
301	3.4.22.6	Chymopapain A	Melon tree	25	1ch., s.	7.2		[735, 736–738]
302		„ B	The same	28 (35.5)	1ch.	6–7		[736–738]
303		„ S	„	24.7	1ch.	6–7		[739]
304	3.4.22.–	Peptidase A	„	24	1ch., s.	5–8	Proteinase Ω	[736, 738, 740, 741]
305		„ B	„	20.7	1ch.	6	Proteinase III	[736, 738, 740]
306	3.4.22.–	Proteinase IV	„	24.5	1ch.	7		[742]
307	3.4.22.8	Clostripain	Clostridium hystolytica	55.5	2ch. (43; 12.5), s.	7–7.2	Activated by Ca	[743, 744]
308	3.4.22.10	Streptococcal proteinase	Streptococcus sp.	32	1ch., s.	7–8		[745, 746, 747]
309	3.4.22.12	γ-Glutamyl hydrolase	Bovine, etc. liver	108	1ch., gl.	4.5	Carboxypeptidase G	[748]
310		The same	Rat small intestine	80	1ch.	4.5		[749, 750]
311	3.4.22.13	Staphylococcal cysteine proteinase	Staphylococcus aureus V8	12.5	1ch.	8.8	Proteinase II	[405]
312	3.4.22.14	Actinidin	Gooseberry Actinidia chinensis	26	1ch., s.	5–6.5		[751, 752]

No.	EC	Enzyme	Source	M_r	Subunits	pH	Notes	Refs
313	3.4.22.15	Cathepsin L	Human liver	30	2ch. (25, 5) gl., s.	6–7	Several forms	[753, 754]
314	„	„	Rat liver and other lysosomes	23–24	1ch., s.	6–7	Also	[755–758, 759–761]
315	3.4.22.–	Homoglobin-hydrolyzing proteinase	Rat liver	20	1ch.	3.5		[762]
316	3.4.22.16	Cathepsin H	Human, rat lysosomes	28	1ch., gl., s.	5.5–6.8	2 Forms	[721, 763–765]
317	3.4.22.17	Calpain I	Different animal tissues	120	2ch. (80, 30) s.	7.5–8	Activated by μM Ca concentration	[766–770, 772, 773]
318		„ II	The same	120	2ch. (80, 30) s.	7.6	Also, but mM Ca	„
319		„ III	„	78–94	(94-cDNA) (2 × 67)	7.2		[771, 774]
320	3.4.22.18	Prolyl endopeptidase	Human placenta	140		6.7		[775]
321		„	Adrenohypophysis	70		7.5		[776]
322	3.4.22.19	Enkephalin-converting enzyme	Bovine adrenal chromaffin granules	220		5.5		[777]
323	3.4.22.20	Dinorphin-converting enzyme	Rat brain			8	Membrane	[778]
324	3.4.22.21	Proteinase ysc E	Sacharomyces cerevisiae	ca. 600	(? × 70)	8.5		[779]
325	3.4.22.22	Proteinase ysc D	„	83		6.5–7		[780]
326	3.4.22.–	Proteinase	Mouse macrophages	21	1ch., s.		Similar to cathepsin H	[781]
327	3.4.22.–	Cathepsin N	Human placenta	34.6	1ch.	3.5	Collagenolytic cathepsin	[757, 782, 783]
328	3.4.22.24	Cathepsin S	Bovine spleen	23–25	1ch.	4–6		[757, 759, 784, 785]
329	3.4.22.–	Cathepsin T	Rat kidney	33–35	1ch.	5.8–6.5		[786, 787]
330	3.4.22.–	Cathepsin J	Human tissues	230		6.2–6.8		[788]
331	3.4.22.–	Cathepsin K	The same	650		6.2–6.8		„
332	3.4.22.–	Aleurain	Barley	37	1ch., s.	3.5		[789]
333	3.4.22.–	Proteinase I	Silkworm eggs	160		3.5		[790]
334		„ II	„	68				„
335	3.4.22.–	Leucoergosin-generating enzyme	Inflammatory rabbit skin	14	Aggregated	7.1		[791]
336	3.4.22.–	Cytosol insulin-glucagon degrading peptidase	Rat tissue	80		7.5	Insulinase	[792, 793]

Table 12. Continued

N	EC Number	Name of enzyme	Source	M_r, kDa	Structural data[a]	pH_{opt}	Comments	References[b]
337	3.4.22.–	Proteinase	Rabbit carcinoma	68	gl.	6.5	Activates coagulation factor X	[794]
338	3.4.22.–	,,	Xenopus laevis skin	30	1ch.	5–6		[795]
339	3.4.22.–	Protease	Serratia marcescens	73		7.5–8	Cleaves IgG	[796]
340	3.4.22.–	Proteinase I	Dictostelium discoideum	36	1ch., gl., s.	4.5	Contains phosphorus	[797, 798]
341	3.4.22.–	Alkaline proteinase	Micropogon opercularis	363		9.1	t_{opt} 60 °C	[799]
342	3.4.22.–	Tissue endopeptidase	Animal mitochondria	56–95	1ch.	7	"PZ"-peptidase; possibly identical to 3.4.22.19	[800–804]
343	3.4.22.–	Neutral endopeptidase	Pituitary cytosol	>100		7–7.5	Cation-sensitive	[805]
344	3.4.22.–	Hydrolase H	Rabbit muscle	340	3 (4 × 51, 72, 92) $\alpha_2\beta\gamma$	7.5–8	Aminoendopeptidase	[806]
345	3.4.22.–	Prorenin-processing protease	Human kidney	27	1ch.	4.8–5.6		[807]
346	3.4.22.–	Macropain	Reticulocytes, etc.	ca. 600	(? × 30)	7.6–9.8	Probably identical to multicatalytic proteinase (N228)	[808–812]
347	3.4.22.–	ATP-dependent protease	,,	ca. 600			See N229	[813, 814]
348	3.4.22.–	Viral proteinases	Comoviruses	24	1ch., s.			[654, 656, 815, 816]
349		The same	Picornaviruses 3C	22	1ch., s.			[654, 816, 817]
350		,,	,, 2A	17	1ch., s.			[656, 818]
351		,,	Nepoviruses	22	1ch., s.			[654, 656]
352		,,	Alfaviruses	26	1ch., s.		Hydrolyzes nonstructural proteins	[819]
353	3.5.1.14	Aminoacylase	Pig kidney	86	(2 × 43)	8–9	Acylase I, dehydropeptidase II	[820, 821]
354	3.5.1.30	5-Aminopentanamidase	Pseudomonas putida	67	1ch.	7.5–8.5		[822]

No.	EC	Name	Source	MW	Chains	pH	Remarks	Ref.
355	3.5.1.31	Formylmethionine deformylase	Small intestine of rats and others	ca. 50		8–9		[823]
356	3.5.1.42	Nicotinamide-nucleotide amidase	Azotobacter viaelandii	26		6.5–7		[824]
357	3.5.1.43	Peptidyl-glutaminase	Bacillis circulans	90		8	Peptidoglutaminase I	[825]
358	3.5.1.44	Protein-glutamine glutaminase	The same	125		7.5	,, II	[825]
359	3.5.1.55	Acetylglutamatdeacylase	Pseudomonas diminuta	300		8		[826]
360		,,	Mycobacterium smegmatis	40–48		5.7	Hydrolyzes long-chain acylderivatives	[827]
361	3.5.1.60	N-Acylethanolamine amidohydrolase	Rat liver			9		[828]
362	3.5.1.–	Ascamycin hydrolysing amidase	Xanthomonas citri	38		7.5–8	Activated by Mn	[829]

Aspartic amide hydrolases

No.	EC	Name	Source	MW	Chains	pH	Remarks	Ref.
363	3.4.23.1	Pepsin A	Human stomach	34	1ch., s.	1.9	4 Forms	[830, 831, 832, 833]
364		,,	Other mammalian stomach	34–42	1ch., s.	2–4	Several forms	[830, 834–836, 837–840]
365		,,	Dogfish stomach	35	1ch.	2–4	3 Forms	[842]
366		,,	Chicken ,,	35	1ch., s.	2–3		[843, 844, 845]
367	3.4.23.2	Pepsin B	Pig ,,	38.6	1ch.	2–4	Gelatinase parapepsin I	[834]
368	3.4.23.3	Pepsin C	Human ,,	31	1ch., s.	2–3	Gastricsin	[846]
369		,,	Stomach of pig and other	32	1ch., s.	3		[841, 847, 848]
370	3.4.23.–	Uropepsin	Human urine	33	1ch., s.	2		[849]
371	3.4.23.4	Chymosin A and B	Calf	30.5	2ch., s.	3.5	2 Forms I, II	[850, 851]
372		,, C	,,	30.5			"Rennin"	[852]
373		The same	Cat	ca. 30	1ch., s.	2.5		[853]
374	3.4.23.5	Cathepsin D	Human lysosomes	42	2ch. (27, 14) gl., s.	4.5–5	3 Forms	[854, 855]

Table 12. Continued

N	EC Number	Name of enzyme	Source	M_r, kDa	Structural data[a]	pH_{opt}	Comments	References[b]
375		The same	Different mammalian organs	40–50	1ch., [pig spleen: 2ch. (35, 15)] gl., s.	3–4.5	Multiple forms	[856–860, 861–863]
376		,,	Rabbit ovary	42	(2 × 20)	3.8		[864]
377		,,	Porcine adrenocortical gland	ca. 65	3ch. (27, 24, 14), gl.		5 Isoenzymes	[865]
378		,,	Chicken muscle	36	1ch.	3.5		[866]
379	3.4.23.-	Cathepsin D-like proteinase	Human and other gastric mucosa	86 (rat)	(2 × 42)	3	"Slow" cathepsin	[867–869]
380		The same	Pig small intestine	60	(2 × 30)	3.2		[870]
381	3.4.23.-	Cathepsin E	Mammalian neutrophiles	ca. 100	(2 × ca. 50) gl., s.	2.5–3.2		[871–874, 875]
382	3.4.23.-	Seminal pepsin	Human seminal plasma	35	1ch.	3	In plasma – proenzyme	[876]
383	3.4.23.-	Proteinase	Pig uterine	41	1ch.	3.5		[877]
384	3.4.23.6	Microbial aspartic proteinases	Aspergillus orysae	39.4	1ch., gl.	3.2–3.6	"Takadiastase"	[878, 879]
385		The same	,, saitoi	34.5	1ch.	2.7	Aspergillopeptidase A	[880]
386		,,	,, awamori	35	1ch., s.	2.3–2.6	Avamorin	[881, 882]
387		,,	,, niger var. macrosporum	136	(2 × 60), gl.	3	Protease A, not inhibited by pepstatin	[883]
388			Penicillium janthinellum	32	1ch., s.	3–4	Penicillopepsin	[884, 885]
389			,, roqueforti	33.4	1ch., gl.	3.5		[886]
390			Rhizopus chienensis	35	1ch., s.	2.9–3.3	Rhizopuspepsin, 3 isoenzymes	[887, 888–890]
391		,,	,, niveus	35	1ch., s.	3		[891]
392		,,	Sacharomyces cerevisiae	45	1ch., gl., s.	2.4–3.3	Proteinase A	[892–894, 895]
393		,,	Candida albicans	45	1ch.	2–3.9		[896]
394		,,	Endothia parasitica	34.5	1ch., s.	2.5	Endothiapepsin	[897, 898]
395		,,	Mucor miehii	38	1ch., gl., s.	4		[899, 900]

	EC	Name	Source	MW	Chain	pH opt.	Notes	References
396		,, pusillus		40	1ch, gl., s.	ca. 5		[901, 902]
397		,,	Myxococcus xanthus	45	1ch.	2.3–2.8		[903]
398		,,	Trichoderma sp.	32	1ch.	2	Thermopsin, t_{opt} 90 °C	[904]
399		,,	Sulfolobus acidocaldarius	45	1ch, gl., s.	5.5		[905]
400		,,	Tetrahymena pyriformis	29.3		3–3.6		[906]
401			Plasmodium berghei			2.2		[907]
402	3.4.23.–	Russulin	Russula	35	1ch.	2.2		[908]
403	3.4.23.–	Protease B	Scytalidium lignicolum	43	1ch., s.	3–3.5		[909, 910, 911]
404	3.4.23.–	"Barrier" proteinase	Saccharomyces cerevisiae	62	gl., s.		Product of BAR1 gene	[912]
405	3.4.23.11	Thyroid aspartic proteinase	Bovine thyroid gland	21	Dimer	3.6		[913]
406	3.4.23.–	Pro-opiomelanocortin converting enzyme	Bovine pituitary	70	gl.	4–5		[914, 915]
407	3.4.23.–	Myelin endopeptidase	Human nervous tissue	10–16	1ch.	4	Membrane	[916]
408	3.4.23.12	Aspartic proteinase of insectivorous plants	Nepenthes, Drosera peltata			2–3		[917]
409	3.4.23.13	Aspartic proteinase of Lotus seeds	Lotus	36.8	1ch.	2–3	Stabilized by Cu	[918]
410	3.4.23.14	Aspartic proteinase	Sorghum vulgare	80		3.6		[919]
411	3.4.23.–	The same	Buckwheat seeds	27.8	1ch.	3.5	2 Forms	[920]
412	3.4.23.–	,,	Wheat seeds	58		4.5		[921]
413	3.4.23.–	,,	Tomato	37	1ch.	2.5–3.5		[922]
414	3.4.23.15	Renin	Human and other kidney	36–44	1ch. (or 2ch.) gl., s.	5.5–7	High molecular form (ca. 65)-complex with renin-binding protein	[923–928, 930–932]
415		The same	Bovine pituitary	36	1ch., s.	7	Do not bind lectins	[927]
416		,,	Mouse submaxillary gland	40				[929]
417	3.4.23.–	Proteinase of erythrocyte membrane (EMAP) ,, I	Human erythrocytes	80	1ch.	2.8–3.9	Similar to cathepsin E	[873, 933–936]
418		,, II	,,	40	1ch.		Also	,,
419		,, III	,,	30	1ch.			,,

Table 12. Continued

N	EC Number	Name of enzyme	Source	M_r, kDa	Structural data[a]	pH_{opt}	Comments	References[b]
420		"	Rat erythrocytes	80–82		3.5–4	Is active up to pH 9	[937]
421	3.4.23.–	γ-Glutamyltranspeptidase (light chain)	Rat kidney	22	1ch.			[938]
422	3.4.23.–	Aspartic proteinases of retroviruses	HIV-I	11	1ch., s.	5.5	Active as dimer	[939, 940–942]
423		The same	HTLV-I	12	1ch., s.			[939, 943]
424		"	HTLV-II	12	1ch., s.			[939, 944]
425		"	Mouse leukemia virus	13.3	1ch., s.			[939, 945]
426		"	Bovine " "	14	1ch., s.			[946]
427		"	Avian sarcoma virus	15	1ch., s.			[947]
428	3.5.1.38	Glutamin-(asparagin)ase	Achromobacter sp.	140	(4×35)	7		[948]
429		The same	Acinoetrobacter sp.	35.5	1ch., s.			[949]

Metallo-amide hydrolases

N	EC Number	Name of enzyme	Source	M_r, kDa	Structural data[a]	pH_{opt}	Comments	References[b]
430	3.4.11.1	Cytosole aminopeptidase	Kidney, liver, etc. (bull, pig, etc.)	270–300	(4×64–66) s.	9–9.5	Leucine aminopeptidase, Zn, activated by Mg	[950–953]
431		The same	Bovine lens	360	(6×54–60) s.	8.5–9	Also	[954, 955, 956]
432	3.4.11.–	"	Sacharomyces cerevisiae	610	(12×51), s.	7–8.5		[957, 958]
433		"	Lactobacillus acidophilus	156	(4×40)	6–7	Zn	[959]
434	3.4.11.2	Microsomal aminopeptidase	Human and other small intestine	210–280	(2×100–140) s.	8	Zn, Aminopeptidase M (N)	[960–964, 965]
435		The same	Human and other kidney, liver	235	(2×120) gl., s.	8	Membrane, Zn	[226, 966–968, 969]
436		"	Bovine, rat brain	115		6.5–7	Cleaves enkephalin, Zn	[970, 971]
437		"	Human, etc. skeletal muscle	100–120		7.3	Mn, dimer, polymer	[972, 973]
438		"	Fibroblasts	150		7.3		[974]

No.	EC	Enzyme	Source	MW	Subunits	Notes	pH	References
439	3.4.11.3	Cystyl-aminopeptidase	Human blood plasma	280	(2×140), gl.	Zn, Oxytocinase	7.3	[975, 976]
440	3.4.11.4	Tripeptide aminopeptidase	Enterocytes and other animal tissues	50–70	1ch., gl.	Zn, Mn	7.8	[977, 978]
441	3.4.11.7	Aspartate aminopeptidase (aminopeptidase A)	Human blood plasma	260	(2×130)	Activated by Ca	6.7–7.2	[976, 979, 980]
442		The same	Pig kidney	ca. 400	4ch. (155, 110, 90, 45) gl.	Also	7.5	[962, 981, 982]
443		„	Rabbit kidney	170	1ch.		7	[983–985]
444		„	Bovine kidney	65	(2×33)	„	7	[986]
445	3.4.11.9	Aminopeptidase P	Pig kidney, small intestine	>1000		Membrane, Mn	7.8	[987–989]
446		„	Escherichia coli	200	(4×50), s.	Mn	8.6	[990, 991]
447	3.4.11.10	Aminopeptidase	Aeromonas proteolytica, etc.	25–32	1ch.	Zn	7–10	[992–995]
448		„	Streptomyces griseus	25	1ch.	Ca, 2 forms	7.5–10	[996, 997]
449		„	Salmonella typhimurium	34	1ch., s.	Ca, peptidase M		[998, 999]
450		„	Bacillus subtilis	46.5	1ch.	Zn, activated by Co		[1000]
451		„	Escherichia coli	87	1ch., s.			[1001, 1002, 1003, 1004]
452	3.4.11.11	Aminopeptidase	Human blood	80–90		Mn, "peptidase a"	8	[1005]
453		„	„ brain	61.5 (in SDS)		Co, "enkephalinase"	7.5	[1006, 1007]
454		„	Rat „	ca. 100	2ch. (62, 66)	Zn, Co, also	7–8	[1008, 1009]
455		„	Bovine „	105	1ch.	Zn; soluble	7.2	[1010]
456	3.4.11.12	Thermophilic aminopeptidase	Bacillus stearo-thermophilica	400	(12×36.5)	Ca (?)	7	[1011]
457		„	„ I	46	(2×23)		6.9	„
458			„ II	47.5	(2×24)		6.4	„
459			„ III	400		Co, t_{opt} 65°C		[1012]
460			Telaromyces duponti	62		Co, Mn		[1013]
461			Streptomyces thermophilica I	450	Tetramer	Zn	7.5	[1014]
462			Altermonas proteolytica II	88	Monomer	Zn	8.2	„
463			„ III	63	(2×33)	Zn, t_{opt} 70°C	9.3	[1015]
464	3.4.11.13	Aminopeptidase	Clostridium hystolytica	340	(6×56)	Mn, Co	8.6	[1016]
465		„	Escherichia coli	323	(6×52)	Mn, Mg	9–11	[1017]

Table 12. Continued

N	EC Number	Name of enzyme	Source	M_r, kDa	Structural data[a]	pH_{opt}	Comments	References[b]
466	3.4.11.14	Aminopeptidase	Human liver	360	2ch. (53, 65) gl.	8	Zn	[1018–1020]
467	3.4.11.-	"	Pig "	96		6.5–7	Co	[1021]
468			Bovine lens	96	1ch.	6	Zn, Cu	[1022]
469	3.4.11.15	Aminopeptidase Co	Yeast	ca. 100			Co	[1023]
470	3.4.13.3	Aminoacyl-histidin dipeptidase	Pig kidney	84	1ch.	7.6	Mn, "Carnosinase"	[1024]
471		"	Rat brain	85	1ch.	7.8	Mn, also	[1025]
472	3.4.13.4	Aminoacyl-lysin dipeptidase	"			6.5	Mn, Mg	[1026]
473	3.4.13.5	Aminoacyl-methylhistidin dipeptidase	Rat tissues					[1027]
474	3.4.13.6	Cysteinyl-glycin dipeptidase	Pig kidney			7.3	Mn	[1028]
475	3.4.13.7	α-Glutamyl-glutamate dipeptidase	The same					[1029]
476	3.4.13.8	Prolyl dipeptidase	"	300	Aggregated	8–7.5	Mn, Prolinase	[1030]
477		The same	Bovine kidney	100	1ch.	8.7	Mn	[1031]
478	3.4.13.11	Microsomal dipeptidase	Human "	130	(2 × 60), gl.	8	Zn	[1032]
479		The same	Pig "	87–93	(2 × 47)	7.6	Zn, Co	[1033]
480		"	Rat lung	150	(2 × 80)	8		[1034]
481		"	Mouse ascyte tumor	85	1ch.	8.2	Zn	[1035]
482			Escherichia coli B	100	(2 × 53)	8	Zn	[1036]
483	3.4.13.-	Dipeptidase	Human pancreas	135	(2 × 68)	9	Zn (2 atoms)	[1037]
484	3.4.13.12	Methionyl dipeptidase (dipeptidase M)	Escherichia coli B	93–95	(2 × 45)	7.8–9	Mn	[1038]
485	3.4.13.13	Homocarnosinase	Pig kidney	57	1ch.	7.2	Co	[1039]
486	3.4.13.14	γ-Glutamyl dipeptidase	The same					[1040]
487	3.4.13.15	N^2-β-Alanyl-arginine dipeptidase	Rat brain	89	1ch.	8.3–9.4	Mn	[1025]

No.	EC	Enzyme	Source	MW (×10³)	Structure	pH	Metal / Notes	Ref.
488	3.4.13.16	Aspartyl-phenylalanine dipeptidase	Pig kidney	130	1ch.	8	Zn, aminopeptidase W	[1041]
489	3.4.15.1	Dipeptidyl carboxy-peptidase I	Human tissue and blood plasma	170	1ch., gl., s.	7.5	Zn, angiotensin converting enzyme, Cl-activated	[1042, 1043–1046]
490		„	Bovine and other kidney, lung	80–190	1ch., gl., s.	7.5	Also	[1047–1051]
491		„	Bovine seminal plasma	460		8.3	Zn	[1052]
492	3.4.15.3	Dipeptidyl carboxy-peptidase II	Escherichia coli	97	1ch.	8.2	Co, Zn	[1053]
493	3.4.17.1	Carboxypeptidase A	Human, bovine, etc. pancreas	35	1ch., s.	7.5–8.5	Zn	[1054, 1055, 1057]
494		The same	Rat skeletal muscle I	39.3	1ch.	7.5–8.5	Zn	[1058]
495		„	The same II	37.8	1ch.			„
496	3.4.17.–	„	Mouse mast cells	36	1ch., s.	7	Zn	[1056]
497	3.4.17.2	Carboxypeptidase B	Human pancreas	34.2	1ch.	7	Zn, 2 forms	[1059]
498		„	Bovine, etc. pancreas	34–37	1ch., s.	7.5	Zn	[1060, 1061, 1062]
499	3.4.17.3	Arginine carboxypeptidase (carboxypeptidase N)	Human blood plasma	280	2ch. (2 × 50, 2 × 90), gl., s.	7.5	Zn, kininase I	[1063, 1064, 1065, 1066]
500	3.4.17.–	Carboxypeptidase	„ urine	75.6 (in SDS)		7	Co	[1067]
501	3.4.17.4	Glycine carboxypeptidase	Sacharomyces cerevisiae			6–6.2	Zn, Co	[1068]
502	3.4.17.5	Aspartate carboxypeptidase	Pseudomonas sp.			6–9	Zn, Co (?)	[1069]
503	3.4.17.6	Alanine carboxypeptidase	Soil bacteria			7–8	Zn	[1070]
504	3.4.17.7	The same	Bacillus subtilis			7	Mn	[1071]
505		Acetylmuramoyl-alanine carboxypeptidase	Bacillus sphaericus	90	(2 × 45)	7.5–8	Zn, Co	[1072]
506	3.4.17.8	D-Alanine carboxypeptidase	Streptom. albus G	22	1ch., s.	7.5	Zn	[1073, 1074]
507	3.4.17.9	Carboxypeptidase S	Sacharomyces cerevisiae			5.5	Zn, Co	[1075]
508	3.4.17.10	Carboxypeptidase H(E)	Human and other brain	50	1ch., gl., s.	5–6	Co	[1076–1078]
509		„	Insulin-secretory granules	55	1ch., gl.			[1079]

Table 12. Continued

N	EC Number	Name of enzyme	Source	M_r, kDa	Structural data[a]	pH_{opt}	Comments	References[b]
510		"	Human liver tumor	55 and 57	1ch.	5–6	Zn	[1080]
511	3.4.17.–	Carboxypeptidase M	Human placenta	62	1ch., gl. s.	7	Co	*[1081]*
512	3.4.17.–	" T	Thermoactinomyces sp.	38	1ch	7–8	Zn, t_{opt} 70 °C	[1082]
513	3.4.17.–	" L	Streptomyces griseus	41.2	1*ch.*, s.		Zn, Ca (2 atoms)	*[1083]*
514	3.4.17.–	" L	Steptomyces spheroides	33	1ch.	7.6	Activated by Ca	[1084]
515	3.4.17.–	" P	Pig kidney I	135	1ch.		Zn, Mn	[1085]
516		" "	The same II	230	2ch. (128, 95)		Zn, Mn	"
517	3.4.17.–		Crayfish Astacus fluviatilis	35–36	1ch., s.	6.5	Zn	*[1086]*
518	3.4.19.–	Pyroglutamylpeptidase II	Guinea pig brain	230			Cleaves thyroliberine	[1087]
519	3.4.24.1	Metalloproteinase	Crotalus atrox I	20	1ch.	7	Zn	[459]
520		"	The same IV	46	1ch., gl.	7	Zn	
521		" "e"	" "e"	25.7	1ch., gl., s.	8–9	Zn, toxine "e"	" *[1088–1090]*
522	3.4.24.–	"	Crotalus horridus horridus	56		8	Zn, activated by Ca	[1091]
523	3.4.24.–	"	" adamanteus I and II	24	1ch.	9–10	Zn (Ca)	[1092]
524	3.4.24.–	"	Trimeresurus flavoviridis	23	1ch., s.		Zn, HS-proteinase	*[1093]*
525	3.4.24.2	"	Sepia gastric juice	21				[1094]
526	3.4.24.3	Collagenase	Clostridium hystolyticum A	105	(4 × 25)	7.5	Zn, Ca-dependent	[1095, 1096]
527	3.4.24.4	Bacterial metalloproteinases	Aeromonas proteolytica	34.8	1ch.	7	Zn	[1097]
528		The same	Aspergillus orysae	41	1ch.	5.6	Co	[1098]
529			Bacillus brevis	35		6.7	Zn, t_{opt} 60 °C, 2 Forms	[1099]
530			" subtilis A	35.1	1ch., s.	6.5–7.5	Zn, Ca	[1100, 1101]
531			" thermoproteolyticus	37.5	1ch., s.	8	Zn, Ca; thermolysin	[1102, 1103]
532		"	" stearothermophilus	36	1ch., s.	8	Zn, Ca	*[1104]*
533		"	" thuringiensis	78		6.5	Zn, "inhibitor A"	[1105]

No.	EC / Name	Source	MW	Chains	pH	Metal / Notes	Ref.
534		„ amyloliquefaciens	33	1ch., s.	8	Zn, Ca	[1106]
535	„	„ mesentericus	33	1ch., s.	8	Zn, Ca	[1107]
536	„	Escherichia coli	110	1ch., s.	7.4	Zn, Co, protease Pi, protease III	[1108, 1109]
537	„	The same	75	1ch.	8.5–9		[1110]
538	„	Escherichia freundii	45.4	1ch.	10	Zn	[1111]
539	„	Legionella pneumophila	38		6.7	Zn	[1112]
540	„	Penicillium roqueforti	20	1ch.	5.5	Co	[1113]
541	„	Pseudomonas aeruginosa	33	1ch., s.	8	Elastase	[1114, 1115]
542	„	The same	48.5	1ch.	7–9	Co, Zn	[1116]
543	„	Serratia sp.	46	1ch., s.		Zn	[1117, 1118, 1119]
544	„	Staphylococcus aureus	28	1ch.	7.4	Protease III	[1120]
545	„	„ cremoris	80	(2 × 40)		Zn	[1121]
546	„	„ faecalis	31.5	1ch.	6–8	Zn	[1122]
547	„	Streptomyces moderatus	21	1ch.	9	Zn	[1123]
548	„	Thermus sp.	21	1ch., gl.	8	Zn, "Caldolysin"	[1124]
549	„	Trichophiton granulosum	34.3	1ch.	9.5	Zn	[1125]
550	3.4.24.5 Neutral proteinase	Bovine lens	250	Dimer	7.5	Ca, Mg	[1126]
551	3.4.24.6 Peptidase A	Agkistrodon venom	22.5	1ch.	8.5	Zn (1 atom), Ca (2 atoms)	[1127]
552	3.4.24.7 Vertebrate collagenase	Human gastric mucosa	38	1ch.	7.5–8.5	Ca	[1128, 1129]
553	„	Human fibroblasts	52	1ch., gl., s.	7	Ca, oncogene induced	[1130]
554	„	Rat uterus	50		7	Ca	[1129, 1131]
555	„	Bovine dental sac	34		7.8	Ca	[1095, 1132]
556	3.4.24.8 Collagenase	Achromobacter iophagus	70	(2 × 35)		Zn	[1133–1135]
557	3.4.24.9 „	Trichophiton schoenlenii	20	1ch.	6.5	Ca, Mg	[1136]
558	3.4.24.– „	Planarian Bipalium kewense	47	(2 × 25)			[1137]
559	3.4.24.– Proteoglycan-degrading proteinase	Human skin	120–150	gl.	7–7.5	Ca	[1138]

Table 12. Continued

N	EC Number	Name of enzyme	Source	M_r, kDa	Structural data[a]	pH$_{opt}$	Comments	References[b]
560		"	Rat tumor cells	30–34	1ch.	7	Ca, Zn	[1139, 1140]
561		"	Synovial fibroblasts and other tissues of rabbit, rat, etc.	21–28	1ch., gl., s.	5–9	"Stromelysin" in [10c] 3.4.24.17	[1141, 1142, 1143]
562		"	The same of human	45	1ch., gl., s.			[1144–1146]
563	3.4.24.–	Gelatinase	Rabbit bone	65	1ch.	7		[1147]
564	3.4.24.10	Keratinase	Trichophiton mentagrophytes	48		8		[1148]
565	3.4.24.11	Membrane metallo-endopeptidase	Human and other kidney, intestinal mucosa	90–93	1ch., gl., s.	6–7	Zn	[1149–1153, 1154–1156]
566		"	Mouse kidney	320	(4×81), gl.	9.5	Zn, Ca "meprin"	[1157, 1158]
567		"	Bovine, pig brain	87–90		7–9	Zn	[1159, 1160]
568		"	Human, rat brain	50–58	1ch.	7	Zn	[1161, 1162]
569	3.4.24.–	Mitochondrial protease	Rat liver	108		8.5–9	Co, Mn	[1163]
570		"	Bovine adrenal cortex	60–70		7.5–8		[1164]
571	3.4.24.–	PABA-proteinase	Human small intestine	200	(2×100)	8.5	Zn	[1165]
572	3.4.24.–	Metalloendoprotease	Cytosol of mammalian cells	300	(2×115)		Zn, Co	[1166]
573	3.4.24.12	Hatching enzyme	Sea urchin	28.5	1ch.	8.2	Ca	[1167]
574		"	Oryzies latipes	24	1ch.	8–9	Zn, 2 Forms	[1168, 1169]
575	3.4.24.–	Proteinase	Ancylostroma caninum	37		9–11	Fe, elestolytic	[1170]
576	3.4.24.–	"	Astacus fluviatilis	22.6	1ch., s.	8	Zn	[1171, 1172]

No.	EC	Enzyme	Source	$M_r \times 10^{-3}$	Subunits	pH	Metal/Notes	Ref.[b]
577	3.4.24.–	Histone-specific protease	Sea urchin	41		7–11	Zn, Co	[1173]
578	3.4.24.13	Immunoglobulin A proteinase	Streptococcus sp.					[1174]
579	3.4.24.14	Procollagen N-proteinase	Bovine aorta	72		7	Ca	[1175]
580		„	Chick embryo tendons	500	4 × (61, 120, 135, 161)	7	Ca, 135 and 161 subunits are active	[1176]
581	3.4.24.–	Procollagen C-proteinase	The same	80				[1177]
582	3.4.24.15	Soluble metallo-endopeptidase	Human kidney	100				[1178]
583		„	Rat	260	2 × (122, 148)	7.5	Ca, Mg	[1179]
584		„	Rat liver cytosol	200	3 × (110, 74, 40)	7.5	Ca	[1180]
585		„	Rat brain	90		7	Somatostatin 28-convertase	[1181, 1182]
586	3.4.24.16	Neurotensin endopeptidase	Bovine	58				[1183]
587		„	Rat, guinea pig brain	70–75		7	Cleaves prooxytocin	[1184]
588	3.4.24.–	Endopeptidase	Rat kidney	220	2 × 2ch. (80, 74)	7.3	Zn	[1185]
589	3.4.24.–	Maize peptidases	Zea mays	83–92		7–8	3 Forms	[1186]
590	3.4.24.–	Chloroplast protease	Pea	180		9		[1187]
591	3.4.24.–	Proteinase	Buckwheat seeds	34	1ch.	8.3	Zn	[1188]
592	3.5.1.26	4-N-(2-β-D-glucosaminyl)-L-asparaginase	Pig kidney	70		5.5		[1189]
593	3.5.1.56	N,N-Dimethylformmamidase	Pseudomonas sp.	250	4 (2 × 15, 2 × 105)	5–6	Fe	[1190]
594	3.5.1.68	N-Formylglutamate deformylase	Rat testes				Mn, Ca, membrane	[1191]
595	3.5.1.–	N-Acyl-L-proline amidohydrolase	Pseudomonas sp.	ca. 600	(? × 55)	6		[1192]
596	3.5.2.6	β-Lactamase II	Bacillus cereus 569/H	35.6	1ch., gl.	6–7	Zn	[1193–1195]
597	3.5.2.11	L-Lysine lactamase	Cryptococcus sp.	250	(4? × 63)			[1196]

[a] Abbreviations: ch. – chain; gl. – glycoprotein; s. – primary structure is known; number and size of chains are given in parentheses, for instance: 2ch. (128, 95) – two chains with M_r 128 000 and 95 000 MW; for subunits, the number of units and their M_r (in parentheses) is given, for instance: (2×35) – two subunits with M_r 35 000 MW each, or $(2 \times 15, 2 \times 105)$ – four subunits: two with M_r 15 000 and two with M_r 105 000 MW.
[b] References given in italics contain structural information.

otherwise absolutely alien to these cells. Numerous representatives of the kingdom of living things – plants and insects – are still poorly investigated as to their content of proteases and peptidases.

According to their location amide hydrolases are distinguished as extra- and intracellular enzymes. The former are secreted by microorganisms into the environment or available in various biological fluids – lymph, blood, gastric juice, etc. The latter are localized in subcellular particles (mitochondria, microsomes, lyzosomes, etc.) and more or less connected with cell membranes. There are also enzymes located in the cell soluble phase – cytosol.

In plants amide hydrolases are found in resting and growing seeds as well as in leaves, stems, roots and different kinds of fruit.

Numerous proteolytic enzymes (peptidases and proteases), especially in animals, are produced by the cell as inactive precursors – zymogens. Zymogen is activated exactly in the place appropriate for its participation in chemical conversion of metabolites (see Chap. 5).

2.3 Characteristics of Certain Types of Amide Hydrolases

Many reviews (e.g. [178–183]) on amide hydrolases describe different aspects of the investigation of this class of enzymes. None of the four groups can be characterized by a common feature, be it the source of isolation, the specificity, i.e., ability to catalyze hydrolysis of amide bonds in endo- or expositions of the polypeptide chain, or the molecular weight, the presence or the absence of subunits. It is safe to state only the lack of exopeptidases among aspartic amide hydrolases. Aspartic amide hydrolases are not found in prokaryotes. On the other hand, exopeptidases are usually represented by metal-containing amide hydrolases. The latter are devoid of virus proteases.

Certain enzyme groups share the following features. All serine amide hydrolases are sensitive to specific inhibitors interacting with an active serine residue. They are diisopropylfluorophosphate (DIP), phenylmethylsulfonyl fluoride (PMSF) and some inhibitors of microbial, animal and plant origin. It is noteworthy that there exist quantitative differences in the mode of action of these or other inhibitors on various enzymes of this group. So, the rate and extent of the activity suppression (inhibition) with DIP of diverse amide hydrolases vary. The activity of some DIP-sensitive serine amide hydrolases is not suppressed by PSMF. Serine protease from the starfish *Dermasterics imbricata* is known to be not suppressed by the protein inhibitor, whereas all trypsin-like proteases, though similar to it, are very sensitive to this inhibitor [184].

The majority of serine amide hydrolases function in neutral or weak alkaline media. However, enzymes with the same properties are found among metal-containing amide hydrolases.

Fairly important enzymes belong to serine proteases (see Table 12). They are proteases of the blood coagulation system, proteases of the complement system, i.e., the system of primary immune response in vertebrates, food digestive enzymes such as enteropeptidase, trypsin, chymotrypsin, etc. There are enzymes of relatively small molecular mass (12 000–15 000 MW) and complex enzymes composed of many subunits (500 000–600 000 MW). Some of them are sensitive not only to DIP but also to reagents of the thiol group.

Cysteine amide hydrolases (see review [185]) are inhibited by the reagents specifically interacting with the SH group of the cysteine residue: N-ethylmaleimide (N-EMI), iodacetate, p-chlormercuribenzoate, etc. Usually, these enzymes are activated with mercaptans and other thiol-containing substances. However, this test is not unequivocal since thiols can react with disulfide bonds in the protein, thus often suppressing the catalytic activity. Many cysteine amide hydrolases function in weak acidic and neutral media (pH 5–7). However, there exist enzymes capable of functioning at weak alkaline pH as well.

Many plant enzymes belong to cysteine amide hydrolases: papain, ficin, bromelain, peptidases A and B, etc. as well as some lyzosomal enzymes: catepsins B, L, H and others. Of much interest are Ca^{2+}-dependent proteases – calpains (see review [186]).

Aspartic amide hydrolases usually function at acidic pH, except for the blood plasma enzyme renin, catalyzing hydrolysis of amides at pH 6–7. The activity of aspartic amide hydrolases is inhibited by diazoketones (e.g., diazooxonorleucine), some epoxycompounds and microbial inhibitor pepstatin. Some enzymes of this group are insensitive to pepstatin. The enzyme from *Scytalidium lignicolum* deserves special attention. Its active site includes the functional glutamic but not an aspartic acid residue. So the old name of this group – carboxylic amide hydrolases – seems more reasonable.

All gastric enzymes of stomach (pepsins), some lyzosomal proteases (cathepsins D and E) and many proteases of the simple fungi belong to aspartic proteases. Proteases of retroviruses, including HIV, are the most interesting representatives of this group.

A large group of amide hydrolases is composed of enzymes containing such metal ions as Zn^{2+}, Co^{2+}, Mn^{2+}. Some of them (e.g., thermolysin) include Ca^{2+}, playing a key role in stabilization of their three-dimensional structure. Several members of this group include Ca^{2+} and Mg^{2+} as the only metals important for catalysis.

All metal-containing amide hydrolases are inactivated by chelating agents such as ethylenediaminetetraacetic acid (EDTA) or o-phenantroline. The inactivation is often reversible, i.e., addition of a metal ion restores the activity.

It is noteworthy that among other amide hydrolase types there are also metal-containing (usually Ca^{2+} or Mg^{2+}) enzymes. However, here the metal presumably is not directly involved in the catalytic event but performs an auxiliary role.

The peptidases in their majority (EC 3.4.11–17) are metal-containing amide hydrolases. A special group is composed of aminopeptidases of microvilly of the small intestine or kidney. These are membrane-bound enzymes located so that almost the whole molecule including the active site is situated on the membrane surface, while a portion of it (4000–5000 MW) forms an "anchor" permeating the lipid membrane.

Many enzymes of thermophilic microorganisms (thermolysin, enzyme of *Bacillus stearothermophilis*, enzyme of *Telaromyces duponti*, etc.) are metalcontaining proteases. The optimum temperature of their functioning is

60–65 °C and they are stable upon heating up to 90–100 °C (see review [187]). Many collagenases and some β-lactamases belong to this group, whereas other enzymes with the same functions compose serine and cysteine groups of amide hydrolayses.

One should point to a very interesting mechanism of degradation of abnormal proteins discovered in both eukaryotic [188, 189] and prokaryotic cells [190, 191]. In reticulocytes this mechanism is coupled with ATP-dependent acylation of these proteins with a special low molecular weight protein – ubiquitin (APF-1) [192, 193]. Acylation proceeds by ε-amino groups of lysine residues of the protein, the formed conjugate being cleaved by specific proteases [194], which seem to enter the groups of serine and cysteine amide hydrolases [189, 195]. No protein like ubiquitin has been found yet in *Escherichia coli*. Apparently, abnormal proteins in prokaryotes degrade by the mechanism of ATP-dependent proteolysis without involvement of ubiquitin-like factors. ATP-dependent La protease plays a key role in this process; its structure was determined in 1988. This enzyme, a product of gene *lon*, is a heat-shock protein and shows both protease and ATPase activities.

Multicatalytic proteases (see review [1961]) – ingensin and macropain – compose another group of proteases with the chymotryptic, tryptic, and elastase specificity. They are made of relatively small subunits aggregated to high-molecular-weight complexes. Alkaline protease from carp muscles [197] can be assigned to this group as well.

A group of nonclassified amide hydrolases (EC 3.4.99) embraces many fascinating enzymes, such as aliphatic amidases of microorganisms [199–202], the tryptophan amidase [203], the theanine hydrolase [204], the pantetheine hydrolyzing enzyme [205], glycopeptidases which chip off a carbohydrate chain in glycoproteins [206–208], and also the amidase acting on cyclic amides – aminocaprolactam [209], keratin [210], and dioxopiperazine [211].

Lysobacter proteases (EC 3.4.99.29–30) [212] should be specially mentioned. No inhibitor specific for the above four types of amide hydrolases suppresses their activity. However, this may be due to some peculiarities of their structure as a whole, rather than the properties of the active sites.

Data in Table 12 have provided another important conclusion: the enzyme function does not predetermine the amide hydrolase group to which one or another enzyme belongs. In other words, one and the same function, e.g., hydrolysis of peptide glucans of the bacterial cell wall, can be performed by the enzymes with different catalytic groups. The opposite statement is also true – the enzymes of the same group can display absolutely different specificity in vitro and, evidently, in the living organism.

2.4 Primary Structure

The primary structure, i.e., the sequence of amino acid residues in the protein polypeptide chain, is known completely or partially for many amide hydrolases (see Table 12, and also [1197]). This information has been rapidly accumulating with the development of the recombinant DNA technique [52], its application promotes a replacement of conventional and rather laborious methods of protein chemistry by those of sequencing of appropriate genes. Here new possibilities have appeared. First, the structures of the mature protein forms and their precursors – pro- and preproenzymes – turned out to be determined (Chap. 5). In particular, it has initiated the understanding of the processes of protein penetration through the cell membranes and a release of enzymes to the extracellular space [1198]. In eukaryotic enzymes as well as in other proteins of eukaryotes quite a small part of genetic information happened to be utilized in the sequence of amino acid residues. Its major part, so-called introns, is not transcribed and there are hypotheses concerning the intron role. The only undoubtful fact is the significance of the intron-exon structure of genes for evolution. Introns may determine the domain organization of the protein molecule [1199, 1200] (see Sect. 2.5).

Structural similarity of the enzymes related in their origin and function was noted in the mid-1960's [1201]. At present, a comparison of structures of proteins from various sources makes it possible to identify philogenetic interrelations between diverse organisms [1202]. Some reliable methods for such a comparison (for example [1203]), which admits application of a one-letter code for amino acid residues (see [1204] and footnote to Table 14), have been proposed.

2.4.1 Serine Amide Hydrolases

Serine amide hydrolases can be tentatively divided into four subgroups: (1) proteases of pancreas, (2) enzymes of blood, (3) enzymes of invertebrates, (4) microbial amide hydrolases.

Within each group a high homology of the primary structure is observed. A structural similarity of the enzymes of different subgroups is less expressed. The choice of a method for comparing the primary structures of different proteins is very important. Evolutionary interrelations of serine proteases are characterized either by the total number of occasional evolutionary changes or by the minimum number of bases differing in DNA coding for these proteins (Table 13).

According to Table 13 the enzymes isolated from different sources (e.g., bovine and porcine trypsin) have minimum variations. For the same enzymes of remote species (e.g., trypsins of mammals and shark) the differences are larger. As to the enzymes of different subgroups (thrombin and enzymes of pancreas), the variations here are even more essential.

Chymotrypsin A and trypsin are the best-studied pancreatic proteases. Chymotrypsin A is known to have several forms developed from zymogen (chymotrypsinogen A) at its activation. Chymotrypsin A$_\alpha$ (or α-chymotrypsin) is a three-chain protein with short A-chain (13 residues) and longer B-(131 residues) and C-(195 residues) chains linked by five disulfide bridges. The relationship between α-chymotrypsin and other chymotrypsins is

Table 13. Comparison of primary structure of vertebrate proteases[a] [1203]

Enzyme	Chymo-trypsin A	Chymo-trypsin B	Porcine elastase	Bovine trypsin	Porcine trypsin	Dogfish trypsin
Chymotrypsin B	138 (62)	–	–	–	–	–
Porcine elastase	723 (192)	723 (192)	–	–	–	–
Bovine trypsin	744 (163)	633 (171)	788 (187)	–	–	–
Porcine trypsin	463 (153)	519 (162)	– (188)	71 (46)	–	–
Dogfish trypsin	767 (171)	753 (172)	978 (192)	358 (110)	353 (104)	–
Thrombin (B-chain)	1006 (197)	1336 (204)	886 (216)	790 (181)	703 (175)	754 (184)

[a] Numerals indicate the total number of random evolutionary hits. In parentheses – the total number of minimum base differences.

shown below [1205]:

```
        1          13   14      15  16      146   147  148   149      245
  H₂N–Cys ——— Leu–Ser–Arg–Ile ——— Tyr–Thr–Asn–Ala ——— AsnOH
```

Chymotrypsinogen A

Breaks   15–16 – π–chymotrypsin,

           13–14; 15–16 – δ–chymotrypsin,

           15–16; 146–147 – μ–chymotrypsin,

           15–16; 148–149 – ω–chymotrypsin,.

           13–14; 15–16; 146–147 – $α_1(χ)$–chymotrypsin,

           13–14; 15–16; 146–147; 148–149 – $α(γ)$–chymotrypsin.

Trypsin exists in several forms as well [292, 1206, 1207]. In addition to single-chain β-trypsin, enzymes with breaks of the polypeptide chain at positions Lys131-Ser132 (α-trypsin), Arg105-Val106 (γ-trypsin) have been found. The latter contains about 10% of the β-form activity.

There is virtually no homologous primary structure in animal and microbial proteases. But there are still small homologous sequences observed in certain regions of the enzyme molecules in animals and some microorganisms. So, typical pancreatic proteases (chymotrypsin, trypsin, etc.) have identical sequences in the vicinity of Ser195 (numeration of the chymotrypsin sequence), His57 and the N-terminal site with such microbial proteases as *Lysobacter* α-lytic protease and protease A of the *Streptomyces griseus* pronase complex. Table 14 shows that extended conservative primary structures are observed in serine proteases of pancreas, blood plasm, lyzosomal serine endopeptidases, collagenases and only some microbial enzymes. Each of them has a branched hydrophobic residue (isoleucine or valine) at the N-terminus, residues Val and Gly at positions 2 and 4, respectively, the same sequence Ala-Ala-His-Cys at positions 55–58 and sequence Gly-Asp-Ser-Gly-Gly-Pro at positions 193–198. Many microbial serine proteases differ much from

Table 14. The sequence of amino acids residues in catalytically significant parts of some serine amide hydrolases[a]

NN in Table 12	Enzyme	The number of residue (numeration by sequence of bovine chymotrypsin A[b])																									
		16	17	18	19	20	55	56	57	58	59	60	100	101	102	103	104	105	190	191	192	193	194	195	196	197	198	199
25	Carboxypeptidase Y	I	I	D	P	K	G	G	H	M	V	P	D	K	D	F	I	C	H	I	A	G	E	S	Y	A	G	H
37	Chymotrypsin A, bovine	I	V	N	G	E	A	A	H	C	G	V	N	N	D	I	T	L	S	C	M	G	D	S	G	G	P	L
53	Trypsin, bovine	I	V	G	G	Y	A	A	H	C	Y	K	N	D	D	I	M	L	S	C	Q	G	D	S	G	G	P	L
64	„ Vespra orientalis	I	V	G	G	T	A	A	H	C	L	V	I	N	D	I	G	I	A	C	H	G	D	S	G	G	P	L
70	Thrombin, human, B-chain	I	V	E	G	Q	A	A	H	C	L	L	N	L	D	R	D	I	A	C	E	G	D	S	G	G	P	F
83	α-Lytic proteinase	I	V	G	G	I	A	G	H	C	G	T	G	N	D	R	A	W	S	C	M	G	D	S	G	G	P	L
87	Subtilisin BPN'	A	Q	S	V	P	N	S	H	G	T	H	V	I	D	S	G	I	A	Y	N	G	T	S	M	A	S	P
106	Protease A Streptomyces griseus	I	A	G	G	E	A	G	H	C	T	S	N	N	D	Y	G	I	A	Q	P	G	D	S	G	G	S	L
115	Proteinase K	A	A	Q	T	N	N	G	H	G	T	H	V	L	D	T	G	I	R	S	I	S	T	S	M	A	T	P
140	Cathepsin G	I	I	G	G	R	A	A	H	C	W	G	Q	N	D	I	M	L	A	F	K	G	D	S	G	G	P	L
159	Coagulation factor XIa, human	I	V	G	G	T	A	A	H	F	Y	G	G	Y	D	I	A	L	A	C	K	G	D	S	G	G	P	L
172	Collagenolytic proteinase Uca pugilator	I	V	G	G	V	A	A	H	C	M	D	S	N	D	I	A	V	T	C	D	G	D	S	G	G	P	L
184	Elastase pancreatic, porcine	V	V	G	G	T	A	A	H	C	L	V	G	Y	D	I	A	L	G	C	M	G	D	S	G	G	P	L
191	Chimase I	I	I	G	G	V	A	A	H	C	K	G	L	H	D	I	M	L	A	F	K	G	D	S	G	G	P	L
195	Thryptase, dog	I	V	G	G	R	A	A	H	C	V	G	G	A	D	I	A	L	S	C	Q	G	D	S	G	G	P	L
210	Factor D, complement system	I	L	G	G	R	A	A	H	C	L	E	D	I	H	L	L	L	S	C	K	G	D	S	G	G	P	L
256	Protease, sobemoviruses						P	H	H	V	W	Y	R	I	D	F	V	L	T	A	K	G	H	S	G	S	Y	F

[a] References, see in Table 12.

[b] One character code [1204]: A-Ala, B-Asx, C-Cys, D-Asp, E-Glu, F-Phe, G-Gly, H-His, I-Ile, K-Lys, L-Leu, M-Met, N-Asn, P-Pro, Q-Gln, R-Arg, S-Ser, T-Thr, V-Val, W-Trp, X-?, Z-Glx.

animal proteases, in these regions, but within each group the sequences are rather conservative. Interestingly, there is a noticeable similarity of microbial proteases, D-alanylcarboxypeptidases and β-lactamases I within the 190–200 region. Though the sequences are not identical here, the substitutions are still equivalent in many cases, i.e., Thr → Ala, Thr → Met, etc. Despite a dramatic difference in amino acid sequences in animal and microbial proteases, these enzymes are similar as referred to the spatial structure of the catalytically significant regions. Apparently, animal proteases are produced in divergent evolution, i.e., originate from serine pro-protease common for all of them that appeared at the early stages of evolution. Microbial proteases as compared to animal proteases exemplify convergent evolution, in whose course various organisms have elaborated a common catalytic mechanism based on different protein structures.

2.4.2　Cysteine Amide Hydrolases

All the cysteine amide hydrolases contain free SH groups of the cysteine residue, its blocking leads to complete loss of the catalytic activity. The thiol group is fairly sensitive to oxidation and even in pure preparations of cysteine enzymes less than one SH group per mole of protein is found due to the presence of the molecules bearing the oxidized group incapable of activation [1208]. Comparison of the amino acid sequences of such plant cysteine proteases as papain, ficin, bromelain and actinidin shows their considerable homology. The similarity of these four enzymes is the highest in the vicinity of catalytically active residue Cys25 and one of the His residues (Table 15).

The compared sequences of plant proteases and microbial enzymes are practically identical. Of utmost interest are data obtained in [720] for cathepsin B from rat liver: the sequence adjacent to the cysteine residue in this enzyme is completely identical to that of the plant enzyme – papain.

Table 15. The sequence of amino acids residues in catalytic significant parts of some cysteine amide hydrolases

NN in Table 12	Enzyme	The number of residue (in sequence of papain)															
		18	19	20	21	22	23	24	25	26	27	158	159	160	161	162	163
292	Cathepsin B1 (light chain)	D	Q	G	S	C	G	S	C	W	A	G	H	A	I	R	I
295	Papain	N	Q	G	S	C	G	S	C	W	A	D	H	A	V	A	A
296	Ficin	Q	Q	G	Q	C	G	S	C	W	A	D	H	A	V	A	L
298	Bromelain	N	Q	N	P	C	G	A	C	W	A	N	H	A	V	T	A
301	Chymopapain A	N	Q	G	A	C	G	S	C	W	A	D	H	A	V	T	A
311	Proteinase Staphylococcus aureus	G	Q	A	A	T	G	H	C	V	A	G	H	A	V	A	A
312	Actinidin	S	Q	G	E	C	G	G	C	W	A	D	H	A	I	V	I
317	Calpain I	C	Q	G	A	L	G	D	C	W	L	G	H	A	Y	S	V

Chymopapain somewhat differs in structure. In total, the homology of the primary structure of cathepsins B and H with papain is 30 and 40% [721], and in the vicinity of the active site, 92 and 75% [1209], respectively.

The structure of Ca-dependent proteases – calpains – is of utmost interest. These proteins consist of two chains (80 000 and 30 000 MW). The heavy catalytic chain includes four domains; as is known, the domain containing residues 100–300 is homologous to other typical cysteine proteases, but the domain containing residues 550–700 has a strong similarity to Ca-dependent proteins, such as calmodulin, troponin C, etc. [772].

Conservativity of the glycine residue at position 23 is of keen interest. In serine proteases the glycine residue is also far from the two residues of the catalytically essential (serine) residue (Table 14). Apparently, glycine plays a crucial role due to its ability to form the so-called β-turn in the polypeptide chain and this results in a certain structure near the active residue of serine and cysteine proteases.

The similarity of cysteine proteases of some viruses and serine proteases deserves special attention [656, 815, 816]. Probably both the enzyme families originate from a common precursor having the proteolytic activity, the active cysteine residue being transformed into the serine residue during evolution (compare [1210]).

2.4.3 Aspartic Amide Hydrolases

In recent years primary structures of numerous animal and microbial aspartic proteases have been deciphered. Their comparison clearly shows the evolutionary relationship of these enzymes. As to proteases of other types, the similarity between the sequences of animal enzymes is more expressed than between the enzymes of animals and microorganisms.

So a comparison of zymogens of aspartic proteases containing 373 residues (the longest of the known sequences – prochymosin) demonstrates 175 identical residues (47%) in all animal proteases and only 63 identical residues (17%) in proteases of animal and microbial origin [1211]. Interestingly, identical residues and sequences are clustered – about two-thirds of similar residues cover only 20% of the polypeptide chain. As for other amide hydrolases, certain regions of the sequence display the highest homology, and can be considered significant for enzyme functioning. The same is true for the N- and C-terminal fragments and the sequences adjacent to Asp32 and Asp215 residues in aspartic proteases (Table 16).

The above regions, especially the sites adjacent to Asp32 and Asp215, are of high internal homology. This is strong evidence in favor of a common precursor of aspartic proteases of animals and simplest fungi [1212], today's proteins being the products of duplication of a gene in their common precursor. Presumably, this precursor was a structural analog of retrovirus proteases of those days. Proteases of retroviruses are similar to one domain of aspartic proteases of animals and active only in a dimeric form [939].

Table 16. The parts of primary structure of aspartic amide hydrolases

NN in Table 12	Enzyme	-2	-1	1	2	3	4	5	6	7	8	30	31	32	33	34	213	214	215	216	217	218	322	323	324	325	326	327
363	Pepsin A, human			V	D	E	Q	P	L	E	N	V	F	D	T	G	I	V	D	T	G	T	G	L	A	P	V	A
364	" porcine			I	G	D	E	P	L	E	N	I	F	D	T	G	I	V	D	T	G	T	G	L	A	P	V	A
364	" bovine			V	S	Q	E	P	L	Q	N	V	F	D	T	G	V	V	D	T	G	T	G	L	A	P	V	A
366	Pepsin, chicken	V	L	T	V	V	T	Y	E	P	M	V	F	D	T	G	I	L	D	T	G	T	G	L	A	K	A	I
368	Pepsin C, human		S	V	T	Y	E	P	M	A	Y	L	F	D	T	G	I	V	D	T	G	T	G	F	A	T	A	A
371	Chymosin A	G	E	V	A	S	V	P	L	T	N	L	F	D	T	G	L	A	D	T	G	T	G	L	A	K	A	I
386	Pepsin *Aspergillus awamori*			V	A	T	N	T	P	T	A	L	F	D	T	G	I	A	D	T	G	T	G	F	A	A	Q	A
388	Penicillopepsin	A	A	S	G	V	A	T	N	T	P	N	F	D	T	G	I	A	D	T	G	T	G	F	A	P	Q	A
390	Pepsin *Rhizopus chinensis*	A	G	V	G	T	V	P	M	T	D	D	F	D	T	G	I	L	D	T	G	T	Q	I	A	P	V	A E
422	Protease HIV-1			P	Q	I	T	L	W	Q	R	L	L	D	T	G	(L	L	D.	T	G	A)[a]	G	C	T	L	Q	F

[a] In dimer

Table 17. The parts of primary structure of metallo-amide hydrolases

NN in Table 12	Enzyme	67	68	68'	69	70	71	72	73	74	259	260	261	262	263	264	265	266	267	268	269	270	271	272	273
493	Carboxypeptidase A, bovine	G	I	–	H	S	R	E	W	I	Y	N	Q	Q	G	I	K	Y	S	F	T	F	E	L	R D
498	" B, "	G	F	–	H	A	R	E	W	I	Y	D	Q	G	I	K	Y	S	F	T	F	E	L	R	D
499	" N, human	G	N	M	H	G	N	E	A	L	Y	L	H	T	N	C	F	E	I	T	L	L	E	L	S C
508	" E, bovine	G	N	M	H	G	N	E	A	V	Y	L	S	S	S	C	F	E	I	T	V	E	L	L	S C
531	Thermolysin	G	G	V	H	I	N	S	G	I	L	S	G	I	D	–	V	V	A	H	E	L	T	H	
553	Collagenase, human fibroblasts	G	L	S	H	S	T	D	I	G	Y	M	R	T	N	P	F	Y	P	E	V	E	L	N	F
565	Membrane metallo-endopeptidase, rabbit	G	G	Q	H	L	N	–	G	I	Y	–	G	G	I	G	M	V	I	G	H	E	I	T	H

2.4.4   Metal-Containing Amide Hydrolases

A group of metal-containing amide hydrolases is rather extended and versatile in size and properties as well as in their functional specificity. Here, as for other enzyme groups, there is observed homology of certain regions in the structures of metal-containing hydrolases fairly remote in origin and specificity (Table 17). First of all, the homology is typical of the metal-binding sites and catalytic groups.

Pancreatic carboxypeptidase A has many forms differing in N-terminal amino acids and one of amino acids at the C-terminus. The C-terminal sequence can have Val or Leu at position 3. Besides, each form differs in length depending on conditions of zymogen activation and isolation techniques [1213]:

$\alpha \rightarrow$ | $\beta \rightarrow$ | $\gamma \rightarrow$ | Val
HAlaArg | SerThrAsnThrPhe | AsnTyrAla– ... –MetGluHisThr– $\genfrac{}{}{0pt}{}{\text{Val}}{\text{Leu}}$ –AsnAsnOH

Carboxypeptidase A includes one disulfide bridge, and carboxypeptidase B three S–S bridges and one free SH group [1214]. Thermolysin made of 316 amino acid residues contains no disulfide bonds. Hydrophobic residues Tyr-Tyr-Tyr-Leu-located at positions 27–30 should be mentioned. A larger cluster of hydrophobic residues (64–69) has been found in the leucine aminopeptidase molecule from bovine eye crystalline lens.

2.4.5   Nonprotein Components

Numerous amide hydrolases, especially intracellular enzymes of eukaryotes and blood enzymes, are glycoproteins (see Table 12). Some plant proteases, for example cucumizine, ficin and bromelain, are glycoproteins as well. Sugar residues bound to a protein are rarely met with at prokaryotes. So, lactamases II from *Bacillus cereus* contain sugars.

Amino sugars and sialic acids occur along with common hexoses. Such residues noticeably affect electrophoretic mobility and chromatographic behavior of the protein but, as a rule, show no influence on catalytic properties.

Oligosaccharide residues are bound to the protein either via the glycosidic bond to hydroxyl groups of serine and threonine residues (O-glycan bonding), or to the nitrogen atom of a side chain of the asparagine residue (N-glycan bonding). The latter usually enters the Asn-X-Thr (Ser) sequence.

Another type of nonprotein component is common for membrane enzymes. This is glycosylphosphatidylinosite functioning as an anchor upon the protein binding to the lipid membrane. In particular, it is found in aminopeptidase P from microvilly of small intestine or kidney [1215].

Some proteases contain phosphate groups. So porcine pepsin A contains a phosphoric acid residue forming the phosphoester bond to the Ser88 hydroxyl group [838]. Splitting-off of the phosphate group (formation of pepsin D) does not influence the enzymatic properties of pepsin [1216]. This group is located in a rather movable enzyme region [1217].

Regarding unusual amino acids, γ-carboxyglutamic acid, found in thrombin [317] and other vitamin K-dependent enzymes [1218] and important in Ca^{2+} binding, attracts major attention.

2.5 Spatial Structure

The spatial (tertiary) structure is the enzyme characteristics essential for understanding its catalytic properties and the mode of action. X-ray analysis provides information on the three-dimensional structure of proteins [1219], a resolution of at least 2.7–2.8 Å being necessary for the detailed structure of the molecule. Quality of a crystallographic model is determined by the refinement factor, the so-called R-factor. $R = \Sigma(|F_0| - |F_c|)/\Sigma|F_0|$, where $|F_0|$ and $|F_c|$ are experimental and calculated factors of structural amplitudes. X-ray analysis with this or higher resolution has been performed for numerous amide hydrolases, some of their zymogens, and a large number of enzyme complexes with substrate analogs (Table 18 and Chap. 6).

Table 18. High resolution X-ray analysis of amide hydrolases and some zymogens

Protein	R-factor[a]	Resolution, Å	References
Serine amide hydrolases			
α-Chymotrypsin A, bovine	0.179	2.8; 2.0; 1.8; 1.67; 1.68	[1220–1226]
γ- „	0.173	2.7; 1.9; 1.6	[1225, 1227, 1228]
Chymotrypsinogen A, bovine	0.173	2.5; 1.8	[1229, 1230]
Trypsin, „	0.193	1.8; 1.5	[1231–1236]
Trypsinogen, „	0.166	1.9; 1.8	[1237–1239]
Elastase pancreatic, porcine	0.169	2.5; 1.65	[1240–1242]
Kallikrein A pancreatic	0.220	2.05	[1243]
Tonin	0.196	1.8	[1244]
Chymase II	0.191	1.9	[1245]
Prothrombin (fragment 1), bovine	0.240	2.8	[1246]
Trypsin *Streptomyces griseus*	0.161	1.7	[1247]
Subtilisin *Bacillus amyloliquefaciens*	0.140	1.8	[1248]
Subtilisin BPN' (NOVO)	0.230	2.8; 2.0	[1249, 1250]
Thermitase	0.240	2.2	[1251]
Protease A *Streptomyces griseus*	0.126	1.8; 1.5	[1252–1254]
„ B „	–	2.8	[1255, 1256]
α-Lytic protease *Lysobacter enzymogenes*	0.131	2.8; 1.7	{1257, 1258}

Table 18. Continued

Protein	R-factor[a]	Resolution, Å	References
Proteinase K	0.167	1.5	[1259]
D-Alanine carboxypeptidase streptomyces sp. R61	–	2.8	[1260]
β-Lactamase Bacillus licheniformis	0.208	2.0	[1261]
Cysteine amide hydrolase			
Papain	0.161	2.8; 1.65	[148, 1262, 1263]
Actinidin	0.171	2.8; 1.7	[1263–1265]
Calotropin DI	–	3.2	[1266]
Aspartic amide hydrolases			
Pepsin A, porcine	0.174	2.7; 2.0; 1.8	[1267–1270]
Pepsinogen	0.173	1.8	[1271, 1272]
Chymosin	0.165	2.3	[1273, 1274]
Renin	0.24	2.5	[1275]
Penicillopepsin	0.136	2.8; 1.8	[1276, 1277]
Protease Endothia parasitica	0.178	3.0; 2.45, 2.1	{1278–1280]
Protease Rhizopus chinensis	0.143	2.5; 1.8	[1278, 1281, 1282]
Protease from retrovirus of avian myeloblastosis	0.26	2.4	[1283]
Protease from retrovirus of Raus sarcoma	0.15	2.0	[1284]
Protease HIV-1	0.184	3.0; 2.8	[1285, 1286]
Metallo-amide hydrolases			
Carboxypeptidase A, bovine	0.190	2.0; 1.54	[149, 1287–1289]
„ B	0.33	2.8	[1290]
Thermolysin	0.213	2.3; 1.6	[1291–1293]
Neutral proteinase Bacillus cereus	0.217	3.0	{1294}
Carboxypeptidase T	0.343	3.0	[1295]
D-Alanyl-D-alanile carboxypeptidase	–	2.5	[1296]

[a] For the best resolution data.

Usually, amide hydrolases contain not so many α-helical regions. These proteins are composed of the β-structure fragments alternating with random sites. The domains – building blocks of these protein molecules – are identified for many enzymes.

As expected, the proteins with similar primary structures have similar spatial arrangement. Therefore, it is reasonable to consider the tertiary structure of amide hydrolases for each enzyme type.

2.5.1 Serine Amide Hydrolases

Serine proteases of animal origin – highly homologous chymotrypsin, trypsin, and elastase – are very close in their tertiary structure (Fig. 14).

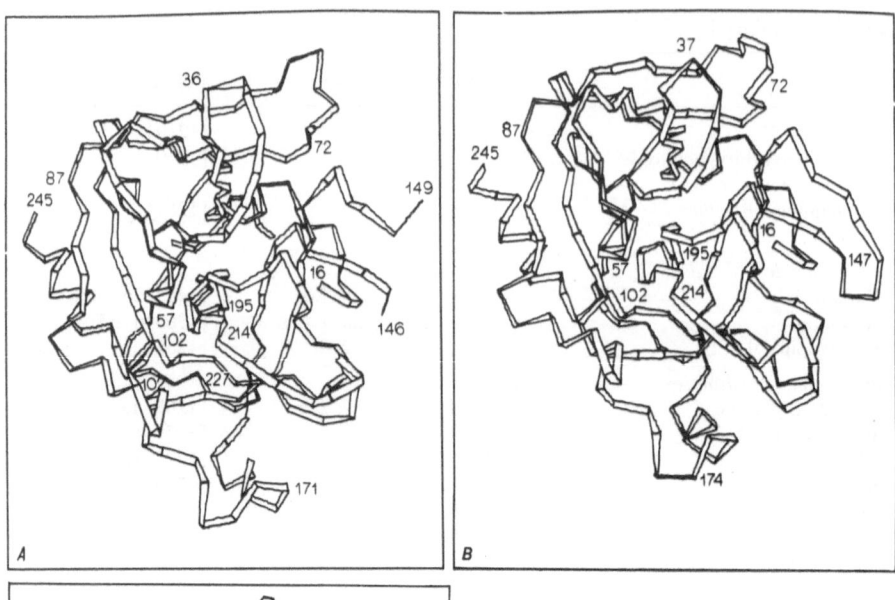

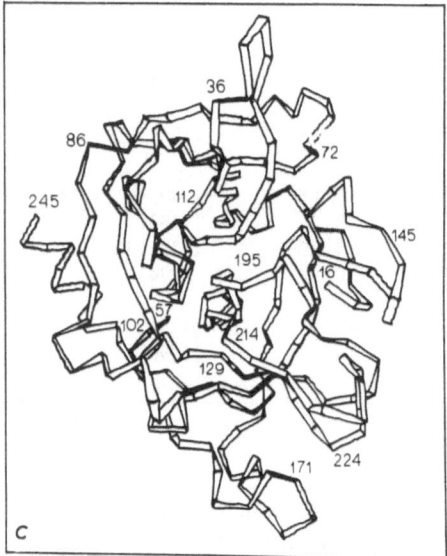

Fig. 14A–C. Comparison of the polypeptide folding in α-chymotrypsin (*A*), trypsin (*B*) and elastase (*C*) [1222]

Let us describe α-chymotrypsin in detail. Its polypeptide chains form a system of antiparallel β-pins linked by a network of hydrogen bonds. Two short α-helical sites are located in the C-terminal region of chain C (residues 232–245) and in the region of residues 165–172. The majority of β-turns are coiled in such a way as to form two cylinders (Fig. 15).

Cylinder I including residues 29–112 is made of six antiparallel chains. Cylinder II is formed by residues 133–230 and also includes six antiparallel fragments of the polypeptide chain. Four of five disulfide bridges enter the cylinder.

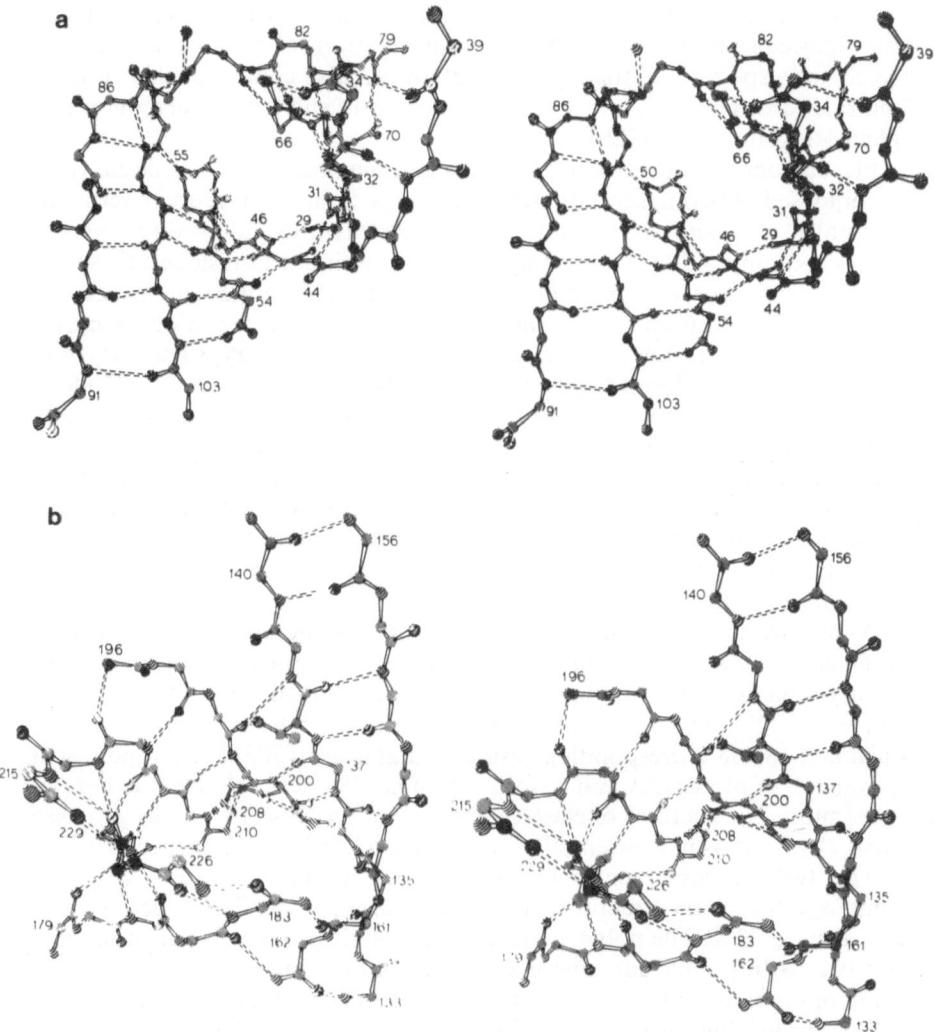

Fig. 15a, b. Stereopairs indicating the structure of "cylinders" I (**a**) and II (**b**) formed by polypeptide chains in α-chymotrypsin [1221]

Side chains of hydrophobic amino acids are oriented inwards, whereas hydrophilic amino acids are located on the cylinder surface. The cylinders touch each other, some residues of hydrophobic amino acids being oriented to the contact area (Trp29, Ile47, Trp51, Val53, Phe89, Ile103, Leu105, and Ile212). The C-terminal α-helix also contacts the two cylinders. About 10% of water molecules bound to the protein are in the contact area. About the same content of water molecules is inside the globule. The remainder of 80% is bound to the protein surface [1225].

The backbone conformation falls into the allowed regions on the diagram of dihedral angles ϕ and ψ [1221].

It is noteworthy that α-chymotrypsin is crystallized as a dimer. Both the dimer molecules have virtually similar conformations of main chains, though side chains, especially in the region of intermolecular contacts, differ in their conformations. The same is true for α-chymotrypsin and crystals of monomeric γ-chymotrypsin [1224, 1226].

Elastase and trypsin are constructed in a similar manner. Both the enzymes are composed of two tightly contacting domains. So in elastase six N-terminal antiparallel rods and six C-terminal rods form two cylinders with internal cavities containing side chains of hydrophobic amino acid residues; these are the hydrophobic core of the molecules. Four disulfide bonds of elastase are involved in stabilization of the cylinder structure and located in places corresponding to the positions of the respective bonds in chymotrypsin and trypsin. The C-helical region formed by residues 234–244 [1240] contacts the cylinders.

Structures of pancreatics kallikrein A [1243], tonin [1244], and chimase II of mast cells [1245] are similar to those of these three enzymes. The major differences here are caused by outer loop structures. The kallikrein fragments surrounding the substrate binding site are more compact than those of trypsin. The same is observed in tonin, which is structurally close to kallikrein. There are variations in structures of the substrate binding sites of these enzymes expressing their different specificity (see Chap. 6).

As shown above, two microbial proteases – α-lytic and protease A from *Streptomyces griseus* – are rather close to animal serine proteases in their primary structures. This similarity is also observed at the level of the tertiary structure [1297]. About 55% of C_α-atoms of α-lytic protease are topologically equivalent to the corresponding residues in elastase [706]. The topological equivalence of protease A and chymotrypsin is 64%. Remarkably, the homology of primary structures of elastase and α-lytic protease, as well as protease A and chymotrypsin, is essentially lower, being 18 and 21%, respectively.

The two microbial enzymes have similar spatial structures (Fig. 16). Homology of primary structures of these proteases is 35%, their topologies are equivalent in 82%. The polypeptide chain of α-lytic protease is coiled so as to form two hydrophobic domains, each including four antiparallel β-loops. The region of domain contact is filled with the side chains of amino acid residues and has rather high electron density. The α-helical site of protease includes only three turns (residues 231–238). Interestingly, one of the proline residues (Pro99a) has *cis*-configuration. The respective residue in proteases A and B of *Streptomyces griseus* occupies the same position.

The tertiary structures of subtilisins BPN' and *Bacillus amyloliquefaciens* [1248–1250] strongly differ from those of the above enzymes. Subtilisins contain seven α-helical segments, each involving 7–10 amino acid residues. Besides, there are six parallel and two antiparallel β-layers. The latter permeate the entire globule and compose the hydrophobic nucleus of the molecule. Many hydrophobic residues in subtilisins contact the surrounding solvent.

Though subtilisins and other serine proteases so far investigated differ in their structure, the catalytically significant regions in all the enzymes of this group with the known structure are rather similar (Sect. 2.6). The structure of serine *D*-alanylcarboxypeptidase differs from those of typical serine proteases

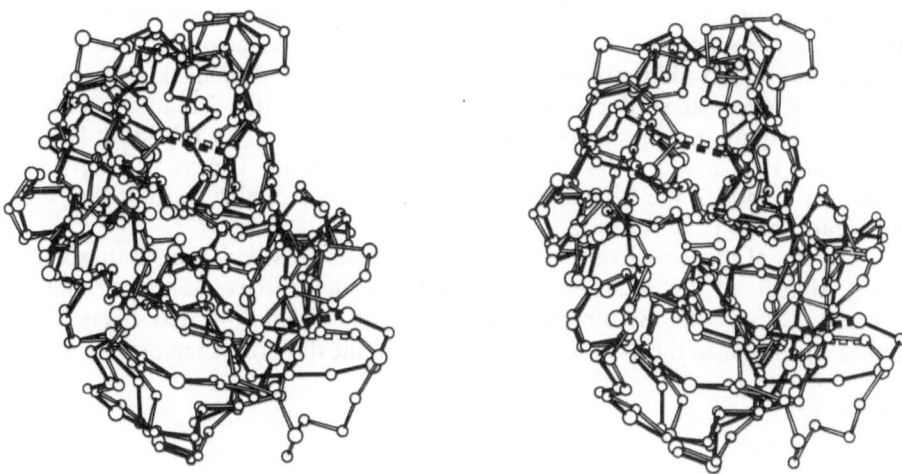

Fig. 16. Stereopair illustrating a similarity of the tertiary structure of α-lytic protease (*light line*) and protease A from *Streptomyces griseus* (*heavy line*) [1257]

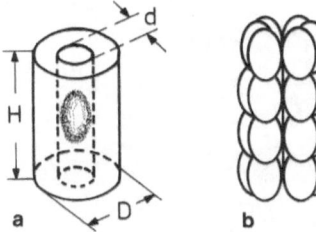

Fig. 17a, b. Structures of multicatalytic proteases; **a** according to electron microscopy. *Size: D = 11 nm, H = 16 nm, d = 4.3 nm;* **b** tentative model of the subunit organization [1298]

[1260]. This enzyme is built of the α/β-cluster and several β-pleated rods. The substrate binding site is located near Ser37, so that the β-lactame ring contacts the hydroxyl group of the serine residue.

According to electron microscopy of multicatalytic proteases, these enzymes appeared [1298, 1299] (Fig. 17) to form a cylindrical complex (110 Å in diameter and 160 Å in height) made of four ring structures.

2.5.2 Cysteine Amide Hydrolases

Of these enzymes, only for papain and actinidin have the spatial structures been found with a fairly high resolution (1.65 and 1.7 Å, respectively). The enzymes have similar three-dimensional structures with clearly expressed two domains. In papain domain L is formed by residues 10–111 and 208–212, and domain R by residues 1–9 and 112–207, i.e., they contain the residues of both the N- and C-terminal sequences. The domains have "nuclea" formed by side chains of nonpolar residues, whereas their contact surface is formed mainly by side chains of polar amino acids and contains a considerable number of bound water molecules. Of three disulfide bridges, two (22–63 and 56–95) are included in domain L, and one (153–200) into domain R (Fig. 18). Secondary structures of papain and actinidin are of type $\alpha + \beta$ (see Fig. 12), domain L containing preferably α-helical conformations and domain R mainly irregular twisted β-sheets. Papain includes three helical regions (24–43, 50–57, 67–78) in L domain, and two helices (117–127, 138–143) in R domain. The mean length of hydrogen bonds in the helical structure is 2.98–2.99 Å which exceeds that for an ideal α-helix (2.87 Å).

The site of the β-structure of domain R consists mostly of four antiparallel rods. Almost all the residues entering the β-structure have dihedral angles $\phi \approx -121°$ and $\psi \approx 143°$, which is typical of right-twisted β-pleated sheets. The four central rods of the β-structure are buried in the domain R region. They are

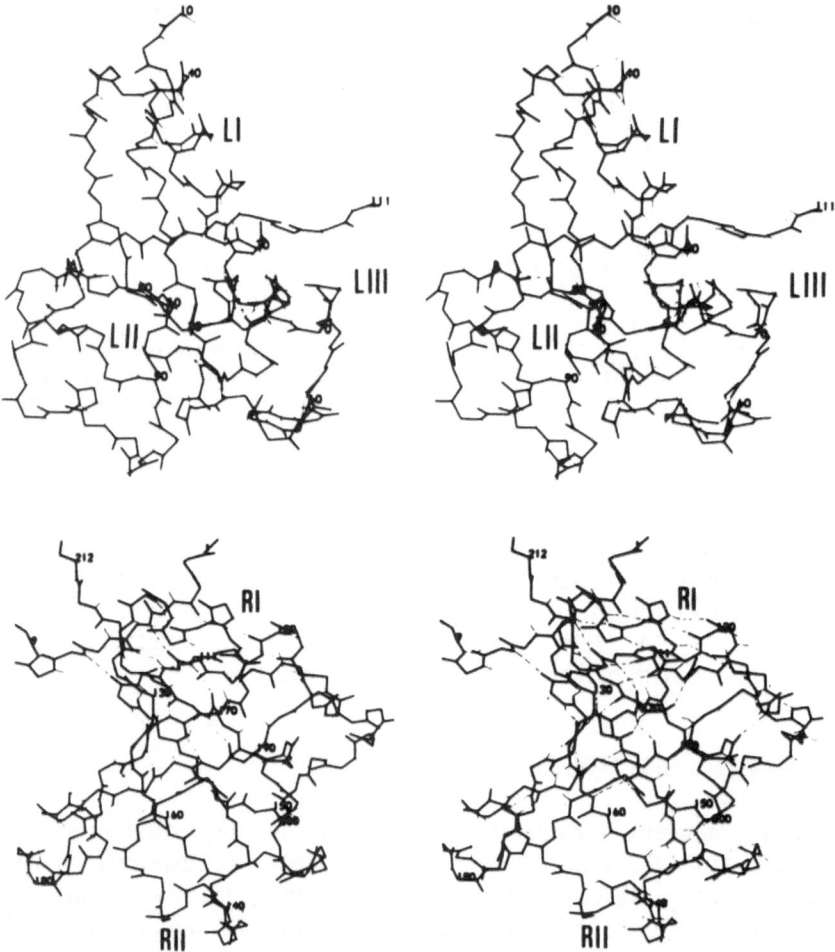

Fig. 18. Stereo image of domains L (N-terminal) and R (C-terminal) of papain [1262]. *Dotted lines* indicate hydrogen bonds between the backbone atoms

built predominantly of nonpolar amino acid residues, the polar residues being located at the rod ends.

As mentioned above, each domain has a hydrophobic nucleus formed by side chains of nonpolar amino acid residues: mainly valine, leucine and isoleucine. As to aromatic residues, they are at the periphery close to its surface or even on the surface, so that a part of the aromatic ring contacts solution. There is virtually no orientation characteristic of the $\pi-\pi$ interactions.

The residues of charged amino acids are on the molecule surface. They are not concentrated but evenly distributed. One charged base group (Lys174) and three acidic groups (Glu35, Glu50 and Glu52) are buried inside the molecule, being inaccessible to solvent. The charges of some groups (not all) are neutralized.

Interestingly, one of the proline residues (Pro155) is in *cis*-configuration in actinidin as well as in papain.

2.5.3 Aspartic Amide Hydrolases

A number of aspartic proteases of animals, microorganisms and viruses have been analyzed by X-ray crystallography with a resolution below 3 Å. In spite of marked differences in amino acid sequences of pepsin and enzymes of microorganisms, their spatial structures appeared to be strikingly similar (for example Fig. 19). The two-domain structures are typical of these enzymes, as in the case of cysteine proteases. Moreover, a thorough study of the pepsin polypeptide chain arrangement reveals [1300, 1301] two "subdomains" with a repetitive spatial structure inside each domain (Fig. 20). This is undoubtedly a result of evolution of aspartic proteases, the key feature of which is gene duplication [1302, 1303].

Aspartic proteases are almost entirely made of β-structures. There are only four short fragments of the distorted α-helix. These are residues 58–62, 137–141, 225–235 and 303–309 (numeration according to the porcine pepsin sequence) in penicillopepsin. β-Structures in these enzymes are formed by rods and "hairpins" linked by short unordered sites. Each "subdomain" is built in a similar way: if one moves from the N- to the C-end of the molecule, first the β-hairpin is located, then a loop, and – further – a short helix followed by the β-hairpin though with opposite orientation as compared to the first spin. This fragment is repeated four times in the enzyme molecule. The nucleus of each domain is formed by a β-layer, which consists of two wide loops entering one another. In general, the domains and enzyme molecules are surprisingly symmetrical.

The renin molecule has an analogous structure. A mean square deviation of the atoms in the main chain from their position in proteases of microorganisms is 0.45 Å [1275].

Retrovirus proteases (aspartic proteases) have low internal homology of the primary structure as compared to aspartic proteases of animals and microorganisms. Their sizes are generally twice as small as those of typical representatives of this group; however, the enzyme active forms are dimers

80 2 Enzymes

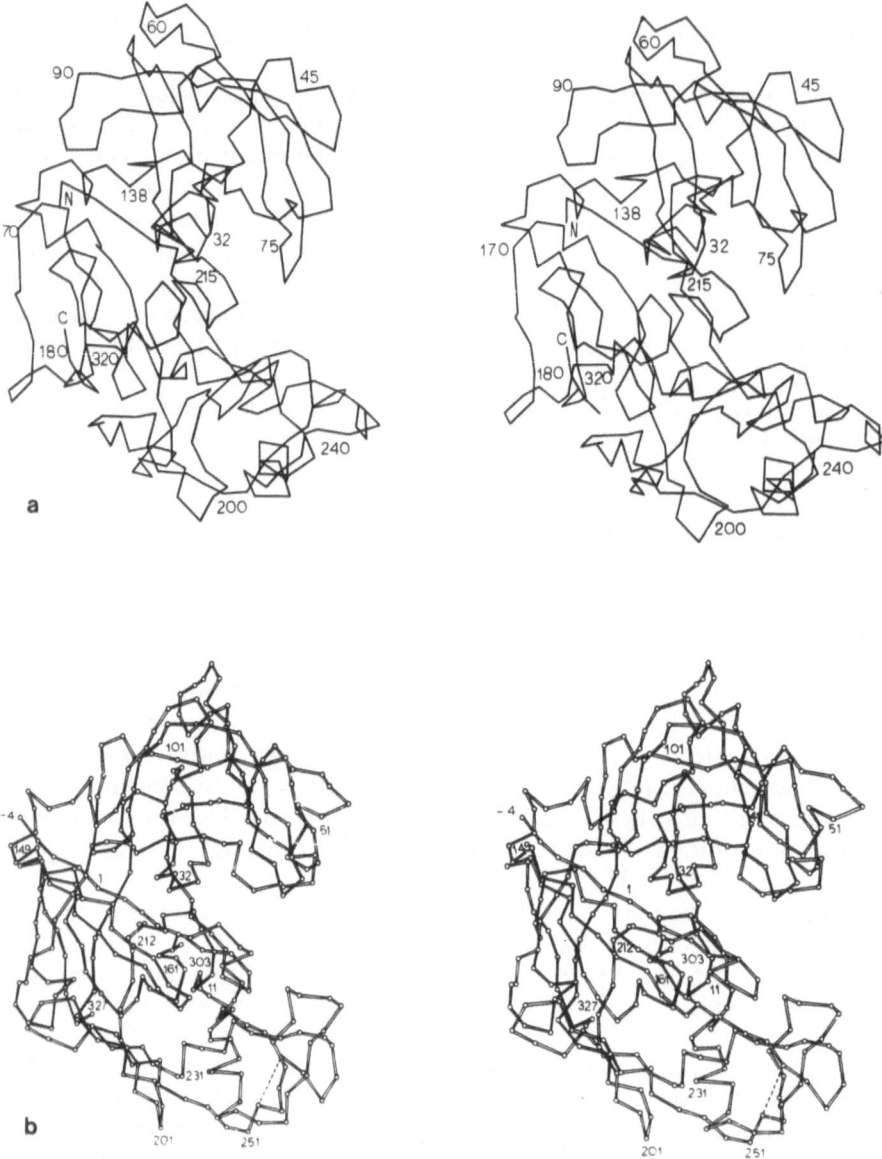

Fig. 19a, b. Stereopairs of the polypeptide folding of porcine pepsin chains (**a**) and penicillopepsin (**b**) [1268, 1276]

[939, 1304]. The dimer structure is similar to the pepsin structure, especially in the vicinity of the active site formed by groups of two protomers. Probably, the N- and C-terminal residues of a protomer molecule play a key role in the dimer stabilization.

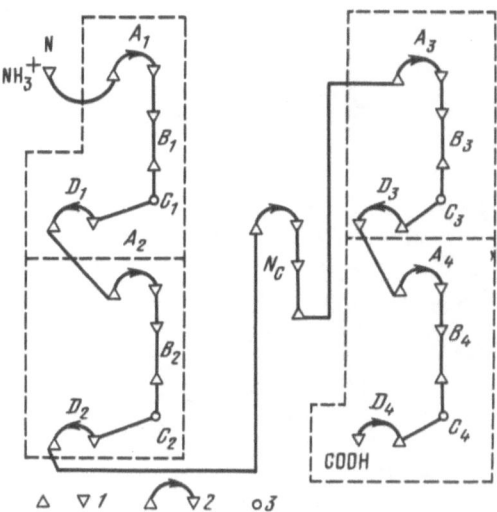

Fig. 20. Alternating of secondary structures in pepsin molecule [1268]. *1* Rods from N- and C-termini; *2* β-pin; *3* α-helix (*dotted line* encounters the structural unit of the molecules repeatedly 4 times)

2.5.4 Metal-Containing Amide Hydrolases

Spatial structures of carboxypeptidase A and thermolysin entering this group have been analyzed with a resolution of about 1.6 Å. A lower resolution has been obtained for carboxypeptidases B and T and for neutral protease from *Bacillus cereus*. As mentioned above, carboxypeptidases and thermolysin do not share so many primary and spatial structure elements [1292]. According to the classification of secondary structures, thermolysin belongs to group $\alpha + \beta$, and carboxypeptidase A – to group α/β. Although with similar molecular weights, these enzymes dramatically differ in molecule size. Carboxypeptidases A has an almost spherical symmetry ($50 \times 42 \times 38$ Å), whereas a thermolysin molecule is an elongated ellipsoid ($64 \times 38 \times 37$ Å).

Thermolysin is a two-domain protein (Fig. 21). The N- and C-terminal domains are connected through the site including residues 153–158. Each domain has virtually spherical symmetry. The α-helical structures compose about 38% of all amino acid residues, the majority of them are at the C-terminal domain. The N-terminal domain is predominantly built of β-pleated sheets including about 70 residues. These sheets, as in α-chymotrypsin, form a "cylinder". The boundary between the two domains is formed mainly by hydrophobic side chains of residues Pro130, Leu133, Val139, Ile188, Val192, Leu202. Some side chains are inside the molecules with the charges neutralized by ion interactions. The high number of salt bridges is typical of thermolysin.

This enzyme contains three to four calcium atoms. Two Ca^{2+} are at a distance of 3.8 Å only. The Ca^{2+}-formed salt bonds include Asp57, Asp138, Asp185, Asp200 as well as Glu177 and Glu190. The calcium atom being nearest to the zinc atom is at a distance of approximately 13 Å. Presumably, two calcium atoms are essential for stabilization of the mutual disposition of the two domains.

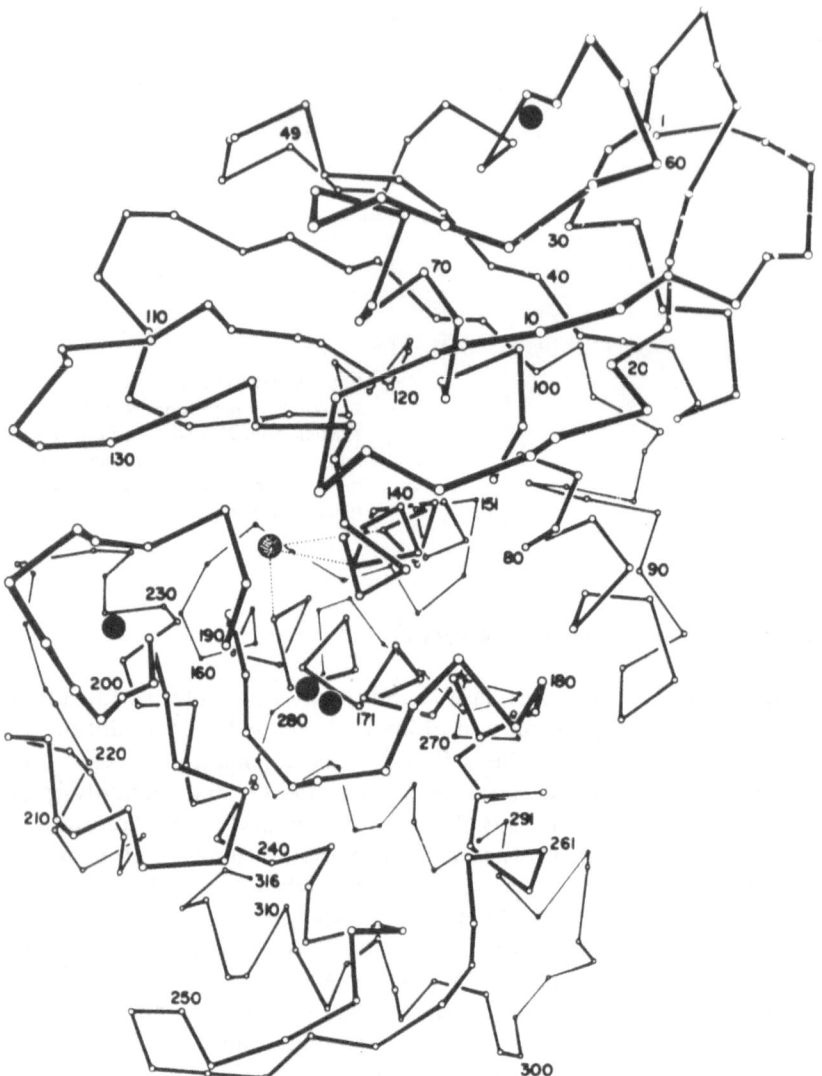

Fig. 21. Conformation of the thermolysin backbone [1291]. *Stippled circle* Zn atom; *solid black circles* Ca atoms

In contrast to thermolysin, carboxypeptidase A (Fig. 22) contains one domain. The total content of α-helical and β-structures is 38% and 17%, respectively. Thus a large portion of the protein molecule structure is unordered.

In the protein amino acid sequence the helical sites alternate with the β-structures and short random fragments. There is an extended region of the unordered structure (residues 122–174) only in the middle of the sequence. The

Fig. 22. Polypeptide chain folding of carboxypeptidase A [149]

β-layers are arranged both as antiparallel and parallel rods. Interestingly, the peptide bond between residues Ser197 and Tyr198 has *cis*-configuration.

In general, the side chains of 164 residues of carboxypeptidase A amino acids are exposed to solution, 78 side chains are embedded to the globule and 65 residues are exposed to the molecule surface only partly contacting a solvent.

As to the structure of neutral *Bacillus cereus* protease, it resembles the thermolysin structure (mean square deviation 0.88 Å), whereas carboxy-peptidases B and T are similar to carboxypeptidase A by their three-dimensional arrangement.

The point worthy of mentioning is the electron microscopy analysis of leucine aminopeptidases from bovine eye crystalline lens and porcine kidney [1305, 1306]. This enzyme composed of six subunits is arranged as shown in Fig. 23.

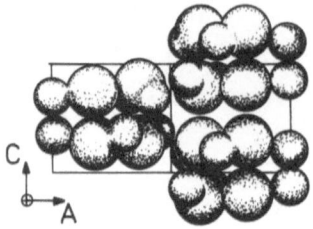

Fig. 23. Model of the subunit organization of hexamer molecule of porcine kidney leucineamine peptidase [1306]. View along axis *b*

2.6 Active Sites

The substrate molecule, or its part which directly interacts with the enzyme during catalysis, is much smaller than the enzyme molecule. Therefore, the events taking place in catalysis occur in a certain enzyme region called the active site. It can be defined as *a set of atoms of the enzyme involved in bonding and chemical conversion of a substrate*. This definition, as any other, is not complete, since it is hardly possible to indicate what atoms participate in catalysis and to explain the meaning of the word "participate". Besides, the catalytic transformation involves different atoms at various stages. At the same time, one can outline the active site region in spatial structures of many enzymes. Within the active site three areas are distinguished: *catalytic site* – a sum of atoms interacting with a cleavable substrate group; *binding site* – a region where attachment of the substrate groups other than the reactive one, occurs; *regulatory site* – an area interacting with external effectors and influencing the state of catalytic and binding sites. A wealth of information on the amide hydrolases active sites has been provided by site-directed chemical modification and kinetic methods; more thorough investigation resulted from X-ray analysis of enzyme complexes with substrate-like ligands. Site-directed mutagenesis has been successfully applied here [1307], that allows substitution of one residue by another. The active sites of representatives of four protease groups are given below. The following chapters considering certain stages of the catalytic process shed light on some details.

2.6.1 Serine Amide Hydrolases

Enzymes of this group are sensitive to inactivation by diisopropylfluoro-phosphate and other phosphoorganic compounds [1308]. Serine proteases interact with diisopropylfluorophosphate forming a phospho-ester bond to the only hydroxyl group, which belongs to the serine residue [1309]. In the α-chymotrypsin sequence, the best-studied enzyme of this group, reactive serine residue occupies position 195 [1310, 1311]. Peptides containing the reactive serine have been isolated from enzymes of the serine proteases group. The amino acid sequence near this residue appeared to be rather conservative (Table 14). Sequence Gly-Asp-Ser-Gly-Gly-Pro remains the same for all pancreatic serine proteases, blood proteases, some microbial proteases, etc. The

same sequence has been discovered in proteases and collagenases of invertebrates [1312–1314]. However, bacterial enzymes such as subtilisin, and fungi
serine proteases have the other sequence – Gly-Thr-Ser-Met-Ala – in the
vicinity of the serine residue modified by phosphoorganic compounds. An
analogous structure has been found for the peptide of the active site of plant
protease cucumizin [1315] and tripeptidyl peptidase II from human erythrocytes [1316]. The Gly193 residue is conservative in a great majority of serine
proteases. The role this residue plays in the enzyme activity is exemplified by
the α-subunit of the nerve growth factor [1317]. This residue is replaced by His
in this subunit, unlike the γ-subunit the α-subunit is inactive. The serine
protease with unique sequence Ser-Ser-Gly is also known [377].

Another group of serine amide hydrolases differing from all those considered above, is composed of D-alanine carboxypeptidases and β-lactamases
I. Sequences Gly-Ser-Val or Ala(Met)-Ser-Met(Thr) are found in their active
sites [257, 258, 1318, 1319]. Serine residues in these enzymes react with
antibiotics of the penicillin group [1320, 1321].

As shown [1322], in α-chymotrypsin p-nitrophenyl ester of acetic acid
acylates the same unique residue Ser195. Several relatively stable derivatives of
α-chymotrypsin and other serine proteases acylated at Ser195 by different
acids have been isolated. Finally, spectral and kinetic studies revealed that
hydrolysis of esters of acylamino acids and their analogs yields relatively stable
acyl derivatives. Diisopropylfluorophosphate and other modifications of
Ser195 completely inactivate the enzyme. Ser195 or, to be more exact, its
hydroxyl group, which enters the active site, is crucial for catalysis.

Imidazole of His57 (in sequence of α-chymotrypsin) is another catalytically
essential group. Its significance has been proved by investigations of pH
dependence in catalysis by serine proteases: a group with $pK_a \approx 7$ functioning
in an unprotonated state is found. Specific compounds, which react with His57,
have been revealed. They are halogenoketones and derivatives of acylamino
acids and peptides [1323]:

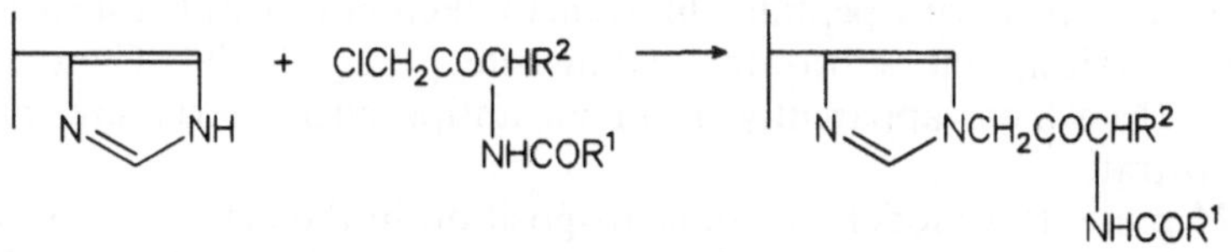

These compounds completely inactivate serine proteases. The amino acid
sequence near His57 also appeared to be fairly conservative for many enzymes
of this group.

Methylation of $N^{\varepsilon 2}$-atom of the imidazole group of His57 residue by
methyl p-nitrobenzenesulfonate has also pointed to a special role this group
plays in catalysis [1324]. It turned out [1325, 1326] that N-methylchymo-
trypsin is practically devoid of catalytic activity, though still being able of
substrate binding. Thus, modification affects the catalytic rather than the
binding site of the active center.

Further prospects in determining the structure of active sites of serine
proteases depend mainly on success in X-ray analysis. The hydroxyl group of

Ser195 and the imidazole ring of His57 in α-chymotrypsin are located near the molecule surface within the low electron density range. In the derivatives acylated at Ser195 (or tosyl derivative) this region is occupied by the side chain (usually aromatic group) of the acyl residue and is called the "tosyl hole". The $N^{\varepsilon 2}$-atom of His57 is located near the O^{γ}-atom of Ser195.

X-ray analysis showed that $N^{\delta 1}$-atom of His57 interacts with the carboxyl group of Asp102 (earlier considered as an asparagine residue). The close disposition of three functional groups Asp102, His57, and Ser195 has evoked a hypothesis about an activation of the Ser195 hydroxyl group by the proton transfer to Asp102 along the so-called "charge relay system" [1327]:

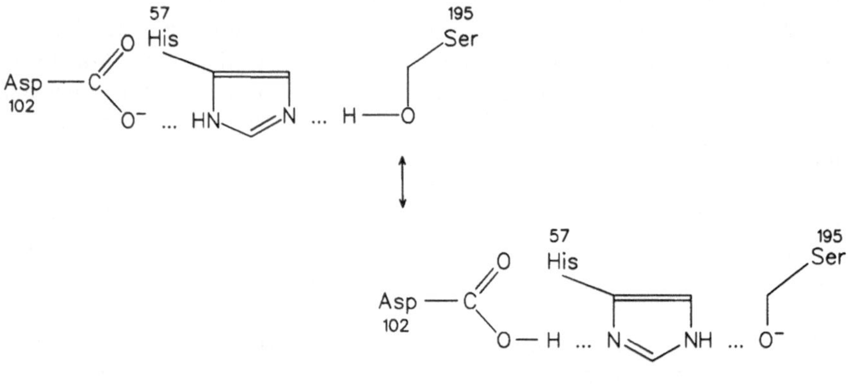

The concept of the charge relay system in the native enzyme has been criticized, at present many investigators regard it as dubious (for details, see Chap. 7).

Research into the crystals of stable O-acyl derivatives of α-chymotrypsin has uncovered the interaction of carbonyl oxygen of the acyl group with NH groups of Ser195 and Gly193 in the enzyme backbone. These groups form an "oxyanion hole" [1328]. Finally, an investigation of the γ-chymotrypsin crystals modified by His57 with peptide chloromethylketones has identified the enzyme groups within the substrate binding site [1329, 1330]. These are residues 214–216 which, apparently, form an antiparallel β-structure with a peptide substrate.

Figure 24 presents a model for atoms disposition in the active site of the α-chymotrypsin analog – protease A from *Streptomyces griseus*. The model is based on high resolution (1.5 Å) X-ray data. It virtually does not differ from the model for α-chymotrypsin gained with resolution 1.68 Å.

"Tosyl hole" is a binding site of the side chain of the substrate amino acid forming a cleavable bond by its carboxyl group. It is a rather narrow cleft with dimensions ($12 \times 6.5 \times 4$ Å) accommodating an aromatic residue of phenylalanine, tyrosine, or tryptophan. This part of the active site is formed mainly by side chains of nonpolar amino acids and peptide bonds of the backbone [190–191, 191–192, 215–216]. The hole includes deep-seated residues determining the enzyme specificity: Ser189 in α-chymotrypsin and Asp189 in trypsin. Residues Gly216 and Gly226 in chymotrypsin and trypsin, respectively, do

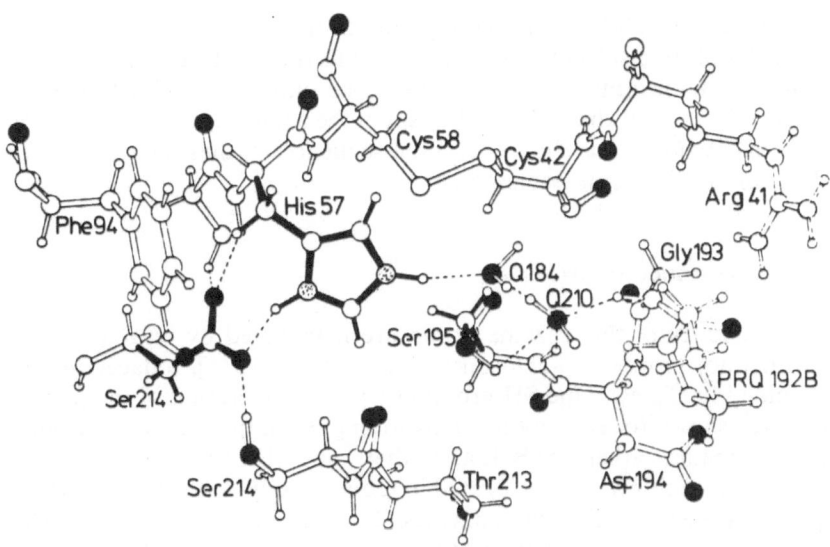

Fig. 24. The active site area of *Streptomyces griseus* protease A [1254]. *Dotted line* shows hydrogen bonds; Q water molecule

not interfere with a substrate, whereas elastase, specific to nonpolar amino acid residues with short side chains, has residues Val216 and Thr226, which prevent bulky side chains of the substrate from entering the cavity. The "tosyl hole" in α-chymotrypsin and related enzymes is screened from the environment by the side chain of Met192.

Studies of the trypsin and chymotrypsin complexes with naturally occurring protein inhibitors of serine proteases were of help in identifying the sites of the enzyme active site responsible for binding the C-terminal fragment of a substrate (leaving group). These regions are disulfide bond 42–58 (chymotrypsin), the area in the vicinity of His40 and Gly193 as well as residue Tyr151 (in trypsin).

Of great importance for the activity of serine proteases is the ion pair formed in the enzyme by carboxylate ion of Asp194 and the ammonium group of Ile16. This ion pair can be considered as a regulatory site, absent in zymogens and formed upon the activation of the latter. Here the enzyme active site is finally formed. pH-Shift to the alkaline region causes disruption of the ion pair and transition of the enzyme to a new inactive conformation.

The above structure of active sites of serine proteases is typical, though with slight variations, of all studied enzymes of this group. The state of the active site in subtilisins and other microbial serine proteases is an example. As indicated, there is no homology in the primary structure of subtilisins and pancreatic enzymes. Also, no similarity in the spatial structure as a whole is observed. However, the structures of the active sites in serine proteases of these two types are rather identical. The major difference between chymotrypsin and subtilisin is that the "tosyl hole" in the former is a narrow pocket with a strictly fixed position of the substrate aromatic group. In the latter the analogous site is a cleft accessible to a solvent from one side. Thus the substrate aromatic ring in a subtilisin complex interacts with the enzyme only by one side [1330].

Another difference is the lack in subtilisin of the acidic residue homologous to Asp189 in chymotrypsin. The N-terminal sequences of these enzymes differ from those of animal serine proteases.

2.6.2 Cysteine Amide Hydrolases

The representatives of this enzyme group are inactivated by reagents interacting with thiol groups – p-chloromercuribenzoate [1331], iodacetate and iodacetamide [1332], etc. One SH group of Cys25 (numeration of the papain sequence) was shown to react upon inactivation [1333]. In numerous enzymes of animals (calpains, cathepsins B, L and others) and plants (papain, actinidin, chymopapain, etc.) this residue is located in sequence Gly-Xaa-Cys-Trp, where residue Xaa varies for different enzymes. In microbial cysteine proteases (*Staphylococcus* protease, clostripain) these sequences are somewhat different [720, 726, 728, 735, 747, 752, 1334–1336]. According to X-ray analysis of papain and actinidin, this residue is in an extended cleft at the boundary of two domains (Fig. 25). It can be transformed to dehydroalanine or the glycine residue with a complete loss of the activity [1337]. The imidazole group of residue His159 is situated close to SH group, its $N^{\delta 1}$-atom forms a hydrogen bond with SH group of Cys25. The second $N^{\epsilon 2}$-atom of His159 forms a hydrogen bond with $O^{\delta 1}$-atoms of Asn182. This bond is in the nonpolar surround and protected from contacting the solvent via the indole ring of Trp184. The role of His159 in catalysis appeared to be initially doubtful, since pH dependence shows the involvement of the unprotonated group with pK_a 4.2 in catalysis. This pK_a value was assigned to the carboxyl group of the aspartic acid residue. However, X-ray analysis shows that the carboxyl group (Asp158) nearest to Cys25 of histidine is at a distance of about 7.5 Å from the SH group.

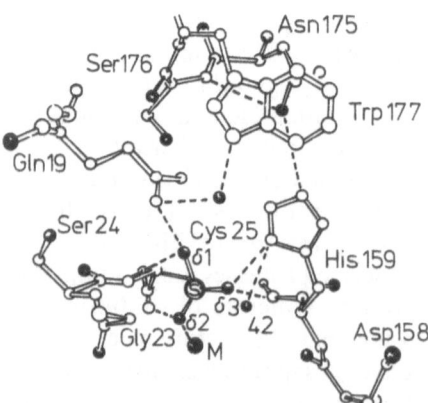

Fig. 25. The active site region of papain [1262] *Dotted line* indicates hydrogen bonds; *M* methanol (crystalline); *42* water molecule

Apparently, the low pK_a value of histidine depends on the interaction with thiolate ion of Cys25 (see Sect. 7.1).

Though there is a formal analogy between system Asn182-His159-Cys25 and the charge relay system in chymotrypsin (Asp102-His57-Ser195), it is evident that cysteine proteases are devoid of such a system, the role of link Asn182-His159 is the holding of the His imidazole ring position.

The active site of cysteine proteases contains also the Gln19 side chain. Its $O^{\varepsilon 1}$-atom forms the hydrogen bond with O^{γ}-atom of Ser183.

The side chains of nonpolar amino acids surrounding the active site form a binding substrate region. A relatively deep "pocket" is distant from catalytically active groups so that it evidently binds substrate residue P_2 (the second from the cleaved bond to the N-terminal side) and is formed by the side chains of Tyr67, Pro68, Val133 and Val157 of the R-domain.

The depth of the active site cleft estimated by spin-labeled papain derivatives [1338] is about 10 Å. In ficin this value is smaller (≈ 8 Å) [1339].

In cysteine proteases there are no groups which could play a regulatory role similar to that fulfilled by Asp194 and Ile16 in α-chymotrypsin.

2.6.3 Aspartic Amide Hydrolases

Aspartic amide hydrolases function in acidic or weak acidic media, therefore it is expected that carboxyls, virtually the only groups in proteins capable of ionization under these conditions, are the catalytically active groups. In fact, some reagents for carboxyl groups completely inactivate aspartic proteases [1340, 1341]. These are diazoketones – derivatives of acylamino acids and peptides [1342, 1343] as well as some epoxycompounds [1344]. Diazoketones in the presence of copper salts react with one certain group in the pepsin-carboxyl group of Asp215. The reaction runs through the formation of carben, which is a reactive part (see Chap. 5).

Analysis of the amino acid sequence adjacent to the modified group in various aspartic proteases shows a considerable conservativity of this region (Table 16).

Epoxycompounds, e.g., p-nitrophenoxypropyloxide react with another carboxyl group belonging to Asp32 in the pepsin sequence [1344]. Besides, under

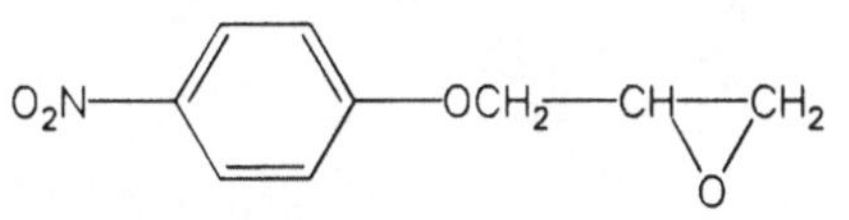

certain conditions this reagent interacts with Asp215 through its secondary hydroxyl group [1345]. The sequence of the residues within the region of Asp32 is also fairly conservative. Of interest is the similarity of the peptide sequences in the region of Asp32 and Asp215 which might occur due to evolutionary peculiarities of the above aspartic proteases.

The catalytically significant residues Asp32 and Asp215 are embedded deeply in the extended cleft formed by two protein domains and are actually crossing the total globule (Fig. 26). Carboxyl groups of these residues are near each other and form a hydrogen bond. Asp32 is buried in the protein globule and forms hydrogen bonds with O^γ-atom of Ser35, with NH group of Gly34 in the backbone, and with a water molecule. The carboxyl group of Asp215 has an identical surrounding. It forms, in addition to the bond with Asp32, hydrogen bonds with O^γ-atom of Thr218, NH group of Gly217 and the same water molecule (in pepsin N355) located between the two carboxyl groups of the active site within the plane of group Asp215. There are at least two water molecules close to these groups, one of which is probably a key element of catalysis (see Chap. 7). Of other functional protein groups within the active site region one should mention the COOH group of Asp304 forming H-bond with the peptide carbonyl of the backbone residue Thr216, ion pair, formed by Lys308 and Asp11, and the hydroxyl group of Tyr75. The latter is in the loop, which partially covers the active site cleft. Upon rotation around bond C_α–C_β the hydroxyl group of this residue and carboxyl groups of Asp215 and Asp32 draw together.

The side chains of Phe189 (in penicillopepsin or Tyr189 in pepsin), Ile213 and Tyr299 (in pepsin Leu299), which can form hydrophobic binding sites of the enzyme, are located within the active site region.

Protease B from *Scytalidium lignicolum*, differing from the above enzymes, is insensitive to the majority of specific inhibitors of aspartic proteases. The glutamic acid residue functions in the active site of this enzyme [1346]. Interestingly, the amino acid sequence near this residue is homologous with

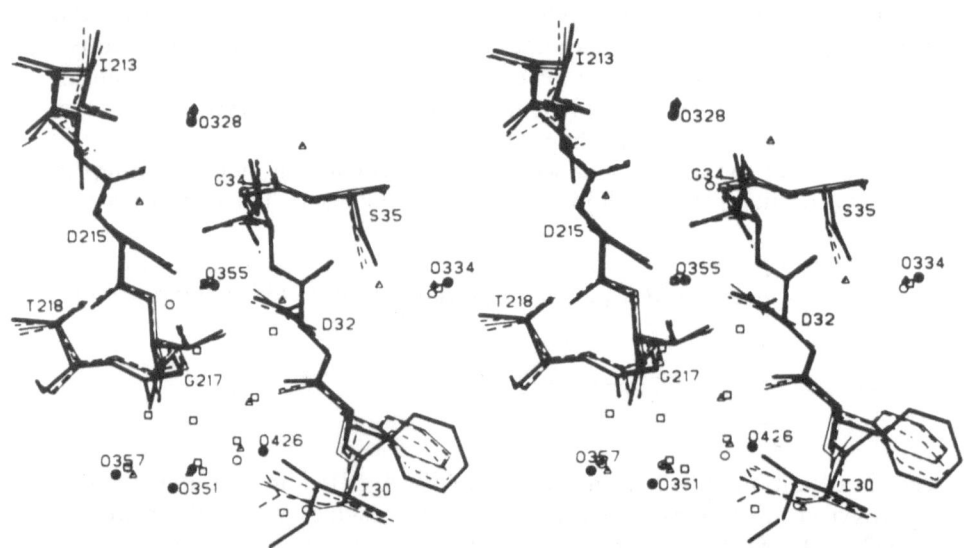

Fig. 26. Superimposed active sites of four aspartic proteases (stereo image). *Solid lines* pepsin; ● water; *thin lines* penicillopepsin; △ water; *short-dotted* lines rhizopuspepsin; □ water; *long-dotted lines* endothiapepsin; ○ water [1270]

the sequence within the region of the active sites of other aspartic proteases:

protease B: Cys—Gln—Thr—Ala—Ile—Leu—Glu—Thr—Gly—Phe

pepsin: (32)   Phe—Thr—Val—Ile—Phe—Asp—Thr—Gly—Ser

         (215)   Ala—Ile—Leu—Asp—Thr—Gly—Thr

Finally, proteases of retroviruses should be mentioned. In the dimeric state they function as aspartic proteases. Residue Asp25 is the catalytically active site in the protease of HIV-1. Apparently, upon dimerization, this residue in each protomer occupies positions corresponding to Asp32 and Asp215 in pepsin [1347].

2.6.4 Metal-Containing Amide Hydrolases

Metal-containing amide hydrolases with the known structure of the active site contain zinc atoms bound to the protein. Despite a big difference in the primary and spatial structures of two main representatives of Zn-dependent proteases – carboxypeptidase A and thermolysin – the structures of their active sites are similar [1348, 1349].

The Zn^{2+}-ion in carboxypeptidase and thermolysin has four coordination bonds. Three of them are formed with functional groups of the protein molecule and one with water molecule. In the metal coordination sphere there are two histidine residues and one glutamic acid residue. However, their mutual disposition in carboxypeptidase A and thermolysin is different (Table 19).

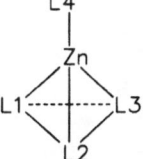

Table 19. Ligands of the Zn atom [1348]

Enzyme	Function	L_1	L_2	L_3	L_4	Proton-acceptor group
Carboxy-peptidase A	Ligand	His196 ($N^{\delta 1}$)	Glu72	His69 ($N\delta^1$)	H_2O	Glu270
	Orienting group	Ser194		Asp142		
Thermolysin	Ligand	His142 ($N^{\epsilon 1}$)	His146 ($N^{\epsilon 1}$)	Glu166	H_2O	Glu143
	Orienting group	Asp170	Asn165			

Along with the groups directly entering the coordination sphere of the zinc atom, there are groups which orient the rings of two ligands relative to the metal atom (Fig. 27). These groups differ in carboxypeptidase and thermolysin. In the close vicinity of the Zn atom (4.2–4.9 Å) the glutamic acid residues, probable proton acceptors, are located.

There is a hydrophobic "pocket" formed by side chains of Tyr192 and Phe279 in carboxypeptidase near the Zn atom. The guanidine group of Arg145 and the phenol group of Tyr248 occupy an adjacent place [149].

In thermolysin the region near the atom is more extended than that in carboxypeptidase. It is formed by aromatic residues Trp115 and Phe114 as well as by a group of hydrophobic side chains Val139, Leu133, Phe130, Leu202 and Val192. An essential difference between this enzyme and carboxypep-tidases is His231 in the active site of the former, which plays an important role in catalysis. This residue is bound to the carboxyl group of Asp226 via its $N^{\delta 1}$-atom which, to a certain extent, resembles the situation in serine proteases.

As in carboxypeptidase, the active site in thermolysin contains the side chain of the arginine residue (Arg203).

Dipeptidylcarboxypeptidase I (angiotensin I-converting enzyme) has the sequence of residues adjacent to catalytically crucial residues His and Glu, similar to that of thermolysin. This enzyme has two potential Zn-binding sites in different domains, but it can bind only 1 g-atom of Zn per mol of protein [1350]. The sequence within the region of active Glu residue in angiotensin-

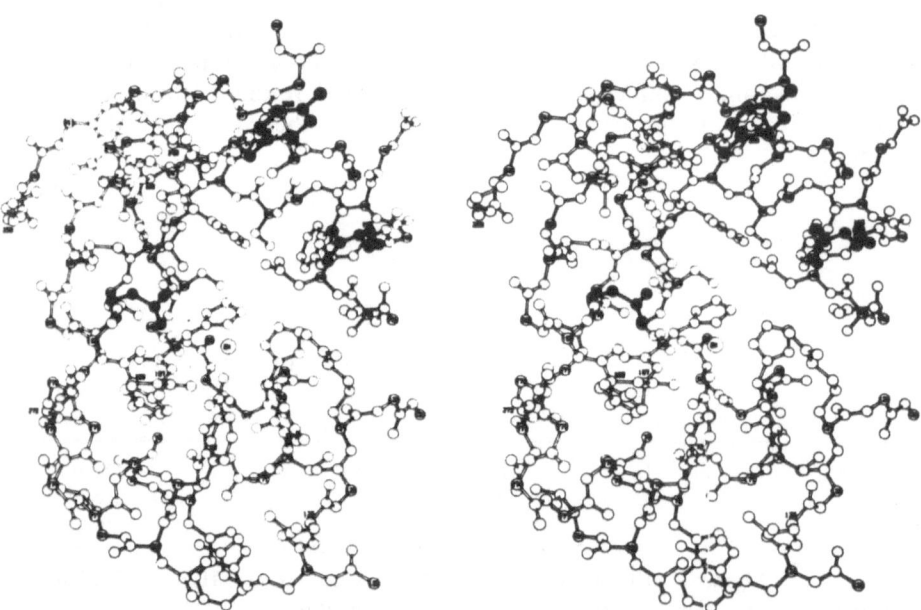

Fig. 27. Stereo image of the active site region of carboxypeptidase A [149] *Black circles* carbon atoms of side chains of catalytically important residues Arg145, Tyr248 and Glu270; *open circles* atoms C, N and O of backbone

converting enzyme has higher homology with the sequence of carboxy-peptidase A (near Glu270) as compared to that of thermolysin [1351].

The glutamic acid residues important for catalysis are revealed by chemical modification and directed mutagenesis in neutral protease from *Aeromonas* sp. [1352] and endopeptidase "24.11" [1353]. Presumably, SH group of the cysteine residue is involved in Zn^{2+}-ion binding in the active site of leucine aminopeptidase [1354, 1355] (however, see [1356]).

2.6.5 Structural Similarity and Differences in Active Sites of the Enzymes of Various Groups

The above data show an extremely high homology in the active site structure within each enzyme group. This is displayed even in the enzymes, the primary and spatial structures of which are markedly distinguished. The structure differences within the group are mainly induced by their various specificities. Still, it should be remembered that only information extracted when studying a few representatives of each enzyme group underlay all these conclusions.

Of interest is the problem of similarity and difference in the structure of active sites of the enzymes of various groups.

Such a comparison between the serine and cysteine proteases [1357], serine and metal-containing [1348], aspartic and metal-containing [1358] enzymes reveals a certain identity in geometry of all four groups of amide hydrolases or, at least, their representatives yielded by the X-ray crystallography studies. The similarity of disposition of the active site groups in trypsin and thermolysin with those of phospholipase A_2 has been established as well [1359].

According to Table 20 superposition of the structures of chymotrypsin and papain active sites results in different positions of functionally significant atoms not exceeding 1 Å.

A comparison of the positions of atoms and groups in trypsin and thermo-lysin (complexes with quasisubstrates) equivalent in functions demonstrates that the active sites of the two enzymes differ much (Table 21).

Proton-acceptor groups His57 and Glu143, respectively, though differing in position, are located at one side of the region, probably, occupied by a cleavable bond.

Table 20. Distances in the active sites of chymotrypsin and papain at structure superposition [1357]

Chymotrypsin			Papain			Distance, Å
His	57	$N^{\delta 1}$	His	159	$N^{\varepsilon 2}$	0.4
His	57	$N^{\varepsilon 2}$	His	159	$N^{\delta 1}$	0.8
Asp	102	C^{γ}	Asn	175	C^{γ}	0.9
Asp	102	$O^{\delta 1}/O^{\delta 2}$	Asn	175	$O^{\delta 1}/N^{\delta 2}$	0.7
Ser	195	O^{γ}	Cys	25	S^{γ}	1.0

Table 21. Comparison of positions of equivalent groups in trypsin and thermolysin [1348]

Group	Trypsin			Thermolysin			Distance, Å
Nucleophile	Ser	195	O^γ	OH_2-Zn			0.6
Anion site	Gly	193	NH	Zn			2.1
Proton acceptor	His	57	$N^{\epsilon 2}$	Glu	143	$O^{\epsilon 2}$	2.1

2.6.6 The Structure of Protease Active Sites in Crystals and Solution

To what extent do data, at least those from X-ray analysis, describe the enzyme structures in solution? Unfortunately, at present there are no techniques which allow spatial positioning of protein atoms in solution precisely. The physico-chemical methods for characterizing the protein molecule either have too low resolution or give a sum of data hardly assignable to certain protein atoms and groups. Therefore, the above question is still unanswered [1360].

As known, the conclusions based on structural analysis of crystals are, as a rule, in line with the protein properties in solution. Hence, in many cases the protein undergoes no significant structural changes during crystallization.

Deviations may arise in the following situations: (1) if crystallization occurs under conditions far from those favorable for enzyme activity; (2) if several protein forms in solution differing in biological activity and in ability of crystallization are in equilibrium; (3) if the packing of the molecules in crystal strongly distorts the molecular structure.

So chymotrypsin crystals for X-ray analysis have been obtained at pH 4.5, i.e., under conditions when the enzyme is practically inactive relative to typical substrates. However, the rate of decay of derivatives of chymotrypsin acylated by Ser195 under these conditions was shown to be equal in crystals and solution [1361].

α- and γ-Chymotrypsins formed upon crystallization under various conditions exemplify interconverting enzyme forms [1362]. In solution one of these forms is converted into the other very slowly. The properties of α- and γ-forms are not identical [1363]. Thus α- and γ-chymotrypsins in solution, presumably, have slightly differing structures. At the same time, comparison of the tertiary structure of crystals of these chymotrypsin forms [1227, 1364] does not reveal any substantial difference between them. However, it is mentioned that γ-chymotrypsin according to the latest data [1228] is, presumably, the complex (or acylenzyme) of protein with the tetrapeptide.

^{15}N-NMR spectra of α-lytic protease in solution and in solid state indicate unimportant differences of these two enzyme states [1365].

Carboxypeptidase A has dramatically distinguished structures in crystal and in solution. In addition to X-ray information available it has been demonstrated that: (1) the character of kinetic dependences of carboxy-peptidase A in crystal and solution varies [1366, 1367]; (2) the surrounding of chromophore labels incorporated to Tyr248 in crystal and in solution differs as well [1368–1370].

The unusual form and the crystallographic characteristics of carboxy-peptidase A crystals subjected to X-ray analysis elicit these distinctions [1371]. Conformations of carboxypeptidase B were shown to be unlike in crystal and solution [1372]. The situation was the same for the enzymes of other classes (for example [1373]).

Thus, one cannot be sure a priori that the protein conformation found in crystal corresponds to it in solution. In each case this problem should be carefully considered. Application of synchrotronic irradiation to protein crystallography seems promising since it allows a study of reactions in crystals within a millisecond period [1374].

2.7 Conformational Mobility of Enzymes

There is a wealth of information in favor of the mobility of groups in protein molecules and variety of conformational protein states (see reviews [1375–1379]). External conditions (pH, temperature, etc.) or attachment of ligands often entails conformational changes. However, even under fixed conditions the proteins exist in a few or many states, which rapidly interconvert. This follows, first, from the experiments on isotope exchange of protons in proteins, which reveals both the rapid and the slow stages; the latter is assigned to the proton exchange inside the globule, its rate being limited by the rate of protein conformational rearrangement [1380]. Secondly, such conformational changes can be detected by using "reporter groups" incorporated into the protein, and when studying spectral or other physicochemical changes in the protein. For instance, the pH-independent conformational transfer with an apparent rate constant of about 5 s^{-1} has been detected in the case of modified carboxypeptidase [1381]. The conformational mobility in proteins has been registered also by the methods of nuclear magnetic resonance of high resolution [1382] according to the position and the form of signals from certain atoms and groups. There are many other methods to identify conformational rearrangements in proteins [1383–1385], but they are outside the scope of this book. Another method should be mentioned here: a possibility in principle of calculating a priori relatively small protein molecules. It provides at once information on the energies of a large set of protein states and, consequently, its conformations [153, 1386], as well as a possibility of computer modeling of the protein mobility by the molecular dynamic methods [1387, 1388].

2.8 Proteolytic Activity of Enzymes Other than Amide Hydrolases. Catalytically Active Antibodies

The ability of enzymes to catalyze hydrolysis of the amide bond has been revealed in the enzymes of subclasses 3.1 and 3.2, namely, in lysozyme [1389,

1390] and cholinesterases [1391, 1392]. Proteolytic activity of lysozyme can be a reason for artifacts on isolation of microbial proteins by enzymatic lysis of bacteria cell walls.

In 1987 two teams simultaneously discovered an ability for specially obtained antibodies to catalyze hydrolytic and some other reactions [1393, 1394].

The method for obtaining these antibodies includes immunization of animals by the so-called "analogs of transition states" of substrates (see Chap. 5). For example, for esters of type I such analogs are phosphonates II [1395]:

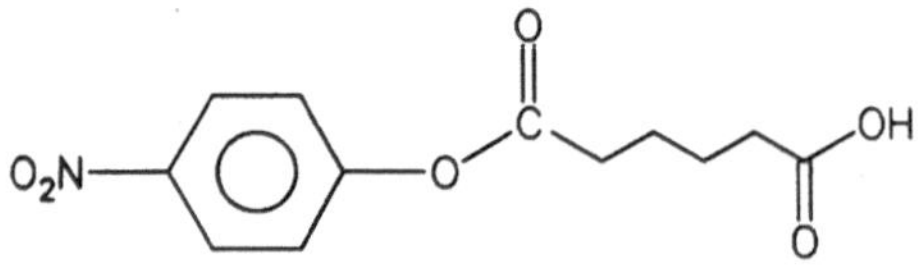

I 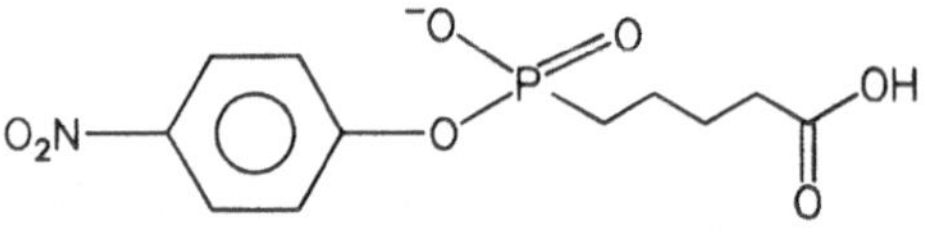   II

It is possible to find clones producing antibodies active in hydrolysis of the corresponding esters and with the properties close to those of enzymes. Particularly, they possess definite stereospecificity and are capable of synthesizing the amide bond [1396, 1397]. The catalytic activity of such antibodies achieves six orders of magnitude [1398]. Antibodies with artificially introduced imidazole groups have been obtained [1399]. However, no antibodies catalyzing hydrolysis of amide bonds have been yet described (see review [1400]).

2.9   Conclusion

Numerous enzymes with diverse properties and structures catalyze hydrolysis of the amide bond. Despite this variety, most of the enzymes of this class, if not all, can be assigned to four main groups differing in the catalytic apparatus structure and, evidently, in the mechanism of action. Each group contains several subgroups, the enzymes of which can be clearly distinguished by their primary and spatial structures from other subgroups of the same group. However, within the active site range, as follows from quite a few data now available, the similarity remains rather strong. Moreover, the enzymes of different groups share some structural features. Presumably, all reflect the unity of functions of amide hydrolases. The observed differences are related, on the one hand, with the evolutionary prehistory of enzymes of this type and, on the other hand, with structural peculiarities of the substrates of these enzymes. Since the structure of the most cleavable amide bond depends slightly on the structure of the rest moiety of the substrate molecule, the most important properties of catalysis of reactions of hydrolytic cleavage of amides should have much in common, and this generality is expressed in the identity of active sites of amide hydrolases.

Chapter 3   Nonenzymatic Hydrolysis. Models

Current views on the mechanisms of enzymatic reactions in many respects have been initiated with an intensive research into nonenzymatic processes catalyzed by relatively simple compounds. First of all, this refers to hydrolytic transformations of the derivatives of carboxylic acids – the object of thorough study for the last decades (see reviews [8, 1401–1406]).

3.1   Thermodynamics

Amide hydrolysis in aqueous solutions at pH close to neutral is, in fact, the sum of three processes:

1. $R{-}CONH{-}R^1 + H_2O \; \overset{K_H}{\rightleftharpoons} \; R{-}COOH + H_2N{-}R^1$

2. $R{-}COOH \; \overset{K_A}{\rightleftharpoons} \; R{-}COO^- + H^+$

3. $R^1{-}NH_2 + H^+ \; \overset{K_B}{\rightleftharpoons} \; R^1{-}NH_3^+$

$$R{-}CONH{-}R^1 + H_2O \; \overset{K_C}{\rightleftharpoons} \; R{-}COO^- + {}^+H_3N{-}R^1 \qquad (1)$$

Thus the equilibrium constant of hydrolysis (K_C) observed in the experiment is equal to

$$K_C = \frac{[R{-}COO^-]\cdot[{}^+H_3N{-}R^1]}{[RCONHR^1]\cdot[H_2O]} = K_H\cdot K_A\cdot K_B , \qquad (2)$$

where K_H is the equilibrium constant in a hydrolytic reaction; K_A is the dissociation constant of the formed acid; K_B is the constant of amine association (protonation). The activity of water is assumed to be a unit in the K_C expression. Substituting K_B by a usually used constant of dissociation of the conjugate base (K_{BH}) we get

$$K_C = K_H\cdot K_A/K_{BH} \qquad (3)$$

or in terms of free energy

$$\Delta G_C = \Delta G_H + \Delta G_A - \Delta G_{BH} \qquad (4)$$

It is possible to calculate the free-energy change of hydrolysis from the experimentally obtained value of the equilibrium constant (and free energy change) at the given pH and temperature, as well as the values of the free-energy change at acid and base ionization.

The free-energy change (ΔG) of the system is usually expressed at pH 7 in biochemical studies. Therefore ΔG should be determined or calculated for this pH. The same is true for the free-energy changes of ionization being calculated for component concentrations at pH 7. These values are related to K_A and K_{BH} through the following expressions:

$$\Delta G_A = - RT\ln\left(1 + \frac{K_A}{10^{-7}}\right) + RT\ln(1 + K_A),\tag{5}$$

$$\Delta G_{BH} = - RT\ln\left(1 + \frac{10^{-7}}{K_{BH}}\right) + RT\ln\left(\frac{K_{BH} + 1}{K_{BH}}\right).\tag{6}$$

According to Table 22 the reactions of hydrolysis are endoergic, i.e., those with absorption of the energy. Ionization of the forming acid and base proceeds with the energy release, so the total change of the standard Gibbs free energy for peptides within pH 4 to 7 tends to zero (Fig. 28).

The free-energy change (ΔG_H) slightly depends on the peptide structure. Major changes of this value are observed for peptides containing a free amino or carboxyl group near a cleavable bond. This occurs due to a distinct change of pK_a of the amino or carboxyl group in free amino acids as compared to their N- or C-substituted derivatives. The amino acid derivatives, such as anilides or esters, also differ much from the substituted peptides by the ΔG_H value.

The free-energy change of hydrolytic reaction (ΔG_H) at about pH 7 is a function of basicity of the leaving group. A few empiric dependences for these values are proposed, for example [1410]:

$$\Delta G_H = 0.3 + 0.6 \cdot pK_{BH}.$$

Interestingly, the coefficient β of variable pK_{BH} is always about 0.6 in these correlative dependences, it being a form of the more general Brønsted equation

Table 22. ΔG_C and ΔG_H values for certain peptides, amides and esters

Compound	ΔG_C, kcal/mol (pH 7, 25 °C)	pK_a	pK_{BH}	ΔG_H, kcal/mol	References
HCO–NH$_2$	−3.54	3.75	9.26	4.1	[1407]
CH$_3$CH$_2$CO–NH$_2$	−2.1	4.87	9.26	4.0	[1408]
CH$_3$CO–NHC$_6$H$_5$	−3.7	4.75	4.58	−0.36	[1409]
AcPhe(NO$_2$)–NH$_2$	−2.05	3.2	9.26	6.2	[1410]
AcPhe–NHC$_6$H$_4$NO$_2$	−3.94	3.3	1.02	1.1	[1411]
AcPhe–GlyNH$_2$	−0.8	3.3	7.93	5.4	[1412]
AcPhe–AlaNH$_2$	−0.91	3.3	8.2	5.8	[1412]
AcPhe–TyrOEt	−0.05	3.3	7.5	5.84	[1413]
Gly–Gly	−4.3	2.35	9.78	3.9	[1414]
AcPhe–OMe	−6.0	3.3	–	1.2	[1415]
CH$_3$CO–OCH$_2$CH$_3$	−4.72	4.75	–	1.6	[1416]

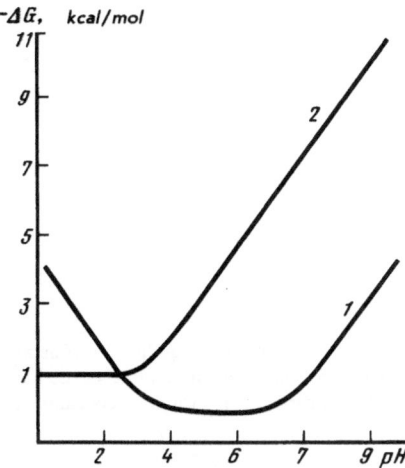

Fig. 28. Dependence of the overall free energy of hydrolysis of ethyl ester of N-acetyl-*L*-phenylalanyl-*L*-tyrosine; *1* (amide bond) and methyl ester of N-acetyl-*L*-phenylalanine; *2* (ester bond) on pH [1413]

(see Sect. 3.4.2). This value is the same for most of the reactions of acyl transfer onto the compounds containing O- or N-atoms (but not S-atom) [1409]. The hydrolysis involves not only processes 1–3 outlined in scheme (1). The nucleophile (water) addition itself is a multistep process.

Amides are resonance-stabilized structures, so nucleophile addition to the carbonyl group leads to loss of the energy of resonance stabilization. Thus, the nucleophile addition can be considered as a combination of at least two processes: electron rearrangement resulting in a bond localization and a loss of resonance stabilization, and the nucleophile addition itself. An attempt [1417] was made to estimate the free energy of the two processes separately by using data on barriers of rotation around the C–N bond and on the nucleophile addition to ketones [1418] with the corresponding corrections concerning inductive and resonance effects of substituents. The results show a change in resonance stabilization as making the main contribution to free energy of the nucleophile addition (Table 23).

The values presented in Table 23 certainly cannot be compared to ΔG_H values in Table 22, as here but a part of the process, which includes only the

Table 23. Change of free energy for a nucleophile addition (ΔG_1) and of free energy for amide resonance stabilization (ΔG_2) [1417]

Amide	Nucleophile	K_1, M^{-1}	ΔG_1, kcal/mol	ΔG_2, kcal/mol
$HCONHCH_3$	CH_3OH	2.2×10^3	−4.6	22.0
$CH_3CONH(CH_3)_2$	CH_3OH	1.2	0.11	18.7
„	CF_3CH_2OH	3.8×10^{-2}	1.96	18.7
„	C_6H_5OH	1.7×10^{-2}	2.44	18.7
„	C_2H_5SH	60	−2.45	18.7
$HCON(CH_3)_2$	CH_3OH	2.2×10^3	−4.6	21.0
$C_6H_5CON(CH_3)_2$	CH_3OH	6.0×10^{-3}	3.07	15.8
$HCONHC_6H_5$	$NH(CH_3)_2$	4.9×10^7	−10.6	17.7

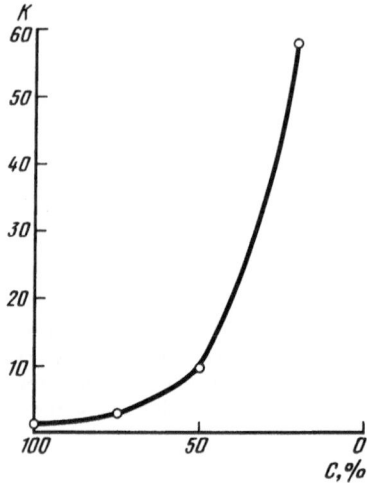

Fig. 29. Change of the apparent equilibrium constant for synthesis-hydrolysis of propionamide vs the concentration of *n*-propanol in water (according to [1408])

formation rather than the breakdown of the intermediate product of the nucleophile addition to a carbonyl carbon atom, is considered:

$$R\!-\!\overset{\overset{\displaystyle O}{\|}}{C}\!-\!NHR^1 + HX \rightleftharpoons R\!-\!\overset{\overset{\displaystyle OH}{|}}{\underset{\underset{\displaystyle X}{|}}{C}}\!-\!NHR^1.$$

This product is usually called a tetrahedral intermediate compound, it will be detailed in Sect. 3.3.2.

Calorimetric studies on hydrolysis of the peptide bond [1419–1422] reveal that the reaction is accompanied with a rather slight negative thermal effect. The effect is larger, when the pK_a value of the hydrolysis products formed is higher. So ΔG_C° for hydrolysis of $BzTyrGlyNH_2$ at bond Tyr-Gly is -1.55 kcal/mol, while for hydrolysis of $BzTyrNH_2$ — 5.84 kcal/mol.

Since ionization of the forming amine is accompanied with a large negative change of enthalpy (-12.4 kcal/mol [1423]), the hydrolysis proceeds with heat absorption which is lower when the pK_a value of the leaving amine is lower.

Estimation of entropy change for the two above compounds yields the values -5 and -12.5 e.u. Ionization of the amino group results in the entropy change close to zero, whereas ionization of the carboxyl group gives $\Delta S^\circ = -22$ e.u. Thus the cleavage of the amide bond is accompanied by a positive change in entropy of about 10–15 e.u.

The equilibrium position is shifted much to synthesis if the medium polarity changes (Fig. 29) [1408, 1424].

3.2 Kinetics

The equilibrium state in the system (7) is reached very slowly. A direct detection of the hydrolysis rate of the peptide bond of N-methylacetamide at neutral pH

$$RCONHR^1 + H_2O \underset{k_a}{\overset{k_h}{\rightleftharpoons}} RCOOH + R^1NH_2 \tag{7}$$

and room temperature [1425] provides the pseudo first-order rate constant equal to 3×10^{-9} s^{-1} ($\tau_{1/2}$ 7 years), while the same constant obtained in base hydrolysis by extrapolation of the concentration of OH$^-$ ions to zero turned out to be only 3.5×10^{-13} s^{-1} [1426].

The rates of the amide hydrolysis are much higher under real conditions, as ions OH$^-$ and H$_3$O$^+$, which catalyze hydrolysis, are present at any pH value in water solution. Thus the constant of the hydrolysis rate (k') observed in the experiment is an effective value and consists of at least three values characterizing the rates of the water-catalyzed hydrolysis (k_{H_2O}), hydrolysis catalyzed by hydroxyl (k_{OH}) and oxonium ion (k_H):

$$k' = k_{H_2O} + k_{OH}[OH^-] + k_H[H_3O^+] \,. \tag{8}$$

In certain pH intervals, where concentrations of hydroxyl or oxonium ions can be neglected, Eq. (8) gives Eqs. (9) and (10):

$$k' = k_{H_2O} + k_{OH}[OH^-] \,, \tag{9}$$

$$k' = k_{H_2O} + k_H[H_3O^+] \,. \tag{10}$$

These equations describe base and acid hydrolyses or, to be more accurate, hydrolyses promoted with ion OH$^-$ and catalyzed with oxonium ion. For obtaining constants k_{H_2O}, k_{OH} and k_H the dependences of the observed rate constant on the concentration of OH$^-$ or H$_3$O$^+$ ions are plotted (Fig. 30). The point, where the straight line is crossing the Y-axis gives k_{H_2O}, and the slope $-k_{OH}$ (or k_H).

Linearity of the plots of the type presented in Fig. 30 indicates the first order in the concentration of OH$^-$ or H$_3$O$^+$ ions. However, the first order cannot be kept in the whole interval of the concentrations, and this indicates a more complicated mechanism of hydrolysis (see Sect. 3.4.2).

The rate constants for base and acid hydrolyses of amides are significantly higher than those of the noncatalyzed process [1429]. Their values for most of the amides are within $1 \times 10^{-4} \div 1 \times 10^{-5}$ M^{-1}s^{-1} (Table 24). So the base

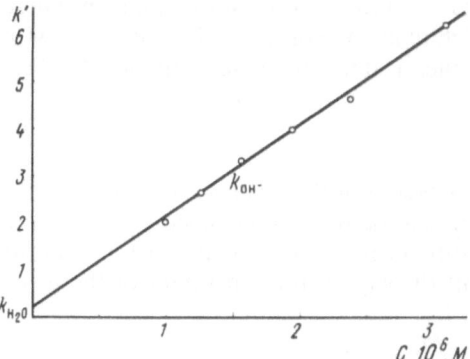

Fig. 30. A typical plot for the observed rate constant of alkaline hydrolysis vs the concentration of hydroxyl ions

Table 24. Constants of basic and acidic hydrolysis of some amides and other derivatives of carboxylic acids (25 °C)

Compound	$k_{OH} \cdot 10^4$, $M^{-1}s^{-1}$	$\Delta H^{\neq}$, kcal/mol	$\Delta S^{\neq}$, e.u.	$k_H \cdot 10^4$, $M^{-1}s^{-1}$	$\Delta H^{\neq}$, kcal/mol	$\Delta S^{\neq}$, e.u.	References
CH_3CONH_2	38	13.3	−32.8	2.1	22.0	−18.4	[1426]
$CH_3CONHCH_3$	3.29	16.6	−29.1	0.145	20.6	−21.6	[1426]
$CH_3CONHC_3H_7\text{-}n$	1.06	16.4	−32.7	0.143	17.6	−31.8	[1426]
$CH_3CON(CH_3)_2$	6.8	15.1	−31.4	0.134	20.5	−22.2	[1426]
$CH_3COOC_2H_5$	980	–	–	1.12	–	–	[1427]
$BzPheOC_2H_5^{a}$	500	–	–	0.033	–	–	[1428]
$CH_3COOC_6H_4NO_2\text{-}m$	5.5×10^4	–	–	0.19	–	–	[1427]
$CH_3COSCH_3^{b}$	400	12.5	–	166	16.5	–	[1401]

[a] in 85% C_2H_5OH (k_{OH}) and in 70% dioxan (k_H).
[b] in 62% acetone.

hydrolysis of esters runs faster than that of amides. At the same time, the acid hydrolysis of the ester bond is by 3–4 ordering of magnitude slower than the base hydrolysis. For amides, the differences in the rate constants of alkaline and acid hydrolyses are slight.

3.2.1 Influence of Structure

The hydrolysis rate of derivatives of carboxylic acids depends on the structure of substituents at the cleaved group. Polar, steric and resonance effects of the substituents influence this dependence (see reviews [1430–1433]). The polar, or inductive, effect embraces all processes through which the substituent can change electrostatic interactions in the reaction center, as compared to a certain substituent taken as a standard. This can be charge separation, electron delocalization, etc. If the substituent is conjugated with the reactive group, such a polar effect is considered as a resonance effect.

The steric effect occurs upon repulsive interactions at the drawing together of two covalently unbound atoms. One distinguishes the primary steric effect by direct repulsion of interacting atoms or groups, and the secondary steric effect by a change of polar and resonance effects as a result of repulsion. There exist a great number of equations which relate one or a few of these effects with the rate and equilibrium constants in chemical reactions. Among them the best known is the Hammett equation:

$$\lg k = \lg k_0 + \rho\sigma , \tag{11}$$

where k is the rate constant of the conversion of the given substance; k_0 is the rate constant for the substance, chosen as a standard, σ the reaction constant of the substituent, a measure of the substituent polarity being independent of the reaction type, ρ the reaction constant displaying the sensitivity of the given reaction to a polarity change.

Taft [1431] proposed another correlative equation that separates the steric and the polar effects (including resonance) of the substituents:

$$\sigma^* = [\lg(k/k_0)_B - \lg(k/k_0)_A]/2.48 , \tag{12}$$

where indices A and B refer to acid and base hydrolyses, respectively, and the value 2.48 is introduced so that σ and σ^* have comparable values.

A set of assumptions underlies this correlation: (1) polar, steric and resonance effects make an additive contribution to free-energy activation; (2) steric and resonance effects make the same contribution to the reactions of base and acid hydrolyses; (3) polar effects of base hydrolysis are much higher than those in the reactions of acid hydrolysis.

Thus, if these statements are true, the ratio of the rates of acid hydrolysis of two compounds, of which one is chosen as a standard, characterizes the sum of steric and resonance effects, and if the conjugation of substituents with the reactive atoms is absent, only steric effects are defined. The expression (12) can be written as

$$\lg(k/k_0) = \sigma^* \rho^* + \delta E_S , \tag{13}$$

where E_S is the steric constant; δ the constant showing the sensitivity of the given reaction to the steric effects, ρ^* the constant describing the sensitivity of the reaction to the polar effects. It is noteworthy that time and again the role of polar effects in hydrolysis of derivatives of aliphatic carboxylic acids seems doubtful, in many works it is evidently overstated [1433–1435].

Data on kinetics of hydrolysis of amides and esters were often subjected to correlative analysis by Eq. (13) or its modified variant (see, for example, [1436–1438]). For alkyl substituents at a carbonyl carbon as well as at a nitrogen atom of the amide group the σ^* value is within the range 0–0.4 and seems slightly effective for the observed rate of hydrolysis. There exists an opinion that σ^* can constitute the steric constant (E_S), i.e., for alkyl groups there exists a relation between σ^* and E_S expressed by the equation [1439]:

$$E_S = 0.88 + 27.78\sigma^* - 1.9(n - 3) ,$$

where n is the number of hydrogen atoms at α-carbon atom.

The steric effect might involve also a resonance component. It has been suggested [1440] that the corrected steric constant $E_S^c = E_S - h(n - 3)$ is used where h is a reaction constant, which accounts for superconjugation, and n is the number of α-H atoms. The number of protons at the sixth carbon atom [1441] or the number of protons in position 7 (H-7) [1437] affecting the steric effect of the substituent was proposed to be taken into account.

Table 25 outlines some correlative equations for hydrolytic reactions.

The reaction rate constant can be often correlated with the basicity of the leaving group or acidity of the acid forming the amide, i.e., the dependence of the Brønsted type equation is preserved [1442, 1443]:

$$\lg k = \lg k_0 + \alpha \cdot pK_a .$$

More informative is analysis of the influence of the structure on the rate constants of separate stages of hydrolytic reaction (see, for example,

Table 25. Correlative equations for hydrolysis of some derivatives of acetic acid [1437]

Reagents	Solvent	Temperature, °C	Correlative equation
$CH_3CONHR + OH^-$	1 n. NaOH	75	$lgk = -1.67 + 0.9E_s^c + 0.056$ (H-7) $R = 0.984;\ s = 0.056$
$CH_3COOR + OH^-$	40% dioxan	35	$lgk = 1.31 + 1.1\sigma^* + 0.78E^c + 0.068$ (H-7) $R = 0.998;\ s = 0.044$
$CH_3COOR + H_3O^+$	14% H_2SO_4	25	$lgk = -1.78 - 0.042$ (H-7) $R = 0.866;\ s = 0.046$
$CH_3COSR + OH^-$	40% dioxan	35	$lgk = 1.14 + 0.66\sigma^* + 0.23E_s^c$ $R = 0.955;\ s = 0.061$

[1444, 1445]). According to the details below, the transfer of the acyl group in the derivatives of carboxylic acids is described as minimal with a two-stage scheme:

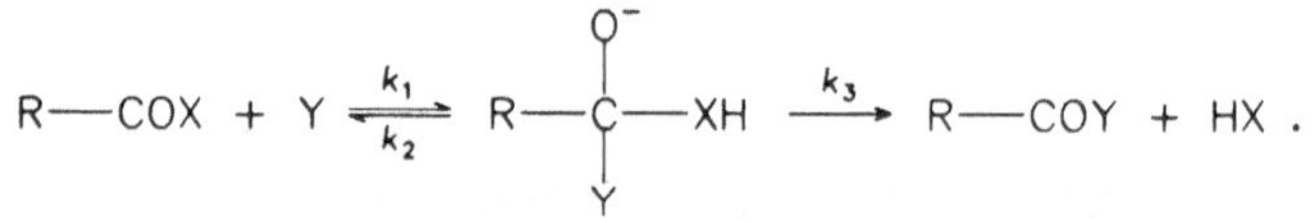

$$R-COX + Y \underset{k_2}{\overset{k_1}{\rightleftharpoons}} R-\overset{\displaystyle O^-}{\underset{\displaystyle Y}{\overset{|}{\underset{|}{C}}}}-XH \overset{k_3}{\longrightarrow} R-COY + HX .$$

The changes in the substrate structure can influence the observed constant, the value k_1 and the ratio k_2/k_3 [8] (Table 26).

We have considered the behavior of the "normal" amides in hydrolysis and acyl transfer reactions. Among substrates of amide hydrolases there exist abnormal amides, i.e., β-lactams entering the bicyclic system of penicillin antibiotics. The peculiarity of the β-lactam group is the lack of resonance interaction of an electron lone pair of the nitrogen atom with π-electrons of carbonyl double bond. This affects markedly the stability of the β-lactam cycle in hydrolytic reactions [1446–1448].

The same is observed for other types of resonance-destabilized amides – 2,2-dimethylquinuclidone (I) [1449], 3,4-dihydro-2-oxo-1,4-ethanoquinoline (II) and 2,3,4,5-tetrahydro-2-oxo-1,5-ethanobenzazepine (III) [1450–1452]. The rate of alkaline hydrolysis seems to increase by 6–8 orders of magnitude, as compared to normal amides [1450].

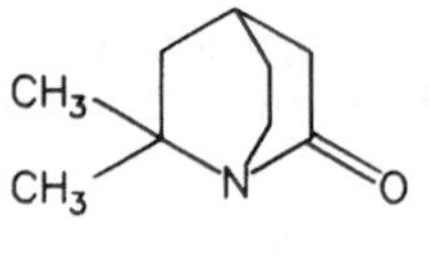

I 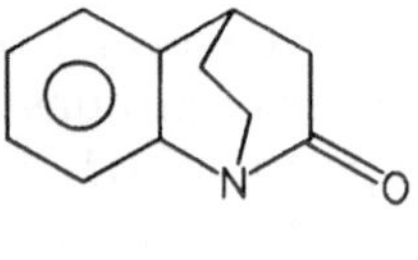   II 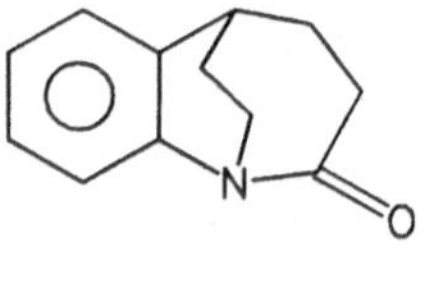   III

Thus resonance stabilization of the hydrolyzed group in derivatives of carboxylic acids makes an essential contribution to the stability of these compounds to hydrolysis. As stated above, the resonance stabilization is a crucial factor upon a change of free energy of nucleophile addition to amides (Table 23). Consequently, the compounds with the α,β-unsaturated bond,

Table 26. Influence of structural changes on the rate of the reaction
RCOX + Y

Increase of electron acceptor properties of substituent in group	k_1	$\alpha = k_2/k_3$	k'
R	Increase	Increase	Increase
X	„	Decrease	„
Y	Decrease	Increase	Decrease

"conjugated" with the amide group, should be hydrolyzed slower than uncon-
jugated amides. However, the resonance effect in hydrolysis (alcoholysis) of
amides turned out to be slight. So the activation energy of alcoholysis increases
by 0.6 kcal/mol upon introduction of α,β-double bond [1453]. A similar effect
at the addition of nucleophile to ketones is +3.1 kcal/mol [1454]. Studies of
acylimidazoles hydrolysis [1455] show that the resonance effect in these
compounds is also less than the expected one.

There are numerous correlations of substituent constants with various
physicochemical properties of carboxylic acid derivatives [1432, 1437] including amide conformations. The value σ^* correlates well with a total charge on
an alkyl group calculated by quantum-chemical methods [1456].

Correlative analysis is widely applied when studying hydrolysis mechanisms, and we intend to employ it to tackle these problems.

3.2.2 Influence of the Medium

The reaction medium can much influence not only the reaction rate [1457],
but also its mechanism [1458, 1459]. As a rule, the rate of amide hydrolysis
decreases with the lowering of the medium's dielectric permeability. This
appears to be due to the fact that hydrolysis is an ion-dipole or dipole-dipole
interaction, while the free energy of electrostatic interaction upon the formation of the activated complex of the reactive atoms is inversely proportional to
dielectric permeability:

$$\Delta G^* = \frac{Z_A Z_B e^2}{Dr^*},$$

where Z_A and Z_B are the ion charges, e the electron charge, D dielectric
permeability, r^* the distance between atoms in the complex. Hence the
dependence of the rate constant on D can be written as [1460]

$$\lg k = \lg k_0 - C \cdot \frac{1}{D},$$

where k_0 is the rate constant at $D \to \infty$ and $C = NZ_A Z_B e^2 / RTr^*$ (N is the
Avogadro number, R the universal gas constant, T the absolute temperature).

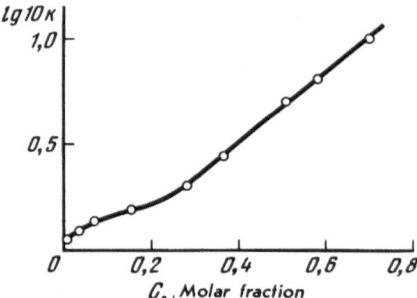

Fig. 31. Rate constant of the ester hydrolysis vs the dimethylsulfoxide concentration in water [1462]

This equation is not true in all cases. The reason for it is different structures of the transition state, specific solvation effects, etc. Thus the rate of ester hydrolysis in the acetone-water mixture linearly depends on the acetone concentration (up to 60%), whereas it increases much in dimethylsulfoxide-water mixtures at high concentrations of the organic solvent (Fig. 31) [1462].

A few correlation equations combining the reaction rate and solvent properties have been proposed [1461].

Ionic strength of the solution can also influence the hydrolysis rate of the carboxylic acid derivatives. These effects are mainly developed at high ion strength due to changes of the coefficients of the activity of the reactants, of solvation in a transition state, etc. Direct effects of ionic strength evident at its low values are usually observed in reactions between ions, which can be described by the Debay-Hukkel equation

$$\ln k = \ln k_0 + 2Z_A Z_B \alpha \bar{\mu} \, ,$$

where $\bar{\mu}$ is the ionic strength and α the constant depending on the solvent and temperature.

Finally, it is interesting to consider the changes of hydrolysis on substitution of water by deuterium oxide [1463, 1464]. The influence of D_2O is stipulated by a change of the rate of deuterium transfer as compared with a proton, as well as different solvating ability of H_2O and D_2O and, at last, by the so-called secondary isotope effects due to a proton change in the substrate by deuterium atoms. Strong influence of D_2O ($k_H/k_D \approx 2$–3) is observed only if the proton transferred at the rate-determining step.

Analysis of isotopic effects is widely applied to study the mechanism of hydrolytic reactions.

3.3 Mechanism

The description of the mechanism of the chemical reaction implies the knowledge of atomic coordinates of the reactive system in each moment of time and energy of the system at every position of reactive atoms. Therefore, the state of each atom is characterized at least by five variables. To describe the process for some reactive atoms in detail, the construction of a multidimensional potential surface, which the reactants pass on their way from the initial state to reaction products, is required. If some intermediate states differing in the energy are identified in this way, we get the picture presented in Fig. 32a. It is possible to design "the reaction pathway" as the projection of the above curve and to obtain the so-called map of alternative pathways [1465], where the orders of a formed and broken bond (Fig. 32b) serve as independent variables.

It is possible to define a few special points on any of the given plots: *1* initial state, *2* transition state, *3* intermediate, *4* final state. Point *2* in current theories of chemical reactions attracts special attention. According to the theory of absolute reaction rates, the reactants in a transition state are in dynamic equilibrium with an initial (and final) state:

$$A + B \xrightarrow{K^{\neq}} (A \dots B)^{\neq}, \quad \text{where} \quad K^{\neq} = \frac{[A \dots B]^{\neq}}{[A][B]}.$$

The standard Gibbs free-energy change corresponding to the K^{+} value is called activation free energy (ΔG^{+}). The frequency of the breakdown of the activated complex $(A \dots B)^{+}$ is equal to that of vibrations of the cleavable bond, i.e., $v = kT/h$, where k is the Boltzman constant, h the Planck constant, T the absolute temperature. At $25\,°C$ $v = 6.2 \times 10^{12}\,s^{-1}$. As

$$[A \dots B]^{+} = [A][B]e^{-\Delta G^{+}/RT},$$

so

$$v = v[A \dots B]^{+} = [A][B]\frac{kT}{h}e^{-\Delta G^{+}/RT},$$

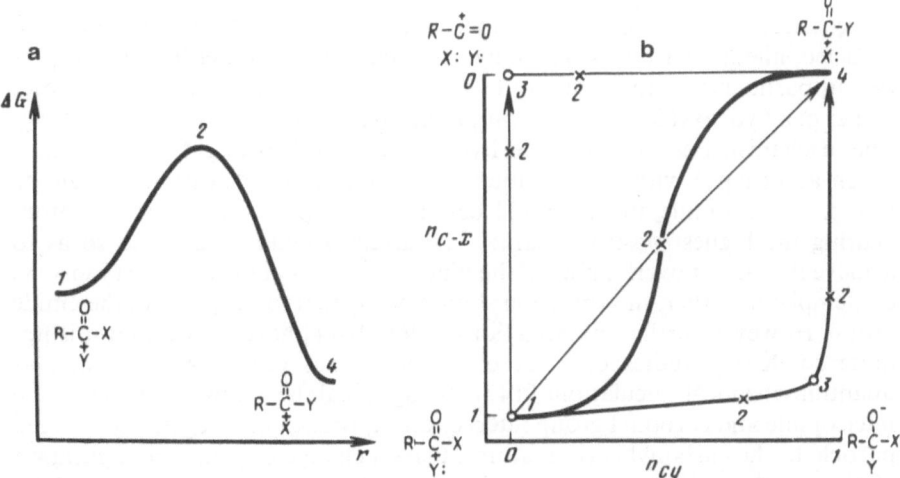

Fig. 32a, b. Representations of reaction pathways: **a** potential energy change along reaction coordinate r; **b** change in the order of cleavable and formed bond *n* (map of alternating pathways) [1465]; *1* original reagents; *2* transition states; *3* intermediates; *4* reaction products

i.e., the lower is the free energy of activation and the higher is the concentration
of the activated complex relative to the molecule concentration in the initial
state, the higher will be the reaction rate (Fig. 32a).

There exists a correlation between the position of the transition state along
the reaction coordinate (χ^{+} – order of the bond being formed), the magnitude
of the energetic barrier (ΔG^{+}), and the difference in the energy of initial
substances and end products (ΔG°), for example [1466, 1467]:

$$\chi^{+} = \frac{1}{2 - \Delta G^{\circ}/\Delta G^{+}} .$$

The transition from the initial state to the final one can proceed via
different transition states. Figure 32b for the reaction

$$RCOX + Y: \rightleftharpoons RCOY + X:$$

presents four pathways [1468]: two of them are dynamically coupled, i.e.,
characterized by a simultaneous change of bond orders C–X and C–Y and
kinetically coupled (passing through one and the same transition state), while
having different vibrational coordinates in these states. The other two pathways are dynamically uncoupled passing through the map angles and including either the formation of the intermediate acylium cation (mechanism S_N1) or
the formation of the tetrahedral intermediate.

Below it will be shown that amides and numerous other derivatives
of carboxylic acids are hydrolyzed via the formation of the tetrahedral
intermediate.

3.3.1 Nucleophilic Attack

Depending on the pH of the solution, OH^{-} ion or a water molecule can be
a nucleophile in hydrolysis. These two particles sharply differ in their properties, in particular by their affinity to proton, i.e., by nucleophilicity. The pK_a
values of a hydroxyl ion and water molecule are 15.74 and -1.74, respectively.
The mechanisms of base and acid hydrolyses of amides might be different.

In accordance with the distribution of the electron density in the amide
group, a nucleophile attack should be directed to the carbonyl carbon atom
bearing the highest positive charge. The attack should be directed so as to
achieve the largest overlapping of the binding orbitals of the reactive atoms. As
contemplated earlier, this direction should be normal to the plane of the amide
group. However, further investigations [1469–1474] have shown that it is not
quite so. X-ray studies of a set of aminoketones [1469, 1470] as well as
quantum-chemical calculations [1471–1475] reveal the following picture of the
nucleophile and carbonyl group interaction. Approach of a negatively charged
particle to the carbonyl carbon atom causes a change of planar configuration
of the amide group, a carbon atom protrudes from the plane formed by
carbonyl oxygen, nitrogen and α-C-atoms. There is observed "pyramidalization" of the amide group (Fig. 33), which increases with a decrease of the

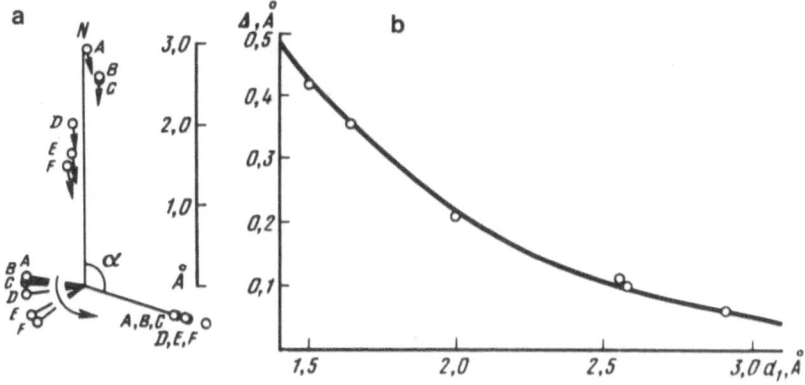

Fig. 33a, b. Attack of nucleophilic nitrogen atom (N) in cyclic aminoketones on the carbonyl carbon atom [1469]. **a** Distances between N and C' and angles C_a–C'–O in aminoketones; different sizes of cycle (A–F); **b** dependence of pyramidalization of carbonyl group (Δ) on distance between nucleophile and carbonyl carbon

distance between nucleophile and the carbonyl C-atom. For ketones the interrelationship of the pyramidalization degree (Δ) and the distance between the carbon atom and nucleophile (d_1) is expressed by an empiric formula

$$d_1 = -0.85 \lg(\Delta/\Delta_t)^2 + 1.48 \text{ Å},$$

where Δ_t is the value of pyramidalization in the tetrahedral structure ($\Delta_t = 0.44$ Å). The angle of the nucleophilic attack remains constant along the whole way of its approach to the carbonyl carbon and is 107°, i.e., very close to tetrahedral.

The same conclusions are obtained [1476] by the concept of "stereoelectronic control", according to which the most stable bond in the tetrahedral compound is that bond which contains the least number of free orbitals located in antiperiplanar conformation to free orbitals of other atoms. This concept turned out to be of benefit for analysis of the breakdown of tetrahedral intermediates and is considered later.

Pyramidalization of the amide group during the nucleophilic attack leads obviously to disturbance of resonance interaction of an electron lone pair of the nitrogen atom and π-electron of the carbonyl group, which provides nucleophile addition. Thus, upon the bond formation the reactive groups seem to adjust to each other, which promotes more complete overlapping of the binding orbitals.

Quantum-chemical analysis of addition of methoxyl ion to N-methylacetamide allows the reaction pathway on the map of the alternative ways to be presented as in Fig. 34 [1477].

The addition of nucleophile results in a decrease of the C–N bond order, and the charge on the nitrogen atom increases. The essential part of double bonding of atoms C and N is lost at early stages of nucleophile addition, when the CH_3O–C bond is still not formed.

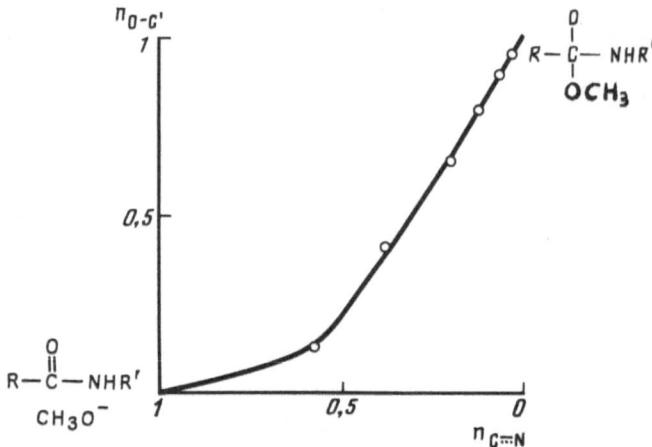

Fig. 34. Change in double bonding in $C-N$ vs the order of $O-C'$ at the methoxy anion interaction with N-methylacetamide (according to [1477])

At acid hydrolysis the events appear to submit the same rule with the only difference that low nucleophilicity of water can hardly induce the noticeable pyramidalization of the amide group. At high concentrations of H^+ ions water addition is facilitated with protonation of the oxygen atom of the carbonyl group (for amides pK_a 0– +1), which decreases resonance stabilization of amide and increases electrophilic ability of the carbonyl carbon. Low rates of acid hydrolysis of esters (Table 24) are possibly stipulated by a lower pK_a value ($-6-$ -7) of these compounds, as compared with amides.

3.3.2   Tetrahedral Intermediates

Nucleophilic addition to the amide group, as already mentioned, decreases the order of the C–N bond and increases the negative charge on the nitrogen atom. One can imagine that the process runs further in the same direction and finally the bond C–N breaks, i.e., the reaction proceeds according to the S_N2 mechanism:

$$N:\ \ \begin{matrix} R \\ | \\ C-NHR^1 \\ || \\ O \end{matrix} \rightleftharpoons \begin{matrix} R \\ | \\ N....C-NHR^1 \\ | \\ O^- \end{matrix} \rightleftharpoons \begin{matrix} R \\ | \\ N-C\ \ \ ^-NHR^1 \\ || \\ O \end{matrix}.$$

However, a leaving of the amino group as an anion is impossible, since the anion is a rather strong base ($pK_a \approx 18$; see Sect. 1.2.3). It is necessary that the leaving group first add a proton. The escape of the amino component of amides as an anion occurs perhaps only in case of very weak basic anilides. So pK_a of p-nitroanilide-anion is equal to 14 [1478], and this group, in principle,

can be cleaved up to protonation. Upon hydrolysis of usual amides protonation precedes cleavage of the leaving group, i.e., there exists an intermediate of
type (IV) – the subject of protonation

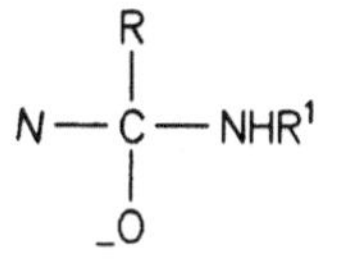

IV

The formation of these tetrahedral intermediates during hydrolysis of
carboxylic acid derivatives has been proved in special cases by their direct
isolation, by physicochemical methods (see review [1479]), as well as by the
studies on the oxygen exchange in the initial substrate by isotope ^{18}O from
$H_2^{18}O$ [1480].

Relatively stable tetrahedral intermediates are formed when adding
alcoholate ions to trifluoroacetylamide or ester of trifluoroacetic acid [1481].
The interaction of acetyl chloride with imidazole derivatives also turns out well
for isolating a stable tetrahedral intermediate [1482]:

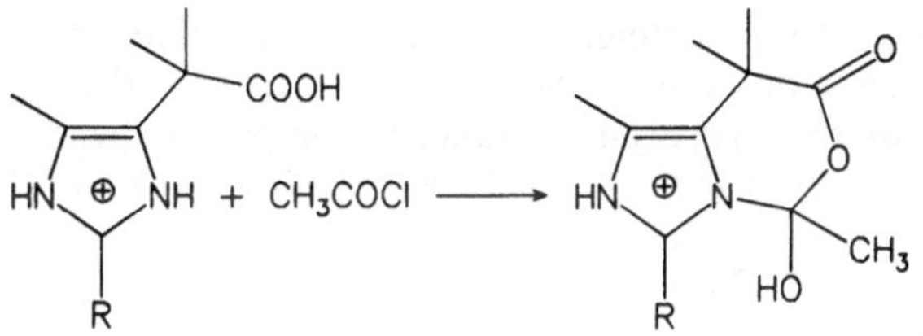

Intramolecular rearrangement of N-(α-hydroxyacyl)-lactams (V) yields
stable compounds of this type – the so-called cyclols (VI) [1483, 1484]:

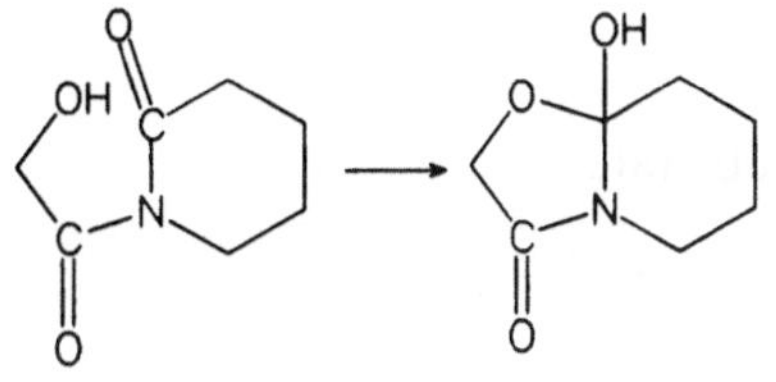

V   VI

An investigation of properties of these compounds [1484] shows that
tertiary hydroxyl has pK_a equal to 10–11. The bonds C–O (cycle) and C–N are
easily cleaved. So the cyclols of type (VI) exist in equilibrium with acyl lactams
(V). However, if the total number of atoms within the cycle is ten or more,
isolation of the corresponding cyclols is impossible because of the formation of

macrocycles (VII) → (VIII):

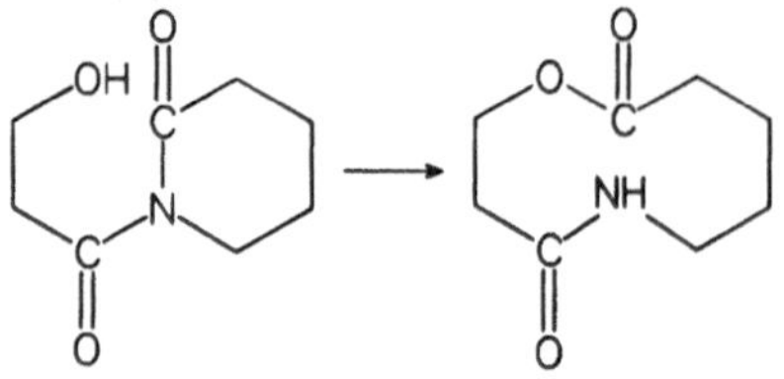

VII   VIII

Cyclols of type (VI) are rapidly dehydrated yielding (IX), and form salts of the corresponding carbo cations in the strong acidic medium (X):

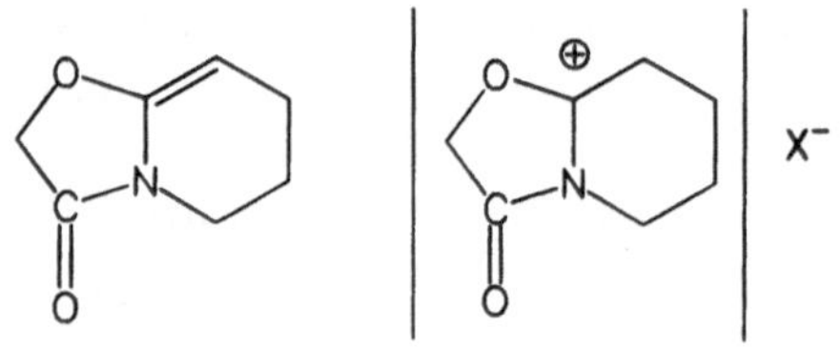

IX   X

The study of oxygen exchange in amides and other derivatives of carboxylic acids provides quantitative characteristics of the stability of tetrahedral intermediates. This exchange is observed if the breakdown rate for the tetrahedral intermediate towards the hydrolysis product is comparable with the rate of its decay in the direction of the initial substance [1480, 1485, 1486]:

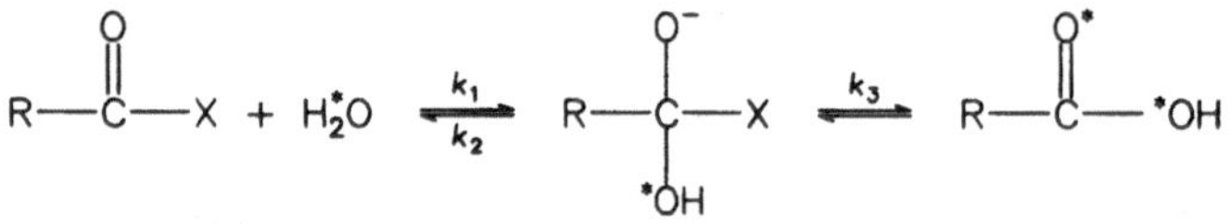

Here the observed hydrolysis rate is

$$k_h = \frac{k_1}{1 + k_2/k_3},$$

and the rate of oxygen exchange in the substrate

$$k_e = \frac{k_1}{2(1 + k_3/k_2)}.$$

The studies on the rate of inclusion of ^{18}O from $H_2^{18}O$ into the substrate (or the rate of ^{18}O label loss in the labeled substrate) yield the value of $k_1 = 2k_e + k_h$ and the ratio $k_2/k_3 = \alpha = 2k_e/k_h$.

It turned out that in primary amides $k_2/k_3 > 1$, i.e., the tetrahedral intermediate compound transforms into the initial amide much faster, while in the case of esters the process proceeds in the opposite way (Table 27). For tertiary amides the value is $k_2/k_3 < 1$.

Table 27. Hydrolysis and ^{18}O-exchange in amides and esters

Compound	$t°C$	$k_h \cdot 10^5$, $M^{-1}s^{-1}$	$k_e \cdot 10^5$, $M^{-1}s^{-1}$	k_1 $M^{-1}s^{-1}$	$\alpha = k_2/k_3$	References
$C_6H_5CONH_2$	40.7	4	76	1.92×10^{-4}	4.22	[1485]
$HCON(Me)CH_2Ph$	27	12	3.97	1.99×10^{-4}	0.66	[1487]
$CD_3CON(Me)CH_2Ph$	40	0.893	0.0862	1.06×10^{-5}	0.19	[1487]
$CH_3CD_2CON(Me)CH_2Ph$	70	2.91	0.767	4.44×10^{-5}	0.52	[1487]
$C_6H_5COOC_2H_5$	25	840	7ʻ4	9.70×10^{-3}	0.18	[1485]

Upon amide acid hydrolysis the product resulting from the addition of water has the following structure (XI)

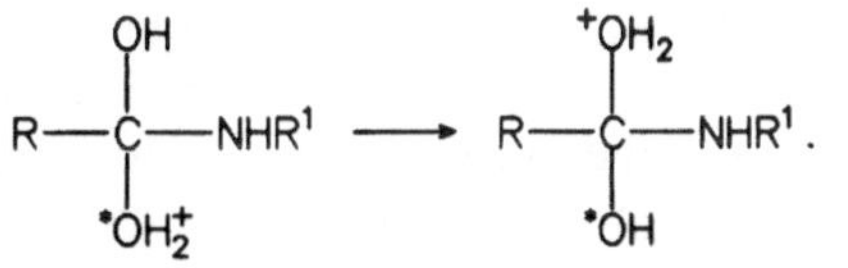

For an oxygen exchange it is necessary to deprotonate the added water molecule and to add a proton to the hydroxyl group (XI) → (XII). However, hydroxyl is a weaker base as compared to the NHR group, and protonation should be mainly directed to that group. Thus, at acid hydrolysis of amides the ratio k_2/k_3 is obviously very low and no oxygen exchange in the substrate is observed [1480]. But at high concentrations of H_3O^+ it is possible to detect this exchange [1488].

3.3.3 Proton Transfer and Product Formation

The formation of the tetrahedral intermediate and its breakdown into the reaction products is accompanied with the proton transfer. This process is exoergic only if it proceeds from a weaker base to the stronger one, i.e., when donor pK_a is lower than pK_a of the acceptor [1489]. In this case the rate of proton transfer is controlled by diffusion. The direct detection of the rate of diffusion proton transfer by the method of laser-induced proton shift [1490] provides rate constants of approximately $2 \div 16 \times 10^{10} \, M^{-1}s^{-1}$. In case of proton collision transfer (direct collision of two molecules) this value is $10^8–10^9 \, M^{-1}s^{-1}$.

At a reverse ratio of the basicity pK_a (donor) $> pK_a$ (acceptor), the proton transfer rate can be found by the equation

$$\lg k = \lg k_0 - [pK_a(\text{donor}) - pK_a(\text{acceptor})] ,\tag{14}$$

where k_0 is the rate constant of the proton diffusion transfer, usually equal to $10^{10} \, M^{-1}s^{-1}$.

Just this situation is observed at acid amide hydrolysis. As mentioned earlier, water is a too weak nucleophile so that a compound of type (XI) could be formed at a high concentration. At the same time, proton removal from the attacking molecule is a thermodynamically unfavorable process, its rate should be characterized by the constant of about $10^{-7}\,M^{-1}\,s^{-1}$, i.e., proton removal at pre-steady state might not provide the observed hydrolysis rates. In this case a concerted process of the proton attack and its removal should occur:

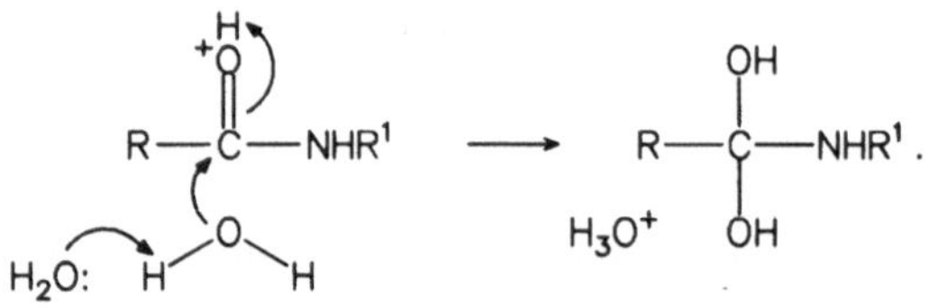

XIII

If so, the formation of the C–O bond increases a positive charge on the oxygen atom of water, i.e., decreases pK_a of the conjugated acid. At the rate constant of hydrolysis of about $10^{-4}\,M^{-1}\,s^{-1}$ the proton transfer from water does not limit the rate of the overall process, if pK_a of the attacking molecule decreases from 15.74 to 12–13.

The hydrogen bonding should precede the proton transfer. A geometry of the system of the hydrogen bond much predetermines the ready proton transfer from one atom to another [1491–1493]. This geometry stipulates also the structure of the transition state. In acid-catalyzed reactions of hydrolysis of carboxylic acid derivatives the transition state has been studied theoretically and experimentally [1494–1498], both linear (XIV) and cyclic (XV) structures being proposed:

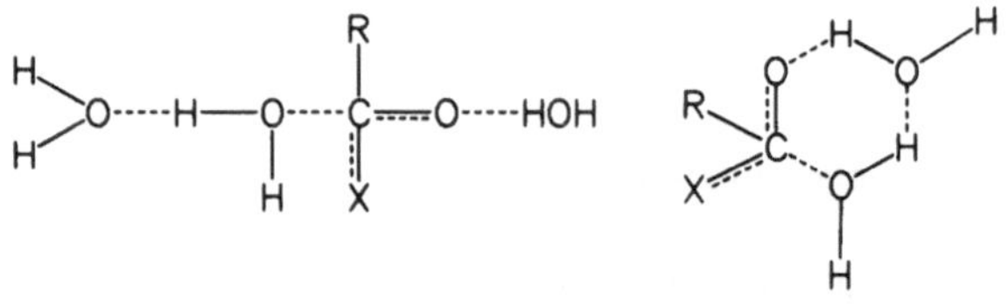

XIV   XV

Protonation of the leaving amino group and deprotonation of the hydroxyl group should precede the decay of the tetrahedral intermediate of type (XIII). Deprotonation occurs because the equilibrium in the reaction is shifted to the

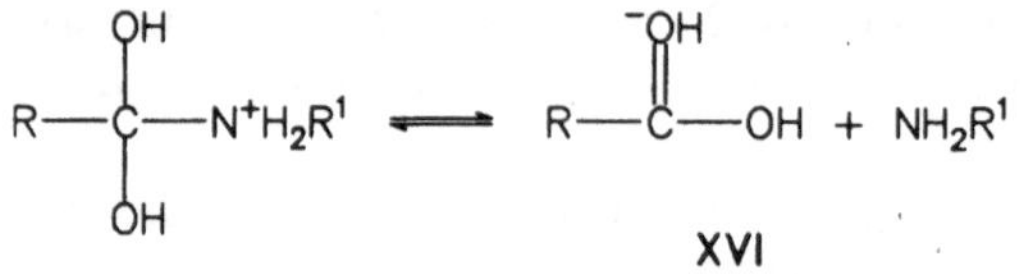

XVI

left due to low pK_a values of the conjugated acid (XVI). In some cases [1499] the transformation of the neutral tetrahedral intermediate (XIII) into zwitterion can be a limiting stage in the process.

When forming the tetrahedral intermediate, pK_a of the amino group increases much, and eventually obtains the value of about 5–7 [1499, 1500]. Thus this group is to be protonated with diffusion-controlled rates actually in the whole pH interval used for hydrolysis. Moreover, it is possible [1501] that sometimes protonation and removal of the amino group occur before the formation of the tetrahedral intermediate, i.e., the reaction runs preferentially by the S_N2 mechanism.

The situation at ester hydrolysis is much more complicated. In this case pK_a of the C–OR group is rather low (about −6) and its protonation might be the rate-determining step. If the R-group is a strong electron acceptor, pK_a of the corresponding alcoholate ion R–O⁻ becomes low [1502], and this group can escape as an anion. Perhaps in case of alkyl substituents the OR group is cleaved off and protonated due to the concerted mechanism.

The proton transfer involvement into the rate-determining step of the process can be established when studying the reaction in deuterium oxide and upon determination of the ratio k_{H_2O}/k_{D_2O} [1503]. Usually for these reactions it is about three. A method for the calculation of the proton number carried at the limiting stage has been proposed [1504]. This allows measuring the reaction rate in mixtures H_2O–D_2O with diverse amounts of deuterium oxide (proton inventory). The rate constant is related to the mole fraction of D_2O (n) and the portion of isotope substitution in the transition state φ_i^T with the ratio (15):

$$k_n = k_0 \prod_1^i (1 - n + n\varphi_i^T) \tag{15}$$

where k_0 is the constant of the reaction rate in H_2O, i the number of transferred protons. For $i = 1$ the plot in the coordinates $k_n \div n$ is linear, and at other values there appeared a deviation from linearity (for example, [1505, 1506]).

It is noteworthy that the theoretical prerequisites of kinetic isotopic effects of the solvent are based on the enthalpy change at isotope substitution. However, it was shown [1507] that the ratio k_{H_2O}/k_{D_2O} can be determined by the translation entropy change due to solvation in a transition state, i.e., this ratio does not obligatorily characterize the proton transfer rate. Besides, the transfer of the proton is rather a complex process involving stages of diffusion, solvent rearrangement, orientation, etc. Therefore, the proton transition in itself can rarely be a slow stage of the process [1508, 1509]. The studies of kinetic isotopic effects of multistage reactions are described in [1510], the relation of kinetic isotope effects and the reaction free energy are outlined in [1511].

Finally, the proton transfer can be realized by a tunnel effect. The corresponding quantum-mechanic theory of the proton transfer kinetics has been developed by Dogonadze et al. [1512].

The breakdown of tetrahedral intermediates is studied by means of model substances of the acetal type (see, for example, [1513, 1514]) and by analysis of the influence of the nucleophile basicity and leaving groups on the reaction rate as well as by data on the influence of isotope substitution on the rate of the process [1515–1517]. The value of the isotope effect imported by a substitution

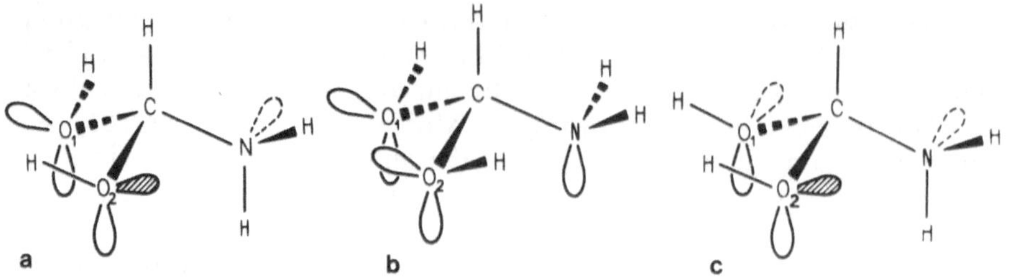

Bond	C–N (a)	C–O$_1$ (a)	C–O$_2$ (a)	C–N (b)	C–O (b)	C–N (c)	C–O$_1$ (c)	C–O$_2$ (c)
Antiperiplanar lone pairs	O$_1$	O$_2$	N	O$_1$, O$_2$	–	–	O$_2$	O$_1$, N
Bond length, Å	1.440	1.414	1.417	1.485	1.406	1.430	1.427	1.447

Fig. 35. Three conformations (a, b, c) of tetrahedral intermediates

of ^{16}O by ^{18}O or ^{14}N by ^{15}N, can be estimated by zero levels of the bond energy by IR spectra of the corresponding compounds. A general conclusion is that at base ester hydrolysis a limiting stage is, as a rule, the formation of the tetrahedral intermediate, and at amide hydrolysis its breakdown.

Experimental and theoretical studies [1470–1472, 1518–1521] show the decay of tetrahedral intermediates to depend on their conformation, i.e., it is subjected to stereoelectronic control (see, however, [1522]). The bond with the minimum overlapping of orbitals of free electron pairs with the orbitals of other atoms is mostly broken. Minimal overlapping is achieved if their relative location is antiperiplanar.

According to Figure 35 the three bonds are equivalent by the number of antiperiplanar orbitals in the conformation (a) of dioxyaminomethane; in conformation (b) a lone pair of the N-atom has two antiperiplanar orbitals of O-atoms, and in conformation (c) the situation is similar for lone pairs of atom O$_2$. Therefore, the splitting of the NH$_2$ group will occur chiefly in conformation (b), while the separation of the OH-group in conformation (c).

Studies on hydrolysis and oxygen exchange in tertiary amides of type RCONH(Me)CH$_2$Ph show that the conformational transitions in corresponding tetrahedral intermediates can have an activation barrier of about 5.5–8 kcal/mol [1487].

3.4 Catalysis

The reactions of hydrolysis of amides and other derivatives of carboxylic acids are accelerated in the presence of different catalysts. Later on the main regulatory mechanisms of homogeneous catalysis of hydrolytic reactions will be considered.

3.4.1 Classification

Homogeneous catalysis according to its catalytic action can be classified as the acid-base and the nucleophile-electrophile one. Upon acid-base catalysis the catalyst, acid or base, promote the proton addition or its release. In the case of nucleophilic or electrophilic catalysis the catalyst forms a covalent inter-mediate with a substrate.

Acid-base catalysis, in its turn, is subdivided into specific and general. At specific catalysis a catalyst totally accepts (base) or releases (acid) a proton at the rate-determining stage of the process. Thus the catalyst does not enter a transition state of the transformed substrate or, if it still does enter, the substrate transformation is not a rate-limiting step:

$$AH + S \xrightleftharpoons{\text{slowly}} A^- + SH, \quad SH \xrightleftharpoons{\text{rapidly}} (SH)^* \longrightarrow \text{products}$$

Upon general catalysis a catalyst always enters a substrate transition state and its breakdown is a rate-determining stage of the overall process:

$$B + HS \xrightarrow{\text{rapidly}} (B\text{----}H\text{----}S)^* \xrightarrow{\text{slowly}} \text{products} + BH$$

$$AH + S \xrightarrow{\text{rapidly}} (A\text{----}H\text{----}S)^* \xrightarrow{\text{slowly}} \text{products} + A^-$$

So catalysis with hydroxonium ion is always specific, whereas catalysis with hydroxyl ion can be specific (since in aqueous solution there is no stronger base than OH^- ion) or nucleophilic.

At nucleophilic (and electrophilic) catalysis, as well as at general ones, a catalyst enters a transition state of the substrate. The difference between these catalysis types is as follows: the attack of nucleophilic catalyst is directed to a carbon atom of the carbonyl group of a derivative of carboxylic acid, while the attack of a general base catalyst to a hydrogen atom (or in the case of general acid catalysis to heteroatom).

Besides these types there are also inter- and intramolecular, mono- and polyfunctional catalyses, depending on molecularity and the number of the catalytic groups involved in reactions.

Thus in the presence of the base with pK_a close to pH of the solution, the expression for the observed hydrolysis rate constant is

$$k' = k_{H_2O} + k_H[H_3O^+] + k_{OH}[OH^-] + k_B[B] + k_{BH}[BH] .$$

It is noteworthy that some cases of acid-base catalysis cannot be distinguished because of the principle of kinetic equivalence. So, for instance, the reactions:

$$A + H_2O + S \longrightarrow \text{products} \quad \text{and} \quad AH + OH^- + S \longrightarrow \text{products}$$

are kinetically equivalent, as $[A^-]$, $[AH]$ and $[OH^-]$ are related by the equation $[A^-] = [HA][OH^-]K_a/K_w$, where K_a is a constant of acid ion-ization, K_w the ionic product of water.

General acid-base catalysis can occur in principle according to concerted

$$B \ HX \quad \overset{}{\underset{}{>}}C{=\!\!=}O \quad \longleftrightarrow \quad BH^+ \ X{-}\overset{|}{\underset{|}{C}}{-}O^-$$

and stepwise

$$HX \quad \overset{}{\underset{}{>}}C{=\!\!=}O \quad \longleftrightarrow \quad H\overset{+}{X}{-}\overset{|}{\underset{|}{C}}{-}O^- \quad \overset{B}{\longleftrightarrow} \quad BH^+ \ X{-}\overset{|}{\underset{|}{C}}{-}O^-$$

mechanisms. Jencks [1523] has formulated the rules by which the concerted acid-base catalysis in aqueous solution can be directed only to the group that undergoes a large pK_a change in the reaction, and will occur if this change transforms a thermodynamically unfavorable proton transfer into the favorable one (see also [1524, 1525]).

3.4.2 Catalytic Efficiency

Efficiency of acids and bases as catalysts of hydrolytic reactions depends on their pK_a values. Brønsted and Pedersen [1526] proposed a simple dependence between the rate constants of the catalyzing reaction and the pK_a value of the acid and base:

$$\lg k_A = \alpha pK_a' + C \qquad \text{and} \qquad \lg k_B = \beta pK_a' + C \,,$$

where k_A and k_B are the rate constants of the reaction catalyzed by an acid or base, α, β and C are the constants. The existence of these correlations (Fig. 36) is quite clear because at acid-base catalysis the action of a catalyst is associated

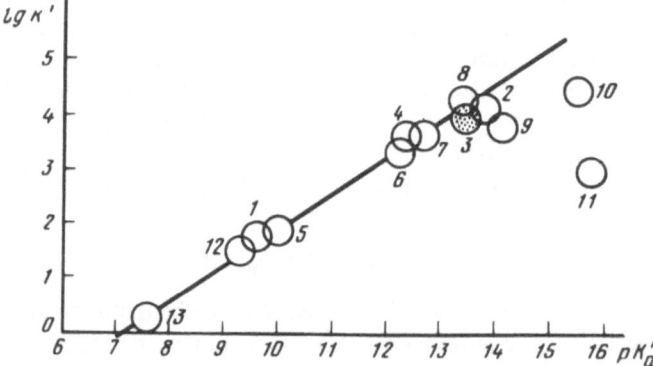

Fig. 36. The Brønsted function for the reaction of oxygen-containing nucleophiles with p-nitrophenylacetate (H_2O, 25–30 °C) [1401, p. 115]; 1 $CCl_3CH(OH)O^-$; 2 $C(CH_2OH)_3CH_2O^-$; 3 $CH_3CONHCH(CH_2O^-)CONH_2$; 4 $F_3CCH_2O^-$; 5 $C_6H_5O^-$; 6 $Cl_3CCH_2O^-$; 7 $F_2CHCF_2CH_2O^-$; 8 $CH{\equiv}CCH_2O^-$; 9 $ClCH_2CH_2O^-$; 10 CH_3O^-; 11 OH^-; 12 p-$ClC_6H_4O^-$; 13 $CH_3OC_6H_4O^-$

with an addition or a release of the proton, and pK_a is a measure of the affinity of the substance to proton. The Brønsted equation is an example of linear free-energy correlations and similar to the equations of Hammett, Taft and others, mentioned in Section 3.2.1.

The physical sense of the Brønsted equation can be illustrated by the following example [1423]. Let us consider the reaction of the proton transfer

$$B: +\ SH \rightleftharpoons S^- + BH\ ,$$

where SH is the substrate, and B the catalyst. The potential energy change depending on the distance between HS and B can be presented with curve *I* (Fig. 37). The potential energy change in the reaction $S^- + BH$ is described with curve *II*. If now the base B is replaced by a weaker base B′, the corresponding curve is shifted to *II′*. The activation energy change (δE^0) in this case is related with the free-energy change of the reaction by expression

$$\delta E^0 = \frac{S_1}{S_1 + S_2}\,\delta\varepsilon_0 = \beta\delta\varepsilon_0\ ,$$

where S_1 and S_2 are the slopes of the corresponding curves in the crossing point (supposed $S_2 = S_{2'}$). The equation is a different representation of the Brønsted equation, from which it is clear that the value β (and thus α) is an energy change level of the system depending on the position of the proton between substrate and catalyst atoms in the transition state of the reaction.

The linear dependence between pK_a and $\lg k$ is true only in a special interval of differences between donor pK_a and proton acceptor pK_a [1489]. This interval corresponds to the mechanisms of general base or acid catalysis. If these differences are very slight, $\alpha \to 1$ ($\beta \to 0$) and specific acid (or base) catalysis is observed. When the differences are great (i.e., transfer occurs from a weak base to the strong one with diffusion-controlled rate), the rate of catalyzed reaction does not depend on the catalyst pK_a. Thus the dependence of the reaction rate on pK_a of catalysts can determine the catalysis type.

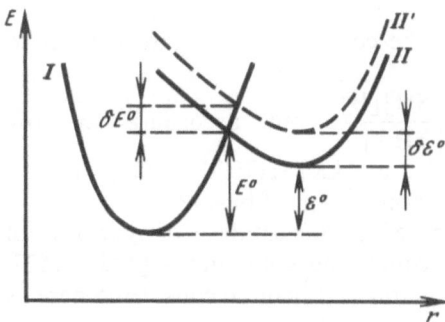

Fig. 37. Interpretation of the Brønsted function [1423, p. 247] (description in text)

3.4.3 Catalysis by a Hydroxyl Group

A hydroxyl group as a free anion or as a part of water, alcohols, and phenols can catalyze hydrolysis of amides and other derivatives of carboxylic acids according to the mechanisms of specific (OH^- ion), general base and nucleophilic catalyses.

Hydroxyl-containing compounds are very weak acids, but their conjugated bases have high pK_a values [117]:

$$R\text{—OH} \overset{K_{a(1)}}{\rightleftharpoons} R\text{—O}^- + H^+ \quad \text{and} \quad R\text{—OH} + H^+ \overset{K_{a(2)}}{\rightleftharpoons} R\text{—OH}_2^+$$

R	$pK_{a(1)}$	$pK_{a(2)}$	R	$pK_{a(1)}$	$pK_{a(2)}$
H	15.74	-1.74	C_2H_5	16.0	-2.0
CH_3	15.0	$-$	C_6H_5	10.0	-6.7

Therefore, a hydroxyl ion in the case of specific catalysis and alcoholates at general base catalysis are proton acceptors. In specific base catalysis by hydroxyl ions the reaction rate is directly proportional to pH of the medium. In principle, OH^- ion can be generated in solution with any base. However, the proton transfer onto a base weaker than hydroxyl occurs at a relatively low rate (Sect. 3.3.3). If the base is involved in the reaction generating the hydroxyl ion, the rate of the process does not depend on the base concentration. Upon the specific catalysis the equilibrium is

$$B + H_2O \overset{K_b}{\rightleftharpoons} BH^+ + OH^- \tag{16}$$

and besides the equilibrium

$$B + HS \overset{K}{\rightleftharpoons} S^- + BH^+ \tag{17}$$

is reached fast.

A deprotonated substrate is slowly transformed into the products of the reaction

$$S^- \overset{k}{\longrightarrow} \text{products} \tag{18}$$

The reaction rate is determined by the conversion rate and is equal to $v = k[S^-]$.

As $K = \dfrac{[S^-][BH^+]}{[SH][B]}$, $[S^-] = \dfrac{K[SH][B]}{[BH^+]}$, but according to (16), $[B]/[BH^+] = [OH^-]/K_b$. Consequently, $v = kK[SH][OH^-]/K_b$, i.e., the reaction rate does not depend on the concentration of base B.

There are quite a few examples of "pure" specific catalysis of hydrolysis of carboxyl acid derivatives by hydroxyl-containing compounds [1426, 1527, 1528]. Base hydrolysis of anilides of aliphatic acids can be considered here as an example [1529]. Its rate is proportional both to the first and the second degree of the concentration of hydroxyl ions (19):

$$k' = k_I[OH^-] + k_{II}([OH^-])^2 . \tag{19}$$

The second member in this equation is evidently associated with catalysis of deprotonation of the tetrahedral intermediate (XVII) by the hydroxyl ion (see, however, [1530]):

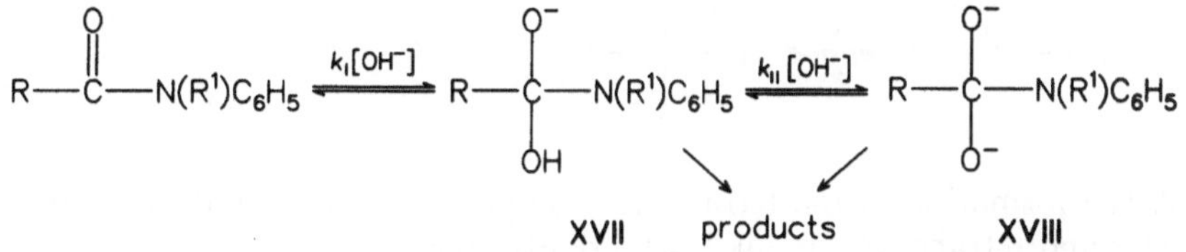

More often, hydroxyl ions and alcoholates serve in hydrolytic reactions as nucleophilic catalysts, which form unstable intermediates when added to a carbonyl carbon atom.

The ability of nucleophilic reagents to be added to the carbonyl carbon atom is proportional in total to their basicity ($\beta = 0.66$). This is evident, for example, when comparing pK_a of these reagents and their reactivity relative to p-nitrophenylacetate (Fig. 36). However, strong basic hydroxy anions deviate from the linear Brønsted dependence, obviously due to solvation effects, and also because a transition state for high reactive compounds can be reached at earlier stages of the process (Hammond postulate) [1531].

Nucleophilic catalysis by the hydroxyl ion, strictly speaking, is not a true catalysis, since the hydroxyl ion is used during hydrolysis. This is a reaction of nucleophile-promoting hydrolysis. The same is true for the alcoholysis of amides. However, ester alcoholysis is a true catalysis, as alcoxide ion formed during the reaction is in equilibrium with alcoxide ion attacking a substrate.

A true catalysis of amide alcoholysis is observed if the forming ester is hydrolyzed to an acid with the rate exceeding that for the ester formation:

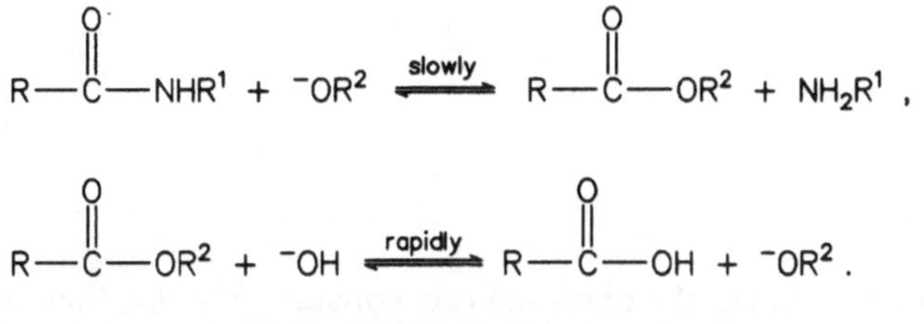

However, alcoholysis of amides runs very slowly and this reaction has been thoroughly studied in the reverse direction, i.e., as aminolysis of esters [1532, 1533].

The efficiency of nucleophilic catalysis with the hydroxyl ion is obvious when comparing the rates of base hydrolysis (see Table 24) of N-methyl-acetamide and the rate of noncatalyzed (catalyzed by water) hydrolysis of this compound. A catalytic efficiency of OH^- ions in concentration 1 M is about 1×10^9 (at $25\,^\circ C$).

Hydroxy anions can act as catalysts by the mechanism of general base catalysis [1401]. Similarly, the conjugate acids (alcohols, phenols, etc.) can function according to the mechanism of general acid catalysis. A few mechanisms and corresponding kinetic schemes are quite possible here.

So a slow step can involve the formation of the tetrahedral intermediate and is accompanied by a proton transfer:

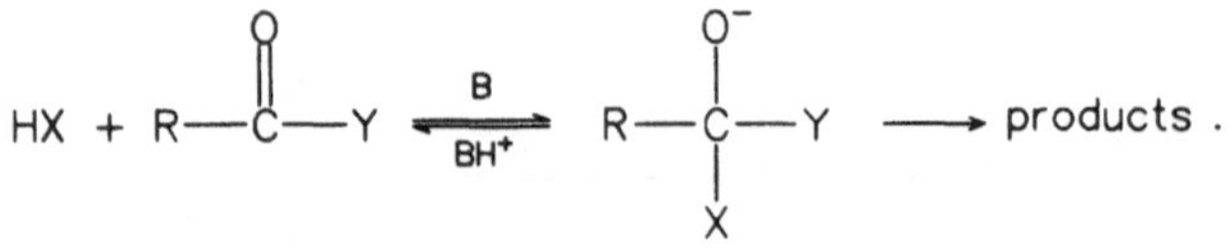

$$HX + R{-}\overset{\displaystyle O}{\underset{\displaystyle}{C}}{-}Y \underset{BH^+}{\overset{B}{\rightleftharpoons}} R{-}\overset{\displaystyle O^-}{\underset{\displaystyle X}{C}}{-}Y \longrightarrow products\ .$$

A fast formation of the tetrahedral compound with the following proton transfer and tetrahedron breakdown are also possible:

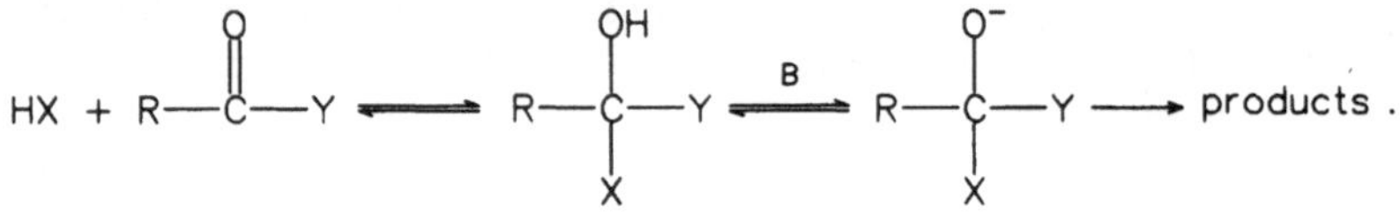

$$HX + R{-}\overset{\displaystyle O}{\underset{\displaystyle}{C}}{-}Y \rightleftharpoons R{-}\overset{\displaystyle OH}{\underset{\displaystyle X}{C}}{-}Y \overset{B}{\rightleftharpoons} R{-}\overset{\displaystyle O^-}{\underset{\displaystyle X}{C}}{-}Y \longrightarrow products\ .$$

For general acid catalysis these mechanisms are possible:

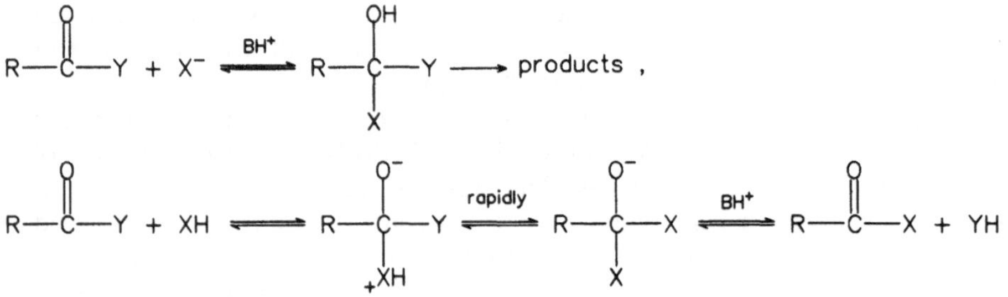

$$R{-}\overset{\displaystyle O}{\underset{\displaystyle}{C}}{-}Y + X^- \overset{BH^+}{\rightleftharpoons} R{-}\overset{\displaystyle OH}{\underset{\displaystyle X}{C}}{-}Y \longrightarrow products\ ,$$

$$R{-}\overset{\displaystyle O}{\underset{\displaystyle}{C}}{-}Y + XH \rightleftharpoons R{-}\overset{\displaystyle O^-}{\underset{\displaystyle +XH}{C}}{-}Y \overset{rapidly}{\rightleftharpoons} R{-}\overset{\displaystyle O^-}{\underset{\displaystyle X}{C}}{-}X \overset{BH^+}{\rightleftharpoons} R{-}\overset{\displaystyle O}{\underset{\displaystyle}{C}}{-}X + YH\ .$$

The type of catalysis is determined by the ratio of the constants of the rates of individual stages

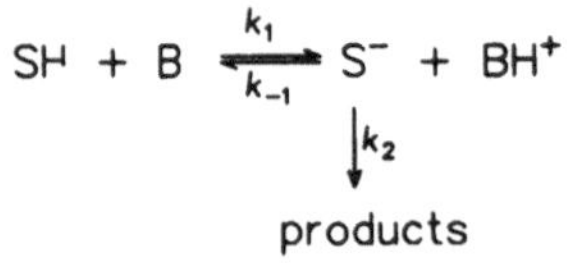

$$SH + B \underset{k_{-1}}{\overset{k_1}{\rightleftharpoons}} S^- + BH^+$$
$$\downarrow k_2$$
$$products$$

If $k_1 \approx k_{-1}$ and $k_2 \gg k_{-1}$, i.e., the observed rate constant $k' = k_1$, then this is general catalysis. If $k_1 \ll k_{-1}$ and $k_2 \ll k_{-1}$, so $k' = k_1 k_2 / k_{-1}$ this is the case of specific catalysis.

Besides, the mechanisms of general catalysis from the point of view of kinetics can be hardly distinguished from those of nucleophilic catalysis, since in all cases both the substrate and catalyst concentrations enter the expression for the rate:

$$v = k[SH][B]\ . \tag{20}$$

Moreover, there are variants when the mechanism of general catalysis is kinetically equivalent to that of nucleophilic catalysis with the participation of the conjugated acid of the catalyst as a general acid catalyst (principle of kinetic equivalence, see Sect. 3.4.1), for example:

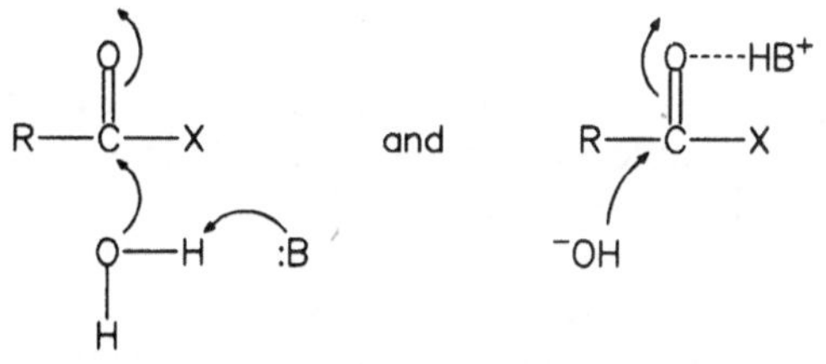

$$\text{and}$$

As a prototropic equilibrium exists

$$B + H_2O \rightleftharpoons BH^+ + OH^-$$

kinetic Eq. (20) is equivalent to the equation

$$v = k[BH^+][OH^-] \,. \tag{21}$$

To distinguish general catalysis from the nucleophilic one is possible if the reaction runs in deuterium oxide (D_2O). Because in general catalysis the rate-determining step includes the proton transfer, a distinctive ($k_{H_2O}/k_{D_2O} \approx 3$) difference in the reaction rate in D_2O as compared to the rate in H_2O is observed. As stated above, an interpretation of the kinetic isotopic effect of the solvent presents certain difficulties (see Sect. 3.3.3).

There are a number of examples of general base and general acid catalysis of hydrolytic reactions of esters and amides [1534–1536], as well as of amino-lysis of esters and thioesters [1537–1539]. In their majority, relatively weak hydroxy anions and other bases serve as catalysts. As to strong bases, they usually function by the mechanisms of nucleophilic or specific (hydroxyl and oxonium ions) catalysis.

3.4.4 Thiol Group

A thiol group as compared to the alcohols has a considerably lower pK_a of 7–9. However, nucleophilicity of the thiolate ion upon addition to the carbonyl carbon is almost the same as of hydroxy anion. This is stipulated by deviation of aliphatic hydroxy anions and hydroxyl ion from the usual Brønsted equation (see Sect. 3.4.3). The nucleophilicity of thiolate ions is higher than that of hydroxy anions comparable to them in basicity. So dianion of o-mercapto-benzoic acid reacts with p-nitrophenylacetate 10–100 times faster than the phenolate ion [1541].

Nucleophilic catalysis by thiolate ions is governed by the Brønsted equation. However, the β value for these anions (0.38) is much lower than for hydroxy anions (0.66). A nucleophilic attack of thiolate anion on carbonyl carbon of the derivatives of carboxylic acids results in the formation of thioesters [1542]:

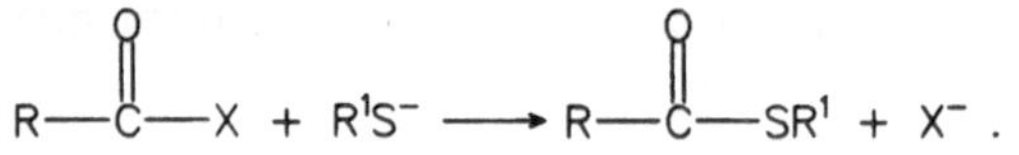

$$R-\overset{O}{\underset{}{C}}-X + R^1S^- \longrightarrow R-\overset{O}{\underset{}{C}}-SR^1 + X^- .$$

The rate constants of base hydrolysis of thioesters are close to the corresponding constants for oxygen esters (Table 24). Therefore, catalytic efficiency of nucleophilic catalysis by thiolate ion of ester hydrolysis is slight.

In the acidic medium thiol esters are hydrolyzed rather slowly and they can be identified by spectrophotometry under these conditions (pH < 7) [1543].

Thiolysis of amides is a very slow process and they almost cannot be directly studied. The reaction of hydrosulfide anion with formamide has been studied by the quantum chemistry and molecular mechanics methods [1544] and compared with the similar reaction of OH^--ion [1457], and vigorous differences in gaseous phase and in solution are found. Experimental data on thiolysis reactions can be obtained when studying reverse reactions of thioesters aminolysis. Knowledge of the equilibrium constant of a reversible reaction

$$RCONHR^1 + R^2SH \overset{K}{\rightleftharpoons} RCOSR^2 + NH_2R^1$$

and the rate of thioester aminolysis facilitate an estimation of the rate of the forward reaction.

According to Table 28 at almost similar pK_a of thiol (7.3) and nitrophenol (7.1), the first is approximately three orders of magnitude more reactive than the second. The thiolysis rate linearly correlates with basicity of the leaving amide group with $\rho = +0.4$. The value of Brønsted coefficients for thiolysis of amides is -0.9, which testifies to the absence of general base catalysis at thiol attack on the amide carbonyl carbon [1533].

Thus the amide thiolysis mechanism involves the attack of thiolate anion on the carbonyl carbon atom, but not the protonated thiol group. In the weak acidic medium the reaction evidently runs according to the mechanism described in [1544, 1545]:

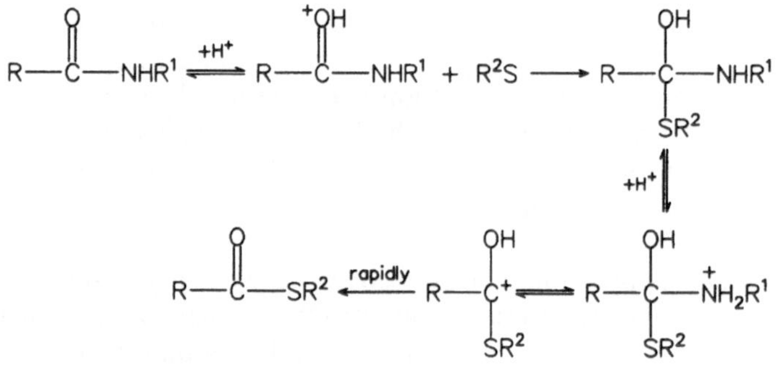

$$R-\overset{O}{\underset{}{C}}-NHR^1 \overset{+H^+}{\rightleftharpoons} R-\overset{^+OH}{\underset{}{C}}-NHR^1 + R^2S \longrightarrow R-\overset{OH}{\underset{SR^2}{C}}-NHR^1$$

$$R-\overset{O}{\underset{}{C}}-SR^2 \overset{rapidly}{\longleftarrow} R-\overset{OH}{\underset{SR^2}{\overset{}{C^+}}} \rightleftharpoons R-\overset{OH}{\underset{SR^2}{\overset{}{C}}}-\overset{+}{N}H_2R^1$$

Table 28. Thiolysis of amides by 2,2,2-trifluoroethanol and amide alcoholysis by p-nitrophenol [1533]

Amide	k_{SH}, $M^{-1}s^{-1}$	$k_{RO}-$, $M^{-1}s^{-1}$	pk_a of leaving group
CH$_3$CONHOH	8.2×10^{-7}	5.7×10^{-10}	6.17
CH$_3$CONHNH$_2$	4.5×10^{-8}	2.8×10^{-11}	8.20
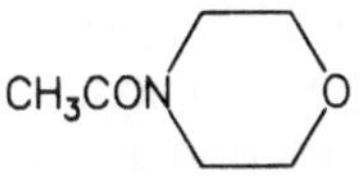 CH$_3$CON⟨hexO⟩	6.0×10^{-10}	8.0×10^{-13}	8.87

3.4.5 Carboxyl Group

Due to low basicity of carboxylate ions (pK_a 3–5), their nucleophilicity is also low. The rate of the catalyzing hydrolysis of p-nitrophenylacetate by acetate anion is approximately 1.7×10^6 times lower than that of the same reaction catalyzed by a hydroxyl ion (see Fig. 36). Therefore, it is unlikely that catalysis with carboxylate ion occurs by the nucleophilic mechanism. Such a type of catalysis is feasible only in the presence of a good leaving group in the cleavable compound with pK_a lower than that of a catalyst.

As shown [1546], hydrolysis of 2,4-dinitrophenylbenzoate catalyzed with an acetate ion proceeds with an intermediate formation of anhydride, and the formed benzoic acid includes up to 75 atomic percent of isotope ^{18}O, if hydrolysis runs in the presence of labeled acetic acid:

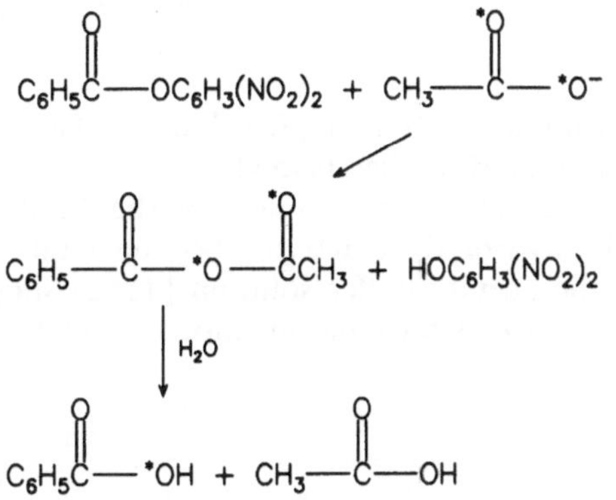

In aprotic solvents a carboxylate ion is also able of displacing an acyl group from nitrophenyl esters forming an anhydride [1547]. A nucleophilic mechanism of catalysis with a carboxylate ion is observed in intramolecular reactions of the transfer of acyl groups (see Sect. 3.4.8).

However, in aqueous solutions hydrolysis of p-nitrophenylacetate probably proceeds by the mechanism of general catalysis induced by a carboxylate ion [1548]:

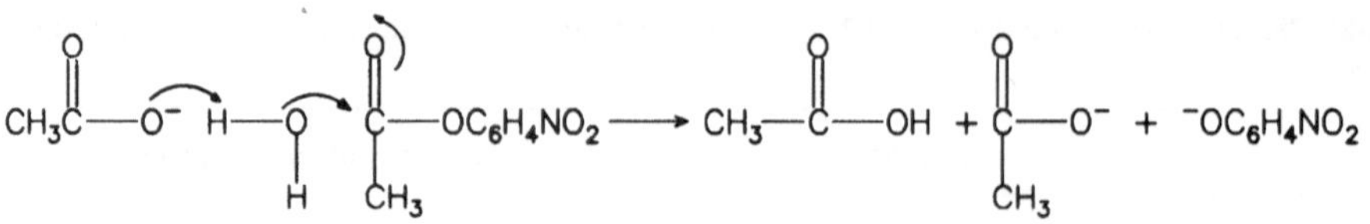

$$CH_3\overset{O}{\overset{\|}{C}}\!\!-\!\!O^- \; H\!-\!\overset{O}{\underset{H}{|}}\;\; \overset{O)}{\underset{CH_3}{\overset{\|}{C}}}\!\!-\!\!OC_6H_4NO_2 \longrightarrow CH_3\!-\!\overset{O}{\overset{\|}{C}}\!\!-\!\!OH \; + \overset{O}{\underset{CH_3}{\overset{\|}{C}}}\!\!-\!\!O^- \; + \; {}^-OC_6H_4NO_2 \; .$$

General base and acid catalysis with a carboxyl group depends both on the catalyst basicity and on the sensitivity of the given reaction to such a type of catalysis, i.e., values of α and β in the Brønsted equation. Though carboxylate ions are weak bases, their catalytic influence on the rate of the reaction at the given pH value can exceed the influence of ions OH^- or H_3O^+, as the concentration of general catalyst can be essentially higher than that of specific ions (see Table 29).

According to Table 29, almost exclusively general acid catalysis with acetic acid is observed in this system at $\alpha = 0.5$.

The efficiency of general acid-base catalysis by a carboxyl group should depend on the direction of proton addition or elimination [1550]. Since *syn*- and *anti*-conformations of the carboxyl group differ much by the energy (about 3.5 kcal/mol in favor of *syn*-conformation [1551]), it is clear from the scheme:

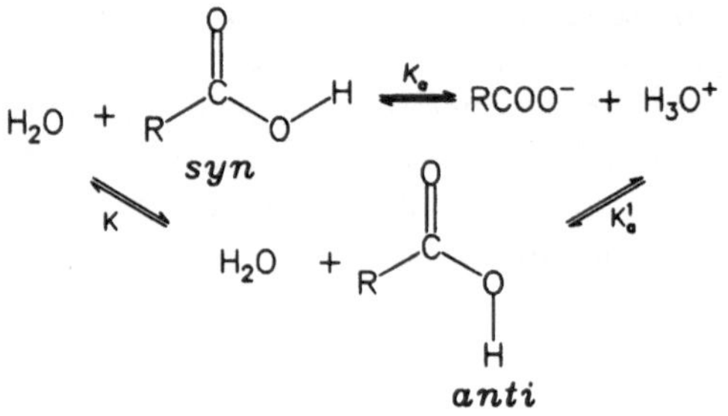

where $K_a' = K_a/K$, that K_a' should be about 10^4 times higher than K_a. This can be distinctly seen upon intramolecular catalysis (see below).

It is rather an intricate task to reveal the mechanism of the catalytic involvement of carboxyl groups in hydrolytic reactions. So the study of hydrolysis of N-butylacetamide in the acetate buffer solution [1552] shows that the reaction has the first order as regards the concentration of acetic acid

Table 29. Part of general catalysis (%) at various α values for the system 0.1 M acetic acid–0.1 M acetate ion [1549]

α	H_3O^+	CH_3COOH	H_2O
0.1	0.002	2	98
0.5	3.6	96.4	0.01
1.0	99.8	0.2	5×10^{-12}

and zero order relative to the acetate ion concentration. Two kinetic equations correspond to these results

$$v = k[HA][Amide] \quad \text{and} \quad v = k'[A^-][AmideH^+],$$

i.e., a catalyst can function according to conserted general acid and nucleophilic mechanisms (XIX), according to the mechanism of the nucleophilic attack of a carboxylate ion on the protonated amide (XX), or the mechanism of general acid catalysis accompanied by the nucleophilic attack by water (XXI)

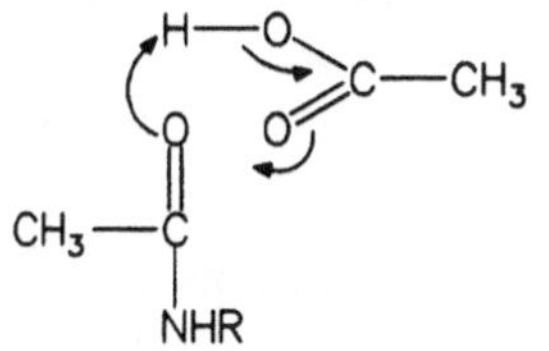

XIX 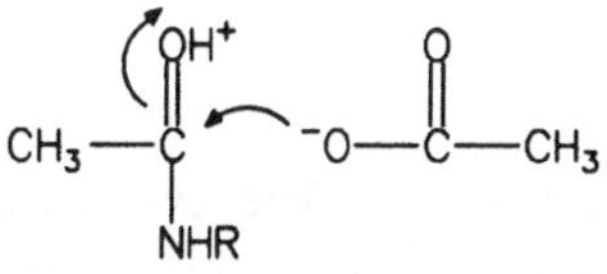   XX

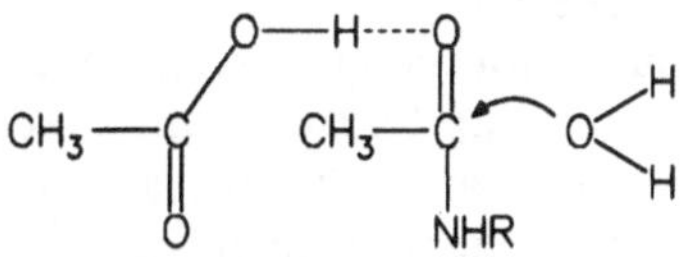

XXI

An attempt to make these mechanisms distinct on the basis of the value of kinetic isotope effect D_2O ($k_{H_2O}/k_{D_2O} = 2$) has failed [1401, p. 123]. The rate constants of hydrolysis of phenylacetate and p-nitrophenylacetate catalyzed by acetate ion obeys the Brønsted dependence on pK_a for the leaving group with $\beta = -0.25$ [1546]. This testifies to rather a low degree of proton transfer in the transition state of the acetate ion catalyzed hydrolysis. More data on intramolecular reactions catalyzed with the carboxyl group are available. These are considered in Section 3.4.8.

3.4.6 Catalysis by Metal Ions

Metal ions carry a positive charge and therefore they can imitate the proton action in hydrolysis of carboxylic acid derivatives, i.e., function as "superacids". In so doing they can indirectly interact with a carbonyl group of the substrate or change the ionization of water bound to them.

As shown [1553, 1554], ions of bivalence metals effectively catalyze hydrolysis of the amino acids ester (Table 30). Here, the metal is coordinated with a free ester amino group and an oxygen atom of the carbonyl group, which results in strong polarization of the latter:

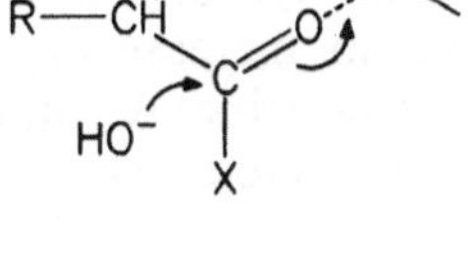

Table 30. Catalysis of hydrolysis of amino acid esters with metal ions, hydroxyl- and hydroxonium ions (25 °C)

Substrate	Catalyst	pH	Solution	k', s^{-1}	k_K, M^{-1}s^{-1}[a]	References
HGlyOMe	Cu^{2+}	7.3	Water	42.5×10^{-3}	2.65	[1553]
HPheOEt	Cu^{2+}	7.3	„	2.67×10^{-3}	0.17	[1554]
HGlyOMe	Co^{2+}	7.9	„	15.6×10^{-3}	0.975	[1553]
HGlyOMe	Mn^{2+}	7.9	„	3.51×10^{-3}	0.22	[1553]
HPheOEt	OH$^-$	7.3	85% Ethanol	5.8×10^{-9}	2.9×10^{-2}	[1428]
HPheOEt	H$_3$O$^+$	7.3	70% Dioxan	1.46×10^{-14}	6×10^{-6}	[1555]

[a] $k_K = k'/[\text{Cat}]$.

According to Table 30 the efficiency of catalysis with metal ions is 10–100 times higher than with ions OH$^-$ and by 5–6 orders higher than catalysis with hydroxonium ions.

The fact that catalysis with metal ions is stipulated by a direct interaction of a catalyst with the reaction center, but is not merely a result of the appearance of a positive charge on the ester amino group follows from the comparison of rate constants listed in Table 30 with that of hydrolysis of betaine ester $^+$Me$_3$NCH$_2$CH$_2$COOR. The latter is 3 orders of magnitude lower than that observed in the presence of copper ion [1556].

Metal ions, evidently, stabilize a tetrahedral intermediate in hydrolytic reactions. The formation of the intermediate is in line with data on an oxygen exchange in esters labeled by carbonyl ^{18}O [1554]:

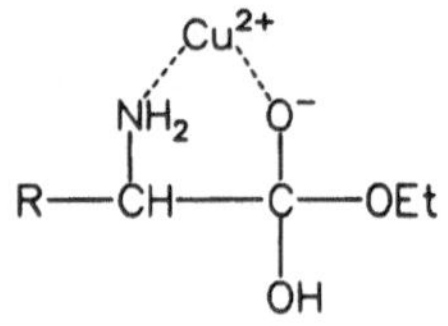

Catalysis of amide hydrolysis at neutral pH, when a rate-determining stage is a breakdown of the tetrahedral intermediate, perhaps occurs due to metal coordination with alcoxyde oxygen (XXII), which promotes protonation of the leaving group and break of the C–N bond [1557]:

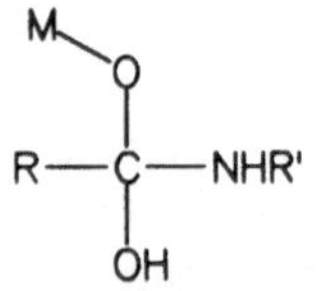

XXII

As known [1558, 1559], metal ions form complexes with hydroxyl ions at neutral pH values. Therefore, a mechanism related to an increase of the hydroxyl ion concentration in the reaction conditions is possible. A similar mechanism was first proposed [1554] for hydrolysis of monoesters of dicarboxylic acids catalyzed with metal ions:

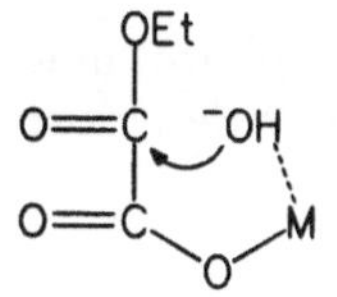

For metal-water complexes pK_a was shown to be close to 7 [1558, 1559]. So pK_a of complex $(NH_3)_5Co^{3+}OH_2$ is equal to 6.6, while pK_a of the $Zn^{2+}OH_2$ complex is about 7.

There are many examples of metal ion catalysis [1560–1564]. One of the most exciting is hydrolysis of glycyl-glycine in the presence of the ethylenediamine-cobalt complex [1565]:

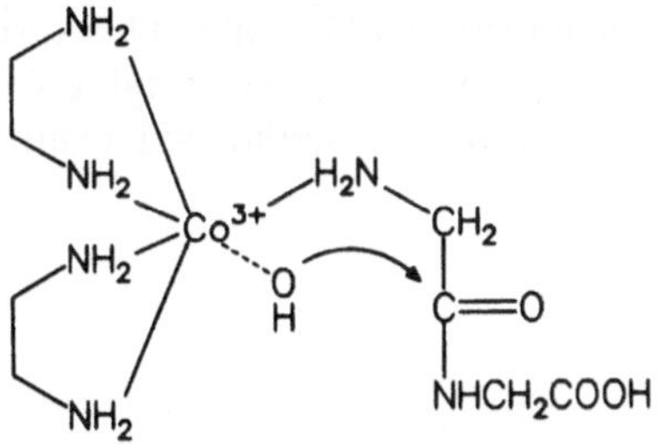

Here dipeptide is hydrolyzed 10^{10} times faster than under the action of the hydroxyl ion.

Substantial accelerations are observed in the complex of bivalence metals of type (XXIII) [1553, 1566] and in other complexes of metals with chelating agents [1567–1569]:

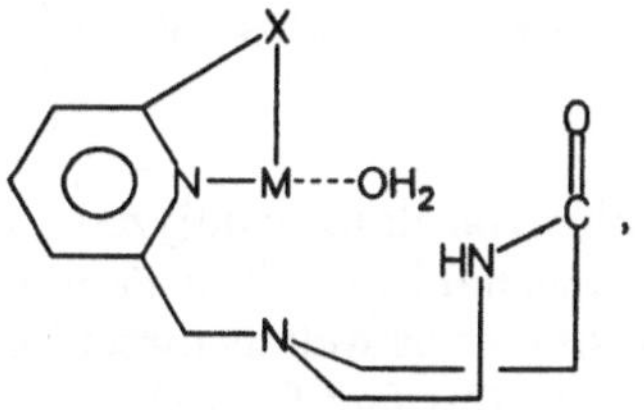

XXIII

where X = COOH, CH_2NMe_2, $CH_2N(CH_2COO^-)_2$.

The cleavage of the lactam cycle in (XXIII) in the presence of Cu^{2+} is enhanced 1.6×10^6-fold, and in the presence of Zn^{2+} 1.9×10^5-fold, as compared with the rate in the absence of metal ions.

It is noteworthy that unlike the esters in case of amides a catalytic effect of noncoordinated metal ions is usually slight. So hydrolysis of glycinamide in the presence of 0.02 M Cu^{2+} is 20 times faster than the spontaneous reaction [1570], despite the fact that copper binds to glycinamide more effectively than to the corresponding ester. It is explained by the firm binding in a ground state in the absence of stabilization of the reaction transition state that must lead to a decrease but not to an increase in rate.

3.4.7   Imidazole Catalysis

Imidazole and amino acid histidine, containing this group are unique acid-base catalysts, since under physiological conditions (pH 7) they exist as acids and as bases:

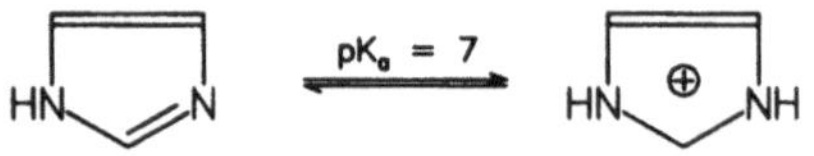

Therefore, imidazole can function as a general base or acid catalyst. Besides, nucleophilicity of the imidazole N-atom is high enough, so that it can attack a carbon atom of the carbonyl group, it being a nucleophilic catalyst.

A typical example of general base catalysis by imidazole is hydrolysis of acetylimidazole in imidazole buffer. As it turned out [1571], the reaction rate enhances at a higher buffer concentration at constant pH value. This result cannot be explained by an addition of imidazole to the carbonyl group, because in this case the breakdown of the tetrahedral intermediate will result in acetylimidazole regeneration.

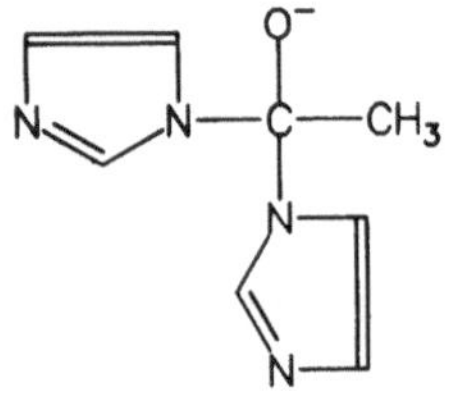

Evidently, imidazole promotes a proton cleavage from the water molecule:

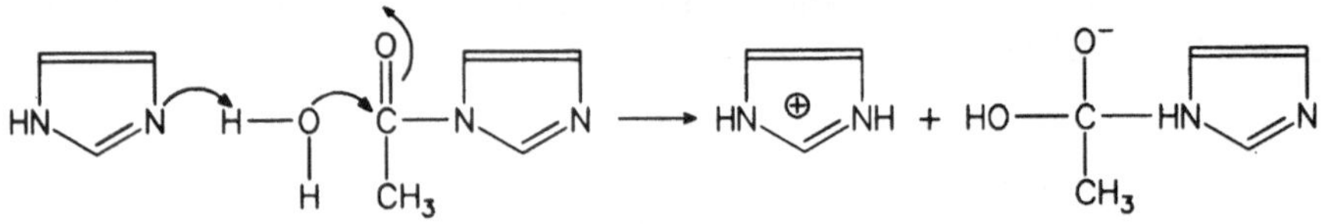

However, hydrolysis of p-nitrophenylacetate catalyzed by imidazole proceeds mainly according to the mechanism of nucleophilic catalysis [1572–1574]. The spectrophotometric studies as well as direct isolation of the intermediate – acetylimidazole – testify to this fact [1575]:

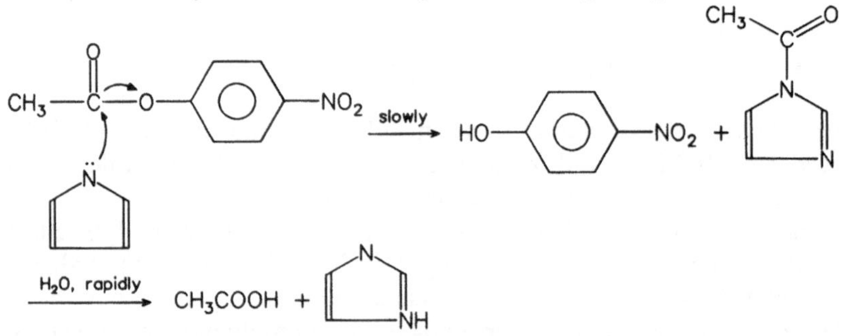

Imidazole catalysis depends on the nature of a leaving group. The compounds, containing weak basic ("good") leaving groups are hydrolyzed in the presence of imidazole chiefly according to the mechanism of nucleophilic catalysis. In case of strong basic leaving groups, general base catalysis by imidazole and general acid catalysis by imidazolium ion are realized.

Catalysis with nonprotonated forms of the substituted imidazoles is governed by the Brønsted linear equation [1576]

$$\lg k = 0.8 pK_a - 4.3 .$$

However, low basic imidazoles (for example, 4-nitroimidazole: pK_a 1.5) stay out of this correlation. This occurs because low basic imidazoles at pH > 7 are partially in an anion form:

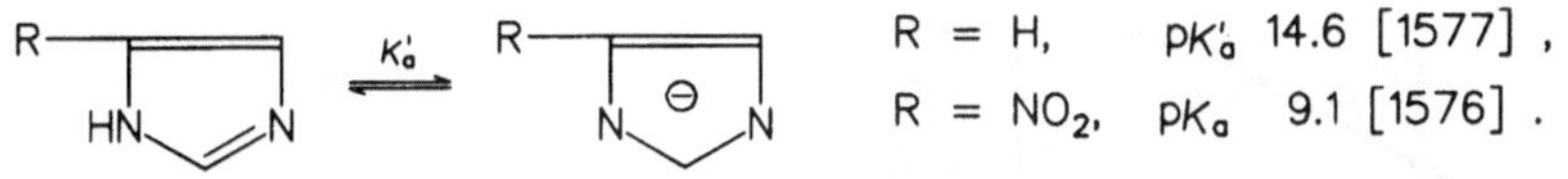

$R = H, \quad pK'_a \; 14.6 \; [1577]$,

$R = NO_2, \quad pK_a \; 9.1 \; [1576]$.

A set of examples is available [1578] when imidazole functions as a nucleophilic and general base catalyst simultaneously. For example, cresylacetate hydrolysis is described by an equation containing a member proportional to the square of the imidazole concentration. Here a kinetic isotope effect of the reaction is equal to 2.2. This is explained by the mechanism [1579]:

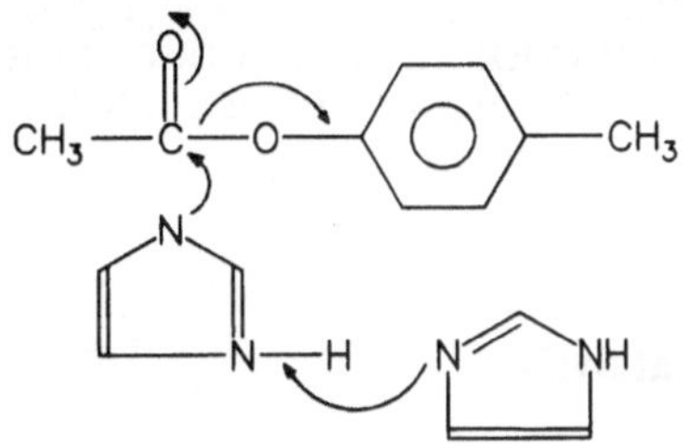

A general equation for the ester hydrolysis rate catalyzed by imidazole can be written as follows:

$$v = [RCO_2R'](k_n[Im] + k'_b[Im] + k_b[Im]^2 + k_A[Im^-]) ,$$

where k_n, k'_b, k_b characterize nucleophilic and general base catalysis, and k_A imidazolyl anion catalysis. A relative contribution of k_n and k'_b is illustrated by the dependence of the reaction rate constants catalyzed with imidazole ($\lg k_{Im}$) on the rate of base hydrolysis ($\lg k_{OH}$) (Fig. 38).

An imidazole protonated form – imidazolium cation – can function as a general acid catalyst [1581, 1582] promoting, for example, a breakdown of tetrahedral intermediates:

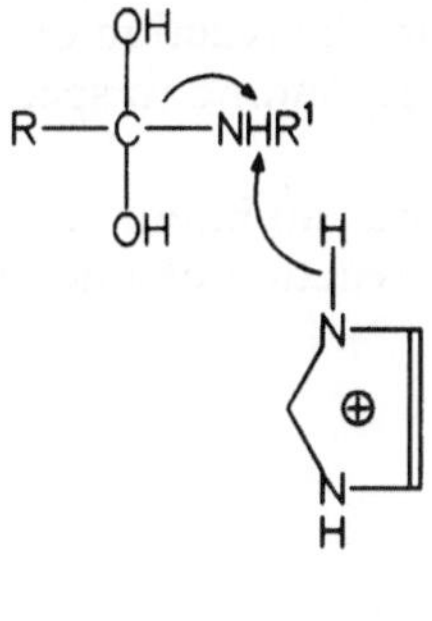

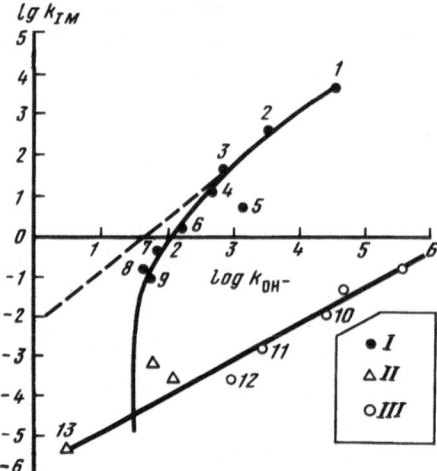

Fig. 38. Effect of the imidazole-catalyzed hydrolysis rate of esters on the base hydrolysis rate ($\mu = 1.0$; 25 °C) [1580]. *I* Reaction of nucleophilic substitution of acetates; *II* general base catalysis of hydrolysis of acetates; *III* general base catalysis of hydrolysis of methyl and ethyl esters. For trifluoroethylacetate the hydrolysis rate was measured in the presence of N-methylimidazole; *1* $(CH_3CO)_2O$; *2* $CH_3COOC_6H_3(NO_2)_2$-2,4; *3* $CH_3COOC_6H_5$; *4* $CH_3COOC_6H_4NO_2$-*m*; *5* $CH_3CON(Me)OCOCH_3$; *6* $CH_3COOC_6H_4Cl$-*p*; *7* $CH_3COOC_6H_5$; *8* $CH_3COOC_6H_4CH_3$-*p*; *9* $CH_3COOC_6H_4OCH_3$-*p*; *10* $CH_3COOCH_3(C_2H_5)$; *11* $CF_3CH_2OCOCH_3$; *12* $HONCOCH_3$; *13* $CH_3COOC_2H_5$

3.4.8   Intramolecular and Polyfunctional Catalysis

A hydrolytic reaction catalyzed with amide hydrolases runs in the enzyme-substrate complex, i.e., by an intramolecular mechanism. The examples of intramolecular mono- and polyfunctional catalysis of organic compounds are the models for this situation. There are many examples of catalysis of this type, most of them are summarized in reviews [1401, 1402, 1583–1585]. Here, I would like to give only a few of them – the most obvious cases in order to compare the efficiency of inter- and intramolecular catalysis.

Hydrolysis of esters and amides of salicylic acid is an example of intra-molecular involvement of a hydroxyl group [1586, 1587]. The pH curve of the hydrolysis rate of *p*-nitrophenyl ester of 5-nitrosalicylic acid is characterized by three regions (Fig. 39): (1) a plateau at pH 6–10 corresponding to water-catalyzed hydrolysis of a neutral molecule; (2) a rise and plateau within the range of weak acid to neutral pH values corresponding to the reaction of the anion form of a molecule; (3) a rise within basic pH corresponding to specific catalysis with the hydroxyl ion.

The reaction within pH 6–10 can be considered either as an interaction of salicylic acid anion with a water molecule (**XXIV**) or as a reaction of a neutral substrate with a hydroxyl ion (**XXV**):

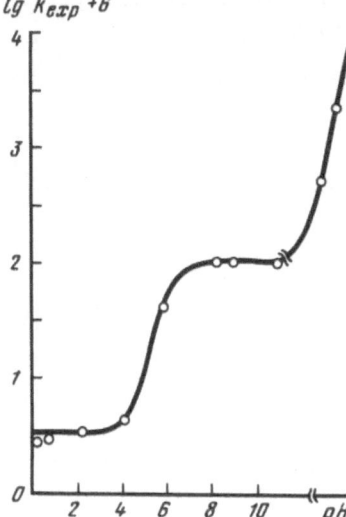

Fig. 39. The rate constant for hydrolysis of *p*-nitro-phenyl ester of 5-nitrosalicylic acid vs pH [1586]

These two mechanisms are kinetically equivalent. However, the hydrolysis rate of the corresponding methoxy anion is 2000 times lower than that of the hydroxy compound. If imidazole or ion N_3^- is used as nucleophile, the transition rates of methoxy- and oxycompounds are becoming almost the same. Hence the reaction of salicylic acid derivatives runs by general base catalysis (mechanism **XXIV**) [1586].

A hydroxyl group can be involved in an acyl transfer also as nucleophile. So imidazole catalyzes the reaction [1588]:

This reaction can be considered as an example of simultaneous intra- and intermolecular catalysis, whereas imidazole, probably, is involved when a tetrahedral intermediate breaks down.

A similar type of intra- and intermolecular catalysis is realized [1589] upon the action of compound (XXVI) on nitrophenylacetate:

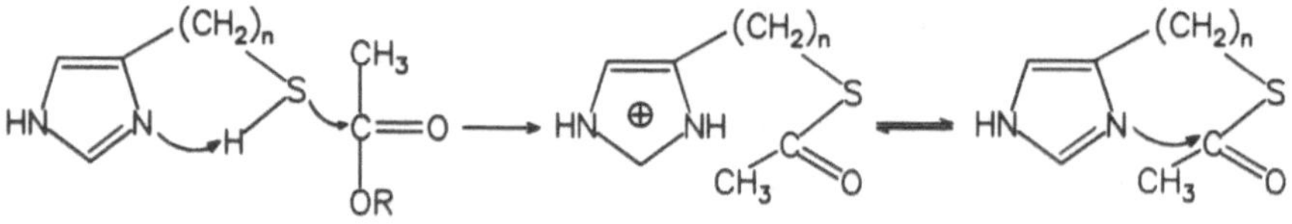

XXVI

n = 1,2

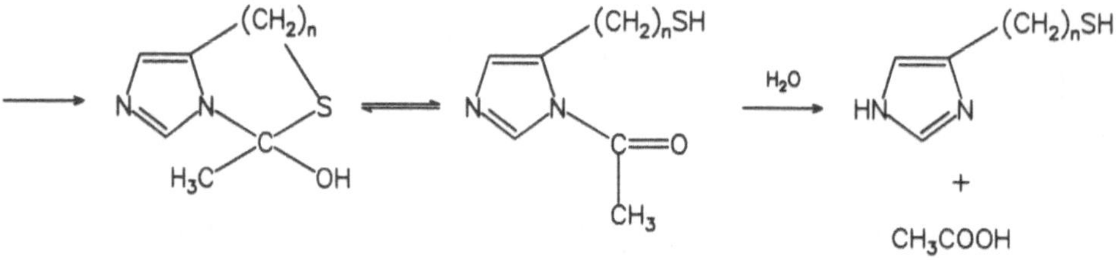

+

CH_3COOH

Interestingly, a thiol group in this example acts as a nucleophilic catalyst, while imidazole as both a general and a nucleophilic one. The reaction rate is only four to five times lower than that of the corresponding reaction catalyzed with α-chymotrypsin.

The best known example of intramolecular catalysis with a carboxyl group is aspirin hydrolysis [1590, 1591]:

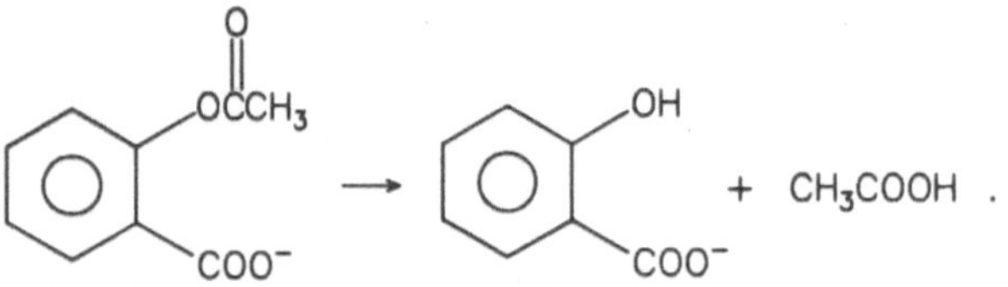

+ CH_3COOH .

This reaction was once considered as an example of a nucleophilic involvement of the carboxylate ion in hydrolysis, because during the reaction in $H_2^{18}O$ there is observed an insertion of oxygen isotope in the carboxyl group of salicylic acid [1590]. However, later on [1591] these data have been disproved. Studies [1590] of hydrolysis of the substituted aspirins reveal the effects of substituents on the properties of carboxylic and ester groups separately ($\rho_{carb} = -0.52$ and $\rho_{est} = 0.96$). This allows the conclusion that the involvement of the carboxyl group in the reaction appears to be due to general base catalysis with a carboxylate ion:

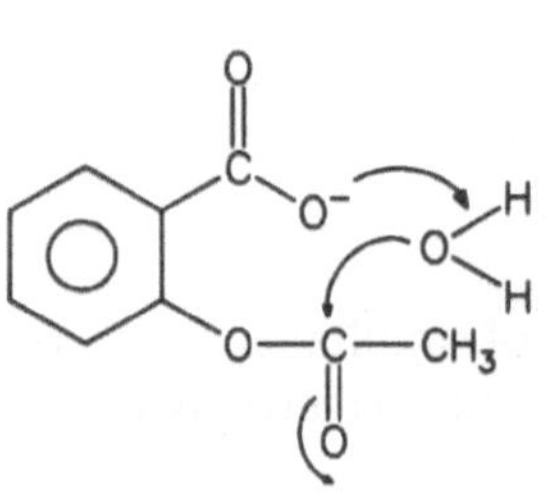

Upon hydrolysis of methyl ester of 2,6-dicarboxybenzoic acid a catalytic involvement of carboxyl groups depends on their ionization [1593]. In monoanion of this compound one group (ionized) acts as a nucleophilic catalyst, while the other as a general acid:

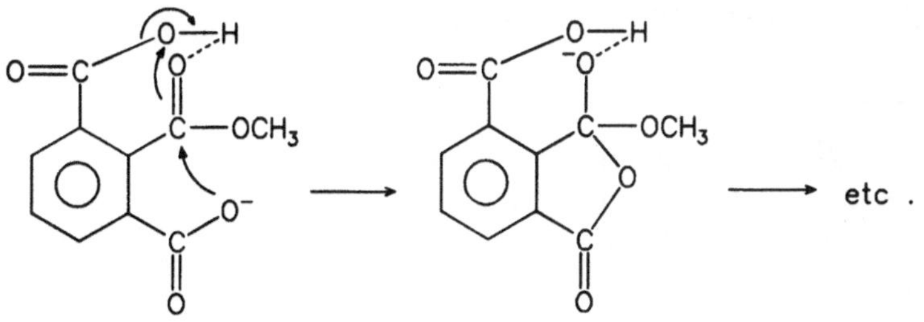

A series of examples of intramolecular catalysis with a carboxyl group is known, where this group functions as a nucleophilic catalyst [1594–1598]. Hydrolysis of 2,3-dimethylmaleamic acids (XXVIIa) is the most effective [1583, 1597]. At pH 3 these compounds are hydrolyzed 10^8–10^9 times faster than N-methylacetamide and 16 000 times faster than a compound free of methyl groups (XXVIIb). However, amide of cyclopenten-1,2-dicarboxylic acid is hydrolyzed slower than compound (XXVIIb):

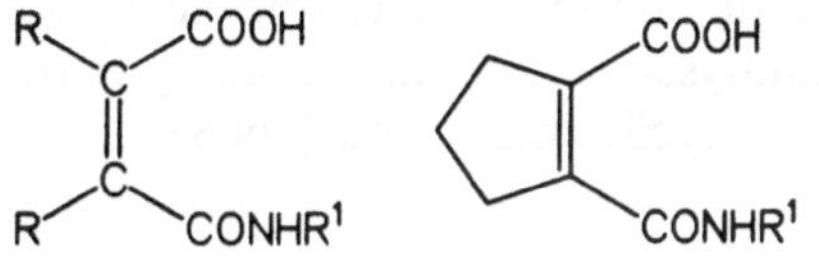

XXVII   **XXVIII**

a: R = CH$_3$; b: R = H

Hydrolysis of amide (XXIX) is an interesting example of nucleophilic catalysis:

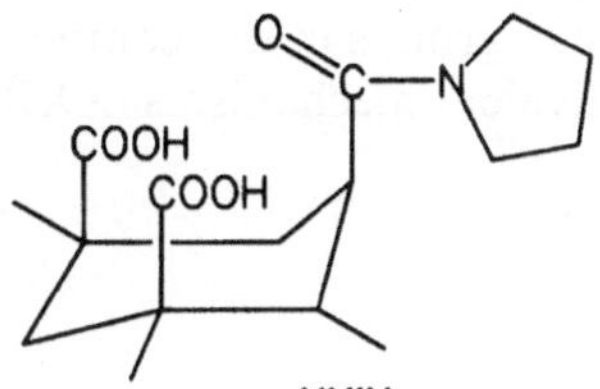

XXIX

In this compound a carboxyl oxygen is located above the amide bond plane at a distance of 2.8 Å, whereas one of the groups provides proton cleavage. An effective molarity in this reaction (see below) exceeds 10^{14} mol/l [1599]. A catalytic role of the carboxylate ion can be related to the cation electrostatic stabilization, for example, to imidazolium involved in the reaction [1600].

The peculiarity of catalysis with a carboxyl group lies in the above difference of basicity of *syn-* and *anti-*conformations. The majority of the models for intramolecular catalysis includes *anti-*conformation of the carboxylate ion. The model for the ion involvement in the proton transfer to (and from) the imidazole residue in a *syn-*conformation (for example, XXX) shows that imidazole pK_a increases here by 1.5 units [1601, 1602]

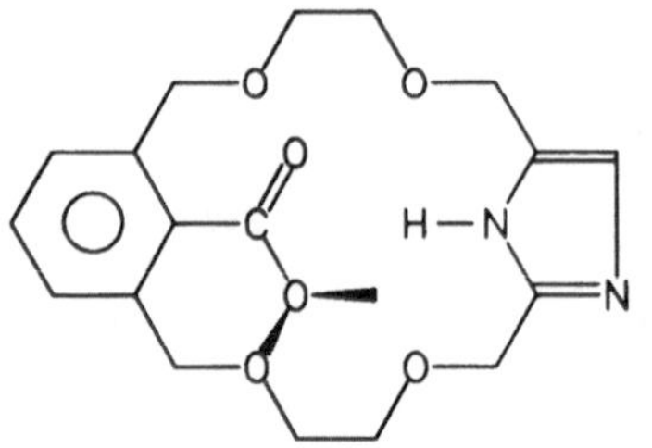

XXX

Many examples of intramolecular catalysis with metal ions are available. Metal can facilitate catalysis of an other, for example, a carboxylic group, thus stabilizing a favorable conformation as well as promoting catalysis with OH^- ions or water [1603–1605].

Intramolecular as well as intermolecular imidazole catalysis occur according to different mechanisms depending on pK_a of the leaving group [1606, 1607]. So hydrolysis of methyl ester of 2-imidazolylbenzoic acid proceeds according to the mechanism of base catalysis with imidazole, and phenyl ester is hydrolyzed with the formation of a tricyclic compound [1608]:

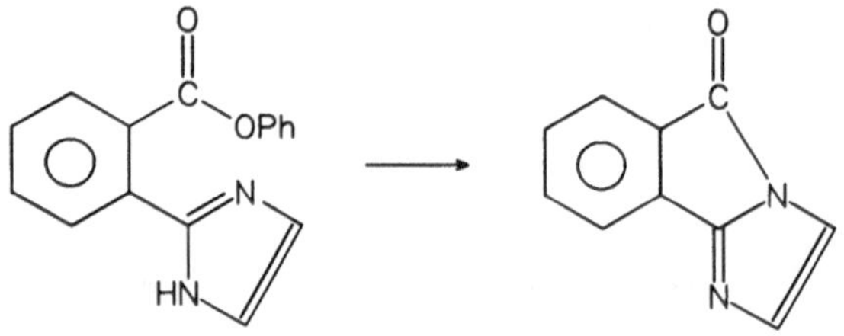

Another example of bifunctional catalysis is hydrolysis of methyl ester of 2,6-dihydroxybenzoic acid [1609]. Phenolic hydroxy groups in this compound cleave the ester bond according to kinetically equivalent mechanisms (XXI) or (XXXII):

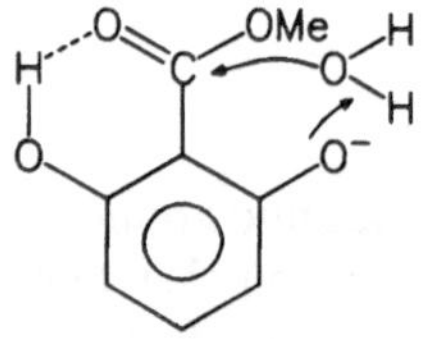

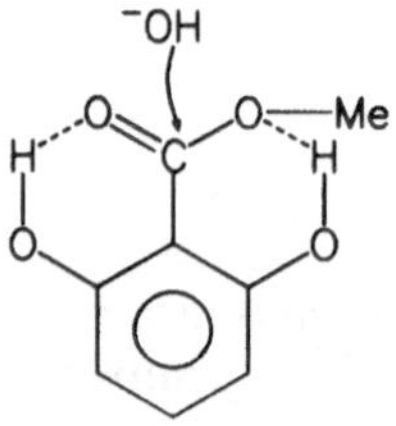

XXXI   **XXXII**

This compound is hydrolyzed 10^5–10^6 times faster than 2,6-dimethyl-benzoate.

A comparison of efficiency of intra- and intermolecular catalyses (Table 31) faces numerous difficulties. First is a possible difference in catalysis mechanisms. So an acetate ion in intermolecular reactions usually functions as

Table 31. Comparison of rate constants of intramolecular (k_1) and intermolecular (k_2) catalysis

Substrate	$k_1 \cdot s^{-1}$	$k_2, M^{-1}s^{-1}$	$k_1/k_2, M$	References
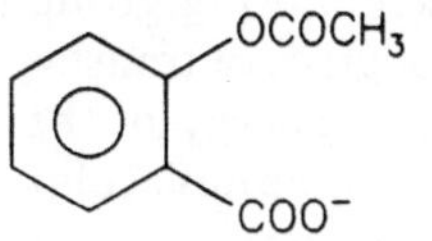 OCOCH₃ / COO⁻	83.8×10^{-6}	–		[1590]
			8.2	
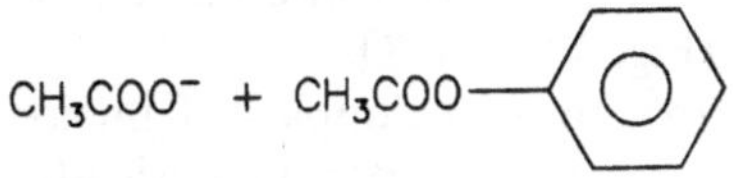 $CH_3COO^- + CH_3COO$—	–	10.2×10^{-6}		[1546]
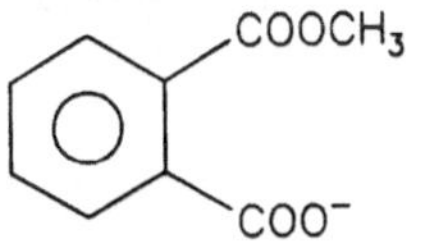 COOCH₃ / COO⁻	5.9×10^{-5}	–		[1611]
			49	
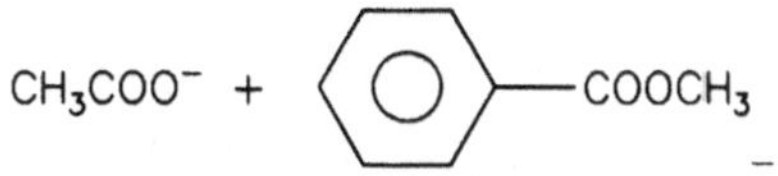 $CH_3COO^- +$ —COOCH₃	–	1.2×10^{-6}		[1611]
—(CH₂)₃COOC₆H₅ / N═ NH	0.043	–		[1612]
			24	
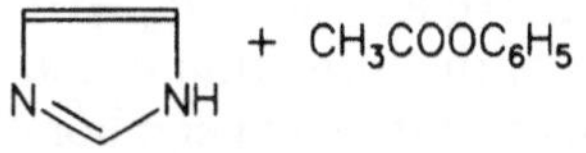 + CH₃COOC₆H₅ / N═ NH	–	0.18×10^{-2}		[1572]
CH₂COOC₆H₄NO₂-p / CH₂COO⁻	0.8	–	2×10^5	[1613]
$CH_3COO^- + CH_3COOC_6H_4NO_2$-$p$	–	4×10^{-6}		[1613]
COOH / COO⁻ OCOCH₃	0.02	–		[1614]
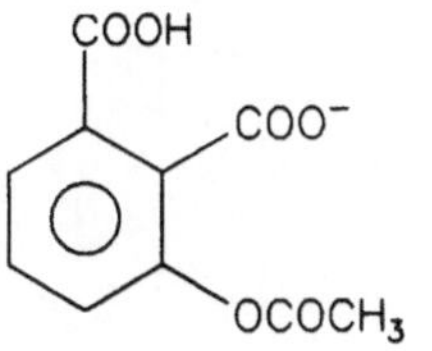			2×10^8	
$C_6H_5OCOCH_3 + C_6H_5COO^-$	–	10^{-10}		[1614]
CH₃ COO⁻ / C ‖ C / CH₃ CONHPr	1.04	–		[1583]
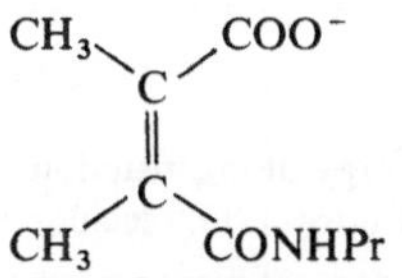			6.5×10^9	
$CH_3CONHBu + CH_3COO^-$	–	1.6×10^9		[1522]

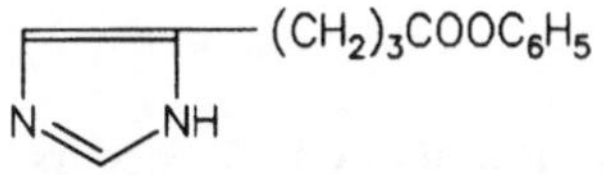

a general catalyst, while in intramolecular ones it can be a nucleophilic reagent. Secondly, it is not so easy to make a comparison under the same conditions. Furthermore, intramolecular catalysis is very sensitive to the substrate conformation [1610]. Finally, these two types of catalysis differ in the order of the reaction.

Usually, the efficiency of intramolecular catalysis is expressed in the effective molar concentration of catalytic groups. For a long time it was held that an effective concentration of neighboring groups cannot exceed 55 mol/l [1615], i.e., water concentration in water. However, according to Page and Jencks [1616, 1617] the maximum effective concentration of the neighboring group can reach 6×10^9 mol/l. This is caused by a limitation of the degrees of translational and rotational freedom of reactants; as a result the free energy of the ground state in intramolecular reactions is higher than that in intermolecular processes. At one and the same level of free energy of the transition state, or close to it, this decreases an activation barrier (Fig. 40) [1578].

The role of entropic factors is clear from the above examples of intramolecular catalysis as well as from the examples of lactonization in sterically restricted acids of type (XXXIII) [1618–1621]:

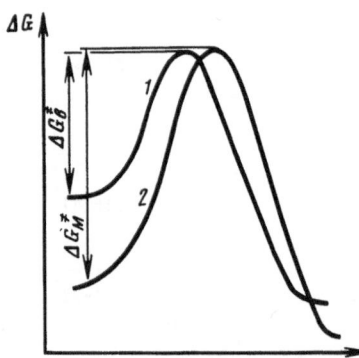

XXXIII

Lactonization of a substance (XXXIII; $R_1 = R_2 = R_3 = CH_3$) occurs $\approx 5 \times 10^{10}$ times faster than that of the compound (XXXIII; $R_1 = R_2 = R_3 = H$) free of methyl groups. As compared with the reaction of the formation of the ester bond from phenol and acetic acid, an effective concentration of neighboring groups here increases approximately by 15 orders of magnitude!

Here a certain freedom of intrarotation of the reacting groups seems possible. In any case, an angular shift approximately by 10° at the transfer from

Fig. 40. Change of the free energy along reaction coordinate (r) for inter- (*1*) and intra- (*2*) molecular reactions

the compound (XXXIV) to (XXXV) does not influence the lactonization rate [1622]:

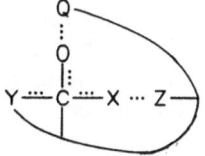

XXXIV   **XXXV**

In real systems, the level of the free energy of the transition state of the inter- and intramolecular reactions remains very seldom the same due to differences in the catalysis mechanism.

The efficiency of polyfunctional intermolecular catalysis is usually a little bit higher than that of monofunctional catalysis with the same compound. An increase of the order of the reaction entails a decrease of the activation entropy of the reaction. Such decrease compensates for a lowering of enthalpy and the activation free energy remains almost the same [1623]. To realize an effective catalysis it is necessary, without changing the reaction order, to increase the number of catalytically functioning groups, i.e., to realize intramolecular polyfunctional catalysis. However, there are still some restrictions due to possibilities of synchronization of an elementary step of the chemical transformation.

Indeed, if a polyfunctional intramolecular reaction runs in one elementary event forming a complex transition state of type this reaction requires synchronic oscillations of a few heavy atoms throughout the reaction coordinate.

A theoretical consideration of this problem [1624–1626] shows that it is necessary to introduce the so-called coefficient of the synchronization (α) in the equation for the rate constant, i.e.

$$k = \frac{KT}{h}\, \alpha e \frac{\Delta G^{+}}{RT}, \quad \text{where } \alpha = \frac{n}{2^{n-1}} \cdot \left(\frac{nKT}{\pi E_c}\right)^{n-1/2};$$

n is the total number of freedom degrees for m of reactive atoms $n = (3m - 6)$, E_c is the activation energy of the synchronization. Figure 41 presents the α values at various n and E_c. It is clear that even for small E_c values at $n = 18$ (i.e., $m = 8$) the α will make an essential negative contribution to the observed rate constant.

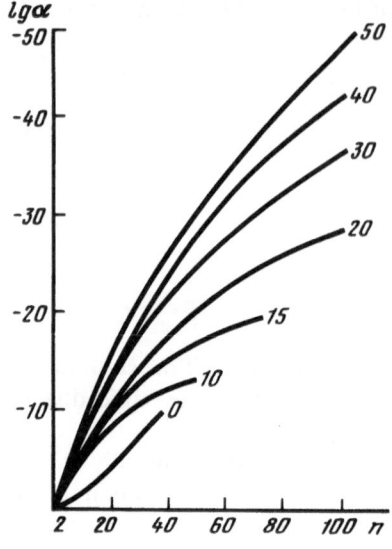

Fig. 41. Synchronization factor (α) vs the number of freedom degrees (n) for elementary acts with various activation energies of the synchronous reaction (indicated in kcal/mol) [1625]

3.4.9   Micellar Catalysis

Another type of catalysis regarding enzyme functioning is micellar catalysis. Lipophilic substances can be concentrated inside a micellar phase, and if it contains functionally active groups, the reaction shall proceed more effectively than in a homogeneous solution. For instance, p-nitrophenylacetate is hydrolyzed faster with alkali or in the presence of ethylamine than p-nitrophenyldecanoate [1627]. However, the hydrolysis rate of an ester bond in mixed micelles of p-nitrophenyldecanoate and decylamine is approximately 100 times higher than that of p-nitrophenylacetate with the same amine (Table 32).

The reasons for the effect of micellar catalysis are becoming quite clear upon its kinetic analysis [1628, 1629]. A substrate in solution containing a micellar phase is distributed between phases

$$S_w \xrightleftharpoons{P} S_m ,$$
(22)

Table 32. Rate constants of hydrolysis of p-nitrophenylacetate and p-nitrophenyldecanoate in mixed micelles

Substrate	Alkali k_1, $M^{-1}s^{-1}$	Ethylamine k_2, $M^{-1}s^{-1}$	Decylamine k_3, $M^{-1}s^{-1}$	k_3/k_2
p-Nitrophenylacetate (XXXVI)	9.45	5.74	39.1	6.8
p-Nitrophenyldecanoate (XXXVII)	1.19	0.19	133	700
k(XXXVII)/k(XXXVI)	0.13	0.033	3.4	100

where P is the equilibrium constant, and indices w and m designate water and micellar phases, respectively, where the following reactions run. In water and micelles the reaction proceeds:

$$nS_w \xrightarrow{k_w} \text{products} \quad \text{and} \quad nS_m \xrightarrow{k_m} \text{products} , \qquad (23), (24)$$

where n is the kinetic order of the reaction.

The rate we observed is

$$k' = \frac{k_m P^n C V + k_w (1 - CV)}{(1 + KC)^n} , \qquad (25)$$

where C is the concentration of the surface-active substance, V is its molar volume, $K = (P - 1)V$. At $k' \gg k_w$ the value of acceleration is equal to

$$(k'/k_w)_{max} = \frac{k_m}{k_w} \frac{(n - 1)^{n-1}}{n^n} P^{n-1} . \qquad (26)$$

According to formula (26), the efficiency is a product of two parameters: k_m/k_w shows an increase of the reaction efficiency due to a change of the medium composition of the factors orientating and stabilizing a transition state, and $(n - 1)^{n-1} P^{n-1}/n^n$ characterizes an effect of the reagent concentration, which depends much on the reaction order. Thus, micellar catalysis might be very effective for the reactions of high order.

Numerous systems containing different surface-active substances as micellar phases and various nucleophilic groups have been studied (see reviews [1628, 1630, 1631]). The acceleration observed reaches 10^4–10^5 [1632, 1633].

Use of enantiomeric micellar catalysts provides modeling stereospecificity [1634, 1635], complex formation, acylation and deacylation of a catalyst [1636, 1637], as well as the other peculiarities of enzymatic catalysis. Catalytic effects are observed not only when esters but also when amides (anilides) are hydrolyzed [1638].

3.4.10 Macromolecular Models of Amide Hydrolases

Polymeric catalysts show many properties typical of the enzymes. The first such catalysts are poly-4-vinylpyridine and poly-N-vinylimidazole [1639, 1640]. Hydrolysis with these polymers of substituted nitrophenylacetates shows typical bell-shaped pH-dependence. Polymers on the basis of polyethylene, containing different catalytically active groups, appeared to be rather effective [1641, 1642]. So the polymers of type (XXXVIII) hydrolyze p-nitrophenylcaproate with the rate constant 6.15×10^4 M^{-1} s^{-1}, a fast acylation of a catalyst with the following slower deacylation being observed. There were proposed polymeric catalysts containing a few different functional groups

[1643]:

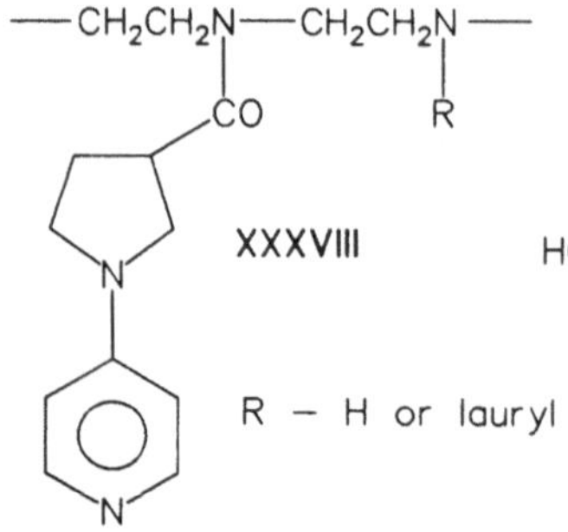

Polyelectrolytes – styrene copolymers with maleimides of histidine and phenylalanine (**XXXIX**) – can form complexes with proteins (casein) and catalyze the amide bond hydrolysis [1644, 1645].

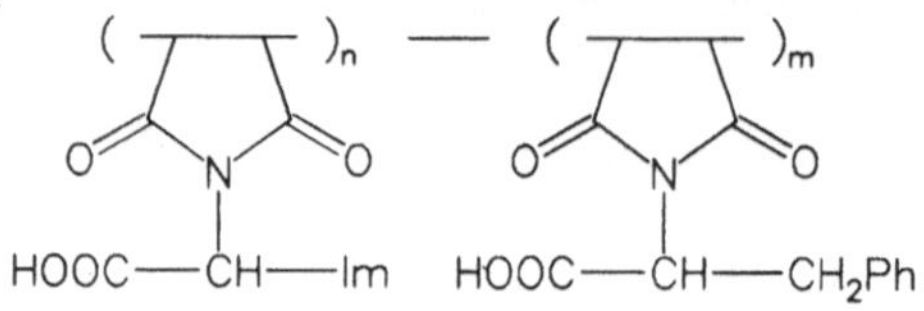

Although they are not referred to macromolecular compounds, they are still close to them by their ability to form complexes with the substrate, crown-esters (**XL**) and the structurally similar compounds compose another type of models of amide hydrolases [1646]:

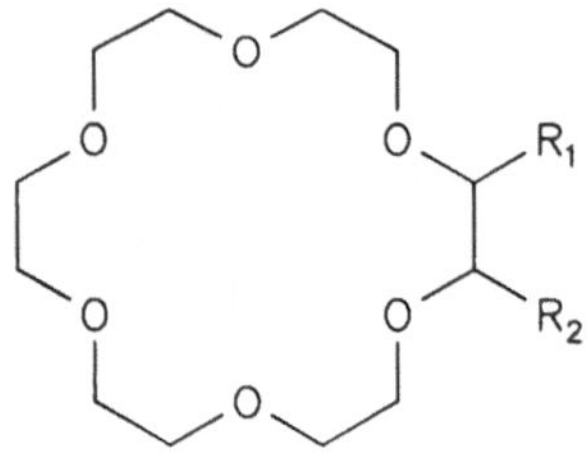

These ethers are capable of binding metal ions and charged groups (for example, ammonium ion), and side chains can contain catalytically active groups. Binding of chiral molecules is stereospecific. Of interest is an example of acylation of macrocycle (**XLI**) with *p*-nitrophenol ester of alanine [1647]. As compared to the transfer of the acyl residue on 3-phenylbenzyl alcohol, its transfer on substance (**XLI**) is accelerated 10^{11} times.

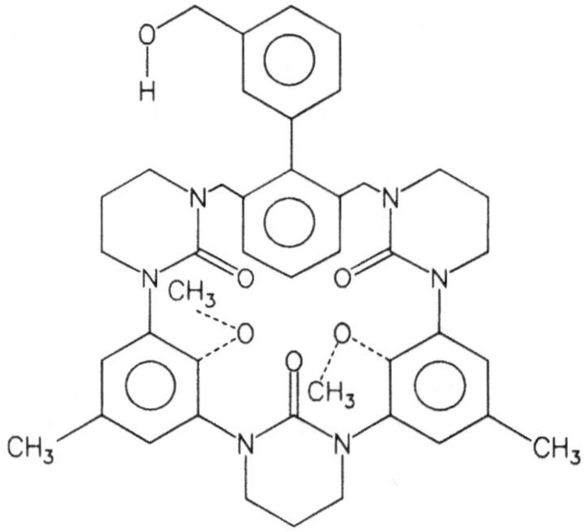

α-, β-, γ-Cyclodextrins (Fig. 42) – cyclic glucose oligomers of 6, 7, or 8 residues, respectively, are another type of artificial catalysts of hydrolytic reactions. They bind small molecules and can be modified by an insertion of

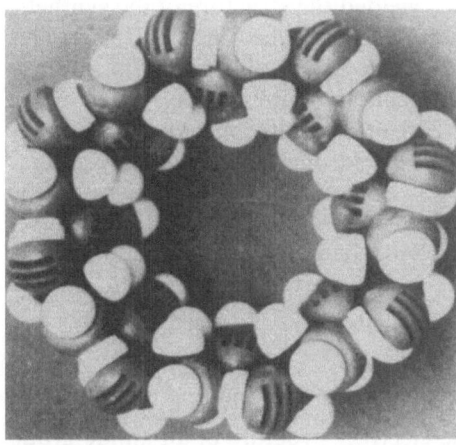

Fig. 42. Model of cyclooctaamilose (γ-cyclo-dextrin) [1648]

functional groups [1402, 1648]. Hydroxyl groups are acylated upon binding of *p*-nitrophenyl esters in the catalyst cavity. The acyl group then is cleaved, evidently, because of neighboring groups that participate in catalysis. The rate of acylation sometimes can be compared with that of *p*-nitrophenylacetate hydrolysis with chymotrypsin [1649]. This model, which adequately mimics its properties, was designed in 1985 (XLII). This is a compound called β-benzim, which catalyzes hydrolysis of *m*-tretbutylphenylacetate with the rate constant close to the enzymatic one [1650]:

XLII

It is noteworthy that β-benzim is more stable to pH and the temperature than chymotrypsin [1651].

An attempt was made to obtain a hybrid protein with the properties of amide hydrolase through expression in *Escherichia coli* of the gene encoding seven repeats of octapeptide Glu-Ala-His-Ala-Ser-Phe-Phe-Phe in the reading frame [1652]. The protein catalyzes hydrolysis of *p*-nitrophenylacetate with the rate exceeding 1000 times that of hydrolysis of this substrate with tripeptide Ser-His-Asp. Analysis of the binding of benzamidine and bovine pancreatic trypsin inhibitor (BPTI) with trypsin has yielded the structure of the synthetic 26-membered peptide, modeling a region of the enzyme binding [1653]. This peptide bound benzamidine and BPTI, reacted with diisopropylfluorophosphate, but did not deploy any catalytic activity. Interestingly, most of the models of enzymes are rather effective as regards *p*-nitrophenyl esters, but, as a rule, inactive as to amide substrate. This "syndrome of *p*-nitrophenyl esters" [1654] is explained, at least in case of cyclodextrins, by the high energetic *S-cys*-conformation of the ester group in acyl-*β*-cyclodextrins. Besides, cyclodextrin O^--anion is a better leaving group than alcoxy and amino groups.

3.5 Conclusion

The aforementioned peculiarities of the electron structure of an amide group – stability and sensitivity to outside effects – govern low rates of noncatalyzed splitting of amides and the existence of numerous types of catalysis, which substantially promotes hydrolysis. Resonance destabilization of amide is the most important and energy consuming process, i.e., these are interactions decreasing the C–N bond order. This is reached either by protonation of a carbonyl group (acid catalysis) or by a charge increase of the molecule attacking carbonyl carbon (base catalysis), as well as by a capability of the amide group to outside plane deformation without large energetic expenditures. Nucleophilic attack in itself is, obviously, a concerted process, i.e., the nearer the negatively charged group approaches to a carbonyl carbon, the more effectively occurs the redistribution of electron density, which decreases resonance stabilization. The nucleophile attack yields, as a rule, an unstable tetrahedral intermediate, its breakdown direction is established both by its conformation and the properties of the substituents related to tetrahedral carbon. At this stage the breakdown towards the products is accelerated according to general acid and base catalysis. The catalysis is very effective if it proceeds by the intramolecular mechanism. Of importance is mutual disposition of the reactants. In most of the favorable cases the acceleration value is close to that observed at enzymatic catalysis.

Chapter 4 Enzyme Hydrolysis. Phenomenology

There exist two peculiarities that distinguish enzymes from nonbiological catalysts – efficiency and specificity. Though biocatalysis has been successfully modeled, still it is impossible to mimic fully these peculiarities. This chapter presents information on the efficiency and specificity of amidehydrolases.

4.1 Enzyme Kinetics

Like any catalyst, the enzyme can work in concentrations much lower than the substrate concentration. Here, three reaction phases are identified: (1) pre-steady state, when the free enzyme concentration falls dramatically (usually to zero at $[S_0] \gg [E_0]$); (2) steady state, when the reaction rate is actually constant in time: (3) post-steady state, when the reaction rate lowers in time due to a great decrease in the substrate concentration (see reviews [1613, 1655–1661]).

4.1.1 Steady-State Kinetics

The enzymic reaction velocity is conventionally measured during the steady-state period. Here, the concentrations of a substrate and intermediates change slightly as compared with the initial one and can be neglected. The steady state is attained at the extent of the substrate conversion being below 10%–15% (Fig. 43).

If the initial substrate concentration grows, in contrast to nonenzymatic reactions the velocity finally reaches the limit, being then independent of the substrate concentration. The enzyme is saturated with a substrate and the reaction rate is maximum. This points to the formation of the enzyme-substrate complex, the latter undergoing chemical transformation

$$E + S \underset{k_{-1}}{\overset{k_1}{\rightleftharpoons}} ES \overset{k_2}{\longrightarrow} E + P .$$ \hfill (1)

The formation of the enzyme-substrate complex is a reversible process, the reagents being kept by the forces of a noncovalent nature. Numerous physicochemical and direct X-ray data in favor of such complexes are available (see Chap. 6).

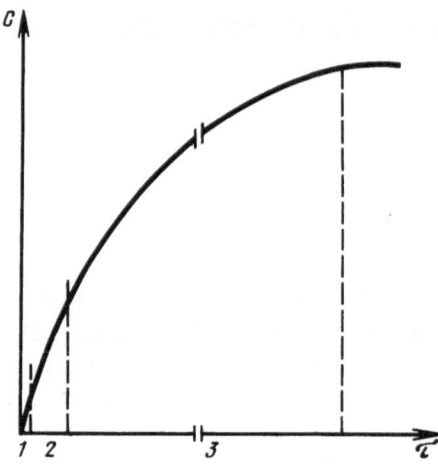

Fig. 43. Dependence of the concentration of the reaction product on time; *1* pre-steady state; *2* steady state; *3* post-steady state

The expression for the substrate conversion rate corresponding to scheme (1) is well-known as the Michaelis-Menten equation [1662]:

$$\frac{d\mathrm{P}}{dt} = v = \frac{k_2[\mathrm{E}]_0[\mathrm{S}]_0}{K_m + [\mathrm{S}]_0}, \tag{2}$$

where $K_m = k_{-1}/k_1$ is the Michaelis constant. To derive this equation a breakdown of complex ES to the initial substrate is supposed to proceed more rapidly than its chemical transformation, i.e., $k_{-1} \gg k_2$. If this assumption is not introduced, $K_m = (k_{-1} + k_2)/k_1$ as Briggs and Haldane have shown [1663]. A more sophisticated situation entails a more complicated expression for K_m. Therefore, the dissociation constant for the enzyme-substrate complex is assumed to be designated as the substrate constant $K_s = k_{-1}/k_1$ [1664, 1665], and in all other cases it is termed K_m (the Michaelis constant) or K'_m, where the apostrophe indicates the experimental (apparent) value. The value of the chemical transformation rate constant can be complex as well. It is a first-order velocity constant, usually called the catalytic constant (k_{cat}). In scheme (1) $k_{cat} = k_2$ and characterizes the number of substrate molecules converted to the product in a time unit (turnover number).

When $K_m \gg [\mathrm{S}]_0$, Eq. (2) is written as

$$v = k_{cat}[\mathrm{E}]_0[\mathrm{S}]_0/K_m. \tag{3}$$

The k_{cat}/K_m value is a second-order velocity constant for the enzyme-substrate reaction.

If $K_m \ll [\mathrm{S}]_0$, the reaction rate is maximum

$$v = V_{max} = k_{cat}[\mathrm{E}]_0, \tag{4}$$

and if $K_m = [\mathrm{S}]_0$, $v = 0.5\,V_{max}$, i.e., the Michaelis constant corresponds to the substrate concentration, when the reaction rate is half a maximum rate.

Here we have considered the simplest scheme of a single-substrate reaction that obeys the Michaelis-Menten kinetics. More general are cases, where several intermediate complexes exist on the pathway from a substrate to a product. For instance, if there are two sequentially formed complexes (due to isomerization of ES to E*S)

$$E + S \underset{k_{-1}}{\overset{k_1}{\rightleftharpoons}} ES \underset{k_{-2}}{\overset{k_2}{\rightleftharpoons}} E^*S \overset{k_3}{\longrightarrow} E + P \;, \tag{5}$$

the rate is expressed as

$$v = \frac{k_3 K_P [E]_0 [S]_0}{K_m + (1 + K_P)[S]_0} \;, \tag{6}$$

where $K_P = \dfrac{k_2}{k_{-2} + k_3}$ and $K_m = \dfrac{k_{-1} + k_3}{k_1}$.

If the enzyme-substrate complex undergoes sequential transformations [1666]

$$E + S \underset{k_{-1}}{\overset{k_1}{\rightleftharpoons}} ES \overset{k_2}{\longrightarrow} EA \overset{k_3}{\longrightarrow} E + P_2 \;,$$
$$+$$
$$P_1 \tag{7}$$

then

$$v = \frac{\dfrac{k_2 k_3}{k_2 + k_3} [E]_0 [S]_0}{K_m \dfrac{k_3}{k_2 + k_3} + [S]_0} \;. \tag{8}$$

The reactions catalyzed by amidehydrolases are in fact single-substrate processes, since a change in water concentration is usually ignored. However, sometimes one should analyze transformations of two substrates catalyzed by the enzymes of this class, particularly reverse reactions of the peptide bond synthesis and reactions of transpeptidation (Sect. 4.8). Two substrates can be attached to the enzyme independently (random mechanism):

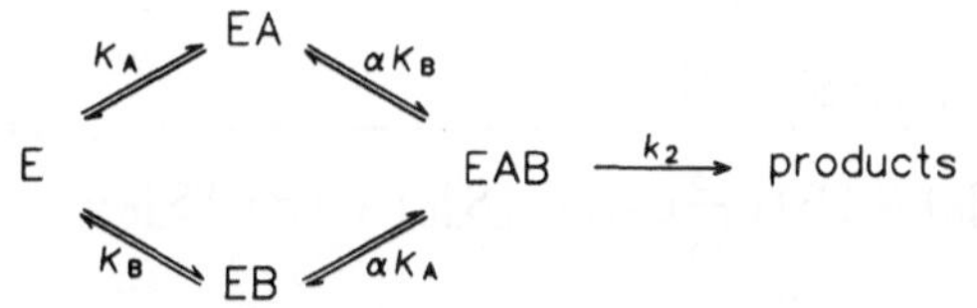

$$
\begin{array}{c}
\quad\quad\; \overset{EA}{\nearrow} \;\; \searrow \alpha K_B \\
\overset{K_A}{} \quad\quad\quad\quad\quad\quad \\
E \quad\quad\quad\quad EAB \overset{k_2}{\longrightarrow} \text{products} \\
\quad \searrow \;\; \nearrow \alpha K_A \\
K_B \; EB
\end{array}
\tag{9}
$$

or sequentially (ordered mechanism):

$$E \underset{}{\overset{K_A}{\rightleftharpoons}} EA \underset{}{\overset{K_B}{\rightleftharpoons}} EAB \overset{k_2}{\longrightarrow} \text{products} \;. \tag{10}$$

Several types of ordered mechanisms (ping-pong, Theorell-Chance and other mechanisms; see [1660]) are distinguished depending on the order of attachment of substrates A and B and the sequence in release of products.

If the equilibrium is quickly reached, i.e., when chemical changes proceed slower than the complex formation, a general expression for the rate of the random double-substrate reaction is as follows

$$v = \frac{k_2[E]_0}{1 + \dfrac{\alpha K_A}{[A]_0} + \dfrac{\alpha K_B}{[B]_0} + \dfrac{\alpha K_A K_B}{[A]_0[B]_0}} \; . \tag{11}$$

For the ordered mechanism (10) the rate is given by equation

$$v = \frac{k_2[E]_0}{1 + \dfrac{K_B}{[B]_0} + \dfrac{K_A K_B}{[A]_0[B]_0}} \; . \tag{12}$$

The proteolytic catalysis of the reaction involving two substrates is exemplified by transpeptidation

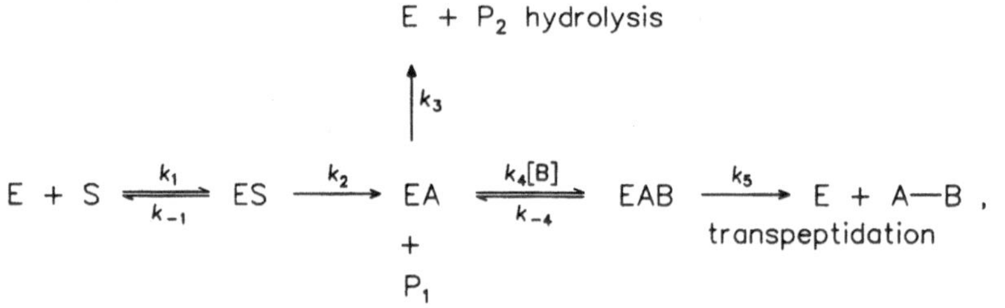

$$\mathsf{E} + \mathsf{P_2}\ \text{hydrolysis}$$

$$\uparrow k_3$$

$$\mathsf{E} + \mathsf{S} \underset{k_{-1}}{\overset{k_1}{\rightleftharpoons}} \mathsf{ES} \overset{k_2}{\longrightarrow} \underset{\substack{+\\ \mathsf{P_1}}}{\mathsf{EA}} \underset{k_{-4}}{\overset{k_4[B]}{\rightleftharpoons}} \mathsf{EAB} \overset{k_5}{\underset{\text{transpeptidation}}{\longrightarrow}} \mathsf{E} + \mathsf{A}{-}\mathsf{B}\ , \tag{13}$$

where B is the acceptor of transpeptidation.

If the concentration of the transpeptidation product can be detected during the initial period of the reaction, the rate of its formation (v_t) is expressed as [1667]

$$v_t = \frac{k_5[E]_0[S]_0[B]_0}{K_T \dfrac{k_3}{k_2}(K_m + [S]_0) + [B]_0 \dfrac{k_5}{k_2}(K_m + [S]_0) + [B]_0[S]_0} \; , \tag{14}$$

where $K_T = \dfrac{k_{-4} + k_5}{k_4}$.

For hydrolysis rate (v_h)

$$v_h = \frac{k_3 K_T[S]_0[E]_0}{K_T \dfrac{k_3}{k_2}(K_m + [S]_0 + [B]_0) + [B]_0 \dfrac{k_5}{k_2}(K_m + [S]_0) + [B]_0[S]_0} \; . \tag{15}$$

The King-Altman method of determinants [1668] as well as various formalized methods based on the theory of graphs [1669–1671] are used to derive the equations of the enzyme kinetics.

4.1.2 Method for Detecting Kinetic Parameters

To measure the magnitudes of kinetic parameters of the equations in Sect. 4.1.1, the initial rates of the substrate conversion with its concentration change are determined. These dependences are usually analyzed graphically, when transforming the rate equation to a linear form. According to the Lineweaver-Burk method [1672] a linear transformation of eq. (2) is

$$\frac{1}{v} = \frac{1}{k_{cat}[E]_0} + \frac{K_m}{k_{cat}[E]_0} \cdot \frac{1}{[S]_0} \,. \tag{16}$$

Plots $1/v$ vs $1/[S]_0$ are presented in Fig. 44a. Intercepts of a straight line on coordinate axes give the reciprocal values of $k_{cat}[E]_0$ and K_m.

Another way to linearize the equations of type (2) has been proposed by Eadie and Hofstee [1673] (Fig. 44b):

$$v = k_{cat}[E]_0 - K_m \frac{v}{[S]_0} \,. \tag{17}$$

The accuracy of the defined kinetic parameters in these and other types of linear presentations of the Michaelis-Menten equation is evaluated in [1674–1677].

The reactions with two substrates are generally analyzed by detection of their rates depending on the concentration of one of the substrates at fixed concentrations of the second substrate. The double reciprocal plots (Fig. 45) often allow distinguishing of the ordered and the random mechanisms, and secondary dependences of the values of intercepted lines on the second substrate concentration give the values of kinetic constants of the bisubstrate reaction.

Practical methods for measuring the initial reaction rates and analysis of kinetic equations are given in many manuals [1613, 1659].

The analysis provides the value of the apparent Michaelis constant, its meaning depends on the kinetic mechanism type (Sect. 4.1.3), and the maximum rate $V_{max} = k_{cat}[E]_0$.

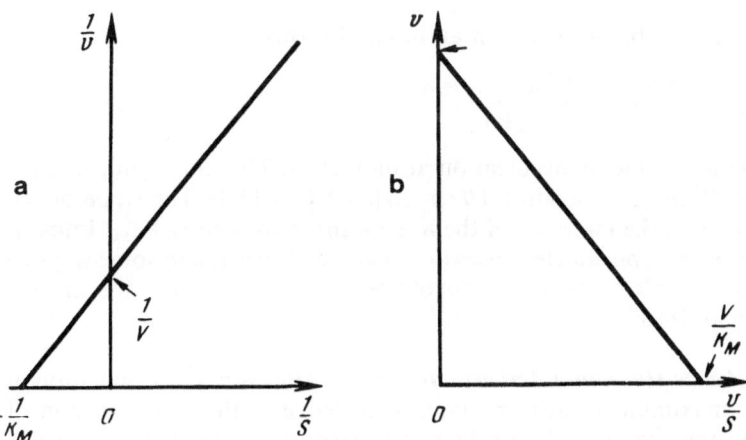

Fig. 44a, b. Detection of kinetic parameters of enzymatic reactions by methods of Lineweaver-Burk (**a**) [1672] and Eady-Hofstee (**b**) [1673]

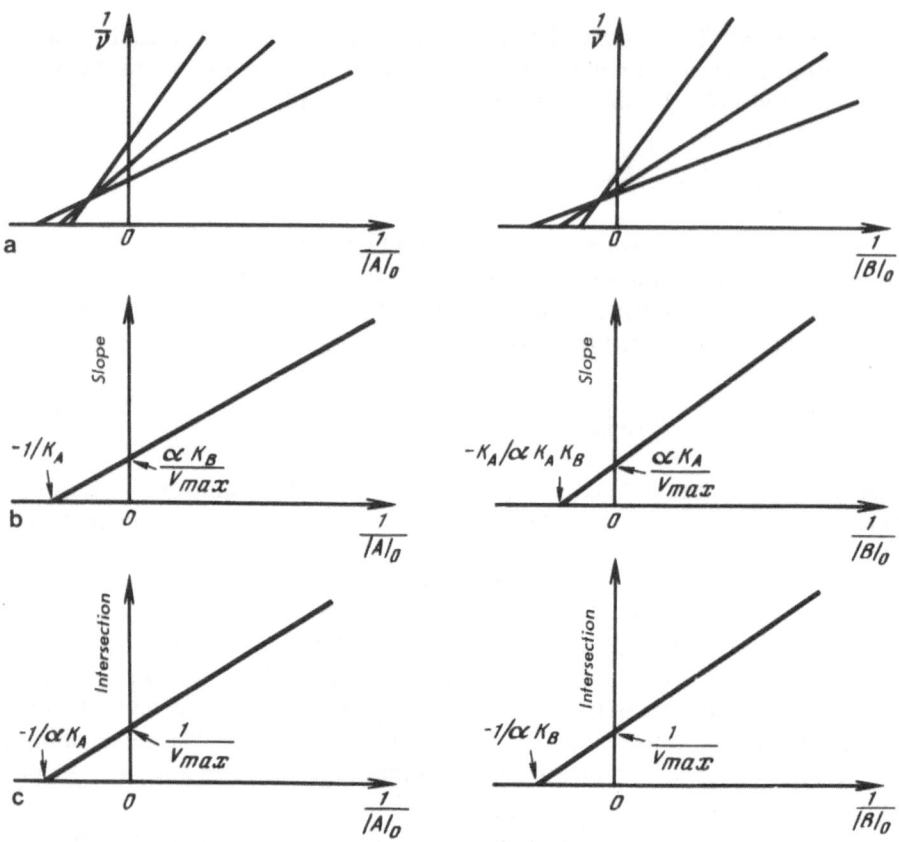

Fig. 45a–c. Detection of kinetic parameters of bi-substrate reactions; **a** primary, **b, c** secondary plots

Equation (2) can be presented in an integral form:

$$\frac{[P]}{t} = V_{max} - \frac{K_m}{t} \ln \frac{[S]_0}{[S]_0 - [P]} ,$$

where [P] is the product concentration at moment t. This equation [1678] is defined by plotting $[P]/t$ against $1/t \cdot \ln [S]_0/([S]_0 - [P])$. The slope of the line is equal to K_m, the intercept of the line on the Y-axis gives V_{max}. Integral equations for more sophisticated cases (for example, [1679]) and some ways to analyzing the complete kinetic curve of the enzymic reaction are as well proposed [1680, 1681].

Assay of the Amide Hydrolase Activity. Active Site Titration. The calculation of k_{cat} from the maximum velocity requires knowledge of the concentration of active sites in the enzyme. The problem has been overcome for quite a few amide hydrolases.

To determine the absolute concentration of the enzyme active sites, several methods have been introduced [1402]. Some of them are based on the application of irreversible (or quasi-irreversible) inhibitors, selectively reacting with active sites of the enzyme studied. If an inhibitor contains a chromogenic or radioactive label, the concentration of the active sites can be calculated when the modified protein is removed and the label content is determined, and of course if the protein molecular weight is known. This has a disadvantage arising from a necessity of a highly selective inhibitor reacting quantitatively with the protein. However, to be sure of this one should obtain a pure protein and study the reaction of modification. Such irreversible inhibitors are described in Chap. 5. Some tightly bound reversible inhibitors, such as *p*-aminobenzamidine, are proposed for titration of trypsin and thrombin [1682].

If the above conditions are followed and the number of active sites in the enzyme molecule is known, the absolute enzyme concentration can be found without labeled inhibitors and isolation of the modified protein. Using inhibitor concentrations certainly lower than the enzyme concentration and detecting the residual enzyme activity, the desired concentration may be obtained (Fig. 46).

Quite a different way for titration of the active site is proposed when the reaction goes stepwise with the accumulation of an intermediate. The primary "burst" of the product is established at the pre-steady-state stage [1208, 1683]. Let us assume that the enzyme-substrate reaction leads to a rapid formation of the first product (P_1) and to a slow breakdown of the intermediate complex (ES') yielding a second product (P_2):

$$E + S \xrightarrow{k_1'} ES' \xrightarrow{k_2} E + P_2 \cdot$$
$$+ \tag{18}$$
$$P_1$$

The curve of the product (P_1) accumulation in time is presented in Fig. 47. Extrapolation of the straight line corresponding to the steady state of the

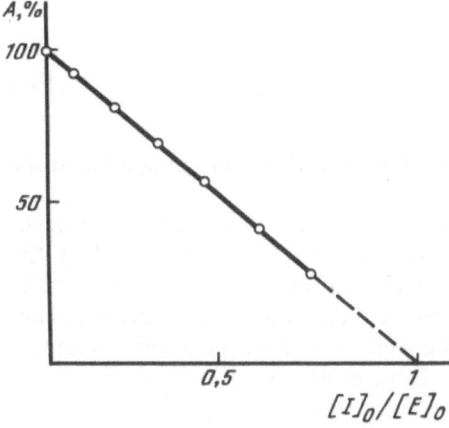

Fig. 46. Titration of the enzyme active site. Effect of activity (A) on the molar ratio of the inhibitor and enzyme

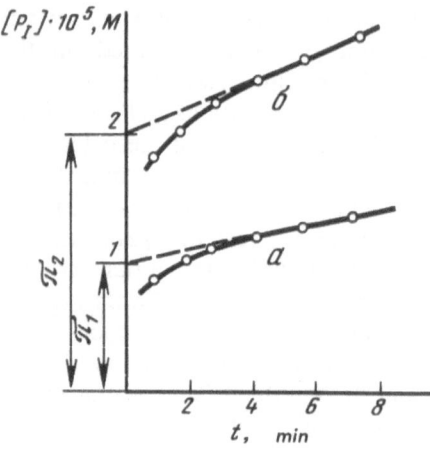

Fig. 47. Kinetics of p-nitrophenol formation upon α-chymotryptic hydrolysis of p-nitrophenylacetate for two enzyme concentrations a and b

process to zero time results in the value of "burst" (π) related to the rate constants by the ratio:

$$\pi = [E]_0 k_1'/(k_1' + k_2)^2 ; \tag{19}$$

if $k_1' \gg k_2, \pi = [E]_0$.

Suitable substrates for the active site titration of serine proteases are trans-cinnamoylimidazole, p-nitrophenylacetate [1208], nitrophenyl esters of N-protected amino acids [1684] as well as amino acid derivatives of rodamine [1685, 1686]. The latter are used to titrate femtomolar concentrations of the enzyme. For serine proteases there is proposed an interesting titrant derived from a bovine pancreatic trypsin inhibitor (BPTI) with the lysine residue of the active site substituted by nitroanilides of various amino acids. Cleavage of nitroaniline leads to regeneration of BPTI inhibitory properties, and the enzyme reaction here is stoichiometric [1687].

The detection of the rate of substrate hydrolysis, when $[E]_0 \approx [S]_0$ is a common method for detection of the absolute enzyme concentration [1659, p. 114].

In accordance with equation

$$\frac{1}{1 + \lambda} = \frac{[S]_0}{\lambda_0 [E]} - \frac{K_m}{[E]_0}, \tag{19a}$$

where $\lambda = 1/(V_{max}/v - 1)$, the plot in coordinates $1/(1 + \lambda)$ vs $[S]_0/\lambda$ has slope equal to $1/[E]_0$.

When the absolute concentration cannot be measured due to whatever reason, it is sufficient to perform the activity assay. It is expressed as the enzyme amount required for conversion of 1 μmol of substrate per 1 min. Numerous substrates and techniques, among them automated procedures are available [1688], which allow detecting very low concentrations of amide hydrolases [1689–1714]. Some of the substrates are listed in Table 33.

Table 33. Some substrates used for sensitivity assay of amide hydrolase activity

Substrate	Enzyme	Method	Sensitivity, ng	References
Sucinylalbumin + fluorescamine	Pepsin	F	1	[1694]
^{125}I-Hemoglobin (sp)	Trypsin	R	0.2	[1700]
^{14}C-Elastin (sp)	Elastase	R	1	„
CM-Lysozyme (cytochrome C)	Trypsin, Chymotrypsin	Ch	2.5–10	[1706]
Casein-β-glycosidase (sp)	„	E	0.3	[1701]
„	Plasmin Bromelain		30	„
Ovomucoid + trypsinogen	Enteropeptidase	Ch	2	[1714]
LeuNH$_2$ + leucine dehydrogenase	Leucine amino-peptidase	E	–	[1702]
^{3}H-BzPhe-Ala-ArgOH	Carboxypeptidase H and B	R	0.01	[1713]
GlnHisPhe$_N$PheAlaLeuNH$_2$	Pepsin, cathepsin cathepsin D	Sp	100	[1703]
ZPheOMC	Chymotrypsin	F	2	[1705]
N-peptidyl-AMC	Coagulation factor, thrombin, kallikrein, urokinase, etc.	F	5–50 <1	[1693, 1695, 1711]
7-(GltPheNH-)4MCL	Chymotrypsin	F	10	[1708]
AntLys-p-NA	Trypsin	F	25	[1709]
Dns-D-AlaGlyPhe$_N$Gly^a	Enkephalinase	F	–	[1704]
ZLysSBzl	Kallikrein, Plasmin, etc.	S	≈4	[1697]
HPhePhe$_T$OH	Leucine amino-peptidase	S	10–300	[1707]

Abbreviations: sp – solid-phase method; Phe$_N$ – p-nitrophenylalanine; OMC – 7-hydroxy-4-methylcumaryl; AMC – 7-amino-4-methylcumary; MCL – 7-amino-4-methylquinolonyl; AA – 2-aminoacrydonyl; Ant – antronoyl-(O-aminobenzoyl); p-NA – p-nitroanilide; SBzl – thiobenzyl ester; Phe$_T$ – thiaphenylphenylalanine; F – fluorescent; R – detection of radioactivity; Ch – chromatographic; E – enzymatic; Sp – spectrophotometric; S – detection of SH-group
^{a} Based on intramolecular extinguishing of fluorescence in substrate.

4.1.3 Meaning and Value of Kinetic Constants

The experiments on steady-state kinetics of substrate hydrolysis and values of molar concentrations of the enzyme active sites provide kinetic constants k_{cat}, K'_m, and their ratio.

In the simple kinetic scheme (2) only the meaning of the k_{cat} and K'_m values are unequivocal – $k_{cat} = k_2$, i.e., k_{cat} is equal to the first-order rate constant for the chemical conversion of the enzyme-substrate complex to reaction products, and K'_m is equal to K_s – the dissociation constant for the enzyme-substrate complex, the latter being true only if $k_{-1} \gg k_2$.

Table 34. Values of the observed kinetic constants at various relationship for the rate constants of individual stages in kinetic scheme (5) [1715]

Case	K_P	k_3	k_{cat}	K'_m	k_{cat}/K'_m
1	$\ll 1$	$\ll k_{-2}$	$k_3 k_2/k_{-2}$	K_s	$k_3 k_2/K_s k_{-2}$
2	$\ll 1$	$\gg k_{-2}$	k_{-2}	K_s	k_{-2}/K_s
3	$\gg 1$	$\ll k_{-2}$	k_3	$K_s k_{-2}/k_2$	$k_3 k_2/K_s k_{-2}$
4	$\gg 1$	$\gg k_{-2}$	k_3	$K_s k_3/k_2$	k_2/K_s

The situation becomes more complicated if several enzyme-substrate complexes are formed on the pathway from a substrate to products. So for scheme (5) the values of kinetic parameters are

$$k_{cat} = \frac{k_3 K_P}{1 + K_P} = \frac{k_2 k_3}{k_2 + k_{-2} + k_3}, \tag{20}$$

$$K'_m = \frac{K_m}{1 + K_P} = \frac{(k_{-1} + k_3)(k_{-2} + k_3)}{k_1(k_2 + k_{-2} + k_3)}, \tag{21}$$

$$\frac{k_{cat}}{K'_m} = \frac{k_3 K_P}{K_m} = \frac{k_1 k_2 k_3}{(k_{-1} + k_3)(k_{-2} + k_3)}. \tag{22}$$

In general, no experimentally determined constant reflects either stability of the enzyme-substrate complexes or the rate of their conversion.

Depending on the ratio of the rate constants of individual steps the experimentally detected kinetic parameters can have the values given in Table 34 (if $k_{-1} \gg k_3$).

For scheme (7):

$$k_{cat} = \frac{k_2 k_3}{k_2 + k_3}, \quad K'_m = \frac{k_{-1} + k_2}{k_1} \cdot \frac{k_3}{k_2 + k_3}, \quad \text{and} \quad \frac{k_{cat}}{K'_m} = \frac{k_1 k_2}{k_{-1} + k_2}.$$

If the slowest step is the formation of the EA complex, i.e., $k_{-1} \gg k_2$ and $k_3 \gg k_2$, then $k_{cat} = k_2$ and $K'_m = k_{-1}/k_1 = K_s$; if the rate-determining stage is the EA breakdown, i.e., $k_{-1} \gg k_2$, but $k_3 \ll k_2$, then $k_{cat} = k_3$, and $K'_m = K_s k_3/k_2$. Independently of the rate-determining step $k_{cat}/K'_m = k_2/K_s$.

The above consideration shows that the experimentally observed constants can refer to different types of the enzyme-substrate interaction depending on the stage with the rate determining the overall process. According to Cleland [1716] the rate of the enzyme process is often limited either by the rate of isomerization of intermediates or by the rate of dissociation of products.

4.1.4 Nonproductive Binding

A substrate can bind an enzyme so that the formed complex is inactive, i.e., incapable of conversion to the reaction products. Such a complex is called nonproductive [1717, 1718], and its formation along with the productive complex results in changing the kinetic parameters of hydrolysis. In fact, for the kinetic scheme

$$E + S \xrightleftharpoons{K_s} ES \xrightarrow{k_2} E + P$$

$$\Big\Updownarrow K_s'$$ $$\tag{23}$$

$$ES'$$

the equation of the product formation rate is

$$v = \frac{k_2 [E]_0 [S]_0}{K_s + (1 + K_s/K_s')[S]_0} . \tag{24}$$

Hence the values of experimentally detected kinetic constants are

$$k_{cat} = \frac{k_2}{1 + K_s/K_s'} \quad \text{and} \quad K_m' = \frac{K_s}{1 + K_s/K_s'} . \tag{25, 26}$$

Thus both constants are lowered by the same value and the change increases with an increase of the ratio K_s/K_s'. Evidently, k_{cat}/K_m' is not altered at nonproductive binding, but remains equal to k_2/K_s. A symbate change of k_{cat} and K_m' at their constant ratio is characteristic of nonproductive binding in a series of similar substrates. Data on the kinetics of hydrolysis of esters and amides of α-chymotrypsin can illustrate this (Fig. 48). Unlike the amides some esters display nonproductive binding [1719].

In the kinetic scheme with several intermediate enzyme-substrate complexes [see Eq. (5), Sect. 4.1.1] another type of nonproductive binding can be

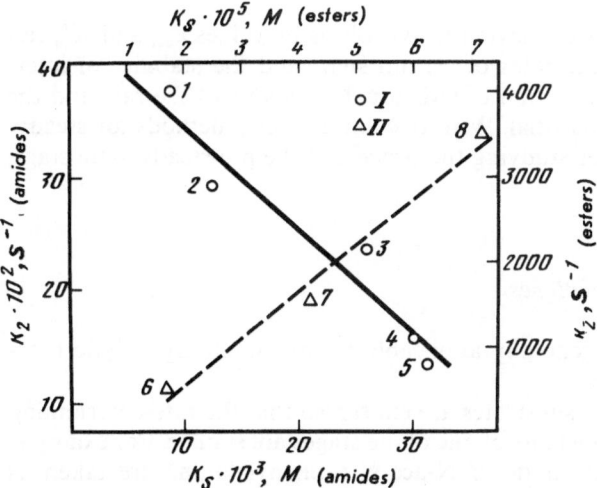

Fig. 48. Effect of k_2 on K_s in a series of amide (*I*) and ester (*II*) substrates of α-chymotrypsin [1719]; *1* 4-PyTyrNH$_2$; *2* 3-PyTyrNH$_2$; *3* ClCH$_2$COTyrNH$_2$; *4* CF$_3$COTyrNH$_2$; *5* AcTyrNH$_2$; *6* CinnTyrOEt; *7* AcLeuTyrOMe; BzTyrOMe (Py, pyridinoyl, Cinn, cinnamoyl)

realized, namely, isomerization of the first complex into the nonproductive one (ES'):

$$\text{E} + \text{S} \xrightleftharpoons{K_s} \text{ES} \xrightleftharpoons[k_{-2}]{k_2} \text{E}^*\text{S} \xrightarrow{k_3} \text{E} + \text{P}.$$

$$K_s' \Big\updownarrow$$

$$\text{ES}'$$

(27)

In this case the kinetic parameters are [1715]:

$$k_{\text{cat}} = \frac{k_3 K_P}{K_P + (1 + K_s')}, \quad \text{and} \quad K_m' = \frac{K_s}{K_P + (1 + K_s')},$$

where $K_P = \dfrac{k_2}{k_{-2} + k_3}$, and $K_s' = [\text{ES}']/[\text{ES}]$.

As in the first case, nonproductive binding does not influence the ratio of kinetic constants.

Finally, when $k_2 \ll k_{-1}$ for reversible reactions

$$\text{E} + \text{S} \xrightleftharpoons[k_{-1}]{k_1} \text{ES} \xrightleftharpoons[k_{-2}]{k_2} \text{E} + \text{P}$$

K_m value towards the substrate formation is equal to $K_{m,P} = k_{-1}/k_{-2}$, i.e., it is not related at all to the product binding (the so-called Van Slyke kinetic constant [1720]).

4.1.5 Detection of Individual Kinetic Constants

The above kinetic analysis of enzymatic hydrolysis by values k_{cat} and K_m' can only catch a glimpse of the rates of certain steps and the stability of intermediates. Therefore, the detection of individual constants of the rate and the equilibrium is especially essential. With this aim in view, methods for steady-state kinetics, as well as for studying the process at the pre-steady-state stage, are applied.

4.1.5.1 Steady-State Techniques

Here, the three methods generally applicable to investigate hydrolytic reactions are considered.

The first: two (or more) substrates are chosen so that the rate-determining steps differ, but the rate constants of one of the stages are similar. For example, if p-nitrophenyl ester and amide of N-acetyl-L-phenylalanine are taken as α-chymotrypsin substrates, the rate-limiting stage of the former includes the breakdown of intermediate acylenzyme, and the second its formation [1721]:

$$E + S \underset{K_s}{\rightleftharpoons} ES \xrightarrow{k_2} EA \ (+P_1) \xrightarrow{k_3} E + P_2 \ .$$

I. $AcPheOC_6H_4NO_2$, $\quad k_2 \gg k_3$;

II. $AcPheNH_2$, $\quad k_2 \ll k_3$.

(28)

Since the two substrates have the same acyl group and differ only in released product P_1, their k_3 values must be similar. To determine k_3 for substrate II, $k_{cat} = k_3$ for substrate I should be obtained. For all compounds derivatives of acetyl-L-phenylalanine k_2 can be found from the ratio

$$k_2 = \frac{k_3 k_{cat}}{k_3 - k_{cat}} \ .$$

Similarly, $K_s = K'_m(k_2 + k_3)/k_3$ can be calculated. This method is of restricted application because certain complications may arise due to nonproductive binding and, besides, at close values of k_3 and k_{cat} it entails a grave error.

The second method implies application of an effector that affects only one step in the process. For instance, there is a compound whose addition influences only the k_2 value [Eq. (28)]. In the presence of this compound the hydrolysis kinetics is described by equation:

$$v^* = \frac{\dfrac{k_2^* k_3}{k_2^* + k_3} [E]_0 [S]_0}{K_s \dfrac{k_3}{k_2^* + k_3} + [S]_0} \ .$$

(29)

Solving the two Eqs. (8) (if $k_2 \ll k_{-1}$) and Eq. (29), it is shown that in coordinates $1/v$ vs. $1/[S]_0$ two lines (corresponding to v and v^*) have intercepts $1/[S]_0 = -1/K_s$ and $1/v = 1/k_3[E]_0$ on coordinates (Fig. 49). Hence, if K_s and k_3 are known, k_2 can be readily found [1722–1724].

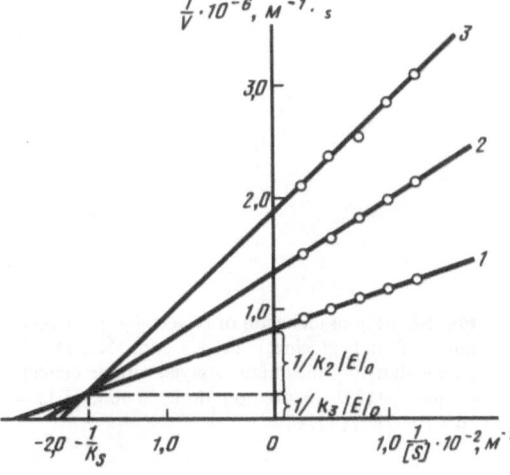

Fig. 49. Detection of individual kinetic constants by added nucleophile [1723]. Hydrolysis of ethyl ester of N-acetyl-L-phenyl-alanine (25 °C; 0.1 N KCl; $[E]_0 = 1.03 \times 10^{-8}$ M) without 1,4-butanediol (1) and in the presence of 0.11 M (2) and 0.22 M (3) 1,4-butanediol

Boric acid, copper ions, alcohols, etc. have been proposed as effectors of α-chymotrypsin [1722, 1723, 1725, 1726].

The third method (Boyer method [1727]) embraces the study of the rate of inclusion of isotope-labeled products into a substrate at the thermodynamic equilibrium of the system.

For scheme

$$
E + S \underset{k_{-1}}{\overset{k_1}{\rightleftarrows}} ES \underset{k_{-2}}{\overset{k_2}{\rightleftarrows}} EA \underset{k_{-3}}{\overset{k_3}{\rightleftarrows}} E + P_2 \atop + \atop P_1 \tag{30}
$$

the equilibrium can be reached by introduction of products P_1 and P_2 in the concentrations that satisfy the equation

$$K_{eq} = [P_1][P_2]/[S] .$$

If P_1 and P_2 contain a radioactive label, the latter is included into the substrate (S). Measurements of the rates of the label appearance in a substrate, depending on diverse equilibrium concentrations of P_1 and S or P_2 and S, yield the curves of the type shown in Fig. 50. The rates of inclusion of P_1 and P_2 (rates R' and R, respectively) are expressed by equations [1721, 1728]:

$$
R = \frac{[E]_0}{\left(\dfrac{1}{k_{-1}} + \dfrac{k_3 + k_{-2}[P_1]}{k_2 k_3}\right)\left(1 + \dfrac{K_1}{[S]} + \dfrac{K_2}{[P_1]}\right)}, \tag{31}
$$

$$
R' = \frac{[E]_0}{\left(\dfrac{1}{k_{-1}} + \dfrac{1}{k_2}\right)\left(1 + \dfrac{K_1}{[S]} + \dfrac{K_2}{[P_1]}\right)}, \tag{32}
$$

where $K_1 = K_s$ and $K_2 = k_2/k_{-2}$. At $[S] \gg K_1$ and $[P_1] \gg K_2$, and if

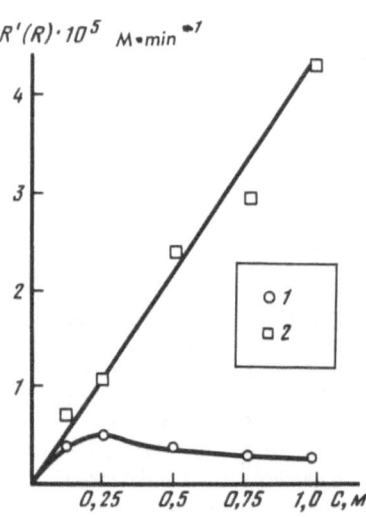

Fig. 50. Rate of inclusion of N-acetyl-*L*-phenylalanine (*1*) and glycineamide (*2*) into N-acetyl-*L*-phenylalanylglycineamide vs glycineamide concentration (pH 8.2; incubation time 30 min; $[E]_0 = 16 \times 10^{-4}$ M) [1729]

$k_{-1} \gg k_2$, $R' = k_2 [E]_0$. If k_2 is known, k_3 and K_s can be found, k_{-2} is calculated from equation:

$$\frac{R'}{R} = 1 + \frac{k_{-2}}{k_3} [P_1] . \tag{33}$$

Finally, k_{-3} can be expressed from the Haldane ratio (Sect. 4.2):

$$K_{eq} = \frac{k_2 k_3}{K_s k_{-2} k_{-3}} .$$

This method was utilized to calculate all individual constants of chymotrypsin hydrolysis of amide of N-acetylphenylalanylglycine [1729–1731], and also to investigate pepsin hydrolysis of some peptides [1732].

4.1.5.2 Methods of Pre-Steady-State Kinetics

These methods are described in numerous manuals [1733–1736], and here only one example is considered. Let us determine individual constants of chymotrypsin-catalyzed hydrolysis of p-nitrophenyl ester of acylamino acid. The reaction proceeds by scheme (7). The rate of the formation of nitrophenol at the pre-steady-state stage is first to be measured, for instance, by a stopped-flow apparatus. The reagents are mixed and a change of the optical density is recorded during several milliseconds. If these experiments are carried out at $[E]_0 > [S]_0$ at various initial enzyme concentrations, the dependence of the product concentration on the time can be obtained for different $[E]_0$. These values are related by expression

$$[P_1] = [S]_0 \left[1 - \exp\left(- \frac{k_2 [E]_0 t}{K_s + [E]_0} \right) \right] . \tag{34}$$

Thus, the effective rate constant of the process is

$$k' = \frac{k_2 [E]_0}{K_s + [E]_0} \quad \text{or} \quad \frac{1}{k'} = \frac{K_s}{k_2} \frac{1}{[E]_0} + \frac{1}{k_2} . \tag{35, 36}$$

Plotting $1/k'$ [calculated from Eq. (34)] on $1/[E]_0$, we obtain a straight line with slope K_s/k_2 and intercept $1/k_2$ on the ordinate axis.

Relaxation techniques (temperature and pH-jump, ultrasonic relaxation, etc.) are widely applied along with the stopped-flow method.

4.1.6 Deviations from the Michaelis-Menten Kinetics

Deviations from the Michaelis-Menten kinetics in reactions catalyzed by amide hydrolases usually appear in the following situations: (1) if a substrate is an effector (inhibitor or activator) as well; (2) if a cooperative binding of a substrate to enzyme subunits occurs; (3) if enzyme molecules interact with each other, thus changing the affinity to a substrate [1737].

Inhibition and activation by a substrate are described by the kinetic scheme

$$E + S \underset{K_s}{\rightleftharpoons} ES \xrightarrow{k_2} E + P$$

$$K_s' \Big\uparrow \tag{37}$$

$$ES_2 \xrightarrow{\beta k_2} ES + P$$

The corresponding expression for the rate of the product P formation [1659] is

$$v = \frac{\left(k_2 + \dfrac{\beta k_2 [S]_0}{K_s'}\right)[E]_0 [S]_0}{K_s + [S]_0 + \dfrac{[S]_0^2}{K_s'}}. \tag{38}$$

A typical plot of dependence $1/v$ vs $1/[S]_0$ for inhibition and activation by a substrate is presented in Fig. 51 [1738].

At high substrate concentrations (left branch) Eq. (38) can be transformed as follows (since $K_s \ll [S]_0$):

$$k' = \frac{v}{[E]_0} = k_2 \frac{1 + \dfrac{\beta [S]_0}{K_0'}}{1 + \dfrac{[S]_0}{K_s'}}, \tag{39}$$

or

$$\frac{1}{1 - k'/k_2} = \frac{1}{1 - \beta} + \frac{K_s'}{1 - \beta} \frac{1}{[S]_0}. \tag{40}$$

Dependence $1/(1 - k'/k_2)$ vs $1/[S]_0$ is a straight line, its slope and intercepts with axes give β and K_s'.

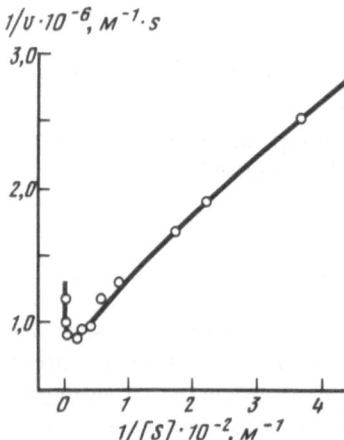

$1/v \cdot 10^{-6}, M^{-1} \cdot s$

$1/[S] \cdot 10^{-2}, M^{-1}$

Fig. 51. Activation and inhibition by substrate. The Lineweaver-Burk plot for hydrolysis of N-carbobenzoxyalanyl-phenylalanine catalyzed by carboxypeptidase A (25 °C; pH 7.5; $[E]_0 = 5.4 \times 10^{-9}$ M) [1738]

Though a few amide hydrolases show cooperativity at the substrate binding, still such types of enzymes exist and their analysis (the simplest case) is
presented here (see review [1737]). The cooperative binding implies the presence of several sites for the substrate binding in the enzyme molecule; binding
to one of these sites increases the affinity of the other to a substrate (positive
cooperativity), or decreases it (negative cooperativity). The enzymes with such
properties are called allosteric enzymes. The plot of the reaction rate vs the
substrate concentration is sigmoidal in case of positive cooperativity (Fig. 52a,
b). The kinetics of allosteric enzymes is analyzed by the Hill equation [1739]:

$$\lg \frac{v}{V_{max} - v} = n(\lg[S]_0 - \lg[S]_{0,5}) \, ,$$

where $[S_{0,5}]$ is the substrate concentration, when $v = V_{max}/2$; n is the Hill
coefficient, it being a measure of cooperativity. If $n = 1$, no cooperativity is
observed, if $n > 1$, the cooperativity is positive, if $n < 1$, it is negative.

The Hill equation is true in a comparatively narrow range of substrate
concentrations and coefficient n is defined within this region from the plot of
$\lg(V_{max} - v)$ vs $\lg[S]_0$ (Fig. 52c) as a slope in the point of 50% saturation (i.e.,
at $\lg(V_{max} - v) = 0$). Other methods to determine the Hill coefficient have been
proposed, when V_{max} is unknown [1740, 1741]. For a dimeric enzyme the
binding of the substrate by one subunit can destabilize the complex of the
other. Here the so-called half of the site reactivity appears [1742].

Deviations of the rate dependence on the substrate concentration from the
hyperbola can arise upon the enzyme oligomerization. This is accompanied
with a change in the activity due to masking some active sites. In the simplest
case, when linear oligomers are formed so that the active sites, except for the
one located at the end of the oligomeric chain, are sterically buried (Fig. 53),

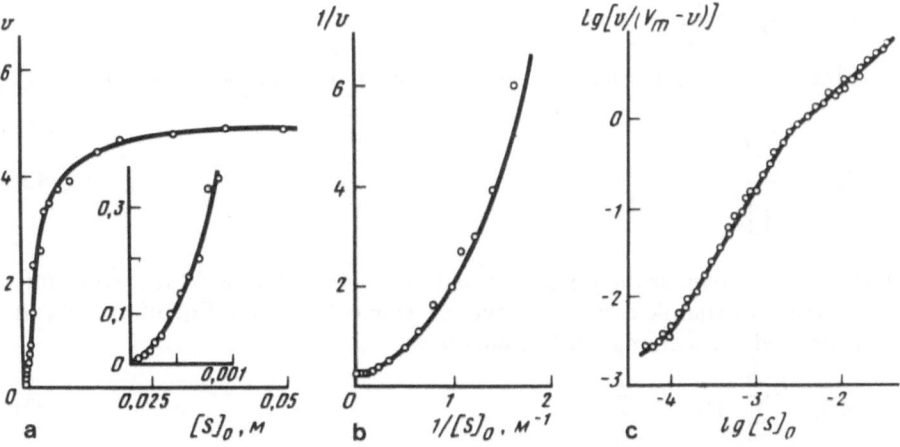

Fig. 52a–c. Kinetic plots for allosteric enzymes. Direct (**a**), double-reciprocal (**b**) and logarithmic
(**c**) scales [1737]

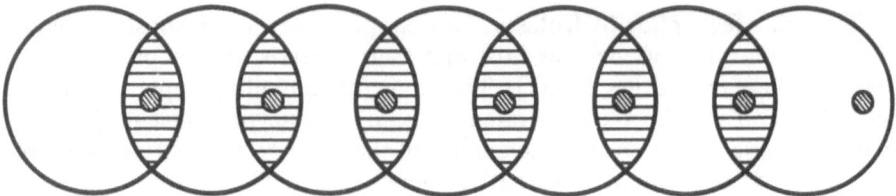

Fig. 53. Enzyme oligomerization with screening of active sites

the activity of the oligomer is related with the enzyme concentration as follows:

$$a = \frac{2a_1}{1 + \sqrt{1 + 4\bar{K}[E]_0}} \, ,$$

where a is the specific activity of an oligomer, a_1 is the specific activity of a monomer, $\bar{K}$ is the association constant for the monomer $\rightleftharpoons N$-mer association (if association constants are equal at each stage) [1743].

4.2 Relationship of Rate and Equilibrium

Like any catalyst, an enzyme does not alter the equilibrium position of the chemical reaction. However, for enzymatic reactions, one point should be taken into account. As a rule, enzymatic catalysis is a multistep process, therefore its equilibrium constant is a complex value including equilibrium constants of certain stages ("microscopic" constants, K_i):

$$E + S \underset{K_1}{\rightleftharpoons} ES \underset{K_2}{\rightleftharpoons} \cdots \underset{K_{i-1}}{\rightleftharpoons} EQ \underset{K_i}{\rightleftharpoons} E + P \, , \qquad (41)$$

$$K_{eq} = \frac{[P]}{[S]} = \prod_1^i K_i.$$

Since for each stage $K_i = k_{-i}/k_i$, expression (41) can be written as

$$K_{eq} = \frac{\prod_{-1}^{-i} k_{-i}}{\prod_1^i k_i} \, . \qquad (42)$$

This ratio is known as the Haldane equation [1744] relating the rate constants to the thermodynamic constant of the reaction equilibrium. Equation (42) can be expressed via a change in free energies

$$\Delta G_0 = \sum_{-1}^{-i} \Delta G^{\ddagger}_{-i} - \sum_1^i \Delta G_i^{\ddagger} \, , \qquad (43)$$

where ΔG_0 is a change of the standard free energy of the reaction, $\Delta G^{\ddagger}_{-i}$ are free energies of the activation of each stage of the reverse reaction, $\Delta G_i^{\pm}$ the same for the forward reaction.

The Haldane ratio shows that forward and reverse reactions proceed via the same transition states, i.e., their pathways are exactly opposite. This is a principle of detailed equilibrium or of microscopic reversibility. It is significant for understanding the reaction mechanisms.

4.3 Specificity

Enzymatic catalysis is highly sensitive to the substrate structure and to the conditions of catalytic reactions. Upon analysis of the substrate specificity of amide hydrolases one should distinguish a few levels.

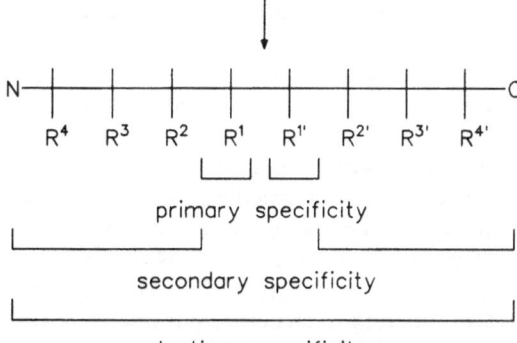

The substrate specificity of proteases (like other depolymerases) can be manifested at the level of the amino acid residue forming a cleavable bond (residues P_1 and P_1' in accordance with the accepted nomenclature [1745]) by its amino or carboxy group; this is the so-called *primary specificity*. The residues remote from the cleavable bond often influence much the ability of substrates to be hydrolyzed in the presence of proteases; this is *secondary specificity*.

In the case of proteins or large oligopeptides the spatial organization of a molecule may be important for positioning a cleavable bond. This is the so-called *tertiary specificity*. Sometimes the catalysis depends on the oligomeric structure of the hydrolyzed protein – *quaternary specificity*.

In accordance with the classification of proteases one can distinguish the levels of *position specificity*, i.e., the enzyme's ability to catalyze hydrolysis of only N- or C-terminal amino acid residues (exopeptidases) or the residues located inside the polypeptide chain (endopeptidases).

There is another level of the substrate specificity – *stereospecificity*, i.e., sensitivity of the amide hydrolases to configuration of asymmetric atoms in the substrate molecule.

To analyze quantitatively the enzyme substrate specificity one should try to choose the best criterion. Since the enzymatic reactions are multistep, a parameter that characterizes the definite stage or a sum of stages in the process is to be a measure of specificity. So for the three-step scheme

$$E + S \underset{k_{-1}}{\overset{k_1}{\rightleftharpoons}} ES \xrightarrow{k_2} EA \xrightarrow{k_3} E + P_2$$
$$\uparrow$$
$$P_1$$

the stages of complex formation ($K_s = k_1/k_{-1}$), enzyme acylation (k_2), or its deacylation (k_3) are considered as those for specificity characterization. Here certain difficulties arise, because the experimentally determined equilibrium or rate constants can be effective values including, for instance, the dissociation constant of a nonproductive complex. Therefore, in many cases the second order rate constant (k_{cat}/K_m) is preferable. It does not depend on the non-productive substrate binding, but can include some kinetic parameters, if the process runs through the formation of several intermediate complexes.

Thus analysis of the specificity requires information on the reaction rates at each stage.

Comparison of the specificity of different enzymes is even more sophisticated. The relative rate of the substrate hydrolysis catalyzed by two enzymes will depend on the choice of a substrate. For one substrate the rates can be equal, for the other, even if of similar structure, absolutely different.

It seems that the most relevant definition of specificity is *the ability of an enzyme to alter its catalytic activity upon changing of the substrate structure*. So enzyme specificity can be defined by a plot shown in Fig. 54, if we have any quantitative characteristics of the substrate structure (Z_s). Here a slope of a curve plotted for many substrates differing in structure parameter Z_s (or the total parameter of such a type) is a specificity measure. Certainly, in this case the choice of substrates is of much importance. If, for example, for analysis of the exopeptidase specificity a series of protected dipeptides is taken, exopeptidase shows no specificity to these substrates.

The utilization of this approach will be demonstrated below. Now let us consider the experimental results on the amide hydrolases specificity.

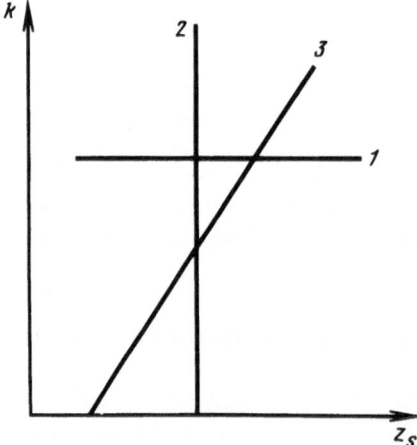

Fig. 54. Rate constant of enzymatic reaction (k) vs substrate structural parameter (Z_s); *1* no specificity; *2* absolute specificity; *3* in-between specificity

4.3.1 General Characteristics

There exists a wealth of information on the amide hydrolases specificity that can be hardly digested. Early reviews are in [178–180, 1746–1748]. Certain enzymes or enzyme groups have been also reviewed [181–183, 185–187, 1749–1752]. An insight into different levels of amide hydrolases specificity will be outlined later.

Already at early stages of the search for and isolation of amide hydrolase it could be assigned either to the group of proteases or of amidases. This determines a substrate type used for testing the enzyme activity during its isolation and purification. If protease is meant, a further characteristic concerns the position specificity, i.e., its assignment to a group of exo- or endopeptidases. No difficulties arise, in principle, since the activity of exoenzymes on the substrates of endopeptidases is generally several orders of magnitude lower than on those with unprotected N- or C-terminal amino acid residues. It is noteworthy that a series of proteases possessing both the exo-and endopeptidase activity have been isolated (see below).

Table 35. Cleavage of insulin B-chain by proteases:

Phe-Val-Asn-Gln-His-Leu-Cys-Gly-Ser-His-Leu-Val-Glu-Ala-Leu-Tyr-Leu-Val-Cys-Gly-Glu-
                      5   10   15   20
Arg-Gly-Phe-Phe-Tyr-Thr-Pro-Lys-Ala
       25   30

Enzyme	Major breaks between residues	Additional breaks between residues	References
Chymotrypsin A$_\alpha$ bovine	16–17; 25–26; 26–27	15–16	[287, 1753]
Chymotrypsin C bovine	4–5; 6–7; 11–12; 15–16; 17–18; 24–25; 25–26	16–17; 26–27	[287]
Trypsin bovine	22–23	29–30	[1754]
Elastase porcine	2–3; 6–7; 9–10; 14–15; 15–16; 18–19; 23–24; 26–27	–	[1755]
Protease I Achromobacter	29–30	–	[587]
Collagenase Entomophtora coronata	11–12; 15–16	–	[474]
Papain	3–4; 13–14; 17–18; 19–20; 25–26; 26–27	6–7; 7–8; 8–9; 14–15	[1756]
Cathepsin L	26–27	3–4; 13–14; 17–18	[1757]
Cathepsin B	5–6; 8–9; 13–14; 14–15; 20–21	1–2; 3–4; 15–16; 17–18	[1757]
Pepsin porcine	16–17; 24–25	10–11; 12–13; 14–15	[1758]
Cathepsin D bovine spleen	1–2; 11–12; 24–25	13–14; 14–15; 23–24; 25–26	[1759]
Cathepsin D erythrocyte	15–16; 24–25	–	[1760]

The determination of a type of amino acids forming a cleavable bond (primary specificity) is the next step in the qualitative characterization of the enzyme specificity. In the case of endopeptidases the B-chain of insulin is often used as a test object (Table 35). These investigations allow the assignment of the enzyme to one of the subgroups with the specificity to side chains of positively charged amino acid residues (of trypsin type), neutral amino acids with a bulky hydrophobic side chain (of chymotrypsin type) or with a small side chain (of elastase type) capable of hydrolyzing the amide bonds near the proline residue (prolidase), etc.

Kinetic studies of hydrolysis of many synthetic and natural substrates provide detailed information on their specificity.

4.3.1.1 Types of Cleavable Bonds

Amide hydrolases of practically all the types can catalyze hydrolysis not only of common peptide bonds, but of the bonds of other types – ester, thioester, anilide, etc. Hydrolysis of esters by various proteases and amide hydrolases has been studied in detail. The ratio between the esterase and amidase activities varies strongly for different enzymes (Table 36): for some serine proteases it reaches four orders of magnitude. The esterase activity of cysteine and aspartic proteases is only several times higher than the amidase activity, whereas the amidase activity of amino- and carboxypeptidases can exceed the esterase activity. Numerous amide hydrolases have no esterase activity at all (for example, [585, 1773]). The esterase and amidase activities can be caused by different bonding types and, probably, catalytic groups of the enzyme. This is true for serine carboxypeptidases (carboxypeptidases W and Y); both activity types of these enzymes are presumably governed by groups with various pK_a values [1774].

Chymotrypsin and trypsin cleave also lactone bonds in dioxomorpholines – derivatives of specific amino- and hydroxy acids [1775–1777]:

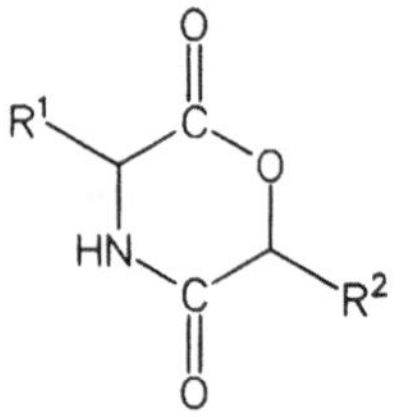

$$R^1 = C_6H_5CH_2;\ NH_2(CH_2)_4,\ etc\ .$$
$$R^2 = H;\ C_6H_5CH_2;\ Me_2CHCH_2\ .$$

The rate of hydrolysis of the compounds by these enzymes is almost the same as that for esters.

It is worthy of note that neither dioxopiperazines [1778], nor cyclopeptides composed of less than six amino acids [1779–1781] are cleaved by chymotrypsin or trypsin. Two more types of ester substrates of chymotrypsin have

Table 36. Comparison of esterase and amidase activities of amide hydrolases

Enzyme	Substrate	k_{cat}, s^{-1}	$K_m \cdot 10^3$ M	k_{cat}/K_m, M^{-1} s^{-1}	$\dfrac{k_{cat}/K_m \text{ (ester)}}{k_{cat}/K_m \text{ (amide)}}$	References
α-Chymotrypsin	AcPheOMe	57.5	1.5	0.38×10^5	3.6×10^4	[1761]
	AcPheNH$_2$	0.039	37.0	1.05		
Trypsin	BzLysOMe	12.9	0.028	4.6×10^5	0.61×10^4	[1762]
	BzLysNH$_2$	0.32	4.25	75.3		
Elastase	AcAla$_3$AlaOMe	120	0.067	17.9×10^5	617	[1763, 1764]
	AcAla$_2$Pro-AlaNH$_2$	6.1	2.1	2.9×10^3		
Subtilisin	AcAla$_3$OMe	2300	1.6	1.44×10^6	1.9×10^3	[1765]
	AcAla$_3$pNA	0.023	0.03	766		
Papain	BzArgOEt	16.1	13	1.24×10^3	4.8	[1766]
	BzArgpNA	0.74	2.86	258.7		
Carboxypeptidase Y	AcPheOEt	120	1.2	1×10^5	189	[250]
	AcPheNH$_2$	5.3	10	530		
Pepsin	ZHisPhe(NO$_2$)PlaOMe	0.13	0.4	325	2.4	[1767]
	ZHisPhe(NO$_2$)PheOMe	0.07	0.52	135		

Table 36. Continued

Enzyme	Substrate	k_{cat}, s^{-1}	$K_m \cdot 10^3$ M	k_{cat}/K_m, M^{-1} s^{-1}	$\dfrac{k_{cat}/K_m \text{ (ester)}}{k_{cat}/K_m \text{ (amide)}}$	References
Carboxypeptidase A	BzGlyGlyPlaOH BzGlyGlyPheOH	500 20	0.33 1.0	1.5×10^6 2×10^4	75	[1768]
Leucine aminopeptidase (lens bovine)	LeuOEt LeuNH$_2$	≈ 25 2×10^4	500 28	≈ 50 7×10^5	$\approx 7 \times 10^{-5}$	[954]
Dipeptidyl carboxypeptidase	BzPhePlaAla BzPhePheAla	18.3 90	0.065 0.065	2.8×10^5 13.8×10^5	0.2	[1769]
Amidase *Pseudomonas aeruginosa*	CH$_3$COOEt CH$_3$CONH$_2$	<1 162	660 0.83	<1.5 1.9×10^5	$<1 \times 10^{-5}$	[1770, 1771]
Penicillin amidase	C$_6$H$_5$CH$_2$COOEt C$_6$H$_5$CH$_2$CONHCH$_2$COOEt	170 47	0.045 0.08	3.7×10^6 0.6×10^6	6.3	[1772]

been described. These are α-substituted acylamino acids (I) [1782] and derivatives of acylcarbazates (II) [1783, 1784]:

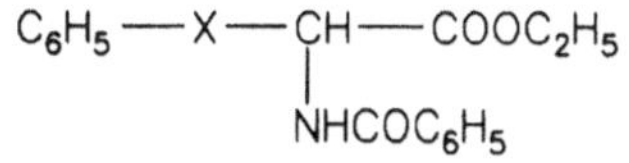

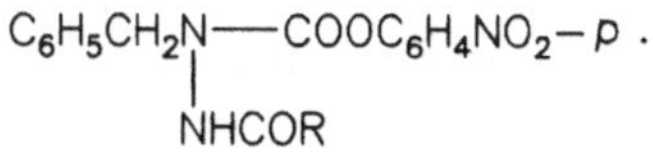

$$C_6H_5 - X - \underset{\underset{NHCOC_6H_5}{|}}{CH} - COOC_2H_5 \qquad C_6H_5CH_2\underset{\underset{NHCOR}{|}}{N} - COOC_6H_4NO_2 - p \; .$$

$$X = O, S, NH \qquad\qquad R = CH_3, \; CH(CH_3)NHCOCH_3 \; \text{or peptide}$$

$$I \qquad\qquad\qquad\qquad\qquad II$$

Carboxypeptidase A effectively cleaves carbamate and carbonate bonds in phenylalanine derivatives:

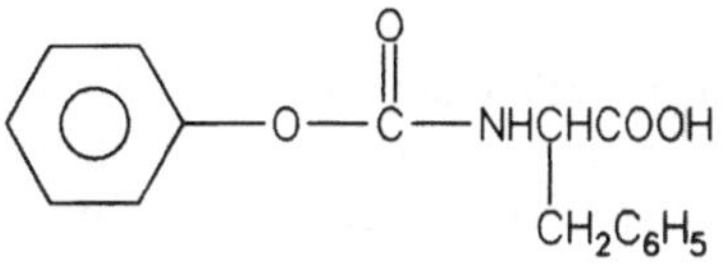

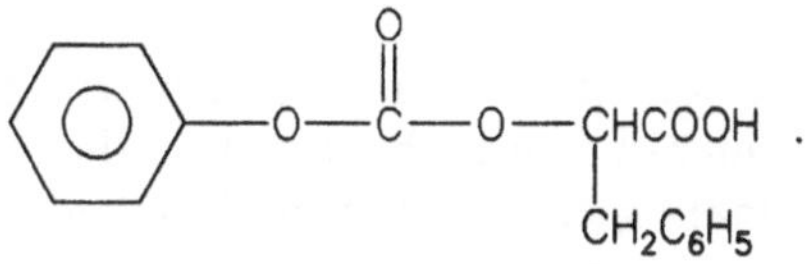

$$III \qquad\qquad\qquad\qquad IV$$

The ratio of the second order constants (IV):(III) = 82 [1785].

Many proteases cleave thioester (V) and thione bonds (VI) [1786–1788]:

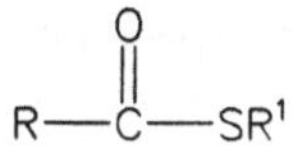

$$R - \overset{\overset{O}{\|}}{C} - SR^1 \qquad R - \overset{\overset{S}{\|}}{C} - OR^1$$

$$V \qquad\qquad VI$$

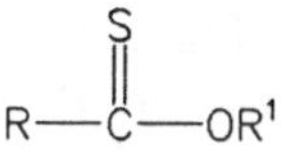

However, thione peptides are inhibitors of leucine aminopeptidase [1789].

p-Nitrophenylthioacetate is hydrolyzed by chymotrypsin with virtually the same velocity as p-nitrophenylacetate [1787]. Hydrolysis of ethyl esters is characterized by higher reactivity of thioesters as compared with oxygen esters [1790–1793].

Of other bonds cleaved by serine proteases one should mention hydrazides, hydroxamates [1794] and p-nitroanilides widely applied to test numerous proteases.

4.3.1.2 Position Specificity

The position specificity of amide hydrolases is clear from their classification number and needs no special consideration except for cases when the enzyme shows a dual specificity or when new data enforce reconsideration of the accepted classification. So it appeared that cathepsin B shows not only the endopeptidase activity, but functions also as dipeptidylcarboxypeptidase [763, 1795–1797]. Another cysteine protease – cathepsin H - functions predominantly as aminopeptidase, but it hydrolyzes p-nitroanilide of benzoylarginine [1798]. Another example is the action of angiotensin converting enzyme on protected oligopeptides. Hydrolysis of the releasing factor of a luteinizing hormone by this enzyme occurs as shown on the scheme [1799]:

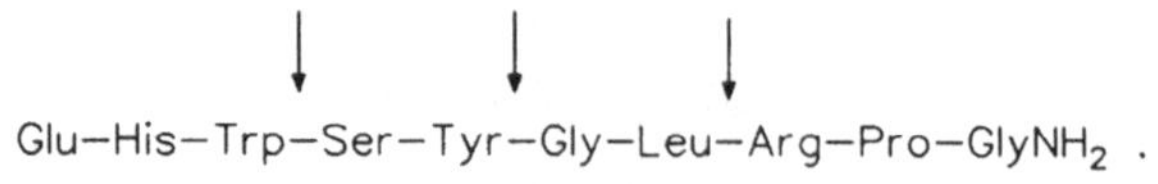

$$\downarrow \quad\quad \downarrow \quad\quad \downarrow$$

Glu—His—Trp—Ser—Tyr—Gly—Leu—Arg—Pro—GlyNH$_2$.

Ability of serine carboxypeptidases to manifest endopeptidase activity has been observed [1800]. Carboxypeptidase Y catalyzes also the hydrolysis of isopeptide bonds formed by the β-carboxy group of aspartic acid in some proteins [1801].

4.3.1.3 Primary Specificity

A variety of investigations deal with the influence of the residues located in the vicinity of a cleavable bond on the rate of the substrate hydrolysis. In fact, the primary specificity is established not only by a position of the certain residue in the polypeptide chain, but also by its contribution to the observed hydrolysis rate. Sometimes the latter plays a crucial role. So trypsin and similar enzymes cleave the bonds formed by the carboxylic group of basic amino acids (Lys, Arg). This factor is the most important for catalysis, whereas in other cases (for instance, for pepsin) the character of residues attached to a cleavable bond can be of less significance than that of remote residues. The primary specificity of papain is determined by the residue at position P_2 (see nomenclature in [1745]) of the polypeptide chain [1802]. To avoid terminological confusion, we consider here primary specificity as a specificity to residues at positions P_1 and P'_1 of the peptide.

In the structure of the amino acid residue three groups essential for specificity can be distinguished: side chain (R), substituents at nitrogen atom (X) and at carboxyl group (Y):

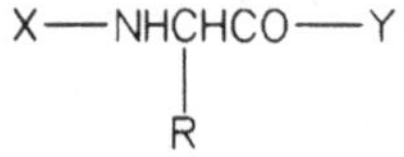

X—NHCHCO—Y
|
R

Structural peculiarities of group R depend on its properties: lipophilicity (hydrophobicity), volume and charge. Group X can be a hydrogen atom or a polypeptide chain, but group Y a polypeptide chain, alkoxy- or alkyl(aryl)amino group or hydroxyl. These properties are decisive for the specificity.

Quite a few proteases possess absolute specificity to a definite amino acid residue. These enzymes are, for instance, 5-hydroxyprolyl peptidase (3.4.19.3) specific to the N-terminal pyroglutamyl residue [704, 1803], some dipeptidases [1024, 1028, 1038], proline specific enzymes [222, 1030]. However, the latter hydrolyze also the bonds including residues of hydroxyproline, N-methyl-alanine, sarcosine, and alanine [1804, 1805].

The majority of proteases show the "group" primary specificity. There can be distinguished enzymes hydrolyzing the amide bonds formed by (a) basic residues of amino acids (Lys, Arg, His), (b) bulky hydrophobic residues (Trp, Tyr, Phe, Leu), (c) aliphatic nonpolar residues (Gly, Ala, Ser, Gln, Asn),

(d) residues containing a carboxylic group in a side chain (Asp, Glu), and
(e) residues with the tertiary nitrogen atom in the backbone (Pro, Hyp), etc.

Apparently, proteases entering the first group (trypsin group) are the most widespread. They include both amino- and carboxypeptidases [682, 1064, 1806], important digestive enzymes – as trypsin itself, and enteropeptidase [1807] enzymes of blood coagulation [313, 1808–1813], complement systems [588, 1814], and numerous though poorly studied enzymes of processing as well [1815, 1816]. The latter include enzymes specific to pairs of basic residues (cleavable bonds formed by such a pair or the bond between this pair and the following residue towards the C-end) [627, 632, 1817]. Trypsin predominantly splits bond Lys-Arg, as compared with Arg-Lys [1818].

Note that some enzymes specific to basic residues at position P_1 effectively cleave the bonds formed by other residues. This refers especially to cysteine proteases – papain [1745, 1802, 1819, 1820], calpains I and II [1821, 1822], etc.

The second group of enzymes embraces proteases with the chymotryptic specificity as well as enzymes with various position specificities – aminopeptidases [985, 176, 1823], carboxypeptidases [250, 252, 1824] and endopeptidases: chymotrypsins from a diversity of sources [1715, 1825–1827]; muscle proteases [1827–1829], digestive enzymes: pepsin [1749, 1750, 1830–1833], gastricsin [1832, 1833], microbial proteases, for example, entering the "pronase" complex [369, 1834], etc. As a rule, these enzymes possess wide specificity and hydrolyze the bonds formed not only by aromatic, but also by different aliphatic amino acid residues. This group involves also ATP-dependent proteases from *Escherichia coli* – proteases La and Ti as well as, presumably, some muscle high-molecular-weight proteases, which compose proteosomes [1835–1837]. Proteases of retroviruses cleave mostly bonds formed by a carboxyl group of phenylalanine and tyrosine; hydrolysis of the bonds containing a methionine residue at position P_1 is observed here [1838, 1839].

Elastases [1840–1845] and the enzymes related to subtilisins [1765, 1846, 1847] hydrolyzing the bonds formed by a carboxygroup of aliphatic residues with a small side chain are the best studied representatives of the third enzyme group. They are less specific and used for a maximum exhausted cleavage of proteins (protease K). This group embraces also collagenases of vertebrates. These enzymes are characterized by the glycine residue at position P_1 of the substrate, and residues Leu, Ile, Phe, Val [1848] at position P_1' [1848]. At the same time, collagenases of *Clostridium* preferably hydrolyze the bonds with residue Gly at the P_1'-position and a bulky hydrophobic residue at position P_1 [1849]. The glycine residue at position P_1 is typical of substrates of dipeptidylcarboxypeptidase (enkephalinase) [1850], and proteinase IV from *Papaya* [1851].

The group of enzymes hydrolyzing bonds formed by carboxyl-containing amino acid residues is comparatively small. It includes proteases from *Staphylococcus aureus* [1852], related to its protease from *Streptomyces thermovulgaris* [1853], aminopeptidase A [985] and other proteases. Interestingly, depending on a composition of the buffer solution, these enzymes can cleave bonds formed by Glu and Asp at P_1 (for example, in Tris-buffer) or only by Glu (in acetate buffer) [1853]. Collagenolytic proteases of invertebrates hydrolyze bond Glu at P_1.

Finally, the enzymes splitting bonds with a proline residue can be specific to the Pro residue at positions P_1 and P'_1 (prolyl- and proline peptidases).

The least expressed primary specificity is observed in the so-called signal peptidases. Among the cleavable bonds there are the bonds formed by the residues of serine, alanine, glycine, isoleucine, glutamine at the P_1-position and isoleucine, asparagine, valine, histidine, etc. at position P'_1. For these enzymes the secondary structure of the signal peptide is apparently more important [1854].

In a series of one-type enzymes the role of the amino acid residue at position P_1 of the substrate for displaying the protease specificity can be characterized by a ratio of the second-order rate constants, when one amino acid is replaced by the other. Table 37 lists data for serine proteases of the chymotrypsin type. The ratio at the substitution of Arg by Lys (in substrates TosGly-Pro-XpNa) is 460 for thrombin, 48 for the protein C of the blood coagulation system [1811], and 49 for factor Xa [1812].

For amidases a separation by the specificity levels has no sense. Modification of substrates of amidase from *Pseudomodas aeruginosa* [1771] shows that, for example, methylation of a nitrogen atom sharply increases k_{cat} and K_m. For acetamide $k_{cat} = 162\ \text{s}^{-1}$ and $K_m = 0.83\ \text{mM}$, and for N-methylacetamide $k_{cat} = 1080\ \text{s}^{-1}$, $K_m = 233\ \text{mM}$, which, apparently, depends on nonproductive binding. In the case of propylamide, methylation increases K_m (7.8 to 248 mM) and lowers k_{cat} (374 to 0.3 s^{-1}).

The specificity of β-lactamases strongly depends on the type of β-lactam antibiotics [1856, 1857]. So β-lactamase from *Escherichia coli* of the wild type hydrolyzes benzylpenicillin (VII), cephalotin (VIII) and cephaloxytin (IX) with the second-order rate constants differing over 10^7-fold.

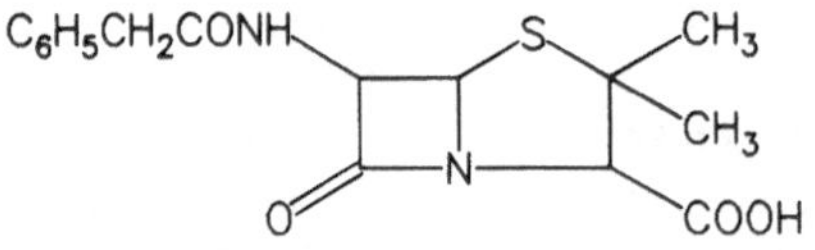

$k_{cat} = 2000\ \text{s}^{-1}$; $K_m = 0.02$ mM

VII

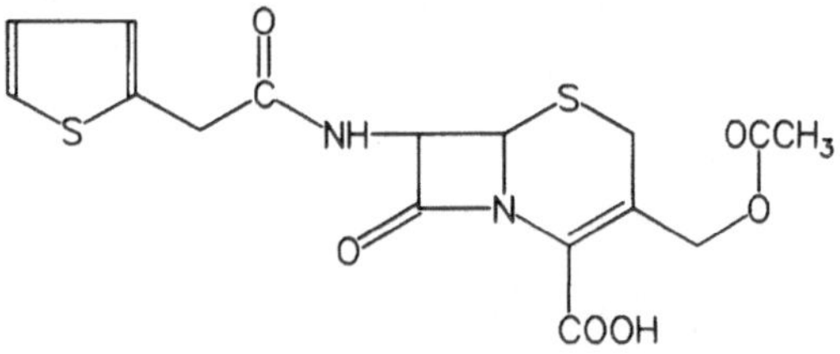

$k_{cat} = 120\ \text{s}^{-1}$; $K_m = 0.2$ mM

VIII

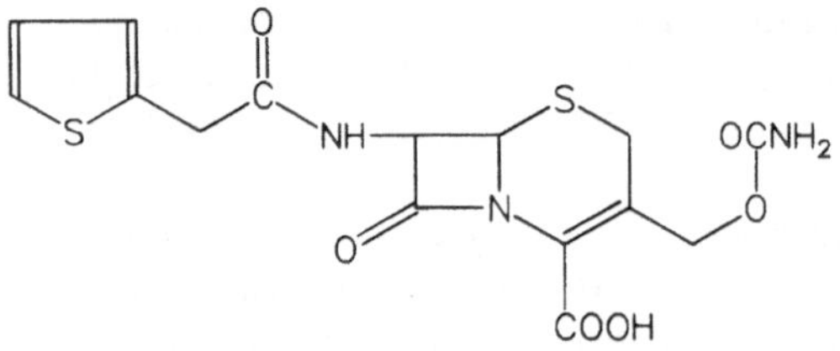

$k_{cat} = 0.004\ \text{s}^{-1}$; $K_m = 0.65$ mM

IX

Table 37. The ratio of second-order rate constants (k_{cat}/K_m) for a series of serine proteases, when replacing one amino acid (A) by the other (B) at position P_1. Substrates BocAlaAla-X-SBzl (according to [1827])

X		$(k_{cat}/K_m)_A/(k_{cat}/K_m)_B$					
A	B	Chymo-trypsin	Leukocyte elastase	Pancreatic elastase	Cathepsin G	Chimase I	Chymase II
Leu	Ala	1428	0.3	5.88	–	–	–
Leu	Met	2.56	2.59	1.35	1.1	0.26	0.26
Met	Ala	1057	1.17	2.32	–	–	–

It is senseless to distinguish some enzymes by the specificity levels because they are highly specific to only one or few substrates. So kidney renin hydrolyzes protein-angiotensinogen by splitting off terminal decapeptide Asp-Arg-Val-Tyr-Ile-His-Pro-Phe-His-Leu [923] and hydrolyzes a small number of synthetic analogs of the N-terminal sequence of angiotensinogen. The enzyme cleaves none of the other compounds investigated.

4.3.1.4 Secondary Specificity

For many proteolytic enzymes the structure of amino acid residues of the substrate remote from a cleavable bond is of much importance. Systematic quantitative studies of secondary specificity have been carried out for α-chymotrypsin [1825, 1858–1861], trypsin [1808, 1862–1864], thrombin, plasmin and other enzymes of the blood coagulation system [1808, 1862, 1865–1868], serine muscle proteases [1828], tonin [1869], kallikrein [1870], elastase [1843, 1871–1874], microbial serine proteases [1846, 1859, 1860, 1875–1877], papain [1819, 1820], pepsin and other aspartic proteases [1082, 1830, 1878–1887], carboxypeptidases [1768, 1806], thermolysin [1888, 1889], aminopeptidases [985, 1890] as well as some other enzymes [448, 1053, 1891–1902].

These investigations have revealed that representatives of distinct groups of proteolytic enzymes are differently "sensitive" to secondary interactions. Trypsin, papain and some other enzymes feel weakly a residue remote from a cleavable bond, whereas pepsin, elastase, etc. change the hydrolysis rate by 4–5 orders of magnitude (Table 38) [1906].

Secondary interactions can differently affect both kinetic parameters k_{cat} and K_m. For pepsin substrates there have been observed relatively small variations in K_m upon the transition from di- to tetra- and pentapeptides, and the k_{cat} values change more than 5 orders of magnitude (Table 39).

At the same time, a change in the length of peptides – substrates of chymotrypsin – entails a change of both k_{cat} and K_m by about 2 orders of magnitude (Table 40).

Table 38. Effect of secondary interactions on proteolysis

Enzyme	Substrate[a]		$\dfrac{k_{cat}/K_m \text{ (II)}}{k_{cat}/K_m \text{ (I)}}$	References
	I	II		
Thermolysin	ZGly-LeuAla	ZAlaGlyGly-LeuAla	0.09	[1888]
Carboxy-peptidase B	AcGly-Arg	Ac(Gly)$_4$Gly-Arg	1.09	[1806]
Trypsin	ZLys-OMe	Z(Ala)$_2$Lys-OMe	3	[1903]
Papain	ZValGlu-LeuGly	Z(Gly)$_2$ValGlu-LeuGly	4.7	[1819]
Postprolin endopeptidase	ZGlyPro-Leu	ZGlyPro-LeuGlyGly	4.8	[448]
Amino-peptidase A	Leu-NH$_2$	Leu-LeuLeu	16	[1890]
Thrombin	BocLys-OMe	Boc(Gly)$_2$Lys-OMe	32.5	[1865]
Chymotrypsin	AcTyr-GlyNH$_2$	Ac(Ala)$_2$Tyr-GlyNH$_2$	192	[1858]
Cathepsin D	ZPhe-PheOpp	Z(Ala)$_2$Phe-PheOpp	>720	[1880]
Pepsin	ZPhe-PheOpp	Z(Ala)$_2$Phe-PheOpp	2000	[1880]
Subtilisin BPN	AcAla-OMe	Ac(Ala)$_2$Ala$_3$-OMe	2275	[1904]
Elastase	AcAla-OMe	Ac(Ala)$_2$Ala-OMe	3875	[1905]

[a] Opp – see Table 39.

Not only the peptide length, but also the character of residues remote from a cleavable bond plays a crucial role in the enzymatic cleavage. A comparison of peptides 18, 24 and 25 (Table 39) shows that substitution of glycine by alanine at position P_2 increases k_{cat}/K_m more than by 1 order of magnitude. At this time, the same substitution at position P_3 enhances the constant only twice. In the case of chymotrypsin substrates the character of the cleavable residue of an amino acid dramatically alters substrate properties of dipeptides – introduction of alanine instead of glycine to P_1 increases k_{cat}/K_m 20-fold, while the substitution by valine virtually does not change the rate constant; addition of proline results in a loss of substrate properties (K_i 75 mM) [1826].

Secondary specificity does not obligatorily attenuates monotonously with an increase of distance from a cleavable bond. In serine proteases – trypsin and chymotrypsin – interactions P_3-S_3 and P_5-S_5 are essential, whereas inter-actions in "even" subsites slightly affect the hydrolysis rate [1908]. Elastase served as an example to show that P_3-S_3 interactions are manifested at enzyme acylation with a substrate [1844]. In the case of this enzyme the primary specificity depends on the substrate length.

Apparently, it is a common fact here that secondary specificity in this locus depends on the character of interaction in the locus or locuses (for example see Table 41).

A dependence of the hydrolysis rate on ionization of the group remote from a cleavable bond has also been observed. So a replacement of the C-terminal carboxylic group in bradykinin by the carboxamide group decreases the velocity of hydrolysis of the Phe5-Ser6 bond 2.5-fold, but it does not affect the cleavage of bond Pro7-Phe8 by endooligopeptidase B [1895]. The role of the

Table 39. Kinetic constants of pepsin-catalyzed hydrolysis of peptides [1878, 1907]

NN	Substrate[a]	k_{cat}, s^{-1}	K_m, mM	k_{cat}/K_m, M^{-1} s^{-1}
1	ZAla-Phe(NO$_2$)Apm	0.002	1.46	1.4
2	ZPhe(NO$_2$)-AlaApm	0.0068	1.30	5.3
3	ZPhe(NO$_2$)-ValApm	0.01	0.78	13
4	ZLeu-Phe(NO$_2$)Apm	0.011	0.73	14.5
5	ZPhe(NO$_2$)-PheArgOMe	0.034	0.92	37
6	ZPhe-Phe(NO$_2$)Apm	0.034	0.88	39
7	AcPhe-PheApm	0.102	2.37	43
8	ZPhe(NO$_2$)-PheApm	0.052	0.74	70
9	HGlyGlyPhe-PheApm	0.95	3.9	245
10	ZGlyProPhe-PheOpp	0.056	0.14	400
11	ZAsnPhe(NO$_2$)-PheApm	0.45	0.84	540
12	ZValPhe(NO$_2$)-AlaApm	0.55	0.96	580
13	ZValPhe(NO$_2$)-PheApm	0.31	0.17	1850
14	AcPhe(NO$_2$)-PheAlaAlaOMe	5.2	0.85	6100
15	ZLeuValPhe(NO$_2$)-AlaApm	4.3	0.59	7300
16	ZAlaAlaPhe(NO$_2$)-PheApm	43.8	1.51	29 000
17	ZGlyHisPhe-PheOpp	15.8	0.44	35 900
18	ZGlyGlyPhe-PheOpp	71.8	0.42	1.8×10^5
19	ZGlyIlePhe-PheOpp	12.6	0.07	1.8×10^5
20	ZPheGlyPhe-PheOpp	24.6	0.11	2.25×10^5
21	ZGlyLeuPhe-PheOpp	134	0.03	4.46×10^5
22	ZAlaGlyPhe-PheOpp	145	0.25	5.8×10^5
23	ZPheGlyGlyPhe-PheOpp	127	0.13	9.76×10^5
24	ZGlyAlaPhe-PheOpp	409	0.11	3.7×10^6
25	ZAlaAlaPhe-PheOpp	282	0.04	7.05×10^6

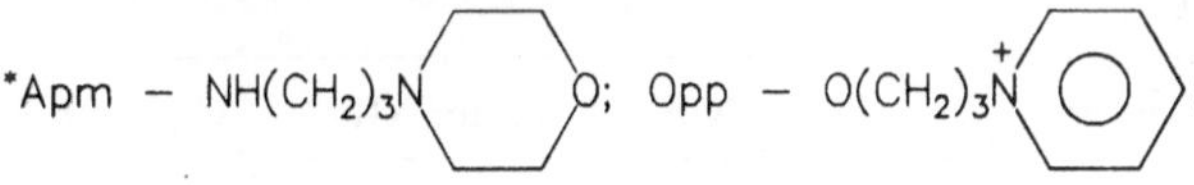

[a] Apm − NH(CH$_2$)$_3$N⬡O; Opp − O(CH$_2$)$_3$N$^+$⬡ ; Hyphen shows place of cleavage

substrate conformational state in its sensitivity to hydrolysis by dipeptidyl-carboxypeptidase has been well illustrated in [1901]. A replacement of the glycine residue at position P$'_1$ of compound X by alanine makes the substrate

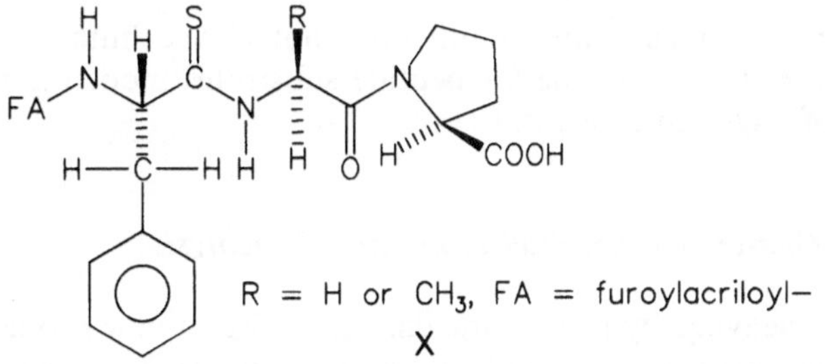

R = H or CH$_3$, FA = furoylacriloyl−

X

insensitive to hydrolysis. NMR analysis of these compounds and the enzyme model has prompted the idea that the methyl group of the alanine residue in a productive conformation of the Michaelis complex runs into a sulfur atom.

Finally, secondary specificity can dramatically change depending on the organism of a species. So human renin hydrolyzes effectively both human and

Table 40. Kinetic parameters of α-chymotryptic hydrolysis of peptides [1825, 1841, 1858]

Substrate[a]	k_{cat}, s^{-1}	K_m, mM	$k_{cat}/K_m \cdot 10^{-3}$, M^{-1} s^{-1}
AcPhe-NH$_2$	0.22	21	0.01
AcPhe-GlyNH$_2$	0.14	14.6	0.0096
AcPhe-AlaNH$_2$	2.8	15.0	0.186
AcProPhe-NH$_2$	0.7	24	0.029
AcAlaProPhe-NH$_2$	2.3	2.2	1.04
AcProAlaPhe-NH$_2$	0.44	18	0.024
AcAlaPhe-GlyNH$_2$	0.355	23.2	0.015
AcProPhe-GlyNH$_2$	0.705	14.9	0.051
ZGlyProPhe-NH$_2$	0.096	2.4	0.04
ZGlyProPhe-GlyNH$_2$	0.08	2.0	0.04
ZGlyProPhe-LeuNH$_2$	0.225	1.5	0.15
ZGlyProPhe-LeuAla	0.35	0.5	0.7
ZGlyProPhe-GlyGly	0.26	1.3	0.2
ZGlyProPhe-GlyLeu	0.68	0.4	1.7
AcProAlaProPhe-NH$_2$	2.8	3.4	0.82
AcProAlaProPhe-AlaNH$_2$	18.7	1.6	11.7
AcProAlaProPhe-GlyNH$_2$	11.0	4.6	2.39
AcProAlaProPhe-AlaAlaNH$_2$	36.6	0.83	44
AcProAlaProPhe-AlaAlaAlaNH$_2$	37	0.82	45
AcProAlaProPhe-PheNH$_2$	8.0	0.21	38.1

[a] Hyphen shows place of cleavage.

Table 41. Ratio k_{cat}/K_m in substrates of type Z-X-A-ArgpNa[a]

A	X$_1$	X$_2$	$(k_{cat}/K_m)_{X_1} : (k_{cat}/K_m)_{X_2}$				
			Thrombin	Factor Xa	Factor XIa	Factor XIIa	C-protein
Gly	Phe	Gly	1.87	1.89	2.7	0.26	3.44
Phe	Phe	Gly	9.2	0.14	0.5	0.82	0.46

[a] According to [1868].

canine angiotensinogen, whereas canine renin does not cleave human angiotensinogen [1800, 1886]. This is true for peptide substrates modeling the N-terminal sequence of angiotensinogens.

4.3.1.5   Tertiary and Quaternary Specificity. Limited Proteolysis

Analysis of the protein cleavage by proteolytic enzymes raises a new problem concerning the spatial structure of substrates (see numerous reviews [1909–1913]). As a rule, native enzymes are more stable to proteolysis than denatured proteins (for instance [1909], Chap. 5). As early as 1952 Linderström-Lung has developed a theory of proteolysis of native proteins proceeding through a stage of their denaturation [1914]. Protease itself facilitates denaturation, i.e., it serves as denaturase. The process runs by "successive" ($k_d < k_h$) or "zipper" ($k_d > k_h$) mechanisms depending on the ratio of the rates of the peptide bond demasking (k_d) and its hydrolysis (k_h). The mechanism type can depend on the nature of a substrate, protease and the reaction conditions, particularly the concentration of hydrolysis products [1915]. These ideas are in line with many experimental results that underlie the methods for detection of protein unfolding rates. It appeared [1916] that the rates of unfolding of lysozyme and RNAase, measured by the rate of proteolysis with different enzymes, coincide with each other and with the denaturation rate independently measured. A higher sensitivity of apoenzymes to proteolysis, as compared to that of holoenzymes [1917, 1918], confirms these concepts.

Meanwhile, more and more information on proteolysis of native proteins has been accumulated. These are the cases of the so-called *limited proteolysis* [1909, 1919] or *processing*: well-known processes of activation of zymogens (see Chap. 5), formation of biologically active peptide hormones, processes of protein transport from cells, etc.

The limited proteolysis sites are determined, first, by the primary and secondary specificity of the hydrolyzed enzyme. So, for example, enteropeptidase with trypsin-like specificity cleaves the bond formed by a lysine carboxylic group in the sequence AspAspAspAspLys of trypsinogens [1920, 1921]. Processing of precursors of peptide hormones usually proceeds by splitting the bond between two residues of basic amino acids or the bond formed by one of the two residues with another amino acid [1815, 1922]. Often the proline residue, which evidently yields a special conformation of a cleavable region, is located near the basic residue [1816, 1923]. The local conformation of a cleavable region is the second essential factor of limited proteolysis. Usually, the bonds in movable loops of the protein molecules or the regions linking the dense subglobules – domains – are involved in cleavage [1917]. For example, pepsin is cleaved by external proteases and autocatalytically within the region of the chain linking the two domains, the 146-membered C-terminal domain being cleaved off [1924]. Processing sites are often located in the so-called "omega loops" – regions of 6- to 16-fold residues – where the distance between N- and C-ends of the loop does not exceed 10 Å [1913].

Analysis of many structures of peptide hormone precursors has revealed that the processing site is located in the region where the β-turn formation is highly probable, and near highly ordered secondary structures [1925]. The secondary structure here is important for the cleavage not only of proteins, but also of short peptides [1926]. Finally, there are two other factors essential for proteolysis of native proteins – accessibility of a cleavable bond and mobility of a cleavable site [1927–1930]. Both the factors can be modeled from X-ray data or experimentally from the values of the temperature B-factor in crystal [1928], a proper coincidence of the proteolysis sites predicted has been experimentally observed. Undoubtedly, an effective cleavage of the protein in the desired site is possible only if all the above conditions are realized.

Limited proteolysis has been studied quantitatively on the example of cleavage of κ_0-casein and other caseins by chymosin and related proteases [1931–1937]. As shown, the cleavage rate of bond Phe-Met in peptide

$$\downarrow$$

His—Pro—His—Pro—His—Leu—Ser—Phe—Met—Ala—Ile—Pro—Pro—Lys—Lys
(98)   (100)   (105)   (110)   (112)

which is fragment 98–112 of κ-casein, exceeds that of the parent protein ($k_{cat} = 68.7$ s^{-1}; $K_m = 0.049$ mM [1933]). The lengthening of peptide 104–111 by residue Leu that results in peptide 103–111 is decisive, since it increases k_{cat}/K_m 330-fold. Sequence 98–102 decreases K_m by over 1 order of magnitude [1931]. These data indicate a predominant role of the secondary, rather than the tertiary specificity in catalysis (see Chap. 6).

The kinetics of hydrolysis of fibrinogen and peptides modeling the proteolysis sites of this protein by thrombin has been studied in detail [1866, 1867, 1938]. At first, thrombin splits off fibrinopeptide A from the α-chain, des-A-fibrinogen is polymerized, then fibrinopeptide B is cleaved off [1938]. The k_{cat}/K_m value for splitting fibrinopeptide A reaches 11.6×10^7 M^{-1}s^{-1}. The rates of limited proteolysis of some proteins are presented in Table 42.

A diversity of examples of limited proteolysis by different enzymes [1909] plays, evidently, a regulatory role.

Let us briefly consider the processing of signal peptides and viral polyproteins. The signal sequence of prokaryotic and eukaryotic proteins is arranged so that there is a cluster of positively charged residues at the N-terminus, the central hydrophobic core (about 12 residues) and the polar C-terminal fragment characterized by the ordered secondary structure. The sequence of three residues A-X-B, where A = Ala, Leu, Val, Ile, and B = Ala, Gly, Ser [1854, 1943], precedes the site of cleavage by the signal protease. The signal protease itself (from *Escherichia coli* [1944]) contains an internal signal peptide (residues 70–76) essential for binding to the membrane.

In the course of maturation many RNA-containing viruses form a polyprotein subjected to processing under the action of the viral protease entering its sequence [654, 1945]. "Cutting" of viral proteins starts with autoprocessing and proceeds strictly by the sites separating the proteins with different functions. These sites can be formed by pairs of amino acid residues, for instance, Gln-Gly, Tyr-Gly and Asn-Ser in picorinaviruses, Phe-Pro, Tyr-Pro and others in retroviruses, etc. In this case the specificity of cleavage remains unclear. Moreover, in retroviruses aspartic protease entering the polyprotein functions as a dimer. The problem of autoactivation of this enzyme remains to be solved.

Intracellular extralyzosomal proteolysis involves the ATP-ubiquitin-dependent proteolysis system [1910, 1912]. The system includes a small protein ubiquitin (9000 MW) and several enzymes-catalyzing attachment of ubiquitin via its C-terminal glycine to the ε-amino group of a cleavable protein, cleavage of the protein in the formed conjugate, and regeneration of ubiquitin. The system hydrolyzes predominantly abnormal cell proteins [1946, 1947]. Recognition of abnormal proteins is proposed to be accomplished through the state of their N-terminal amino group. Acetylation of this group prevents the protein hydrolysis by the ubiquitin-dependent system [1948]. In addition, the

Table 42. Kinetic parameters of limited proteolysis of some proteins

Enzyme	Substrate	pH	k_{cat}, s^{-1}	$K_m \cdot 10^{-3}$, M	k_{cat}/K_m, M^{-1} s^{-1}	References
Chymotrypsin	Pancreatic trypsin inhibitor	7.5	1.4×10^{-8}	2×10^{-6}	7	[1939]
Trypsin bovine	The same	7.5	8.7×10^{-10}	8.7×10^{-11}	1×10^4	[1939]
The same	Soybean trypsin inhibitor	8	2.5×10^{-6}	1.25×10^{-9}	2×10^6	[1939]
"	Trypsinogen bovine	8.1	2.5×10^{-3}	0.4	6.25	[1940]
"	" porcine	8.1	5.5×10^{-3}	0.48	11.4	[1940]
"	Chymotrypsinogen A	–	0.18	1.1	163	[1940]
"	" B	–	1.6	0.57	2.8×10^3	[1940]
Trypsin 1 *Dermasteria imbricata*	Pancreatic trypsin inhibitor	8.2	0.0016	1.6×10^{-7}	1×10^6	[1939]
The same	Soybean trypsin inhibitor	8	0.12	3×10^{-4}	4×10^5	[1939]
Kallikrein human tissue	High-molecular kininogen	"	0.43	0.005	8.6×10^4	[1941]
The same	Low-molecular "		1.83	0.0125	14.6×10^4	[1941]
Enteropeptidase, bovine	Trypsinogen	5.6	1.48	7×10^{-3}	2×10^5	[1920]
" porcine	"	5.6	4.8	7×10^{-2}	6.8×10^4	[1942]
Chymosin	κ-Casein	6.6	68.9	0.05	1.4×10^6	[1934]
"	β-Casein	6.6	0.56	0.007	0.8×10^5	[1934]
Renin	Angiotensinogen	≈6	50	1×10^{-3}	5×10^7	[923]

ATP-dependent, though ubiquitin-independent, system of proteolysis involving high-molecular-weight proteases (proteosomes) presumably functions in the cell [1949].

Many examples illustrate the protein quaternary structure's influence on the resistance to proteolysis. So yeast hexokinase is stable to proteolysis with yeast proteases, but at dissociation of the dimer the formed monomers easily split off the fragment significant for dimerization [1950]. The same is observed for phosphorylase A under the action of trypsin and glucose, when the tetrameric protein dissociates to dimers [1951]. Another example of the quaternary specificity in proteases is the action of trypsin on tetramer $(\alpha_2\beta_2)$ of tryptophan synthase [1952]. It appeared that β-subunits are not subjected to splitting in the tetramer, but if the enzyme dissociates, they are cleaved by trypsin. α-Subunits are step-by-step cleaved by the enzyme so that they lose their activity. Within the intact enzyme only a first hydrolysis step is realized without a loss of the tryptophan synthase activity. Thus, the quaternary structure of the protein-substrate can to a large extent change the pathway of hydrolytic cleavage due to masking of certain proteolysis sites or due to conformational changes in these positions.

4.3.2 Quantitative Dependences of Substrate Structures and Their Reactivity

The aforementioned regularities are qualitative and make no prediction on the substrate reactivity in reactions of the enzymatic hydrolysis. Quantitative analysis of specificity will be considered below.

4.3.2.1 Statistical Analysis of Specificity

One of the methods for a quantitative assay of specificity of proteolytic enzymes is statistical analysis of the cleavage frequency of various peptide bonds in peptides and proteins [1953–1956]. Original data can be found in the literature on structural analysis of protein sequences (for example, see [1197]).

The method embraces the calculation of the distribution frequency of each amino acid in every position of a cleavable sequence and analysis of reliability of the deviations from the random distribution. The example of the calculation of the specificity of porcine pepsin [1955] relative to residue Ile at position P_1 illustrates this.

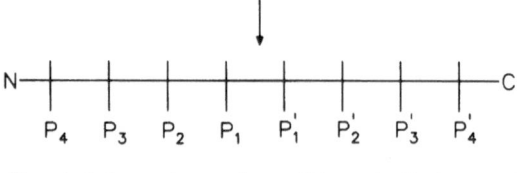

The total number of peptides studied   500
The total number of Ile residues   372
The number of Ile residues at position P_1 3

The distribution frequency for Ile at position P_1: $A_{1,\text{Ile}} = (3/372) \times 100 = 0.81$. The mean distribution frequency for Ile at position P_1: $A_1 = \sum_1^{20} A_{1,j}/20 = 4.99$, where $A_{1,j}$ is the distribution frequency of each of 20 amino acids at position P_1. Deviation from the random distribution: $K_{1,\text{Ile}} = 0.81/4.99 = 0.16$. The boundary of the confidence interval $\Delta\chi_1 = \varepsilon\sigma$, where $\varepsilon = 2$ (for $\alpha = 0.95$) and $\sigma \approx S_n$ is the mean square deviation $K_{i,j}$ from the unity; $\Delta\chi_1 = -0.84$. The specificity index $S_{1,\text{Ile}} = -(1 - \Delta\chi_1)/K_{1,\text{Ile}} = -5.25$.

The calculations yield "specificity indices" ($S_{i,j}$), which characterize preferableness ($S_{i,j} \geqslant +1$), undesirability ($S_{i,j} \leqslant -1$), and indifference ($+1 > S_{i,j} > -1$) of the enzyme to a certain amino acid in the given position of the cleavable sequence.

For pepsin substrates the specificity indices for P_1-position of the most and the least preferential residues are Leu $+2.46$, Phe $+2.28$, Trp $+1.67$, Glu $+1.49$, Ile -5.25, Arg -6.23.

The analysis of specificity indices provoke interesting conclusions. First, at position P_1' the specificity indices correlate with hydrophobicity of side chains (see Sect. 1.4), whereas at position P_1 and others such correlation is lacking. Secondly, at position P_1 isoleucine and positively charged amino acid residues are undesired, whereas Pro is absolutely undesired (the specificity index for this residue is $-\infty$, i.e., it does not occur in any investigated peptide). Of much interest are mean absolute values of the specificity indices for each position. This value characterizes the maximum positive and negative deviations from the random distribution of amino acid residues at the certain position, i.e., "rigidity" of requirements of the enzyme for the structure of the side chain. The value is maximum for position P_1, considerably deviates from unity at positions P_3, P_2 and P_1', P_2', and is close to unity for the other positions (Fig. 55). So pepsin is specific to the structure of the residue forming a cleavable bond by its carboxyl group; the total length of the substrate binding site of the pepsin active site is 6–7 amino acid residues. An analogous situation is observed for chymosin [1932].

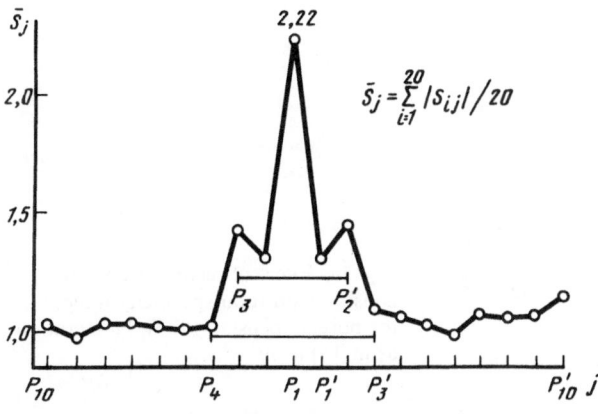

Fig. 55. Effect of mean specificity index (S_j) of pepsin on position of amino acid residues relative to a cleavable bond [1955]

The specificity indices calculated in such a way probably reflect a contribution of each amino acid residue to the total velocity of the certain bond cleavage by pepsin. This is confirmed by a good correlation between the total specificity indices for synthetic peptides and rate constants (k_{cat} and k_{cat}/K'_m) of their hydrolysis catalyzed by this enzyme (Fig. 56) [1907].

The linear dependence of kinetic constants on overall specificity indices is described by formulae

$$\ln k_{cat} = 13.75 \pm 0.67 \ln\left(\sum S_{ij} + \sum \bar{S}_j\right) - 31.34 \pm 1.75 \,, \tag{44}$$

$$\ln \frac{k_{cat}}{K'_M} = 16.13 \pm 1.08 \ln\left(\sum S_{ij} + \sum \bar{S}_j\right) - 36.77 \pm 2.82 \,. \tag{45}$$

No correlation between the total specificity indices and K'_m is observed.

Correlations in Fig. 55 illustrate the dependence of the substrate structure on the hydrolysis rate presented generally in Fig. 54. Evidently, the slope of plot $\ln k_{cat}/K'_m$ vs. $\ln\left(\sum S_{ij} + \sum S_j\right)$ characterizes the pepsin specificity.

Another approach to the statistical analysis of the pepsin specificity [1956] is based on the calculation of a probability cleavage of the given bond, expressed as a ratio of the number of cleavable bonds of the certain type to the

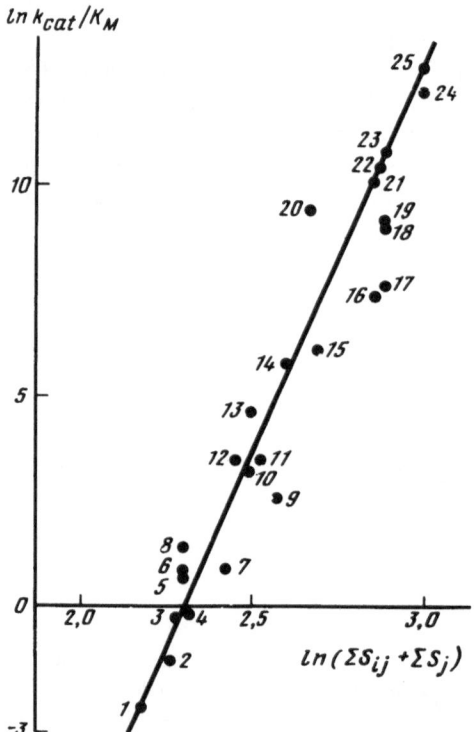

Fig. 56. Effect of second-order rate constants on total specificity indices for porcine pepsin [1907]. Numeration as in Table 39

total number of such bonds. For instance, in 177 peptides and proteins studied (containing in total 6910 peptide bonds) the alanine residue is met 502 times and in 82 cases this residue forms a cleavable bond by its carboxyl group. Probability of cleavage is $82/502 = 0.16$.

Though statistical specificity indices have no certain physical meaning and are simply empiric parameters, correlations of types (44) and (45) are of practical value. They afford a reliable calculation of constants of enzymatic hydrolysis of any amino acid sequences a priori. From the specificity indices there has been proposed the pepsin substrate Pro-Thr-Glu-Phe-Phe(NO_2)-Arg-Leu – one of the most rapidly hydrolyzed by this enzyme ($k_{cat}/K_m \approx 5.2 \times 10^6 \, M^{-1} s^{-1}$) [1883].

Statistical methods do not always provide good predictions of hydrolysis rates. Reasons for it are: first, the original data on the cleavage of peptides and proteins extracted from structural studies are not quantitative and the observed results can strongly depend on experimental conditions. Secondly, if the primary specificity (for instance, at residue P_1) mainly contributes to the hydrolysis rate, variations of remote amino acid residues weakly manifest themselves in experiments on the structure determination, but can markedly affect the initial rates obtained in kinetic experiments. Finally, when the substrate structure influences both the k_{cat} and K'_m, the original data can reflect changes of either the maximum velocity or ratio k_{cat}/K'_m depending on the $[S]_0/K'_m$ ratio. Apparently, these factors are responsible for a poor correlation of total specificity indices and kinetic constants of hydrolysis of synthetic peptides (see Table 40) in case of peptide substrates of α-chymotrypsin.

Another procedure for statistical analysis of specificity is based on application of the Free and Wilson method [1957]. Any variable factor connected with the structure can be represented as a sum of additive contributions of certain structural elements

$$k_n = \sum_i a_{i,n} + \mu \tag{46}$$

if

$$\sum_n a_{i,n} = 0 , \tag{47}$$

where n is the substrate molecule in the given series of substrates, i is the position of the residue in a sequence (P_1, P_2, etc), a is the contribution of a residue to measured parameter k, μ is the total contribution for the whole series of substrates. If there is a series of substrates chosen properly, one can obtain the values of contributions of each amino acid in every position to the measured kinetic constant by solving the system of equations (46) and (47). In principle, this contribution does not depend on a substrate accommodating the given residue, but only on the type of amino acid and its position in the sequence. Thus, the values of kinetic constants can be predicted for each peptide from the calculated contributions. This method has been utilized to analyze the specificity of subtilisin, trypsin, thrombin, and some other proteases [1958–1961].

4.3.2.2 Linear Free-Energy Type Relationship

The above correlations have a serious disadvantage. They do not operate with any parameters characterizing substrate physicochemical properties, though the latter govern the substrate's ability for interaction with the enzyme.

Correlations of the linear free-energy type for enzymatic hydrolysis have been obtained for a definite number of substrates of some proteases. Esters of acylamino acids, which are the substrates of α-chymotrypsin, trypsin and some other enzymes, have been studied in detail [1962–1964].

Influence of the Side Chain Structure. Table 43 shows that the second-order rate constant (k_2/K_s) for chymotrypsin-catalyzed hydrolysis enhances with an increase of the side chain size, though there are some exceptions (derivatives of valine, isoleucine and some tryptophan derivatives). The hydrolysis rate for these substrates correlates with hydrophobic properties of side chains (Fig. 57); however, sterical factors play a certain role both in binding and in catalysis.

There is also observed a linear correlation between hydrophobicity of side chains of substrates and rate constants of acylation and deacylation of chymotrypsin [1965, 1966].

Sterical effects of the side chain are manifested in rate constants of acylation [1967, 1968] and deacylation [1969] of α-chymotrypsin by esters of aliphatic acids, while the slope of plots k_2/K_s and k_3 against E_s is about 1. Analogous dependences have been described in [1726, 1970–1974]. Along with the hydrophobicity index the value of molecular refraction can be employed [1975].

If sterical effects are not taken into account, for methyl esters of N-acetylamino acids a correlation equation [1723] can be written as follows:

$$\lg k_2/K_s = (2.2 \pm 0.1)\pi - 0.9 \pm 0.05 , \tag{48}$$

Table 43. Kinetic constants of α-chymotrypsin-catalyzed hydrolysis of N-acetylamino acid esters (pH 7.8; 25 °C; 0.1 M KCl) [1723][a]

N	Substrate	k_2, s^{-1}	k_3, s^{-1}	$K_s \cdot 10^3 \mathrm{M}$	$k_{-2}, \mathrm{M}^{-1}\mathrm{s}^{-1}$	$k_2/K_s, \mathrm{M}^{-1}\mathrm{s}^{-1}$
1	AcGlyOMe	0.109	0.298	702	0.15	0.155
2	AcAlaOMe	2.27	5.67	1270	7.6	0.018
3	AcButOMe	8.81	1.68	417	3.1	21.1
4	AcValOMe	0.98	0.21	500	0.15	1.96
5	AcNvlOMe	55.8	9.3	100	14.3	558
6	AcNleOMe	103	19.1	34.4	34.4	2990
7	AcIleOMe	1.0	0.18	100	0.09	10
8	AcPheOMe	796	111	7.63	207	1.04×10^5
9	AcTyrOEt	5000	200	17.2	67	2.9×10^5
10	AcTrpOMe[b]	730	29	2.52	–	2.9×10^5

[a] Constants are designated in line with kinetic scheme (30).
[b] pH 7; 25 °C; 3.17% CH_3CN [1761].

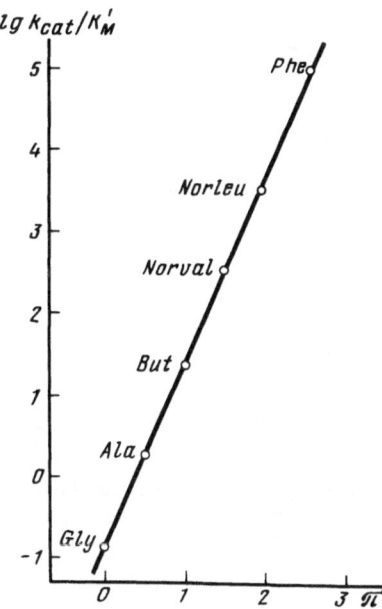

Fig. 57. Effect of the second-order rate constant for reaction of chymotrypsin with methyl esters of N-acetylamino acids on hydrophobicity of the substrate side chain (pH 7.8; 25 °C) [1965]

The equation shows that the specificity of the substrates depends exclusively on the side-chain hydrophobicity. The situation for the substrates with short or branched side chains, for example for nitrophenyl esters of aliphatic carboxylic acids is quite different. In this case the hydrophobicity of a substituent is weakly expressed and inductive and sterical effects make the major contribution to the rate of the enzyme acylation [1964, 1972]:

$$\lg k_2/K_s = (2.18 \pm 0.17)\sigma^* + (0.53 \pm 0.1)E_s + (3.46 \pm 0.17) \,. \tag{49}$$

Unlike nonenzymatic catalysis, for enzymatic catalysis the correlation equations require introduction of the so-called parameter of specificity, that expresses the conformity of the substrate and the enzyme active site [1976]. This is the π value in case of chymotryptic hydrolysis of esters. The dependence of the rate constant of deacylation of p-nitrophenyl esters of aliphatic (and arylaliphatic) acids on inductive, sterical and hydrophobic effects (Fig. 58) in coordinates $\lg k_3 - 2.29\sigma^* - 0.72E_s$ against π passes through the maximum indicating the limited sizes of the active site. In sum, the dependence is characterized by three regions of the curve: (a) the site independent of the alkyl chain hydrophobicity typical of the substrates with the short or branched acyl group; for these substrates the deacylation rate is determined by inductive and sterical effects; (b) the site of linear dependence with a slope about the unity for nonbranched substrates with the number of methylene chains equal to or exceeding three, where the rate depends on the hydrophobicity of a substituent; (c) the site with a negative slope for substrates with the number of CH_2 groups exceeding six, where the limited length of the substrate-binding region of the active site is observed.

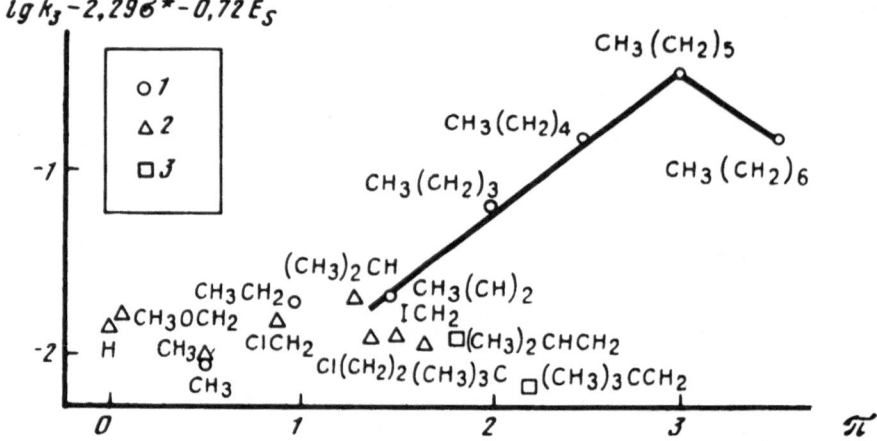

Fig. 58. Effect of rate constants of deacylation of acylchymotrypsins–derivatives of aliphatic carboxylic acids – on the hydrophobicity constant for apolar fragments of the acyl group; *1* [1976]; *2* [1969]; *3* [1972]

Amides of N-acetylamino acids fall outside of the mentioned correlations, however, only a few data are available on the influence of the substrate side chain in the hydrolysis of such amides.

The situation is more intricate in the case of trypsin. The second-order rate constants (k_2/K_m) for the reaction of trypsin with esters of acylamino acids containing a short side chain (glycine, alanine, butyric acid) are close to the corresponding constants of their chymotryptic hydrolysis (Table 44). The same is observed for the rate constants of deacylation of various acylchymotrypsins and acyltrypsins [1977] (compare with Table 43). However, the rate of tryptic hydrolysis dramatically rises upon introduction of a positive charge into the side chain sufficiently remote from C_α-atom.

Table 44. Kinetic parameters of tryptic hydrolysis of esters of acylamino acids [1977–1979]

Substrate	k_2, s^{-1}	k_3, s^{-1}	$K_s \cdot 10^3$, M	k_2/K_s, $M^{-1}s^{-1}$
AcGlyOEt	0.032	0.1	830	0.04
AcAlaOEt	0.25	–	36	6.9
AcButOMe	7.6	1.52[a]	0.47	16
AcPheOMe	55	173	110	5×10^2
AcTrpOMe	–	31	–	–
AcTyrOEt	36	193	47	7.6×10^2
TosOrnOMe	54	5.4	16	3.4×10^2
BzLysOMe	5.9×10^3	22.3	4.9	1.2×10^6
BzArgOMe	2×10^4	23.3	2.2	4.6×10^6

[a] For benzoyl derivative.

Thus, at the acylation stage the trypsin specificity depends on the hydrophobic interaction of the substrate side chain with the enzyme and on the charge, the latter increasing the acylation rate 10^2–10^3-fold. As to the deacylation stage (k_3), the charge interaction is nonessential here: k_3 values for acylchymotrypsins and acyltrypsins are rather close.

As to p-nitrophenyl esters of carboxylic acids, trypsin is less sensitive to the acyl group hydrophobicity. The corresponding correlation equation for these substrates [1980]

$$\lg \frac{k_2}{K_s} = (0.34 \pm 0.06)\pi - (4.4 \pm 0.1) \tag{50}$$

has a coefficient at π twice as low as the coefficient for hydrolysis of such substrates by α-chymotrypsin (0.74) [1970]. Correlation equations of this type have been proposed for papain substrates [1981] and thermolysin [1982].

Another example of correlation of the substrate structures and their reactivity is hydrolysis of peptides by carboxypeptidase A. For substrates of series of Z-X-Phe and Z-Gly-X, where X is a variable amino acid residue, there is found a linear dependence $\lg K_m$ vs hydrophobicity of the side chain of residue X (Fig. 59a). However, the catalytic constant (k_{cat}) for these compounds is virtually independent of the hydrophobic side chain (Fig. 59b).

The dependence of the k_{cat}/K_m value on the size of the side chain of p-nitrophenylcarboxylates has been studied for elastase as well [1983]. It appears to be similar to the dependence observed in case of chymotrypsin (Fig. 58) except for its maximum corresponding to the butyric acid ester.

Resonance effects of acylgroups in reactions of enzymatic hydrolysis should be taken into consideration. In general, a resonance effect of a substituent

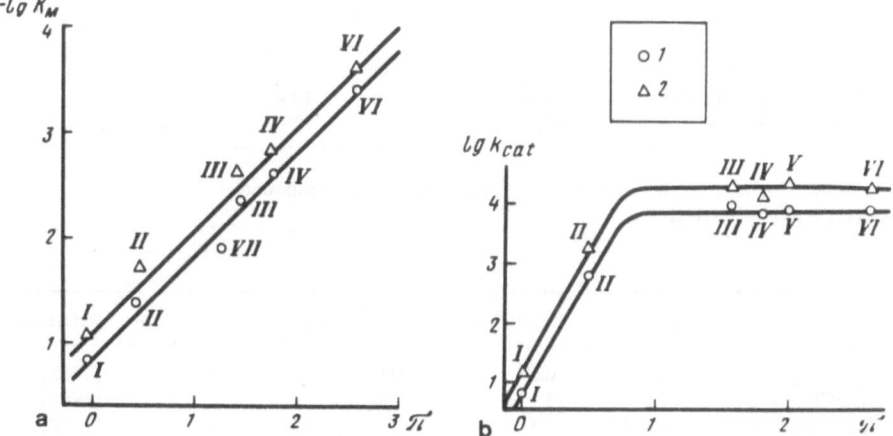

Fig. 59a, b. Effect of binding (**a**) and catalytic (**b**) constants of hydrolysis of substrates ZGlyGlyX in the presence of Zn^{2+} (*1*) and Co^{2+} (*2*) carboxypeptidase A on Ganch hydrophobicity constants [1738]. Substituent (X): Gly (*I*); Ala (*II*); Nva (*III*); Leu (*IV*); Nle (*V*); Phe (*VI*); Val (*VII*)

should decrease the reactivity of carboxylic acid derivatives, thus stabilizing resonance form.

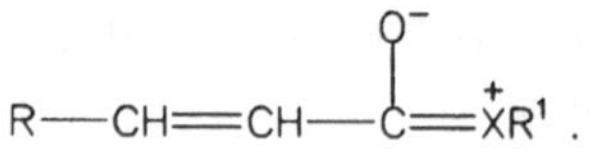

$$R—CH{=}CH—\overset{\overset{\textstyle O^-}{\textstyle \|}}{C}{=}\overset{+}{X}R^1 \ .$$

A few data on chymotryptic hydrolysis of derivatives of cinnamic acid show that this is really observed at the stage of enzyme deacylation rather than acylation (Table 45).

Acylenzymes containing unsaturated acyl groups strongly differ in their spectral properties from the corresponding model compounds – a red shift observed on the absorption maximum [1984–1986].

Influence of the Acylamino Group. Changes in the character of the acylamino group in chymotrypsin substrates of the type

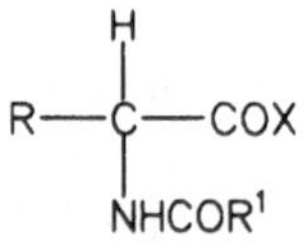

$$R—\overset{\overset{\textstyle H}{\textstyle |}}{\underset{\underset{\textstyle NHCOR^1}{\textstyle |}}{C}}—COX$$

also effect the observed values of kinetic constants. Removal of this group decreases the k_2 and k_3, whereas the K_s is practically the same.

The substitution of group NH by the oxygen atom or its methylation also results in a sharp decrease in catalytic constants (Table 46).

The substitution of the acetylamino group in the methyl ester of acetyl-L-phenylalanyl-L-phenylalanine by the acetoxy group completely deprives the obtained compound of the ability to be hydrolyzed in the presence of pepsin,

Table 45. Constants of alkaline and chymotryptic hydrolyses of cinnamic acid derivatives [1987]

Substrate	k_2, s^{-1}	k_3, s^{-1}	$k_{OH}, \mathrm{M}^{-1}\mathrm{s}^{-1}$
$C_6H_5–CH{=}CH–COOMe$	0.028	0.0125	1.19
$C_6H_5CH_2–CH_2COOMe$	0.02	0.18	7.7

Table 46. Kinetic parameters of chymotryptic hydrolysis of compound $C_6H_5CH_2CH–COOC_2H_5$
           |
          X

X	k_{cat}, s^{-1}	$K'_m \cdot 10^3, \mathrm{M}$	k_2, s^{-1}	k_3, s^{-1}	$K_s \cdot 10^3, \mathrm{M}$	References
H	0.15	0.69	0.67	0.2	3.0	[1965]
$OCOCH_3$	0.6	23	–	–	–	[1988]
$N(CH_3)COCH_3$[a]	0.023	7.5	0.026	0.25	8.4	[1989]
$NHCOCH_3$	96.5	0.9	800	110	7.6	[1965]

[a] Derivatives of tyrosine.

however, the value of the constant of pepsin inhibition by this substance coincides with K_m of the original substrate [1990].

A negative role of the acetoxygroup can be "compensated" by secondary interactions. So carboxamidomethyl ester of acetylglycyl-L-phenyllactic acid

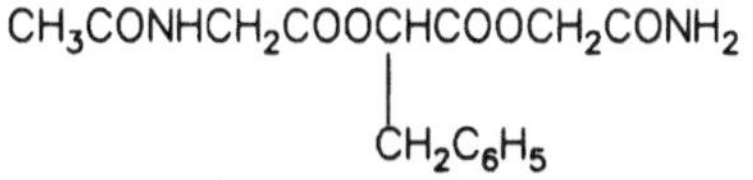

$$CH_3CONHCH_2COOCHCOOCH_2CONH_2$$
$$\overset{|}{CH_2C_6H_5}$$

is hydrolyzed by chymotrypsin with a higher rate than methyl ester of N-acetylphenylalanine ($k_{cat} = 76\ s^{-1}$, $K_m = 0.5\ mM$) [1991].

The acylamino group makes virtually a constant contribution, independently of the acylamino acid type, to the chymotrypsin-induced acceleration of the reaction. At the deacylation stage, the ratio

$$lgk_3(NHCOCH_3) \approx lgk_3(H) + 2.4 \tag{51}$$

is satisfied [1992], i.e., the rate constant of deacylation increases by about 250 times upon transition from the substrate free of acylamino groups to those containing these groups.

With an increase in the hydrophobicity of R^1 in the acylamino group, K_s and k_2 decrease so that the corresponding correlation equations have practically the same angular coefficient, but with the opposite sign [1964, p. 135]:

$$lgk_2/K_s = (0.6 \pm 0.1)\pi + const, \tag{52}$$

$$lgk_2 = -(0.7 \pm 0.1)\pi + const. \tag{53}$$

At the same time the constant of deacylation rate (k_3) does not completely depend on the hydrophobicity of residue R^1 [1970].

A symbate change of k_2 and K_m evidences an enhancement of the degree of nonproductive binding at the hydrophobicity increase of the acylamino group (see Sect. 4.1.4).

Analysis of the temperature dependence of constants k_3 for a series of chymotrypsin substrates containing and lacking the acylamino group has provided interesting data [1992]. According to Fig. 60, hydrolysis of substrates with similar side chains, but of different acylamino groups, proceeds with virtually the same enthalpy of activation, though with dramatically differing entropies of activation. Apparently, the acylamino group makes an entropy contribution to the acceleration of hydrolysis of specific substrates of chymotrypsin.

Substituents at the α-C-Atom. Substitution of the hydrogen atom at α-C-atom of amino acid derivatives by bulkier groups provokes a complete loss of substrate properties [1993, 1994]. However, relatively small groups introduced to the α-position still do not confine hydrolysis [1966].

Leaving Group and the Type of Cleavable Bond. The character of the leaving group is decisive for the reactivity of derivatives of carboxylic acids relative to

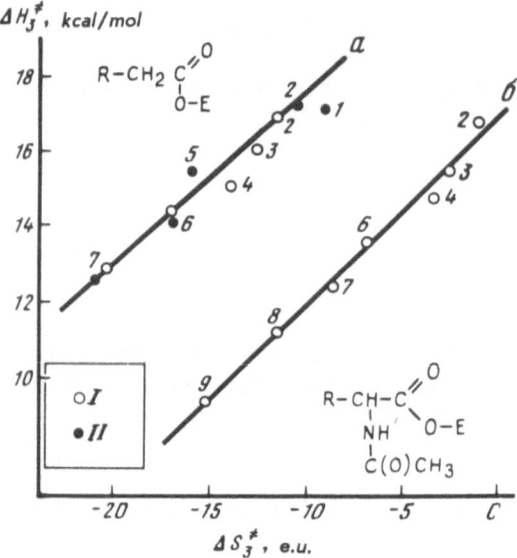

Fig. 60. Effect of activation enthalpy of deacylation of acylchymotrypsins – derivatives of N-acetylamino acids (*a*) – and normal aliphatic carboxylic acids (*b*) on activation entropy [1992]; R–CH₃ (*1*); C₂H₅ (*2*); *n*-C₃H₇ (*3*); *n*-C₆H₁₃ (*4*); H (*5*); *n*-C₄H₉ (*6*); *n*-C₅H₁₁ (*7*); CH₂C₆H₅ (*8*); CH₂C₆H₄OH-*p* (*9*); reaction rates are detected by pH-stat (*I*) and spectrophotometric (*II*) methods

the given nucleophile. In enzymatic catalysis the role of the basicity of the leaving group can be masked by its interaction with the enzyme. For many nonspecific and the so-called semispecific substrates (esters, anilides and some other derivatives of acylamino acids) the dependence of the rate of enzymatic catalysis on pK_a of the leaving group has been thoroughly investigated, and in the case of catalysis by chymotrypsin [1478, 1534, 1996–2002] and pepsin [2003–2005] contradictory data have been obtained. The linear dependence for pK_a of the leaving group on the rate constant of enzymatic hydrolysis (k_{cat}) is observed only in a relatively narrow interval of pK_a values (Fig. 61). All in all, no dependence is traced for the chymotryptic catalysis [2000, 2001].

The hydrophobicity influence of the leaving group on the chymotrypsin-catalyzed alkylacetate hydrolysis has been studied in the reaction of the acetyl group transfer to alcohols during hydrolysis of *p*-nitrophenylacetate [2006]. The transfer rate and hydrophobicity correlate with the angular coefficient close to the unity only for the limited number of alcohols (from ethanol to butanol). The rest of the alcohols follow correlation dependences with other angular coefficients (see also [2007]).

What underlies a considerable difference in the rates of hydrolysis of amides and esters upon catalysis by serine proteases? It appeared [2008–2010] that the rate of enzymatic hydrolysis of various derivatives of the same acylamino acid is linearly related with the value of the standard free energy of the reaction (Fig. 62). This dependence fits the known Hammond postulate (see

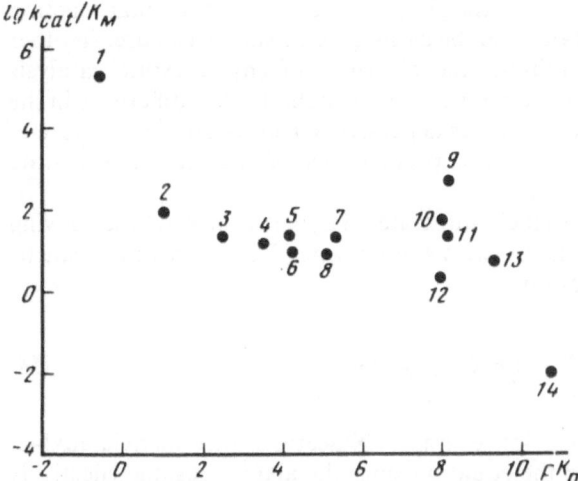

Fig. 61. Second-order rate constant against pK_a of leaving group upon α-chymotryptic hydrolysis of substrates of type AcTyrX [2001]. Substituent (X): OC_2H_5 (*1*), $NHC_6H_4NO_2$-*p* (*2*), $NHC_6H_4NO_2$-*m* (*3*), NHC_6H_4Cl-*m* (*4*), NHC_6H_4Cl-*p* (*5*), $NHC_6H_4OCH_3$-*m* (*6*), $NHC_6H_4OCH_3$-*p* (*7*), $NHC_6H_4CH_3$-*p* (*8*), $NHCH(CH_3)CONH_2$ (*9*), NHOH (*10*), $NHCH_2CONH_2$ (*11*), $NHNH_2$ (*12*), NH_2 (*13*), $NHCH_3$ (*14*)

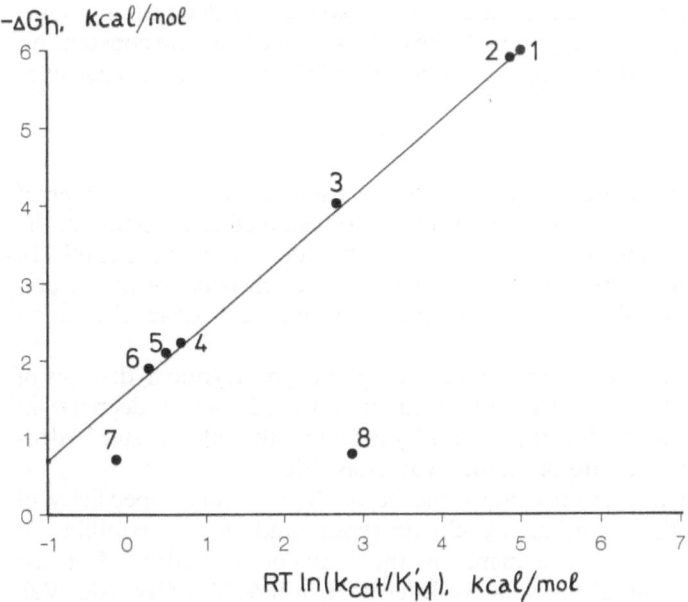

Fig. 62. Changes in standard free energy and second-order rate constant at chymotryptic hydrolysis [2009]. Substrates of type $AcPhe(NO_2$-*p*)-X. Substituent (X): OC_2H_5 (*1*), OCH_3 (*2*), $NHC_6H_4NO_2$-*p* (*3*), NHC_6H_5 (*4*), NH_2 (*5*), $NHNH_2$ (*6*), $NHCH_2CONH_2$ (*7*), $NHCH(CH_3)CONH_2$ (*8*)

Sect. 3.4.3): the higher the energy of the ground state in a unimolecular reaction is, the more the transition state resembles this ground-state structure. In other words, under the same conditions the efficiency of chymotryptic catalysis increases with the lowering of the cleavable bond stability. The difference in the rates of hydrolysis of esters and amides depends exclusively on the structure of the cleavable bond, but not on the peculiarities of the enzyme-substrate interaction.

This is true for "semispecific" substrates, i.e., substrates whose leaving groups do not interact markedly with the enzyme. This correlation can be described by empirical equation

$$-\Delta G_0 = 0.931 \; RT \ln \frac{k_{cat}}{K'_M} + 1.616 \quad (r = 0.999) \; . \tag{54}$$

However, specific substrates, for example N-acetyl-p-nitrophenylalanyl-L-alanineamide, do not fit into the equation since the alanine residue effectively interacts with the enzyme active site [2011].

There is little information on hydrolysis rates of various derivatives of acylamino acids by other enzymes. For carboxypeptidase A the dependence of the rate on the free energy is apparently linear [2012], whereas for pepsin the difference in the hydrolysis rates of amide and ester (depsipeptide) substrates is small [2013].

A study on the influence of the leaving group structure on the chymotryptic catalysis ([1405], p. 388) has demonstrated, that changes in pK_a are accompanied by alteration of the reaction mechanism and the rate-determining stage. Therefore, within the wide pK_a interval the dependence of the rate constant on the basicity of the leaving group has a complicated nonlinear character (Fig. 63).

Secondary Specificity and Substrate-Structure. There are a small number of publications which quantitatively follow the relationship of the structure of the residues remote from the cleavable group and the rate of enzymatic catalysis. At the qualitative level the dependence of the rate of cleavage of the insulin B-chain as well as trypsinogen by carboxylic protease from *Aspergillus saitoi* has been investigated [880].

Interesting data on the secondary specificity of chymotryptic hydrolysis of ester and amides of tripeptides have been obtained [2014]. A decrease in hydrophobicity of the residue at position P_2 enhances the rate of ester hydrolysis and diminishes the rate of amide hydrolysis (Fig. 64).

There has been also studied an influence of the secondary specificity of acyldipeptides on their binding by chymotrypsin and on the equilibrium position in acylation of the enzyme by these dipeptides [2015]. For the compounds of the general formula Ac-X-TrpOH, where X = Gly, Ala, Val, Leu and Phe, the binding by chymotrypsin ($-\Delta G_a$) turned out to be linearly increased with an increase of π for residue X, whereas the equilibrium position for acylation does not depend on this residue hydrophobicity.

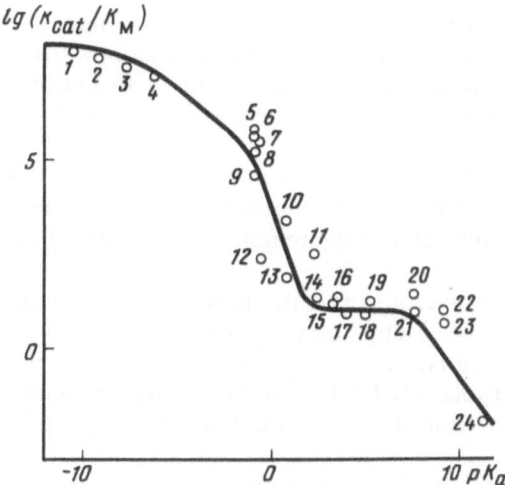

Fig. 63. Effect of lg(k_{cat}/K_m) on pK_a of leaving group in reactions of chymotrypsin with specific substrates [1405]. *1* ZTyrOC₆H₄NO₂-*p*; *2* ZTyrOC₆H₄COMe-*p*; *3* ZTrpOC₆H₄Cl-*p*; *4* ZTrpOC₆H₄OMe-*p*; *5* AcTyrOMe; *6* AcTrpOMe; *7* AcTrpOEt; *8,9* AcPheOMe; *10* BzTyrNHC₆H₄NO₂-*p*; *11* BzTyrNHC₆H₄NO₂-*m*; *12* BzTyrNHC₆H₄NO₂-*o*; *13* AcTyrNHC₆H₄NO₂-*p*; *14* AcTyrNHC₆H₄NO₂-*m*; *15* AcTyrNHC₆H₄Cl-*m*; *16* AcTyrNHC₆H₄Cl-*p*; *17* AcTyrNHC₆H₄OMe-*m*; *18* AcTyrNHC₆H₄CH₃-*p*; *19* AcTyrNHC₆H₄OMe-*p*; *20* AcTyrGlyNH₂; *21* AcPheGlyNH₂; *22* AcPheNH₂; *23* AcTyrNH₂; *24* AcTyrNHCH₃

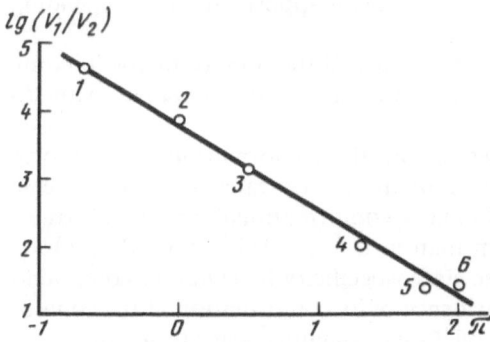

Fig. 64. Relative rates of chymotryptic hydrolysis of esters (V_1) and amides (V_2) on hydrophobicity of side chain at position P₂ for compounds of type RTyrOEt and RTyrNH₂ [2014]. Substituent (R): Ac (*1*), GlyGly (*2*), GlyAla (*3*), GlyVal (*4*), GlyLeu (*5*), GlyNle (*6*)

4.3.3 Stereospecificity

The majority of amide hydrolases show stereospecificity, i.e., they catalyze the hydrolysis of only one enantiomer of an asymmetric substrate *L*(*S*) or *D*(*R*):

L(S)—form   D(R)—form

Evidently, the stereospecificity is manifested when a substrate interacts with an enzyme by at least three groups around the asymmetric atom [2016]. The word "interaction" does not always mean binding, but may indicate sterical hindrances for the entering of a group into the certain region of the enzyme active site.

Proteases can be distinguished by their primary and secondary stereospecificity in line with the influence of the stereochemistry of the amino acid residue forming a cleavable bond, and the residue remote from this bond on the hydrolysis rate.

The primary stereospecificity of numerous proteases is not observed in the substrate binding. So the K_i (K_m) of α-chymotrypsin complexes with D-enantiomers of acylamino acid derivatives is even lower than the K_m of the complexes with L-enantiomers (Table 47) [2017, 2018]. A change in the configuration of the residues set far from the cleavable bond in the peptide chain does not affect the binding.

The alteration of the configuration of α-C-atom of N-acylamino acids influences the rate constant of deacylation of the corresponding acylchymotrypsins and the second-order rate constant (k_{cat}/K_m). Depending on the character of the leaving group, the effect of the substrate stereochemistry can be essential (for amides, methyl esters) and comparatively small (for nitrophenyl esters) [2019].

Table 47 shows, that increasing hydrophobicity of the substrate side chain enhances the ratio of the constants of L- and D-isomers. But it decreases with an increase of the hydrophobicity of a substituent in the N-acylamino group (Fig. 65). Thus D- and L-enantiomers behave in an opposite manner depending on the substituent nature.

Evidently, the ratios result from an increase of the nonproductive binding upon the enzyme interaction with D-enantiomers, as compared with L-compounds.

If the substrate structure changes so that the conformation of its D-form meets the requirements of productive binding, substrates with the reverse stereospecificity can be obtained. So many conformationally rigid substrates possess a higher reactivity in D-form than in L-form (Table 48) [2020, 2021].

As shown [2021, 2022, 2026], the stereospecificity inversion of compound (1) (Table 48) occurs due to the coincidence of its conformation in the equatorial form [2027] with that of N-acetyl-L-phenylalanine ester (Fig. 66).

Table 47. Kinetic parameters of chymotryptic hydrolysis of p-nitrophenyl esters of benzyloxycarbonylamino acids [2017]

Substrate	π	$\dfrac{K_m(L)}{K_m(D)}$	$\dfrac{k_3(L)}{k_3(D)}$	$\dfrac{k_{cat}/K_m(L)}{k_{cat}/K_m(D)}$
ZAlaONp	0.5	2.41	33.8	14.1
ZButONp	1.0	0.9	76.8	83.8
ZNvlONp	1.5	0.84	54.7	64.6
ZLeuONp	2.42	1.98	75.5	38.1
ZPheONp	2.65	1.26	216	171

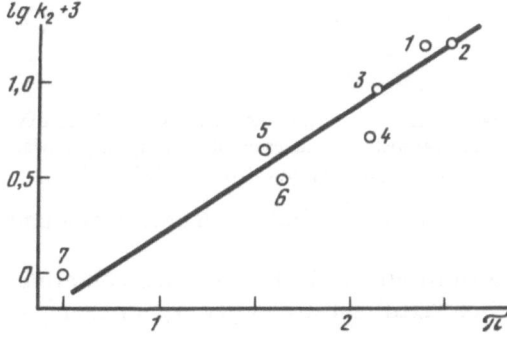

Fig. 65. Effect of acylation rate constants in reactions of chymotrypsin with methyl esters of N-acyl derivatives of *D*-alanine on hydrophobicity of the substrate N-terminal group [1970]. Acyl groups: tetrahydrofuryl (*1*), thiophenyl (*2*), benzoyl (*3*), *o*-aminobenzoyl (*4*), nicotinyl (*5*), isonicotinyl (*6*), acetyl (*7*)

Table 48. Reverse stereospecificity of chymotrypsin to the conformationally rigid substrates (pH 7.8)

N	Compound	Configuration	Substituent[a] X	k_{cat}, s^{-1}	$K'_m \cdot 10^3$, M	References
1		*D*	OMe	20.0	0.98	[2022]
		L	OMe	0.12	12.7	[2021]
		D[b]	ONp	2.6	0.018	[1443]
2		*D*	OMe	56.0	2.5	[2022]
		L	OMe	1.8	28.0	[2021]
3		*S*	ONp	3.3	0.0045	[2023]
		R	ONp	0.072	0.0033	[2023]
4		*D*	OMe	0.006	11.2	[2024]
		L	OMe	0.002	52.3	[2024]
5		*D*	OMe	0.22	14.3	[2025]
		L	OMe	0.024	14.1	[2025]

[a] ONp – *p*-nitrophenyl ester.
[b] pH 6.

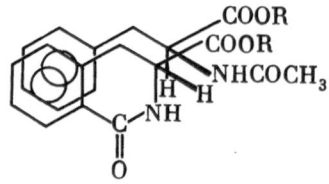

Fig. 66. Tentative positioning of *D*-3-carboethoxy-dihydroisocarbostyryl and ester of N-acetyl-*L*-phenylal-anine in the α-chymotrypsin active site [2029]

Subtilisin behaves similarly relative to the hydrolysis of this compound [2028].

Conformational analysis of the substrates with the restricted conformation mobility as well as other types of "fixed" substrates [1775], for example,

gives a "picture" of the conformation of typical substrates of α-chymotrypsin in the enzyme active site [2029].

The effect of the secondary specificity on kinetic parameters of enzymatic hydrolysis is usually weaker than that of the primary specificity and depends on the distance between the residues in *D*-configuration and the cleavable bond (Table 49). It is noteworthy that stereospecificity of proteases practically disappears if the reaction proceeds in nonaqueous solvents [2032].

Proteases are specific not only to the configuration of the asymmetric center, but also to the geometry of the peptide bond or double bond in the substrate. So prolidase and other proline-specific enzymes have been shown to cleave only the *trans*-form of the amide bond.

As to chymotrypsin and trypsin, they are capable of splitting the substrates with *cis*-proline residues at positions P_2 and more distant from the cleavable bond; however, trypsin does not hydrolyze the substrates with *cis*-Pro at position P'_2 [2035, 2036].

The deacylation rate of *trans*-cynnamoylchymotrypsin is higher than that of its *cis*-isomer [2037, 2038], which makes this enzyme applicable in nonsilver photography [2039].

A series of enzymes do not express stereospecificity. For example, yeast asparaginase II hydrolyzes both *D*- and *L*-asparagine [658].

Conformational Specificity. The peptide sensitivity to the action of proteolytic enzymes can be determined by its conformation.

This specificity type can play a crucial role in the hydrolysis of native proteins with a low conformational mobility. If the substrate is conformationally mobile and the transition between the conformations (S_1 and S_2) is determined by equilibrium constant K

$$S_1 \overset{K}{\rightleftharpoons} S_2, \text{ where } K = [S_2]/[S_1] ,$$

the rate of hydrolysis of one of conformers (for instance, S_2) is

$$v = \frac{k_2[E]_0[S]_0}{K_s(1 + K)/K + [S]_0}, \text{ where } [S]_0 = [S_2] + [S_1] .$$

Table 49. Effect of D-configuration of the amino acid residue remote from the cleavable bond on kinetic parameters of hydrolysis

Enzyme	Substrate[a]	$\dfrac{k_{cat}}{K_m}$, $M^{-1}s^{-1}$	$\dfrac{(k_{cat}/K_m)_L}{(k_{cat}/K_m)_D}$	References
α-Chymotrypsin	AcLeuTyr=OMe	2.9×10^6	152.6	[2030]
	Ac-D-LeuTyr=OMe	1.9×10^4		
Trypsin	BzPheValArg=pNA	2.96×10^5	3.14	[1808]
	Bz-D-PheValArg=pNA	9.42×10^4		
„	BocAlaLys=OMe	1.2×10^6	17.1	[1863]
	Boc-D-AlaLys=OMe	7×10^4		
Thrombin	BocAlaLys=OMe	4.2×10^4	751	[1863]
	Boc-D-AlaLys=OMe	55.9		
Elastase	AlaAlaPhe=AlaAlaAla	3.6×10^4	3.27	[1871]
	D-AlaAlaPhe=AlaAlaAla	1.1×10^4		
Subtilisin BPN'	ZAlaAlaLeu=NH$_2$	1.14×10^3	82.9	[1846]
	Z-D-AlaAlaLeu=NH$_2$	13.75		
Papain	AcPheGlu=pNA	2.04×10^3	330	[2031]
	Ac-D-PheGlu=pNA	5.9		
Thermolysin	ZGlyLeu=GlyAla	3.08×10^4	16.5	[1888]
	ZGlyLeuGly=D-Ala	1.86×10^3		
Prolidase	ZGlyPro=Ala	1.95×10^5	32	[1893]
	ZGlyPro=D-Ala	6.08×10^3		

[a] = Denote place of cleavage.

Thus, the more the equilibrium shifts to the nonhydrolyzed conformer, the higher is the K'_m value, but the conformational equilibrium does not affect k_{cat}.

4.3.4 Relationship Between Catalytic and Michaelis Constants

The amide hydrolase specificity can be displayed through the values of maximum velocity ($V_{max} = k_{cat}[E]_0$) and binding (K_s). Besides, the rate constant of one of the stages (as well as k_{cat}) can increase in a series of substrates, and the dissociation constant of the enzyme-substrate complex (K_s) decreases, i.e., the principle "better binding – better catalysis" works [1962]. The nonproductive binding can cause a decrease in k_{cat} and K_m ("better binding – worse catalysis").

Analysis of such dependences for many substrates of some proteases has shown [2040] that there is a definite interrelation between the k_{cat} and K'_m clearly developed for substrates of the same type (for example, peptides), which have a similar or the same group inducing primary enzyme specificity [1715]. For many substrates this dependence is expressed in an increase in k_{cat} which is accompanied only with slight changes in K'_m until the k_{cat} exceeds the value characteristic of the given enzyme, then k_{cat} tends to the constant value, and K'_m decreases. In other words, first the specificity is manifested via the maximum rate and then via the substrate binding (Fig. 67).

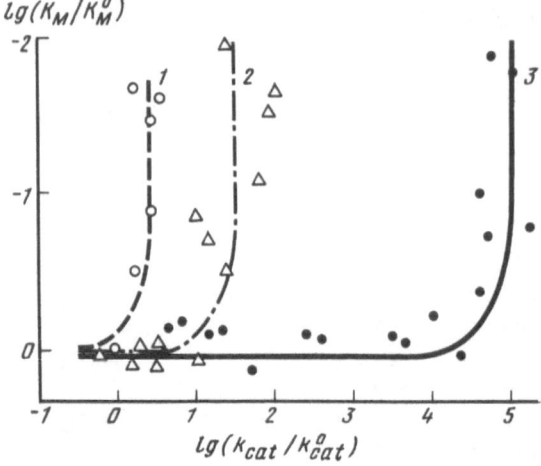

Fig. 67. Relationship between relative values of K_m and k_{cat} for substrates of carboxypeptidase (*1*), chymotrypsin (*2*) and pepsin (*3*) [2040]. Substrate: *1* Ac-X-Phe [1738]; *2* Ac-X-Phe-Y [1858, 1859]; *3* Z-X-Phe(NO$_2$)Phe-Y [1878, 1907] (X amino acid residue or peptide; Y C-protective group or peptide)

Such a relationship between the k_{cat} and K'_m has been, presumably, developed during evolution. To provide the maximum velocity of the substrate hydrolysis at the constant and maximum value of k_{cat}/K'_m and the constant substrate concentration, high K'_m values are favorable (at $K_m > [S]_0$) [1613, p. 325]. Here the maximum rate is reached by increasing k_{cat}. However, the value cannot, evidently, exceed the certain limit depending on the structure of a cleavable bond and on the type of the catalytic apparatus of the enzyme. Therefore, further enhancement of the rate is achieved by a decrease of K'_m, i.e., by strengthening the substrate binding. The consequences of the observed ratios k_{cat} and K'_m will be considered in detail in Chap. 8.

Interestingly, the similar ratios of kinetic constants are observed not only for hydrolytic, but also for other enzyme types [2041, p. 72].

The concept "better binding – better catalysis" is fulfilled for some proteases with the substrates differing by the P$_1$ residue. So, for esters of acylamino acids a linear dependence [2042] linking the k_2 and K_s in the chymotryptic hydrolysis has been proposed:

$$\lg k_2 = C + \rho_a \sigma_x + \frac{\phi_a}{\phi_b} pK_s \, ,$$

where C is the constant, ρ_a and σ_x are the parameters of the Hammet equation, and ϕ_a and ϕ_b are the values of the substrate sensitivity to the hydrophobicity change (π) at the stage of acylation and binding.

4.4 Efficiency

If the enzyme specificity is manifested through the rates of the conversion of a series of substrates, its efficiency is evaluated by comparing these rates with those in the model nonenzymatic reaction. Here several questions arise.

First, this is the problem of a choice of a model reaction. Kosover [2043] has proposed an idea about a congruent model reaction, i.e., such a nonenzymatic reaction as proceeds via intermediates analogous or similar to those of the enzymatic reaction. Thus choice of the model implies the knowledge of the mechanisms of both enzymatic and model reactions. Still, it is not always possible. For hydrolytic reactions catalyzed by four known groups of amide hydrolases the most fitting congruent models are the reactions catalyzed by low-molecular-weight compounds, whose nature is like that of the catalytically active enzyme groups, i.e., ions $R-S^-$ and $R-O^-$, carboxylate ion and metal ion. Here the problems still exist, for example, concerning the concentration of the nucleophile active form in the enzyme.

A second problem is a choice of a kinetic parameter to be compared. The model and enzymatic reactions differ in their molecularity. The former are second-order reactions, whereas the rate of the enzymatic process depends only on the concentration of the enzyme-substrate complex. It is better, evidently, to compare constants of the same order, for example, the second-order rate constant of the model reaction (k_{OH}, k_{SH}, etc.) and the second-order rate constant (k_{cat}/K'_m) of enzymatic reaction. At the same time, it seems interesting to compare the values of the second-order rate constant for a model reaction and k_{cat} of the enzyme. This gives the effective concentration of the substrate, when the rate of its conversion to a product is equal to the enzymatic reaction rate.

It is noteworthy that the values of k_{cat}/K_m, k_{cat} and K_m can be effective, i.e., they do not characterize the rate or the equilibrium at any stage of the process, being just a combination of the rate constants of certain stages (see Sect. 4.1.3). Besides, the rate constant of the model reaction can be effective as well, if there exists the equilibrium between several substrate forms differing in their reactivity. This problem is outlined in Chap. 8.

Finally, it is not easy to choose a substrate when detecting the efficiency of the enzymatic catalysis. The rate of model reactions only slightly depends on the substrate structure (for similar cleavable bonds), whereas the rate of

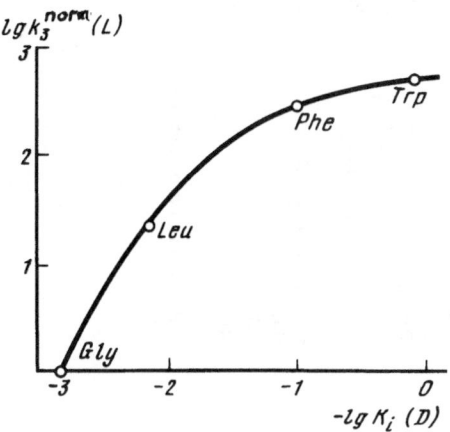

Fig. 68. Dependence of normalized rate constants of deacylation of acylchymotrypsins on the affinity of the corresponding amides of N-acetyl-*D*-amino acids to enzyme [3405]

Table 50. Efficiency of some amidehydrolases

Enzyme	Substrate	k_{cat}, s^{-1}	$K_m \cdot 10^3$, M	$k_{cat}/K_m \cdot$ M^{-1}s^{-1}	$\dfrac{k_{cat}^{\,a}}{K_m k_n}$	References
Chymotrypsin	FaTrpOMe	27.1	4.1×10^{-3}	6.5×10^{7}	1.3×10^{9}	[2044]
	AcTrpONp	30.5	2×10^{-3}	1.5×10^{7}	2.7×10^{6}	[1761]
	AcProAlaProPhePheNH$_2$	8.0	0.21	3.8×10^{4}	1.15×10^{8}	[1841]
Subtilisin Carlsberg	AcGlyGlyPhe (NO$_2$)OMe	1.35×10^{5}	48.5	2.8×10^{6}	5.5×10^{7}	[2045]
Thermomikolin	Ac(Ala)$_3$OMe	2.05×10^{3}	1.5	1.36×10^{6}	2.7×10^{7}	[381]
Papain	ZGlyValGluLeuGly	8.9	2.9	3.1×10^{3}	5.1×10^{12}	[1819]
Pepsin	ZAlaAlaPhePheOpp	282	0.04	7×10^{6}	4.8×10^{11}	[1878]
Co^{2+}-carboxypeptidase A	ZGlyGlyPhe	283	0.3	9.4×10^{5}	3.1×10^{10}	[1738]
Carboxypeptidase B (human)	BzAlaArg	426	0.086	4.9×10^{6}	1.65×10^{11}	[2046]
β-Lactamase	Benzylpenicillin	2000	0.02	1×10^{8}	1.1×10^{9}	[1856]
Renin	Angiotensinogen	50	0.001	5×10^{7}	3.4×10^{12}	[923]

[a] Rate constants of nonenzymatic hydrolysis (k_n, M^{-1}s^{-1}) are given in Tables 26 and 29, and also in References [1533, 1570].

enzyme reactions can alter by 5–7 orders of magnitude with a change of the structure of the region even if remote from a cleavable bond.

As to proteolytic enzymes having proteins as their natural substrates, the latter should be used to determine the efficiency. However, the hydrolysis rate of one amide bond in such a polymer, as in the nonenzymatic reaction can hardly be detected in the enzymatic reaction as well. Table 42 shows several examples of these reactions. For many proteins the hydrolysis rates are low.

Experimental material of vast extent has been accumulated on the synthetic peptide hydrolysis. The efficiency is reliably defined from these data at "saturating specificity", i.e., the substrates converted by the given enzyme with the maximum velocity are applied. Note that each series of the substrates has its own "saturation" level. For example, the deacylation rates of acylchymotrypsins tend toward the limit in a number of acyl residues with the increasing hydrophobicity of the side chain and reach the maximum value in the case of N-acetyltryptophanylchymotrypsin (Fig. 68).

A comparison of the rates of amide hydrolase-catalyzed hydrolysis shows that the efficiency of these enzymes is 10^7–10^{12} (Table 50).

The question arises, what is the limit reaction rate at enzymatic hydrolysis? It is clear, evidently, that in the simplest case of the two-stage enzymatic process

$$E + S \; \underset{k_{-1}}{\overset{k_1}{\rightleftharpoons}} \; ES \; \overset{k_2}{\longrightarrow} \; E + P$$

the overall reaction rate depends on k_1 and k_2 and cannot exceed any of them. If k_2 increases, $k_{cat}/K_m = k_1 k_2/(k_{-1} + k_2)$ tends to k_1 and is equal to this value at $k_2 \gg k_{-1}$ [2047]. Hence, k_{cat}/K_m cannot exceed k_1. In turn, the k_1 value is limited by the diffusion rate and for protein-ligand interactions it is not over $1 \times 10^9 \, M^{-1} s^{-1}$ [1733]. Thus the second-order rate constant cannot exceed this value.

Tables 42 and 50 list the rates of some amide hydrolase-catalyzed reactions approaching the diffusion-controlled limit. Direct evidence in favor of the diffusion control of the enzymatic reaction rate has been obtained [2048, 2049]. The dependence of k_{cat}/K_m of the β-lactamase-catalyzed hydrolysis of benzylpenicillin $(2.35 \times 10^7 \, M^{-1} s^{-1})$ and cephaloridin $(1 \times 10^4 \, M^{-1} s^{-1})$ against the viscosity of the medium shows that the viscosity increase (i.e., decrease of the diffusion rate) affects the hydrolysis of the rapidly hydrolyzed substrate, but not the "slow" substrate. The direct measurement of k_1 of these substrates has yielded the value $1.7 \times 10^7 \, M^{-1} s^{-1}$.

4.5 Comparison of Amide Hydrolase Efficiency and Specificity

Amide hydrolases are 10^7–10^{12} times as effective as the respective model catalysts. The efficiency of serine proteases tends to the lower limit of this interval, whereas that of aspartic and metal-containing enzymes to the upper limit. Despite comparatively low k_{cat}/K_m for papain – a representative of cystein proteases – its efficiency is one of the highest due to a low value of the catalytic efficiency of thiols in amide hydrolysis. Interestingly, the efficiency of serine proteases slightly depends on the type of a cleavable bond.

The deviation observed for *p*-nitrophenyl ester of acetyltryptophan is stipulated by the rate-determining stage of deacylation of acylchymotrypsin common for this compound, and the corresponding methyl ester. At this stage the cleavage of *p*-nitrophenyl esters is not an adequate model reaction.

The efficiency of *β*-lactamase as well as other proteases is within the same range, though model *β*-lactames are hydrolyzed in nonenzymatic reactions 100 times faster than common amides.

Since the experiments do not ensure that the chosen substrates are the best and the models actually congruent, the comparison of the efficiency of amide hydrolases is of a relative character. Still, it is clear that amide hydrolases are rather effective enzymes, their efficiency may depend on the association and dissociation stages of the corresponding complexes with substrates or products. For example, catalase, usually considered as the most effective enzymes $(k_{cat}/K_m = 4 \times 10^7 \text{ M}^{-1} \text{ s}^{-1}$ [2050]), is only 5×10^3 times as effective as model catalyst hematin [2051].

On distribution curves of the k_{cat}, k_{cat}/K_m and k_1 values plotted for numerous enzymes [2041, p. 70], the experimental points for "good" substrates of amide hydrolases lie near the maxima (Fig. 69).

The comparative specificity of different enzymes can be evaluated by comparing probabilities of splitting of the amide bonds in proteins and peptides [1956], i.e., the ratio of the number of cleavable bonds to the total number of bonds in the sequences under study. Among the enzymes investigated in such a way, chymotrypsin and pepsin show the highest specificity:

Enzyme	Probability	Enzyme	Probability
Chymotrypsin	0.14	Elastase	0.24
Pepsin	0.15	Papain	0.26
Thermolysin	0.18		

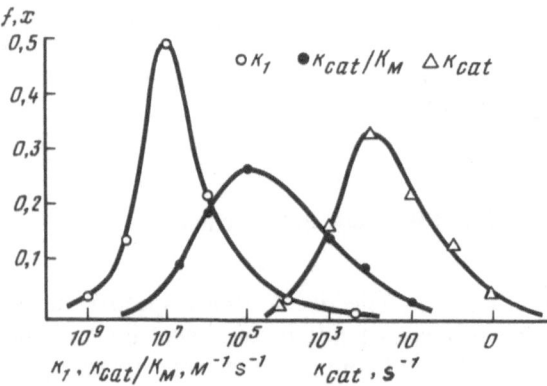

Fig. 69. Distribution functions [$f(x)$] of rate constants of rate-determining stages of enzymatic reactions (k_{cat}), rate constants of the formation of enzyme-substrate complexes (k_1), and second-order rate constants (k_{cat}/K_m) [2041]

However, these evaluations should be treated carefully, since they do not take into account the variable distribution frequency of one or another amino acid in the protein sequence.

As mentioned in Sect. 4.3.1.4, the proteases substantially differ as regards the influence of secondary interactions on catalysis. This refers not only to the enzymes of various groups, but to proteases of the same group, for example, chymotrypsin, subtilisin BPN', or elastase. From this point of view the last two are more specific than chymotrypsin, which contradicts the above data [1956].

Unfortunately, a comparison of the specificity of different amide hydrolases independently of a series of the chosen substrates is impossible by the criteria demonstrated in Sect. 4.3. There are many publications that compare the hydrolysis rates of the same substrates or several related substrates by different enzymes, which belong to related types (for example, [419, 772, 1827, 1833, 1847, 1848, 1863, 1868, 1904, 2052–2059]). However, the specificity evaluation remains, to a large extent, intuitive.

4.6 Effect of Amide Hydrolases Modification on the Activity

Enzyme modification can substantially alter the efficiency and specificity. This often promotes the understanding of the role these or other regions of the enzyme molecule play in the catalysis. There are three modification types: (1) proteolytic modification, i.e., splitting off (or attachment) of a certain fragment of the polypeptide chain; (2) chemical modification incorporation of alien groups and fragments into the protein, and (3) site-directed mutagenesis – a substitution of some amino acid residues by others. The latter has been developed due to recombinant RNA molecules techniques, and seems promising for the relationship of the enzyme structure and activity.

4.6.1 Proteolytic Modification

First of all, let us consider naturally occurring modifications, namely zymogens, various active enzymes appearing due to non-unequivocal activation or the action of foreign proteases.

Zymogens of some serine proteases – chymotrypsinogen and trypsinogen – turned out to possess the pronounced catalytic activity [2060–2063] (Table 51). The table shows that the activities differ by 3–4 orders of magnitude, mainly because of lowering the catalytic constant. With an increase of the ionic strength of the solution the trypsinogen activity rises due to a decrease in K_m [2062].

The active forms of pepsinogen formed during its activation at the pH shift from the neutral to acidic regions have been unraveled [2064, 2065].

As shown above, the activation of zymogens often proceeds by an obscure pattern, while diverse enzyme forms different in structure of the N-terminal part or in the number and the structure of polypeptide chains appear. A variety

Table 51. Activity of zymogens [2061] and corresponding serine proteases

Protein	BocGlyONp[a]			ZSerGlyGlyONp[a]		
	k_{cat}, s^{-1}	$K_m \cdot 10^3$, M	k_{cat}/K_m, M^{-1} S^{-1}	k_{cat}, s^{-1}	$K_m \cdot 10^3$, M	k_{cat}/K_m, M^{-1} S^{-1}
Chymotrypsinogen	0.0077	2.7	2.85	0.0037	7.1	0.52
Chymotrypsin	1.2	1.3	923	0.32	0.47	680
Trypsinogen	0.0036	3.4	1.05	0.004	6.3	0.63
Trypsin	1.2	1.45	827	0.47	5.8	81

[a] *p*-Nitrophenyl ester.

of forms of active proteases can be discriminated by their kinetic characteristics [292, 2066]. Chymotrypsin forms with differently cleaved C-chains (δ-chymotrypsin, α_1-chymotrypsin) differ from α-chymotrypsin by K_m relative to ester substrates in the alkaline region [2066]. Trypsin forms α, β and γ are shown also to be distinguished by kinetic parameters regarding the same substrate [1207].

An analogous picture is observed for thrombin. The prothrombin activation by factor Va (in the presence of Ca^{2+} and phospholipid) yields α-thrombin, which undergoes autolysis with the formation of β- and γ-thrombins. Besides, other forms (ε, β_T, etc.) can be formed. They all differ in their catalytic properties [2067]. The β-chain of α-thrombin isolated under special conditions preserves up to 5% of the catalytic activity [2068].

The renin activation process is rather complicated. Prorenin, inactive protein with M$_r$ $\approx$ 43 kDa, is rapidly activated with splitting off bond Arg63-Ser64, which forms a single-stranded enzyme. The latter is capable of slow conversion to the double-stranded form with M$_r$ 36 500 MW by cleavage of residues 355–356 [2069].

When the disulfide bond is reduced, the double-stranded enzyme, enteropeptidase, forms a catalytically active light chain possessing the trypsin-like specificity to trypsinogen activation. However, the activation in this case proceeds slowly and only at high pH values. Interestingly, the soybean trypsin inhibitor, uncapable of suppressing the native enteropeptidase, reacts effectively with the light chain [2070]. Proteolytic cleavage of enkephalinase [2071], calpains [2072] and plasmin (formation of "microplasmin" [2073, 2074]) slightly affecting the enzyme properties has been also observed.

The subtilisin action on thermolysin is an interesting example here [2075]. In this case the N-terminal tetrapeptide splits off and cleavage by bond Thr224-Gln225 occurs. The two chains of M$_r$ 24 000 and 10 000 MW are closely associated. Thermolysin S (analogous to RNAase S) keeps 3% of the native enzyme activity.

4.6.2 Chemical Modification

There is a wealth of information on the chemical modification of enzymes, among them amide hydrolases (see reviews [163, 2076–2080]).

The chemical modification of amide hydrolases usually leads to a fall of their catalytic activity when K'_m increases or k_{cat} decreases. However, some modifications do not cause a decrease in the substrate-binding features of the enzyme. So methylation of the catalytically significant residue His57 in chymotrypsin [1324] does not affect the ability to bind substrates and inhibitors [1325, 1326, 2081, 2082]. The catalytic activity of (His57)-methylchymotrypsin is preserved, though it decreases by about 5 orders of magnitude [1325]. The substrate binding is preserved in anhydrochymotrypsin. In some cysteine proteases the thiol group in the cysteine residue of the active site can be replaced by a hydroxyl group, and in serine proteases serine hydroxyl by the thiol group, artificial enzymes being so obtained [2083]. It appears that thiolsubtilisin hydrolyzes p-nitrophenyl esters with higher k_{cat}/K_m than the original subtilisin [2084]. Selenious subtilisin has been also obtained [2085]. This artificial enzyme is by 4 orders more active in relation to acyltransfer reactions than subtilisin itself.

The chemical modification usually changes the ratio between the peptidase and esterase enzyme activity. So azocoupling of α-chymotrypsin with the substrate diazoanalog occurring by residue Tyr146 leads to a high rise of the

$$C_6H_5CONHCHCOOCH_3$$

amidase activity (at SucGly$_3$Phe$p$NA), as compared to the esterase activity (at AcTyrOEt) – the ratio of the second-order rate constants for the ester and amide changes from 300 (for unmodified enzyme) to 50 (for the modified enzyme) [2086].

Quite opposite changes are observed [2087] at azocoupling (with diazo-1H-tetrazole) and nitration (by tetranitromethane) of carboxypeptidase A. Diazocoupling occurs at residue Tyr248 and results in an enhancement of the esterase activity, a small decrease of the amidase activity, and the following nitration of residue Tyr198 does not affect the esterase activity, though essentially decreases the amidase activity.

Chemical modification also enhanced the amidase activity of carboxypeptidase A and thermolysin [2088], however the specific modification of Tyr248 in carboxypeptidase A entails a complete loss of the amidase activity, while preserving the esterase activity [1260].

The alteration in the trypsin specificity is observed at the enzyme ethylation by triethyloxonium fluoroborate [2090]. The obtained derivative has lost its ability to hydrolyze cation substrates and to bind specific inhibitors, but kept and even increased the ability for binding and hydrolyzing neutral substrates. An increase in the chymotrypsin hydrophobicity enhances k_2 2.3-fold [2091].

The modification can be directed toward a change in stability and solubility of the enzyme in organic solvents. So the modification of surface amino groups of chymotrypsin and trypsin by polycarboxylic acids dramatically enhances their stability to heat denaturation [2092], and the incorporation of polyethyleneglycol esters into the chymotrypsin molecule yields the preparations soluble in trichloroethane [2079].

In the case of metal-containing amide hydrolases of much interest are the changes of kinetic parameters depending on the character of the metal in the active site. Zn^{2+} ion in carboxypeptidase A can be replaced by Co^{2+} ion [2093]. Here the enzyme is strongly activated: the k_{cat}/K_m value for hydrolysis of ZGly-Gly-Phe rises from $4 \times 10^5 M^{-1} s^{-1}$ for Zn^{2+}-enzyme to $9.4 \times 10^5 M^{-1} s^{-1}$ (due to k_{cat}) [1738]. The same activation has been observed for aminopeptidase from *Bacillus subtilis* [2094] and *Aeromonas* sp. [2095].

Usually, carbohydrates entering the amide hydrolases do not affect their specificity or catalytic properties. However, there are exceptions. Various yeast carboxypeptidases Y, differing by the carbohydrate composition, have different kinetic characteristics [2096].

Many proteins are glycozylated at the β-turns of the polypeptide chain [1917, 2097–2099]. Just these sites are most affected in the limited proteolysis reactions. Probably, this is the reason why glycosylation protects the proteins from unspecific cleavage by proteases. The concept has been experimentally confirmed in some cases [2100].

It should be stressed, that various forms of subunit amide hydrolases possess their own specificity. Amide hydrolase from *Bacillus stearothermophylus* API consists of 12 subunits (α and β) entering the enzyme forms in a different relationship. The peptides containing residues Asp and Glu are hydrolyzed only by complete enzyme forms, whereas isolated α- and β-subunits hydrolyze neutral peptides [1011].

4.6.3 Site-Directed Mutagenesis

There are known natural mutant forms of some amide hydrolases with the distinct activity [1195, 2101–2103]. For point mutations the properties of enzymes, where one amino acid is changed by another, can be compared. However, these possibilities became common practice only after the site-directed mutagenesis techniques have been developed (see reviews [2104–2105]). Since then an avalanche-like accumulation of data on the site-directed mutagenesis is observed [2106–2110]. This novel trend, named protein engineering of enzymes, is successfully applied to reveal the structure-functional relationship in enzymes.

The method has been widely used in subtilisin aimed at obtaining the enzyme resistant to oxidation [2111] and to heat denaturation [2112–2115], changing the enzyme pH optimum, determining the contribution of certain residues within the active site to the enzyme specificity [2117–2121], as well as at supporting the mechanism of stabilization of the transition state [2122].

The oxidation of Met222 results in a pronounced fall of the subtilisin activity. This residue has been substituted by 19 other amino acids. It appears, that the replacement by neutral amino acids only slightly influences the activity, whereas the incorporation of charged amino acids decreases the activity 200–300-fold. Enzymes containing the neutral amino acid residues other than Met222 are resistant to oxidation. Attempts have been undertaken to enhance subtilisin thermostability by incorporation of one or two disulfide bonds into position 22–87 or 24–87. As it turned out, this provides no increase in the thermo- and autolytic stability [2112]. The formed S–S bond is exposed to the surface and has an extremely strained conformation [2113]. Extended investigations of various types of S–S-containing substilisins reveal no relationship between availability and position of disulfide bridges and enzyme stability [2115].

In subtilisin BPN′ the replacements of residue Asp99 by Ser or Glu156 by Ser, etc. shift pK_a of His64 0.08 to -1.0 [2116]. The substitution of Asn155 by Leu decreases the activity as much as 200 times, though it does not affect the K_m. This fact favors the hypothesis of stabilization of the substrate transition state by the enzyme [2122] (see Chap. 7 and 8).

The role of the residues within the substrate-binding cavity of subtilisins BPN′ and from *Bacillus licheniformis* in displaying the substrate specificity has been investigated. The residues at positions 166, 156, 169 and 217 have been substituted. Residue Gly166 has been replaced by the residues of various hydrophobicity; as a result mutant subtilisins with a changed specificity have been obtained. So the incorporation of Ile into position 166 increases k_{cat}/K_m as much as by 1 order of magnitude for the substrates with Ala at the P_1-position, and decreases by as much as 3 orders of magnitude for the substrates with Phe and Tyr [1209]. The simultaneous substitution of three residues Glu156, Gly169 and Tyr217 in subtilisin BPN′ by the typical residues of enzyme from *Bacillus licheniformis* (Ser156, Ala169 and Leu217) yields a mutant with the specificity close to the latter [2120]. Finally, the substitution of Pro172 and Gly131 within the Ca^{2+}-binding site by the Asp residue increases six times the binding of this ion [2121].

Many substitutions have been performed in rat trypsin [2119, 2123–2126]. The replacement of Gly216 and Gly226 by Ala changes the enzyme affinity to Arg- and Lys-containing substrates [2123, 2124]. A lower catalytic activity is probably associated with the deformation of the enzyme oxyanion cavity [2124]. The replacement of the substrate-binding residue Asp189 by Lys results in a loss of the enzyme's ability to bind and convert the substrates with Lys or Arg at P_1, but the protein acquires a weak chymotryptic activity. The side chain of residue Lys189 in the mutant enzyme is presumably directed outside the substrate-binding cavity [2126].

Mutation of zymogen of aspartic protease – prochymosin – [2127] within the processing site (27–31) showed, that this enzyme, and perhaps other aspartic proteases, are formed through splitting off directly at bond Phe42-Gly43, rather than via the cleavage of peptide 1–27 (see Chap. 5). For another aspartic protease – renin – the replacement of Ala317 by Asp entails a shift of pH optimum by 0.5 units to the acidic region [2128]. This implies that this residue plays a definite role in the renin activity at the neutral pH optimum in contrast to other aspartyl proteases.

The literature contains data on the site-directed mutagenesis of neutral endopeptidase (3.4.24.11) [2129], viral proteases [2130, 2131] and β-lactamases [2132–2134]. The substitution of Ser70 in β-lactamase (RTEM) by

Cys yields thiol-β-lactamase partially active to the benzylpenicillin hydrolysis, and sensitive to p-chloromercuribenzoate – a typical inhibitor of cysteine amide hydrolases.

Finally, for carboxypeptidase A the Tyr248 replacement by the phenylalanine residue does not change k_{cat}, but slightly increases K_m [2135]. Since the role of the general-acid catalyst in the carboxypeptidase A-catalyzed reactions has been definitely ascribed to the phenol hydroxyl of Tyr248 (see Chap. 7), such result implies a reconsideration of existing opinion.

4.7 Synthetic Activity of Amide Hydrolases

Since at neutral pH values the equilibrium constant of reaction is about a unity

$$RCONHR^1 + H_2O \xrightleftharpoons{K_P} RCOO^- + {}^+H_3NR^1$$

(see Sect. 3.1), on sufficient concentrations of the reagents proteolytic enzymes can catalyze the synthesis of the amide bond. This possibility was first demonstrated in 1937 [2136]. Later on interest in the preparative enzymatic synthesis of peptides has increased sharply due to prospects of obtaining in preparation some valuable peptides and proteins (see review [2137]).

α-Chymotrypsin [2138–2141], ficin [2142], thermolysin [2141, 2143–2145], carboxypeptidase Y [2146, 2147], microbial metal-containing proteases [588] and other enzymes are used as catalysts. The application of immobilized enzymes [2148, 2149] as well as modified enzymes having a lower amidase activity, but preserving the esterase activity, namely thiolsubtilisin [2150] and $N^{\varepsilon 2}$-methyl(His57)-chymotrypsin [2151], gives fairly good results.

Enzyme specificity in the synthesis accords well with their specificity in hydrolytic reactions [2139, 2152]. The same is true for pH dependences of the peptide synthesis [2151]. If serine proteases are utilized as catalysts, with the ester they form rapidly the acylenzyme, and then the acylgroup is transferred to amine:

$$RCOOET + E\!-\!\!-OH \longrightarrow RCOO\!-\!\!-E \ ,$$

$$RCOO\!-\!\!-E + H_2NR^1 \longrightarrow RCONHR^1 + HO\!-\!\!-E \ .$$

The kinetics of the synthesis under these conditions has been thoroughly investigated and the way for determination of maximum yield proposed [2140, 2143, 2152, 2153].

Ionization of hydrolysis products contributes much to a change of the standard free energy of the reverse synthesis-hydrolysis reaction of the amide bond. Therefore, the synthesis is favorable under conditions when two reagents are not ionized. Under common conditions this is hardly achievable. Elegant methods have been proposed to overcome this difficulty. So the synthesis can be carried out by means of acylamino acid ester and the amino component at pH $\approx$ 10, when the latter is not ionized.

Another method for suppressing ionization of reagents implies that the reaction should run in the water-organic solvent. This changes much the equilibrium constant [1424, 2145, 2154–2158]. The method can be successfully applied for synthesizing not only peptides, but also esters of acylamino acids [1424], whose equilibrium in the aqueous medium essentially shifts to hydrolysis. Note that subtilisin preserves its catalytic activity in nonaqueous dimethylformamide. The reaction in this solvent between the esters of carboxylic and acylamino acids and carbohydrates yields acylmono sugars in high amounts [2159]. The change of the water portion in an organic solvent by ethylene glycol, formamide, etc. ensures success for the segment peptide synthesis [2160].

Methods for the peptide synthesis by proteases appeared to be of value when preparing biologically significant peptides – enkephalins [2161], angiotensin II [2162], secretin [2163] – as well as when modifying the proteins – soybean trypsin inhibitor [2164] and ribonuclease A [2165].

A most striking procedure is the obtention of human insulin from porcine insulin by protease I from *Achromobacter* [588, 2166]. This enzyme selectively splits off Ala30 in the insulin B-chain. A treatment of des(Ala30)-insulin by the excess of *t*-butyl ester of threonine in the presence of the same enzyme and organic solvent leads to human insulin with a yield of 85% after removal of the *t*-butyl group.

4.8 Unusual and Side Reactions Catalyzed by Amide Hydrolases

Numerous amide hydrolases catalyze not only the hydrolysis and synthesis of the amide bond, but other reactions too, of which transpeptidation is the most interesting.

4.8.1 Transpeptidation

There are two types known of the transpeptidation reactions: the acyl transfer type [2167]

$$RCONHR^1 + R^2NH_2 \xrightleftharpoons{\text{E}} RCONHR^2 + R^1NH_2, \tag{56}$$

and the amine transfer type [2168]

$$RCONHR^1 + R^2COOH \xrightleftharpoons{\text{E}} R^2CONHR^1 + RCOOH. \tag{55}$$

The original amide is usually called a donor of transpeptidation, the compound to which the donor fragment transferred is named an acceptor of transpeptidation. One of the products of the substrate (donor) hydrolysis can serve as an acceptor, for example, [2169]

$$2\text{Leu}-\text{TyrNH}_2 \longrightarrow \text{Leu}-\text{Leu}-\text{TyrNH}_2 + \text{TyrNH}_2.$$

The reaction of acyl transpeptidation is catalyzed by numerous serine and cysteine proteases [2170–2172]. Only once has it been reported that α-chymotrypsin can catalyze the reaction of amine transfer as well [2173]:

2ZPhe–Leu $\longrightarrow$ ZPhe–Leu–Leu + ZPhe .

Aspartic proteases catalyze mainly the reactions of amine transfer [2174–2176]. However, if a compound with a free amino group operates as a donor, acyl transpeptidation is also observed [2169, 2177, 2178]. The pepsin-catalyzed amine transfer runs, predominantly, with the donors containing the free carboxyl group at the residue transferred. For a long time pepsin has been proposed to be uncapable of catalyzing the transfer of the residue protected by the carboxylic group [2176], but later [1667] the low yields in the reaction, as compared to the transfer of the free amino acid, have been found to depend on the lower ratio of the rates of transpeptidation and hydrolysis in the latter case (Table 52).

Metal-containing endopeptidases of the thermolysin type catalyze amine and acyl transpeptidation [2173], and aminopeptidase (leucin aminopeptidase) acyl transfer [2179].

It is still unclear whether carboxypeptidases can catalyze the reactions of transpeptidation. The exception is D-alaninecarboxypeptidases [256, 2180, 2181] catalyzing reaction of the type

R–D–Ala–D–Ala + H_2N–R' $\longrightarrow$ R–D–AlaNHR' + D–Ala .

The most important phenomenological distinctions between transpeptidation and the reaction of the peptide synthesis are that the former runs only in the presence of a substrate, whereas resynthesis of the peptide can proceed at the presence of the products of the peptide hydrolysis and the catalyst.

Yields of the transpeptidation products are much higher than the steady-state yields of the products of resynthesis, if the reaction is carried out at low reagent concentrations. In fact, the yield of the synthesis product is determined by the equilibrium constant in reaction

$$R\text{---}COOH + R^1NH_2 \underset{}{\overset{K_P}{\rightleftharpoons}} RCONHR^1 + H_2O ,$$

$$\text{where } K_P = \frac{[RCOOH][R^1NH_2]}{[RCONHR^1]} .$$

Table 52. Kinetic parameters for reaction of transpeptidation, catalyzed by pepsin (pH 4.6; 37 °C); acceptor ZPhe (NO$_2$)OH [1667]

Donar	$k_5 \cdot 10^2$, min^{-1}	K'_T, mM[a]	$\dfrac{k_5\,(1,3)}{k_5\,(2,4)}$	$\dfrac{k_5}{k_{cat}}$
1. AcPheTyr	3.1	0.156	–	0.019
2. AcTyrTyr	2.93	0.424	1.06	0.046
3. AcPhePheApm[b]	4.65	1.7	–	7.6×10^{-3}
4. GlyGlyPhePheApm	4.89	0.17	0.95	8.6×10^{-4}

[a] $K'_T = (k_{-4} + k_5)\, k_3/k_2 k_4$ (see scheme 13).
[b] Apm – N-(3-morpholino)-propylamide.

Upon equilibrium at $K_P \approx 1$ and $[RCOOH] = [R^1NH_2] = 0.01$ M the peptide concentration is 0.0001 M, i.e., the yield is 1%. In transpeptidation reactions at the donor and acceptor concentrations 0.01 M the yield of the transpeptidation product can reach 100%. This is quite understandable, since for reaction

$$RCONHR^1 + R^2NH_2 \xrightleftharpoons{K'_P} RCONHR^2 + R^1NH_2 ,$$

$$\text{where } K'_P = \frac{[RCONHR^2][R^1NH_2]}{[RCONHR^1][R^2NH_2]}$$

the equilibrium constant (K'_P) always tends to unity, and the equilibrium product concentration

$$[RCONHR^2] = \sqrt{[RCONHR^1][R^2NH_2]K'_P}$$

at the same substrate and acceptor concentrations is equal to that of original compounds. In fact, transpeptidation yields are much lower, because of parallel hydrolysis of the substrate and transfer products.

The enzyme specificity in transpeptidation is analogous to that in hydrolysis [2182]. Any enzyme modification decreasing the activity relative to hydrolysis proportionally diminishes the activity in transpeptidation reactions. Evidently, catalysis in both cases depends on the functioning of the same groups of the active site.

Application of a chromogenic acceptor affords detection of the initial rates of transpeptidation reactions by the amino transfer type [1667]. Table 52 shows that the transpeptidation rate consant [kinetic scheme (13), Sect. 4.1.1] is similar for different donors containing the same transferable group (the same acceptor). So the transfer occurs for the complex common for different donors.

4.8.2 Oxygen Exchange in the Carboxylic Group of Acylamino Acid

If acylamino acid is incubated with amide hydrolases of pepsin, chymotrypsin, carboxypeptidase and other enzyme type in heavy oxygen water ($H_2^{18}O$), a rapid exchange of oxygen in the substrate carboxylic group is observed [2183, 2184]:

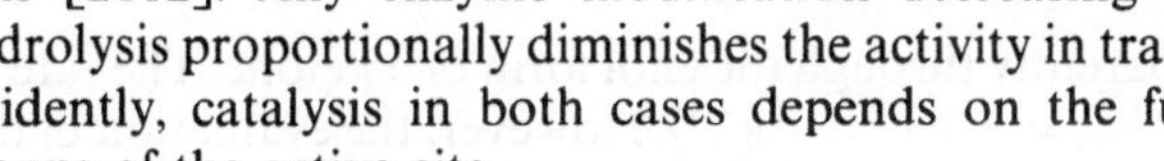

$$R-C\diagup\!\!\!\!\!\!\!^{O}_{OH} + H_2^{18}O \xrightleftharpoons{E} R-C\diagup\!\!\!\!\!\!\!^{^{18}O}_{^{18}OH} + H_2O$$

The enzyme specificity in this reaction is also analogous to that observed for the transpeptidation acceptor and inhibition of hydrolytic reactions by the products [2185].

The rate constants of an isotope exchange are usually close to those of hydrolysis of the substrates containing the same acylamino acid. So the oxygen exchange rate in acetylphenylalanine catalyzed by pepsin is characterized by constants (pH 2) $k_{cat} = 0.014$ s^{-1}, $K'_m = 8.4 \times 10^{-3}$ M [2186] and hydrolysis of AcPhe-Phe-$k_{cat} = 0.015$ s^{-1}, $K'_m = 0.16 \times 10^{-3}$ M [2187]. The K_i value for

AcPhe is 23×10^{-3} M [2187], i.e., it is close to K'_m in isotope exchange reactions.

The isotope exchange is supposed to occur as a result of the synthesis-hydrolysis reaction of acylamino acid with some contaminant peptides contained in the enzyme preparations or (in the case of carboxypeptidase A) with the products of the acyl group cleavage [2188]. However, the oxygen exchange is catalyzed by the immobilized enzyme (pepsin) carefully washed out of the admixtures of low-molecular-weight compounds and not subjected to autocatalytic transformations during its incubation [2189].

4.8.3   Enolization of Ketones, α, β-Elimination, etc.

Carboxypeptidase A catalyzes the hydrogen exchange in the α-CH$_2$ group of ketones – analogs of this enzyme substrate [2190–2192]:

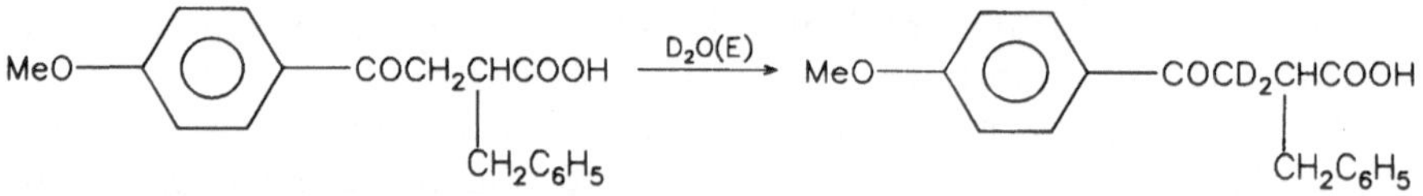

The reaction runs apparently through the enol form of a ketone. The rate of this reaction is fairly low ($k_{cat} = 3.7 \times 10^{-4}$ s^{-1}). However, the evaluation of the carboxypeptidase A catalytic effect here shows, that the enzyme accelerates it as much as 10^4–10^5-fold as compared with the model reaction of acetone enolization in the presence of the acetic acid.

Carboxypeptidase A [2193, 2194] as well as dipeptidylcarboxypeptidase [2195] catalyze the reaction of α, β-elimination in substituted γ-keto acids:

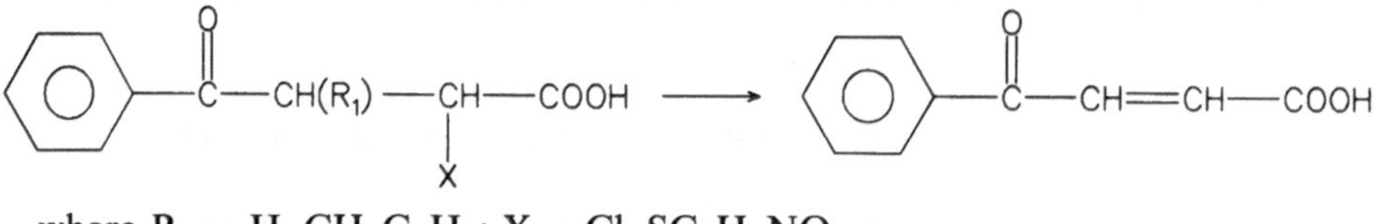

where R$_1$ = H, CH$_2$C$_6$H$_5$; X = Cl, SC$_6$H$_4$NO$_2$-p.

Elimination of HCl proceeds with the second-order rate constant $k_{cat}/K_m = 3300$ M^{-1} s^{-1}, whereas the alkaline-catalyzed process runs with $k_{OH} = 0.18$ M^{-1} s^{-1} [2194].

Finally, in 1955 the chymotryptic hydrolysis of the C–C bond in ethyl ester of 5-(p-hydroxyphenyl)-3-ketoglutaric acid with the formation of p-hydroxyphenylpropionic acid was reported [2196]. However, no support for this result followed.

4.9   Conclusion

Numerous amide hydrolases, except for a few enzymes, catalyze one reaction – hydrolytic cleavage of the amide bond. The diversity of these enzymes is mainly caused by distinctions in the specificity displayed at different levels of the structural organization of substrates. Besides, the amide hydrolases specificity can manifest itself at different stages of the catalytic process – the substrate binding and its chemical transformation. A change of the substrate structure generally increases k_{cat} without any influence on K'_m and only when the catalytic constant is maximum for the given enzyme (or the given series of substrates) does the Michaelis constant decrease. For a relatively simple series of substrates the specificity can be described quantitatively by the ratios of the type of linear dependences of the free energy, in more complicated cases by empiric correlations from the statistical analysis of the specificity. However, the specificity of a variety of amide hydrolases cannot be analyzed in such a way. First of all, this concerns some enzymes with the specificity approaching the absolute value (for example, renin). The efficiency of amide hydrolases changes within a wide range depending on the substrate type. At the same time, some enzymes have the substrates hydrolyzed with rate constants close to those of diffusion controlled processes. Thus the efficiency of the enzymatic hydrolysis seems to be independent of the peculiarities of the chemical mechanisms of catalysis.

Chapter 5   Regulation and Effect of External Factors

The activity of amide hydrolases can be regulated at their biosynthesis as well as at the post-translational level. Here we consider only a second type of regulation, describing possible changes of this enzyme group activity both in vivo and in enzymological experiments in vitro.

The problems of regulation of the amide hydrolases activity are raised in many reviews [6, 2197–2199]. A conventional way of regulation especially characteristic of animal proteases is the activation of zymogens – inactive precursors of proteolytic enzymes. Active enzymes formed upon limited proteolysis by zymogen can serve as activators of other zymogens, thus providing a "cascade" of sequential activation.

The activity of enzymes including amide hydrolases is altered with a change of the medium – concentration of hydrogen ions, ionic strength, and the microenvironment. The latter is of special importance for membrane-bound amide hydrolases. The temperature of the medium where they function is also very substantial.

Another effective way of regulation of the amide hydrolases activity is an influence of the inhibitors. Later on, the action of natural and synthetic inhibitors will be discussed.

Finally, proteolytic enzymes are capable of self-inactivation, i.e., of autolysis.

These effects yield a very complex and, evidently, fine-tuned system of regulation of amide hydrolases activity in the organism.

5.1   Zymogens and Their Activation

It became clear especially from gene structural analysis that the majority of proteins are synthesized by cells as precursors, which then "mature", being transformed into biologically active proteins.

Amide hydrolases are no exception. Precursors usually contain the so-called signal sequence at the N-terminus of the polypeptide chain consisting of 15–30 amino acid residues (see review [2200]). The sequence in various proteins actually has no homology. However, three regions can be here identified: *n*-region, which involves different residues, but as a whole is characterized by a positive charge; then *h*-region (about 10–30 residues), rich in hydrophobic amino acid residues and forming a hydrophobic core; and *c*-region usually uncharged but involving polar residues. The region cleaved by signal proteases corresponds structurally to the so-called rule " -1, -3", the gist of it being as follows: site -1 is occupied by Ala residue in preference, in site -3 the aromatic or charged residue is absent [1854].

A signal peptide foregoes a prosequence of a different size. Thus for chymotrypsinogen it includes 15 residues [2201], while for subtilisin from *Bacillus subtilis* 77 residues [2202]. This sequence is already bound to the sequence of the mature protein. In the case of chymotrypsinogen, after cleavage of the proregion (at Arg-Ile bond) inactive zymogen is formed, but in the case of subtilisin an active enzyme.

Trypsin, formed from trypsinogen under the action of enteropeptidase (enterokinase) and activating most zymogens, is essential for activation of numerous enzymes from the pancreas (Fig. 70) [5, 2203].

The activation of bovine and some other trypsinogens occurs in the presence of Ca^{2+} ions and is accompanied by cleavage of the N-terminal hexapeptide [2204]:

$$\downarrow$$

ValAspAspAspAspLys—IleVal . . .

Octapeptide of porcine trypsinogen splits [2205]:

$$\downarrow$$

PheProThrAspAspAspAspLys—IleVal . . .

In this process enteropeptidase is about 2000 times as active as trypsin which is also capable of activating trypsinogen.

Four residues of aspartic acid typical of an activation peptide of trypsinogen allow the N-terminal fragment to be localized on the globule surface and prevent masking of the cleavable bond. The presence of many charged amino acid residues in the activation peptide is characteristic of numerous zymogens.

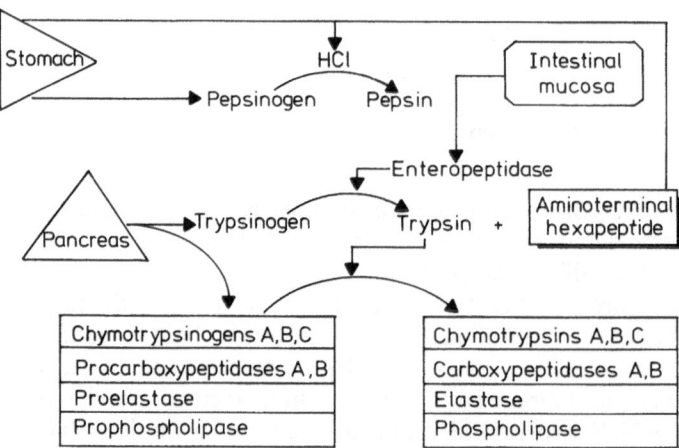

Fig. 70. Scheme for activation of zymogens of gastric enzymes [5]

The trypsin action on chymotrypsinogen A results in a breakdown of bond Arg15-Ile16, an active chymotrypsin A_π is formed [2206]. This enzyme form is very unstable and transforms into δ-, $\chi(\alpha_1)$- and α-chymotrypsins by autolysis [1205, 2206, 2207] due to sequential cleavages of bonds Leu13-Ser14, Tyr146-Thr147 and Asn148-Ala149. The formation of minor forms of μ- and ω-chymotrypsins has been also established [1205].

In the case of chymotrypsinogen B along with the release of dipeptide Ser14-Arg15 the three amino acids Asn147-Ala148-Leu149 split, being catalyzed by chymotrypsin [2208]. In porcine chymotrypsinogen C and shark cation chymotrypsinogen the first bond breaks in sequence Arg15-Val16 [2209, 2210].

Prothrombin, the precursor of one of the key enzymes of the blood coagulation system, thrombin, is formed in liver as a proprotein of 43 residues of a signal (leader) peptide [2211], its biosynthesis depends on vitamin K in the organism (see review [2212]). The bovine prothrombin is a glycoprotein (M_r 70 kDa), made up of 582 amino acid residues forming one polypeptide chain [317]. This protein contains ten residues of unusual γ-carboxyglutamic acid [2213], which may play a crucial role when binding Ca^{+2} ions by prothrombin. Its activation occurs in vivo under the action of the coagulation factors Xa and Va, of Ca^{+2} and phospholipid (prothrombinase) [2212]. The Arg273-Thr274 bond is cleaved and glycopeptide (M_r 33 000 MW) called intermediate protein II is formed. At a second stage the Arg323-Ile324 bond splits and double-stranded active thrombin (M_r 39 000 MW) is formed [2214, 2215].

Active factors Xa and Va are in turn the products of an activation cascade (Fig. 71), which starts with an activation of the Hageman factor (factor XII)

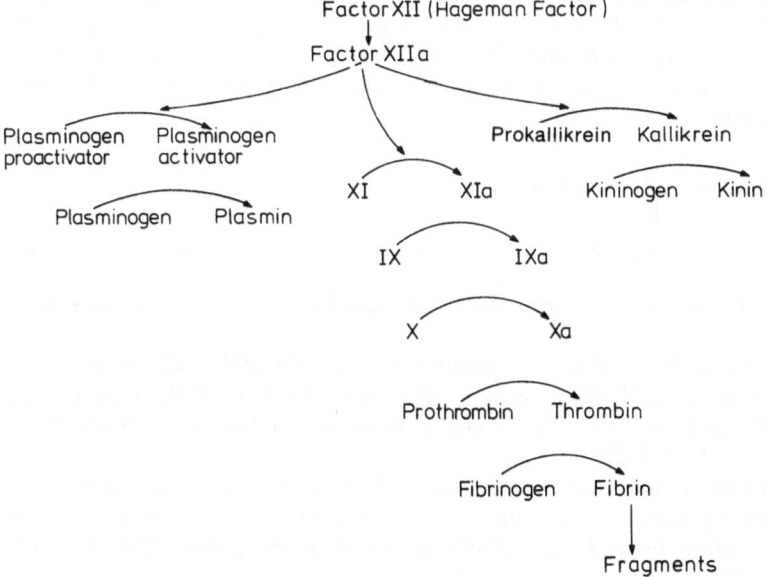

Fig. 71. Cascade processes induced by the Hagemann factor

initiating sequential activation in the blood coagulation system [417], in the kininogen-kinin system [2216], in the system of fibrinolysis (plasminogen activation) [2217] as well as in the system of complement activation [2218]. The latter can be autocatalytically activated [2219, 2200].

Activation of factor XII initiating cascade proceeds possibly autocatalytically and is intensified by the charged surfaces (glass, kaolin, dextransulfate). Factor XIIa (active) can catalyze transformation of prokallikrein into kallikrein. The latter accelerates transformation of XII into XIIa in the presence of dextransulfate 11 000 times, as compared to conditions when no charged component is observed in the system [2221].

Another substantial serine protease acrosin is formed from the precursor through cleavage of the N- and C-terminal peptide containing a large number of proline residues [2222]. The same is observed at aqualysin [2223], neutrophile elastase and cathepsin G [2224] formation. The activation peptide of the latter contains only two residues, namely Gly and Glu. The precursors have been found at collagenases [2225], plant serine carboxypeptidases [2226] and other enzymes as well. Interestingly, the presequence is essential for proper folding of the protein chain of certain microbial proteases – subtilisin, [2227] and α-lytic protease [2228]. Another example may illustrate an unusual mechanism of activation, i.e., activation of the nerve growth factor [2229]. This protein contains one atom of zinc that suppresses an autocatalytic activation. In diluted protein solutions, zinc dissociates and the proenzyme is activated.

No precursor has been found for cysteine enzymes except for protease from *Stretococcus* sp. [745, 746, 2230]. An intracellular form of zymogen of the enzyme contains a free SH group, which transforms into methyldisulfide ($-S-S-Me$) when transporting to extracellular space. The reduction of this bond leads to activation and autoprocessing of the proform (34 000 MW) into an active enzyme (28 000 MW). Crystalline papain after isolation shows a very low catalytic activity, which is manifested only in the presence of thiol compounds, reduction agents or the CN^- ion [1331]. According to [2231, 2232], a catalytically active cysteine residue is blocked by a disulfide bond in the inactive papain:

$$E-S-S-CH_2CH-COOH \, .$$
$$\quad\quad\quad\quad\quad\; | $$
$$\quad\quad\quad\quad\quad NH_2$$

The other cysteine proteases have presumably inactive precursors of a similar type.

Autocatalytic proteolysis of double-stranded (80 000 and 26 000 MW) proenzyme of calpain II occurs at a small subunit, as a result the fragment of M_r 9000 MW splits off and the enzyme activatable by low concentrations of Ca^{+2} is formed [2072, 2233].

Proenzymes of mammalian aspartic proteases are stable in neutral and weakly alkaline medium, being subjected to an autocatalytic activation, when pH decreases up to 2–4 [2234]. Virtually about 40 amino acid residues split off at the N-terminus of the zymogen molecule.

The mechanism of pepsinogen activation is rather complicated. Biosynthesis of this zymogen proceeds in special cells of the stomach wall [2234] as a preproenzyme containing 14–15 residues of the leader sequence [2235, 2236]. A conformational change of pepsinogen in the acidic medium entails the formation of a catalytically active zymogen – δ-pepsinogen [2064, 2086, 2237–2239]. Depending on pH, an intramolecular process of cleavage of the activation peptide [2240] or intermolecular activation [2064, 2234] can take place. These two processes apparently compete, and pH predetermines which reaction prevails (Fig. 72).

An activation peptide can be cleaved at one or two stages in zymogens of various aspartic proteases. Thus, porcine pepsinogen is activated mostly in two stages: first bond Leu16-Ile17 splits, then bond Leu44-Ile45 [2065, 2241–2244]; in the chicken pepsinogen and chymosin the hydrolysis product is also cleaved first at bond Phe27-Leu28, then bond Leu42-Thr43 (chicken zymogen) or Phe42-Gly43 (chymosin) breaks [2127, 2245, 2246]. It is noteworthy that cleavage by bond 27-28, at least for prochymosin, is not obligatory. At pH 4.5 prochymosin, with this sequence changed by site-directed mutagenesis, is normally activated, however no activation is observed at pH 2 [2127]. In human pepsinogen A bond Leu23-Lys24 is cleaved intramolecularly, while cleavage of bond Leu47-Val48 is intermolecular [2247]. The region of the initial cleavage generally depends on the structure of the activation peptide and the specificity of the enzyme formed. The existence of basic amino acid residues in position P_3 of the activation peptide obstructs a breakdown in this region [2246] (Table 53). Interestingly, peptide 1–16 effectively inhibits the pepsin activity [837, 2248–2252]. A series of the enzymes of microorganisms evoke activation of pepsinogens [2253].

Aspartic protease – cathepsin D is also synthesized as a precursor of M_r 53 000 MW [1168, 2254–2258], which step by step converts into an intermediate of M_r 47 000 MW, then into a double-stranded form (31 000 and 13 000–14 000 MW) (in bovine proenzyme due to a release of dipeptide SerSer [2259]). Activation occurs inside lysozomes and is accompanied by ATP

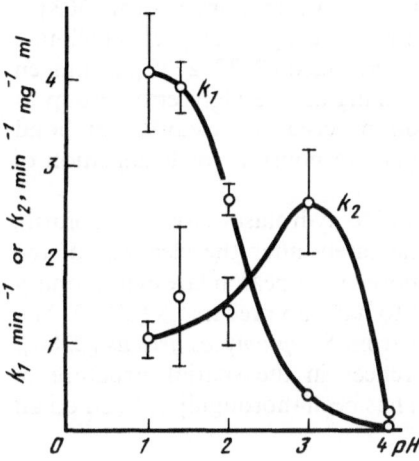

Fig. 72. Pepsinigen activation. pH vs first-(k_1) and second-(k_2) order rate constants (*vertical lines* mean square deviation) [2240]

Table 53. Sequence within cleavable regions in zymogens of aspartic proteases [2246]

	10	15	20	25	30	35
Porcine pepsinogen A	RKKS	LRQNL↓	IKDGK	LKD↓ F*		LKTHKHNPAS
Bovine pepsinogen A	KKKS	LRQNL↓	IENGK	LKE↓ F*		MRTHKYNLGS
Calf prochymosin	KGKS	LRKAL*	KEHGL*	LED* F↓		LQKQQYGISS
Chicken pepsinogen	KGKS	LRKQL*	KDHGL*LED* F↓			LKKHPYNPAS
Human progastricsin	KFKSI	/LRETM*KEKGL* LED* F↓				LRTHKYDPAS
Human pepsinogen A	RKKS	LRRTL*	SERGL*	LKD↓ F*		LKKHNLPARK
Monkey pepsinogen III3	RKKS	LRRNL*	SEHGL*	LKD↓ F*		LKKHNLNPAS
Bear pepsinogen	KKKS	LRKVL*	KEHGL* LKD↓ F*			LKKHSPNPAS

Full arrows show major sites of cleavage; dashed arrows indicate minor cleavage sites; asterisks show scissile bond not cleaved due to basic residue in P_3.

expenditure, obviously for increasing the concentration of H^+ ions [2258]. Cysteine protease seems to be involved in activation of procathepsin D.

Kidney renin is also synthesized as a preproenzyme ($M_r \approx 50000$ MW), containing 20 residues of the leading peptide and 46 residues of the prosequence [2260]. As reported, a high molecular inactive form of this enzymes has been revealed [2261–2264]. This form is supposed [2265] to be a renin complex with an acid-labile inhibitor, though this is perhaps prorenin [2263, 2266]. Prorenin is activated in blood plasma in two stages with the formation of the activated intermediate due to a conformational change induced by H^+ ions and sequential proteolytic cleavage at bond Arg-Leu under the action of the plasma enzymes – kallikrein or plasmin [2267].

Activation of procarboxypeptidase A is more complex [2268, 2269]. Zymogen consists of three subunits, of which only one subunit – subunit I – is zymogen of carboxypeptidase A, and the other two – chymotrypsinogen C and an inactive form of elastase-like enzyme – protease E [503, 2270]. Procarboxypeptidase A with molecular mass about 45000 MW [2271] (the other form in pig is of 71000 MW [2272]) is activated by trypsin much faster when isolated from the complex (as a succinylated derivative). Herewith about 100 amino acid residues [2273–2276] split off, yielding a mixture of quite a few enzyme forms differing in their N-terminal amino acids (Sect. 2.4.4). Interestingly, procarboxypeptidase of shark pancreas is in a monomer form [303].

Information on processing of the membrane enzyme of prepenicillinase from *Bacillus licheniformis* is worthy to be mentioned [2277, 2278]. Zymogen and the enzyme itself are lipoproteins containing an acetylglyceride group in position Cys27 (of proenzyme). Activation proceeds by cleavage of bond Gly26-Cys27, where a lipid residue is perhaps responsible for localization of the protein chain in the membrane.

As mentioned above, many microbial amide hydrolases have no proform, i.e., true zymogen. Activation occurs immediately after the cleavage of the leader (signal) sequence. Zymogens are supposed to appear at late evolutionary stages [4]. So trypsinogen in starfish seems to have no precursors [2279]. The same is true for chymotrypsin-like enzymes from *Streptomyces griseus* [2280].

It is interesting to consider the differences in the spatial structure of enzymes and their zymogens. This question has been thoroughly studied on an example of the α-chymotrypsin-chymotrypsinogen A [1229, 2281–2283]. Essential changes occur in the vicinity of cleavable bonds Ile16 and Arg145. Residue Ile16 is of a crucial value for arranging the enzyme active site, because it forms a salt linkage by its amino group and a carboxyl group of residue Asp194, exposed in zymogen into the region of the substrate-binding protein site. Residues 189–193, forming this substrate-binding site also significantly change. Chymotrypsinogen activation provokes two crucial events; these are the formation of the substrate-binding site and the so-called oxyanion hole due to a change of the orientation of the carbonyl group in residue Gly193 [1229]. As to a catalytic apparatus, the "charge relay system" formed by residues Asp102-His57-Ser195, in zymogen it is formed more distinctly than in the enzyme [1229]. That is because chymotrypsinogen shows a certain catalytic activity. Chymotrypsinogen and trypsinogen also slowly react with diisopropylfluorophosphate [2284] at a catalytically active serine residue and bind competitive inhibitors [2285]. Trypsinogen forms a stable complex with pancreatic trypsin inhibitor [1233, 2286]. As shown by [2287], peptides (for example, IleVal) induce a trypsinogen transition into the conformation typical of trypsin, i.e., these peptides are capable of replacing the zymogen N-terminal fragment and form the binding active site. A similar mechanism is supposed to come into effect upon the activation of plasminogen by streptokinase [2287, 2288].

Thus, at least for trypsin and chymotrypsin, the major difference between zymogens and active proteases arises from their conformations. This is supported by the enzyme chemical modification (for example, [2289, 2290]) as well as by physicochemical studies [2291–2293].

Note that the difference in zymogen and enzyme conformations affects, evidently, mostly the activity with regard to the specific substrates and less the activity with respect to "poor" substrates [2290, 2294].

5.2 The Influence of pH

Amide hydrolases are active within an interval of approximately 10 pH units. For each enzyme group the activity range is localized narrower. However, there are enzymes functioning in the region far from the pH optimum typical of most of the representatives of this group.

The pH-dependence of kinetic constants of the enzymatic hydrolysis, especially for the rate constants of individual stages, provides information on the character of the enzyme groups involved in the substrate binding and its transformation, as well as on the reaction mechanism (see reviews [1613, 1659, 1661]).

5.2.1 The Kinetics

A change in the rate and equilibrium constants of the enzyme-substrate complex depends on a change of the ionization of the groups involved in these interactions. A pH change results in protonation or deprotonation of at least two ionized enzyme groups involved in catalysis:

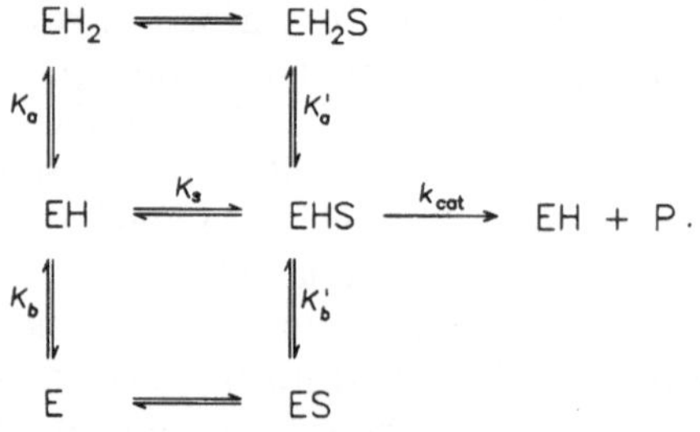

$$
\begin{array}{ccc}
\text{EH}_2 & \rightleftharpoons & \text{EH}_2\text{S} \\
K_a \downarrow & & \downarrow K_a' \\
\text{EH} & \overset{K_s}{\rightleftharpoons} \text{EHS} & \overset{k_{cat}}{\longrightarrow} \text{EH} + \text{P}\,. \\
K_b \downarrow & & \downarrow K_b' \\
\text{E} & \rightleftharpoons & \text{ES}
\end{array}
\tag{1}
$$

If the three enzyme forms (E, EH, and EH_2) can bind to the substrate, but the chemical transformation occurs only in one complex (for example, EHS), the equation of the reaction rate is

$$
v = \frac{(k_{cat}/f')[\text{E}]_0[\text{S}]_0}{K_m f/f' + [\text{S}]_0}\,,
\tag{2}
$$

where $f = 1 + [\text{H}^+]/K_a + K_b/[\text{H}^+]$ and $f' = 1 + [\text{H}^+]/K_a' + K_b'/[\text{H}^+]$, i.e., the so-called Michaelis pH-functions [1661, p. 138].

Thus, the kinetic constants derived experimentally are related with the hydrogen ions concentration by the following equations:

$$
k_{cat} = \frac{\tilde{k}_{cat}}{1 + [\text{H}^+]/K_a' + K_b'/[\text{H}^+]}\,,
\tag{3}
$$

$$
K_m = \frac{\tilde{K}_m(1 + [\text{H}^+]/K_a + K_b/[\text{H}^+])}{1 + [\text{H}^+]/K_a' + K_b'/[\text{H}^+]}\,,
\tag{4}
$$

$$
k_{cat}/K_m = (\tilde{k}_{cat}/\tilde{K}_m)\frac{1}{1 + [\text{H}^+]/K_a + K_b/[\text{H}^+]}\,,
\tag{5}
$$

where $\tilde{k}_{cat}$ and $\tilde{K}_m$ referred to the pH-independent values. The equations show that in the simplest case of the pH dependences for k_{cat} and K_m Eqs. (3) and (4) express ionization of the groups in the enzyme-substrate complex, and the corresponding dependence for k_{cat}/K_m-ionization of the free enzyme groups.

At high concentrations of hydrogen ions the members including K_b and K_b' in Eqs. (3)–(5) can be obviously ignored. The logarithms of the equations for k_{cat}, K_m and k_{cat}/K_m are

$$
\lg k_{cat} = \lg \tilde{k}_{cat} - \lg(1 + [\text{H}^+]/K_a')\,,
\tag{6}
$$

$$
\lg K_m = \lg \tilde{K}_m + \lg(1 + [\text{H}^+]/K_a - \lg(1 + [\text{H}^+]/K_a')\,,
\tag{7}
$$

$$
\lg k_{cat}/K_m = \lg(\tilde{k}_{cat}/\tilde{K}_m) - \lg(1 + [\text{H}^+]/K_a)\,.
\tag{8}
$$

At $[H^+] \ll K_a$ (and K_a') the relative constants are pH-independent, whereas at $[H^+] \gg K_a$ (K_a')

$$\lg k_{\text{cat}} = \lg \tilde{k}_{\text{cat}} - pK_a' + pH \;, \tag{9}$$

$$\lg \frac{k_{\text{cat}}}{K_m} = \lg \frac{\tilde{k}_{\text{cat}}}{\tilde{K}_m} - = pK_a + pH \;. \tag{10}$$

As to the K_m, it does not depend on pH at $K_a = K_a'$. If we follow $K_a \ll [H^+] \ll K_a'$ or $K_a' \ll [H^+] \ll K_a$, Eq. (7) is simplified to equations similar to (9) and (10). So it seems possible to determine the values of pK_a and pK_a' when plotting the logarithm dependences of kinetic constants on pH (Fig. 73). The values pK_b and pK_b' are defined similarly [2295, 2296].

Other ways to determine these values have been proposed in [2297, 2298]. In more intricate cases, for instance at four ionic states of the enzyme E, EH, EH_2, and EH_3, the Michaelis pH-function is:

$$f'' = 1 + \frac{[H^+]}{K_b} + \frac{([H^+])^2}{K_a K_b} + \frac{K_c}{[H^+]} \;, \tag{11}$$

where K_c, K_b and K_a are the ionization constants of EH, EH_2 and EH_3, respectively. The pH dependence of $\lg k_{\text{cat}}$ here is a number of fragments with slopes $+2$, $+1$, 0 and -1.

The compounds carrying ionogenic groups are the substrates of many amide hydrolases. If the substrate ionization is taken into account, rather complex pH-dependences for kinetic constants occur. When dissociation constants of the free enzyme and of the enzyme-substrate complex are the same

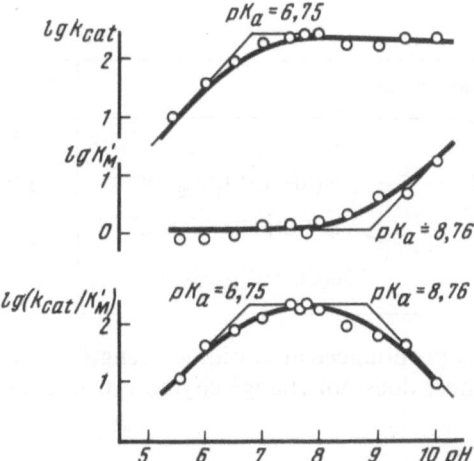

Fig. 73. pH vs kinetic constants for α-chymotryptic hydrolysis of N-acetyl-L-tyrosine ethyl ester (25 °C; $\mu = 0.1$)

and the enzyme binds only one substrate form

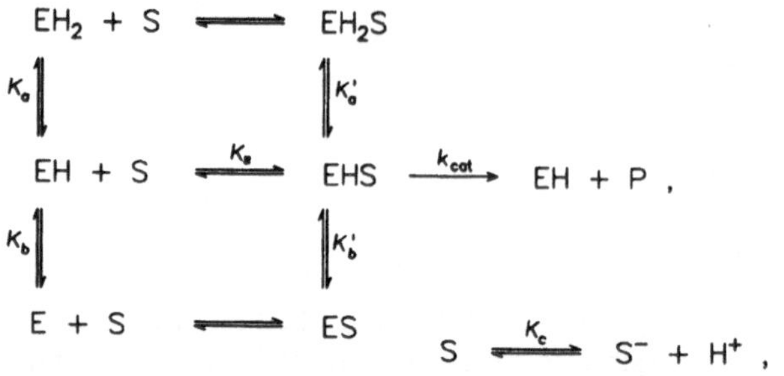

$$
\begin{array}{ccc}
EH_2 + S & \rightleftharpoons & EH_2S \\
K_a \big\updownarrow & & \big\updownarrow K_a' \\
EH + S & \xrightarrow{K_a} EHS & \xrightarrow{k_{cat}} EH + P \;, \\
K_b \big\updownarrow & & \big\updownarrow K_b' \\
E + S & \rightleftharpoons & ES \qquad S \xrightleftharpoons{K_c} S^- + H^+ \;,
\end{array}
\tag{12}
$$

the reaction rate is written as:

$$
v = \frac{\dfrac{\tilde{k}_{cat}}{1 + [H^+]/K_a + K_b/[H^+]}[E]_0[S]_0}{\tilde{K}_s(1 + K_c[H^+]) + [S]_0}.
\tag{13}
$$

5.2.2   Interpretation of Ionization Constant Values

To derive the equations for kinetic parameters vs pH the following assumptions are made: (1) groups behave as perfectly titrated acids and bases, whereupon a change in the ionization state of other ionogenic groups, free of the
substrate binding and its transformations, does not influence ionization of
these groups; (2) only one enzyme form is active; (3) the proton transfer runs
faster than the chemical processes; (4) the rate-limiting step of the overall
process is the same at all pH values; (5) the enzyme is stable at all pH [2299].
    Let us consider whether the conditions are feasible. The protein contains a
high number of ionogenic groups located usually on its surface. Ionization of
the groups may influence the pK_a's of these groups involved in the substrate
binding and catalysis [2300]. The pK_a's of catalytically active chymotrypsin
groups and its derivatives with various surface charges are listed below [1613,
p. 177]:

Enzyme	Type of modification	pK_a
Chymotrypsin	–	7.0
Succinylchymotrypsin	$E-Lys(NH_3^+) \longrightarrow E-Lys(NHCOCH_2CH_2COO^-)$	8.0
Ethylene diamine chymotrypsin	$E{\overset{\textstyle Asp(COO^-)}{\underset{\textstyle Glu(COO^-)}{<}}} \longrightarrow E{\overset{\textstyle Asp(CONHCH_2CH_2NH_3^+)}{\underset{\textstyle Glu(CONHCH_2CH_2NH_3^+)}{<}}}$	6.1

Effects of the surface charge are mostly pronounced at low ionic strength of the
solution. Note that succinylation almost does not change chymotrypsin conformation [2301].

Usually only one enzyme form is active. Still, sometimes two or more maxima are observed on the curves of the pH-dependence, which can be interpreted as a result of the activity of quite a few ionic forms. This may be associated with the presence of another enzyme in the preparation active at other pH or with the activation processes by anion (for example, chloride), etc.

The condition for fast establishment of prototropic equilibrium is, as a rule, met. The most intricate question concerns the constancy of the rate-limiting stage.

For the simplest scheme

$$E + S \underset{k_{-1}}{\overset{k_1}{\rightleftharpoons}} ES \overset{k_2}{\longrightarrow} E + P \,,$$

following the Briggs-Haldane kinetics ($k_{-1} \approx k_2$), rather than the Michaels kinetics ($k_{-1} \gg k_2$), a certain problem might arise due to a change in the relative magnitude of k_{-1} and k_2 at different pH. It turned out [2297, 2298, 2300, 2302] that the pK_a of *free* chymotrypsin for a fast cleavable substrate – *p*-nitrophenyl ester of acetyl-*L*-tryptophan – is 0.3 unit lower than the pK_a from the hydrolysis of ethyl ester or amide of acetyl-*L*-phenylalanine (by definition pK_a of free enzyme should be independent of the substrate type) (Fig. 74).

This is because at high pH for *p*-nitrophenyl ester $k_2 > k_{-1}$, whereas at lower pH k_2 becomes lower than k_{-1}, i.e., at high pH $k_{cat}/K_m = k_1$, and at low pH $k_{cat}/K_m = k_2/K_s$.

The experimental value of the ionization constant (K_k) here is

$$K_k = K_a \frac{k_2 + k_{-1}}{k_1} \,, \tag{14}$$

where K_a is a true ionization constant of the enzyme group.

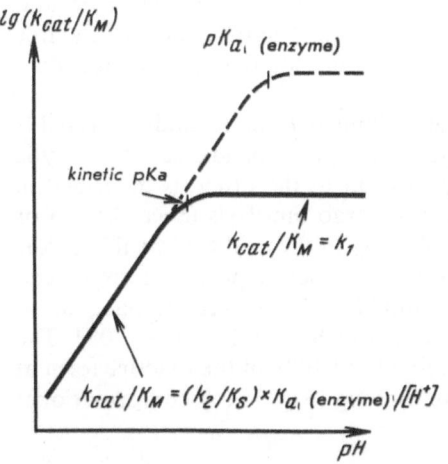

Fig. 74. Plot of differences between true and "kinetic" pK_a by the Briggs-Haldane mechanism followed by a change in rate-determining stage with pH

In the case of three-stage scheme:

$$E + S \underset{k_{-1}}{\overset{k_1}{\rightleftharpoons}} ES \xrightarrow{k_2} EP_2 \xrightarrow{k_3} E + P_2$$
$$+$$
$$P_1$$

a change in pH alters the rate-limiting stage. If the two stages are controlled by groups with different pK_a's (or pK_a of groups changes, when converting from one stage to another), the pH-dependence of k_{cat} and K_m seems very sophisticated. However, a change in the rate-limiting stage should not perturb the pH dependence for k_{cat}/K_m.

If one or more enzyme-substrate complexes are formed along the pathway from the substrate to the product, for example:

$$E + S \underset{k_{-1}}{\overset{k_1}{\rightleftharpoons}} ES \underset{k_{-2}}{\overset{k_2}{\rightleftharpoons}} E^{\bullet}S \xrightarrow{k_3} E + P ,$$

the kinetic parameter will be

$$\frac{k_{cat}}{K_m} = \frac{k_3 K_P}{K_s} , \tag{15}$$

where $K_P = \dfrac{k_2}{k_{-2} + k_3}$.

The pK_a value of the free enzyme can also depend on the substrate type, if a change in pH influences differently the K_P for various enzyme-substrate complexes.

The pK_a from the pH dependence of k_{cat} and K_m might differ from the true pK_a values of ionization of the groups of the enzyme-substrate complex, if the major portion of the substrate is bound nonproductively. In this case the pK_a (pK_a') is [2000]

$$pK_a' = pK_a - \lg \frac{1 + K}{1 + K'} , \tag{16}$$

where K and K' are the equilibrium constants between productive (HES and ES) and nonproductive (HES' and ES') complexes. The pK_a value of the free enzyme (from the pH-dependence of k_{cat}/K_m) is not altered upon nonproductive binding.

Many amide hydrolases are unstable at limit pH values under reversible and irreversible conformational changes. The pH-dependence of catalytic constants, when passing from one enzyme form to the other, is described in [2303]. Besides, proteolytic enzymes can undergo autolysis (Sect. 5.11). For instance, most aspartic proteases are stable only in the acidic or weakly acidic medium, being irreversibly inactivated at pH > 6. Serine proteases are reversibly inactivated in the strongly acidic medium [1375]. A reversible inactivation of chymotrypsin has been thoroughly studied at high pH [2304–2308]. The solution of the enzyme appeared to contain about 10% of the inactive form at pH 6–8, losing the ability for a substrate binding (Fig. 75). A conformational

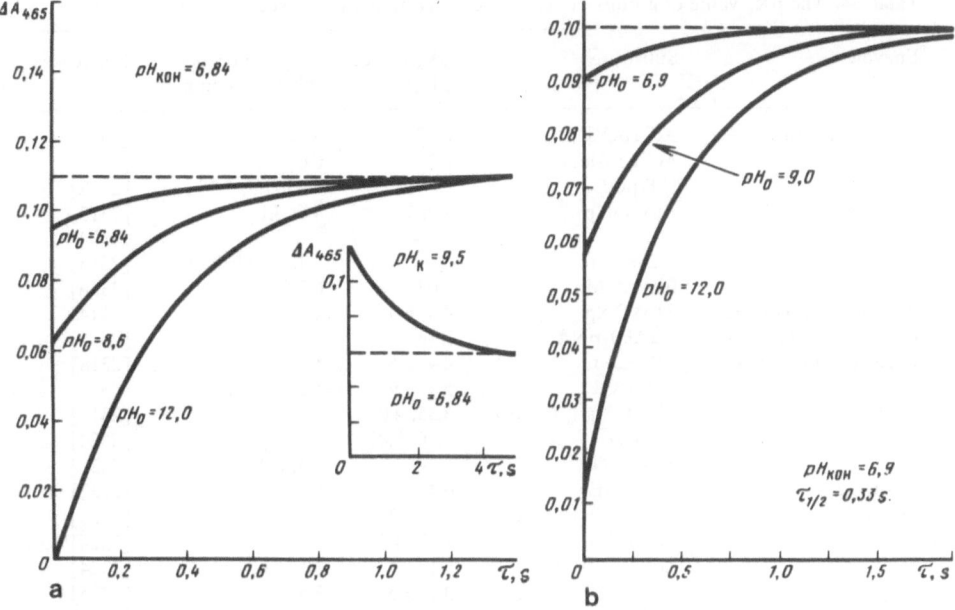

Fig. 75a, b. Kinetics of proflavin binding by chymotrypsin at various pH illustrating the presence of inactive enzyme from [2306]: **a** α-chymotrypsin (*insertion:* proflavin solution, pH 9.5, added to the enzyme, pH 6.84); **b** δ-chymotrypsin. The enzyme was incubated in buffer with pH_0, mixed with proflavin in concentrated phosphate buffer with pH_{KOH}; difference absorbance A_{465} was measured. *Dotted line* indicates rapid association of proflavin with the enzyme active form. Kinetic curves show the rate of conversion of the inactive form to the active

transition of the active form of chymotrypsin into inactive is stipulated by a breakdown of salt bridge Ile16-Asp194. The Ile16 amino group has pK_a 7.8, in the active form the value is much higher (≈ 10). The substrates and substrate-like inhibitors shift equilibrium to the active form.

The rate of transition from an active conformation into the inactive one is relatively small ($k = 0.6$ s^{-1} at pH 6.3 and 25 °C [2306]), therefore it can effect the experimental kinetic parameters of the substrate hydrolysis.

Other examples of conformational changes from different pH can be found in studies on bromelain [2309], clostripain [743], carboxypeptidase Y [2310], etc.

Thus pH-dependence often provides apparent pK_a's of the enzyme groups to these or the other functional groups, their assignment seems rather a complex task. The more so as pK_a values in the protein can strongly differ from those in model compounds.

5.2.3 Experimental Data

According to Table 54, the substrate binding, as a rule, slightly changes the pK_a of catalytically active groups of amide hydrolases. For free enzymes pK_a's

Table 54. The pK_a value of groups of active sites of certain amide hydrolases

Enzyme	Substrate[a]	pK_a of free enzyme	pK_a of enzyme-substrate complex	References
α-Chymotrypsin	AcTrpOEt	6.77	6.86	[2311]
	AcPhe-AlaNH$_2$	6.8	6.6	[2298]
	HTrpOEt	–	5.45; 7.5; 7.25	[2312]
β-Trypsin	ZLysONp	4.5; 7.1	4.6; 7.1	[2313, 2314]
	HLysOMe	–	5.42; 6.57; 7.25	[2315]
	BzAlaOMe	7.08	6.7	[2316]
β-Kallikrein, pancreas	ZLysONp	4.4; 6.9	4.4; 6.9	[2314]
Elastase, leukocyte	Ac(Ala)$_3$pNA	7.16	6.91	[2317]
Carboxypeptidase Y	ZPheLeu	4.4; 6.5	5.4; 7.7	[2318]
	AcPheOEt	5.9; 8.9	5.2; 9.1	„
Papain	ZLysONp (pH < 6)	3.35; 4.0	–	[2319]
	ZAlaOP	3.56; 8.07	–	[2320]
	BzArgpNA	–	4,3; 8,2	[2321]
Clostripain	BzArgOEt	6.4; 8.35	8.25	[2322]
Pepsin	ZPhe(NO$_2$)-PheApm	2.9; 4.2	4.3	[2323]
	AcPhe-TyrNH$_2$	1.17; 4.35	1.35; 4.15	[2324]
	AcPhe-TyrOH	1.17	1.12; 3.7	[2324]
	ZHisPhe-PheOEt	3.7; > 5	3.5;5.2	[2325]
Chymosin	ZHisPhe-TrpOEt	2.9; 5.4	–	[2326]
Carboxypeptidase A	ZGlyGly-Leu	6.19; 9.11	–	[2327]
	NBz-PlaOH	9.4	6.28; 9.3	[2328]
Thermolysin	FALeuNH$_2$	6.2; 7.8	–	[2329]
Leucineaminopeptidase (lens)	HLeupNA (Mg$^+$)	8.8	–	[2330]
Aminopeptidase *Aeromonas*	HLeupNA	4.9; 7.55; 10.0; 10.7	5.3	[2331]
β-Lactamase II	Benzylpenicillin	5.6; 7.6	5.25; 8.2; 9.8	[2332]
β-Lactamase *Pseudomonas*	Cephalosporin	5.5; 9.8	4.2; 8.3; 10.7	„

[a] ONp – *p*-nitrophenyl ester; OP – Phenyl ester; pNA – *p*-nitroanilide; Apm – N-(3-morpholine)-propylamide; NBz – *p*-nitrobenzoyl; FA – furylacryloyl; the pointer shows on cleavable amide bond.

depend little on the substrate type, with the exception of a few cases, which can be hardly explained by different assay conditions. Pepsin is one of them. So in the case of the free enzyme the pK_a values differ much for the substrates containing the His residue and neutral substrates. The same is observed for peptides GlyGlyPhePheOEt, BzLysPhePheOEt, etc. [1878, 2333] as well as for peptides with the aminopropylmorpholine C-protection group [2323]. None of these compounds alter ionization within the pH values under study.

The pK_a's of pepsin as well as of some other aspartic proteases apparently do not characterize the ionization of the active site groups [2334].

The pK_a differences for an uncomplexed pepsin and the substrates containing a free or a protected C-terminal carboxyl group depend on the ionization of the latter at pH $\approx$ 3.5 [2335–2337]. An increase in the dissociation constant of the enzyme-substrate complex is observed in this case [2338]. It is note-worthy that the phosphate group in pepsin influences the pK_a's of the enzyme [1216, 2339]. The pK_a's of the enzyme complex with AcPheTyrOMe in dephosphorylated pepsin are 2.1 and 4.6, while for the native enzyme they are 1.6 and 4.1 [1216].

Table 54 presents one pK_a value for free chymotrypsin, though in fact another value is observed between 8.8–9.0, which depends on a conformational transition into an inactive form discussed earlier. Interestingly, δ-chymotrypsin is more stable in alkali medium than α-chymotrypsin [2066, 2306].

The studies performed within a wide pH range have shown that a change in the rate-limiting stage occurs quite often. When hydrolyzing p-nitrophenyl esters of acylamino acids by trypsin [2313] and pancreatic kallikrein [2340] at acidic pH an overall rate is limited by deacylation, at pH > 4.4–4.8 by acylation. For leukocyte elastase [2317] and papain [2341] a slow stage is acylation at acidic pH, an increase of the basicity makes deacylation of the enzyme such a stage.

Complications might occur when defining the pK_a of catalytically active groups, due to wrongly chosen conditions. For example, determination of the pK_a of chymotrypsin at low pH (5–6) and at high enzyme concentrations entails errors as these conditions are much favorable for the enzyme dimer-ization affecting the k_{cat} value [2298]. The pK_a of the Cys25 residue of papain [2342–2347] depends on whether its SH group is incorporated into a salt bridge with His159 or not. At the same time, pK_a of the His159 residue also depends on the presence or absence of the salt bridge. Kinetic studies provide no distinction between two states bearing the same charge:

$$R—S^- \dots HIm^+ \quad И \quad R—SH \dots Im \,.$$

Therefore physicochemical and chemical modification methods are widely applied to tackle the problem of ionization of catalytically active papain groups. NMR spectroscopy [2348–2352], Raman spectroscopy [2353–2355], differential UV-spectroscopy [2356, 2357] as well as fluorescent analysis [2356, 2358] are of much value among physicochemical methods for determin-ing the pK_a of catalytically important enzyme groups.

Utilization of chemical modification methods has been described [2359–2363]. Influence of pH on the conformational state of the enzyme can be revealed from X-ray analysis of crystals grown in media differing in basicity [2352].

5.3   The Influence of Ions and Ionic Strength of the Solution

There should be distinguished a specific and an unspecific action of ions on the enzymes. The specific action is the action of only one type of ions (anions or cations), whereas the unspecific action is caused by ions of various nature due to changes of ionic strength of the solution and usually related to ion lyotropic properties. Many amide hydrolases contain constitutive ions being prosthetic groups. Their influence is outlined in other sections.

5.3.1 Specific Ions

A number of cations and anions can activate or inhibit amide hydrolases. Ca^{2+} and Mg^{2+} ions are the most widespread activators. Ca^{2+} ions bind to such serine proteases as chymotrypsin, trypsin, elastase, etc. They usually do not influence the activity of these enzymes, but stabilize them as regards an autolytic cleavage [2364–2371]. Below are listed the association constants of some serine proteases with Ca^{2+} ions [2368–2371]:

Enzyme	$K \cdot 10^{-4}, M^{-1}$*	Enzyme	$K \cdot 10^{-4}, M^{-1}$*
α-Chymotrypsin	0.92	Thermomycolin	50, 0
Porcine trypsin	22.0	Subtilisin BPN'	10^6
Bovine „	11.0	Thermitase	10^8
Elastase	21.0	Proteinase K	10^3
* The values are for the most strongly bound Ca^{2+}-ion			

Studies of Ca^{2+} binding are properly performed by replacing this ion by certain ions of lanthanides (Tb^{3+}, Ln^{3+}, Gd^{3+}) and the following recording of fluorescence of the added ions (Tb^{3+}) or a shift of lines in NMR spectra [2372–2374]. These studies and X-ray analysis have revealed [1233, 2375] that Ca^{2+} in trypsin binds in the vicinity of residues 70–80 forming a polypeptide loop, here Glu80 serves as an ion acceptor. Residue Ile is located in this position in chymotrypsin, which accounts for the lower affinity of Ca^{2+} for this enzyme. Subtilisins and related enzymes bind up to 2–4 Ca^{2+} ions with a different stability degree.

It has been found that Ca^{2+} activates cationic chymotrypsin and trypsin from sheep pancreas [276], acrosin [335], thermomycolin [381], and other serine proteases.

Among cysteine proteases, along with the enzymes activated by Ca^{2+} ions such as clostripain [743], there is a group of proteases (calpains), which display the absolute Ca^{2+}-requirement, i.e., to be almost inactive in the absence of these ions [2376–2379] (see Table 12). Myofibrils of smooth muscles contain these proteases. They are not lyzosomal enzymes and, apparently, are involved in catabolism of proteins of muscle tissue.

There exist at least two types of calpains – calpain I activated by micromolar concentrations of Ca^{2+} (physiological concentration is ≈ 10 μM[2380]) and calpains II that require millimolar concentrations of this ion. The two types consist of two subunits – catalytic (about 80 000 MW) and regulatory (28 000–30 000 MW). As shown, both heavy and light subunits contain four potential Ca^{2+}-binding sites each [886, 892], while there are two types of the regions – with high and low affinity to Ca^{2+} [2381]. The regions are accommodated in the C-terminal domain IV of the four-domain heavy subunit, and in domain II also the C-terminus of the two-domain light subunit. Ca^{2+}-binding sites have a typical "E–F" structure [2382]: Asp(Glu)-(Xaa)-(Asp)Glu-(Yaa)$_{6-8}$Glu. The catalytic site of calpains resides in domain II of the heavy subunit. The domain can interact with domain IV, this results in derepression or activation of the enzyme, as in the case of activation by calmodulin. An interaction of this domain with the light subunit is another possible mechanism of the activation. Note, autolysis of the light subunit enhances the concentration of Ca^{2+} necessary for the activation of calpain I [2383, 2384]. It is still obscure, whether this phenomenon and the existence of two types of calpains are related. The repression-derepression mechanism probably functions due to the presence of an endogenic inhibitor and the activator of calpain in the protein cells [2385, 2386]. The activator (M_r 40 000 MW) acts as an antagonist to the inhibitor, thus enhancing the enzyme affinity to Ca^{2+} 100 times [2387]. Ca^{2+}-ions can also influence association of protomers in oligomeric enzymes [2388].

The effect of Ca^{2+} on the properties of metal-containing endopeptidase thermolysin has been studied in detail [2389–2301]. Ca^{2+}-ions are of importance for maintaining thermostability of this enzyme [2390]. Thermolysin binds four ions of Ca^{2+} ($K_{1,2} = 2.8 \times 10^9$ M^{-1} and $K_{3,4} > 10^6$ M^{-1}) [2392]. Two ions are located close to each other inside the molecule near the protein residues Asp138, Glu177, Asp185, Glu190 and Asp191, the third, and possibly the fourth, ion reside close to the surface within the residues 196–200. The ion nearest to the surface is apart from the enzyme active site at a distance of 13 Å [2393, 2394]. Ca^{2+} ions also activate aminopeptidase A [979], collagenase [1095] and some other metal-containing proteases.

Many Zn^{2+}-containing aminopeptidases are activated by Mg^{2+} or Mn^{2+} [951, 954]. Mg^{2+} ions bind to leucineaminopeptidase from bovine lens with K_D 2×10^{-6} M. An increase of k_{cat} at hydrolysis of $LeuNH_2$ 1.1×10^5 to 1.19×10^6 min^{-1} without changing the K_m is observed [954]. The activation of leucine-aminopeptidase evolves in time (3–6 h, at 30 °C). Anions display a synergistic effect during metal-cation activation [2395, 2396].

Presumably Na^+- and K^+-ions act very specifically on neutral protease from bovine pituitary [805] and alkaline protease from carp muscles [197], thus inhibiting these enzymes.

Ions of heavy metals act specifically on cysteine amide hydrolases. This was established by Krebs as early as in 1930 [2397]. Mercury ions are the most active inhibitors of cysteine proteases. Such metals as Cu^{2+} or Ag^+ inhibit certain serine proteases. They selectively obstruct acylation of serine proteases by specific and unspecific substrates, but show no influence on deacylation of acylenzyme [2398]. Chymotrypsin is also inhibited by ions Pb^{2+} [2399], Zn^{2+}, Ni^{2+}, and Cd^{2+} [2400].

Not only cations, but anions too specifically influence the activity of amide hydrolases. It is noteworthy that Cl^- activates some amide hydrolases – dipeptidylcarboxypeptidase (angiotensin converting enzyme) [2401], formylmethionineaminopeptidase [823], arginylaminopeptidase [682, 2402], cathepsin C [2403] and yeast aminopeptidase I [957]. In the first case Cl^--ion affects V_{max} and K_m, the action appearing to depend on the type of substrate

used [2404]. As shown [2405], Cl⁻ changes spectral characteristics of dipeptidylcarboxypeptidase, which is perhaps associated with the enzyme conformation. This ion interacts with the lysine residue [2406]. Sometimes it acts as an
allosteric activator (see Sect. 5.10) and can be replaced by Br⁻, I⁻, CNS⁻, etc.
[957]. The specific action of phosphate ions on certain amide hydrolases
should be mentioned. So phosphate ion evokes significant changes in the
specificity of *Staphylococcus* protease [1852]: the enzyme in phosphate buffer
splits bonds formed by Asp or Glu residue, and in ammonium buffer only the
bonds formed by glutamic acid. Polyphosphates (including ATP) increase the
catalytic activity of cathepsin D [2407]. This effect can be partially explained
by a change of ionic strength in solution. Inorganic phosphate also affects the
glutaminase activity [2408, 2409], providing the allosteric activation for this
enzyme (see also Sects. 5.9 and 5.10).

5.3.2 The Effect of Ionic Strength of the Solution

Addition of salts may alter enzyme catalytic properties due to an enhancement
of ionic strength of the solution [1404, Chap. 7; 2410]. This has been thoroughly studied for serine proteases – chymotrypsin and trypsin [2411, 2415].
A change of the NaCl concentration has entailed a proportional increase of
k_{cat} and decrease of K_m when hydrolyzing ethyl ester of acyl-*L*-tyrosine by
chymotrypsin (Fig. 76) [2415]. Interestingly, the ionic strength is more effective
for the chymotryptic hydrolysis of the trypsin-specific substrates like BzArg-
OEt (k_{cat}/K_m is eight times higher) as compared with hydrolysis of the
chymotrypsin-specific substrates (three times); the pH-dependence of kinetic
constants actually does not change due to an increase of ionic strength at pH
6–10. At more acidic pH values (<5) a reverse of the effect of ionic strength on
k_{cat} (by hydrolysis of AcTyrOEt) for chymotrypsin is observed. In the case of
trypsin, the effect of ionic strength is opposite, i.e., an increase in the salt
concentration entails a lowering of k_{cat}/K_m (for BzArgOEt twice) and an
increase of K_m [2415]; a pH change within 5–7 displays no effect. In the case of
salt concentration for chymotrypsin this influences mainly the rate constant of
the enzyme acylation (k_2) and less the rate constant of deacylation (k_3).

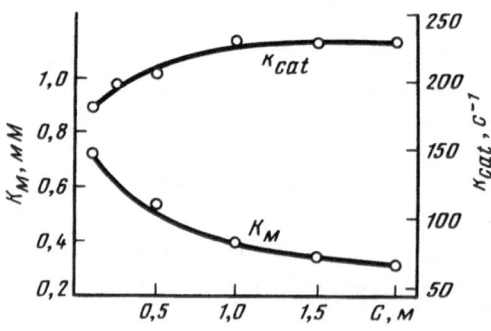

Fig. 76. Kinetic constants of
chymotryptic hydrolysis of N-
acetyltyrosine ethyl ester vs NaCl
concentration (pH 7.9; 25 °C) [2415]

Various cations align, according to their effect, identically to the Hoffmeister lyotropic row:

$$Na^+ > Li^+ > K^+ > NH_4^+ > Me_4N^+ ,$$

and anions are in the following row:

$$SO_4^2 > Cl^- > Br^- > AcO^- > NO_3^- > I^- > ClO_4 > CNS^- .$$

The influence of salt on the dissociation constant of the enzyme-substrate complex, at least for chymotrypsin, can be explained by the so-called effect of salting off, i.e., by the influence on the substrate activity (γ_s) at constant enzyme (γ_E) and the enzyme-substrate complex (γ_{ES}) activity [2416]. This is true at relatively high salt concentrations. When passing from low salt concentrations to high (>0.5 M), the enzyme conformation evidently changes [2416, 2417], which is manifested in a parallel change of absorption at 282 nm and of catalytic activity [2416].

Carboxypeptidase A, unlike chymotrypsin, is inhibited by anions, the efficiency of inhibition here decreases:

$$PO_4^{2-} > SO_4^{2} > N_3^- > Cl^- .$$

The phosphate and chloride ions are supposed to interact with Arg residue in the substrate-binding site [2418]; ion N_3^- interacts with the Zn-atom [2419]. It is noteworthy that a high concentration of Zn^{2+}-ions suppresses the activity of carboxypeptidase A [2420] and some other Zn-containing proteases.

5.4 The Effect of Organic Solvents

Organic solvents can influence the catalytic activity of enzymes including amide hydrolases by changing the dielectric permeability of the solution as well as the coefficient of the substrate distribution between phases of the solvent and the protein and also by specific competition with the substrate for the active site. Besides, the solvents containing nucleophilic functional groups can compete with water in hydrolytic reactions. This part deals only with the first two effects.

A change in the dielectric permeability influences the ionization of buffer components and pK_a of the surface protein groups (see review [2421]). This in turn results in a change of pK_a of groups masked inside the globule, including the groups of the enzyme active site. A change of dissociation constants of the carboxyl and amino groups depending on the amount of an organic solvent in water may reach a few orders of magnitude (Fig. 77) [2422, 2423].

A variety of data are available regarding the influence of the dielectric permeability on catalytic characteristics of enzymes (see reviews of earlier works [2424, 2425, Chap. 15]). So the linear correlation of the rate of the trypsin-catalyzed hydrolysis of BzArgOEt and a reverse value of dielectric permeability ($1/D$) has been found for a series of water-ogranic solvents mixtures [2426]. A change in the medium composition influences k_{cat} only

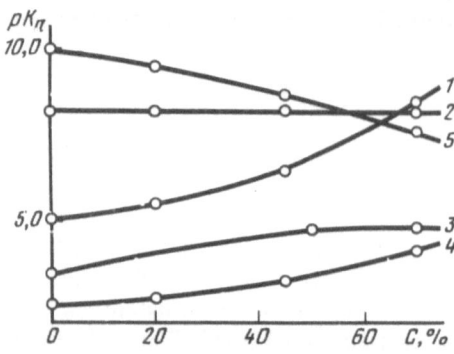

Fig. 77. pK_a of inorganic groups of some compounds vs dioxane concentration in water; *1* acetic acid; *2* Tris; *3* benzoylarginine; *4* glycine (COOH); *5* glycine (NH$_3^+$)

slightly, but alters much the K_m [2423, 2427, 2428]. The pH-dependence of k_{cat} in water and water-organic mixtures are almost the same [2423]. Some proteases were shown to lose amidase, but to maintain esterase activity in water-organic solvents [2429]. Studies on the influence of organic solvents on the kinetics of the enzymatic hydrolysis are perturbed due to specific effects [2427, 2430] we have discussed earlier, as well as due to a narrow range of changes of dielectric permeability mostly used. Studies on hydrolysis of AcTrpOEt and BzArgOEt catalyzed by chymotrypsin and trypsin, respectively, in a large interval of the values of dielectric permeability have revealed no correlation of $1/D$ and k_{cat} and K_m values. In systems where electrostatic enzyme-substrate interactions prevail changes of dielectric permeability do not influence K_m. At the same time, an appropriate correlation of lgk_{cat}/K_m and the value of free energy of the substrate transfer from water into the mixed solvent is observed. The organic solvent effects K_m only when hydrophobic interactions make their contribution to the formation of the enzyme-substrate complex [2431]. At high concentrations of the organic solvent the studies become complex due to processes of denaturation. The reaction performed with the suspended enzyme in a dry organic solvent runs without complications.

As shown [2032, 2432–2435], chymotrypsin, subtilisin and a series of other enzymes maintain their activity relative to such reactions as, for example, transesterification. Here unusual effects are observed: a loss of the enzyme enantioselectivity [2032], an ability for discrimination between polar and hydrophobic substrates [2432], a perturbation of specificity to competitive inhibitors [2433] as well as the appearance of ligand-induced "memory" of the enzyme during its preliminary treatment by a ligand [2435]. The enzymes become more stable to temperature and storage [2433].

5.5 The Kinetic Isotope Effect of the Solvent

Studies of the kinetic isotope effect of the solvent yield important information on the reaction mechanism and the structure of a transition state (see reviews

Table 55. Influence of D_2O on enzymatic hydorlysis

Enzyme	Substrate	pH	$\dfrac{k_{cat}\ (H_2O)}{k_{cat}\ (D_2O)}$	$\dfrac{K'_m\ (H_2O)}{K'_m\ (D_2O)}$	References
α–Chymotrypsin	$CH_3COOC_6H_4NO_2$-p	7.43	2.40	–	[2441]
	AcTyrOEt	8	1.8	–	[2311]
	AcTrpNH$_2$	8	1.9	–	[2311, 2441]
Trypsin	$CH_3COOC_6H_4NO_2$-p	7.5	1.38	–	[2441]
	BzArgOEt	7.5–8.5	3.03	2.8	[2441, 2443]
	BzPheValpNA	7.44	4.3	–	[2441]
Elastase	$CH_3COOC_6H_4NO_2$-p	7.5	2.45	–	"
Thrombin	BzArgOEt	8.0	2.92	–	"
Papain	Cinnamoylimidazole	6.5–8.5	3.35	–	[1984]
Pepsin	AcPheTyrOMe	2.5	1.05	–	[2444]
	(Gly)$_3$Phe(NO$_2$)OMe	4	2.0	1	[2445]
Carboxypeptidase A	BzGlyPhe	7.47	1.33	1.55	[2446]
Asparaginase (*Escherichia coli*)	Asn	7.12	2.93	–	[2447]
Gultaminase (*Escherichia coli*)	Gln	5.5	1.8	–	"

[1405, 2436]). For enzymatic reactions in D_2O there occurs a change of kinetic parameters not only due to changes of the proton transfer at the rate-limiting stage, but also due to changes of the protein solvation, conformation of its molecule, etc. A replacement of H_2O by D_2O results in a shift of the pH value of the solution so that pH and pD have an empirical relationship

$$pH = pD + 0.4 .$$

At the same time the pK_a of ionized groups is higher in D_2O. These two effects can compensate each other.

Table 55 shows that a kinetic isotope effect is 2–3 for the majority of amide hydrolases under study. Contradictory data have been obtained for pepsin catalysis, where certain substrates do not alter k_{cat} in D_2O, while the others display an almost usual isotope effect.

The shift of pK_a of catalytically active groups is 0.5 ± 0.2 units. The effect of D_2O on K_m develops so that the value decreases, i.e., the enzyme-substrate complex (when $K_m = K_s$) becomes more rigid. This depends at least partially on an increase of hydrophobic interactions in D_2O.

Studies on the dependence of the reaction rate on the portion of D_2O in H_2O (proton inventory) have found a wide application [1504, 2437, 2442]. They allow a conclusion about the number of protons being transferred in a transition state. Figure 78 presents a typical dependence.

The investigation of the temperature dependence of the kinetic isotope effect provides important information. As shown [2443, 2448], upon hydrolysis of ester substrates by chymotrypsin and trypsin the energy of activation remains the same in D_2O and H_2O, the kinetic isotope effect is totally stipulated by a change of the pre-exponential term. This is interpreted in favor of a quantum-mechanical model of the proton transfer for the enzymatic catalysis, which includes proton tunneling (Chap. 3).

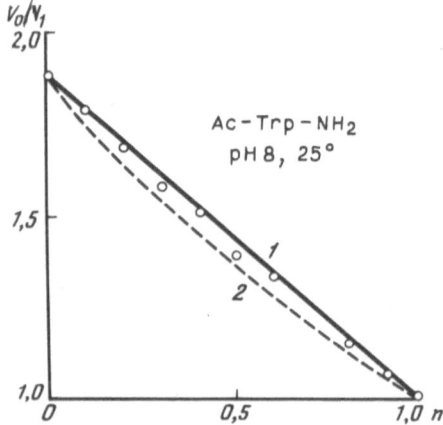

Fig. 78. Typical dependencies of the ratio of rates of enzymatic reactions in water (v_0) and in mixture D_2O–H_2O (v_1) on portion (n) of D_2O in H_2O [2439]; *1* one-proton catalysis; *2* catalysis with transfer of several protons

5.6   The Effect of Temperature and Pressure

5.6.1   Theory

As for every chemical reaction, the dependence of the rate constant on the temperature in the reactions catalyzed by amide hydrolases is written by the Eyring-Polany equation (see reviews [1661, 2449]):

$$\ln k = \ln \frac{kT}{h} + \frac{\Delta S^+}{R} - \frac{\Delta H^+}{R} \cdot \frac{1}{T}, \tag{17}$$

where k is the Boltzman constant, T is the absolute temperature, h the Planck constant, R the universal gas constant, S^+ the entropy and H^+ the enthalpy of activation.

The temperature also influences the dissociation constant of the enzyme-substrate complex. In accordance with the Vant-Hoff equation

$$\ln \bar{K}_s = \frac{\Delta S_0}{R} - \frac{\Delta H_0}{R} \cdot \frac{1}{T}, \tag{18}$$

where $\bar{K}_s = 1/K_s$, and ΔS_0 and ΔH_0 are the entropy and enthalpy of the complex formation.

The second-order rate constant k_2/K_s is thus related to the temperature by expression

$$\ln \frac{k_2}{K_s} = \ln \frac{kT}{h} + \frac{\Delta S^+ - \Delta S_0}{R} - \left(\frac{\Delta H^+ - \Delta H_0}{R} \right) \cdot \frac{1}{T}. \tag{19}$$

Plotting the logarithms of kinetic constants against the reverse temperature, values ΔS^+, ΔS_0, ΔH^+ and ΔH_0 may be defined.

The peculiarities of the temperature dependence of the enzymatic reactions, as compared with the nonenzymatic ones, are as follows: (1) the values of enthalpy and entropy of the enzymatic process seem sensible only if the corresponding kinetic constants are related to the elementary step of chemical transformation of the substrate or to the true equilibrium constant (this is also true for a nonenzymatic reaction, but in this case it is rather easy to check how the indicated requirement is realized); (2) the dependence of kinetic constants on the temperature described by Eqs. (17)–(19) is followed for enzymatic reactions within a relatively narrow range of the temperatures determined by the freezing point of the solution and the enzyme stability to the heat denaturation; (3) the temperature can change pK_a of catalytically active enzyme groups, thus altering the reaction rate; (4) the Arrhenius plots for enzymatic reactions very often show break; the reasons are discussed below.

A suggestion was made [2450] that the Eyring-Polany equation and the empirical Arrhenius plot

$$\ln k = \ln A - \frac{\Delta E}{RT} \tag{20}$$

are inapplicable to the biopolymer reactions. The major argument here is that:
(a) ΔH^+ and ΔE are independent values, and the expression

$$\Delta H^+ = \Delta E - RT \tag{21}$$

shows that there exists a certain temperature T, when dependences $\ln k$ on $1/T$
described by Eqs. (17) and (20) have an identical slope; (b) the constancy of ΔS,
ΔA, ΔH, and ΔE in enzymatic reactions is apparent, which is stipulated by the
tightness of the temperature interval of measurements; (c) a change of temperature, even in the narrow interval, alters the biopolymer structure; (d) the
formation of the enzyme-substrate complex results in substantial rearrangements of the protein molecule.

Despite the soundness of this criticisms, the measurements of the kinetic
constants depending on temperature are often applicable to characterize
enzymatic reactions; they seem sensible if substrates of one type are compared.
Besides, of wide use are direct calorimetric methods for measuring heat effects
of the reactions freed of the numerous drawbacks listed here. The range of
temperatures suitable for measurements with use of cryoenzymological
methods is extending (Sect. 5.7).

The equilibrium of a reversible reaction and, consequently, its rate may
alter when the pressure changes. Experimentally, this can be observed at
pressures over 1000 atm. Studying the influence of the pressure on the rate
constants of enzymatic reactions, important information on their mechanism
can be received (see Sect. 5.6.2). Measurement of the pressure influence on K_m
yields a calculation of a volume change (ΔV) at the formation of the Micheails
complex; data on a change of the rate constant apparently provide the value of
the activation volume (ΔV^+) [2451]:

$$\ln k = \ln k_0 - \frac{\Delta V^+}{RT} P \, ,$$

where k_0 is the rate constant at $P = 0 \div 1$ atm.

The plots of the logarithm of the constant on the pressure give a straight
line (if ΔV^+ does not depend on the pressure) with a slope $-\dfrac{\Delta V^+}{RT}$. The
activation volume is related to the entropy of the activation and characterizes
a structural change of the transition state as compared with the ground state of
the reacting system.

5.6.2   Experimental Data

Analysis of data from Table 56 reveals that the enzyme specificity can be
determined by a change of enthalpy and entropy of the activation. A relative
contribution of each parameter depends on catalyst type and a series of the
substrates chosen. So the rate of deacylation of N-acylaminoacylchymo-
trypsins is established mainly by a change of enthalpy of activation (Fig. 60).
For the corresponding enzyme derivatives acylated by aliphatic acids a change

Table 56. Thermodynamic activation parameters of enzymatic hydrolysis

Enzyme	Substrate	Kinetic parameter	$\Delta H^{\pm}$, kcal/mol	$\Delta S^{\pm}$, e.u.	References
α-Chymotrypsin	$CH_3COOC_6H_4NO_2\text{-}p$	k_2	20.5	+13.5	[2452]
		k_3	15.0	−18.0	[2452]
		k_3	9.7[a]	−15.9[a]	[1987]
	AcPheOMe	k_2	16.9	+11.7	[2453]
		k_3	10.1	−14.8	[2453]
	AcTrpOMe	k_3	12.3	−10.0	[2454]
	Furoylacryloylimidazole	$k_2{}^b$	7.8	–	[2455]
		$k_3{}^b$	15.0	–	[2455]
		k_3	19.4	−5.8	[2456]
Trypsin	TosArgOMe	k_3	5.4	−31.4	[2457]
		k_3	9.77	−17.2	[2458]
Elastase	$CH_3COOC_6H_4NO_2\text{-}p$	k_2/K_s	10.2	−16.1	[2459]
	ZGlyONp	k_2/K_s	8.6	−9.98	[2459]
	BocAlaONp	k_2/K_s	8.53	−7.49	[2459]
Subtilisin	AcPheOMe	k_2/K_s	3.83	−23.9	[2019]
Pepsin	ZPhe(NO₂)PheApm	k_{cat}	7.8	−39.3	[1907]
	ZLeuValPhe(NO₂)-AlaApm	k_{cat}	3.4	−44.6	[1907]
Carboxypeptidase A	ZGlyGlyPhe	k_{cat}	12.6	−6.7	[2327]
	ZGlyGlyVal	k_{cat}	15.1	0	[2327]
Thermolysin	FAGlyPheAla	k_{cat}	8.6	−22.9	[2460]
		k_{cat}/K_m	1.5	−26	[2460]
Neutral proteinase *Bacillus subtilis*	FAGlyLeuAla	k_{cat}	3.8	−37.2	[2460]
		k_{cat}/K_m	5.7	−14.5	[2460]

[a] In Tris-buffer.
[b] Calorimetric data.

of entropy of the activation of deacylation occurs, while the enthalpy changes very little [1992]. Enthalpy "control" is observed at pepsin catalysis, whereas for elastase the contribution of entropy prevails.

The concept that the specificity is defined by lowering the number of rotational degrees of freedom for releasing a "good" substrate, i.e., one determined mainly by entropy of a ground state [1987], appears to be wrong. Data that underlie the concept are obtained when the buffer components (Tris) strongly influence the deacylation rate [1992, 2453].

Many experiments on the temperature dependence of kinetic constants show the break on the Arrhenius plots [1907, 2452, 2457, 2461, 2462]. These occur due to certain reasons [2463]: (a) a change of the rate-limiting step; (b) conformational changes of the enzyme; (c) temperature-dependent changes of ΔH or ΔS.

In the first case, if the reaction runs according to a routine three-stage scheme for serine and cysteine proteases

$$E + S \underset{k_{-1}}{\overset{k_1}{\rightleftharpoons}} ES \overset{k_2}{\longrightarrow} EA \overset{k_3}{\longrightarrow} E + P_2 ,$$
$$+$$
$$P_1$$

the transition from the rate-limiting step of acylation to that of deacylation must result in a break on graphs $\ln k_{cat} \div 1/T$, but the graph $\ln k_{cat}/K_m$ against $1/T$ may remain linear.

$$\text{E} + \text{S} \xrightleftharpoons[k_{-1}]{k_1} \text{ES} \xrightleftharpoons[k_{-2}]{k_2} \text{E}^{\bullet}\text{S} \xrightarrow{k_3} \text{E} + \text{P}$$

In the second case the two plots can give a break, because $k_{cat}/K_m = k_2 k_3 / K_s(k_{-2} + k_3)$ depending on the ratio between k_3 and k_{-2}, so the magnitude of the second-order rate constant is k_2/K_s or $k_2 k_3/K_s k_{-2}$. This is probably the case most often observed [1907, 2452, 2461, 2462, 2464].

The conformational changes that entail a change of the kinetic parameters of the enzymatic reaction have been registered in a number of cases (see [1661, 2449]). Various conformers show, as a rule, distinctive thermodynamic characteristics of binding and substrate transformation. For example, chymotrypsin changes its state at hydrolysis of p-nitrophenylacetate [2452], which is followed by the break on graphs $\ln k_2 \div 1/T$ (at 20.9 °C) and $\ln K_s/k_2 \div 1/T$ (at 20.1 °C). The values of $\Delta H_2^{\ddagger}$ and $\Delta S_2^{\ddagger}$ at low temperatures are 20.5 kcal/mol and $+13$ e.u., while at high temperatures -0.34 kcal/mol and -0.55 e.u. The activation parameters of deacylation do not depend on the temperature. However, for some p-nitrophenyl esters of 5-alkylfuroyl-2-carboxylic acids the dependence of $\ln k$ on $1/T$ is nonlinear, whereas for others the Arrhenius plots trace no break [2462]. For hydrolysis of AcTrpOEt (at pH 8) the transition between two forms of α-chymotrypsin (A_b and A_f) is observed at 25 °C [2461], while between two states in pepsin at 34 °C for hydrolysis of AcPheTyr [2465], and at 20 °C for hydrolysis of ZLeuValPhe (NO$_2$)AlaApm [1907]. Thus the temperature of the conformational transition depends much on the substrate structure.

It seems interesting to compare the parameters of the activation for thermostable thermolysin and closely related neutral protease *Bacillus subtilis*, which is more labile. The protease has higher enthalpy of activation (by k_{cat}/K_m) and less negative entropy than thermolysin; this is perhaps caused by a loose structure of the neutral protease [2460].

Finally, at high values of specific heat capacity change (ΔC_p)ΔH_0 and ΔS_0 are nonlinear temperature functions [2457, 2463, 2466], as

$$\ln K_s = \Delta H_0 \frac{1}{T} + \Delta S_0 + \frac{\Delta C_p}{R} \cdot \ln T .$$

The advantages of calorimetric measurements of heat effects of the reaction at constant temperature are evident. The heat effects of chemical stages are not often defined (for example, [2455]) and their comparison with data cn the temperature dependence of kinetic parameters seems yet difficult. Of more use are calorimetric measurements for determining enthalpy and entropy of equilibrium processes (association of the enzyme with the substrate and inhibitor) [2467, 2468]. Data obtained by the methods for the calorimetry and temperature dependence of dissociation constants of the enzyme-ligand complexes differ much. This is clear from the comparison of parameters ΔH_0 and ΔS_0 for chymotrypsin (Table 57).

Table 57. Thermodynamic parameters of complex formation of chymotrypsin (pH 7.8)

Ligand	Method	ΔH_0, Kcal/mol	ΔS_0, e.u.	References
Proflavin	Kinetic	−8.0	−6	[2469]
	Calorimetric	−11.3	−18	[2468]
Ac-L-Trp	Kinetic	−9.06	−21	[2469]
	Calorimetric	−21.2	−61	[2468]
Ac-D-Trp	Kinetic	−10.0	−23	[2469]
	Calorimetric	−19.0	−52	[2468]

Interestingly, the binding of the ligands (Table 57) by modified chymotrypsin preparation – $N^{\varepsilon 2}$-methylhistidine-57-α-chymotrypsin or anhydrochymotrypsin is characterized approximately by 10 kcal/mol less enthalpy and by 30 e.u. higher entropy [2469]. Thermodynamic parameters of binding are closely related to the values typical of transfer of aromatic compounds from water into nonpolar phase ($\Delta H_0 \approx 0$). Similar parameters are observed for binding hydrophobic compounds to trypsin [2470] and substrate binding to pepsin [1907].

As for catalytic constants, there occurs a nonlinear dependence of binding constants on temperature (see, for example, [2471]).

Measurements of the dependence of catalytic constants of amide hydrolases on the pressure are not very numerous (for example, [2460, 2472, 2473]). Changes of the volume of the substrate binding by thermolysin and neutral protease from *Bacillus subtilis* can be given as examples, in the case of the first thermostable enzyme they are $-20 \div -30$ cm^3 mol^{-1}, and in the second case they are close to zero [2460]. The ΔV^{+} value of chymotryptic hydrolysis of di- and tripeptide p-nitroanilides affords an assumption about the number of hydrogen bonds formed by the substrates in the transition state, where these results accord well with crystallographic data [2335].

5.6.3 Compensation Effect

We have already encountered the compensation effect in considering specific substrates of chymotrypsin, though it is inherent for many other processes studied [2450, 2474–2476].

A thermodynamic compensation effect is a linear dependence of changes of enthalpy (ΔH_i) and entropy (ΔS_i) in a series of similar processes that occur at a constant temperature, i.e.

$$\Delta H_i = \alpha + T_C \Delta S_i ,\tag{22}$$

where α and T_C are the constants, T_C having a temperature dimension and being termed the compensation temperature. So free energy in a number of the transformations under study remains constant. The coefficient

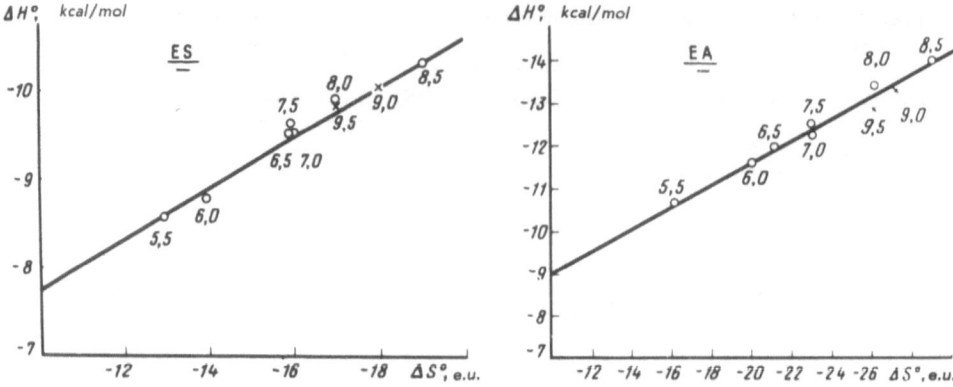

Fig. 79. Compensation plots of the formation of the Michaelis complex (*ES*) and acylenzyme (*EA*) during α-chymotryptic hydrolysis of N-acetyl-*L*-tryptophan ethyl ester at various pH [2476]

$\alpha = \Delta H_i = T_C \Delta S_i$ is a change of the free energy at the compensation temperature.

The compensation effect develops in many enzymatic reactions, for example, in chymotrypsin-catalyzed hydrolysis of ethyl ester of acetyl-tryptophan at different pH (Fig. 79) and many other cases.

Two types of compensation effects should be distinguished. The first, the so-called complete compensation effect, has a temperature of compensation close to that of the experiment. This type of effect is probably an artifact caused by close values of measured parameters (equilibrium and rate constants) within a relatively narrow range of temperatures, which inevitably leads to errors. Since the detected value, for example, the equilibrium constant

$$K = e^{\Delta S_i/R}\, e^{\Delta H_i/RT}\,, \tag{23}$$

is the product of two factors, compensation of ΔS_i and ΔH_i is always observed at $K \approx$ const.

However, there are compensation effects with such temperature of compensation as does not coincide with the temperature of measurements, and with a slight scatter of points. The explanation of these processes [2476] relates a compensation with a change (deformation and breakdown) of the structure of water near the reacting macromolecule. This point of view is not supported by some investigators, who regard that compensation effects of any type result from a "wrong application of certain equations of equilibrium thermodynamics" [2450, p. 120]. Thus, the nature of the compensation effect is an open question.

5.6.4 The Temperature Stability of Amide Hydrolases. Thermostable Enzymes

As already mentioned, the enzymatic reactions follow the Arrhenius equation (and other thermodynamic dependences) in a rather narrow interval of temperatures. Usually, at temperatures over 40–50 °C the reaction rate decreases much and the dissociation constants of the enzyme-ligand complexes increase. This is stipulated by phenomena of enzyme heat denaturation (Fig. 80) [2477, 2478].

Here the problems of the protein denaturation outlined in a series of comprehensive reviews [168, 169, 2479] are not considered. We cover only certain cases regarding amide hydrolases. The temperature stability of most animal amide hydrolases is confined within the range (40–50 °C) stated above. Plant enzymes, as a rule, are more stable (denaturation temperature is about 50–60 °C) [2480]. A large number of amide hydrolases produced by thermophilic microorganisms are known whose stability to temperature is rather high (Table 58) [2481–2483].

Thermostability of metal-containing proteases depends on the presence of metal in buffer solution. Thus, *Bacillus stearothermophillus* aminopeptidase stable in solutions containing Ca^{2+} is inactivated without this ion at 80 °C for 1 h by 65% [1011]. Thermolysin is very stable in the presence of Ca^{2+} ions [381]. Many inhibitors including substrate-like compounds [168, 2484] as well as ATP sometimes [2485] display a stabilizing action on amide hydrolases.

Another inactivation property typical of only proteolytic enzymes is an autolytic cleavage occurring effectively when the temperature increases (Sect. 5.11).

It is noteworthy that still no generally accepted concepts exist about the nature of thermostability of some proteins, eventhough intensive theoretical and experimental studies are carried out [2496–2500]. Two factors apparently dominate: these are hydrophobic interactions and conformational entropy of unfolding of the globule. Some attempts are made to enhance enzyme stability through the methods of site-directed mutagenesis [2501].

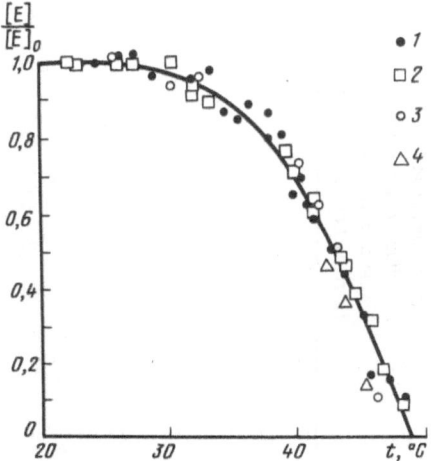

Fig. 80. Change of relative concentration of native chymotrypsin depending on temperature (30 min incubation) [2478]. Detection by proflavin binding (*1*), by initial rate of hydrolysis of N-succinyl-*L*-phenylalanyl-*p*-nitroanilide (*2*), by initial rate of hydrolysis of *p*-nitrophenylacetate (*3*), by changing the value of trough on ORD curve of chymotrypsin at 233 nm (*4*)

Table 58. Certain thermostable amide hydrolases

Enzyme	Source	Temperature of incubation, °C	Time of incubation, min	Residual activity %	Temperature optimum °C	References
Cucumisin	Melon Cucumis melo	70	10	≈90	75	[436]
Neutral protease	Bacillus subtilis	50	15	100	57	[2486]
Thermolysin	Bacillus thermoproteolyticus	80	60	50	–	[1102]
α-Lytic protease	Lysobacterium enzymogenus	50	480	30	–	[2487]
Aminopeptidase	Bacillus stearothermophillus	80	300	100	90	[1011]
"	Telaromyces duponti	65	240	50	65	[1012]
Thermomycolin	Mabranchea pulchella	73	110	50	–	[381]
Protease	Aspergillus ochraceus	55	10	90	40	[2488]
"	Aspergillus oryzae	50	60	100	58	[2489]
"	Cephalosporium acremonas	60	10	90	40	[2490]
"	Melanocarp. albomyces	60	20	100	55	[2491]
"	Mucor pusilus	55	15	100	55	[2492]
"	Myriococcus albomyces	65	15	95	55	[2493]
"	Penicillium duponti	65	15	100	70	[2494]
"	Penicillium roqueforti	80	20	90	50	[1113]
"	Torula thermophila	60	60	100	55	[2495]
Thermitase	Thermoactionmyces vulgaris	55	120 (CaCl₂)	120	80	[394]

The temperature stability of amide hydrolases and many other enzymes can be increased through immobilization, i.e., upon chemical or sorption binding to insoluble carrier (see review [2502]). As an example, entrapment of α-chymotrypsin in the structure of cross-linked polyacrylamide gel can be listed. An enzyme preparation enduring boiling in water has been obtained [2503].

5.7 Cryoenzymology of Amide Hydrolases

If the top level of the temperature stability of enzymes can be set aside by immobilization methods, the bottom range associated with the freezing point of solutions can be surmounted when using the mixture of water and organic solvents with a low freezing point. These methods, which underlie cryo-enzymology, are described in a series of reviews and monographs [2421, 2504–2506].

The investigation of enzyme hydrolysis at negative values of temperature has been first performed in 1964 [2507] with the use of methanol-water mixture. Thorough studies on the properties of a diversity of mixtures at low temperatures provided further recommendations for conditions when the dielectric permeability of the mixture corresponds to the value typical of water solutions at room temperature [2421].

Concentrated solutions of NaCl [2508, 2509], ammonium acetate [2510] and polyelectrolytes [2511] are also applied in cryoenzymological research.

Reactions catalyzed by chymotrypsin [2512–2515], trypsin [2516–2519], papain [2520–2522], carboxypeptidase A [2508, 2509, 2523, 2524], pepsin [2525], penicillopepsin [2526], leucineaminopeptidase [2527–2529] as well as β-lactamases [2510, 2530] have constituted the cryoenzymological studies.

First, the dependence of catalytic constants on the temperature appeared often to be governed by the Arrhenius equation within $-50- -60\,°C$. Secondly, at low temperatures it is possible to slow certain stages of the process and thus to identify some intermediates in catalysis. At pH 5.5 and $-40\,°C$ acylchymotrypsin formed at hydrolysis of p-nitrophenyl ester of acetyltryptophan has been isolated in dimethylsulphoxide-water mixture (65:35) [2512]. On the other hand, in the case of pepsin, penicillopepsin or leucineaminopeptidase no covalent intermediates have been found [2525–2527]. Proof of the formation of an intermediate anhydride during carboxypeptidase A catalysis at low temperatures has been reported in [2524]. Moreover, spectral studies [2508, 2509] provide evidence of the absence of the intermediate, at the same time testifying to the formation of a second enzyme-substrate complex (Chap. 6). A similar two-stage complex formation has been found for leucineaminopeptidase [2528, 2529], and for trypsin isomerization into a less catalytically active form [2518] has been discovered but not the formation of the tetrahedral intermediate, as earlier supposed [2531].

The cryoenzymological technique is now widely applied to X-ray analysis [2532–2534] and to NMR [2535], which affords the study of the structure of enzyme complexes with true substrates.

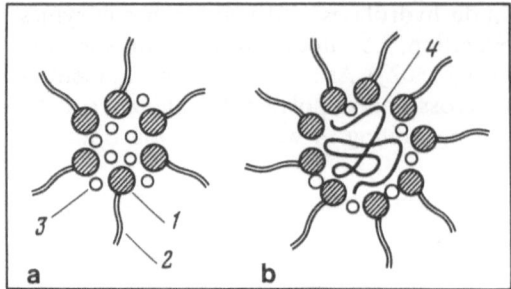

Fig. 81. Scheme for reverse micelle of surfactant in organic solvent (*a*) and the enzyme incorporated into micelle (*b*); *1* ionic (polar) and *2* hydrocarbon group in the surfactant molecule; *3* counterions or solubilized water molecules; *4* enzyme molecule

5.8 Micellar Enzymology

Lately, there appeared new interesting possibilities for studying enzymatic reactions at low temperatures [2516, 2536]. They embrace a creation of reversed micelles in an organic solvent (Fig. 81). This is possible when a surface-active substance is introduced into the system (for instance, sodium *bis*-(2-ethylhexyl)-sulfosuccinate [2536] or lecithin [2537]).

The major principles of the micellar catalysis are presented in Sect. 3.4.9. Reactions in micelles are widely used at room temperatures (see review [1631] and the literature cited). The catalytic activity under these conditions depends on the ratio of water/surfactant and has the maximum. It is characterized by the optimum ratio between the diameter of the internal cavity and the size of the protein molecule. The efficiency of the catalysis, as compared with water solutions, decreases much steeper for more specific substrates than for simple acylamino acid derivatives [2538].

The advantages of reversed micelles are the optical transparency of solutions, their low viscosity and low freezing temperatures.

5.9 Inhibitors

This section deals with the action of organic substances suppressing the enzyme catalytic activity on amide hydrolase. The action of inorganic ions has been considered in Sect. 5.3.

5.9.1 The Kinetics of Inhibition

All inhibitors can be divided into two large groups – reversible and irreversible inhibitors. The first usually form noncovalent bonds with the enzyme, their action can be reversed, for example, by dilution of the mixture or by separation of an inhibitor by gel filtration. While the second react with the enzyme and form covalent bonds, their action cannot be reversed by common approaches. However, sometimes it is difficult to separate a noncovalently bound substrate from the protein and vice versa, a covalent bond is broken very readily.

A reversible inhibitor can interact both with a free enzyme and an enzyme-substrate complex (see reviews [1661, 2425]):

$$
\begin{array}{ccccc}
E + S & \underset{}{\overset{K_s}{\rightleftharpoons}} & ES & \overset{k_2}{\longrightarrow} & E + P \\
\big\downarrow K_i & & \big\downarrow \alpha K_i & & \\
EI + S & \underset{}{\overset{\alpha K_s}{\rightleftharpoons}} & ESI & \overset{\beta k_2}{\longrightarrow} & EI + P
\end{array}
\qquad (24)
$$

An overall equation for the reaction rate corresponding to the kinetic scheme (24) is

$$
v = \frac{k_2 \dfrac{\alpha K_i + \beta [I]_0}{\alpha K_i + [I]_0} [E]_0 [S]_0}{\alpha K_s \dfrac{K_i + [I]_0}{\alpha K_i + [I]_0} + [S]_0} .
\qquad (25)
$$

The type of an inhibition is defined by magnitudes of coefficients α and β [2539].

A *linear competitive inhibition* is characterized by values $\alpha \to \infty$, whereas β has no physical meaning. So Eq. (25) is transformed into

$$
v = \frac{k_2 [E]_0 [S]_0}{K_s \left(1 + \dfrac{[I]_0}{K_i}\right) + [S]_0} .
\qquad (26)
$$

The K_i value is often determined by the Dixon method [1661, p. 350] plotting the $1/v$ vs $[I]_0$ at a few $[S]_0$ values or $1/v$ vs $1/[S]_0$ at some $[I]_0$ values (Fig. 82).

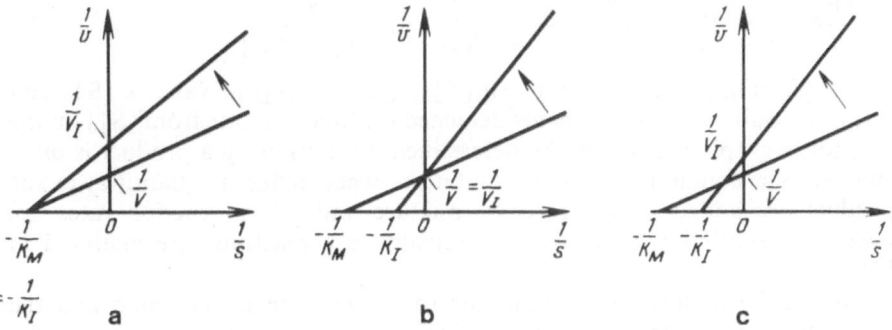

Fig. 82. Determination of the inhibition constant by the double-reciprocal plot method; *a* noncompetitive inhibition; *b* competitive inhibition; *c* inhibition of the mixed type

A linear noncompetitive inhibition ($\alpha = 1$, $\beta = 0$) is described by the rate equation

$$v = \frac{k_2/(1 + [\mathrm{I}]_0/K_i)[\mathrm{E}]_0[\mathrm{S}]_0}{K_s + [\mathrm{S}]_0} . \tag{27}$$

The K_i value is found by the same approaches as for competitive inhibition.

In a mixed-type inhibition (α and $\beta = 1$) the coordinates of intercepts on the graphs $1/v \div 1/[\mathrm{S}]_0$ are given by equations [2539] being independent of K_i:

$$\frac{1}{[\mathrm{S}]_0} = \frac{1}{K_s}\frac{\beta - 1}{\alpha - \beta} \quad \text{and} \tag{28}$$

$$\frac{1}{v} = \frac{1}{V_{\max}}\frac{\alpha - 1}{\alpha - \beta} . \tag{29}$$

If K_s and $V_{\max}$ are found in experiments without an inhibitor, we can find α and β from Eqs. (28) and (29).

Another approach to find α, β and K_i is transformation of k_{cat} and K'_m in Eq. (25) to:

$$\frac{1}{k_{\mathrm{cat}}/(k_2 - 1)} = \frac{\alpha K_i}{\beta - 1}\frac{1}{[\mathrm{I}]_0} + \frac{1}{\beta - 1} , \tag{30}$$

$$\frac{1}{K'_m/(K_s - 1)} = \frac{\alpha K_i}{\alpha - 1}\frac{1}{[\mathrm{I}]_0} + \frac{1}{\alpha - 1} . \tag{31}$$

These two graphs provide the three constants.

Inhibition by a substrate has been considered in Sect. 4.1.6.

Inhibition by a product of the reaction is analyzed at high degrees of the substrate conversion, when the steady-state condition is not followed. Therefore, for the kinetic scheme

$$\mathrm{E} + \mathrm{S} \xrightleftharpoons{K_s} \mathrm{ES} \xrightarrow{k_2} \mathrm{E} + \mathrm{P} , \quad \mathrm{E} + \mathrm{P} \xrightleftharpoons{K_P} \mathrm{EP} \tag{32}$$

an integral expression for the rate is used:

$$\frac{[\mathrm{P}]}{t} = \frac{V_{\max}}{1 - K_s/K_P} - \frac{K_s(1 + [\mathrm{S}]_0/K_P)}{t(1 - K_s/K_P)} \cdot \ln\frac{[\mathrm{S}]_0}{[\mathrm{S}]_0 - [\mathrm{P}]} . \tag{33}$$

By plotting $[\mathrm{P}]/t$ against $1/t \cdot \ln[\mathrm{S}]_0/([\mathrm{S}]_0 - [\mathrm{P}])$ at various $[\mathrm{S}]_0$ and reciprocal values of the fragments detached on X-coordinate from $[\mathrm{S}]_0$ or the secondary graph, the K_P can be determined. Inhibition by a product is often studied as a common reversible inhibition, when different quantities of the product are added to the reaction mixture and initial reaction rates are measured. Inhibition schemes for multisubstrate reactions are analyzed in [2540].

When the affinity of the inhibitor to the enzyme is very high and the concentrations of the enzyme and inhibitor used in the experiment for determination of K_i become comparable, admission on the constancy of the inhibitor concentration ($[\mathrm{I}] \gg [\mathrm{EI}]$) is inapplicable. So for noncompetitive inhibition the K_i value can be detected from the equation [2541]:

$$\frac{[I]_0}{1 - \alpha} = K_i \cdot \frac{1}{\alpha} + [E]_0 \,,$$

where $\alpha = v_i/v$, i.e., the ratio of the rates in the presence and absence of an inhibitor. The slope of the plot $[I]_0/1 - \alpha$ against $1/\alpha$ will be equal to K_i. There exist other types of analysis of these systems [2425, p. 80; 2542].

Complex cases of mixed types of inhibition are proposed [2543] to be analyzed by a method of vector $K'_m - V'_{max}$ coordinate system. The dependences of apparent values of K'_m and V'_{max} for diverse concentrations of the inhibitor are plotted, the position and the value of the straight line (vector) predetermine the type of inhibition. The approach is also applicable to the study of enzyme activation.

It is often useful to consider interrelationships of two inhibitors of one and the same enzyme. Two inhibitors are independent of each other if they can bind simultaneously to the enzyme active site. Interdependent inhibitors displace each other [2544]. It is possible to distinguish these types of inhibition when plotting v/v_i vs $[I_1]_0$ at the constant concentration of the second inhibitor (Fig. 83). For interdependent inhibition

$$v/v_i = 1 + \frac{[I_2]_0}{(K_i)_2} + \frac{[I_1]_0}{(K_i)_1} \,,$$

and for interindependent inhibitors

$$v/v_i = \left(1 + \frac{[I_1]_0}{(K_i)_1}\right) \left(1 + \frac{[I_2]_0}{(K_i)_2}\right).$$

Irreversible Inhibition. Two types of irreversible inhibitors should be distinguished: with and without affinity to the enzyme. In the first case the reaction of the inhibitor and the enzyme is described as the second order reaction or, if the inhibitor is in high excess, as a reaction of the pseudo-first order (see review [2545]).

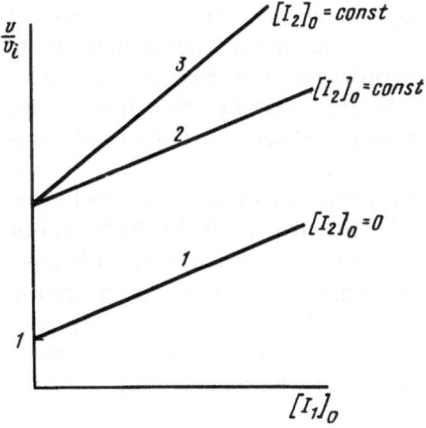

Fig. 83. Interrelationship of the action of two inhibitors on the enzymatic reaction: *1* the lack of the second inhibitor; *2* interdependent; *3* independent action of the second inhibitor

Affinity inhibitors form first an enzyme complex, then the enzyme-inhibitor complex is converted into irreversibly inactivated enzyme:

$$E + I \underset{}{\overset{K_i}{\rightleftharpoons}} EI \xrightarrow{k_i} E—I \cdot \tag{34}$$

When equilibrium is established faster than the chemical inactivation, the rate of a change of the enzyme activation is written by Eq. (35) similarly to the Michaelis-Menten equation:

$$v = -\frac{d[E]}{dt} = \frac{k_i[E]_0[I]_0}{K_i + [I]_0} \cdot \tag{35}$$

It is possible to determine k_i and K_i when measuring the initial inactivation rates at various concentrations of $[I]_0$ and plotting $1/k_{exp}$ vs $1/[I]_0$, where $k_{exp} = k_i[I]_0/K_i + [I]_0$ [2546]. The theory and methods for determination of the rate constants of irreversible inhibition are discussed in [2547, 2548].

A few approaches have been proposed to determine the number of functionally important enzyme residues from data of their chemical modification [2549–2551]. As shown in [2550], the enzyme residual activity (α) and the number of residues (i) important for activity are related to the equation:

$$\alpha^{1/i} = \frac{N}{P} \cdot \chi - \frac{N - (P + S)}{P} \, ,$$

where N is the total number of the residues of the given type in the protein, P is the number of slowly modified residues, among which i residues are essential for the activity, and S is the number of quickly modified residues inessential for the activity.

5.9.2 Unspecific Inhibitors

A large group of organic compounds suppressing the amide hydrolases activity can be called unspecific inhibitors, as their structure differs much from that of substrates, and their action is stipulated by certain general physicochemical properties. These inhibitors can suppress the activity of numerous amide hydrolases, their active sites display hydrophobic characteristics or carry a charge. Besides many other problems of the chemistry of amide hydrolases, their inhibition has been studied on the example of serine proteases chymotrypsin, trypsin, elastase, etc.

Aromatic compounds are the first to be referred to unspecific inhibitors, their action on chymotrypsin has been studied as early as 1963 [2552]. About 130 compounds considered as derivatives of benzene, pyridine, naphthalene, etc. have been tested. These compounds are characterized by the inhibition constant of the chymotryptic catalysis (1×10^{-2}–1×10^{-3} M). Cyclohexanol binding is much weaker ($K_i = 29$ mM [2553]) when comparing with phenol ($K_i = 3.5$ mM [2554]); this testifies to a planar character of the enzyme binding site.

There exists a certain dependence between the area of the inhibitor surface and stability of the enzyme-substrate complex (Fig. 84) [2413, 2555]. However, their hydrophobicity plays a crucial role in binding (Fig. 85) [2554, 2556]. Polar substituents in aromatic compounds slightly influence the efficiency of the complex formation. Presumably, the enzyme "extracts" only a hydrophobic skeleton of the molecule, while polar regions remain hydrated. Similar dependences are observed for aliphatic alcohols [2553, 2556] and other compounds. Binding of alcohols is characterized by a significant positive change of the entropy in the system [2556, 2557].

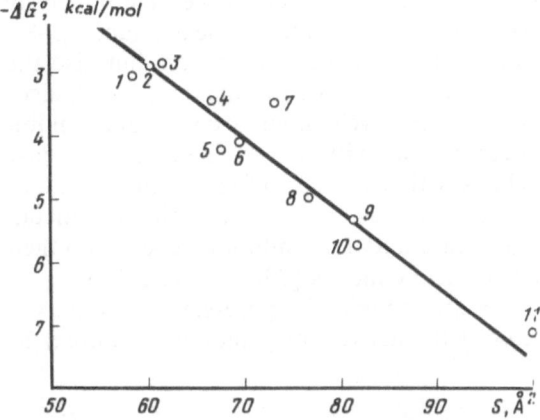

Fig. 84. Free energy of complex formation of aromatic compounds with chymotrypsin vs their surface area [2413, 2555]: *1* pentan; *2* benzene; *3* cyclohexane; *4* toluene; *5* chlorobenzene; *6* nitrobenzene; *7* ethylbenzene; *8* indene; *9* naphtalene; *10* azulene; *11* antracene

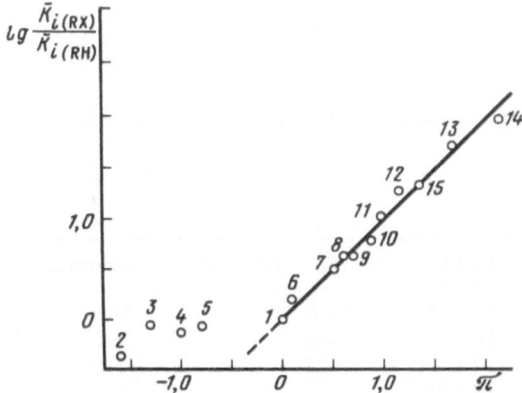

Fig. 85. Relative association constants of phenol derivatives (RX, where R = C_6H_4OH) against substituent hydrophobicity. Substituent (X): H (*1*); 4-NH_2 (*2*); 4-$CONH_2$ (*3*); 4-OH (*4*); 3-OH (*5*); 3-OMe (*6*); 4-Me (*7*); 3-Me (*8*); 2-Cl (*9*); 2-Br (*10*); 4-Cl (*11*); 4-Br (*12*); 2,4-Cl_2 (*13*); 2,4-Br_2 (*14*); β-naphthol (*15*)

The enzymes, the complex formation of which involves electrostatic interactions, are effectively inhibited by organic compounds containing a charged group. This may be exemplified by data on inhibition by the charged compounds of trypsin and a set of trypsin-like proteases [2558, 2559] (Table 59).

Some dyes compose a special group of inhibitors, effectively bound to a series of proteolytic enzymes (Table 60).

An essential change of absorption or fluorescence spectra upon binding to the enzyme is the peculiarity of these compounds. This spectral shift is interpreted [2567] as a consequence of a change in dielectric permeability of the environment during the complex formation of a dye and the protein. The properties of the active site of chymotrypsin are closely related to those of cyclohexane. However, a spectral shift of the dye may not be determined by dielectric permeability, but mostly by a refractory index of the medium [2568].

Not only serine proteases undergo the unspecific inhibition, but also the representatives of other groups of amide hydrolases. So dioxan [2336], alcohols [2569], phenols and other aromatic and cycloaliphatic compounds inhibit pepsin activity. Alcohols suppress the activity of leucineaminopeptidase [2570], and their efficiency correlates with the length of an aliphatic chain.

Chelating agents can be also considered as unspecific inhibitors of metal-containing amide hydrolases, since their ability for inhibition depends on their common property of binding ions of heavy metals [2389, 2486, 2571].

Univalent cations in certain cases are capable of suppressing the activity of proteolytic enzymes. So inhibition of the activity of glandular kallikrein by

Table 59. Inhibition of trypsin-like enzymes by amidines [2558]

Enzyme	K_i, μM		Enzyme	K_i, μM	
	Benzamidine	5-Amidinoindole		Benzamidine	5-Amidinoindole
Bovine trypsin	35.2	29.1	Kallikrein	933	1.2
Bovine thrombin	332	7.7	Acrosin	4.0	0.5
Human thrombin	235	7.9	Plasmin	207	112
Human factor X_a	182	5.45			

Table 60. Dyes–inhibitors of proteases

Dye	Enzyme	K_i, M	λ_{max} of complex	References
Proflavin	Chymotrypsin	2.5×10^{-5}	465	[2560, 2561]
	Trypsin	3.4×10^{-5}	469	[2560]
Thionine	Chymotrypsin	45×10^{-5}	625	[2560]
	Trypsin	3.4×10^{-5}	622	[2562]
Acryflavin	Pepsin	1.2×10^{-4}	476	[2563]
Brilliant green	Chymotrypsin	–	420; 625	[2564]
Rhodamine 6G	Chymotrypsin	0.8×10^{-5}	542	[2565]
Bieberich purple	Chymotrypsin	–	420; 625	[2566]

Na^+- and K^+-ions in physiological concentrations [2572] as well as their inhibition of the activity of multicatalytic proteases (proteosomes) have been noticed [2573].

The type of inhibition depends not only on the structure of the inhibitor, but sometimes on the substrate structure. It was shown as an example [2574] that pepsin inhibition by ethanol is purely competitive as regards hydrolysis of $ZPhe(NO_2)AlaAmp$ and noncompetitive as to $ZAlaPhe(NO_2)Amp$ hydrolysis.

Binding of inhibitors often results in conformational rearrangements of the protein. This is clear from the changes of their spectral and other physicochemical characteristics [1375], in changes of pK_a of ionogenic groups accompanied by the proton binding or its release [2575–2577], as well as in a higher stability of the enzyme complexes with certain inhibitors to denaturation, as compared with free proteins [2578, 2579].

Binding of inhibitors also depends on an oligomeric state of the enzyme [2580].

5.9.3 Specific Reversible Inhibitors

To this type of inhibitors can be referred the substrates, products of the enzymatic reaction, and their analogs.

Substrate inhibition has been thoroughly studied on the example of carboxypeptidase A [1738, 2581–2583]. The kinetic scheme for catalysis by this enzyme is described by Eq. (37), the reaction rate by Eq. (38) (Sect. 4.1.6).

At low substrate concentrations the reaction obeys the Michaelis-Menten kinetics. A second molecule of the substrate binds to the enzyme on an increase of $[S]_0$, thus activating hydrolysis ($\beta > 1$). Further increase in the substrate concentration results in binding of a third molecule, which inhibits the enzyme. The β value for diverse substrates is within 2–4.5, while the K_s' is by 1 order of magnitude higher than that of the K_s. The inhibition is characterized by K_s'' (17–30 mM), except for ZGlyPhe, where the constant is $K_s'' = 0.3$ M [1738]. Interestingly, after acylation of carboxypeptidase by N-acetylimidazole the enzyme cannot be inhibited by the substrate [2584].

The hydrolysis products, as a rule, inhibit the enzymes. Analysis of the inhibition type by the product reveals an important conclusion about the mechanism of the enzymatic reaction [2585]. However, we run into many problems here. As shown [2175, 2585, 2586], at pepsin-catalyzed hydrolysis of the compounds such as AcPhePheOMe an acyl fragment (AcPheOH) shows some properties of a noncompetitive inhibitor, whereas a second product (HPheOMe) is a competitive inhibitor (at pH 2). This suggests a conclusion about an ordered dissociation of products from the enzyme complex

$$E + S \rightleftharpoons ES \longrightarrow EPQ \longrightarrow EP \longrightarrow E + P$$
$$+$$
$$Q$$

(where P = HPheOMe and Q = AcPheOH) as well as about the existence of the intermediate aminoenzyme, to which a covalent nature has been ascribed. At pH ≈ 4 a competitive inhibition by acetylphenylalanine is observed, which testifies to a random mechanism, when pH increased:

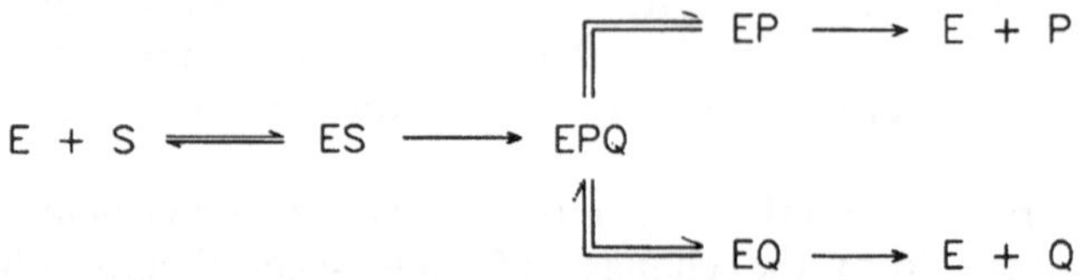

$$\begin{array}{ccc}
 & EP \longrightarrow E + P \\
E + S \rightleftharpoons ES \longrightarrow EPQ & \\
 & EQ \longrightarrow E + Q
\end{array}$$

Opposite results have been obtained [2587] on the inhibition by acetyl-phenylalanine of pepsin-catalyzed hydrolysis of AcPheTyr within pH 2 ÷ 4.5. This entails assumption of the ordered mechanism at low pH and a possibility of the transition to the random mechanism, when the group with $pK_a ≈ 3$ is being ionized.

As shown [2588], the differences in the inhibition type might occur if an inhibitor binds not only to a free enzyme, but also to the enzyme-substrate complex. At least for pepsin the results of inhibition by the product provide no answer on the type of regularity of the hydrolysis mechanism.

Rather effective inhibitors are obtained, when the two products of the enzyme hydrolysis (so-called bi-product analogs) are combined in one and the same structure. For example, L-benzylsuccinate (I) inhibits carboxypeptidase with $K_i = 0.45 \mu M$ [2589], thus partially reproducing the structure of hippuric acid and phenylalanine:

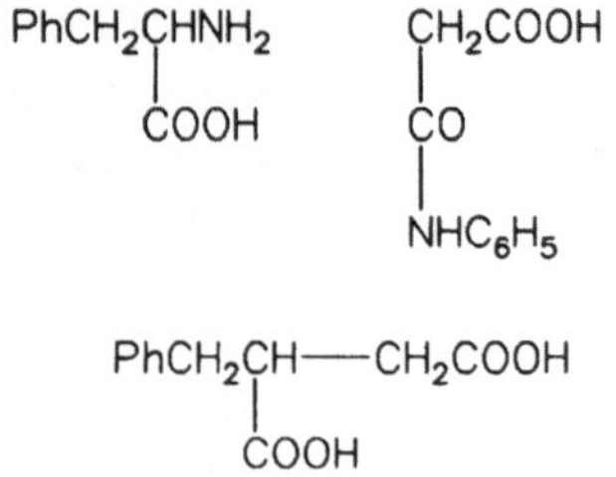

$$\begin{array}{ll}
PhCH_2\overset{|}{C}HNH_2 & CH_2COOH \\
\underset{}{\overset{|}{C}OOH} & \overset{|}{C}O \\
 & \overset{|}{N}HC_6H_5
\end{array}$$

$$PhCH_2\overset{|}{C}H-CH_2COOH$$
$$\overset{|}{C}OOH$$

I

Compounds (II) and (III) appeared to be highly effective inhibitors of carboxypeptidases B and N, respectively [2590, 2591]:

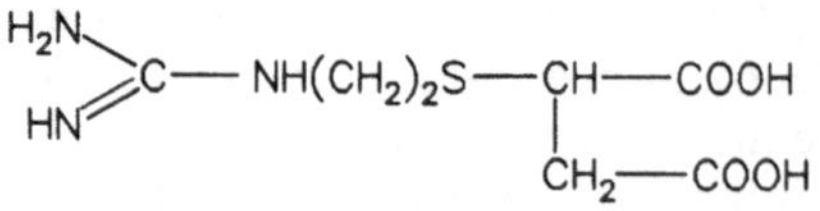

$$\begin{array}{l}
\underset{HN}{\overset{H_2N}{>}}C-NH(CH_2)_2S-\overset{|}{C}H-COOH \qquad K_i = 4 \mu M , \\
\qquad\qquad\qquad\qquad CH_2-COOH
\end{array}$$

II

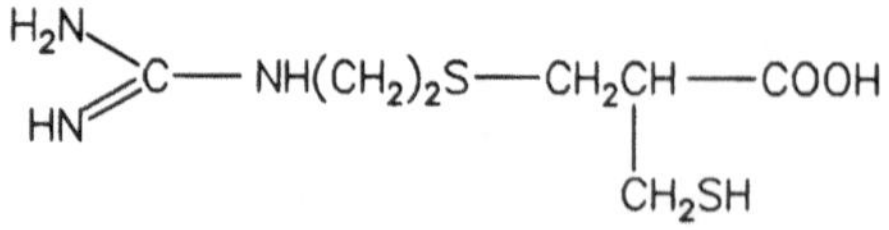

$$\begin{array}{l}
\underset{HN}{\overset{H_2N}{>}}C-NH(CH_2)_2S-CH_2\overset{|}{C}H-COOH \qquad K_i = 0.002 \mu M . \\
\qquad\qquad\qquad\qquad CH_2SH
\end{array}$$

III

There have been proposed inhibitors of a similar type [2592] for a series of serine proteases of the trypsin group (plasmin, kallikrein, thrombin, etc.) as well as for some metal-containing proteases [2593, 2594].

A large number of works deal with studies of an inhibiting action of the substrate analogs. A comparison of the inhibition constants of chymotrypsin by amides and esters of acetyl-D-amino acids has revealed slight differences in the K_i of amides and esters; the K'_m of the corresponding substrates differs much, while K_i and K'_m of amides are rather close [1666]. For esters a distinct difference is observed between K_i and K'_m. Hence, K'_m of amides coincides with K_s – dissociation constant of the Michaelis complex. For esters, K'_m is an effective constant. The equality of K'_m and K_s is observed also at the pepsin catalysis [2338, 2444, 2595]. In the case of pepsin the K_i of neutral inhibitors does not depend on pH [2338], but, for example, in thermolysin such dependence does not exist for charged substrates [2596].

In most cases the inhibiting ability of substrate analogs of proteases is established not only by the primary, but, to a large extent, by a secondary specificity, i.e., by the structure of groups far from the cleavable bond in the proper substrate [2597–2602]. Inhibition of cathepsin D and pepsin by peptides containing the PhePhe bond usually cleavable by these enzymes may be an interesting example here (Table 61).

L-L-configuration of amino acids that form a potentially cleavable bond shows the most effective inhibiting action, whereas D-D-enantiomer inhibits cathepsin D by approximately $\approx 10^3$ worse.

However, for α-chymotrypsin the structure and configuration of secondary amino acid residues import slight changes to the inhibition constant [2597, 2599]. The inhibition of hydrolysis of p-nitroanilide of succinylphenylalanine by chymotrypsin and of tripeptide substrate ValTyrGly by inhibitor GlyTyr-D-Ala proceeds differently [2597].

The influence of secondary interactions is observed in a series of similar thermolysin inhibitors [2603]:

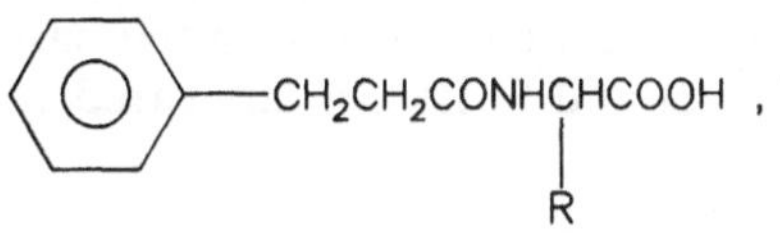

$$\text{IV} \qquad \text{V}$$

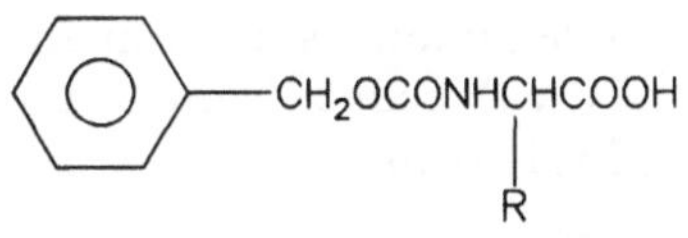

where R = H, CH$_3$, CH(CH$_3$)$_2$, CH$_2$CH(CH$_3$)$_2$, CH(CH$_3$)CH$_2$CH$_3$.

Table 61. Inhibition of cathepsin D and pepsin [2601]

Inhibitor	K_i, μM	
	Cathepsin D	Pepsin
pGlu-D-PheProPhePheVal-D-Leu	0.031	0.47
D-PheProPhePheVal-D-Leu	0.52	5.2
D-PhePro-D-Phe-D-PheVal-D-Leu	650	–
D-PheProPhePheVal	No inhibition	

If the K_i values of compounds IV correlate with hydrophobicity of group $R (\lg K_i = -0.737\pi - 1.8)$, there is observed a correlation for inhibitors of type V with steric effects of this group ($\lg K_i = 0.88 E_s - 3.086$), but not with hydrophobicity. These compounds distinguished by a replacement of the CH_2 group by O-atom, are perhaps bound differently in the enzyme active site. A replacement of the NH group in residue P_1 of the pepsin substrates by oxygen atom [1990] or a replacement of the CO group by the CH_2 group in the P_1'-residue of these substrates [2604] transforms them into the enzyme competitive inhibitors, whereas the K_i values appeared to be very close to the K_m' values.

Similar enzymes can also display quite a different sensitivity as regards the inhibitors. Thus m-$CF_3C_6H_4COAlaAla$-pNA inhibits leukocyte elastase with $K_i = 4 \times 10^{-6}$ M, but does not influence pancreatic elastase [2605].

A large group of effective inhibitors of serine proteases is represented by fluorinated peptide-ketones of type [2606–2608]:

$$\text{Acyl} - \text{NH} - \underset{\underset{R^2}{|}}{\text{CH}} - \text{CONH} - \underset{\underset{R^1}{|}}{\text{CH}} - \text{COCF}_2\text{X} \ ,$$

where R^1 and R^2 are the residues of specific amino acids for S_1- and S_2-binding sites of the enzyme, and X is a fluorine atom or a chain specific to the S'-binding sites. The majority of these inhibitors bind in the active site slowly and tightly ($K_i \approx 1 \times 10^{-9}$ M), thus forming semiketals with serine hydroxyl:

$$\text{E} - \text{O} - \underset{\underset{O^-}{|}}{\overset{\overset{R}{|}}{\text{C}}} - \text{CF}_3 \ ,$$

i.e., they are similar to the so-called "transition state analogs" (Sect. 5.9.7). Similar inhibitors have been proposed for aspartic protease renin [2609].

The same "pseudo-irreversible" inhibitors of cysteine proteases [2610] are peptide–nitriles, for example, $AcPheLeuC \equiv N$ ($K_i = 5.8 \times 10^{-6}$ M) forming intermediates of type R–C(=NH)–S–E with the enzyme SH group.

Reversible inhibitors of renin and other aspartic proteases on the basis of compounds with reduced amide groups (VI) and (VII) also appeared to be effective [2611–2614]:

$$\text{R} - \text{CH}_2\text{NH} - \text{R}' \qquad \text{and} \qquad \text{R} - \underset{\underset{OH}{|}}{\text{CH}}\text{CH}_2\text{NH} - \text{R}' \ .$$

$$\qquad\quad \text{VI} \qquad\qquad\qquad\qquad\qquad\qquad \text{VII}$$

The activity of collagenase and some metal-containing proteases is suppressed by inhibitors of a similar type [2615–2617].

Analogs of the substrates of aspartic proteases – the phosphine acids derivatives of peptides – show a high efficiency [2618], for example:

$$\text{IvaVal} - \text{NHCH} - \overset{\overset{\displaystyle ^-\text{O} \quad \text{O}}{\diagdown \diagup}}{\text{P}} - \text{CH}_2\text{CO} - \text{AlaXaa}$$

The structure data of certain proteases underlay the development of effective inhibitors of metal-containing enzymes. They act due to introduction into the molecule of the group forming a coordination bond to a metal atom [2619–2631]. The most effective and applicable are captopryl (VIII) and thiorphan (IX):

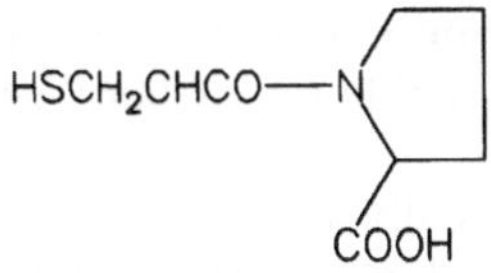

VIII 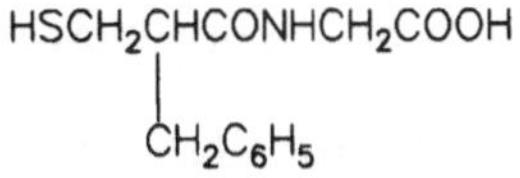   IX

Captopryl is used to suppress dipeptidylcarboxypeptidase (angiotensin converting enzyme) in vivo. This enzyme is suppressed with $K_i = 4.5 \times 10^{-12}$ M by a substrate analog – [des-Pro^3]-bradykinin [2632].

Specific substrate-like inhibitors containing a dye group (dansyl, acrydine, etc.) are applicable to introduce reversible reporter groups into amide hydrolases [1881, 2633–2635].

5.9.4 Irreversible Group-Specific Inhibitors

We have already discussed the representatives of these inhibitors (Sect. 2.6). Here they will be described in detail. Information on many of the inhibitors can be found in reviews [2545, 2636–2638].

Serine Amide Hydrolases. The derivatives of phosphoric acid, certain chloromethylketones and sulfonylfluorides are the most important group-specific inhibitors of serine amidehydrolases.

The derivatives of phosphoric acid of a general formula react with serine

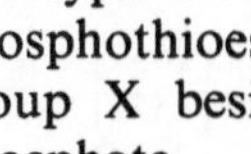

enzymes and phosphorylate a catalytically active serine residue (in chymotrypsin Ser195). The best-known inhibitor is diisopropylfluorophosphate ($R_1 = R_2 = i$-Pr, X = F) [1308]. The earlier works on the action of inhibitors of this type are summarized in review [175]. Phosphoester [2639, 2640] or phosphothioester [2641] as well as other groups [2642, 2643] can be used as group X besides the halogene atom. The reaction of diisopropylfluorophosphate with chymotrypsin runs with the rate constant $k = 2.7 \times 10^3$ M^{-1} min^{-1}. Trypsin reacts with this substance much slower ($k = 0.3 \times 10^3$ M^{-1} min^{-1}) [2284]. These rates are higher by 4–5 orders of magnitude than the reaction rate with a model compound – N-acetyl-L-tyrosinamide. For various alkylphosphates there have been found structure-functional correlations of inhibiting properties [2644, 2645]. A phosphoester group can be removed, when treated by hydroxylamine and other nucleophilic agents with partial regeneration of the enzyme activity [2646, 2647].

The serine residue selectively reacts also with phenylmethanesulfonvlhalides [2648–2650] and diphenylcarbamoylchloride [2651], but not all serine amide hydrolases are sensitive to alkylphosphates and these irreversible inhibitors.

The action of chloroketones – the derivatives of amino acids and peptides – on serine amide hydrolases has been widely studied (Table 62; earlier works see in [2652]).

Chloroketones react with atom $N^{\varepsilon 2}$ of a catalytically active residue of histidine (His57 in chymotrypsin) [2660, 2661], though data are also available on alkylation of other residues, in particular Met192 [2662]. These reagents are quite suitable to incorporate reporter labels into the active sites of serine proteases [2663].

The reaction of chloroketones with enzymes follows kinetic scheme (34) with the formation of an intermediate enzyme-inhibitor complex (with a model compound by acetylhistidine chloroketones react by 6 orders of magnitude slower [2664]).

Besides chloroketones other derivatives of peptidylketones type $R–COCH_2X$ ($X = COCH_3$, $OCOCH_3$, F, etc.) react with this group of the enzymes, whereas the efficiency of inhibition is related to the inductive constant σ_1 of group X by a linear ratio [2665]:

$$-\log K_i = 7.2\sigma_1 + 2.1 \ .$$

Table 62. Rate constants of the reaction of chloroketones with amide hydrolases

Enzyme	Inhibitor	$k_i/[I]_0$, $M^{-1}s^{-1}$ (Scheme 34)	References
Chymotrypsin	ZPheCH₂Cl[a]	73 (pH 7)	[2653]
"	ZGlyLeuPheCH₂Cl	3.0 (pH 5.8)	[2654]
Trypsin	TosLysCH₂Cl[a]	7.0	[2653]
"	D-ValPheLysCH₂Cl	1.08×10^{4}[b]	[2655]
Pancreatic elastase	ZPheCH₂Cl	0.05	[2653]
The same	Suc(Ala)₂ProValCH₂Cl	73	[2656]
"	Tfa(Ala)₂AlaCH₂Cl	60	[2656]
Leukocyte elastase	MeOSuc(Ala)₂ProValCH₂Cl	922	[2654]
Cathepsin G	ZGlyLeuPheCH₂Cl	260	"
Plasmin	D-ValPheLysCH₂Cl	2.38×10^{4}[b]	[2655]
Human thrombin	D-ValLeuLysCH₂Cl	2.54×10^{4}[b]	"
Bovine factor Xa	D-ValPheLysCH₂Cl	117[b]	"
Kallikrein	ProPheArgCH₂Cl	7.48×10^{4}[b]	[2657]
Postproline protease	ZGlyGlyProCH₂Cl	109	[2553]
Enkephalinase	TyrGlyGlyPheLeuCH₂Cl	(16 nM)[c]	[2658]
Human skin chymase	ZPheCH₂Cl	18.3	[2659]

[a] Not a group-specific inhibitors, react with papain with $K_i/[I]_0$ equal to 50 (ZPheCH₂Cl) and 46 (ToSLysCH₂Cl) [2653].
[b] k_i/K_i in $M^{-1}s^{-1}$.
[c] IC_{50} by contraction of smooth muscles (guinea pigs).

The reaction with haloidmethylketones runs perhaps via the formation of hemiketal with serine hydroxyl of the enzyme and with the following intra-molecular alkylation of the histidine residue [2665, 2666].

Note that chloroketones react with thiol groups [2667], therefore they can inactivate cysteine amide hydrolases. Thus their use as the only group-specific inhibitors does not provide unambiguous information on the nature of the modified enzyme.

Fluoromethylketones react slower with serine and cysteine proteases than the corresponding chloroketones [2668, 2669]. Fluoroketones produce a hydrate form faster than chloroketones, but react slower with the groups of the active site [2670]. The inhibitors apparently bind to the enzyme in keto-, but not in a hydrate form, the formation of hemiketal occurs only in the active site.

α-Bromacetamides, derivatives of aromatic amines, are effective inhibitors of serine proteases [2671]. Thus bromacetylisoindoline reacts with methionine of the chymotrypsin active site with the second-order rate constant of the $k_i/K_i = 147\ \mathrm{M}^{-1}\,\mathrm{min}^{-1}$.

There have been proposed sulfonates of type [2672]:

$$R-CH(COOCH_3)NHCOSO_3^- \ ,$$

reacting with serine proteases with the cleavage of bisulfite and alkynyl esters of aromatic acids [2673]:

$$R-C_6H_4COOC\equiv CCH_3 \ .$$

Phenylpropynal reacting with the arginine and lysine residues is proposed as a specific irreversible inhibitor of β-lactamase:

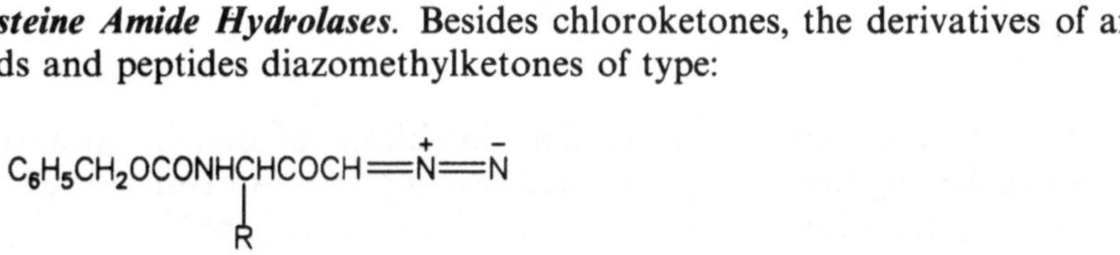

Cysteine Amide Hydrolases. Besides chloroketones, the derivatives of amino acids and peptides diazomethylketones of type:

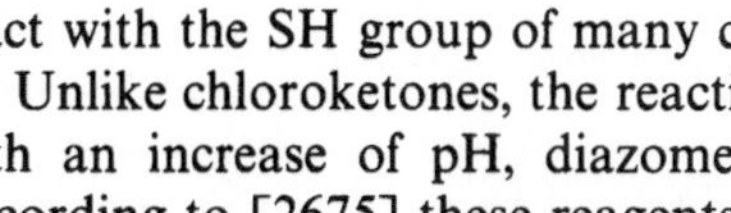

$$C_6H_5CH_2OCONHCHCOCH=\overset{+}{N}=\overset{-}{N}$$
$$| \atop R$$

react with the SH group of many cysteine amide hydrolases [2675–2677].

Unlike chloroketones, the reaction rate of which with SH group enhances with an increase of pH, diazomethylketones effectively react at low pH. According to [2675] these reagents attack an uncharged sulfhydryl group. It was shown [2678] that the pH-dependence of the reaction can be explained by rate-limiting protonation of intermediate thiolhemiacatal (VIII) with the following three-center rearrangement:

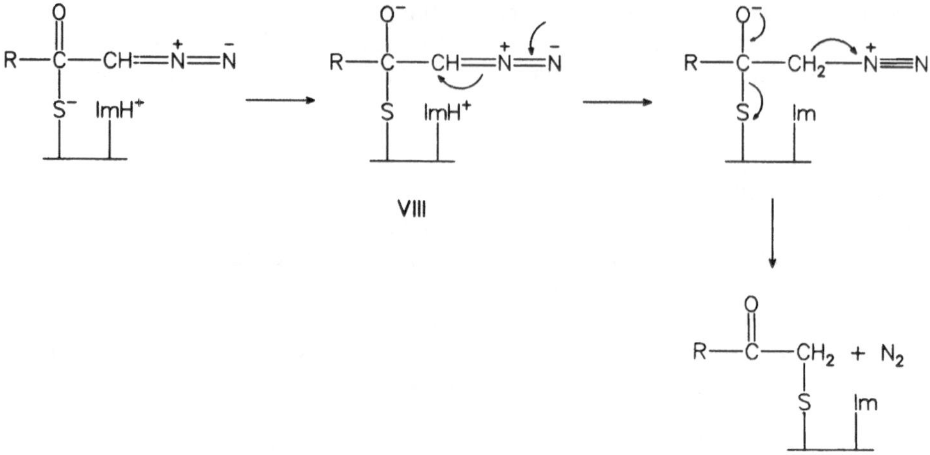

VIII

The reaction of diazomethylketones with cysteine amide hydrolases runs effectively according to scheme (34) (Sect. 5.9.1). The second-order rate constant of the cathepsin B reaction with ZPheAlaCHN$_2$ is $k_i/K_i = 7.5 \times 10^4 \, M^{-1} \, min^{-1}$ [2679], and with ZPheTrpCHN$_2$: $8.5 \times 10^6 \, M^{-1} \, min^{-1}$ [2680]. These compounds do not react with low-molecular thiols.

1,3-Dibromacetone reacting simultaneously with the SH group and the histidine residue of the active site can be referred to group-specific inhibitors [2681]:

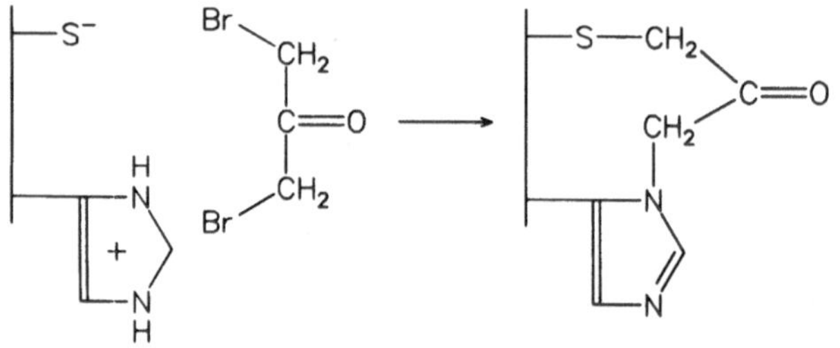

There exist other specific irreversible inhibitors of cysteine proteases [2682–2684]. Among them compound (IX) reacting with cathepsin B with the rate constant $k_i/K_i = 2.6 \times 10^6 \, M^{-1} \, s^{-1}$ is most effective [2682]:

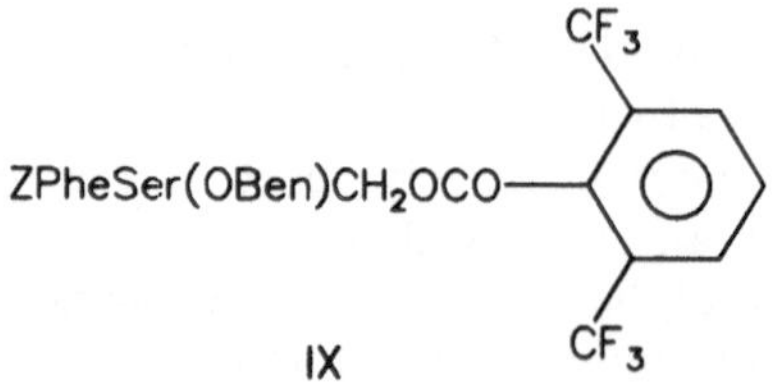

IX

Aspartic Amide Hydrolases. In 1965 it was found [1340] that two to three carboxyl groups of pepsin are subjected to esterification by diphenyl-diazomethane with a partial loss of the enzyme's catalytic activity. The search for more specific diazocompounds has resulted in obtaining a great number of reagents specifically inhibiting aspartic proteases, when reacting with a catalytically important carboxyl group (Asp215 in pepsin) [1341–1343, 2685–2688]. Among them methyl ester of diazoacetylnorleucine is well known

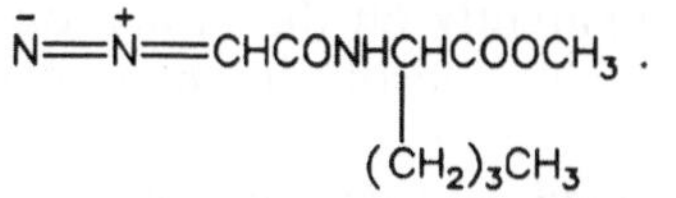

$$\bar{N}{=}\overset{+}{N}{=}CHCONH\underset{\underset{\displaystyle (CH_2)_3CH_3}{|}}{C}HCOOCH_3 \ .$$

These compounds react with proteases in the presence of Cu^{+2} ions. The reaction mechanism presumably consists first of the formation of a complex of carben with Cu^{+2} ion [1343, 2689]:

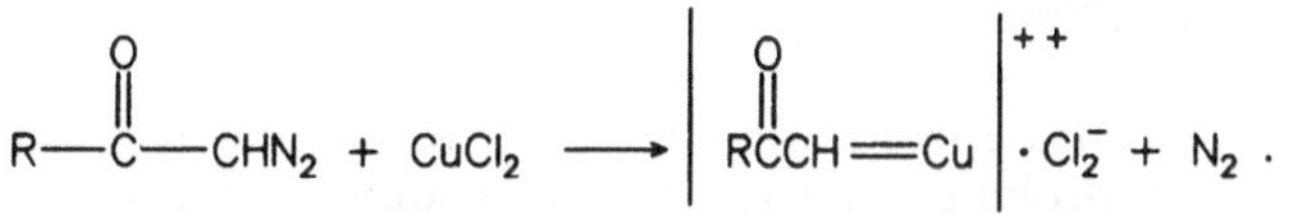

$$R{-}\overset{\overset{\displaystyle O}{||}}{C}{-}CHN_2 + CuCl_2 \ \longrightarrow \ \left| \ R\overset{\overset{\displaystyle O}{||}}{C}CH{=}Cu \ \right|^{++} \cdot Cl_2^- + N_2 \ .$$

Since a carben complex bears a positive charge, it probably binds to an ionized carboxyl group. The studies of the pH-dependence of inactivation [1342, 1343] have revealed that a protonated group with $pK_a \approx 5$ is involved in the process. Thus, the reaction runs with the protonated COOH group of the Asp215 residue when a neighboring ionized carboxyl group is also involved. Neither denatured pepsin nor pepsinogens reacts with diazocompounds [1341, 2685]. They were used to introduce EPR labels into the active site of aspartic proteases [2690].

Epoxides like 1,2-epoxy-3-(*p*-nitrophenoxy)-propane are the other group-specific reagents for aspartic proteases [2691]:

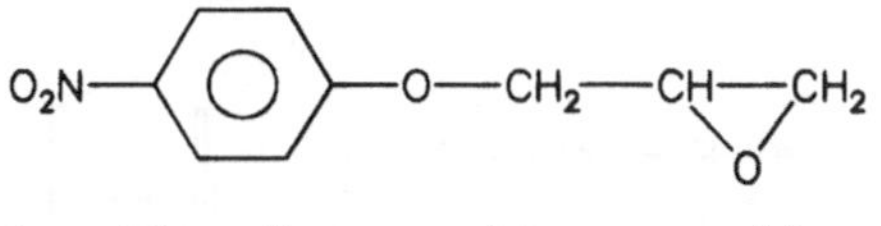

$$O_2N{-}\langle O \rangle{-}O{-}CH_2{-}\underset{\underset{\displaystyle O}{\diagdown \diagup}}{CH}{-}CH_2$$

In acid medium epoxides react with carboxylate groups [2692]. As shown [2693], this reagent forms an ester bond with the Asp32 residue in pepsin. Besides, under certain conditions group Asp215 and residue Met180 are also affected.

X-ray studies of penicillopepsin inhibited by epoxycompounds have revealed [1345] that carboxylate-ion of Asp32 attacks the epoxide C1-atom, while the reaction with Asp215 runs at the C2-atom.

Metal-Containing Amide Hydrolases. For this enzyme group chelating substances (ethylene diamine tetraacetic acid; 1,10-phenanthroline) binding the metal ion are mostly specific reagents (Sect. 5.9.2). Besides, reversible inhibitors described in Sect. 5.9.3 react very specifically. Group-specific irreversible inhibitors for these enzymes have not been found. Information is available only on the action of some inhibitors on certain representatives of metal-containing amide hydrolases. So the action of chloroacetylhydroxamate of methyl leucine ester and corresponding nitrile on thermolysin has been described [2694] as

$$
\underset{\underset{CH_2CH(CH_3)_2}{|}}{ClCH_2CON\overset{\overset{OH}{|}}{CH}COOCH_3}
\qquad\qquad
\underset{\underset{CH_2CH(CH_3)_2}{|}}{ClCH_2CON-\overset{\overset{OH}{|}}{CH}CN}
$$

The action of these compounds is based apparently on the formation of chelate with Zn^{2+} and on esterification of residue Glu143 in thermolysin [2695].

Haloidketones, the derivatives of unprotected leucine, act on leucine aminopeptidase as reversible inhibitors [2696].

5.9.5 "Syncatalytic" Inhibitors

Certain substrates and substrate-like compounds react with amide hydrolases so that they undergo a hydrolytic cleavage and inactivate the enzyme at the same time. Such compounds can be called "syncatalytic" inhibitors. They are also termed "suicide substrates".

An example of these compounds are 5-acylhydroxyoxazoles [2697]:

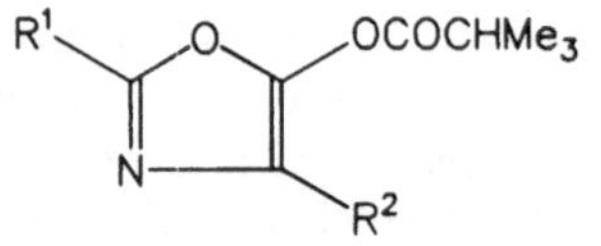

X: R^1 = Ph,
   R^2 = $-(CH_2)_3NHC(=NHNO_2)NH_2$;

XI: R^1 = Ph, R^2 = Me .

These compounds acylate the catalytically active serine residue and form a stable pivaloyl-enzyme, when interacting with thrombin, chymotrypsin or trypsin (but not with plasmin):

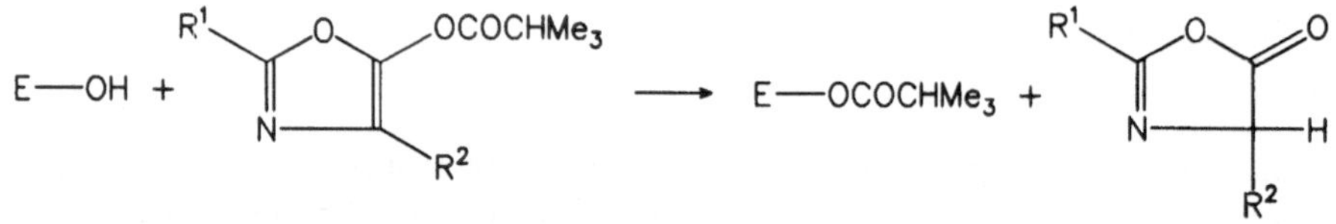

Acylation of thrombin by compound XI proceeds with the rate constant $k_i/K_i = 2.6 \times 10^4\ M^{-1}\ min^{-1}$.

Another type of the compounds inhibiting elastase, chymotrypsin, trypsin, and cathepsin G embraces compounds (XII) and (XIII) [2698]:

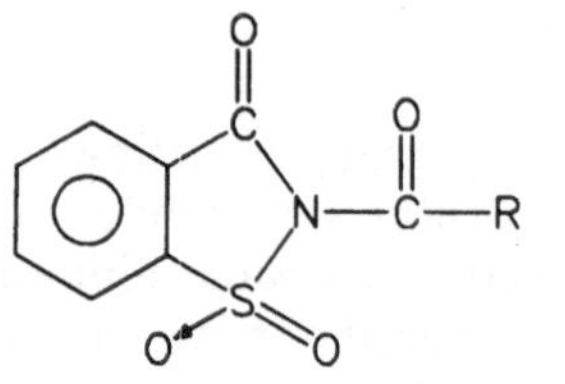

XII 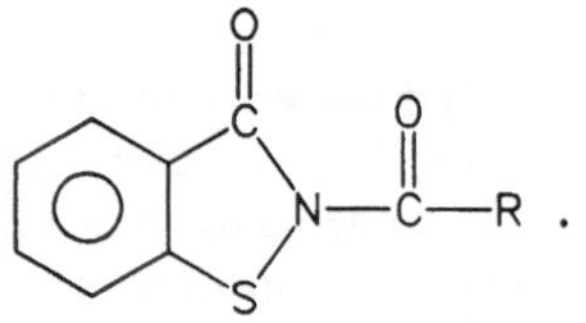   XIII

Their action also provides for acylation of the enzyme. A similar mechanism of inactivation is postulated as regards benzoxazinones (XIV) [2699], protected pyrones (XV) [2700], isocoumarins (XVI) [2701] and enollactones (XVII) [2702]. The rate of deacylation of trypsin acylated by compound (XVI) [X = Y = H, R = (CH$_2$NH$_3^+$)] is only 5.3×10^{-4} s^{-1}. X-ray analysis of the complex of compound (XV) with chymotrypsin reveals [2700] that an inhibitor forms a salt bridge with imidazole of His57 and, apparently, protects an approach of the water molecule upon deacylation.

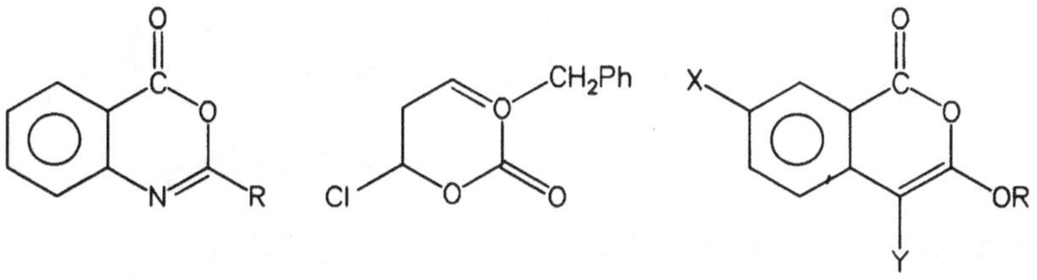

XIV   XV   XVI

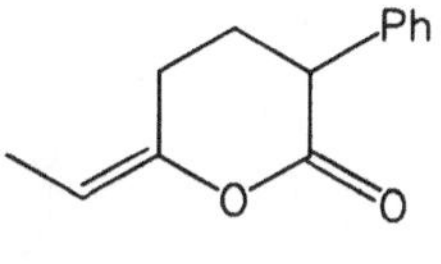

XVII

Elastase effectively reacts with carbamates of type XVIII [2703] via intermediate (XIX):

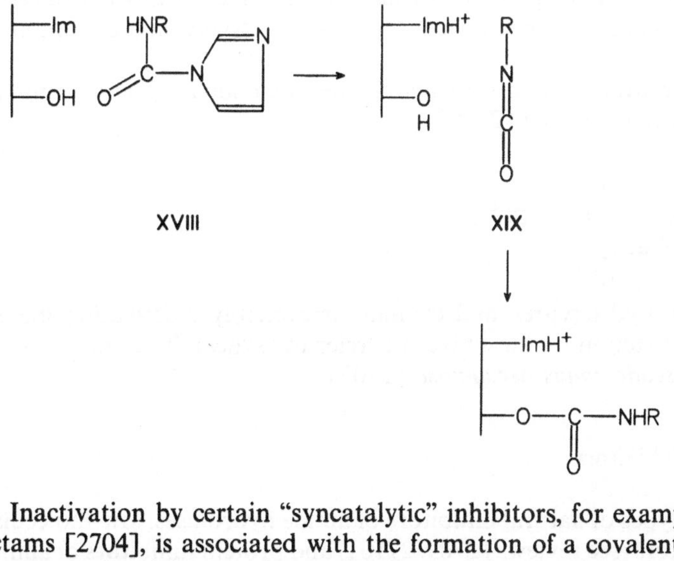

XVIII   XIX

Inactivation by certain "syncatalytic" inhibitors, for example haloidenollactams [2704], is associated with the formation of a covalent bond with an enzyme nucleophil occurring at the stage of the intermediate acylenzyme.

Another type of inhibitors is based on the intermediate formation of carbonium ions [2705]:

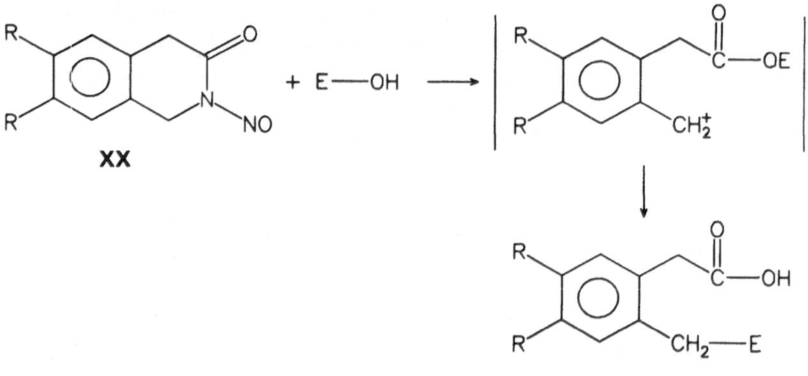

Compound XX (R = OMe) inhibits chymotrypsin by 98% for 5 min.

Nitriles of type (XXI) inactivate carboxypeptidase A via the stage of a formation of intermediate ketimine (XXII) [2706]:

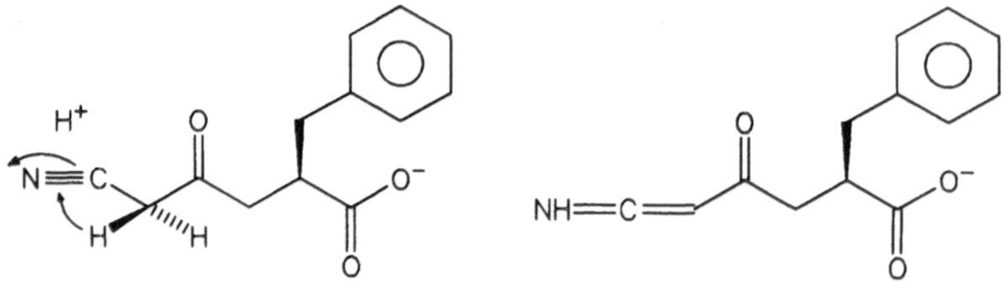

Many antibiotics of the penicillin group inactivating D-alanine carboxypeptidases and β-lactamases due to acylation of the catalytically active serine residue by the β-lactam carbonyl group pertain to "syncatalytic" inhibitors [256, 263, 2707].

Besides, 6-β-brompenicillic acid inactivates β-lactamase I and forms an acyl derivative of the enzyme [2708]

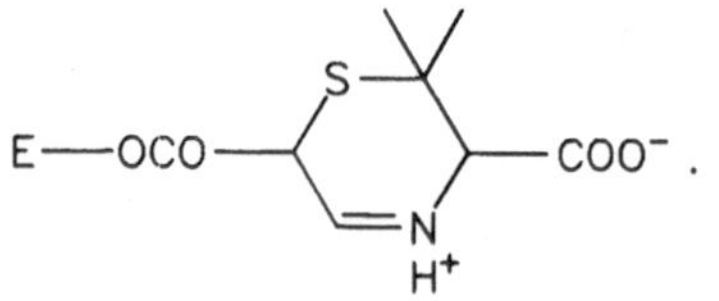

Finally, urea, hydroxyurea and cyanate, irreversibly inactivating the enzyme due to the reaction by the active site, refer to "syncatalytic" inhibitors of amidase from *Pseudomonas aeruginosa* [2103].

5.9.6   Natural Inhibitors

There exist two types of natural inhibitors of amide hydrolases: low-molecular, microbial inhibitors (see reviews [2709, 2710]) and protein inhibitors of animal or plant origin [2711–2714].

Microbial Inhibitors are produced by a diversity of *Aspergillus* strains. These compounds effectively and specifically inhibit proteases of various classes so that they can be referred to a number of group-specific inhibitors. They contain very unusual amino acid derivatives:

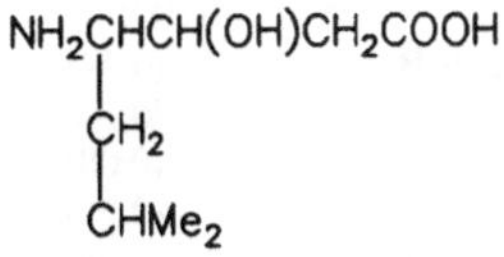

NH$_2$CHCH(OH)CH$_2$COOH

CH$_2$

CHMe$_2$

statin [S,S] (Sta)

NH$_2$CHCH(OH)COOH

CH$_2$

CHMe$_2$

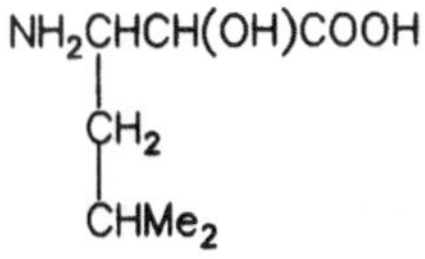

desmethylenestatin [R,S] (Dst)

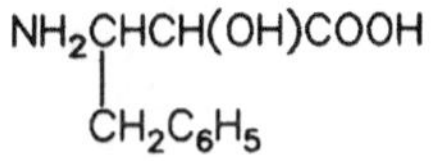

NH$_2$CHCH(OH)COOH

CH$_2$C$_6$H$_5$

4−phenyl−3−amino−2−

hydroxybutyric acid [R,S]

(Pab)

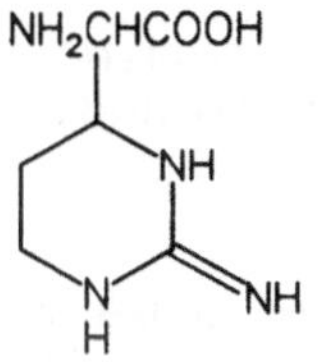

NH$_2$CHCOOH

NH

N

H

NH

α−[2−iminohexahydro−

4−pyrimidyl]−glycine [S,S]

(Pgl)

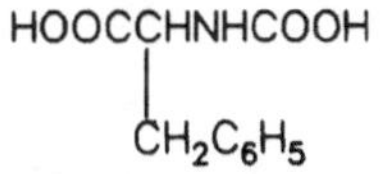

HOOCCHNHCOOH

CH$_2$C$_6$H$_5$

N−carboxy−

phenylalanine [S] (Cph)

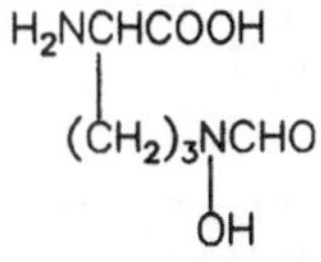

H$_2$NCHCOOH

(CH$_2$)$_3$NCHO

OH

5−(formylhydroxyamino)−

2−amino−valeric acid

[S] (Oav)

Table 63 presents the efficiency of a number of such inhibitors as regards amide hydrolases of different groups [2709, 2710, 2715–2722].

Interesting structure-functional studies have been made for a series of microbial inhibitors of amide hydrolases. So the action of various analogs and derivatives of pepstatin [2723–2727] has been investigated. Esterification of the C-terminal group appeared not to influence its inhibitory properties. However, acylation or removal of the β-hydroxy group in the amino acid residue of statin decreases the activity approximately 1000 times [2724]. Some fragments of pepstatin are still strong inhibitors ($K_i \approx 3 \times 10^{-6}$ M) [2725]. Their inhibitory activity is supposed to be stipulated by the similarity of the pepstatin structure with that of a transition state in the reactions catalyzed by aspartic proteases (about analogs of transition states see Sect. 5.9.7) [2725]. Identical concepts envisage phosphoamidone [2728]. Pepstatins with distinct acyl groups (R = *i*-Valeryl, Acetyl, Lactyl) inhibit aspartic proteases in a different way [2729]. Protease of the HIV-1 virus is slightly suppressed by

Table 63. Microbial inhibitors of amide hydrolases

Name	Structure[a]	Inhibited enzyme	K_i, nM or (I_{50} μg/ml)
Leupeptin	RCOLeuLeuArgH	Trypsin	130.00
		Plasmin	(100)
		Cathepsin B	(0.44)
		Acrosin	70
		Papain	(0.08)
Antipain	N-CphArgLeuArgH	Trypsin	(0.26)
		Cathepsin A	(1.2)
		" B	(0.6)
		Plasmin	(93)
		Papain	(0.16)
Chymostatin	NCphPglXaaPheH	Chymotrypsin	(0.15)
	Xaa = Leu(A), Ile(B), Val(C)	Cathepsin A	(62.5)
		" B	(2.6)
		" D	(49)
		Papain	(7.5)
Elastatynal	N-ClePglGlnAlaH	Elastase	240
Pepstatin	RCOValStaAlaStaOH	Pepsin	1
	(R = i-Pr, Ac, Lac)	Cathepsin D	10
		Renin	0.13
		Gastricsin	100
		Protease HIV-1	20
E-64	EpsLeuAbg	Cathepsin D	(240)
E-64c	EpsLeuDmpa	" L	(110)
Lochistatin	EtOEpsLeuDmpa	Papain	–
Bestatin	PabLeuOH	Aminopeptidase B	(160)
		Leucineamino-peptidase	(32)
Phosphoamidone	6-dManPLeuTrpOH	Thermolysin	28
Amastatin	DstValValAspOH	Aminopeptidase A	(0.54)
		Leucineamino-peptidase	(0.5)
Arphamenin	HArgCH$_2$CH(CH$_2$C$_6$H$_4$OHp)COOH [S, R]	Aminopeptidase B	(0.002)
		Carboxypeptidase A	(5.2)
Foroxymitin	Oav-Ser-dKOav	Dipeptidyl-carboxypeptidase	(7.0)

[a] Abbreviations see above, and also: ArgH – arginal; NCph – in the formation of the amide bond HN–COOH group is involved; PheH – phenylalaninal; NCle – N - carboxy - [S] - leucine; AlaH - alaninal; Abg - 4 - guanidino - 1 - aminobutane; Dmpa - 3,3 - dimethyl - 1 - aminopropane; 6 - dManP - 6 - deoxymannopyranosyl - 1 - phosphate; dkOav - diketopiperasin of Oav.

isovalerylpepstatin ($K_i \approx 1 \times 10^{-4}$ M) [2730], whereas acetylpepstatin is very effective ($K_i \approx 20$ nM at pH 4.7) [2731].

Different derivatives of the epoxy inhibitor of cysteine proteases E-64 have been studied [2720, 2732–2735] as regards the inhibition rate of papain and cathepsins B and H. The latter reacts with E-64 much slower than the first two enzymes. Investigation of various stereoisomers and analogs of bestatin [2717, 2736, 2737] have shown the importance of configuration of the C3-atom in the residue of 3-amino-2-oxy-4-phenylbutyric acid. The configuration of the rest of asymmetric atoms of bestatin influences the activity very little.

From the *Stachybotiys complementi* culture [2738, 2739] the interesting inhibitor (K-67) suppressing the complement system during the activation of the C5-factor has been isolated:

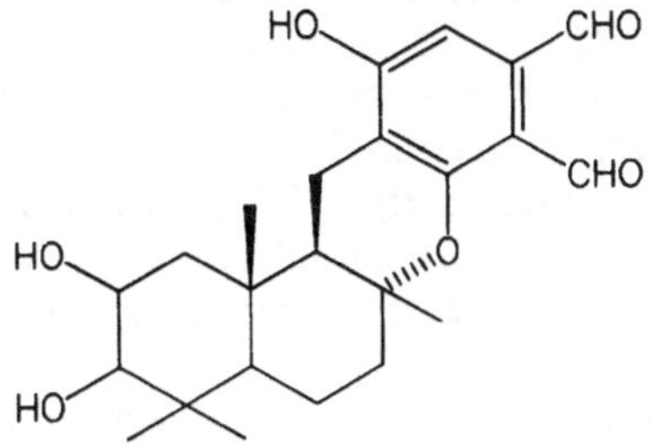

K-76

Cases when some proteases are insensitive to group-specific microbial inhibitors should be registered. So *Scytalidium lignicolum* proteases and protease A from *Aspergillus niger*, being aspartic amide hydrolases as regards their properties, are insensitive to pepstatin [909, 2740]. Besides, inversion of the leupeptine properties relative to the activity suppression of cathepsin B is observed upon changing the in vitro experiments to those on animals. Here leupeptine appeared to be an effective stimulator of cathepsin B [2741].

Protein Inhibitors. Biological liquids and tissues of animals and plants contain a large number of protease inhibitors. These are the proteins or glycoproteins of M_r 3000 to 700 000. However, most of the protein inhibitors studied contain 60–70 amino acid residues. The number of inhibitors known at present is very high. Information on this point is available [2711–2714, 2742–2758].

Along with the inhibitors of a wide spectrum of action, such as α_2-macroglobulin [2759], there are many suppressors which show a certain group-specificity.

Widespread are the inhibitors of serine proteases. They are classified into at least ten families [2749]. In these are involved the inhibitors of the family of the Kunitz pancreatic trypsin inhibitor [2711, 2712] as well as of the pancreatic secretory inhibitor (Kasal) [2745], a subtilisin inhibitor from actinomycetes [2760], etc.

The blood plasma is the richest source of the protein inhibitors of serine proteases. The content of inhibitors in plasma reaches 10% of the total content of the proteins [2758]. α_1-Antitrypsin (or α_1-inhibitor of proteases) shows the highest concentration in plasma [2761]. This protein (M_r 52 000 MW) inhibits a series of serine proteases, but it suppresses mainly the activity of leukocyte elastase [2762]. Its complete amino acid sequence [2763] and spatial structure [2764, 2765] are known. Plasma also contains antithrombin III [2743], α_1-antichymotrypsin [2751, 2766], α_2-antiplasmin [2767, 2768], the inhibitor of the protein C [2769] of the complement C1-factor [2742, 2770], as well as other inhibitors of the complement system [2771, 2772].

This group of inhibitors of a middle size has been called "serpins" (from *Serine Proteinase Inhibitors*) [2758]. Some inhibitors from other sources, for example insects, are also referred to "serpins" [2773].

Inhibitors from human and animals pancreas are mainly the basic Kunitz trypsin pancreatic inhibitor and a Kasal trypsin secretory pancreatic inhibitor mentioned above. These are small proteins (58 and 56 amino acid residues, respectively) suppressing the activity not only of trypsin, but also of some other serine proteases. The first has been structurally studied in detail (Chap. 6). The primary structure of the second has been clarified lately [2774]. The Kunitz inhibitor is contained not only in pancreas, but also in spleen [2775] and plasma [2776]. Effective inhibitors of serine proteases have been found in the spermatic liquid [2744, 2777] and cow colostrum [2746].

The inhibitors of serine proteases of a wide spectrum of action are ovomucoids from hen and turkey eggs and from snail as well [2714, 2778–2780]. Ovomucoid of the Japanese snail is a three-domain protein (186 residues). One of its domains (domain III – 56 residues) on being isolated suppresses the activity of serine proteases of animals and microorganisms. Its spatial structure has been deciphered [2778].

A rich source of inhibitors of serine proteases is a secretion of the medical leech. Hirudine, a protein of 76 amino acid residues, is rather active as regards thrombin [2747, 2781–2784]. The other inhibitor – eglin – is specific as regards elastases and cathepsin G [2755]. An effective inhibitor of factor Xa – antistazin [2785, 2786] – has been found in the other leech (*Haementaria officinalis*). This protein containing 20 residues of hemicystine consists of 119 amino acid residues.

Snake venom (*Russel viper, Vipera ammodytes*, etc.) contains a large number of inhibitors of trypsin, chymotrypsin, kallikrein and plasmin [2664, 2787].

Other inhibitors, nexins produced by fibroblasts and glial cells, are close to serpins and suppress the activity of thrombin and urokinase [2788].

Finally, diverse inhibitors of serine proteases are contained in plant seeds, these are the seeds of soy [2788] and lima beans [2789], barley [2790], squash [2791], etc., as well as in microorganisms, for example the inhibitor of subtilisins from *Streptomyces* [2792], and also the inhibitor of plasminogen activator from *Escherichia coli* [2793].

The activity of cysteine proteases is subjected to regulation by many specific inhibitors [2714]. Of much value are the inhibitors of tissues and biological liquids of mammals. They are divided into two large groups: low molecular, which involve cystatins and stephins, and high molecular – mainly kininogens [2794].

Cystatins are the proteins containing about 115 residues of amino acids and two disulfide bonds close to the C-terminus of the molecule. This group embraces mostly studied cystatins 1 and 2 from the egg-white of hen [2795–2797] and a set of analogous proteins from tissues and plasma in human and rat, colostrum, snake venoms, etc. [2798–2803]. Egg cystatins 1 and 2 differ in the presence of the phosphate residue bound to Ser80 in cystatin 2 [2804]. Stephins are the proteins of about 100 amino acid residues and, as a rule, free of S–S bridges [2805–2808].

Lately, it has been found out that both low- and high-molecular forms of kininogens are effective inhibitors of cysteine proteases containing two to three regions similar by structure to cystatins [2794, 2809–2811]. Evolutionary relationship of the three groups of inhibitors are illustrated by Fig. 86 [2794, 2812].

All these inhibitors suppress the activity of papain, ficin, cathepsins B, L and H, dipeptidylpeptidase I in picomolar concentrations [2714, 2813, 2814], but do not influence calpains. The latter are suppressed by the specific inhibitor of 718 amino acid residues and containing four domains, each of them interacting with calpains [2815, 2816].

Inhibitors of cystein proteases are widespread in plants [2817–2819].

Less information is available on the protein inhibitors of aspartic and metal-containing proteases. Pepsin and carboxypeptidases are suppressed by an inhibitor for helminths (*Ascaris*) [2820, 2821]. A renin-binding protein (M_r 56 000 MW), endogenic inhibitor of kidney renin, is known [2822].

Inhibitors of cathepsin D, carboxypeptidase A, and thermolysin have been found in potatoes [2823, 2824] and tomatoes [2825]. These are small proteins of 38–39 amino acid residues. Thermolysin inhibitor from *Streptomyces* contains 102 residues [2826]. The protein of M_r 36 000 MW inhibiting collagenase has been isolated from rabbit bone tissue [2827]. This protein is similar to the tissue inhibitor of metallo-proteinases [2828].

The blood plasma inhibitor of wide action – α_2-macroglobulin – occupies a special place [2750, 2757, 2759]. This is a tetrameric protein (M_r 725 000 MW) consisting of two noncovalently bound dimers, each containing two chains bound by a disulfide bond. It inhibits almost all types of proteases [2750, 2759, 2829].

Actually, all protein inhibitors contain conservative regions, in their sequence being their active sites (Table 64).

There are inhibitors of serine proteases which influence only, or virtually, the enzymes of the trypsin group, and those specific to chymotrypsin-like

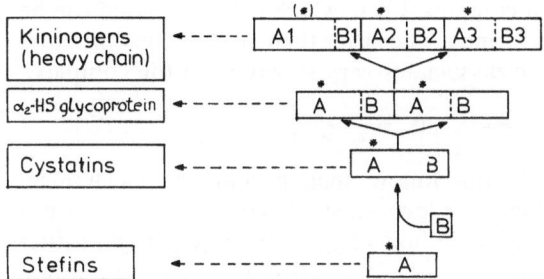

Fig. 86. Evolutionary interrelation of inhibitors of animal cysteine proteases; *A* N-terminal fragment of "internal long repetition" of heavy chain of human kininogens; *B* corresponding; *C* terminal fragment *Asterisks* indicate the regions for potential active sites [2749, 2812]

Table 64. Amino acid sequence within the active sites of certain protein inhibitors [2745, 2758, 2812, 2815, 2830]

Inhibitor	Sequence within active site[a]
	⬇
α_1-Antitrypsin	ProMet-SerIleProPro
α_2-Antiplasmin	Ser-ArgMetSerLeu
Antithrombin III	GlyArg-SerLeuAsnPro
α_1-Antichymotrypsin	LeuLeu-SerAlaLeuVal
C1-Inhibitor	AlaArg-ThrValLeuVal
Kunitz's pancreatic inhibitor	CysLys-AlaArgIle
Kazal secretory inhibitor	TyrLys-IleTyr
3rd Domain of turkey ovomucoid	ProArg-AspTyr
Russel inhibitor of adder venom	CysArg-GlyHis
Trypsin inhibitor from soybeans	ThrLys-SerAsnProPro
Cystatins	GlnVal-ValAlaGly
Calpains inhibitor	ThrIle-ProProXTyrArg

[a] The arrowhead shows on cleavable or potentially cleavable amide bond.

proteases. However, so-called multivalent, or "two-head", inhibitors suppressing the activity of trypsin and chymotrypsin are widely spread.

Of crucial importance for the activity of protein inhibitors is a relatively small region in the sequence covering the residues of basic amino acids (Arg, Lys) for the inhibitors of tryptic enzymes, or the residues of aliphatic amino acids (Leu, etc.) for chymotrypsin-inhibiting proteins. The two sequences enter the single molecule in multivalent inhibitors. Essential homology is observed in the sequence close to the active sites of inhibitors from various sources, but residues Lys and Arg can vary even for one inhibitor from different animal species.

The set of semisynthetic pancreatic trypsin inhibitors (BPTI) has been obtained. In these inhibitors the Lys15 residue is changed by Gly, Ala, Val, etc. [2831]. Such replacement increases the affinity of the inhibitor to elastase.

There has been proposed [2749] a standard kinetic mechanism of inhibition true for most of the inhibitors of serine proteases and for a series of other protein inhibitors. It involves first the formation of the noncovalent complex EI being isomerized into the tight complex (C), in which a definite bond can be cleaved, and an "acyl enzyme" is formed. The latter then deacetylates entailing a "modified" inhibitor (I*), which dissociates very slowly from the complex:

$$E + I \rightleftharpoons EI \rightleftharpoons C \rightleftharpoons EI^* \rightleftharpoons E + I^*.$$

However, many inhibitors do not follow such a standard mechanism [2757]. The pancreatic trypsin inhibitor forms a stable compound with a noticeably distorted (pyramidalized) amide bond Lys15-Ala16 (Chap. 6), rather than a covalent bond with the serine residue of the active site. This is confirmed by an ability of the modified enzymes (anhydrotrypsin, N-$^{\varepsilon 2}$-methylhistidine-57-chymotrypsin) for the effective binding of the inhibitor by the active site [2081, 2832].

Of much importance is segment Gly9–Ala10 in four cystatins besides a nonconservative region in Table 64. The segment presumably might be cleaved by an enzyme [2796], though according to X-ray analysis this bond is located far from the thiol group of the active site [2797]. On the other hand, the site-directed mutagenesis in the relation of sequence GlnValValAlaGly does not influence the binding of cystatin A by cysteine proteases [2833].

A covalent bond with the enzyme is formed by such inhibitors as α_2-macroglobulin [2834]. To explain the inhibitory properties of this protein a hypothesis of "trapping" has been proposed [2835]. It suggests that the enzyme-inhibitor binding is accompanied by the cleavage of the inhibitor and substantial conformational rearrangements, which entail free SH groups being masked in the native α_2-macroglobulin as thioester with Glu residues. Carboxyl groups of Glu form then a covalent bond with an inhibited enzyme [2759, 2836–2839]:

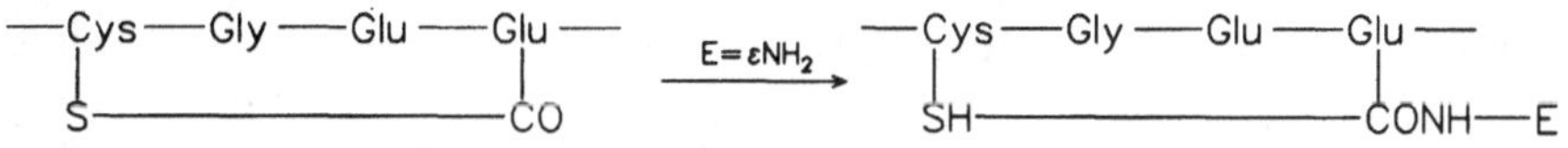

$$\begin{array}{ccc}
\text{—Cys—Gly—Glu—Glu—} & & \text{—Cys—Gly—Glu—Glu—} \\
| \qquad\qquad\qquad\qquad | & \xrightarrow{\text{E}=\varepsilon\text{NH}_2} & | \qquad\qquad\qquad\qquad | \\
\text{S————————CO} & & \text{SH————————CONH—E}
\end{array}$$

Binding of the inhibitor of carboxypeptidase A from potatoes to the enzyme causes the cleavage of the C-terminal amino acid of the inhibitor [2840]. However, the shortened inhibitor is still capable of binding to the enzyme. The modification of the active site of the Glu270 residue in carboxypeptidase A prevents binding of the inhibitor from *Ascaris* [2841, 2842].

5.9.7 Inhibitors – "Transition States Analogs"

As early as 1946 Pauling [2843] stated that the substrate in the transition state of the enzymatic reaction must bind to the enzyme more effectively than in the ground state. Only in the 1970's has been this idea developed [2844, 2845] and attempts taken to evolve stable inhibitors, which mimic the transition state (see review [2846]). These first inhibitors of amide hydrolases – alkylboronic acids – were obtained and studied in 1968 [2847]. Lately, numerous derivatives of boric acids including peptidylboronic acids have been synthesized [2848–2853]. These compounds inhibit not only serine proteases, but also lactamases [2854, 2855], aminopeptidases [2856] and some esterases [2857, 2858].

Though the derivatives of boric acids – analogs of unspecific substrates of serine proteases – are very weak inhibitors (Table 65), peptidylboronic acids inhibit in nanomolar concentrations.

Alkylboronic acids are confirmed by X-ray [2874–2877] and ^{11}B-NMR [2878] studies form a complex with serine proteases structurally identical to tetrahedral intermediates (Fig. 87).

Table 65. Inhibition of amide hydrolases by certain analogs of transition states

Enzyme	Comp. No.	Inhibitor	K_i, nM	References
α-Chymotrypsin	1	AcLeuPheCH$_3$	2.5×10^4	[2859]
	2	CH$_3$(CH$_2$)$_5$B(OH)$_2$	2.7×10^6	[2847]
	3	C$_6$H$_5$(CH$_2$)$_2$B(OH)$_2$	9×10^4	[2848]
	4	C$_6$H$_5$CONHCH$_2$B(OH)$_2$	5×10^3	[2849]
	5	C$_6$H$_5$CH$_2$CH(NHCOCH$_3$)B(OH)$_2$	2×10^3	[2850]
	6	C$_6$H$_5$CH$_2$CH$_2$CHO	3×10^5	[2860]
	7	MeOSuc (Ala)$_2$ProNHCH(R)B(OH)$_2$		
	8	Substance 7, R=CH$_2$C$_6$H$_5$	0.16	[2852]
Leukocyte elastase	9	„ R=CHMe$_2$	0.57	„
Chymase	10	Substance 8	58	[2853]
α-Lytic protease	–	„	68	[2861]
Mesenterico-peptidase	11	CH$_3$(CH$_2$)$_7$B(OH)$_2$	1.58×10^6	[2862]
	12	C$_6$H$_5$(OH)$_2$	0.63×10^6	„
Subtilisin BPN′	–	Substance 11	1.5×10^6	[2863]
	–	„ 12	6.2×10^6	[2864]
Penicillinamidase	13	CH$_3$(CH$_2$)$_4$B(OH)$_2$	830	[2865]
Trypsin	–	Substance 1	89×10^6	[2866]
	14	NH$_2$(NH=)CNHC$_6$H$_4$CH$_2$COCOOH	650	[2867]
Pancreatic elastase	–	Substance 8	0.25	[2852]
Asparginase Escherichia coli	15	HOOCCH$_2$CH(NH$_2$)CHO	3.6	[2868]
Papain	16	C$_6$H$_5$CONHCH$_2$CHO	2×10^3	[2869]
	17	AcPheNHCH$_2$CHO	5.2	„
	18	ZLeuNleH[a]	138	[2870]
Calpain I	–	Substance 18	52	„
Carboxypeptidase A	19	HOCCH$_2$CH(CH$_2$Ph)COOH	480	[2871]
Leucineamine-peptidase	20	NH$_2$CH(CH$_2$Ph)BO$_2$R′[b]	60	[2872]
Renin	21	BocPheValPheCF$_2$CH$_2$NH-i-Bu[a]	3.5	[2873]

[a] $[I]_{50}$.
[b] R′ – pinacolic ester.

So the complexes of these inhibitors (XXIV) really remind one, due to their structure, of a transition state or a tetrahedral intermediate (XXIII):

XXIII   XXIV   XXV

However, according to data on ^{15}N-NMR, phenylboronic acid forms a covalent bond with histidine, but not with serine of the active site of α-lytic protease (XXV) [2879]. Apparently, such a structure is typical for the complexes with nonspecific substrate analogs.

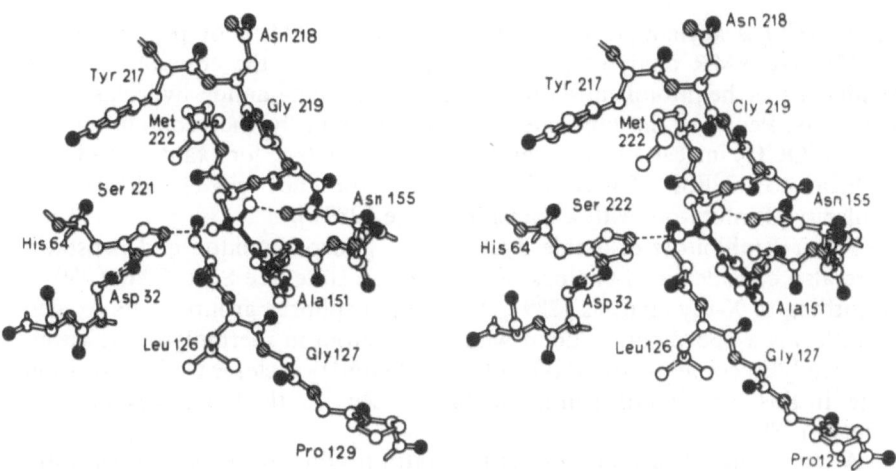

Fig. 87. Stereopair of the active site region of subtilisin BPN' in complex with phenylboronic acid (*solid line*) [2875]

Binding of alkylboronic acids to enzymes is accompanied by conformational rearrangements, which are detected by spectral changes in the protein [2880].

The complex formation of alkylboronic acids with some enzymes is gradual, first a coordination bond with imidazole of the active site (EI) probably occurs, then the serine residue is involved (EI*) [2861, 2864, 2880]:

$$E + I \underset{}{\overset{K_i'}{\rightleftharpoons}} EI \underset{k_{-2}}{\overset{k_2}{\rightleftharpoons}} EI^* \,,$$

where apparent constant of inhibition

$$K_i = K_i'\left(1 + \frac{k_2}{k_{-2}}\right).$$

The $K_i' = 0.6$ mM and $k_2/k_{-2} = 9.25$ have been defined for the reaction of phenylboronic acid with subtilisin BPN' [2864], whereas the binding of compound 8 (Table 65) to α-lytic protease is characterized by the $K_i' = 6.4$ nM and $k_2/k_{-2} = 9.6$ [2861].

Aminoacyl- and peptidylaldehydes [2860, 2870, 2881–2883] including native aldehydes – leupeptine, antipain, chymostatin, elastatinal – are considered as "transition states analogs" for amide hydrolases (see Sect. 5.9.6). The inhibiting ability of aldehydes depends on the formation of hemiacetals or thioacetals with catalytically active groups of serine or cystein proteases [2872, 2884, 2885], and is accompanied by conformational rearrangements in the protein [2886].

In Sect. 5.9.3 I have already mentioned as inhibitors peptidylfluoromethylketones of a general formula:

$$R–COCF_2–X \,,$$

where RCO is acylamino acid or peptide, and X is H, F or peptide. X-ray [2887] and NMR analyses [2888, 2889] reveal, that the compounds form hemiketals or hemithioketals with serine and cysteine amide hydrolases, respectively. Peptides, the substrate analogs, act similarly; they contain a keto-group $COCH_2$ instead of a cleavable peptide group (see, for example, [2890]). A linear correlation of $\lg K_i$ with $\lg K_m/k_{cat}$ (but not $\lg K_m$) indicates the similarity of inhibitors with a transition state [2891].

Natural inhibitors of amide hydrolases – phosphoamidon and pepstatin – are also considered as analogs of a transition state (see Sect. 5.9.6) [2891]. According to X-ray analysis [2892, 2893] a phosphorus group via its oxygen atom forms a coordination bond with a zinc atom in thermolysin; however, a thermolysin complex with $(HO)_2 P(O)NHLeuNH_2$ is closer to the transition state. In this case two oxygen atoms bind to Zn, i.e., the latter is pentacoordinated [2893].

A tetrahedral β-carbon atom of the statin residue in the pepsin-pepstatin complex [2725] mimics the transition state (for details, see Chap. 6). The complex formation of pepstatin and pepsin proceeds stepwise [2894].

Are the inhibitors – "transition states analogs" – real analogs of the substrate transition state in hydrolysis? In my opinion, this term seems rather doubtful.

Binding of boric acid to, for example, chymotrypsin and its binding to a model compound – amide of salicylic acid – are approximately the same ($\Delta G_0 = -1.4$ and -2.1 kcal/mol, respectively [2848]), so the binding of boric acid is even more effective in the model experiment. At the same time, a free energy of the transfer of the phenylethyl group from water to octanol (-4.2 kcal/mol [1971]) is equal to that of the transfer of this group from water into the cavity of the enzyme active site. Thus the overall change of the free energy upon the binding of phenylethylboronic acid to the enzyme (-5.6 kcal/mol) is not higher, but even lower, than the free energy of binding of this compound to the model substance and the energy of transfer to nonpolar surrounding (-6.3 kcal/mol). A high affinity of the "transition state analogs" is determined by additional ($\Delta G_0 = 3.7$ kcal/mol [2468]) interactions in the enzyme active sites compared, for example, with ester of hydrocinnamic acid. The same can be true for aldehydes, which form a covalent bond with a catalytically active enzyme group. As to pepstatin, its binding to pepsin ($\Delta G_0 = -12.4$ kcal/mol) is getting worse when hydroxyl groups (imitating, as supposed [2725, 2895], the structure of the transition state) are removed, and decreases the free energy by only 3.2 kcal/mol [2724]. This may be attributed to the hydroxyl formation of weak hydrogen bonds into enzyme. In the best case, this type of inhibitors might be termed an analog of the tetrahedral intermediate.

Alkylboronic acids appeared to be of use for analyzing the topography of the active sites of serine proteases [2847, 2862, 2863]. The study of the K_i dependence on the length of the hydrocarbon chain yields information on the distance between the catalytic and sorption sites as well as on the stretch of the latter (Fig. 88).

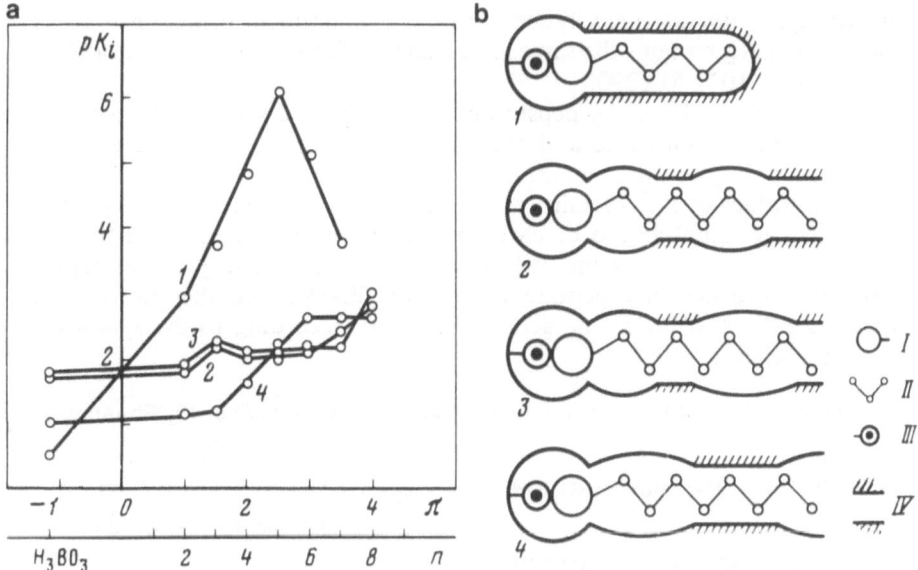

Fig. 88a, b. Association constants for alkylboronic acids vs length of the alkyl chain (and hydrophobicity) (**a**) and scheme of topography of amide hydrolases active sites (**b**); *1* penicillinamidase; *2* mesentericopeptidase; *3* subtilisin; *4* chymotrypsin; *I* $B(OH)_2$; *II* alkyl chain; *III* catalytic enzyme groups; *IV* hydrophobic site

5.10 Activation

A series of organic compounds activate the enzymatic hydrolysis. Substrate-like compounds and those that mimic certain substrate fragments may be referred to these activators. An ability for activation with an excess of the substrate can be traced for some enzymes (see Sect. 5.9.3). Besides, metal ions are important from the viewpoint of their activation. As an example, of much interest is here the plasma protein C; its activity shows linear dependence on the concentration of Na^+-ions of the second power [2896, 2897].

The kinetic activation is described by the same generalized Eq. (25) (Sect. 5.9.1) as the action of inhibitors at $\beta > 1$ and $\alpha \leqslant 1$. At $\alpha > 1$, depending on the ratio of α and β, inhibition or activation is observed.

Catalysis by carboxypeptidase A is a typical example of the substrate activation, where ES_2 complex is transformed into the reaction products two to five times faster than complex ES [1738]. Substrate activation has been also noticed at catalysis by leukocyte elastase [2898] and by some other enzymes.

Introduction of a series of peptides not being hydrolyzed by carboxypeptidase A also enhances the catalysis by this enzyme [2899]. Such peptides show higher affinity than the substrates to the enzyme, while the EA complex (where A is an activator) binds much better the substrate than a free enzyme.

Aspartic proteases, pepsin and penicillopepsin, act identically [2900–2904]. For penicillopepsin the LeuTyrNH$_2$ hydrolysis is increased by nonhydrolyzed peptide LeuGlyLeu, whereas enhancement of k_{cat} ($\beta \approx 10$) is observed, but K'_m remains the same. The enzyme affinity for the activator (K_A) is equal to 2×10^{-3} M [2900].

A thorough study of the pepsin activation [2901–2904] has shown that depending on the substrate and the activator structure the activation can proceed according to noncompetitive ($\alpha = 1$, $\beta > 1$) and mixed (α and $\beta > 1$) types. The values of β in certain cases reach 70. The presence of at least one unprotected terminal amino or carboxy group in the substrate and the activator appeared indispensable. It has been established [2902] that at pepsin catalyzed hydrolysis of dipeptide HPhe(NO$_2$)PheAmp in the presence of tripeptide ZPheAlaAlaOH the activation occurs according to the synthesis-hydrolysis scheme:

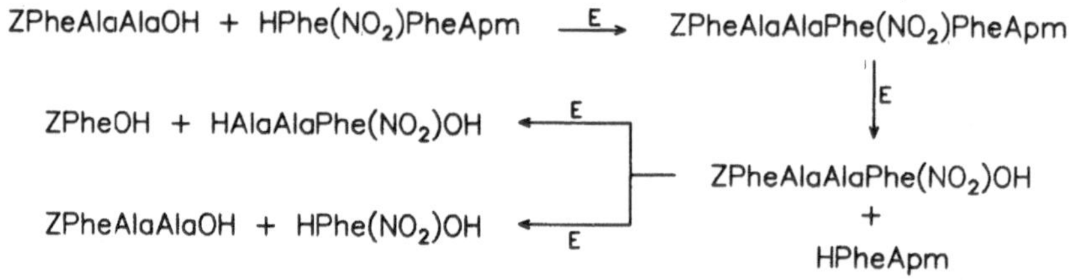

ZPheAlaAlaOH + HPhe(NO₂)PheApm ──E──▸ ZPheAlaAlaPhe(NO₂)PheApm

ZPheOH + HAlaAlaPhe(NO₂)OH ◂──E──

ZPheAlaAlaOH + HPhe(NO₂)OH ◂──E──

                                                        │ E
                                                        ▼

ZPheAlaAlaPhe(NO₂)OH
+
HPheApm

Similar data have been obtained for other peptides [2903, 2904]. Synthesis-hydrolysis apparently is of importance for activating hydrolysis by other enzymes.

Binding of the activator by aspartic proteases entails certain conformational rearrangements in the enzyme registered by a spectral change of the circular dichroism [2900].

A purely sorption type of the activation is observed upon trypsin-catalyzed hydrolysis of short-chain esters of acylamino acids in the presence of alkylammonium ions [2905–2907]. In this case, at the constant length of the substrate side chain the shorter is an alkyl chain of the activator, the more effective will be the catalysis. Alkylammonium ions perhaps interact with the trypsin cation site, thus providing a right orientation of the substrate or inducing conformational changes in the enzyme, which promote the catalysis.

Data are available [2908] on the existence of the protein activator ($M_r \approx 20\,000$ MW) of cathepsin D from bovine spleen, which increases enzyme activity as regards the protein substrates by 50–300%. The enzyme is activated also by ethyl ester of glycine [2909], phospholipids [2910] and ATP [2911, 2912]. ATP-dependent protease La from *Escherichia coli* is activated by DNA [2913].

5.11 Allosteric Effectors

Conforming to the meaning of the word "allosteric" ($\alpha\lambda\lambda o\zeta$ – other, $\sigma\tau\varepsilon\rho\varepsilon o\zeta$ – space) some of the above activators as well as noncompetitive inhibitors of the enzyme may be referred to this type. So salts of trialkylammonium derivatives of azobenzene have been discussed as allosteric activators [2914]:

$$R \cdot \underset{\underset{CH_3}{|}}{\overset{\overset{CH_3}{|}}{N}} \cdot CH_2 \overbrace{\hspace{1cm}} -N{=}N- \overbrace{\hspace{1cm}} CH_2 \cdot \underset{\underset{CH_3}{|}}{\overset{\overset{CH_3}{|}}{N}} \cdot R \cdot$$

They enhance the chymotrypsin catalyzed hydrolysis of anilides of N-acylamino acids. However, these compounds presumably prevent nonproductive binding typical of anilides, since no activation is observed when esters, hydroxyamides of acylamino acids, and proteins as well are utilized as substrates.

Properties of the trypsin allosteric effector have been found in dansyl-arginine. This compound binds to the enzyme within the active site (at $K_m = 6.7 \times 10^{-3}$ M) and outside (at $K_A = 4.8 \times 10^{-4}$ M).

More typical is the allosteric activation of carboxypeptidase B as regards its unspecific substrates of type ZGlyPheOH, when β-phenylpropionic acid is added [2916]. At high concentrations of NaCl, already without effectors the S-shape dependence of the hydrolysis rate of the marked substrates on their concentration is observed. The activator at low concentrations increases the hydrolysis rate, while at high concentrations the rate decreases. These facts have been interpreted as a result of the presence of two binding sites of hydrophobic compounds in the enzymes, whereas the state of one of them – catalytic – depends on the ligand binding in the second site.

Compounds such as lysin, 6-aminohexanoic acid, etc. act similarly as regards the plasmin-catalyzed hydrolysis of benzoylarginine ester [2917].

The rate dependence on the substrate or effector concentration typical of allosteric enzymes can be observed when the enzyme undergoes slow conformational transitions, for example the alkaline conformational transition in chymotrypsin (see Sect. 5.2.2) [2918], or when the enzyme is capable of reversible association [2919]. The latter case has been studied in detail on the example of glutaminase from pig kidney [2920–2922]. Depending on the buffer solution (Tris, phosphate, or phosphate-borate) the enzyme exists in various oligomeric forms with diverse catalytic activity. A phosphate-borate form has apparent molecular weight of about 1.5×10^6, and forms the double-helix type structure [2922]. The enzyme is allosterically activated by the dye – bromthymol blue [2920].

Cl^--Ion is an allosteric effector of dipeptidylcarboxypeptidase (angiotensin converting enzyme) [2923, 2924]. For ATP-dependent La-protease ATP functions as an allosteric activator [2925].

Protease from sheep erythrocytes is a true allosteric protease [435]. This enzyme (M_r 340 000 MW) consists of six identical subunits. The kinetics of this enzyme-catalyzed hydrolysis of tripeptide GlySerAla (but not GlyMetAla!) is cooperative with the Hill coefficient $n = 1.8$. A close value of the Hill coefficient has been found [957] for the yeast aminopeptidase I. This enzyme (600 000 MW) is composed of 12 subunits activated by Zn^{2+} and Cl^- (or Br^-). The halogen anions act so that a positive enzyme cooperativity decreases by the substrate or Zn^{2+} ions, making the rate dependence on the concentration close to hyperbolic. The action of cathepsin C is similar [2926].

Lastly, of interest are the allosteric action of the C1q-factor of the complement system on the activation of C1s-C1r-factors [2927], and the identical action of streptokinase on the plasminogen activation [2928].

5.12 Autolysis

The enzyme autolytic cleavage is a property typical of proteases only. This embraces a protease catalyzed hydrolysis of amide bonds in the molecules of the same enzyme subjected to partial, or complete, denaturation. The kinetics of the enzyme inactivation due to autolysis is described by the scheme:

$$E_N \xrightleftharpoons{K_D} E_D; \quad E_D + E_N \xrightarrow{k_i} \text{products}$$

where E_D is a denatured form of the enzyme, and the relative kinetic expression for the inactivation rate is [2478]:

$$\frac{d[E]_N}{dt} = k_i K_D ([E_N])^2 .$$

The initial rate of denaturation is shown to be

$$v_0 = \lim_{t \to 0} \frac{1}{t} \cdot \ln \frac{[E]_0}{[E]_t} ,$$

where $[E]_t$ is the E_N concentration in the t moment. From the dependence $\frac{1}{t} \cdot \ln \frac{[E]_0}{[E]_t}$ on t, when extrapolating to $t = 0$ we can get the value of $k_0 = k_t K_D$, i.e., the value of the effective rate constant of denaturation. For α-chymotrypsin these values appeared to be equal to 9 $M^{-1} s^{-1}$ at 40 °C and to 100 $M^{-1} s^{-1}$ at 49.5 °C (pH 7.6) [2478]. According to the equation of the inactivation rate, the latter depends on the enzyme initial concentration of the second power, therefore the autolysis rate enhances quickly with an increase of the concentration of protease solutions. An appreciable autolysis of trypsin occurs even at the storage of the lyophilized enzyme [292]. It is extremely effective in the pH-optimum of the enzyme action. Sometimes autocatalytic inactivation is observed at pH differing from the optimal [2929]. Even at purification of trypsin on sulfoethyl Sephadex at pH 2.6 a distinct autolysis is observed [2930]. The pepsin autolysis proceeds at its crystallization [2931], and in the solution under the action of various effects (Fig. 89) [2932]. The autolysis rate is different in the enzymes of diverse groups. Thus exopeptidases (carboxypeptidase A, etc.) are very stable as regards self digestion [1213]. Moreover, various isoenzymes of chymosin are distinctive by the autolysis rate [2933, 2934].

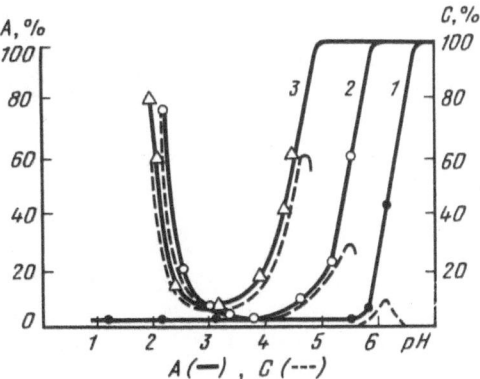

Fig. 89. pH-Dependence of pepsin activity (A) and concentration of the pepsin autolysis products soluble in trichloroacetic acid (C) [2932]; *1* in aqueous solution; *2* in 8 M urea; *3* in 4 M guanidinium chloride

Autolysis can enhance due to diverse compounds. So dithiothreitol and dithioerythritol stimulate papain autolysis, while mercaptoethanol does not influence the process [2934]. Chymotrypsin autolysis is promoted by the Evans blue dye [2935].

As supposed earlier, autolysis can result in the fragments of proteases keeping the catalytic activity (see, for example, [2936]). Really autocatalytic cleavage often leads to active forms (for instance, conversion of δ-chymotrypsin into α-chymotrypsin). But deep autolysis inevitably entails the enzyme inactivation.

5.13 Conclusion

The catalytic activity of amide hydrolases is altered due to different factors. Two of them are the most important: a change of the charge of the enzyme ionized groups, especially of the groups of the active site, and conformational rearrangements of the protein globule. At the activation of zymogens the formation of the catalytically active protein is mainly related to a change of the conformation, which results in the formation of the substrate binding active site. The change in pH, temperature, ionic strength of the solution entails local or more general conformational rearrangements of the enzyme molecule. Thus the conformational lability is presumably indispensable for the activity and regulation of amide hydrolases.

A great number of specific inhibitors allow not only alteration of the amide hydrolases activity, but also investigation of the structure of the active sites including interlocation of functionally important groups. This is of much value for crystallographic research into the enzymes.

Chapter 6   Enzyme-Substrate Complexes

In the foregoing, we have discussed the structure and the properties of the partners invovled in the enzymatic hydrolysis of the amide bond as well as the major experimental facts concerning the process itself. Here and in the next chapter we consider the events occurring in catalysis, analyzing these from the standpoint of the structure of the interacting substances. Unfortunately, at present this can be illustrated by only a few examples in which, due to advances in protein crystallography, a detailed picture of the events can be appreciated.

The concept of the formation of the enzyme-substrate complex underlies modern enzymology. This has been checked and confirmed by numerous direct and indirect methods, most of which have been discussed earlier and to which we will return later.

6.1   The Forces of Interaction

6.1.1   Dispersion Forces

An approach of two atoms to one another yields an interaction of two types – repulsion forces, which specify the interaction of electron shells, and attraction or dispersion forces [2937]. The latter obey a law of dependence proportional to the 6th power of the distance between the atoms, whereas the repulsion forces are proportional to the 12th power of the distance. The overall potential energy of interaction is written by the equation

$$U = A/r^{12} - B/r^6 , \tag{1}$$

where A and B are the constants (Fig. 90).

Attraction forces are initiated by mutual induction of the so-called instant dipole moment in neutral atoms or molecules due to local fluctuations of electron density. The distance corresponding to the minimum of the potential energy is termed Van der Waals radius of the atom (Table 66).

In complex molecules, in proteins mostly, the distances may be a little bit shorter [2940]. The energy of dispersion interaction is very small (Table 66) [2939]. However, many interacting atoms in the protein molecule or within the contact area of the enzyme and substrate make a great contribution to an overall energy of binding (see Sect. 6.3).

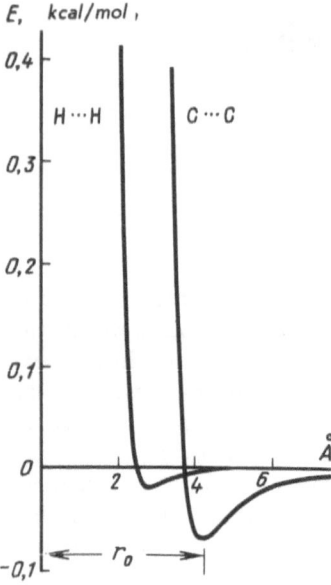

Fig. 90. Potential energy of interaction between hydrogen atoms and two (tetrahedral) carbon atoms in protein molecule; r_0 Van der Waals interatomic distance

Table 66. Van der Waals radii of atoms and groups; the energy of interaction between similar atoms [2938, 2939]

Atom or group	Radius, Å	Energy, kcal/mol
H	1.4	0.0186
O (in OH)	1.5	
O (in CO)	1.7	0.0519
O^- (in COO^-)	1.8	
C (trigonal)	1.7	0.4519
S	1.8	
N (amide)	2.0	0.1366
N^+ (ImH^+)	1.55	
CH (tetrahedral)	2.1	0.0679
CH (aromatic)	1.8	

6.1.2 Hydrophobic Interactions

Hydrophobic interactions arise from rearrangement of water hydrogen bonds in the vicinity of a nonpolar hydrophobic group.

As a result, a local order of the solvent increases, and this entails a decrease of the entropy in the system. To compensate an unfavorable change in the entropy, water displaces a hydrophobic molecule or a group into the hydrophobic region of the protein. Thus hydrophobic interactions develop (see reviews [2941–2944]).

Free energy of hydrophobic interaction can be detected from the equilibrium constant for the distribution of the given substance between water and a certain solvent, for example, n-octanol [2942]:

$$\Delta G_h = -RT\ln \frac{[A]_{oct}}{[A]_w} .$$ (2)

The interaction is characterized also by the Hansch constant of hydrophobicity:

$$\pi = \lg([A]_{oct}/[A]_w) .$$ (3)

The values of ΔG_h and π for amino acids are listed in Table 9.

Thorough analysis of the dissolving of hydrocarbons in water has revealed a high contribution of enthalpy to free energy of hydrophobic interactions [2945], i.e., for one CH_2 group $\Delta H_0 = -0.57$ kcal/mol, while $T\Delta S = +0.34$ kcal/mol. Hydrophobicity changes with a change of the surface area of the molecule interacting with water [2943, 2944]. As established empirically, a change of the energy of hydrophobic interaction per $1\,A^2$ of the area is 20–25 cal/mol.

The microenvironment of the active sites of many enzymes is less polar and shows less dielectric permeability than water, as follows from spectral changes of dyes upon their adsorption by enzymes [2946]. Dielectric permeability of the microenvironment of the chymotrypsin active site is evaluated by $\varepsilon < 10$, i.e., close to dielectric permeability of octanol. However, hydrophobicity of amide hydrolases active sites appears often to be higher than that of the model environment – octanol. Section 4.3.2.2 contains a series of examples on the correlation of the reaction rate constants and the hydrophobicity of substrate side chains with a slope over a unit. The same dependence is observed for the K_i values of certain inhibitors (for instance, of substituted formanilides [2000]). The thing is that hydrophobic interactions in water arise only if nonpolar compounds are incorporated, whereas the hydrophobic cavity exists in the enzyme even without a hydrophobic ligand. Therefore, depending on the structure of the ligand and the enzyme, hydrophobic interactions give a greater advantage in energy than the transfer of the substance from water into octanol.

6.1.3 The Hydrogen Bond

The characteristics of hydrogen bonds for a set of model amide-containing compounds have been described in Sect. 1.2.3 (see also review [72]). Note that the energy of a hydrogen bond dramatically decreases in polar media. The contribution of an uncharged hydrogen bond to the energy of the ligand-protein complex formation is evaluated as 0.5–1.8 kcal/mol [2947].

Enthalpy makes a preference contribution to the free energy of the formation of the hydrogen bond (see Table 4). This is true for the compounds forming one hydrogen bond, as only translational degrees of freedom change here. For the compounds with two or many hydrogen bonds the rotatory

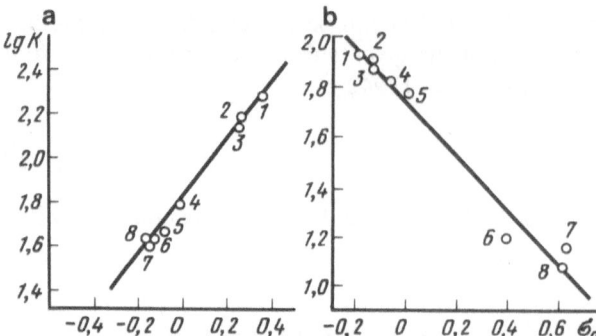

Fig. 91a, b. Effect of substituents on association constants due to hydrogen bonds in phenols (a) and pyridines (b) in CCl_4 [2950]. Substituents in phenol (a); *1* m-Cl; *2* p-I; *3* p-Cl; *4* H; *5* m-CH$_3$; *6* p-CH$_3$O; *7* p-t-Bu; *8* p-CH$_3$; in pyridines (b); *1* 4-t-Bu; *2* 4-CH$_3$; *3* 4-C$_2$H$_5$; *4* 3-CH$_3$; *5* H; *6* 3-Br; *7* 3-CN; *8* 4-CN

degrees of freedom disappear during the bond formation and the entropy contribution increases. A compensation dependence of ΔH_0 and ΔS_0 at the formation of a hydrogen bond is shown to exist [2948].

The strength of the hydrogen bond depends on the pK_a of the compounds involved in the bond formation and thus on the electron properties of substituents [2949, 2950], whereupon electron-acceptor substituents, for example, in phenols, enhance the H-bond strength (Fig. 91).

6.1.4 Electrostatic Interactions

These interactions are the most long-range at the formation of enzyme-substrate complexes, since their energy is inversely proportional to the first power of the distance between ions (see review [2410], and Chap. 7).

The ions are solvated in aqueous solution and the energy of solvation is very high. So for the carboxyl group in water it is about 65 kcal/mol [2939]. When an ion pair appears between the charged groups of the protein and substrate, solvating water is released, thus increasing entropy in the system. However, the water molecules inside the protein globule or within protein-protein contacts can serve as "bridges" of electrostatic interactions and make a contribution to the stabilization of the complexes [2951].

It is not so easy to evaluate the contribution of ionic interactions to an overall energy of the complex stabilization. As shown [2306], for example, an ion pair localized inside the globule of α-chymotrypsin and formed by a carboxylate ion of the Asp194 residue and ammonium group of residue Ile16 makes a contribution (about 3 kcal/mol) to the stabilization of the enzyme structure.

It seems rather difficult to define this value when comparing K_s or K_i of substrates and inhibitors, because the energy of the interaction is partially consumed on a conformational change of the enzyme. So the K_s for the substrates of trypsin is equal to 0.1 and 262 mM, respectively, i.e., the charge in

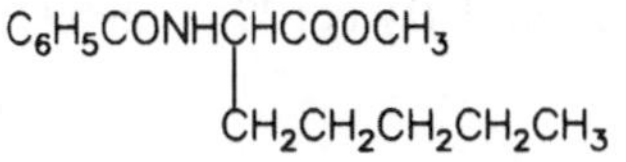

$$C_6H_5CONHCHCOOCH_3$$
$$CH_2CH_2CH_2CH_2CH_3$$

$$C_6H_5CONHCHCOOC_2H_5$$
$$CH_2CH_2CH_2NHC(\!=\!\overset{+}{N}H)NH_2$$

I 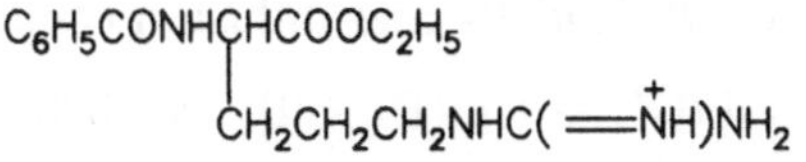   II

compound II provides a negative contribution to the binding. However, the difference in free energies of activation (by k_2/K_s) for the pair AcLysOMe and AcHepOMe is 5.9 kcal/mol [1979], which is mostly assigned to an ionic interaction. This agrees well with the results on the trypsin site-directed mutagenesis [2952]. A substitution of Asp189 in the substrate-binding pocket of the enzyme by the Ser residue decreases k_{cat} as regards Lys- and Arg-containing substrates by 5 orders of magnitude, whereas K_m increases it only two to six times. The interaction of the substrate-charged group with the carboxylate ion of residue Asp189 in trypsin provides a contribution of about 4 kcal/mol to the free energy of binding.

6.1.5   Other Types of Interactions

Donor–acceptor π–π interactions as well as the formation of charge-transfer complexes are possible between aromatic cycles and between these cycles and an amide group. The studies on model complexes of such a type are presented in [2953–2955]. In intermolecular charge-transfer complexes in organic solvents the association constant is very low (0.4–1.2 M^{-1}), and the enthalpy of the complex formation is -0.5 to -2.5 kcal/mol.

Theoretical conformational analysis has revealed [143] possible interactions of aromatic rings of phenylalanine and the peptide bond in peptides like AcPhePheNHMe. The enthalpy of this interaction is about -4 kcal/mol.

As shown [2956, 2957], the aromatic groups may interact with the S–S bond and sulfur atoms in proteins. The energy of the benzene – dimethylsulfide interaction is -3.3 kcal/mol:

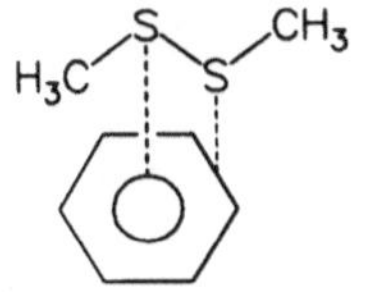

6.2   Kinetics of the Complex Formation

6.2.1   Theory

The motion of the reactant molecules in solution is stipulated by the diffusion rate depending on the molecule size. For uncharged molecules of a spherical form the rate constant of the diffusion-controlled reaction of the two molecules with radii R_1 and R_2 and diffusion coefficients D_1 and D_2 is [2958, 2959]:

$$k_d = 4\pi(R_1 + R_2)(D_1 + D_2)N/1000 ,\qquad (4)$$

where N is the Avogadro number. Since the diffusion coefficient is expressed as

$$D = \frac{kT}{6\pi\eta R} ,\qquad (5)$$

where η is the viscosity of the solution, the equation for the rate constant can be written as:

$$k_d = \frac{2RT(R_1 + R_2)}{3000\eta R_1 R_2} .\qquad (6)$$

Electrostatic interactions (in dilute solutions) can be considered from the Debye–Huckel equation [2960]:

$$\lg \frac{k_d^e}{k_d} = Z_1 Z_2 \sqrt{\mu} ,\qquad (7)$$

where Z_1 and Z_2 are the charges of interacting particles, μ is the ionic strength.

This equation is applicable at relatively high ionic strength ($\mu > 0.01$ [2961]) and for the protein solutions close to the isoionic point, where a net charge of the protein globule can be ignored. An overall charge of the protein globule, not only a local charge within the radius of collision, can influence the rate of diffusion-controlled reactions far from the isoionic point. This results in a rather complex equation [2962], the plot of which for the certain case ($R_1 = 25$ Å and $R_2 = 3$ Å) is depicted in Fig. 92 [2963]. The modified Eq. (4), considering these and some other factors, is presented in [2964].

These data promote estimation of the rate constant of diffusion-controlled reactions as $k_d = 10^8$–10^{10} M^{-1}s^{-1} [2965].

The viscosity of many solvents is temperature-dependent [2966]:

$$\eta = Ae - \frac{\varphi}{RT} .\qquad (8)$$

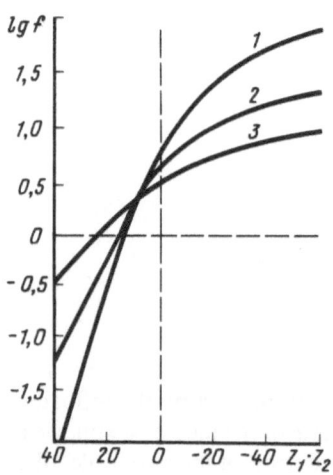

Fig. 92. Efficiency factor $\mathfrak{f} = k_d^e k_d$ vs product of protein and substrate charges (influence of total charge of protein globule). Ionic strength: *1* 0 M; *2* 10^{-3} M; *3* 10^{-2} M NaCl [2963]

Placing the expression to Eq. (6) and taking the logarithm we get:

$$k_d = \ln A + \ln T - \frac{\varphi}{RT}. \tag{9}$$

Hence the dependence of k_d on the temperature is defined mainly by a change of viscosity. This fact directly correlates with the influence of the solution viscosity on catalytic constnats of enzymatic reaction (see Sect. 4.4). The φ value as well as free energies of activation of diffusion-controlled reactions are within 1–5 kcal/mol.

6.2.2 Experimental Data

According to Table 67 the association rate is often comparable to diffusion-controlled values, though certain distinctions are observed for some

Table 67. Rate constants of the formation and breakdown of complexes of amide hydrolases with ligands:

$$E + L \; \underset{k_{-1}}{\overset{k_1}{\rightleftharpoons}} \; EL$$

Enzyme	Ligand[a]	k_1, M^{-1}s^{-1}	k_{-1}, s^{-1}	References
Chymotrypsin	FATrpNH$_2$	6.2×10^6	2.7×10^3	[2305]
	AcPhe-pNA	1.5×10^7	4×10^3	[2047]
	AcTrp-ONp	1.3×10^8	3.5×10^3	[2967,
		(6×10^7)	(6×10^4)	(2297)]
	CF$_3$CO-D-TrpOH	1.5×10^7	–	[2968]
	Proflavin	1.5×10^8	2.15×10^3	[2969]
	Rhodamine 6G	1.2×10^2	3×10^{-3}	[2565]
	BPTI	1.2×10^8	8.3×10^3	[2970]
	Chymotrypsin	3.7×10^3	0.68	[2971]
Trypsin	BzArg-pNA	$\approx 10^6$	–	[2972]
	SBTI	8.2×10^6	(2.2×10^4)	[2973,
		(6.8×10^6)		(2974)]
	BPTI	1.1×10^6	6.6×10^{-8}	[2974]
	LBTI	1.4×10^6	–	[2973]
	OTI	2.7×10^6	–	„
Anhydrotrypsin	BPTI	7.7×10^5	8.5×10^{-8}	[2974]
	SBTI	4×10^6	1.4×10^{-3}	„
Subtilisin BPN′	SSI	3.1×10^6	–	[2975]
Papain	ZLys-pNA	2.7×10^6	0.27×10^3	[2522]
Pepsin	DnsPhe-PheOpp	$>1 \times 10^6$	–	[2976]
	Pepstatin	7×10^5	–	[2977]
Thermolysin	Phosphoamidon	$\approx 5 \times 10^{-3b}$	–	[2978]

[a] FA – furylacryloyl; Bz – benzoyl; Dns – dansyl; pNA – p-nitroanilide; ONp – p-nitrophenyl ester; Opp – 4-pyridine-propyl ester; BPTI – bovine pancreatic trypsin inhibitor; LBTI – lima bean trypsin inhibitor; SBTI – soybean trypsin inhibitor; OTI – egg white ovomucoid; SSI – protein subtilisin inhibitor.
[b] Rate constant of the enzyme inactivation of pseudo-first order (s^{-1}); evaluation from the plot [2978].

complexes, which cannot be explained by a contribution of electrostatic interactions. The low rates for the interaction of rhodamine 6G or dimerization of chymotrypsin apparently depend on the limiting stage of the protein conformational change [2965]. The rates of conformational changes are within $10^2-10^4\,\mathrm{s}^{-1}$ [2979].

Another reason for the lowered association rates, as compared with the diffusion rate, can be associated with its stepwise processing. The details are discussed later.

Only a few data are available on the temperature dependence of the association rate. When subtilisin BPN′ associates with the protein inhibitor a change in the free energy of activation we observe almost depends on the enthalpy of activation ($\Delta G^{+} = 8.91$ kcal/mol and $\Delta S^{+} = 2.4$ e.u.) [2975].

6.2.3 Multistage Complex Formation

More and more information appears in favor of protein-ligand interactions in two or more stages. It is quite natural, because the interaction is accompanied with processes of desolvation of the reactive molecules, with the formation of numerous noncovalent bonds, and often with conformational rearrangements of both the protein and the ligand. The processes seem highly improbable during a diffusion-controlled reaction.

As above, the rate observed in case of the two-stage complex formation can be lower than that of diffusion-controlled processes.

In fact, in the process

$$E + L \underset{k_{-1}}{\overset{k_1}{\rightleftharpoons}} EL \underset{k_{-2}}{\overset{k_2}{\rightleftharpoons}} EL^* \tag{10}$$

the rate constant of the EL* complex formation, if $[E]_0 \ll [S]_0$, is written as

$$k = \frac{\dfrac{k_1}{k_{-1}} k_2 [S]_0}{1 + \dfrac{k_1}{k_{-1}}[S]_0} + k_{-2}\,.$$

If the breakdown of the EL* complex can be ignored, then at low concentrations of the ligand ($k_1/k_{-1}[S]_0 < 1$) the apparent second-order rate constant of the complex formation is:

$$k = \frac{k_1}{k_{-1}} k_2 = \frac{k_2}{K_s}\,, \quad \text{where } K_s = \frac{k_{-1}}{k_1}\,.$$

Thus, even if the first stage proceeds with the diffusion-controlled rate ($k_1 = 1 \times 10^9\,\mathrm{s}^{-1}$) and K_s value is within 1×10^{-3} M, the second-order rate constants equal mostly to $1 \times 10^6-1 \times 10^7\,\mathrm{M}^{-1}\mathrm{s}^{-1}$ (see Table 67) may result from the rate constant of about $10^3-10^4\,\mathrm{s}^{-1}$ of the second stage (k_2). These values are much lower than the desolvation rates of model compounds ($10^8-10^{11}\,\mathrm{s}^{-1}$ [2980, 2981]) and they are within the range typical of the rate

Table 68. Rate constants of stages of the EL* complex formation and falling (scheme 10)

Enzyme	Ligand[a]	Conditions t °C; pH	k_2, s^{-1}	k_{-2}, s^{-1}	References
Chymotrypsin	FATrp-NH$_2$	15; 6.7	1.5[b]	30	[2982]
	Dns-D-TrpOEt		19	9	[2983]
	Proflavin	–; 8.4	500	7000	[2984]
Trypsin	Dns-Arg-OMe	21; 7.1	≈3	0.6	[2985]
	Leupeptin		166	0.0017	[2986]
	STI		1580	750	[2987]
Subtilisin BPN'	PhB(OH)$_2$	15; 6.5	2420	262	[2864]
	m-NO$_2$C$_6$H$_5$B(OH)$_2$	” ”	332	24.6	
	SSI	” ; 7	774	–	[2975]
Papain	Mns-G-V-E-L-G	25; 6.5	4000	15	[1820]
	Mns-G$_2$-V-E-L-G		2000	15	
	ZLyspNA	–30; – (60% Me$_2$SO)	0.91	1 × 10^{-5}	[2522]
Pepsin	Pepstatin	25; 5	600	0	[2977]
Complement factor C1s	C1-inhibitor	–	3.5 × 10^{-4}	–	[2988]
Carboxypeptidase A (arsanylazoTyr248)	HGly-PheOH	25; 8	3250	100	[2989]
	HGly-TyrOH	” ”	4030	271	”
	HGly-IleOH	” ”	1970	895	”
	HPheOH	” ”	1156	1746	”
Thermolysin	P-Leu-TrpOH	–	900	2.5 × 10^{-3}	[2990]
	Tallopeptin	–	70	6.3 × 10^{-4}	[2990, 2991]

[a] Abbreviations, see Table 67; STI – Streptomyces typsin inhibitor; Mns – mansyl; P – (OH)$_2$P(=O)–.
[b] See, however, [2515]

constants of conformational changes [1735, 2979]. If $k_1[S]_0/k_{-1} \gg 1$, $k = k_2 + k_{-2}$ and the rate constant may be k_2 (if $k_{-2} \ll k_2$) or k_{-2}.

Unfortunately, only a few data are available on the values of k_2 and k_{-2} for typical substrates of amide hydrolases (Table 68). These investigations have often been performed with bad substrates or inhibitors. In addition to data from Table 68 there have been obtained the qualitative evidence for the stepwise complex formation with amide hydrolases [2972, 2976, 2992–2997] as well as data on the enzymes of other classes and nonenzymatic proteins (see, for example, [2998–3003]).

Structural differences in complexes EL and EL* are often obscure. So when subtilisin interacts with arylboronic acids sound data [2864] are available that the first complex is formed by sorption, whereas the second is a product of the salt formation of the residue of boric acid and the hydroxyl group of the active site serine residue:

$$E\text{---}OH + PhB(OH)_2 \rightleftharpoons E\text{---}OH \cdot PhB(OH)_2 \rightleftharpoons [E\text{---}O\bar{B}(OH)_2Ph] \cdot H^+.$$

Often there are observed the enzyme conformational changes (for example, with arsanylazotrosine carboxypeptidase A), a gradual displacement of the dye (pepsin-pepstatin), or a change in fluorescence of the substrate (papain, trypsin).

In any case, the first stage of interaction is an incomplete "tuning" of the enzyme and ligand. For true amide hydrolase substrates the final complex (EL* or ES*) must be apparently a productive enzyme-substrate complex. The first complex can result from "anchoring" of the substrate by the enzyme at the region of the highest specificity, for example at the region of the primary specificity. One can imagine that at the second stage secondary interactions (secondary specificity) are realized, which can be accompanied or followed by conformational changes of the enzyme-substrate complex. Section 8.9 shows how these concepts contribute to understanding the specificity and efficiency of catalysis by amide hydrolases.

6.3 Thermodynamics

The stability of the enzyme-substrate complex depends on a change of the free energy of the complex formation

$$\Delta G_0 = -RT\ln\bar{K}_s \,,$$

where $\bar{K}_s = 1/K_s$ is the association constant of the enzyme and ligand

$$E + S \xrightarrow{\ K_s\ } ES \,, \quad K_s = \frac{[E][S]}{[ES]} \,.$$

As above, the Michaelis constants determined in the kinetic experiment are often not identical to the dissociation constants of enzyme-substrate complexes. The latter can be defined by an independent method, for instance by the methods of equilibrium dialysis, gel filtration, via a change of spectral characteristics of the enzyme or substrate during their interaction [1661]. The choice of the conditions is always of importance, when no chemical transformation of the substrate occurs.

If the enzyme-substrate interaction is stepwise as in the previous section, the experimental association constant (K'_s) can be an effective value. If $[E]_0 \ll [S]_0$, for scheme

$$E + S \xrightleftharpoons{K_s} ES \xrightleftharpoons{K^*_s} ES^*$$

this value is

$$K'_s = \frac{K_s}{1 + K^*_s}, \quad \text{where } K^*_s = \frac{[ES^*]}{[ES]}.$$

When equilibrium shifts much to $ES^*(K^*_s \gg 1)$ $K'_s = K_s/K^*_s$, if it shifts to the complex ES $(K^*_s \ll 1)$, then $K'_s = K_s$.

Thus the change of the free energy depending on the equilibrium state demonstrates either an overall change in the system resulting in the formation of the ES* complex, or only a partial change related to the formation of the ES complex.

All this initiates an interest in the pepsin substrate binding. For this enzyme no kinetically essential stages have been found between the Michaelis complex and the stage of a chemical transforamtion [1749, 3004], i.e., $K'_m = K'_s$. Many substrates of this enzyme have almost the same values of K'_m equal to 1×10^{-3} M at significant differences in the k_{cat} values (see Table 39), i.e., despite a substantial structural diversity, the observed stability of their complexes with the enzyme is practically the same. The temperature dependence of the K'_m [1907] unravels a slight increase in enthalpy of association and a compensating positive change in entropy upon transfer from "slow" substrates to the "quick" ones. In sum, the substrate binding is stipulated by an essential positive change in entropy, which can be interpreted [1375] as a consequence of the desolvation process.

This accords well with the concept about a step-by-step complex formation, when $K^*_s < 1$. Since the substrates studied have one and the same group within the primary specificity region [(Phe(NO$_2$) or Phe], we can suggest that in complex ES the interactions of only this substrate fragment with the enzyme are realized. So the rest of secondary interactions must be realized in the productive complex (ES*) rapidly transformed into reaction products. The differences in the conversion rate measure secondary interactions. In Section 8.8 we will analyze again such an "accumulation" of the free energy for lowering the activation barrier of the chemical stage of the reaction. Here we consider a contribution of secondary interactions to catalysis exemplified by two compounds [1878]:

ZGlyGlyPhePheOpp with $k_{cat}/K_m = 1.78 \times 10^5$ M^{-1}s^{-1} and

ZAlaAlaPhePheOpp with $k_{cat}/K_m = 7.05 \times 10^6$ M^{-1}s^{-1}.

The ratio of these constants corresponds to a change of the free energy of activation by 2.2 kcal/mol, which is close to the contribution of hydrophobic interactions (see Sect. 6.1.2) of two methyl groups reacting with subsites S_2 and S_3 of the pepsin active site. But, if specific interactions (for example, electrostatic) proceed in these subsites (or in others), the contribution to the secondary interactions can be sufficient, and the situation can be realized, when $K_s^* > 1$ [3005].

The contribution of various interactions to the free energy, enthalpy as well as entropy of the complex formation is different and depends not only on the type of interaction, but also on the ligand structure and on the energetics of the binding-associated processes – a change of solvation, pK_a of the protein groups, conformational state and changes of the enzyme.

As a rule, hydrophobic interactions cause very small changes in enthalpy and are controlled mainly by entropy of association (see, for example, [2470, 3006]). The hydrogen binding is accompanied by a change in enthalpy.

Desolvation processes result in a positive change in entropy due to an increase of the number of the translational degrees of freedom of the system, when water molecules withdraw. The contribution of various factors to the binding energy of inhibitors by thermolysin is examined in detail in a series of publications [2596, 3007].

Desolvation seems important for interactions of amide hydrolases with protein inhibitors. Often such an interaction is accompanied by positive alterations of entropy [3008, 3009] (however, see [3010]). In the case of small ligands, of much value are the factors inducing a negative change in entropy (Table 57).

The energy of the protein-ligand interaction is often calculated by the methods of molecular mechanics or quantum chemistry (see, for example, [153, 3011–3015]). The values of enthalpy of binding here strongly depend on the method of calculation (see also Sect. 6.6).

An overall heat effect of the ligand binding depends on pH of the medium and, apparently, on the ionic state of protein groups under the given conditions. This is evident, for instance, when comparing enthalpy of the complex formation of a series of inhibitors with chymotrypsin at pH 7.8 and 5.6 [2467, 2468].

To detect the contribution of conformational changes of the protein to thermodynamic parameters of the complex formation is very difficult. Of much interest are here data [2469, 3016, 3017], where the thermodynamics of association of various ligands with native chymotrypsin and trypsin and these enzymes modified at the groups of the active site (His57 and Ser195) has been compared.

First, binding of specific substrate-like inhibitors is characterized by a higher negative change in entropy than that of unspecific inhibitors. Secondly, dehydration of Ser195 or methylation of His57 show an essential positive change in entropy of association, which, apparently, is explained by a higher mobility of ligands in the complexes with modified proteins and by the absence of ligand-induced conformational changes in the proteins [2469]. Note that the ligand binding to a monomer form of chymotrypsin is more effective, as compared to a dimer [3018].

Thus the changes in the free energy, enthalpy and entropy of association depend on a set of factors, which sometimes cannot be taken into account.

In Sect. 6.6 we try to evaluate the overall energy of the enzyme-substrate binding from X-ray data on interactions within the complex.

6.4   Conformational Changes

The binding of substrates and inhibitors to enzymes often entails conformational changes of the protein molecules, which can be registered by spectral methods (changes in UV-spectra, in spectra of fluorescence and circular dichroism, etc.), by thermochemical methods (by breaks on Arrhenius plots) as well as by X-ray methods. The changes may be very slight and concern only local regions of the protein molecule, especially the orientation of the side chains of its amino acid residues (see, for example, [1375, 1378, 3019–3022]), and also can be associated with the changes in mutual orientation of domains in proteins [3023].

The dynamic mobility of proteins is the inherent property of the portein molecule being independent of the ligand. A change in conformation induced by the ligand proceeds within the states also allowed for a free protein molecule. Calculations, for example, by the method of atom-atomic potentials [3024] provide evidence that the side chain of residue Ser195 in α-chymotrypsin is fixed in position with $\chi_1 = 90°$ due to a hydrogen bonding to the water molecule. When the substrate binds and water is displaced, the side chain of this residue gains actually a fully free rotation around bond C_α–C_β within $\chi_1 - 120°$ to $+120°$ (Fig. 93). The global minimum of the energy of the residue (at $\chi_1 = 90°$) is only 2–3 kcal/mol lower than that of the energy of other conformations.

Nevertheless, since the amino acid conformation in proteins coincides with those of isolated residues, incorporation of a relatively small ligand can hardly promote substantial changes.

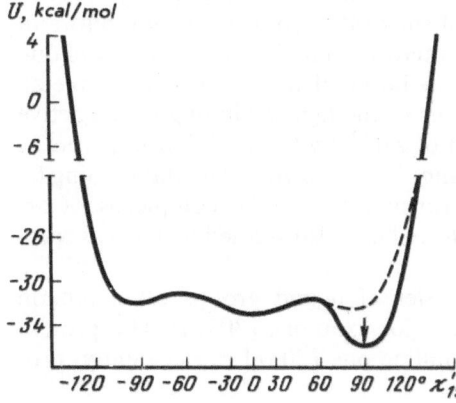

Fig. 93. Dependence of conformational energy of the Ser195 side chain in α-chymotrypsin on the angle of rotation around bond C_α–C_β (χ^1_{195}) [3024]

The study of CD spectra of the complexes shows that the ligand binding usually alters ellipticity at absorption of aromatic fragments, but does not affect the area of absorption of peptide groups (see, for example, [3025–3027]). However, some pH-dependent conformational transitions may be an exception. So, ligands (indole, acetyltryptophanamide) can shift the equilibrium between active and inactive conformations of chymotrypsin at pH > 8 to the active form. This is accompanied by a change of dispersion of optical rotation of the protein or the pK_a of group Ile16, that results in absorption of protons from medium [2307, 3028]. The kinetics of such transitions is described in detail in [3029]. Identical substrate-induced changes have been observed in bromelain [2309].

Crystallographic studies of amide hydrolases complexes with the substraste-like compounds have identified numerous conformational changes, the character of which will be thoroughly discussed in the next section.

Incorporation of "reporter" labels is an effective way for recording conformational changes. Such labels can be chromophore groups [3030], spin labels [3031], fluorescent labels [3032], etc. A proper example illustrating utilization of such an approach is the study of carboxypeptidase A modified by azocoupling at the Tyr248 residue with diazoarsanilic acid [1964, loc. cit.]:

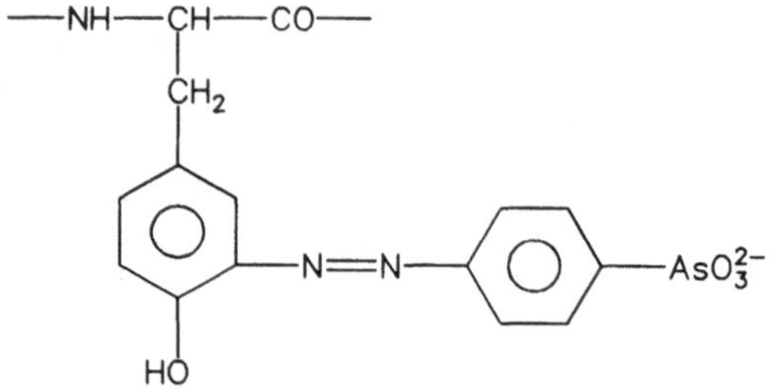

The arsanylazo group is coordinated in the molecule of carboxypeptidase A with Zn^{2+}, which results in an intensive band at 510 nm and the corresponding band in CD spectra. The binding of substrates, competitive inhibitors of the enzyme, and of certain compounds initiates an elimination of the band due to conformational changes that break a bond of an azo group to metal. Investigations have revealed some regions of the ligand binding in the active site of carboxypeptidase. A substitution of Zn^{2+} by the Co^+-ion in carboxypeptidase A allows a detection of the change in the enzyme at ligand binding by the method of magnetic circular dichroism [3033]. The complexes of arsanylazo-(Tyr248)-carboxypeptidase A have been also studied by Raman spectroscopy [3034, 3035].

Interestingly, free rotation and transfer of ligand groups often remain possible in the enzyme-ligand complexes (for example, [3036]). The protein surrounding, as mentioned for the crystalline one [3037] can evidently, promote a free rotation.

6.5 Structure of the Complexes

The crystallographic studies yield detailed information on the structure of the enzyme-substrate complexes. There are data on the structure of a large number of complexes of amide hydrolases with substrate-like compounds, "quasi-substrates" and inhibitors, i.e, the substances practically not transformed in the course of crystallographic experiment. Besides, the structures of a series of

Table 69. X-ray analysis of complexes and amide hydrolases derivatives

Enzyme	Ligand or derivative[a]	Resolution Å	R-Factor	References
α-Chymotrypsin	For-*L*-TrpOH	2.0	–	[1329]
	p-CH$_3$C$_6$H$_4$SO$_2$-E	2.0	–	[1221]
	Ind-CH=CHCO-E	2.5	–	[1328]
	p-RC$_6$H$_3$(OH-*o*)CH=CHCO-E	1.9	0.203	[3038]
	CH$_2$=CHCH=C(CH$_2$Ph)CO-E	,,	0.195	[2700]
	C$_6$H$_5$CH$_2$CH$_2$B(OH)$_2$	1.8	0.2	[3039]
	SLPI	2.3	0.19	[3040]
	BPTI (model)	2.0	–	[3041]
	OMT-3	1.8	0.168	[3042]
Trypsin	Benzamidine	1.9	0.23	[1233]
	BPTI	,,	,,	[3043, 3044]
	SBTI	2.6	–	[3046]
	PSTI	1.8	0.195	[3095]
	APPA	1.4	–	[3045]
Pancreatic	TfaLysAlaNHC$_6$H$_4$CF$_3$-*p*	2.5	0.21	[3047]
elastase	AcAlaProAla	1.65	0.184	[3048]
	BocValNHC$_6$H$_3$(*m*-Cl)(*o*-COO-E)	1.76	0.172	[3049]
	ThrProVal-N-Me-LeuTyrThr	1.8	0.190	[3050]
	AcAlaProValCF$_2$CONH(CH$_2$)$_2$C$_6$H$_5$	1.78	0.16	[2887]
Leukocyte	MeOSucAlaAlaProValCH$_2$-E	1.84	0.164	[3051]
elastase	OMT-3	1.8	0.21	[3052]
Thrombin	*D*-PheProArgCH$_2$-E	1.9	0.171	[3053]
Kallikrein A	BPTI	2.5	0.23	[3054]
α-Lytic	BocAlaProNHCH(CHMe$_2$)B(OH)$_2$	2.0	0.138	[2876]
protease	MeOSucAlaAlaProNHCH(R)B(OH)$_2$	2.0–	0.141–	[3055]
	[R=Me, CH$_2$Ph, etc.]	2.15	0.147	
Subtilisin Carlsberg	Eglin C	1.2	0.178	[3056]
Subtilisin	CI-2	2.1	0.193	[3057]
BPN′	SSI	2.6	0.34	[3058]
	BzArgOH	2.5	–	[3059]
	ZGlyGlyTyrOH	,,	–	[1330]
	AcAlaGlyPheCH$_2$-E	,,	–	[3060]
	PheAlaLysCH$_2$-E	,,	0.229	[3061]
	C$_6$H$_6$B(OH)$_2$	,,	–	[2875]
Streptomyces	Chymostatin	1.8	0.123	[3062]
griseus	AcProAlaProPheOH	,,	0.133	[3063]
protease A	AcProAlaProTyrOH	,,	0.122	,,
	AcProAlaProPheCHO-E	,,	0.142	[3063, 3064]

Table 69. Continued

Enzyme	Ligand or derivative[a]	Resolution Å	R-Factor	References
Streptomyces	BocAlaGlyPheCH$_2$-E	2.8	–	[3065]
griseus protease B	OMT-3	1.8	0.125	[3066]
Thermitase	Eglin C	1.98	0.165	[3067]
Proteinase K	ZAlaAlaCH$_2$-E	2.2	0.22	[3068]
Papain	E-64	2.4	0.233	[3069]
	E-64c	2.5	0.269	[2735]
	Cystatin (model)	2.0	0.19	[2797]
Pepsin	HPhe (I$_2$)TyrOMe)	3.0	–	[3070]
Penicillopepsin	IvaValValStaOEt	1.8	0.131	[3071, 3072]
	IvaValValLysStaOEt	„	0.144	[3072]
Endothiapepsin	ProThrGluPheCH$_2$NHPheArgGlu	2.0	0.2	[2612]
	BocPheHisNHCHCH(OH)NCOLysPhe $\quad$ $\|$ $\qquad$ $\|$ $\qquad$ PhCH$_2$ $\qquad$ CHCH$_2$Me$_2$	1.8	0.16	[3073]
	BocHisProPheHisStaLeuPheNH$_2$	2.2	0.17	[3074]
Rhizhopus	Pepstatin	2.5	0.265	[3075]
pepsin	*D*-HisProPheHisPheCH$_2$NHPheValTyr	1.8	0.147	[3076]
Carboxy-	BzPheOH	1.8	0.187	[3076]
peptidase A	F$_3$CCOCH$_2$(CH$_2$Ph)COOH	1.7	0.172	[3078]
	HGlyTyrOH (at −9°C)	1.6	0.178	[3079]
	„ (with apoenzyme)	2.3	–	[1288]
	BocPheCH$_2$CH(CH$_2$Ph)COOH	1.75	0.236	[3080]
	ZGlyPO$_2$NHPhe	2.0	–	[2628]
	ZAlaNHCH(CH$_3$)P(OH)(=O)CH (CH$_2$Ph)COOH	„	0.193	[3081]
	PCI	2.5	0.196	[3082]
Thermolysin	ZPheOH	2.3	–	[3083]
	C$_6$H$_5$CH$_2$CH(COOH)CH$_2$COOH	2.3	–	[3084]
	C$_6$H$_5$(CH$_2$)$_2$COPheOH	2.3	–	[3083, 3085]
	ValTrp	–	–	[3086]
	Ph(CH$_2$)$_2$CH(COOH)ValNHCH (COOH)CH$_2$Trp	1.9	0.171	[3087]
	E-CH$_2$CON(OH)LeuOMe	2.3	0.17	[2695]
	HSCH$_2$CH(CH$_2$Ph)COGlyOH (*S*)	1.7	0.183	[2629]
	HSCH$_2$CH(CH$_2$Ph)NHCOCH$_2$COOH (*R*)	„	0.187	„
	(HO)$_2$P(O$_2$)NHLeuNH$_2$	1.6	0.179	[2893]
	ZGlyPLeuLeuOH	„	0.177	[2728]
	ZPhePLeuAlaOH	1.7	0.17	„
	Phosphoamidon	2.3	–	[3085]
D-Ala-*D*-Ala- peptidase	Cephalosporin C	„	0.106	[3088]

[a] Ind – indolyl; BPTI – basic pancreatic trypsin inhibitor; OMT-3 – third domain of turkey ovomucoid; SBTI – soybean trypsin inhibitor (Kunitz); PSTI – pancreatic secretory trypsin inhibitor; SLPI – leucocyte secretory protease inhibitor; APPA – *p*-amidinophenyl-pyruvic acid; CI2 – chymotrypsin inhibitor 2 from barley grain; SSI – *Streptomyces* subtilisin inhibitor; E-64 and E-64c see Section 5.9.6; Sta – statin; Lysta – 4,8-diamino-3-hydroxyoctanoic acid; PCI – inhibitor of carboxypeptidase from potato; GlyP and PheP – phosphorus analogs of Gly and Phe [NHCH$_2$(R)PO$_2$].

amide hydrolases derivatives, which mimic various steps of the enzymatic catalysis, are also known (Table 69).

Note that the conclusions regarding the structure of the complexes of enzymes with true substrates should be interpreted very carefully.

Here we raise some general questions regarding the structure of amide hydrolases complexes, and then will illustrate certain examples on the structure of the complexes of representatives of four groups of the enzymes under study.

6.5.1 Conformation of the Substrate in the Active Site

Complementarity of the interacting regions of the reactants seems indispensable for the formation of a stable enzyme-ligand complex. It can be defined [3089] as a presence in such regions of a mutually complementary Van der Waals shell (Fig. 94), so that external parts of one molecule fit internal parts of the other.

The Van der Waals shell is fixed in conformationally rigid ligands, however for most of the substrates of amide hydrolases both the conformation and thus the type of shell can change within a certain range. Hence a question arises: to what extent does the conformation of these substrates in solution conform to that in the active sites of enzymes?

The conformational mobility of peptides is within relatively narrow ranges – it seems possible within energetically allowed regions shown on Ramachandran maps (see Sect. 1.4.2). Conformational changes in the active sites of amide hydrolases at substrates binding are also usually slight and local. Therefore, they are expected to be inessential at the enzyme binding.

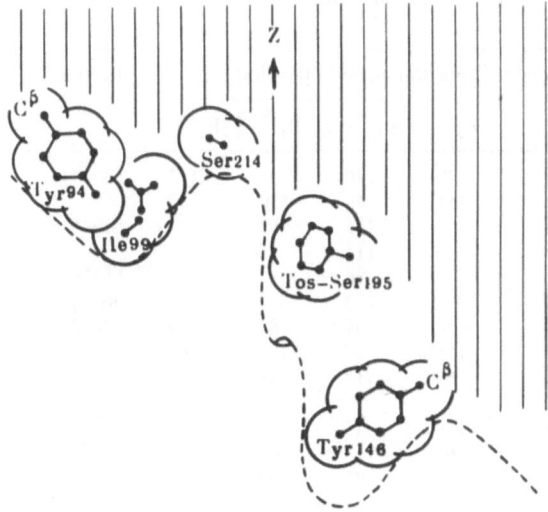

Fig. 94. Part of Van der Waals shell of the α-chymotrypsin molecule [3089]

The identity of conformations of N-acetyl-*L*-phenylalanine derivatives in solution and in the active site of chymotrypsin has been postulated [2029] from the comparison of the kinetics of the substrate hydrolysis of this enzyme with that of a limited conformational mobility (Fig. 66). To prove an extended conformation of side chain of the Lys residue in complexes with trypsin a similar approach has been applied [3090].

The conclusion has found confirmation in the studies [3091] on NMR complexes of acetyl- and trifluoroacetyltryptophan with chymotrypsin, Raman spectra of thioacylpapain and other cysteine proteases [3092, 3093] as well as by calculations of the substrate conformation by atom-atomic potentials [3024, 3094]. Lastly, the crystallographic analysis of chymotrypsin complexes with substrate-like inhibitors has also provided confirmation of this conclusion (Table 70). The differences observed in the conformation of the substrate backbone sometimes are related to the peculiarities of the protein crystal packing. Thus, in crystals of the complex of chymotrypsin and formyltryptophan the orientation of the ligand apparently is slightly disturbed due to the obstruction caused by the second enzyme molecule contacting the substrate-binding region [1329].

The ligand conformation corresponds to the regions allowed for angles φ and ψ on Ramachandran maps (compare Table 70 and Fig. 7). Distinct differences here can be observed for side chains, but the variation in energies of the conformations is usually small. So the aromatic ring of formyltryptophan upon binding turns by about 60° relative to the optimum conformation in solution [2029]. For the pepsin substrates the optimum conformation in solution is that where the two aromatic rings of dipeptide AcPhePheNHMe do not interact with each other and with the peptide bond [143]. Data on the conformation of dipeptide HPheTyr(I_2)OMe in the active site of pepsin

Table 70. The conformation of the backbone chain of some substrate analogs in the active sites of amide hydrolases

Enzyme	Ligand[a]	Angle	Dihedral angle in position				References
			P_2	P_1	P_1'	P_2'	
Trypsin	BPTI	φ	-70	-120	-94	-112	[3044]
		ψ	154	49	170	79	
	SBTI	φ		-135	-89		[3046]
		ψ		-115	85		
Chymotrypsin	For-*L*-TrpOH	φ	$-$	-86	$-$	$-$	[1329,
		ψ	$-$	3	$-$	$-$	3059]
Leucocyte elastase	OMT-3	φ	-76	-99	-82	-104	[3052]
		ψ	155	29	156	106	
Streptomyces griseus protease B	OMT-3	φ	-69	-117	-83	-99	[3066]
		ψ	159	45	153	115	
Subtilisin Carlsberg	Eglin C	φ	-61	-115	-96	-117	[3056]
		ψ	143	43	169	110	

[a] Abbreviations, see Table 69.

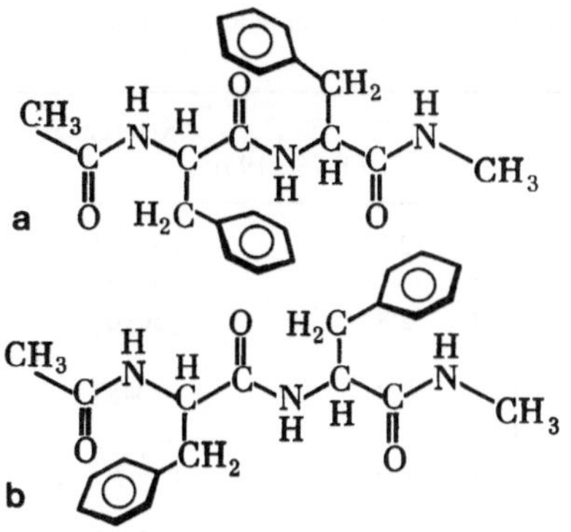

Fig. 95a, b. Conformation of AcPhePheNHMe [143]; a conformation with parameters: χ_1^I 180°, χ_1^{II} − 60° and ΔG_{tot} = 0.6 kcal/mol; b conformation with χ_1^I − 60°, χ_1^{II} 180° and ΔG_{tot} = 4.2 kcal/mol

[3070] show that the aromatic cycles accept an unfolded conformation (Fig. 95). The difference in overall energies of the two conformations is about 4 kcal/mol. However, this energy is obviously compensated in the complex due to hydrophobic interactions of the aromatic rings.

Sometimes a distinct deviation of the conformation of the amide bond from the planar structure is observed in the regions of the secondary specificity. For example, in complexes of endothiapepsin with the analogs of the substrate containing His residue in position P_2, the amide bond P_2–P_1 has $\omega = -168°$ [3073]. This deviation is 1–2 kcal/mol less efficient as compared with the planar structure.

6.5.2 The State of the Cleavable Bond

As shown above, the amide bond in solution has a planar configuration. However, the turn around bond C–N by about 10°–15° as well as a slight deviation of the carbon carbonyl atom from the plane formed by atoms C_α, O and N are not associated with high energetic expenditures (Sect. 1.2.1).

Studies on the trypsin complex with the bovine pancreatic inhibitor have revealed [3043, 3044] that the amide group of the Lys15-Ala16 of inhibitor is deformed so that the carbon carbonyl atom protrudes from the plane formed by C_α-, O- and N-atoms (pyramidalization). Further analysis of the structure of the enzyme-inhibitor complexes of serine proteases has shown that this thoroughly analyzed situation is not unique (Table 71) [3045].

It is noteworthy that in the free bovine pancreatic trypsin inhibitor χ_C for bond Lys15-Ala16 is close to zero [3045].

In complexes with protein inhibitors the amide group of the inhibitor (Lys15-Ala16) is supposed [3043] to have a true tetrahedral structure forming a covalent bond with the hydroxyl group of the serine residue of the enzyme active site. However, further investigations performed, mainly by ^{13}C-NMR methods, when using anhydroenzymes, have disclosed [2832, 3097, 3098] that the bond is not formed, and the development of the carbonyl tetrahedral state is apparently associated with the deformation of the bond in the complex.

Table 71. Outside planar deformation of the amide bond in complexes of some amide hydrolases and their zymogens (designation of angles, see p. 7)

Complex	Bond	ω_1	χ_c	r C'–O$^\gamma$ (Ser195)	References
Trypsin-BPTI	Lys15-Ala16	163.9	−12.9	2.7	[3043]
Trypsinogen-SPTI	Lys15-Ile16	172.8	−14.5	–	[3095]
Trypsinogen-BPTI (Arg15)-IleVal	Arg15-Ala16	165.6	−10.4	2.8	[3096]
Trypsinogen-BPTI	Lys51-Ala16	162.8	−18.5	2.9	[3045]
Anhydrotrypsin-BPTI	,,	159.5	−10.3	–	,,
Subtilisin Carlsberg Eglin C	Leu45-Asp46	158.0	−10.1	2.81	[3056]
Leukocyte elastase OMT-3	Leu18-Glu19	172.0	−5	2.6	[3052]

Qualitative data on this fact are also available for other enzyme – quasi-substrate complexes [3049, 3054, 3057, 3087].

This suggests that the active sites at least of a series of proteases fit the substrate amide group deformed so as to approach the tetrahedral state. Numerous data on the binding of ketoanalogs of the substrates and peptide-aldehydes as tetrahedral hemiacetals and hemithioacetals (ketals) also bear witness of this fact [3045, 3051, 3060–3062, 3068, 3078, 3080, 3099].

This distortion is supposed to remain also in acylenzymes, which is confirmed by Raman spectra of chromogenic acylchymotrypsins [3100, 3101].

Note that not all complexes of protein inhibitors and proteases have deviations of the amide group of the "reaction site" as regards the planarity [3041, 3067]. Moreover, in certain acylenzymes, for example in acylpapains, double bonding of a cleavable single C–S bond enhances [3093, 3102].

Though the level of distortion of the amide bond in crystals of amide hydrolases complexes with substrate-like inhibitors depends on the method employed to refine crystallographic parameters [3066], the distortion presumably, evokes no doubt. This might be of importance for understanding the efficiency and specificity of amide hydrolases [1715, 3103, 3104] (for details see Chap. 8). The distortion mechanism of catalysis is observed also for peptidyl-proline *cis-trans* isomerase [3105].

6.5.3 Location on the Enzyme and Interactions

Here we discuss the details of the location of substrate-like compounds in the active sites of some enzymes from X-ray data.

Serine Proteases. Ample data are available on the structure of the complexes of serine proteases with quasi-substrates, products of hydrolysis, as well as with inhibitors (Table 69). The interaction of the aromatic ring of the substrate in position P_1 was first described in [1329] on the example of the complex of chymotrypsin and formyl-*L*-tryptophan, and secondary interactions when using the complexes of trypsin with the pancreatic trypsin inhibitor [3043, 3044] as well as the complexes of protease A from *Streptomyces griseus* with tetrapeptides like AcProAlaProTyrOH [3063] (Fig. 96).

The binding region of the side chain of residue P_1 (S_1-region) at chymotrypsin is restricted "below" by the chain of residues 213–220, and "above" by residues 190–192. Deep inside this "hydrophobic pocket" there is located a side chain of the Ser189 residue, while the side chain of residue Met192 is outside.

The aromatic group of quasi-substrates of chymotrypsin is sumberged in this pocket so that its plane is parallel to that of amide bonds Ser190-Cys191 and Trp215-Gly216. C_β-Atom of the substrate contacts the solvent according to NMR data [3106].

Residues Lys and Arg of the trypsin substrate are positioned similarly. However, the presence of Asp189 (instead of Ser189) in trypsin-like enzymes distinguish them noticeably from the enzymes of the chymotrypsin group. The charged groups of the substrate form a salt bridge with a carboxylate ion of this residue [1235].

What distinguishes the S_1-region of subtilisin from the relative region in chymotrypsin is its higher accessibility to the solvent so that the aromatic group bound here by one of its sides is exposed into the solvent [3060]. The same is observed in subtilisin complexes with macromolecular inhibitors, where side chains of hydrophobic residues (Pro42, Leu45, Leu47 in egline C) contact partially the solvent [3056, 3107]. Finally, for pancreatic elastase [3108] in positions 216 and 226 there are located residues Val and Thr (Asp in leukocyte elastase) instead of residues Gly (in chymotrypsin and trypsin). These residues prevent the binding of bulky groups important to the elastase specificity for the side chains of residue Ala in P_1.

The location of the aromatic group in locus S_1 of protease A is actually identical to that of the tryptophan residue in chymotrypsin [3063].

The tyrosine residue in the complex protease A and product (AcProAla-ProTyrOH) forms a few hydrogen bonds (Fig. 96). This is, first, a hydrogen bond with the carbonyl group of residue Ser214 (3.03 Å). Its significance for catalysis has been postulated time and again. However, the bond has neither been found in the formyltryptophan – chymotrypsin complex [1329] nor in complexes with trypsin [1231]. Calculation of the substrate conformation in the enzyme active site by the method of atom-atomic potentials has not revealed it either [3024]. But this bond is observed in serine proteases alkylated by peptide chlormethylketones as well as in the complexes of these enzymes with the proteinacious inhibitors [3040, 3043, 3061].

The carboxyl group in the tyrosine residue forms three hydrogen bonds: *a*) with protonated $N^{\varepsilon 2}$-atom of His57 (2.82 Å); *b*) with NH of the peptide group of residue Gly193 (2.81 Å) and *c*) a weak bond (3.2 Å) with NH group of Ser195.

The first bond mimics the proton transfer onto the leaving group in the substrate, while the second and third being typical of carbonyl groups of substrates and acylenzymes compose a binding region termed an "oxyanion hole" [1328, 3060].

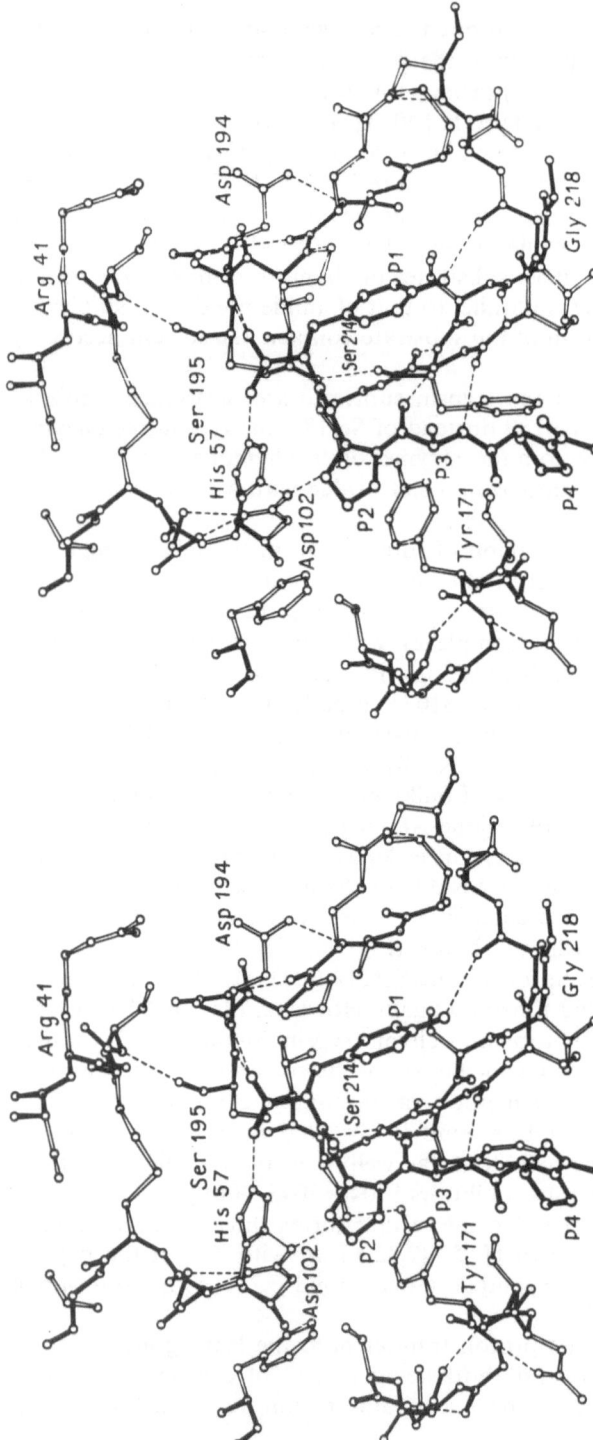

Fig. 96. Stereopair of the AcProAlaProTyrOH disposition in the active site of *Streptomyces griseus* protease A [3063]

Hydrogen bonds with NH groups of residues Gly193 and Ser195 are observed actually in all complexes of serine proteases with protein substrate-like inhibitors [3021, 3044, 3049, 3066, 3109]. As described above, the amide group of the "reaction site" is pyramidalized in many such complexes. This lowers the distance between the O^γ-atom of Ser195 and the carbon carbonyl atom of the potentially cleavable bond by up to 2.6–2.9 Å, as compared with the sum of Van der Waals radii, but it is greater than the length of the covalent bond. The O^γ-atom of Ser195 is located approximately at the same distance from $N^{\varepsilon 2}$-atom of His57.

Residues P_2–P_4 of the ligand form an antiparallel β-folded structure with enzyme chain 214–218, whereupon Gly216 by its NH and CO groups binds to CO and NH of the P_3 residue. The same interaction is observed with peptide chloromethylketones and protein inhibitors in α- and γ-chymotrypsin, elastase, subtilisin, protease B from *Streptomyces griseus* as well as trypsin [1242, 1330, 3044, 3046, 3061, 3065]. In trypsin, however, due to deletion in position 218, the CO group of Gly216 is oriented so that it cannot accept a hydrogen bond with the substrate, though NH of Gly216 still enables the accepting of the hydrogen bond with the P_3-residue.

The quasi-substrate forms many hydrophobic contacts in the active site of the enzyme. In the complex AcProAlaProTyrOH with protease A over half (44) of the contacts fit the Tyr residue in position P_1, which interacts with residues Ala192, Gln192A, Pro192B, Gly215, and Thr226. The indole ring in the ForTrpOH-chymotrypsin complex forms 41 contacts with the enzyme. Residue Pro (P_2) forms 12 contacts with residues His57, Tyr171 and Ser214 and, as mentioned above, with carbonyl oxygen of Ser214. Residue Ala (P_3) shows no interaction with the enzyme, except for the above two hydrogen bonds, and residue Pro (P_4) gives 12 contacts with residues Val169, Asn170, Tyr171 and Gly216. Further lengthening of the peptide chain yields additional residues exposed into solution, but does not affect their cleavage rate [3110].

The position of the residue in the leaving group of substrate (P'_1) has been analyzed from data on the structure of the complexes of the pancreatic tryptic inhibitor with trypsin and chymotrypsin (Fig. 97). Residue Ala16 in this inhibitor interacts by its side chain with sulfur atom of the disulfide bond Cys42-Cys58 as well as with residue His57 and O^γ-atom of Ser195. C^α-Atom of the P'_1-residue and its carbonyl group contact the side chains of residues Ser195 and Met192 [2011]. The authors [2011] suppose that some of the contacts make a negative contribution to a total energy of binding, but these unfavorable interactions are eliminated in the transition state of the reaction. The binding region of the P'_2 residue is formed in chymotrypsin by the "cavity" between residues His40 and Gly193. The most essential conformational changes, occurring when the substrate-like inhibitors bind, might be traced by the interaction of protease A from *Streptomyces griseus* with tetrapeptide AcProAlaProTyrOH [3063]. These are: *a*) C^β- and O^γ-atoms of Ser195 move by 0.21 and 0.46 Å to the peptide; a similar, but much stronger, shift has been observed [3111] in the complex trypsin and trypsin pancreatic inhibitor; *b*) an imidazole ring of His57 shifts by 0.2 Å (by $N^{\varepsilon 2}$-atom) to the peptide; *c*) the shift of a carbonyl group of Ser214 is by 0.24 Å in the same direction.

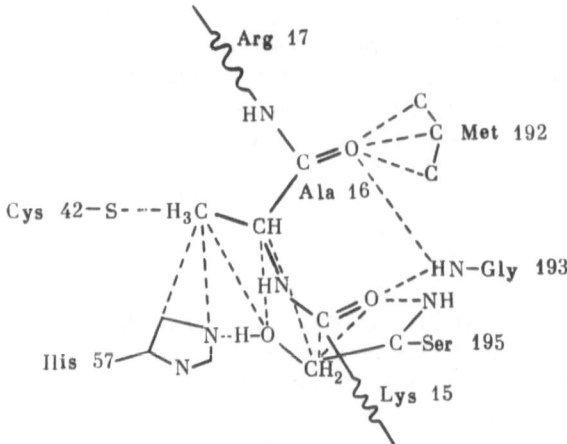

Fig. 97. Interaction of amino acid residues of bovine pancreatic inhibitor (bond *Lys15-Ala16* is shown) with groups of chymotrypsin active site [2011]

Besides, using the chymotrypsin-formyltryptophan complex [1329], there is observed a turn of the side chain of residue Met192 around C_α–C_β bond by 120°. In this case, the chain of Met192 has moved from the solution-exposed state to deeper layers of the globule so that it protects the substrate side chain from contacting water. Residue Met192 seems to be of importance, which is confirmed by kinetic [3112] and physicochemical [2292, 3113, 3114] studies.

Conformational changes in serine proteases occur when binding ligands are shown to be slight and concern the side chains of the enzyme amino acid residues. The exception is only for, apparently, subtilisin BPN' where, as indicated [3060], the backbone within residues 125–127 (subtilisin numeration) shifts by 0.5 Å, approximately.

X-ray studies of peptides, ketoanalogs of the substrates and derivatives of boric acids [2876, 3039, 3051, 3062] yield a concept regarding the position of tetrahedral intermediates in the active sites of serine proteases (Fig. 98).

Investigations of chymotrypsin complexes with different inhibitors by NMR methods [3115–3118] confirm the character of interactions, observed for the enzyme crystals in general. Note that the crystals of chymotrypsin and protease A have been obtained at pH ≈ 4, i.e., when the enzymes lose their activity. However, for protease A they still hydrolyze the peptide substrate [3063], while chymotrypsin crystals obtained at higher pH [1222, 3119] do not differ much from the crystal of the "acid form".

According to X-ray analysis and data on NMR spectra of complexes with *L*- and *D*-trifluoroacetyl-*p*-fluorophenylalanine [3116, 3117] and trifluoroacetyltryptophan [3015], the reasons for primary stereospecificity of chymotrypsin are the orientation of the acyl group of *D*-isomer in the complex to pair Ser195-His57, while a proper orientation of *L*-isomer allows this group to form a bond with the CO group of Ser214.

Calculations by the methods of molecular mechanics have revealed that in the tetrahedral intermediate *L*-AcTrpOH is more stable (by 9 kcal/mol) than the *D*-form [3015]. Hence the stereospecificity is observed in the transition state, but not in the ground one. Identical calculations of complexes BPTI with

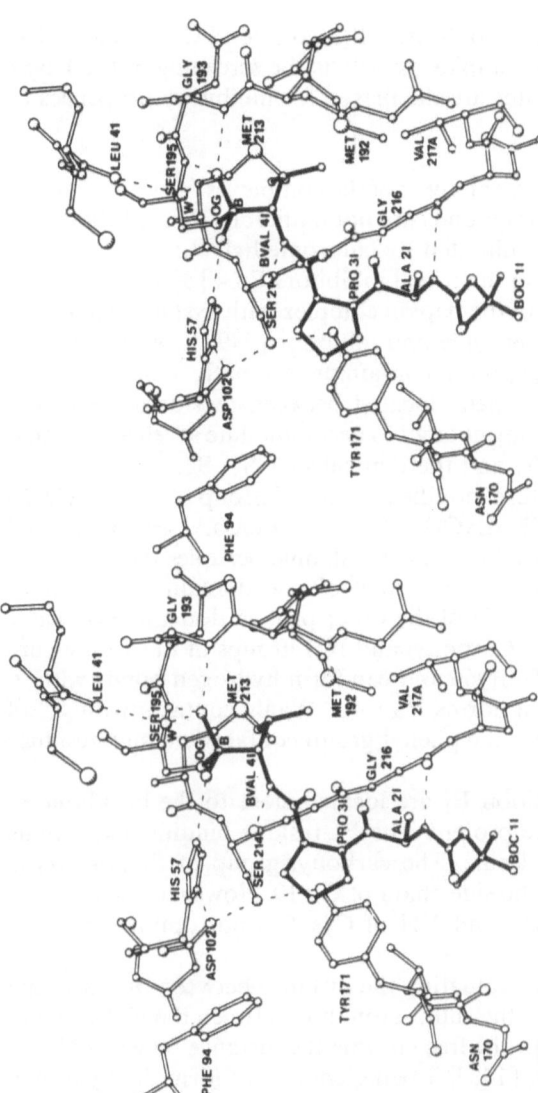

Fig. 98. Stereo image of complex BocAlaProNHCH(CH$_2$CHMe$_2$)B(OH)$_2$ (shown by *full bars*) in the active site of α-lytic protease [2876]

trypsin [3094, 3120, 3121] have confirmed the concept [1235, 3095] that rigidity of the enzyme-inhibitor complex as well as the screening of the His57 residue in this complex from water are the reason for inhibitory properties of this and analogous inhibitors.

Cysteine Proteases. Data on X-ray analysis of the complexes of cysteine amide hydrolases and substrate-like compounds are not numerous. There is information on the structure of papain inhibited by chloromethylketones [3122], on the complexes of the enzyme with microbial inhibitors E-64 [3069] and E-64c [2735], as well as on the model of the papain complex with cystatin of chicken egg (from X-ray analysis of the enzyme and inhibitor [2797]). As mentioned above, papain is specific to peptides containing aromatic amino aicd in position P_2. Its active site is a "cleft" located between two domains of the protein molecule. The "cleft" is long enough to accommodate seven amino acid residues – four in subsites S_1-S_4 and there in subsites $S'_1-S'_3$.

Analysis of the papain structure modified at the SH group of residue Cys25 by chloromethylketones $ZPheAlaCH_2Cl$ and $AcAlaAlaPheAlaCH_2Cl$ (Fig. 99) has detected that the residue of phenylalanine occupies the S_2-region formed by residues Tyr67, Pro68, and Trp69 of one domain and residues Phe207, Ala160, Val133, and Val157 of the other protein domain. The water molecule in this case is replaced. Carbonyl and NH groups of this residue are oriented to peptide bond Gly65-Gly66 and can form hydrogen bonds with it. The C_β-atom of the Phe residue forms Van der Waals contacts with C^δ of Pro68 and C_β of Ala160, whereas the phenyl group contacts, mainly, residues Val133 and Val157.

Amino acid residues in position P_1 are located close to the backbone of residues Gly23 and Ser24 of the protein. The P_1-residue binding also entails a replacement of the water molecule. The carbonyl group of P_1-position is directed to the amide group of the side chain of Gly19. However, the location of groups $CO(P_1)$, $N^{\varepsilon 2}$ of Gln19 and NH of Cys25 is not optimal here for hydrogen bonding.

Data available do not allow evaluating the distance between the carbonyl C-atom of the true substrate and the sulfur atom nor between the $N^{\delta 1}$-atom of His159 and the leaving group. In the free enzyme the distance between $N^{\delta 1}$ of His159 and S^γ of Cys25 is 3.4 Å [148], it being enough to form the hydrogen bond between them.

The binding of epoxy inhibitors E-64 and E-64c to papain is identical [2735, 3069]. These compounds form covalent bonds with the sulfur atoms of Cys25, which attacks the C2-atom of the epoxy cycle, whereupon the inversion of the configuration of this atom is observed. The carboxyl group of the inhibitor forms H-bonds with the amide group of Gln19 and NH of the backbone of Cys25, as well as in the case of E-64c with the $N^{\varepsilon 2}$-atom of the His159 residue.

The P_3-residue of chloroketones as well as the terminal isopropyl (in E-64c) and the guanidine group (in E-64) bind close to the side chain of Tyr67, and the Ala residue in this position contacts the phenol ring of Tyr61 and the C_α-atom of Gly65.

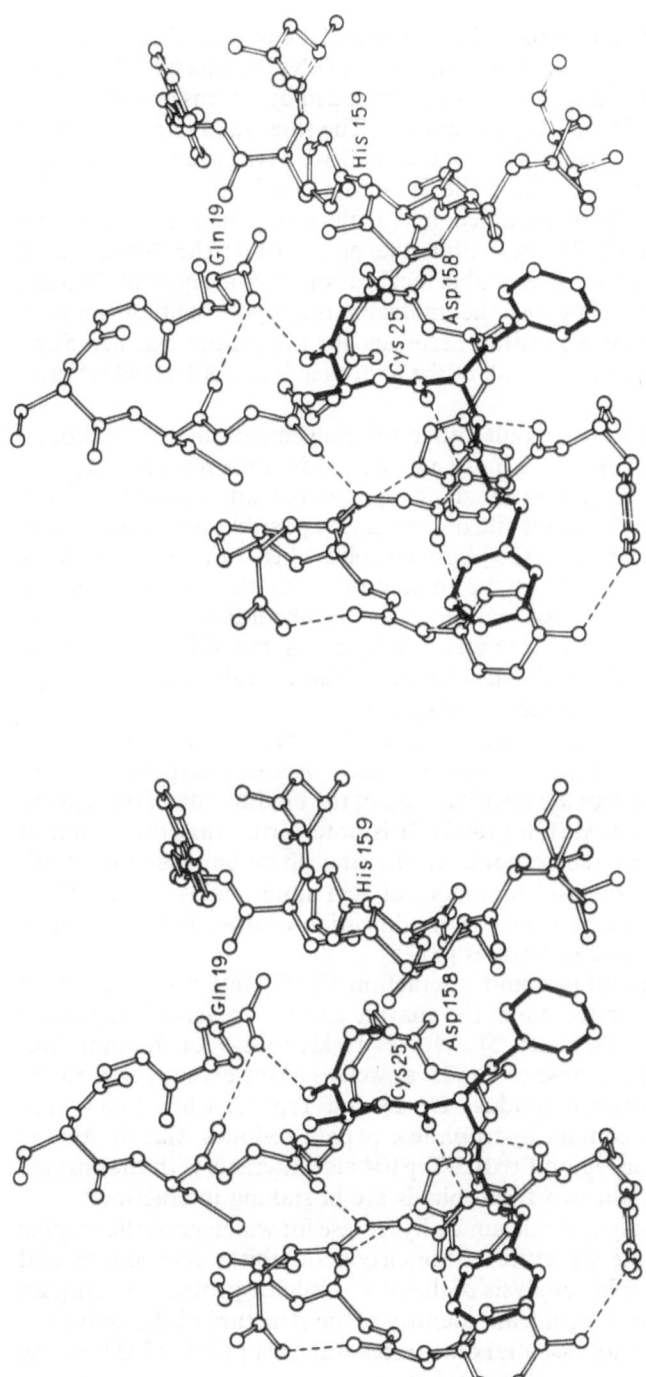

Fig. 99. Stereo image of the papain active site modified by benzyloxycarbonyl-*L*-phenylalanyl-*L*-alanylchloroketone (*full bars*) [3122]

The inhibitors binding to papain induce essential conformational changes in the protein. For chloroketones the sulfur atom of Cys25 shifts by 0.9 Å inside the globule of the protein; the process is accompanied by a transfer of atoms of the backbone of the Ser24 and Gly43 residues. The side chain of Ser176 turns around bond $C_\alpha–C_\beta$ through $\approx 120°$ so that its hydroxyl group is located at a distance of a hydrogen bond from the $O^{\varepsilon 1}$-atom of Gln19.

In the free enzyme the protonated imidazole group of His159 and the ionized sulfur atom of Cys25 are within the plane, while the $N^{\varepsilon 2}$-atom of His159 appears to be far from a probable localization of N-atom of the leaving group of true substrates. However, the imidazole ring upon binding can turn through $\approx 30°$ and occupy a position necessary for the proton transfer. Such a turn is not accompanied by a break of the hydrogen bond of His159 $N^{\alpha 2}$ and CO of Asn175.

Binding of the inhibitors entails other rearrangements in the backbone within Arg59-Pro68, Ser21-Gly23 and Lys156-His159, thus reaching the amplitude of 1 Å. These conformational changes make an approach of the substrate or inhibitor to the active site of the enzyme possible. Model studies of the papain-substrate complex [3123] have established that the substrate leaving group can be unfavorable at the interaction with the α-CH_2 group of His159, but the interaction is not realized in the tetrahedral intermediate. The unfavorable interaction is the stronger the larger is the side chain of the P_2-residue, which accounts for the dependence of the k_{cat} value (but not of K_m) on hydrophobicity of this residue in substrates.

Studies on the model and data on X-ray low resolution (6 Å) for the complexes of papain with quasi-substrates AlaAlaPhe(I-p) Arg and BocPhe(I-p) Arg [3124] provide identification of the binding site of the leaving group within Trp177 and Ala136 [1802]. It is noteworthy that the region of residue Trp177 apparently corresponds to the site, where unproductive binding of some substrates, like p-nitroanilides of acyl amino acids, occurs. Such a conclusion has been reached because modification of this residue enhances the papain activity to these substrates [3125].

Modeling of the cystatin-papain interaction [2797] has revealed a high complementarity of the active site of the enzyme and the "reaction" site of the inhibitor. This site (see also Sect. 5) involves residue Gly9 of the inhibitor, which must be located here close to Cys25, as well as a conserved region 53–57 buried in the "cleft" between residues Cys25 and Trp177, while Leu54 and Val155 of the inhibitor contact hydrophobic papain residues Ala136, Ala137 and Gly23. The inhibitor region Pro103-Trp104 also interacts with the enzyme within Trp177, whereupon two tryptophans are in staking interaction.

Papain now is the only cysteine amide hydrolase for which crystallographic data are available on the structure of the enzyme-inhibitor compounds and complexes. Crystallographic analysis of the other cysteine protease – actinidin in a free-state – indicates a significant identity of the structure of the active site with that of papain, though essential differences still exist [1264, 1265]. So the sulfur atom of the cysteine residue catalytically active in actinidin is distant from the plane of the imidazole cycle by 2.7 Å and cannot form the H-bond with the $N^{\delta 1}$-atom of the latter at such orientation. However, as in papain, the imidazole ring can turn through 35° so that the sulfur atom falls within the same plane with the imidazole cycle. Such a position is stabilized by its location in the plane parallel to that of the indole cycle of Trp177. Papain in solution has two forms, differing in the orientation of the imidazole cycle relative to residues Asn175 and Asp158 (forms "up" and "down") [2243]. At the same time, ficin exists, probably, in one conformation ("up"). Apparently, actinidin also relates to the type of ficin, but not of papain.

Differences in catalytic characteristics of actinidin and papain have been also interpreted [3126] as a result of those in the distribution of the electrostatic field in the active site of enzymes, but this interpretation has met with objections [3127]. The characteristic differences from papain are observed for bromelain as well. The investigation of Raman spectra of dithioacyl enzymes has uncovered [3092] that the substrate in the active site of bromelain has a conformation other than that in papain.

Aspartic Proteases. X-ray data are available for the complex of pig pepsin with dipeptide – methyl ester of phenylalanyldiiodtyrosine at a resolution of 3 Å only [3070, 3128]. However, for the very similar microbial aspartic proteases, along with the results of low resolution X-ray analysis (2.8–3.7 Å) [1281, 1345, 3129], data appeared, which allow a thorough description of the substrate binding in the active site of the enzymes [2612, 3071–3073, 3076].

The active site of aspartic proteases, as for papain, is located in the "cleft" formed by two domains of the protein molecule. Catalytically active groups Asp32 and Asp215 enter various domains, but in the spatial structure they approach at ≈ 2.6 Å (between oxygen atoms). The groups are deep in the "cleft".

The binding of substrate-like inhibitors can be illustrated by the complex of heptapeptide (D-HisProPheHisPheCH$_2$NHPheValTyrOH) reduced at a cleavable bond with rhizopuspepsin [3076], as well as by the complex of structurally related peptide (H256) (ProThrGluPheCH$_2$NHPheArgGluOH) with endothiapepsin [2612]. These inhibitors are incorporated into the active site of the extended conformation (Fig. 100) so that the carbon atom, which mimics carbonyl group of the substrate, is located near the catalytically active residues Asp32 and Asp215 (numeration of pepsin). The NH group of Phe residue in P$_1'$-position is at a distance of 2.83 Å from the O$^{\delta 2}$-atom of Asp215 [3076]. The orientation of the carboxyl group of the cleavable bond can be defined from an analysis of the structure of aspartic proteases complexes with pepstatin and its analogs [3071, 3072, 3075], peptides like R–CH(OH)CH$_2$COR' (where R and R' are the amino acid residues) [3073, 3130], as well as from the models [2334, 3131]. Groups Asp32 and Asp215 in free enzymes form a rather symmetrical structure (Fig. 101). The carbonyl group is apparently located so that the water molecule bound to four oxygen atoms of Asp32 and Asp215 is replaced by a carbonyl oxygen atom of the substrate [3131]. The side chain of the P$_1$-residue interacts with the hydrophobic residues in enzyme – Ile30, Tyr75, Phe111 and Ile120. Group NH of this residue forms the H-bond with the carbonyl group of Gly217. As set forth in Chap. 4, aspartic proteases can be divided into two groups by their specificity

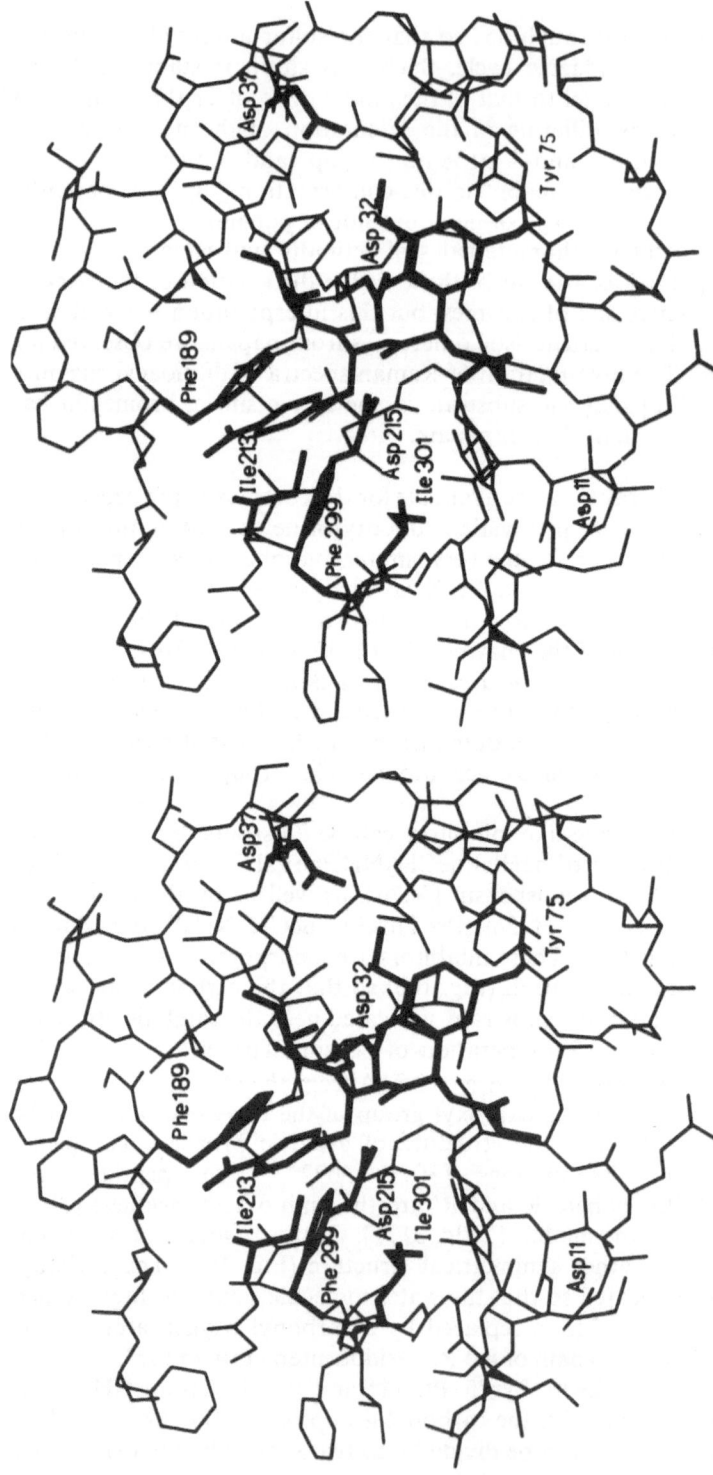

Fig. 100. Stereo image of pencillopepsin active site bound to tetrapeptide AlaLysPhe(NO₂)Ala (*bold lines*) [3129]

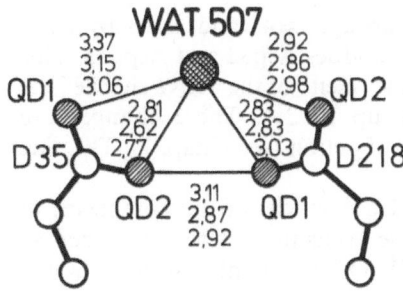

Fig. 101. Distances between oxygen atoms of carboxylic groups of active sites of aspartyl proteases (numeration of risopuspepsin) and water molecule. *Numerals* (from *top to bottom*) for endothiapepsin, penicillopepsin and risopuspepsin [3076]

to the P_1-residue: a pepsin group involving pepsins from diverse sources, chymosin, renin, etc. specific for aromatic and bulky hydrophobic residues in this position, and the group of penicillopepsin (endothiapepsin, rhizopuspepsin, etc.) with the expressed lysine specificity [2334].

An essential difference between the two enzyme groups is residue Asp114, which is typical of all the enzymes under study capable of hydrolyzing the substrates containing the lysine residue in P_1. Their lysine ε-NH_2 group is located near residue Asp114 and interacts with it, thus providing obviously the H-bond via a bridge molecule of water [2334]. Note that pepsin also has the cation-binding site positioned within a secondary enzyme – substrate interactions in regions S and S′ [3005].

Models and X-ray analysis yield zones of secondary interactions for the "ideal" (according to indices of specificity, see Sect. 4.3.2.1) substrate of pepsin listed in Table 72 [1270].

Substrate NH and CO groups in positions P_2, P_1', and P_2' form hydrogen bonds with the appropriate partners of the enzyme backbone and carboxyl groups of the aspartic acid residues. The N-terminal fragment of the peptide substrate is apparently more mobile in the complex than the C-terminal fragment according to EPR spectra of the spin-labeled pepstatin analogs [3132]. As to conformational rearrangements of the enzymes upon binding of substrate-like inhibitors, the process is accompanied by insignificant changes within active sites. However, as shown above for pepsin [1270] and for endothiapepsin [3073], a slight but very important change in the interlocation of two domains – N- and C-terminal – is observed. The C-terminal domain

Table 72. Pepsin residues contacting side chains of "ideal" substrate HsAnIleGluPheTyrValLeuOH

Substrate		Pepsin residues
P_4	Asn	S_4	Thr12, Ser219, Leu220, (Glu244, Met245)
P_3	Ile	S_3	Glu13, Ile30, Phe111, Phe117, Ser219
P_2	Glu	S_2	Thr218, Thr222, Glu287, Met289, Gly76, Thr77
P_1	Phe	S_1	Ile30, Asp32, Tyr75, Thr77, Phe111, Leu112, Phe117, Ile120, Gly217
P_1'	Tyr	S_1'	Tyr189, (Gln191), Ile213, Val291, Thr293, Leu298, Ile300
P_2'	Val	S_2'	Gly34, Ser35, Asn37, Ile73, Tyr75, Ala130
P_3'	Leu	S_3'	Ile128, Gly188, Tyr189

(residues 193–298) fully turns (as "solid") through $\approx 4°$ and shifts by 0.3 Å along the axis passing through C_α-atoms of residues Asp32 and Asp215. This distinctly changes a configuration of the whole "cleft" of the active site, i.e., the distance between some residues varies by up to 2 Å. These changes are supposed to be essential for the mechanism of action (see Chap. 7). The side chains of certain residues shift by 2.5–3.5 Å.

The above experimental data have yielded a model of the complexes of aspartic proteases with the substrate in the transition state (or rather as tetrahedral intermediate [3130]) (Fig. 102). Of much interest here is the interaction of one of the oxygen atoms of the tetrahedron with the aromatic cycle of Tyr75 residue, which can contribute to the energy of binding by as much as 5 kcal/mol [3133].

Metal-Dependent Proteases. Known are the structure of the complexes of carboxypeptidase A with a series of substrate-like inhibitors and the structure of the enzyme-inhibitor complexes of thermolysin (see Table 69). The complex of carboxypeptidase A with GlyTyr has been studied in detail. The binding of this dipeptide to the enzyme seems to reflect a productive binding of the true substrate. However, further investigations [3079] have revealed that the substrate replaces water from a coordination sphere of Zn^{2+}-ion, i.e., it binds unproductively. Moreover, the study of this complex has contributed much to understanding the structure and mechanism of action of carboxypeptidase A.

Glycyltyrosine binds to enzyme near Zn^{2+} so that the aromatic side chain of the ligand is placed into the "hydrophobic pocket", formed by the side chains of residues Ile243, Ile247, Tyr248, Gly253, Ile255 and Thr268. The side chain of residue Tyr248 is located in the plane almost normally to the substrate tyrosine ring plane (Fig. 103).

A carboxylate ion of the substrate forms a salt bridge with the guanidine group of residue Arg145, and the carbonyl group of the amide bond coordinates with Zn^{2+}. In the true substrates this group directs, apparently, to guanidine group of Arg127 residue [3081]. The charged group at high pH (in nonprotonated form) yields a coordination bond with Zn^{2+} [3134], which entails the replacement of water.

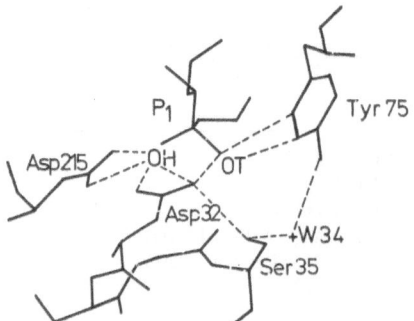

Fig. 102. Structure of tetrahedral intermediate in the active site of aspartic proteases; *OH* and *OT* oxygen atoms of a cleavable substrate group. *Dotted lines* hydrogen bonds. Interaction of one of oxygen atoms (*OT*) with aromatic cycle Tyr75 is shown [3130]

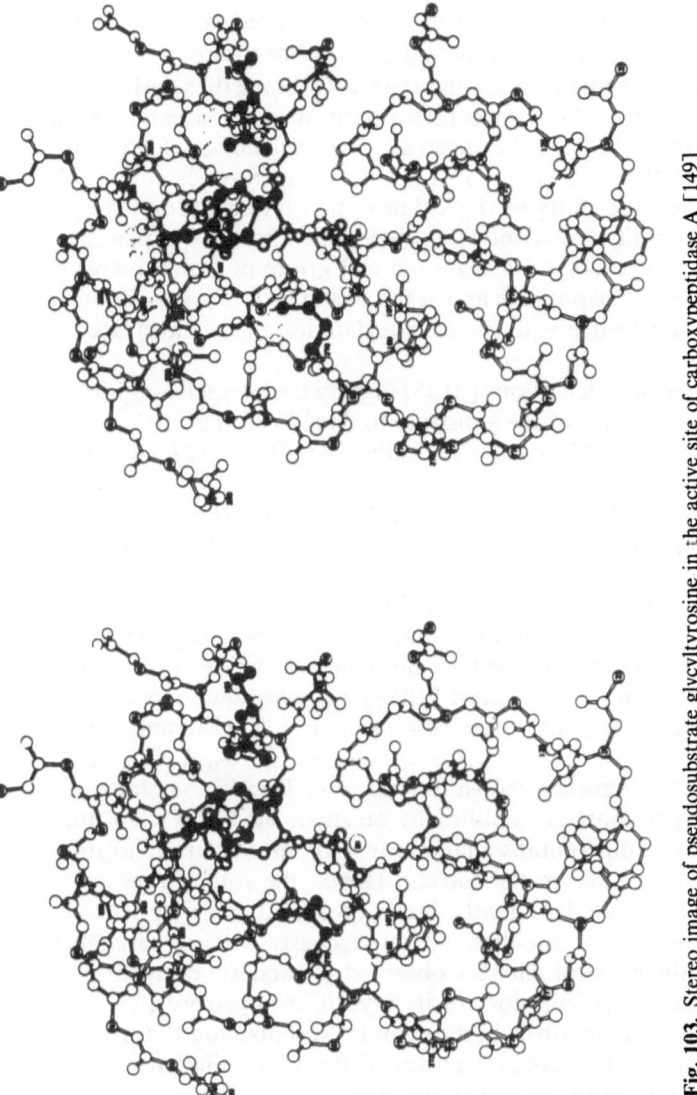

Fig. 103. Stereo image of pseudosubstrate glycyltyrosine in the active site of carboxypeptidase A [149]

These interactions are accompanied by pronounced conformational re-arrangements in the enzyme active site: (a) a guanidine group of Arg145 shifts by 2 Å to the ligand due to a turn around the C_β–C_γ bond; (b) a carboxyl group of Glu270 also shifts by 2 Å and occupies a position, when one of the oxygen atoms is remote by 3.6 Å from the substrate carbonyl carbon; (c) a change in the orientation of the side chain of Tyr248 is the most essential. Phenol hydroxyl of this residue shifts by 12 Å due to a rotation around the C_α–C_β bond and a slight shift (1–2 Å) of the polypeptide chain. In a new position phenol hydroxyl of Tyr 248 is 3 Å from the NH group of the cleavable bond. The shift of Tyr248 seems possible as a result of the shift of Arg145 during the formation of the salt bridge with the carboxylate ion of the substrate by this residue.

Quantum-chemical calculations [3135] display the energetic importance of the mechanism of "sliding" of the substrate to Arg145 so that the carboxylate ion alternatively forms salt bridges with residues Arg71, Arg127, and then with Arg145.

Analysis of the models has revealed that at the substrate binding in the active site of carboxypeptidase A an essential distortion of the planar structure of the amide bond seems possible. This fact is of much value, when formulating the enzyme action mechanism [3136, 3137].

The structural modeling of the complex of carboxypeptidase A with peptide ZAlaAlaTyr [3138], as well as the structural analysis of the enzyme complex with potato protein inhibitor [3082], has promoted identification of the secondary binding regions within residues Arg71, Tyr248 and Phe279.

All this has initiated an acute discussion (see [1964, Chap. 1; 3139]) especially concerning data on the shift of residue Tyr248. As shown [1370, 3140], in carboxypeptidase A modified by arsanylic acid at Tyr248 the sub-strate binding (in solution) entails a label transition into external layers of the protein molecule accessible to the solvent. Hence, the role of this residue in catalysis seems disputable. Eventually, it can be substituted by phenylalanine residue by the method of site-directed mutagenesis without changing the activity. Apparently, spectral changes observed for arsanyl derivatives of the enzyme depend on the peculiarities of its crystal arrangement [1371, 3035]. Moreover, Tyr248 is still involved in the complex stabilization thus resulting in hydrogen bonding to the carboxyl group of the C-terminal residue of the substrate and the NH group of the P_1-residue.

A hot discussion has arisen about the number of the substrate binding sites in carboxypeptidase A. This is caused by a variety of spectral characteristics of the enzyme complexes with esters and amides [3141]. However, most of the investigators share the opinion that the differences in the behavior of esters and amides appear, because carbonyl of these substrates binds either to Zn^{2+} or to the guanidine group of Arg127. Of much importance might also be the difference in the rate-limiting stage of the hydrolysis [3077, 3142].

Studies of ketoanalogs of the substrates [3078, 3080] as well as of phospho-amidates [2628, 3081] in the complexes with carboxypeptidase have provided the structural view of tetrahedral intermediates in catalysis. The two oxygen atoms of tetrahedron enter the coordination sphere of the Zn^{2+} ion, and bind to residues Glu270 and Arg127. The scheme of the tetrahedral intermediate is presented in Fig. 104.

Despite the significant differences in the structure of carboxypeptidase A and thermolysin the orientation of substrate-like inhibitors in the active sites of the two enzymes has much in common. Structural investigation of the thermolysin complexes with β-phenylpropionyl-L-phenylalanine, benzyloxycarbonyl-L-phenylalanine, L-benzylsuccinic acid, phosphoamidon and other ligands [2628, 2728, 2891, 3083–3087] has clarified the structure of the enzyme-substrate productive complex (Fig. 105). The aromatic side chain of residue P'_1 of the substrate binds in hydrophobic pocket of the enzyme formed by residues Val139, Leu133, Phe130, Leu202, Gly189, Val192 and Ile188. The carboxyl group of P'_1-residue forms a hydrogen bond with the guanidine group of Arg203, but groups NH and CO of the P'_2-residue of the substrate form hydrogen bonds with the amide group of the side chain of Asp112. Residue P_2

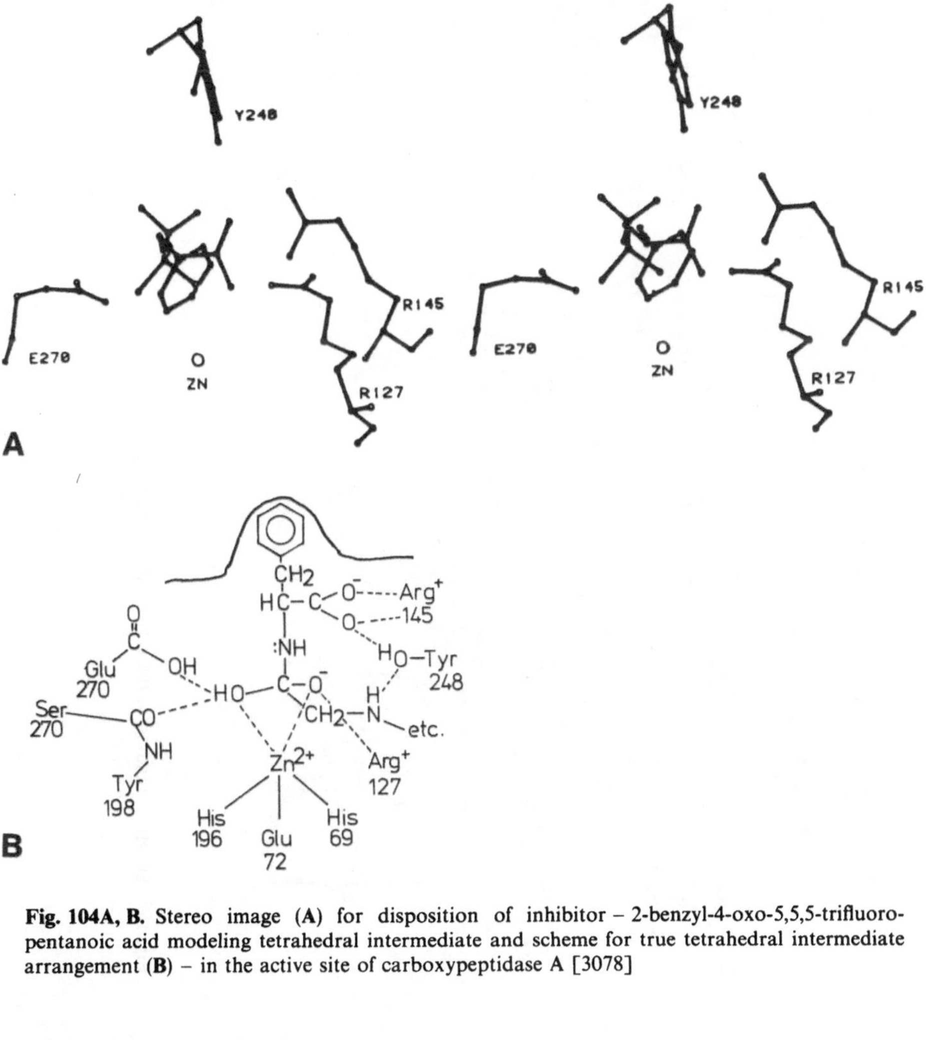

Fig. 104A, B. Stereo image (A) for disposition of inhibitor – 2-benzyl-4-oxo-5,5,5-trifluoropentanoic acid modeling tetrahedral intermediate and scheme for true tetrahedral intermediate arrangement (B) – in the active site of carboxypeptidase A [3078]

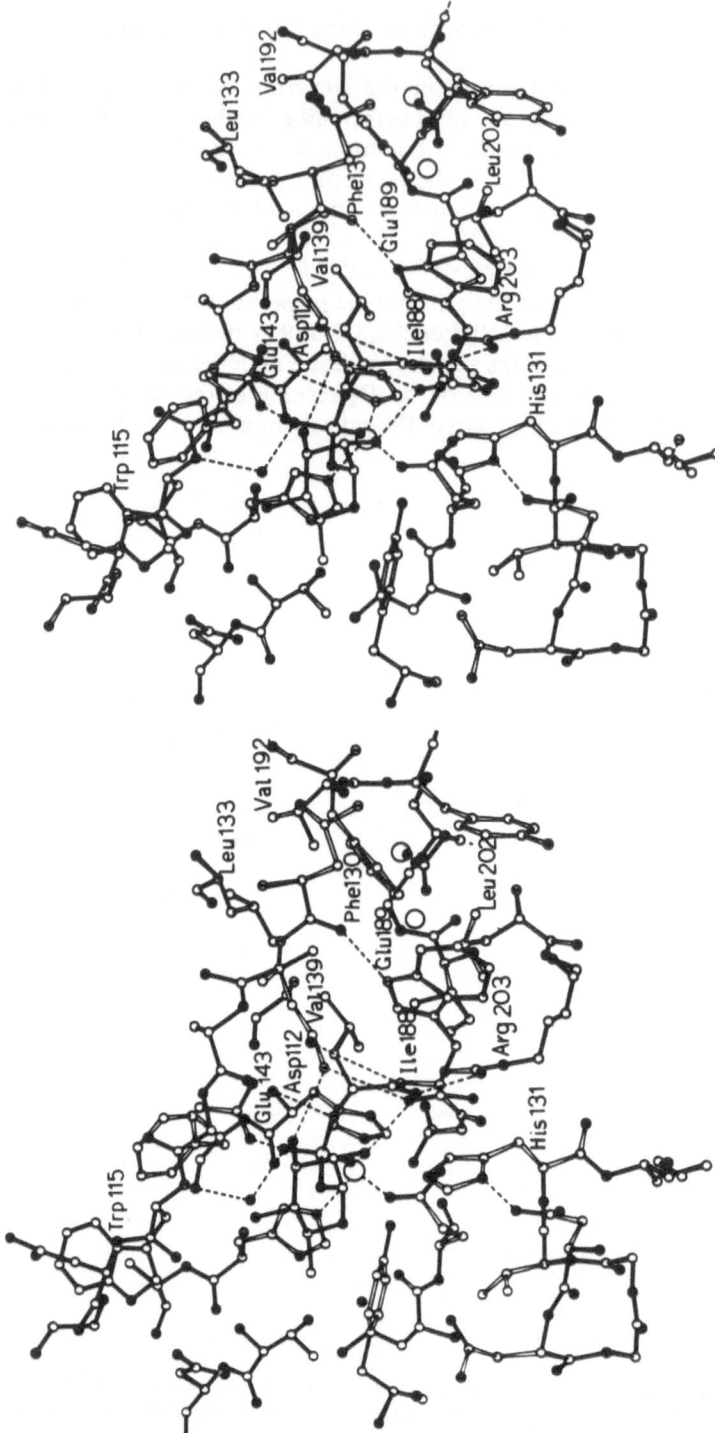

Fig. 105. Substrate binding in the active site of thermolysin [3085]

binds to the backbone of Trp115 and forms the region of the β-structure. The substrate cleavable bond is localized near Zn^{2+} and its carbonyl group occupies the position of the fourth ligand of the metal atom. The carboxyl group of Glu143 appears to be about 4 Å apart from the carbon carbonyl atom, the water molecule being close to carboxyl. Finally, the $N^{\varepsilon 2}$ atom of His231 is near the NH group of P'_1-residue of the substrate. Conformational changes, resulting from the binding of phosphoamidon, are restricted by tenth parts of an Å, and invade the side chains of Asn112, Leu202 and Glu143.

The direct surrounding of the Zn^{2+} ion in thermolysin and carboxypeptidase A is identical [1348, 2628]. Moreover, the same surrounding is observed in some dehydrogenases.

Of crucial value is the fact that the nearest surrounding of metal atom involves the substrate carbonyl group and the water molecule, i.e., zinc is presumably in the pentacoordination state.

6.6 Evaluation of the Overall Energy of the Enzyme-Substrate Interaction

The evaluation of the energy of the enzyme-substrate interactions and its comparison with the observed values seem rather exciting, and can be done only roughly. Data on the structure of the complex AcProAlaProTyrOH in the active site of protease A from *Streptomyces griseus* reveal [3063] that this tetrapeptide forms 78 Van der Waals contacts ($r < 4$ Å) with the enzyme and seven hydrogen bonds. Taking an average value for the energy of the Van der Waals contact as -0.05 kcal/mol and the energy of the hydrogen bond as about -3 kcal/mol (see Table 66), we get an overall energy of the interaction of, approximately, -25 kcal/mol. The evaluation of the interaction of the P_1-residue in carboxypeptidase A [3143] has yielded the value of -15 kcal/mol. The total energy of nonvalent interactions in acylenzyme – indolylacryloylchymotrypsin [3024] is about -30 kcal/mol. Electrostatic interaction of the carboxylate ion and guanidine group in carboxypeptidase contributes about -11 kcal/mol [3135]. Lastly, Van der Waals interactions in subsite D of the active site of lysozyme with the glycopyranoside cycle is -14 kcal/mol [2939].

The equilibrium constants of the complex formation are about $1 \times 10^{-3} - 1 \times 10^{-4}$ M for all the compounds mentioned. This corresponds to a change of the free energy by 4.2–5.5 kcal/mol. The enthalpy value of the complex formation, for example acetyltryptophan with chymotrypsin, is about -9 kcal/mol [2469]. Thus the energy of the complex formation seems much lower than that obtained by calculation, when all the interactions of the protein and ligand are summarized.

Here, apparently, no attention is paid to the energy expenditure on the conformational changes of the protein and substrate, many other factors are hardly accounted for. These expenditures seem to be of importance for catalysis.

6.7 Conclusion

The above information sets forth the following substantial characteristics of the enzyme-substrate interactions: (a) the enzyme-substrate binding is often stepwise, a fast diffusion-controlled stage of association preceding by a slow stage of "tuning" of the reactants; (b) a distortion of the structure of the planar amide bond during hydrolysis, resulting in a nonplanar bond ("pyramidalization"), seems possible; (c) the complex formation, as a rule, is accompanied by conformational rearrangements of the protein which is usually local; (d) the enzyme-substrate complex is stabilized by numerous noncovalent bonds, overall energy of which exceeds the value of the free energy yielded from the values of association constants of the enzyme and substrate.

Chapter 7 Chemical Transformation
of the Substrate

The formation of the productive enzyme-substrate complex provides the next step of the catalytic event – substrate chemical transformation to the reaction products. Its efficiency is governed by the state of the enzyme catalytic groups within the complex, their disposition relative to a cleavable bond, as well as the state of the substrate scissile bond. Some reviews [9, 1405, 1406, 1613, 3144–3149] summarize information on the mechanisms of the catalytic event at this stage.

7.1 The State of Catalytically Active Groups

Effective hydrolysis of amides and other derivatives of carboxylic acids implies the system's including a nucleophile that directly attacks a carbonyl carbon of the compound hydrolyzed; a general (or specific) base catalyst to release proton from a water molecule or another nucleophile as well as a general acid catalyst stabilizing intermediate structures and facilitating release of the leaving group. Evidently, the groups bearing the negative charge or having excessive electron density can be nucleophiles and general base catalysts, whereas electrophilic groups with a readily dissociating proton can act as general acid catalysts.

The groups involved in the catalysis are well identified now for all amide hydrolase types (see Sect. 2.6). Yet their state and role in catalysis are argued in many cases.

For serine proteases the hydroxyl group of Ser195 (according to chymotrypsin numeration), the imidazole group of His57, and the carboxyl group of Asp102 residues are the major catalytically active groups. Besides, amide groups of the protein backbone – NH groups of Ser195 and Gly193 – are also essential for the catalysis.

For cysteine enzymes the SH group of Cys25 (numeration according to papain) and the imidazole group of His159 are analogs of the two former groups. Cysteine proteases are presumably devoid of the group similar to Asp102, however, the carboxyl group of Asp158 may take part in the catalysis [3150]. The group of the Gln19 side chain and the NH group of the Cys25 backbone are analogs of the other two groups.

Aspartic proteases contain two catalytically active residues – Asp32 and Asp215 (according to pepsin).

The catalysis by metal-containing exopeptidases (carboxypeptidases) envisages the zinc atom as well as groups Glu270 and Arg127 (numeration by bovine carboxypeptidase A). Endopeptidases of this type have the carboxyl group of Glu143 residue and instead of the arginine the residue is His231 (thermolysin numeration).

Besides, high resolution X-ray analysis has revealed bound water molecules in the active sites of many amide hydrolases.

The residues strongly discriminated by pK_a values operate as catalytically active groups, thus differing in their ability of ionization within the pH region optimum for the enzyme action.

The ionization state of the active groups depends on their surrounding in the protein molecule. Analysis of the pH-dependence of catalysis (see Sect. 5.2), chemical modification [2344, 2359, 3151–3156], physicochemical investigations [2345, 2347, 2349, 2350, 2355, 3157–3162] as well as theoretical calculations [3163–3165] allowed a reliable detection of pK_a of the catalytically active groups in amide hydrolases (Table 73).

Data from Table 73 need some comments. The basicity of the serine hydroxyl in chymotrypsin, trypsin, etc. seems undetectable, since the proteins cannot be titrated up to pH $\approx$ 14. The pK_a here is derived from that of the hydroxyl group in N-acetylserineamide. It has been assumed [3166] that pK_a of the serine hydroxyl is much lower (≈ 8.0) due to the vicinity of the positively

Table 73. The pK_a and ionization state of active groups of some amide hydrolases

Group	Enzyme	pK_a	pH_{opt}	Quota of charged form in pH_{opt}	References
$-CH_2OH$	Chymotrypsin, etc.	≈ 13.5	8.0	3.2×10^{-6}	[1527]
$-CH_2SH$	Papain	4.1[a]	6.0	0.99	[2345]
	Chymopapain A	6.8	7.2	0.71	[3153]
	Chymotrypsin	6.8	8.0	0.06	[3160]
	Trypsin porcine	5.0	8.0	1×10^{-3}	[3160]
	α-Lytic protease	5.7	8.0	5×10^{-3}	[3160]
	Papain	4.26[a]	6.0	2×10^{-2}	[3169]
	Thermolysin	7.8	8.0	0.39	[1527, 2329]
$-COOH$	α-Lytic protease (Asp102)	4.5	8.0	1	[2350]
	Papain (Asp158)	4.2	6.0	0.98	[3150]
	Pepsin (Asp32)	1.5	2.5	0.91	[2324, 3170]
	„ (Asp215)	4.7	2.5	6.3×10^{-3}	[2324, 3170]
	Carboxypeptidase A (Glu270)	6.0	8.0	0.99	[3171]
	Thermolysin (Glu143)	6.2	8.0	0.98	[1527]
$Zn^{2+}\ldots OH_2$	Carboxypeptidase A	≈ 7	8.0	0.9	[1566]
	Thermolysin	≈ 7	8.0	0.9	[1566]
CONH	All enzymes	$\approx 18^b$	–	–	[71]

[a] The pK_a values of papain catalytic groups strongly depend on pH and the SH group state.
[b] For ionization $CONH \rightleftharpoons CON^- + H^+$.

charged groups, for example, arginine. However, crystallographic data do not support this.

The ionization state of Cys25 and His159 in papain and some other related cysteine enzymes is very intrinsic. The pK_a value of each of this group depends on the ionization state of the other group. It is suggested that papain has the following system of prototropic equilibria:

$$
\begin{array}{ccc}
 & \text{Im—E—SH} & \\
 \nearrow^{K_1} & & \searrow^{K_{12}} \\
\text{ImH}^+\text{—E—SH} & & \text{Im—E—S}^- \\
 \searrow_{K_2} & & \nearrow_{K_{21}} \\
 & \text{ImH}^+\text{—E—S}^- &
\end{array}
$$

with pK_1 4.26, pK_2 3.34, pK_{12} 7.55, and pK_{21} 8.47. This may be caused by two pH-dependent conformational states of the enzyme (see Sect. 5.2.3) which yields the effect of the electrostatic field created by the group with pK_a 5–6 remote from the catalytic site [3167]. In papain ionization virtually does not affect the activity, but in cathepsins B and H and in actinidin this influence is fairly high. The carboxyl group of Glu52 residue is apparently such a group in actinidin [3167].

Experimental ionization constants are related with the above microscopic constants as follows [3168]:

$$
K_1' = K_1 + K_2 \quad \text{and} \quad K_2' = \frac{K_{12}K_{21}}{K_{12} + K_{21}}.
$$

The pK_a values are ascribed to many enzyme groups only from the pH-dependences of the kinetic constant of hydrolytic reactions. As stated in Sect. 5.2.2, such assignments seem difficult.

The pK_a values have been assigned to the carboxyl groups of Asp32 and Asp215 (1.5 and 4.7, respectively) from pH-dependence of their chemical modification with epoxides and diazo compounds [1343, 2693] as well as from pH-dependence of pepsin catalysis. Chemical modification involves complicated processes, so these data may not unequivocally characterize the ionization state of the groups under study. Moreover, if the striking symmetry of disposition and interaction of carboxyl groups of these enzymes is taken into account (Sect. 6.5.3), the pK_a values are supposed to be rather close and the ion assymmetry can appear only within the substrate complex. Although, in general, quantum-mechanical analysis [3165] supports these views, it seems probable to revise the assignment of the pK_a's.

According to Table 73, the candidates for nucleophilic groups of many enzymes are either within the pH optimum of the enzyme action at very low concentration of the charged form, for instance serine hydroxyl, or, if fully ionized, they belong to rather weak bases with a low nucleophilicity (for example, COO^- group in aspartic proteases).

Apparently, mechanisms exist for enhancing the nucleophilicity (or concentration) of the group or the reactivity of the substrate with regard to the weak nucleophile.

7.2 The Charge Relay System

In 1969 an idea regarding the abnormally high reactivity of the serine hydroxyl in chymotrypsin was put forward [1327]. Crystallographic data as well as a thorough study of the enzyme primary structure show the three functional groups – Asp102 (earlier considered as Asn), His57 and Ser195 – to be closely located in the enzyme spatial structure, thus forming the system linked by hydrogen bonds:

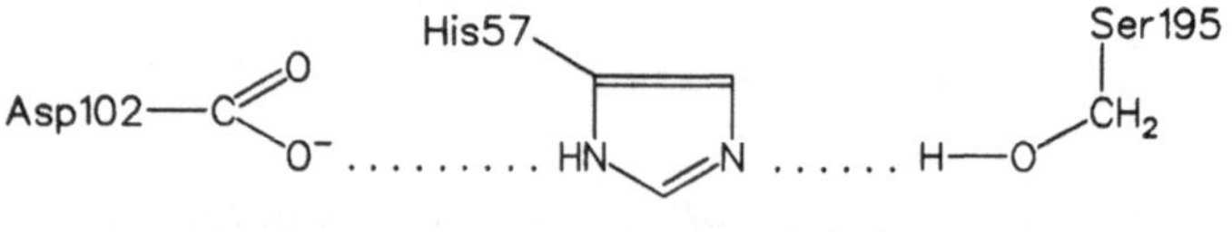

$$\text{Asp102}-\text{C} \overset{\text{O}}{\underset{\text{O}^-}{}} \ldots\ldots\ldots \underset{\text{His57}}{\text{HN}=\text{N}} \ldots\ldots \text{H}-\text{O}-\overset{\text{Ser195}}{\text{CH}_2}$$

Since the Asp102 carboxyl group is in the hydrophobic environment and screened from the solvent, its pK_a is proposed to be abnormally high, therefore the bond network can be isomerized:

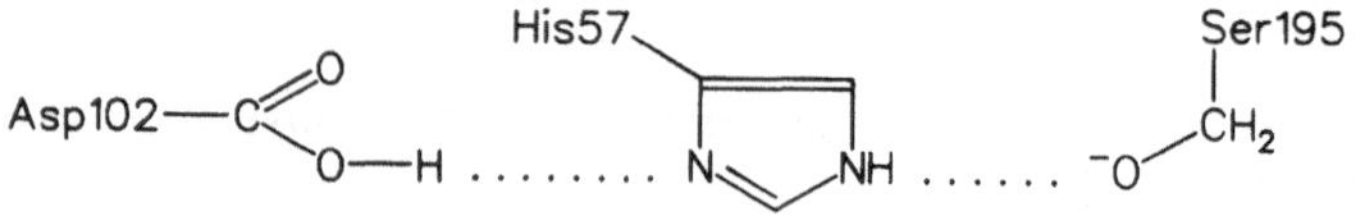

$$\text{Asp102}-\text{C} \overset{\text{O}}{\underset{\text{O}-\text{H}}{}} \ldots\ldots\ldots \underset{\text{His57}}{\text{N}=\text{NH}} \ldots\ldots {}^-\text{O}-\overset{\text{Ser195}}{\text{CH}_2}$$

An analogous bond network has been postulated for other serine proteases [1231, 1240, 3063, 3172]. In the neutral medium the charge can be transferred from the carboxyl group to the serine hydroxyl, the latter acquiring an extremely high nucleophilicity.

Data showing that destruction of the bond network in serine proteases by methylation of $N^{\epsilon 2}$-atom of His57 in chymotrypsin decreases the catalytic activity over five orders of magnitude could testify in favor of such a bonding system [1325]. Despite slight changes in the active site, His57-methylchymotrypsin differs in many respects from the native enzyme [3173], for instance in its resistance to denaturation [3174]. Thus the reason for the low catalytic activity of this derivative can be other than the destruction of the "charge relay system". Moreover, in active chymotrypsinogen and the alkaline form of chymotrypsin the groups of this system are positioned as in the native enzyme [2290].

Close values of pK_a of imidazole and Asp102 and proper disposition of the atoms in the system for stable hydrogen bonding are necessary conditions for the charge transfer.

A variety of methods – difference titration under enzyme denaturation conditions [3175], IR- [2355] and NMR-titration [2349, 2350, 3157, 3158, 3176] – have been employed to find out the true pK_a of the groups of the charge relay system. The results are rather contradictory. It has been suggested [2355, 3176] that these data accord with the existence of the charge relay system and lead to a hypothesis that the enzyme group ionized at pH $\approx$ 7 belongs to Asp102, while His57 is not protonated up to pH 2.

So another scheme including the "charge relay system" in catalysis has been proposed [3176]:

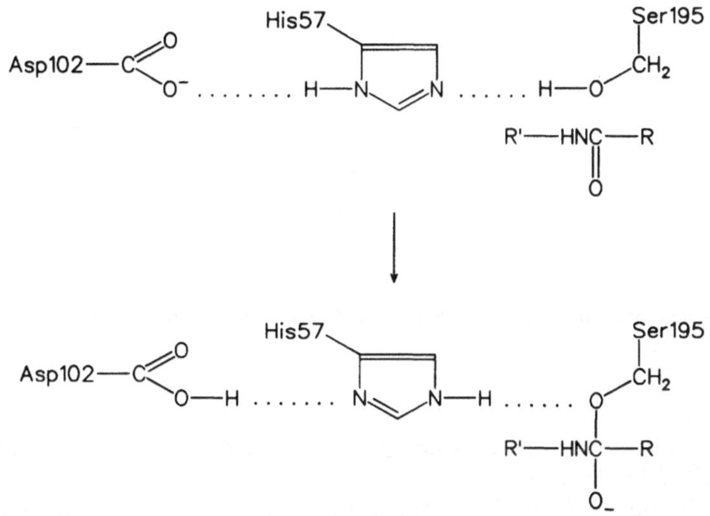

His57   Ser195

Asp102—C(=O)O⁻ .... H—N⟨imidazole⟩N ...... H—O—CH₂

R'—HNC(=O)—R

↓

His57   Ser195

Asp102—C(=O)O—H .... N⟨imidazole⟩N—H ...... O—CH₂

R'—HNC—R
       |
       O₋

It implies the concerted transfer of two protons at the rate-determining stage, which is not contradicted by "the proton inventory" method (measurement of the dependence of the kinetic isotope effect of solvent in mixtures H_2O/D_2O of various composition; see Sect. 3.3.3) [3177], indicating a "multiproton" catalysis [3178]. However, catalysis of this type is also observed for ($N^{\varepsilon 2}$-methylhistidine)-chymotrypsin free of the charge relay system [3179].

For α-lytic protease (containing only one histidine) pK_a of His57 has been shown to be about 7. This is in line with the results on low field NMR signals of chymotrypsin and other serine proteases [2349, 3157] as well as with studies on C2-proton imidazole resonance of trypsin [3158]. Structural analysis of protease A from *Streptomyces griseus* has disclosed that the carboxyl group of Asp102 forms the maximum possible hydrogen bonds (four bonds) and therefore it cannot have high pK_a.

X-ray data have been revised many times. New interpretation of the original data [1327] shows [1249] that in native α-chymotrypsin, γ-chymotrypsin, trypsin, subtilisin BPN' and elastase the serine hydroxyl is outside the plane of the imidazole ring and the O^γ-atom of Ser195 sets itself apart from the $N^{\varepsilon 2}$-atom of His57 by 3.5–3.7 Å, deviating thus from the disposition required for the hydrogen bonding by 2.5–3.5 Å (Fig. 106). Though Ser195 in serine proteases can freely rotate around C_α–C_β-bond by about 120° (Sect. 6.5.3), sometimes it approaches the $N^{\varepsilon 2}$-atom of His57 so as to form a proper hydrogen bond [1229].

The above distance, later revised for chymotrypsin [1226], trypsin [1243], kallikrein [1243] and protease A from *Streptomyces griseus* [3063], is 2.7–2.9 Å.

Finally, neutrongraphic analysis of trypsin and its complex with the inhibitor imitating the tetrahedral intermediate testifies to the lack of proton transfer

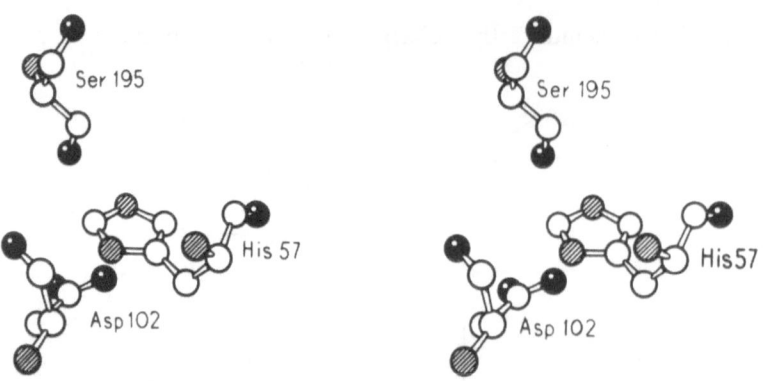

Fig. 106. Stereo image for location of residues Ser195, His57 and Asp102 in the active site of bovine α-chymotrypsin [1249]

to Asp102 [3182], though there evidently exists a hydrogen bond between Ser195 and His57 [3182, 318].

The substitution of Asp102 by Asn [3815] in trypsin by site-directed mutagenesis yields a sharp (10^4-fold) fall of the activity in the neutral medium. The structures of mutant and native enzymes are almost the same [2124].

Thus, at least the free enzyme and apparently its enzyme-substrate complex are free of the "charge relay system", the catalytic "triad" can be shown by scheme:

$$\text{Asp102}-C\overset{O}{\underset{O^-}{\diagdown}}\ldots\ldots\ldots\overset{\text{His57}}{HN\diagdown\diagup N}\ldots\overset{\text{=}}{\ldots}\ldots H-O\diagup\overset{\text{Ser195}}{CH_2}$$

Quantum-chemical calculations [3186–3191] have also confirmed high barriers for the proton transfer from Ser195 to His57. Chymotrypsin chemical modification [3152] and the solvent effect on the deacylation rate of acyl-chymotrypsins [3192] accord well with the concept about Ser195 functioning in the nonionized form.

A role of the hydrogen bond between Asp102 and His57 residues is to fix the imidazole ring orientation and to distribute the electron density within the ring so as to provide the π-tautomer structure favorable for catalysis [2350, 3193]

$$\underset{\pi}{\overset{R\diagdown}{HN\diagdown\diagup N}}\quad\longleftrightarrow\quad\underset{\tau}{\overset{R\diagdown}{N\diagdown\diagup NH}}\text{ ,}$$

as well as a positive charge stabilization on the imidazolium cation at further steps [2189].

Cysteine proteases have no analogs of the "charge relay system". As mentioned (see Sect. 5.2.3), the state of catalytically active groups in papain matches the ion pair structure

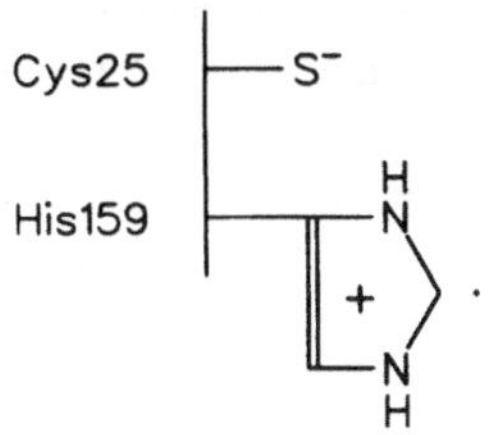

Cys25 ├── S⁻

His159 ├── (imidazolium ring structure with N, H and + charge)

The imidazole orientation is stabilized by the hydrogen bond of the $N^{\varepsilon 2}$-atom with the Asn175 carbonyl group [3122].

The nucleophilicity of thiolate ions is high enough to react with the substrate carbonyl group. However, study of the model reactions (see Sect. 3.4.4) shows that thiols are rather inert in reactions with amides. Certain difficulties may arise in describing the following catalysis stages.

The carboxyl groups of Asp32 and Asp215 of aspartic proteases are within the same plane forming a stable hydrogen bond. Besides, each oxygen atom is linked by the hydrogen bonding network with other amino acid residues and solvent molecules (see Fig. 100). So the carboxyl group of Asp32 binds to the Ser35 hydroxyl group and the NH group of Gly34, and Asp215 to the hydroxyl of Thr218 and the NH group of Gly217. Besides, the oxygen atoms of the carboxyl groups of Asp32 and Asp215 are located at a distance close enough for the hydrogen bonding of two water molecules [3131, 3194].

As shown above, the symmetry in the disposition of the carboxyl groups implies a similarity of their pK_a values. However, the bell-shaped pH-dependence for the enzymatic catalysis and some other facts bear evidence to the difference in the pK_a of Asp32 and Asp215.

The reason for it may be associated with two closely located negtative charges resulting from a removal of the proton between these residues. To withdraw the charges from each other requires higher energy owing to the hydrogen bonding network. The situation resembles that observed for conformationally fixed dicarboxylic acids [2574, 3165, 3194].

Zn^{2+}-Ion plays a key role in catalysis of metal-containing proteases. The ion can coordinate with four to six ligands [3195] including a water molecule. Dissociation of the water within the coordination compound presumably proceeds at neutral pH (see Sect. 5.2.3). If the water molecule attacks the substrate, the concentration of OH^--ions is virtually equal to the enzyme concentration within the pH optimum. However, nucleophilicity of the metal-bound OH^- ion should dramatically decrease (as its pK_a).

Thus none of the amide hydrolases types considered is expected to possess high nucleophilic catalytic groups that attack the substrate carbonyl. The efficiency of the attack must be achieved by other mechanisms.

7.3   The State of the Cleavable Substrate Group

An increase in the reactivity of the cleavable bond relative to weak nucleophiles is possible if the level of the resonance stabilization of this bond decreases. This, in turn, occurs when the amide bond coplanarity is disturbed. There are two ways for the disturbance of the amide group coplanarity – "bending" and "twisting":

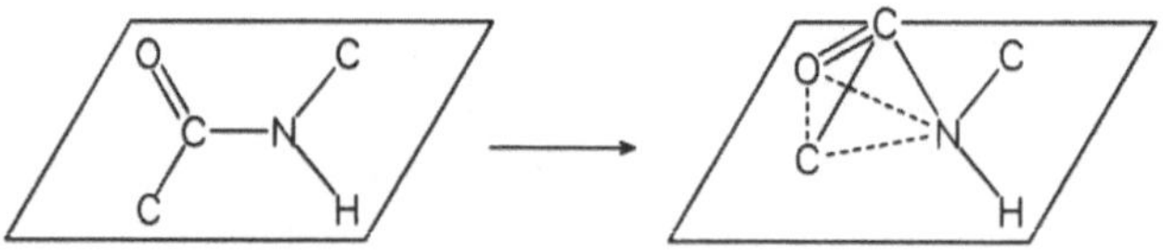

Bending

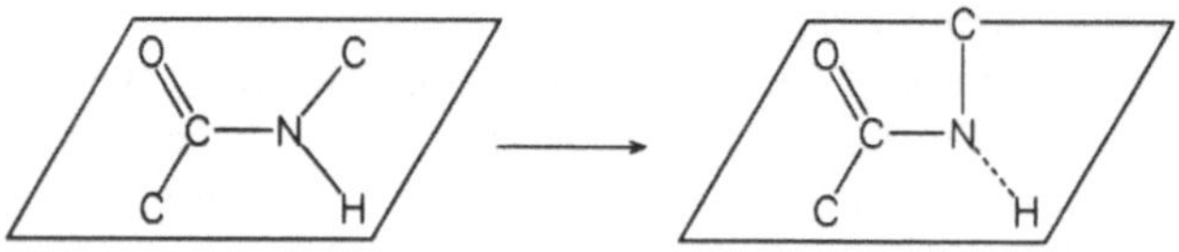

Twisting

Both these ways entail a change in coupling the electron lone pair of the nitrogen atom and π-electrons of the carbonyl group. In the first case, the carbonyl C-atom extrudes from the plane formed by atoms O, C_α, and N, in the second it remains in this plane. These deformations, if not large, do not require a high energy expenditure. Even ethylene "twisting" around the C=C double bond by 20° is associated with an energy expenditure of about 2.5 kcal/mol [3196].

The problem of resonance destabilization of the amide group in free peptides and enzyme-substrate complexes has been already discussed (Sects. 1.2.1, 3.3.1 and 6.5.2). It is noteworthy that the bending degree depends on a distance between the nucleophile and the carbonyl carbon of the amide [1469, 3197] and can be essentially stabilized by secondary enzyme-substrate interactions.

Thus the amide bond coplanarity seems to be disturbed in the enzyme complexes with the substrates and, as a result, the energy of its resonance stabilization falls.

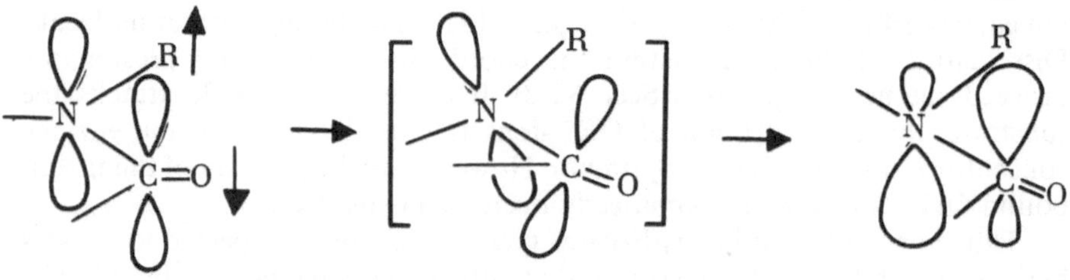

Fig. 107. Change of hybridization in the amide group upon rotation around bond C–N [3136]

The amide group "twisting" should change hybridization of the carbonyl carbon and nitrogen atoms (Fig. 107) [3136], which makes the directions of the nucleophile or proton attachment nonequivalent.

7.4 Covalent or General Base Catalysis?

The catalytically active nucleophilic group of the enzyme can be a covalent catalyst if it directly attacks the carbonyl carbon of the substrate:

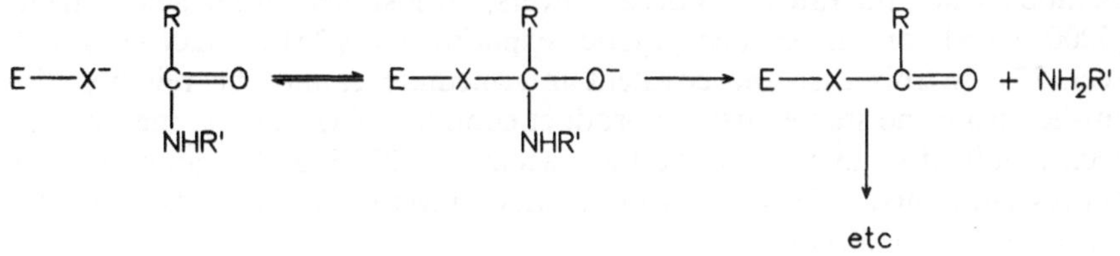

$$
\text{E—X}^- \quad
\begin{array}{c} R \\ | \\ C{=}O \\ | \\ NHR' \end{array}
\;\rightleftharpoons\;
\text{E—X—}
\begin{array}{c} R \\ | \\ C{-}O^- \\ | \\ NHR' \end{array}
\;\longrightarrow\;
\text{E—X—}
\begin{array}{c} R \\ | \\ C{=}O \end{array}
+ \; NH_2R'
$$

$$\downarrow$$

$$\text{etc}$$

(1)

or a general base catalyst, when facilitating the proton withdrawal from the water molecule:

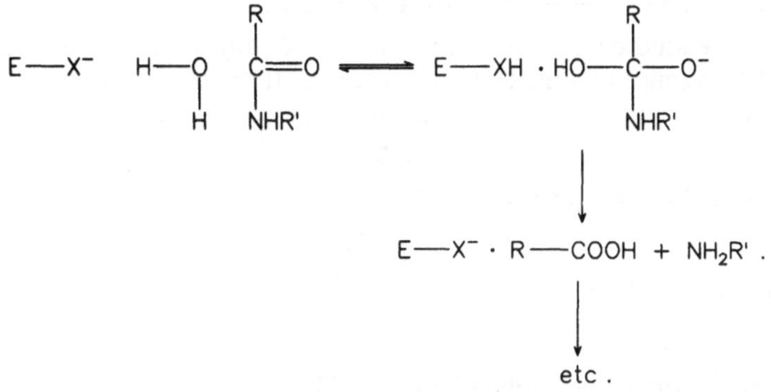

$$
\text{E—X}^- \quad \text{H—O}
\begin{array}{c} \\ | \\ H \end{array}
\begin{array}{c} R \\ | \\ C{=}O \\ | \\ NHR' \end{array}
\;\rightleftharpoons\;
\text{E—XH} \cdot \text{HO—}
\begin{array}{c} R \\ | \\ C{-}O^- \\ | \\ NHR' \end{array}
$$

$$\downarrow$$

(2)

$$\text{E—X}^- \cdot \text{R—COOH} + NH_2R' \;.$$

$$\downarrow$$

$$\text{etc .}$$

According to scheme (1), the acyl fragment of the substrate formed after the breakdown of the tetrahedral intermediate remains covalently bound to the enzyme, whereas scheme (2) shows no covalent bond. Depending on the ratio of the formation and decay rates, the intermediate acylenzyme can be accumulated or not during the reaction.

If the acylenzyme is accumulated in the system, the covalent catalysis is readily identified. A general approach is to measure the catalytic constants for a series of substrates with an identical acyl fragment and different leaving groups. In this case (see Sect. 4.1.1) the reaction proceeds by the three-stage scheme:

$$
E + S \underset{}{\overset{K_s}{\rightleftharpoons}} ES \xrightarrow{k_2} EA \xrightarrow{k_3} E + P_2,
$$
$$
\begin{array}{c} + \\ P_1 \end{array}
$$

where EA is the acylenzyme, $k_{cat} = k_2 k_3 / k_2 + k_3$ and at $k_2 \gg k_3 k_{cat} = k_3$. For the whole substrate series k_{cat} is to be the same [1666, 1721]. According to Table 74 the k_{cat} values are constant for some acylamino acid esters – chymotrypsin substrates.

For various amides of acylamino acids k_{cat} is much discriminated, since $k_2 \gg k_3$ is not fulfilled.

This method has been used for numerous serine and some cysteine proteases [1766, 3198, 3199]. As noted, in some cases the investigations do not reveal the character of the rate-determining stage [1427].

Sometimes the intermediate acylenzyme can be detected by spectral methods, when the rate of its decay lowers, for instance, in the acidic medium [3200–3205] or in cryoenzymatic experiments [2512, 3206–3208] (see Sect. 5.7). In this case the covalent intermediate cannot be reliably distinguished from the stable enzyme-product complex. The spectral methods have been modified so as to increase their accuracy [3205, 3209], among them of fluorescent analysis, for identifying the acylenzyme in β-lactamase I catalyzed processes [3210, 3211].

The so-called method of trapping, i.e., the formation of a new substrate derivative from the acylenzyme by the acyl fragment transfer onto a nucleophilic acceptor, is successfully applied:

$$E\text{---}X\text{---}COR + HYR^1 \longrightarrow E\text{---}XH + R^1 YCOR .$$

These reactions are studied by measuring the k_{cat}/K_m ratio for the substrate hydrolysis in the presence of a nucleophilic acceptor. If the transfer occurs, the second-order rate constant depends on the acceptor concentration:

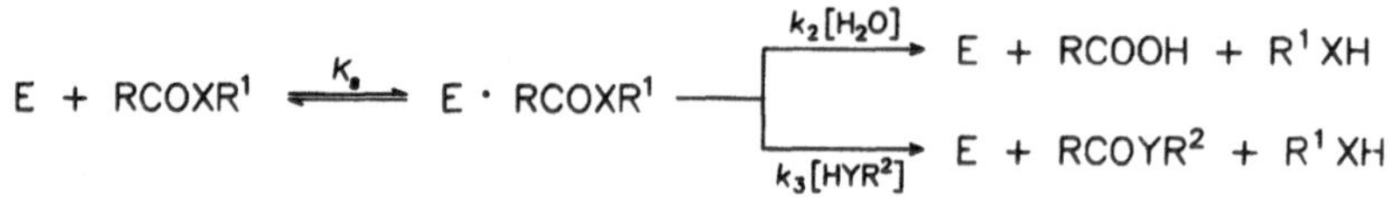

$$E + RCOXR^1 \underset{}{\overset{K_s}{\rightleftharpoons}} E \cdot RCOXR^1 \longrightarrow \begin{cases} \xrightarrow{k_2[H_2O]} E + RCOOH + R^1 XH \\ \xrightarrow{k_3[HYR^2]} E + RCOYR^2 + R^1 XH \end{cases}$$

$$\frac{k_{cat}}{K_m} = \frac{k_2[H_2O] + k_3[HYR^2]}{K_s} .$$

1,4-Butandiol is a proper nucleophile here [1723].

Table 74. Kinetic constants of chymotrypsin-catalyzed hydrolysis of acylamino acid derivatives [1761]

Substrate[a]	k_{cat}, s^{-1}	K_2, s^{-1}	k_3, s^{-1}
AcTrpOMe	27.7 ± 0.5	730	29
AcTrpOEt	26.9 ± 0.5	700	28
AcTrpONp	30.5 ± 1.2	38 300	30.5
AcTrpNH$_2$	0.026	0.026	30
AcPheOMe	57.5	280	72
AcPheOEt	63.5	530	72
AcPheONp	77.0	23 700	77
AcPheNH$_2$	0.039	0.039	72

[a] ONp – p-nitrophenyl ester.

This method has been utilized to identify the acylenzyme at the chymo-tryptic hydrolysis of benzoylglycine esters (HYR^2-hydroxylamine) [3198, 3212]. However, its application for proving the acylenzyme formation at the chymotryptic hydrolysis of amides of acylamino acids has met with certain difficulties (compare [3213] and [3214]). Unequivocal evidence in favor of the acylenzyme formation appeared only in 1973 [1412]. The same methods have been also used to state the intermediate formation of acylenzymes upon catalysis by β-lactamase [1856], D-alanylcarboxypeptidase [2180, 3215], and cathepsin B [3216]. For identification of the rate-limiting stage it is suggested [3217] to determine the dependence of k_{cat} and k_{cat}/K_m from temperature. These two parameters alter identically with temperature changes, when the enzyme acylation is of a slow stage.

The method of trapping, in particular its version transpeptidation [2169, 2177, 3218–3221], has convinced investigators that the acylenzyme (anhydride E–CO–O–COR) and the "aminoenzyme" (E–CONHR^1) are involved in hydro-lysis of peptides by pepsin and other aspartic proteases. The formation of the "aminoenzyme" has been also confirmed when studying initial rates of trans-peptidation of the substrates with the same transferable group (see Sect. 4.8.1). Later, the covalent intermediates have been show not to have been formed at the catalysis by pepsin. Consequently, the covalent and stable noncovalent intermediates cannot be discriminated by this method.

Another method, though with the same shortcoming, implies a study of the order of the products release in the hydrolytic reaction by the character of inhibition by products (see Sect. 5.8.3). It has been employed to analyze the mode of action of pepsin [2175, 2587, 2604]. However, the results have been criticized [2588].

Attempts have been undertaken to ground the formation of covalent intermediates from the linear free-energy relationships [2010, 2012, 2013]. This analysis is applied only to the systems with known standard changes in the free energy, but it is not devoid of arbitrary assumptions.

Lastly, the influence of enzyme modification on amidase and esterase activities can procure information on the mechanism of amide and ester hydrolysis. However, this method cannot supply us with unequivocal results (compare [3222] and [3223]).

In my opinion, the most reliable methods for identification of a catalysis type are those based on the inclusion of heavy oxygen from $H_2^{18}O$ into the substrate, transpeptidation or hydrolysis products.

In fact, the reaction proceeding by mechanism (1) involves water only at the stage of the acylenzyme hydrolysis, and by mechanism (2) at the first chemical stage of the process. It is important that the reaction runs by the mechanism of the acyloxygen cleavage established for model systems (see Sect. 3.3) and for ester hydrolysis by some proteases [3224].

The ^{18}O incorporation into a substrate can be studied as well in a steady-state regime as [3225–3228] under equilibrium conditions [1729]. In the first case, label ^{18}O can be included into the substrate only if water is involved at the first stage, i.e., if the reaction follows mechanism (2). A positive result evidences mechanism type (2), while a negative one does not provide any conclusion, since no inclusion into a substrate can follow from the unfavorable ratio of the rates of formation (k_2) and decay (k_3) of the tetrahedral intermediate:

$$E\!-\!X^- \quad H\!-\!\overset{\bullet}{\underset{H}{O}} \quad \underset{NHR'}{\overset{R}{\underset{|}{\overset{|}{C}}}}\!\!=\!\!O \quad \underset{k_2}{\overset{k_1}{\rightleftharpoons}} \quad E\!-\!XH \cdot H\overset{*}{O}\!-\!\underset{NHR^1}{\overset{R}{\underset{|}{\overset{|}{C}}}}\!-\!O^- \quad \overset{k_3}{\longrightarrow} \quad products \;.$$

If the reaction proceeds through the acylenzyme formation and the system is in thermodynamic equilibrium:

$$E + RCONHR^1 \;\rightleftharpoons\; E \cdot RCONHR^1 \;\rightleftharpoons\; E\!-\!COR \;\overset{H_2\overset{\bullet}{O}}{\rightleftharpoons}\; E + RCO^*OH \;,$$
$$+$$
$$R^1NH_2$$

the rate of the ^{18}O incorporation into the substrate from $H_2{}^{18}O$ cannot exceed the rate of the acyl product inclusion, and the rate of R^1NH_2 inclusion can be considerably higher, since the $R'NH_2$ and ^{18}O inclusion occur independently.

If the acylenzyme is not formed:

$$E + H_2\overset{*}{O} + RCONHR^1 \;\rightleftharpoons\; E \cdot RCONHR^1 \cdot H_2\overset{*}{O} \;\rightleftharpoons\;$$

$$E \cdot RCO^*OH \;\rightleftharpoons\; E + RCO^*OH \;,$$
$$+$$
$$R^1NH_2$$

the inclusion of ^{18}O into the substrate should run with the rate equal to that at the $R'NH_2$ inclusion. The rates of inclusion both of radioactive-labeled products and ^{18}O from heavy oxygen water into the substrate at various but equilibrium concentrations of the substrate and products (the Boyer method, see Sect. 4.5.1) show that in the course of the chymotrypsin-catalyzed hydrolysis of AcPheGlyNH$_2$ the reaction runs through the acylenzyme.

Thus, if mechanism (1) works, the incorporation of ^{18}O from $H_2{}^{18}O$ in the products of acyl transfer or transpeptidation is improbable. If the reaction proceeds according to mechanism (2), the inclusion is observed. Table 75 lists data [2040, 2189, 2574, 3227–3232] provided by the study of hydrolysis catalyzed by some amide hydrolases. Serine and cysteine proteases function by covalent catalysis (1), and aspartic and metal-containing proteases use general base catalysis (2) at the stage of a water molecule attack of a substrate.

It is well known (see Sect. 4.8.2) that many amide hydrolases catalyze an oxygen exchange in the acyl product of the hydrolytic reaction. This is accepted by some enzymologists [2183, 3233, 3234] as evidence for the acylenzyme formation, while others [2188] believe that this exchange can be a result of synthesis (from acylamino acid and peptides – products of autolysis or release of an acyl group) of a novel peptide and its further hydrolysis. In fact, such enzymes as pepsin and leucineaminepeptidase functioning by mechanism

Table 75. Inclusion of ^{18}O to transpeptidation products [2040, 3229, 3230]

Enzyme	Examined reaction	Form of analyzed product[a]	Exchange, %
Chymotrypsin	$2LeuLeuNH_2 \rightarrow (Leu)_3 + LeuNH_2$ $+ NH_3$	Tfa(Leu)$_3$OMe	0
Papain	$ZGlyNH_2 + LeuGly \rightarrow ZGlyLeuGly$ $+ NH_3$	ZGlyLeuGlyOMe	0
	$ZGlyOMe + LeuGly \rightarrow ZGlyLeuGly$ $+ MeOHZPheOMe$	ZGlyLeuGlyOMe	0
	$+ HLeuNH_2 \rightarrow ZPheLeuNH_2$ $+ MeOHZPheLeuNH_2$		0
Carboxypeptidase Y	$ZPheAlaOH + HLeuNH_2 \rightarrow$ $ZPheLeuNH_2 + Ala$	ZPheLeuNH$_2$	0
Pepsin	$2LeuTyrNH_2 \rightarrow (Leu)_2 + 2TyrNH_2$	TMS(Leu)$_2$OTMS	56 ± 10
Leucine-aminopeptidase	$2LeuNH_2 \rightarrow LeuLeuNH_2 + NH_3$	LeuLeuNH$_2$ ^{18}O-LeuLeuNH$_2$	46 ± 5 55 ± 5^b
Thermolysin	$2LeuLeuNH_2 \rightarrow (Leu)_2 + 2LeuNH_2$	Tfa(Leu)$_2$OMe	47 ± 5

[a] Tfa – trifluoroacetyl, TMS – trimethylsilyl; the substances are studied by mass spectrometry; water containing 60–80 atomic % is used.
[b] ^{18}O-LeuNH$_2$ and H$_2^{16}O$ are applied.

(2) enable to catalyze the ^{18}O exchange in acylphenylalanine and leucine:

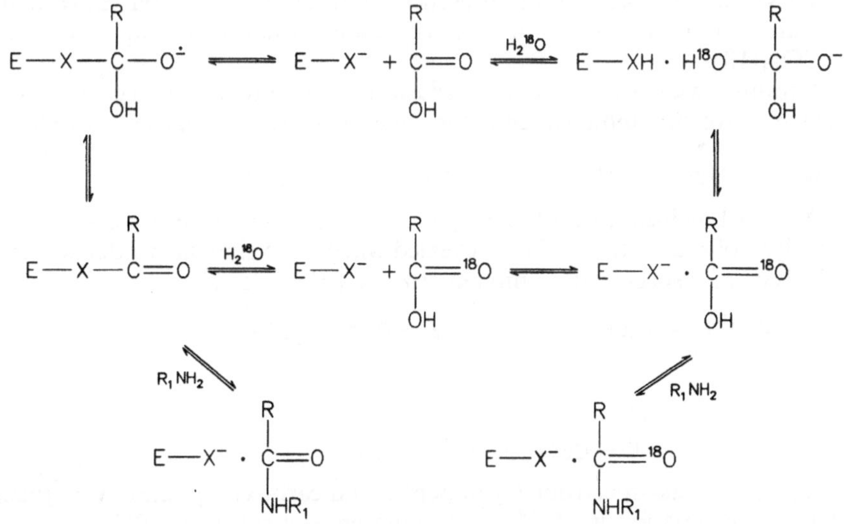

The scheme indicates that an addition of the amine product (R_1NH_2) should accelerate the ^{18}O exchange in the acyl product only if the reaction follows mechanism (2) due to parallel synthesis-hydrolysis of the peptide. According to Table 76, this mechanism governs catalysis by pepsin, leucineaminopeptidase, and carboxypeptidase A.

Table 76. The rate of oxygen èxchange catalyzed by pepsin, leucineaminepeptic ase and carboxypepetidase A [2189, 3228, 3235]

Enzyme	Substrate	Experimental V_{max} exchange, $M \cdot min^{-1}$	Calculated value of exchange rate due to synthesis-hydrolysis[a], $M \cdot min^{-1}$
Pepsin	AcPheOH	2.5×10^{-5}	–
	AcPheOH + HPheAlaOMe	5.6×10^{-5}	7.8×10^{-5}
Leucineamino peptidase	HLeuOH	1.6×10^{-3}	–
	HLeuOH + NH$_4$Cl	1.8×10^{-3}	8.3×10^{-6}
	HLeuOH + HLeuNH(CH$_2$)$_2$OH	3.5×10^{-3}	1.2×10^{-3}
Carboxypeptidase A	AcPheOH	3.3×10^{-6}	–
	AcPheOH + HPheOH	11.6×10^{-6}	9.9×10^{-6}

[a] Calculated by formula (1) in [2188].

Evidently, the contribution of synthesis-hydrolysis may be observed only if its overall rate is comparable with the rate of intrinsic exchange. This is illustrated by system HLeuOH + NH$_4$Cl, where the synthesis-hydrolysis rate is fairly low, and in which the acceleration by ammonia does not exceed experimental error.

These methods witness the presence or absence of the acyl enzyme, still they are not fit to identify the intermediate formation of "aminoenzyme" (E–CONHR).

For this reason, the lack of the inclusion of ^{18}O into carboxyl groups of the enzyme active site upon the aminoenzyme hydrolysis should be proved;

$$E\!-\!CO^{*}NHR^{1} + H_{2}^{*}O \longrightarrow E\!-\!CO^{*}OH + NH_{2}R^{1}$$

No label inclusion into the enzyme in the course of the pepsin-catalyzed hydrolysis of substrates has been revealed when analyzing the products of mild hydrolysis of a specifically inhibited enzyme [2189, 2574]:

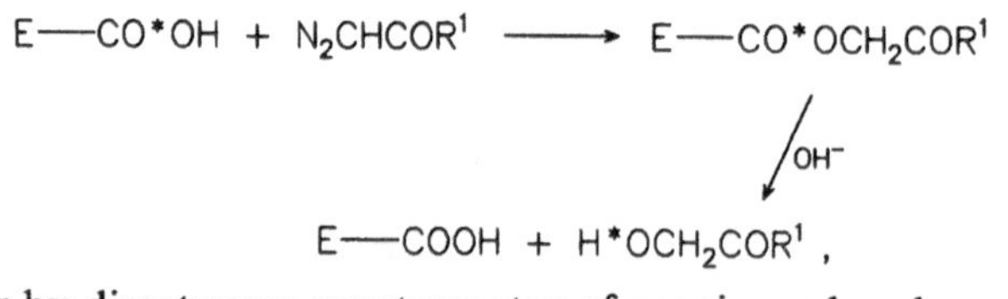

$$E\!-\!CO^{*}OH + N_{2}CHCOR^{1} \longrightarrow E\!-\!CO^{*}OCH_{2}COR^{1}$$

$$\Big/ OH^{-}$$

$$E\!-\!COOH + H^{*}OCH_{2}COR^{1} ,$$

nor by direct mass-spectrometry of pepsin and carboxypeptidase A fragments after their incubation in H$_2$^{18}O in the presence of substrates [3236].

Identical methods are applied to elucidate the mode of action of aminoacylase from pig kidney [3237], asparaginase [3238], and the enzymes of other classes [3239].

Thus, the catalytically active hydrozyl group of Ser195 of serine proteases is the group attacking the scissile bond of amide and ester substrates. The same is true for the Cys25 SH group of cysteine enzymes.

In the case of aspartic proteases a water molecule is an attacking nucleophile. Nucleophilic catalytic groups of these enzymes serve as general base catalysts for proton release from water molecules.

The results of the ^{18}O inclusion into the transpeptidation product catalyzed by pepsin have been argued [3240] and explained by a complicated sequence of the synthesis-hydrolysis reaction. However, general base catalysis at the stage of nucleophile attack is commonly recognized today.

As to carboxypeptidase A numerous data are available in favor of general base catalysis of the attack by water bound to Zn^{2+} at the hydrolysis of amide and ester substrates [3241–3244]. Nevertheless, data of cryoenzymology (see Sect. 5.7) testify to the intermediate formation of the anhydride between an acyl fragment of the substrate and the carboxyl group of Glu270 residue [2523, 3206, 3245–3248]. General base catalysis in case of leucineaminepeptidase and thermolysin [2695] is no doubt.

The involvement of the phenol hydroxyl group of the tyrosine residue as a nucleophile, which directly attacks the substrate carbonyl and forms the ester bond with acyl fragment of the substrate (acylenzyme), has been postulated for aminopeptidase M [3249].

There are some data on covalent catalysis at the pepsin hydrolysis of sulfide esters [3221, 3250, 3251]. However, they need verification.

It is unlikely that the carboxyl group behaves as a covalent catalyst. Though models exist where the carboxylate ion fulfills this function (see Sect. 3.4.5), these are hindered by relatively stable cyclic anhydrides. Comparison of a tetrahedral intermediate formed by a carboxyl group and a hydroxyl ion shows that the breakdown of the former in the direction of initial substances should be considerably more effective (low basicity of COO^- group):

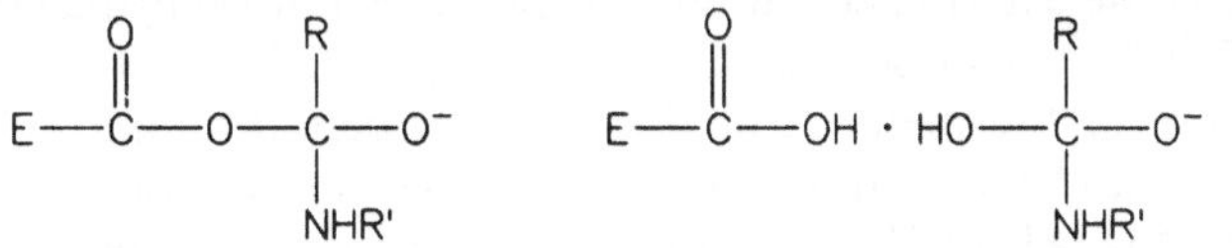

$$E-\underset{\underset{\displaystyle O}{\|}}{C}-O-\underset{\underset{\displaystyle NHR'}{|}}{\overset{\overset{\displaystyle R}{|}}{C}}-O^- \qquad E-\underset{\underset{\displaystyle O}{\|}}{C}-OH \cdot HO-\underset{\underset{\displaystyle NHR'}{|}}{\overset{\overset{\displaystyle R}{|}}{C}}-O^-$$

7.5 Electrophile

Electrophilic catalysis is one of the ways to enhance the reactivity of carboxylic acid derivatives in hydrolytic reactions. It involves the interaction with the oxygen atom of the carbonyl group and initiates redistribution of the electron density so that the positive charge at the carbon atom increases:

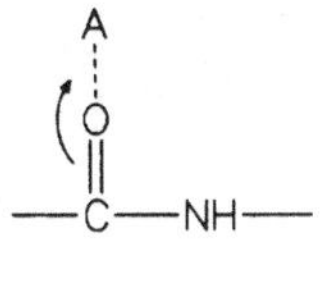

$$\left(\begin{matrix} A \\ \vdots \\ O \\ \| \\ -C-NH- \end{matrix}\right.$$

In the case of serine proteases (see Sect. 6.5.3) the carbonyl oxygen in the enzyme-substrate complex occupies the so-called oxyanion hole forming hydrogen bond with the NH group of Gly193 (≈ 2.8 Å in chymotrypsin) and the NH group of Ser195 (3.2 Å). These bonds per se cannot effectively polarize the substrate carbonyl group. According to quantum-chemical calculations, the charges on the amide group atoms remain actually unchanged at the hydrogen bonding by the carbonyl oxygen. Only protonation of the latter changes the charges on atoms C, N, and O as well as the orders of bonds C=O and C–N [3252].

The role of these bonds becomes more important upon the formation of the tetrahedral intermediate with the carbonyl oxygen acquiring the full negative charge. A charge appearing in the nonpolar surrounding of the enzyme active site is extremely unfavorable. However, the two dipoles N–H in the vicinity of the charge compensate much for this energy loss [3253–3257].

It is suggested that the NH group of Cys25 and the amide group of the Gln19 side chain play the same role [3258]. It is noteworthy that a comparison of the hydrolysis rates of oxygen (RC(=O)OR') and thione (RC(=S)OR') esters by serine and cysteine amide hydrolases [3259, 3260] provoke questions on the role of these interactions in stabilization of the tetrahedral intermediate by cysteine enzymes. Though serine proteases are unable to hydrolyze thione esters because the size of a sulfur atom prevents a proper accommodation of the thiocarbonyl group in the oxyanion hole, cysteine enzymes readily hydrolyze both the ester types.

In metal-containing enzymes Zn^{2+} ion apparently serves as an electrophile. This ion in carboxypeptidase A and other similar enzymes can include not only water molecules in its coordination sphere, but also the oxygen atom of the substrate carbonyl group [3244, 3261, 3262]. The carbonyl group enters apparently the external coordination sphere of a metal displacing the water molecule. This is evident because Co^{3+} carboxypeptidase A with the octahedral coordination of the metal is unable to bind peptides and esters [3263]. For thermolysin and dipeptidyl carboxypeptidase the substrate binding proceeds similarly [3244]. In the case of carboxypeptidase A and thermolysin, the $N^{\epsilon 2}$ atom of histidine residues (His69 and His142, respectively) forms the H-bond with aspartic acid residues (Asp142 and Asp170), when the former are bound to Zn^{2+}-ion. It can influence nucleophilicity of water associated with Zn^{2+} ion [3264].

The role of electrophile of aspartic proteases is ascribed to the hydrogen atom between the carboxyl groups of Asp32 and Asp215 [1345]. So this atom affects the cleavable bond if it forms not only hydrogen bond with the substrate oxygen atom, but also protonates it. However, the distance between the proton and carbonyl oxygen of the substrate is presumably too large to form a stable hydrogen bond. A theoretical calculation [3265] from crystallographic data provides no evidence for a possible protonation of a cleavable group oxygen even in a tetrahedral intermediate. According to X-ray analysis, the shortest distance between the inhibitor hydroxyl oxygen modeling the tetrahedral state and aspartic acid oxygens is 2.7 Å [3073].

7.6 Nucleophilic Attack. Elementary Step

As mentioned in Sect. 7.2, the catalytically active hydroxyl group of serine proteases in the productive enzyme-substrate complex is in a neutral form and cannot be ionized by proton transfer to imidazole of His57, since it has no "charge relay system". Though the carbonyl group forms hydrogen bonds with NH groups of Ser195 and Gly193, these bonds polarize the carbonyl only slightly. The catalysis efficiency cannot be high in such a system. Apparently, the major factor of the effective reaction is a disturbance of the resonance stabilization of the cleavable bond. There are solid arguments for suggesting that in the ground state of the reacting system (in the productive enzyme-substrate complex) the resonance stabilization of amides is disturbed, mainly owing to bond pyramidalization (see Sect. 6.5.2).

At a certain degree of initial pyramidalization (bending) determined by the structure of the enzyme-substrate complex, further decrease of the resonance stabilization of amide (and increase of the bending degree) depends mainly on the distance between the substrate carbonyl carbon and O^{γ}-atom of serine hydroxyl (see Sect. 3.3.1). Any thermal fluctuation approaching these atoms increases the bending degree of the amide group. It enhances the positive charge on the carbonyl carbon and decreases π-orders of C–N and C=O bonds. On the other hand, the fluctuation weakens O–H bond in the attacking alcohol group, thus intensifying its nucleophilicity.

Such fluctuation "automatically" leads to the O–C bonding, i.e., to the tetrahedral intermediate. The decisive factors here are the state of the amide bond before the reaction and the initial distance between O^{γ} (Ser195) and C' (carbonyl). The substrate specificity can be determined by these two factors; for the most specific substrates the bending degree and O^{γ}–C' distance are optimum *in the ground state of the system*.

The above ideas on the elementary step of the nucleophilic attack in serine proteases conform to the quantum-chemical calculations [1477, 3266, 3267]. The changes in the bending degree of the amide group depending on the distance between different nucleophiles and the carbonyl carbon have been calculated by the semi-empiric method CNDO/2 (Fig. 108). Extrapolation of the straight lines to distances exceeding 2.5 Å shows that even at these distances there is observed an inductive effect of charged nucleophiles on the hybridization state of the carbonyl C-atom, leading to the energetically preferable and partially pyramidalized structure of the amide group. In the case of the alcohol hydroxyl at a distance corresponding to a sum of Van der Waals radii of alcohol oxygen and carbonyl carbon (about 3.2 Å; Table 66), the bending degree is still very low ($\Delta_C < 0.007$ Å). However, in the case of "good" substrates the interactions of nonreacting groups of the substrate and the enzyme may result in shortening this distance in the productive enzyme-substrate complex; it can be shorter than the sum of Van der Waals radii (compare [3268]).

Figure 109 shows a change in the potential energy of the system, when methoxy anion and alcohol hydroxyl intract with the amide bond of

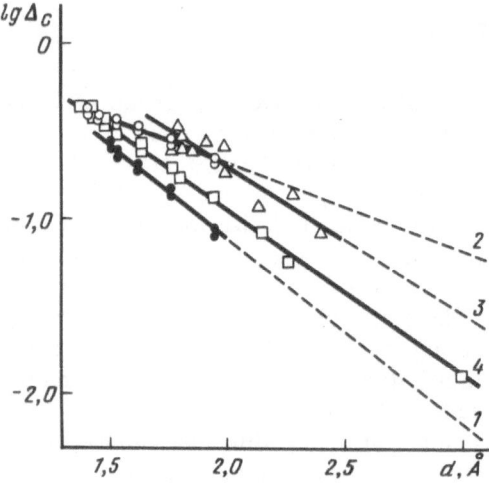

Fig. 108. Dependence of pyramidalization degree (Δ_c) of the amide group in N-methylacetamide on distance between nucleophile and carbonyl carbon [1477]; *1* methanol; *2* methoxyanion; *3* methylthiolate anion; *4* water + formiate ion

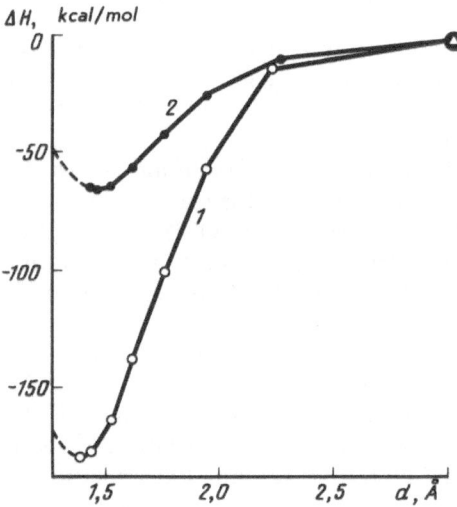

Fig. 109. Change of potential energy at the attack of carbonyl carbon in N-methylacet-amide with methoxyanion (*1*) and methanol (*2*) [1477]

N-methylacetamide. It is noteworthy that the calculations model an approach of reacting atoms so that the p_z-orbital of a nucleophile is directed normally to the amide group. Here a decrease in the system energy due to the overlapping of bonding orbitals of the reacting atoms is observed at distances of about 2–2.3 Å. The reaction with methoxy anion ends with the covalent bond formation, whereas in the case of methanol the decrease in the system energy is much less, and the equilibrium degree of bending is 0.22 Å (in tetrahedron $\Delta_C \approx 0.47$ Å). The order of the bond between the alcohol oxygen and carboxyl carbon is only ≈ 0.5 in the minimum of the potential curve. This corresponds to the O–C' distance equal to ≈ 1.6 Å, and the equilibrium structure here is represented by the formula:

$$
\begin{array}{c}
CH_3 \\
| \\
HO^{\delta+} \\
\vdots \\
CH_3\!-\!\!\overset{|}{\underset{\underset{O^{\delta-}}{\|}}{C}}\!-\!NHCH_3
\end{array}
$$

A further movement of the system along the reaction coordinate seems impossible without proton withdrawal. However, there are prerequisites for proton acception by a proper base, because the state of the attacking group is close to that of the oxonium cation. According to X-ray analysis of the trypsin-pancreatic trypsin inhibitor complex [3044], the distance between the serine hydroxyl and carbonyl carbon of Lys15 is also greater than the covalent bond. Moreover, the study of the trypsin and chymotrypsin complexes with protein inhibitors [3044] has shown the distance between O^{γ}-atom and $N^{\varepsilon 2}$-atom of His57 in the enzyme-"pyramidalized" substrate complex to be close enough for hydrogen bonding, which provides further proton transfer to the imidazole residue.

Thus the nucleophilic attack on the substrate carbonyl carbon is a "self-developing" process facilitating further interaction of reactants at any movement along the reaction coordinate. It is noteworthy that isolation of intermediates on the pathway from the ground state of the system to the transition state has no sense if the process is an elementary step occurring in the fluctuation time, i.e., during $10^{-12} - 10^{-13}$ s. Therefore, its efficiency in this case should depend on the initial (ground) state of the system only – proximity of a nucleophile and carbonyl carbon, as well as the extent of destabilization of the scissile group. This is evidently the major difference between enzymatic and nonenzymatic reactions; to keep reactants in the position favorable for the nonenzymatic reaction is impossible because of the movement of free thermal molecules. So the enzyme creates a "frame" or a "corridor" directing the elementary step along the reaction pathway.

The fact that many enzymatic reactions proceed with diffusion-controlled rates (see Sect. 4.4) is proof of the rather low activation energies of the *elementary step* of the chemical conversion itself. Presumably, for "ideal" substrates the activation energy of this event is always lower than that of the diffusion-controlled reactions (≈ 5 kcal/mol).

Other investigators share the same opinion concerning the mechanism of nucleophile attack for serine proteases [1330, 3044, 3045, 3060, 3062, 3103, 3104]. It is interesting that the β-lactam ring in benzylpenicillin by β-lactamase from *Escherichia coli* is cloven with the second-order rate constant $(k_{cat}/K_m) \approx 1 \times 10^8$ M^{-1}s^{-1}, i.e., with the rate close to that of the diffusion-controlled process. However, its efficiency ($k_{cat}/K_m k_n = 10^9$; Table 50) does not exceed that of other amide hydrolases. The β-lactam ring in penicillins is actually devoid of the resonance stabilization (see Sect. 1.3.1), and the mechanisms of the cleavable bond destabilization presumably do not function. There is another important factor providing the efficiency: orientation, i.e., the distance between the enzyme nucleophile and carbonyl carbon, which is to be optimum for effective catalysis. The hydrolysis rate of other substrates non-specific to this enzyme is low, if this condition is not complied with.

The attack of the substrate by the thiolate ion in cysteine proteases occurs apparently similarly. "Artificial" enzyme thiolsubtilisin is inactive relative to specific substrates, but the deacylation rate of arylacryloyl derivatives (obtained upon acylation at the SH group by arylacryloyl imidazoles) is higher than that of acylsubtilisins [3269]. This is explained by a lower nucleophilicity of the thiolate ion as compared to the hydroxy anion; however, the reasons may be more intricate [3270].

In aspartic proteases a water molecule is a nucleophilic agent. Extremely low nucleophilicity of water virtually excludes an attack of the resonance-stabilized amide group. At the same time, ionization of water by the proton transfer to the carboxylate ion is a thermodynamically unfavorable process occurring with a very low rate. Therefore, the proton transfer and the nucleophilic attack must be synchronous in the single elementary step.

The situation here is in principle analogous to that for serine proteases, the only difference being that imidazole is a more effective acceptor of the proton than the carboxylate ion. The value of the synchronization coefficient (see Sect. 3.4.9), which limits the number of possible simultaneous shifts of heavy atoms in the single elementary step, should be always taken into account. This magnitude seems hardly detectable for such enzymatic processes, since nothing is known about the activation energy of the synchronous process. At $E_c = 0$ the maximum number of atoms in the synchronous elementary step is not higher than eight [3271, 3272].

There is another possible way based on the study of the general base catalysis in model compounds [3273–3275], when the nucleophilic attack and the proton transfer pass via two subsequent transition states (Fig. 110). The rate-determining water attachment promoted by destabilization of the scissile group is not coupled with the proton transfer to a general base. A first transition state in this case is characterized by hydrogen bonding to a general base. Further moving of a nucleophile along the reaction coordinate entails a second transition state, when the formation of the normal nucleophile bond to the carbonyl carbon is completed, and the proton moving also along the reaction coordinate (since it is bound to the general base) is halfway to the acceptor. The reaction ends with a rapid proton transfer to the general base. This mechanism correlates well with the kinetic isotope effect of the solvent, since its essential characteristic is the hydrogen bonding of the nucleophile to the general base.

If the concept that water coordinated with Zn^{2+}-ion is a nucleophile in metal-dependent proteases and the carbonyl carbon of the substrate enters the metal coordination sphere is true, no problem associated with the nucleophile

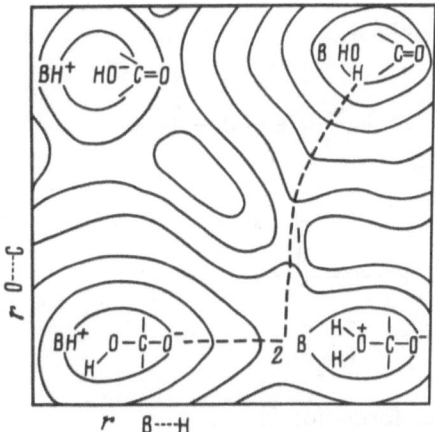

Fig. 110. Motion of proton along the potential energy surface at general base catalysis of water attachment to the carbonyl group [3275]

ionization and the carbonyl group polarization appears. Moreover, it is quite understandable why bending of the amide group is the enzyme-inhibitor complexes of metal-containing proteases is not so important as bending to serine enzymes.

Quantum-chemical studies [3276] show that the zinc atom is substantial not only in polarization of the carbonyl group, but also in stabilization of the tetrahedral intermediate, when it is neutralizing the negative charge and providing more stable bonds to the central carbon atom of tetrahedron.

7.7 Tetrahedral Intermediate

The nucleophilic attack on the substrate carbonyl carbon yields the formation of tetrahedral intermediates, identified upon nonenzymatic hydrolysis mainly from the experiments on the ^{18}O inclusion in the substrate from heavy oxygen water (see Sect. 3.3.2).

The mechanism of the S_N2 type is alternative and does not include the tetrahedral adduct formation:

To make a choice between these mechanisms is rather difficult. So attempts to find out the ^{18}O inclusion into the substrate in the course of the chymotryptic catalysis, which could evidence the formation of the tetrahedral intermediate, have failed [3225]. However, the negative result can only indicate an unfavorable ratio of rate constants k_{-2} and k_3:

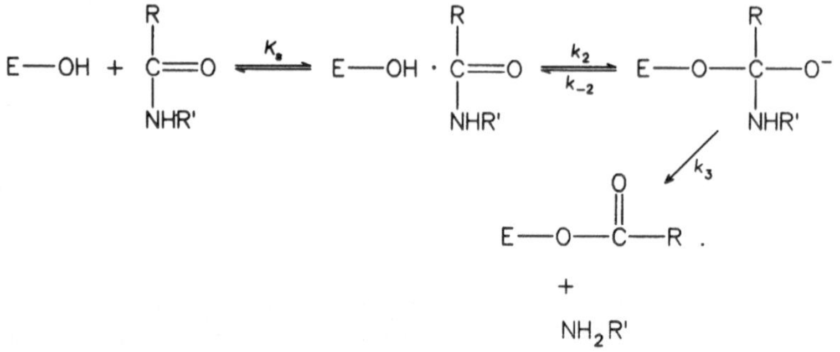

Here the rate constant for the acylenzyme formation is

$$k_{ac} = \frac{k_2}{1 + k_{-2}/k_3} \qquad (3)$$

and at $k_3 \gg k_{-2} k_{ac} = k_2$, i.e., no exchange is observed.

In the leucineaminopeptidase and pepsin-catalyzed reactions much of the ^{18}O included into the substrate remained unhydrolyzed [3228, 3231]. The study of this reaction in detail testifies to incorporation as a result of trans-peptidation, i.e., transfer of the substrate acyl fragment to the amine product formed during hydrolysis:

$$E + RCONHR' + H_2^*O \longrightarrow E \cdot RCO^*OH + R'NH_2 ,$$

$$E \cdot RCO^*OH + NH_2R' \longrightarrow E \cdot RC^*ONHR' .$$

This has been proved by comparing the inclusion degree of ^{18}O- and ^{15}N- (or ^{14}C-) labeled products into the original substrate [2040, 3277]. Thus neither positive nor negative results on the oxygen exchange in the substrate evidence pro et con the formation of a tetrahedral intermediate.

The S_N2 mechanism including the proton transfer to the leaving group in the pre-transition state was proposed for chymotrypsin as early as 1967 [3278–3280]. It envisages the involvement of the imidazole of the histidine residue in the proton transfer from the serine hydroxyl to the amide group nitrogen. The reaction rate in this case should depend on pK_a of the leaving group.

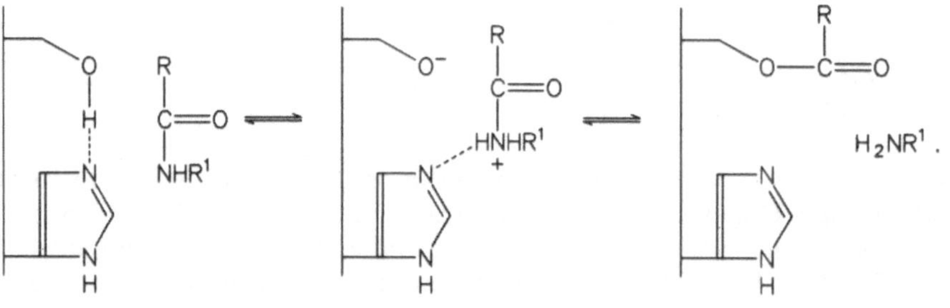

In fact, electron donor substituents in the aniline ring of substituted anilides of N-acetyltyrosine increase the catalyic constant ($\rho \approx -2.0$) [1996, 1998, 3278, 3281]. However, detailed investigations [2001] of the compounds for which pK_a values of leaving groups vary within a very wide range show that actually no dependence is observed (see Sect. 4.3.2.2). The dependence is evidently an artefact stipulated by narrow limits of pK_a values and a high degree of nonproductive binding of anilides [2000].

The concept on the pre-transition state of the proton transfer has been sharply criticized [1999, 2302, 3282, 3283]. This mechanism does not account for the lack of the influence of the released group on K_m, besides the rates of the proton transfer (10^{13}–10^{14} s^{-1}) by this mechanism are unachievable. In the light of current views on serine proteases free of the charge relay system, this concept is found to be unsound.

Evidence for the tetrahedral intermediate formation upon the amide hydrolases catalysis has been obtained [1515, 3284–3286] on studying the kinetic isotope effect by incorporation of ^{15}N or ^{18}O into the leaving group of the substrate (see review [1405]). The prime conclusion here is that at hydrolysis of ester substrates of chymotrypsin the formation of the tetrahedral intermediate is a rate-limiting stage, whereas at hydrolysis of amides the breakdown of this intermediate is the rate-determining step [3287, 3288]. In papain hydrolysis of benzoylarginineamide the isotope effect (^{14}N/^{15}N) correlates with the above suggestion [3289]. Indeed, for substrates with leaving groups of the low basicity (esters, anhydrides) the probability of protonation of two potentially leaving groups

$$-CH_2O-\overset{\overset{\displaystyle O^-}{|}}{\underset{\underset{\displaystyle R}{|}}{C}}-OR$$

is virtually the same. The k_{-2}/k_3 value [formula (3)] is high, and the tetrahedron decay in the acylenzyme direction can be slow (k_3).

For aliphatic amides (peptides)

$$-CH_2O-\overset{\overset{\displaystyle O^-}{|}}{\underset{\underset{\displaystyle R}{|}}{C}}-NHR'$$

differences in pK_a of leaving groups are high and the rates of formation (k_2) and decay (k_3) of the tetrahedral intermediate are comparable. This correlates well with the dependence of the acylation rate (and the rate of the acyl group transfer in the acylenzyme to the nucleophilic acceptor) on pK_a of leaving groups (or the pK_a acceptor) [1533, 1989, 3290–3292].

The study of the reactions of peptide-aldehydes, ketones and inhibitors – "analogs of transition states" (see Sect. 5.8.7) – with amide hydrolases has provided the arguments about the formation of tetrahedral intermediates

[3293–3298]. Their formation has been identified in cryoenzymatic experiments from kinetic and spectral data [3299–3302], their interpretation seems, however, doubtful [3303].

The fact itself that a tetrahedral intermediate is formed in catalysis and, particularly, its kinetic significance remain open to discussion. So it has been suggested [1501] that hydrolysis of peptides (but not esters or anilides) catalyzed by serine proteases proceeds by the mechanism of the S_N2 type. This conforms to the isotope effect of the chymotryptic hydrolysis of 4-methoxyphenylformiates [3304], indicating the carbonyl intermediate hybridization in the transition state between sp^2 and sp^3.

An essential increase in pK_a of the leaving group at the tetrahedral configuration of the carbonyl carbon (see Sect. 3.3.3) implies the proton transfer from imidazole to the leaving group before the formation of the tetrahedral intermediate, since at early stages pK_a values of imidazole and the leaving group are leveled off. However, ^{13}C NMR of the trypsin-halogen methylketones complexes [1160, 3305] and pH-dependence of inhibition of chymotrypsin by these compounds [3298] show that pK_a of imidazole in hydrated complexes modeling the tetrahedral intermediate is not lower than 10.

A tetrahedral intermediate, if formed, is stabilized by numerous interactions with the enzyme. Their energies for the derivatives of N-acetyl-L- and D-tryptophan have been calculated by the methods of molecular mechanics (Table 77). The interactions in the oxanion hole of the enzyme contribute much to tetrahedron stabilization [3306, 3307]. If the tetrahedral intermediate models the reaction transition state, the difference -7.7 kcal/mol between L- and D-forms explains the difference in the rates of the enzyme acylation by stereoisomeric derivatives of acetyltryptophan.

Indirect evaluations of the equilibrium constant at the tetrahedral intermediate formation give $\approx 10^{-6}$ [3254]. However, they can understate the tetrahedron stability, especially for specific substrates.

Table 77. The energy of various interactions of tetrahedral compound formed by AcTrpX and chymotrypsin [3306]

Interaction	ΔH_o, kcal/mol		Difference $(L–D)$
	L-form	D-form	
Binding energy	19.5	19.5	0
O$^-$-atom with NH-Gly193 and CO-Met192	−5.3	−4.7	−0.6
O$^-$-atom with NH-Ser195 and CO-Asp194	−5.2	−3.2	−2.2
O$^-$-atom with CO-Ser214	0.6	0.8	−0.2
NH-Ac with NH-Gly193 and CO-Met192	−0.1	0.2	−0.3
NH-Ac with NH-Ser195 and CO-Asp194	−0.1	0.1	0.2
NH-Ac with CO-Ser214	−1.8	0	−1.8
Other Van der Waals interactions	−113.6	−111.0	−2.6
All unbinding interactions	−125.5	−117.8	−7.7
All binding and unbinding interactions	−106.0	−98.3	−7.7

7.8   Proton Transfer and Decomposition of Tetrahedral Intermediate

A tetrahedral intermediate (or $S_N 2$ reaction) can collapse only if the leaving amine attaches the proton. High pK_a values of the conjugate base prevent releasing the amine as an anion:

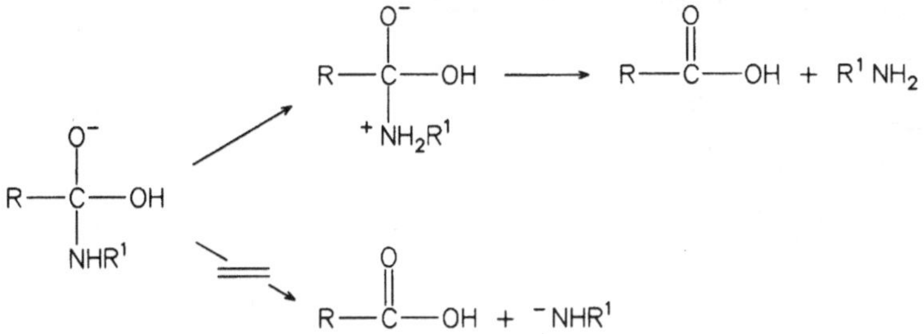

The decomposition of tetrahedral intermediates under nonenzymatic conditions is catalyzed by general acids. The same catalysis type is observed in case of enzymatic hydrolysis.

A proton donor of serine and cysteine amide hydrolases is a conjugate acid of imidazole (His57 in chymotrypsin – imidazolium cation [2350]. Consequently, synchronous attack by the hydroxyl and the proton transfer an imidazole with a further displacement of a proton from the imidazolium ion to the leaving group can be assumed:

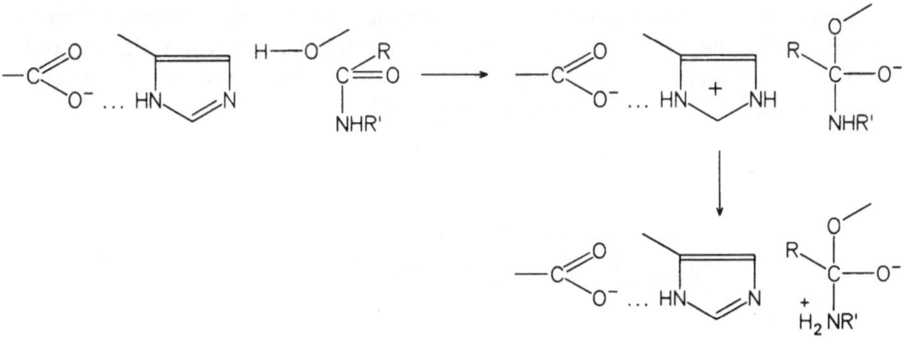

The question arises whether the proton can be transferred directly from the serine hydroxyl to the leaving group. As shown above, it is impossible in the pre-transition state (see Sect. 7.7), mainly because of a large difference in pK_a of alcohol hydroxyl and amide nitrogen. The transfer rate is in line with Eq. (14) (Sect. 3.3.3).

$$lgk = 10 - [13.5 - (-2.5)] = 5.5, \quad \text{i.e. } k = 3 \times 10^{-6} \text{ M}^{-1}\text{s}^{-1}.$$

It is evident that pK_a of the leaving group increases with the tetrahedral intermediate formation. The attack of the alcohol group of the substrate carbonyl carbon cannot result in covalent bonding without a proton release (see Sect. 7.6); the bond length is equal to the equilibrium O–C distance ≈ 1.6 Å. This distance corresponds to the pyramidalization degree of the amide group $\Delta_c \approx 0.23$ Å. Here the calculated pK_a of the leaving group is about 6.5 [1500].

Conventional values of pK_a of imidazole underlie the concept [1790, 3024, 3309, 3310] on the synchronous proton transfer from the serine hydroxyl to the leaving group through the transition state:

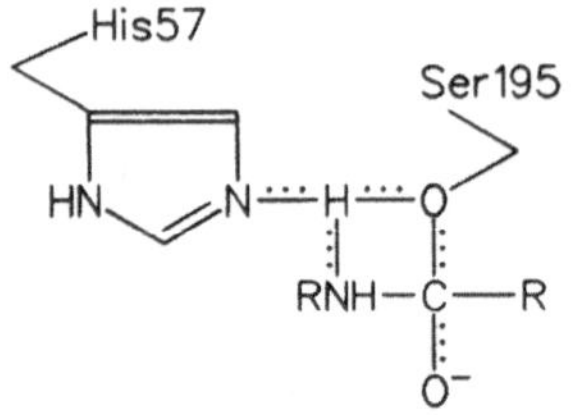

However, for sterical reasons a four-membered cyclic transition state seems improbable.

If pK_a of imidazole in the tetrahedral intermediate increases by two-three units [1160, 3028, 3160, 3298, 3305] its formation can be complete. According to X-ray analysis of serine proteases complexes with inhibitors – "transition state analogs" [3308] – the distance between the nitrogen atoms of imidazole of the leaving group in the tetrahedron is still too large for an effective proton transfer from O^γ-atom of Ser195 to $N^{\varepsilon 2}$-atom of His57 is accompanied by a dramatic fall in the system energy, and the sequential proton release envisages the overcoming of the high activation barrier.

There are two feasible pathways in the process: (1) conformational fluctuations within the active site, which draw together the proton donor and acceptor [3308], and (2) involvement of the intermediate acceptor-donor and – water molecule – in the proton transfer [3311]:

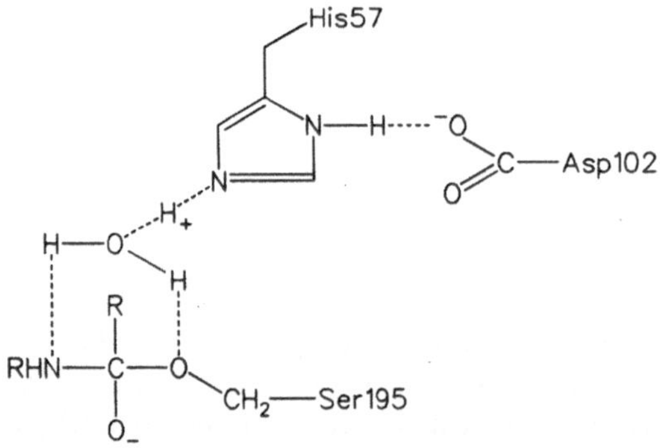

The quantum-chemical analysis finds the two ways possible.

Stability of the tetrahedral intermediate depends, on the one hand, on its "solvation", i.e., on the interaction of the oxoanion with dipoles of NH bonds (Gly193 and Ser195 in chymotrypsin), on the other hand on the efficiency of the general acid catalysis of its decay. The solvation efficiency has been studied on the models – products of interaction of serine proteases with peptidoaldehydes [9, 2860] – and expressed by the value of about 7 kcal/mol [9]. However, the major part of this energy (≈ 5 kcal/mol) is referred to the intramolecular character of the solvation. The kinetic study of deacylation of N-acetylphenylalanyl-(methyl-$N^{\varepsilon 2}$-His57)-chymotrypsin yields an "effective concentration" of general acid (His57 imidazole) in the enzyme, which is equal to $\approx 2 \times 10^6$ M [3312]. Hence, the efficiency of general acid catalysis by enzymes is very high and exceeds much that reached by model systems.

It has been also postulated that the proton transfer in enzymatic hydrolysis reactions occurs presumably via tunneling (see Sect. 5.5). This accords with the temperature dependence of the solvent isotope effect.

At hydrolysis of esters and anilides the pK_a values of the leaving group remain low for the tetrahedral intermediate as well, and the nucleophilic attack on such compounds may end with their formation. In the case of esters a release of the leaving group as an alcoholate ion ($pK_a \approx 15$) and its protonation in the "post-transition" state are probable.

Some investigators believe that the tetrahedron breakdown is stereoelectron-controlled (see Sect. 3.3.3) [1476, 1518, 1519, 3313]. In brief, the concept implies the conservation of the bond configuration during the acyl transfer. In other words, if the most resistant ester with Z-configuration is hydrolyzed, the acylenzyme should have Z-configuration also. This occurs only if the tetrahedral intermediate conformation provides the maximum of antiperiplanar orbitals in a cleavable group relative to the remaining two tetrahedron groups (Fig. 35). The situation here resembles, perhaps, that in Fig. 111 and envisages a stage of isomerization of tetrahedral intermediates.

For papain the formation of the tetrahedral intermediate upon amide hydrolysis has been derived from the value of the ^{15}N-isotope effect. The reasons for the distinction between papain and serine proteases in this case are as follows. First, pK_a of His159 sharply decreases, if a sulfur atom is uncharged (see Sect. 7.1), which leads to low concentration of the imidazolium ions in the pH optimum (pH ≈ 6). Secondly according to analysis of model complexes of papain and substrate (see Sect. 6.5.3) the contacts of the leaving group with α-CH_2-group of His159 are unfavorable. This "pressure" is eliminated in the tetrahedral complex. Therefore, even in the ground state of an enzyme-substrate complex a scissile bond is strongly disturbed so as to resemble the tetrahedron (high bending degree), and the attack by a strong nucleophile (thiolate ion) ends with a quick tetrahedron formation.

There are few data on the catalysis of protonation of the leaving group in aspartic amide hydrolases. The involvement of the Tyr75 hydroxyl group has been assumed in the case of penicillopepsin [1345]. It is noteworthy that the proton transfer from the tyrosine phenol group in the acidic medium is possible only by the estafette mechanism [3314]:

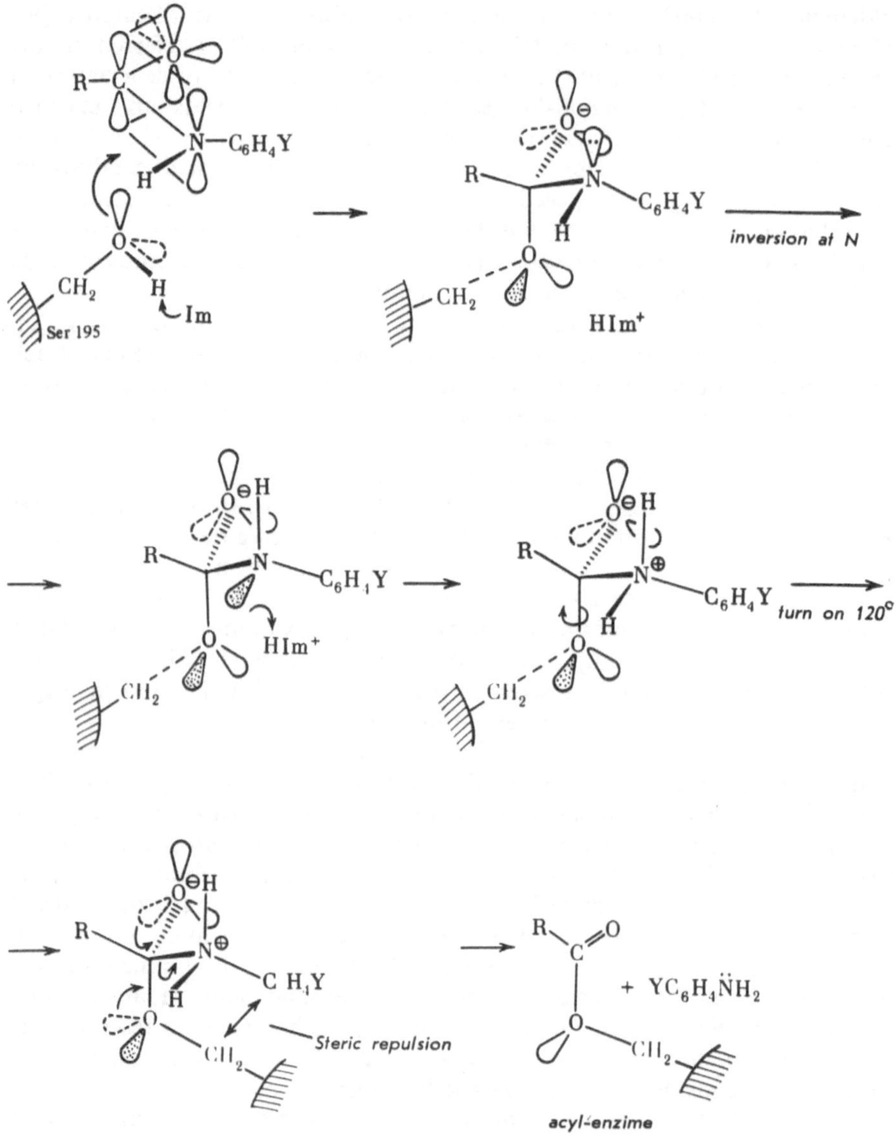

Fig. 111. Scheme for chymotrypsin acylation including isomerization of tetrahedral intermediates [1406]

According to X-ray data, the structures of enzymes of this type is free of AH groups, except for Asp215 which has been earlier proposed to be only an electrophilic catalyst polarizing the substrate carbonyl group. However, the pH-dependence ($pK_{ES} \approx 4.5$) confirms participation of Asp215 in the proton transfer.

The study of the quasi-substrate – pepsin complex shows that Tyr75 is located far from the ligand binding site [3128]. The Asp215 position corresponds to its role as a general acid catalyst at protonation of the leaving group.

It is just the way how this group serves evidently as a proton donor to the leaving group [1715]. This can occur by concerted general base – general acid catalysis of the proton transfer from the hydroxyl of the tetrahedral intermediate to Asp32 and from Asp215 to the leaving group [1715, 3315] (see Sect. 7.10). The other mechanism [3131] is the proton transfer from one hydroxyl group of the tetrahedron to the leaving group including protons from the medium (Fig. 112). According to the quantum-chemical model [3265], the proton transfer from Asp215 to the leaving group can be accompanied with a turn of the carboxyl group by $\approx 180°$. This is probably reached by moving the domains relative to each other (see Sect. 6.5.3).

In metal-containing amide hydrolases – carboxypeptidase A, thermolysin and dipeptidyl carboxypeptidase – the breakdown of tetrahedral intermediates is catalyzed by the carboxyl group (Glu270 and Glu143 in carboxypeptidase A and thermolysin, respectively) [3136, 3137, 3316, 3317]. As mentioned above, the involvement of Tyr248 of carboxypeptidase A in this process has not been confirmed yet [149]. The carboxyl group functions as a general base promoting ionization of the Zn^+-bound water and as a general acid transferring protons to the leaving group. In thermolysin one of the Glu143 oxygens is at a distance of 3.2 Å from the nitrogen atom of the leaving group [3316]. The involvement of the two atoms of the tetrahedron oxygen sheds light on the

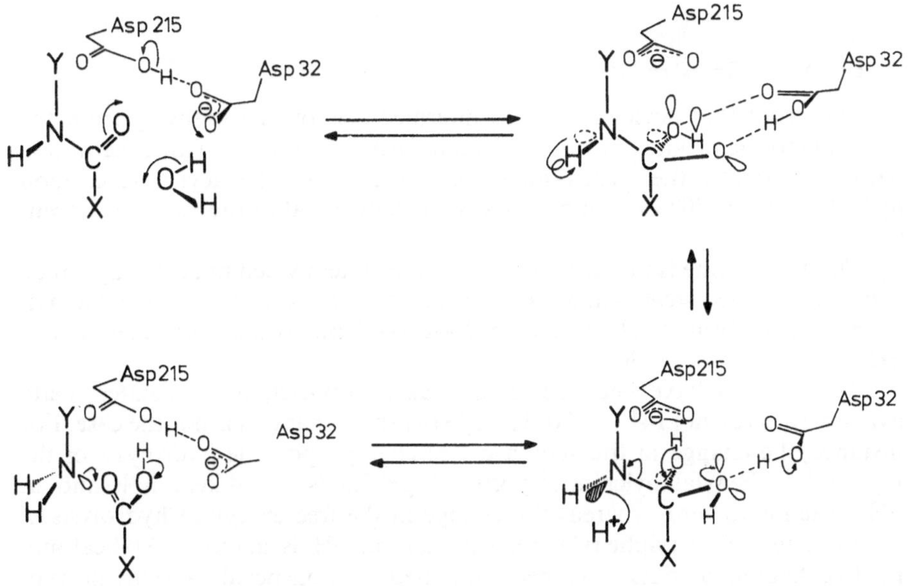

Fig. 112. Tentative mechanism of action of aspartic proteases including protonation of the substrate leaving group via proton transfer from the hydroxyl group of tetrahedral intermediate [3131]

methanol's disability to substitute water in the reactions catalyzed by metal-containing proteases [3139]:

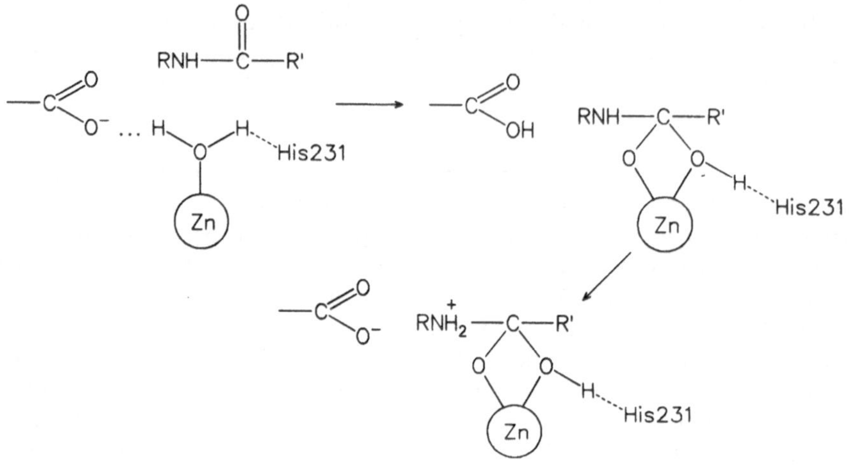

7.9   Acylenzymes and Enzyme-Product Complexes

Decomposition of tetrahedral intermediates or reactions of the S_N2 type in case of cserine and cysteine amide hydrolases yields acylenzymes

The proof that acylenzymes are intermediates on the pathway from the substrate to the product has been given in Sect. 7.4. If the enzyme is saturated with a substrate, the steady-state concentration of the acylenzyme upon hydrolysis of specific ester substrates is virtually equal to the enzyme concentration.

The acylenzyme is not accumulated in the system when hydrolyzing amide substrates. The steady-state concentration of N-acetyl-L-phenylalanyl-chymotrypsin formed at hydrolysis of N-acetyl-L-phenylalanylglycineamide at pH 5.5 is of 0.15% [1730].

Acylenzymes have been considered as energy-rich intermediates. Data available at present [1412, 1730, 3290, 3318] show that this is not the case. For instance, the change in the standard free energy upon the hydrolysis of the N-acetyl-L-phenylalanylchymotrypsin at pH 7.3 is −3.4 kcal/mol, and at pH 5.5 tends to zero, whereas the change in the free energy of hydrolysis of methyl ester of acetylphenylalanine at neutral pH is about −6.0 kcal/mol [1415]. Analogous data have been obtained for nonspecific acylchymotrypsins. So furoylchymotrypsin is by about 3 kcal/mol as stable as O-furoyl-N-acetylserineamide [3290], and hyppurylpapain is more stable (by ≈8 kcal/mol), as compared to the corresponding thiol esters [3318].

The acylenzyme formation is an exothermic reaction: for acylchymotrypsin $\Delta H_0 = 7\,\mathrm{kcal/mol}$ [3319] at pH 8.0. It was suggested that chymotrypsin is acylated by nitrophenylacetate at first through the formation of the derivative acetylated by the His57 imidazole isomerized into O-acetyl(Ser195)-chymotrypsin [3320]. Two stages deacylation has been postulated for acetylchymotrypsin. However, these data were later disproved [3321].

As mentioned in Sect. 4.3.2.2, a compensation for the enthalpy and entropy of activation of deacylation of acylenzymes is observed with $T_c = 420\,^{\circ}\mathrm{K}$ [3322], which is presumably explained by solvation of the transition state.

Though the free-energy change at the thioester hydrolysis is somewhat higher (by 2–3 kcal/mol) than that of oxygen esters, the rates of their alkaline hydrolysis differ slightly. Deacylation of unspecific acylpapains occurs more rapidly than that of the corresponding acylchymotrypsins, but specific acylchymotrypsins are hydrolyzed noticeably quicker, as compared with acylpapains [3323].

Deacylation of acylenzymes poses the same problems as acylation, but more urgently. Since a water molecule is a nucleophilic agent, the effective general base catalysis or destabilization of the acylenzyme ester group can be a driving force of effective deacylation.

Early acylchymotrypsin studies have already revealed a spectral anomaly of chromogenic unspecific acylchymotrypsins [1985], which could evidence the distortion of the ester group. This has been supported by X-ray data [1328] on β-(3-indolyl)-acryloylchymotrypsin (see Fig. 113):

Analysis of this shows that the ester group is essentially distorted: it is pyramidalized ($\Delta_c = 0.3\,\text{Å}$), and dihedral angle ω is only -122°.

Raman spectra of a series of acylenzymes [3100, 3101] testify that the frequencies of the ester group are close to those of the aldehyde carbonyl. Similar conclusions have been derived from NMR spectra of selenium-containing acylchymotrypsins [3324]. A correlation between the carbonyl group polarization in acylchymotrypsins and a rate of their deacylation is observed [3325].

Thus the ester bond is destabilized at least for investigated acylenzymes and, apparently, for others too. The ester group bending changes a charge on the carbonyl carbon atom and the bond order between the oxygen atom of water and carbonyl carbon (Fig. 114). The Δ_c value (0.3) is close to that of pyramidalization of the totally tetrahedral carbon atom. Probably, such a value is not realized during hydrolysis of specific chymotrypsin substrates. Nevertheless, the ester bond destabilization in the acylenzyme can be one of the reasons for high deacylation rates (cf. [3326]).

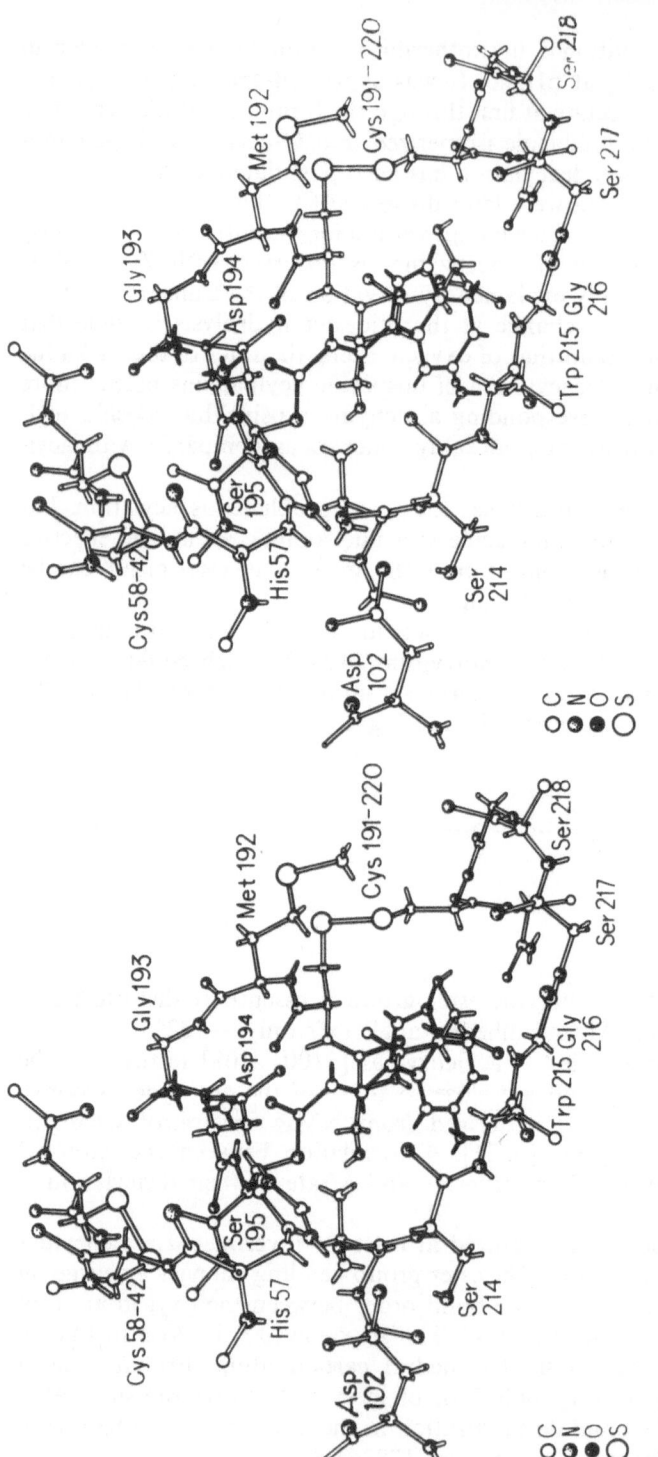

Fig. 113. Stereo model of β-(3-indolyl)-acryloyl-α-chymotrypsin [1328]

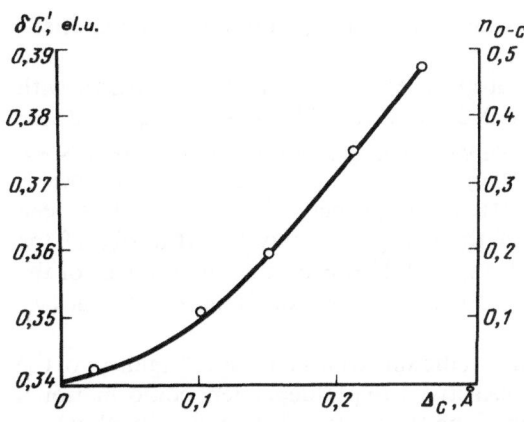

Fig. 114. Dependence of charge on carbonyl carbon in methylacetate and bond order between the carbon atom and oxygen atom of the attacking water molecule on the pyramidalization degree of the ester group (calculated by CNDO/2 by S.L. Alexandrov)

The ester group in indolylacryloylchymotrypsin has a somewhat distorted Z-configuration (Fig. 113). This contradicts a proposal [1985, 3327] on E-conformation of this bond in the derivatives of substituted acryloylchymotrypsins based on spectral changes at the enzyme acylation.

A distance between $N^{\varepsilon 2}$-atom of His57 and the carbonyl carbon atom of the acylenzyme is 4.1 to 4.8 Å from the X-ray [1221, 1328, 3328] and conformation [3024] analyses. The water molecule in indolylacryloylchymotrypsin is located so as to form two hydrogen bonds: 1) to $N^{\varepsilon 2}$-atom of His57 and 2) to carbonyl carbon of the ester bond. Such water position is not productive. In acylenzymes subjected to rapid hydrolysis the water should occupy quite the other position, possibly close to that of the leaving group in the enzyme-substrate complex. If so, the distance between $N^{\varepsilon 2}$-atom of His57 and the water oxygen atom is to be 2.8–3.0 Å. It corresponds to the presence of the proper hydrogen bond and meets the condition of the efficiency of general base catalysis of water attachment to the carbonyl group of the acylenzyme.

The general base catalysis of deacylation correlates with steric effects of an acyl group [1967]. It is noteworthy that the reactivity of a water molecule participating in deacylation of acylenzymes, which contain the specific acyl group, is close to that of the hydroxyl ion. This conclusion is made [3329] from the available data on deacylation rates and alkaline substrate hydrolysis. However, the question about the reasons for such high reactivity remains: either they are a result of peculiarities of the water structure in the active site of acylenzyme, or of some properties of the cleavable bond. This problem requires additional study.

A nucleophilic attack of the acylenzyme ester bond by the water molecule yields the tetrahedral intermediate. Reaction of the $S_N 2$ type is improbable here because of low pK_a of the leaving group – serine hydroxyl (pK_a R–$\overset{+}{O}H$–R ≈ -1). The formation of the tetrahedral intermediate is confirmed by the dependence of the deacylation rate on pK_a of substituted acylchymotrypsins (substituent in acyl group) [3330].

Serine hydroxyl can be protonated in the "post-transition" state, when the tetrahedral intermediate decays, i.e., serine hydroxyl, being a better leaving group than the hydroxyl of the attached water molecule (pK_a 13.5 and 15.5, respectively), can be withdrawn as an anion and is protonated by the His57 imidazolium cation.

Investigation of the kinetic isotope effect of a solvent in mixtures with different H_2O and D_2O ratios [2439, 3177, 3178, 3331] has given grounds for the proposal that deacylation of unspecific acylchymotrypsins is a one-proton transfer ($k_H/k_D \approx 2.5$), whereas deacylation of high specific acylenzymes is accompanied by two-protons transfer. For proper substrates this has been associated with the functioning of the charge relay system (transfer of the second proton between His57 and Asp102). However, the involvement of the proton forming the hydrogen bond in the oxyanion hole is more probable (see also [2438]).

Serine proteases acylation with specific substrates increases "rigidity" of the protein molecule and decreases its sensitivity to pH-dependent conformational transitions [3332]. The side chains of the P_1-residue of the substrate play here a crucial role. Their interaction with the binding region in the active site (S_1) includes a conformational change of the enzyme that facilitates further reaction. This has been demonstrated [3333, 3334] on a series of the so-called reverse trypsin substrates of type

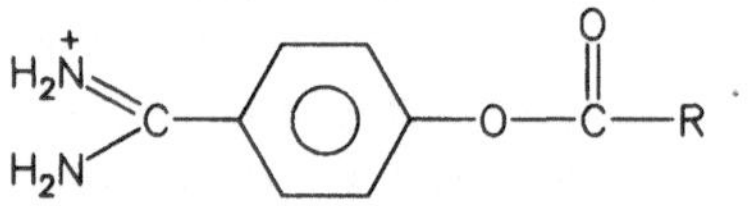

Cleavage of these substrates yields the acylenzymes with unspecific group COR, where $R=CH_3$, C_2H_5, C_3H_7, etc. An addition of amines to the acyl-enzyme essentially increases their deacylation rate, whereupon fluorescent labeled groups R show that the conformational change enhances the acyl group's accessibility to water.

In the case of cysteine proteases the acylenzyme deacylation follows the same mechanism as for serine proteases, though possibly with two major distinctions. First, pK_a of His159 in papain (≈ 4.2) is essentially lower than pK_a of His57 in chymotrypsin (≈ 7), therefore the water proton transfer to imidazole rate is expected to be 3 orders of magnitude lower in cysteine proteases as compared to serine enzymes. This can be possibly compensated by higher destabilization of the ester group in thioesters. However, according to Raman spectra of dithioacylpapains [3102, 3335], bond C–S is shortened due to interaction of the enzyme sulfur atom with the nitrogen atom of residue P_1. Secondly, the tetrahedral intermediate breakdown in the case of cysteine enzymes proceeds without protonation of the leaving group. A release of group $-CH_2S^-$ immediately regenerates the enzyme active form.

Acylenzymes of cysteine proteases exist in two conformations – A and B, the latter being predominant [3336]:

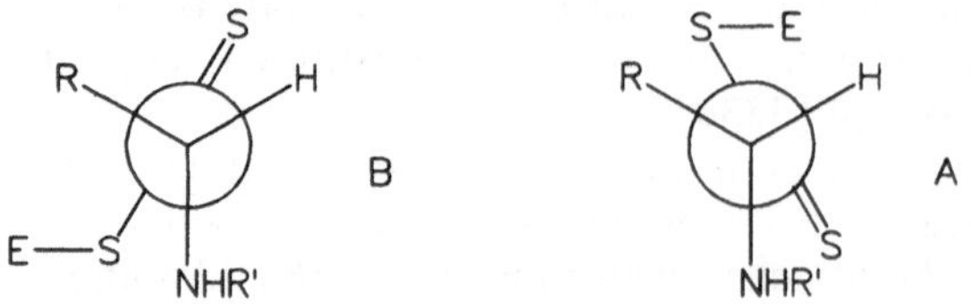

Apparently, acylchymotrypsins exist in two forms as well [3337]. The dynamics of acylenzymes of serine proteases has been investigated in [3338–3340].

The enzyme-product complex resulting from deacylation should dissociate yielding an acyl product and regenerating the free enzyme. Though the affinity of hydrolysis products to the enzyme is usually close to the substrate affinity, this does not affect much the hydrolysis rate in the region of moderate degrees of substrate transformation, since the concentration of the free product remains low. However, at high concentration of products the rate falls due to competitive inhibition by the product. Dissociation of the enzyme and substrate (see Sect. 6.2.3). This is considered below in detail. Besides, since the free energy of the acylenzyme hydrolysis at acid pH values is low, deacylation under this condition proceeds slowly, and the product accumulation can reverse the reaction yielding the acylenzyme [2003, 3341]. In fact, X-ray analysis of complexes of *Streptomyces griseus* protease A with the product shows [3063, 3069] that the product carboxyl group forms the hydrogen bond to the imidazolium cation of His57. This bond stabilizes the complex, and its decomposition (at neutral pH) promotes the product dissociation.

As shown above (see Sect. 7.4), pepsin, leucineaminopeptidase, thermolysin and carboxypeptidase A (amide substrates) function by the type of general base catalysis at the stage of the nucleophilic attack of the carbonyl group of cleavable compounds. This evidently excludes the formation of acylenzymes. As to ester substrates hydrolysis by carboxypeptidase, the cryoenzymatic experiment provides data interpreted as evidence in favor of the intermediate acylenzyme – anhydride, formed by enzyme and substrate carboxyl groups. However, they need verification (see Sect. 7.4).

At the carboxypeptidase A catalyzed hydrolysis of the peptide substrates at least two enzyme-substrate complexes are formed (see Sect. 6.2.3). In the case of ester (depsipeptide) substrates the second complex is the complex of the enzyme with hydrolysis products, i.e., in this case the product dissociation is the rate-determining stage [3242]. Apparently, in the case of metal-containing amide hydrolases the substrate carboxyl group remains in the metal coordination sphere after hydrolysis and only later is it replaced by a water molecule. The mechanism of cleavage by thermolysin is presented in Fig. 115 [3316].

The absence of covalent intermediates in the cases just described poses a question regarding the acyl transfer and transpeptidation mechanisms. For a long time the fact of catalysis of such reactions by amide hydrolases of this type served as major evidence for the formation of intermediate covalent acylenzymes, and in the case of aspartic proteases these are also "aminoenzymes" [1749, 1750, 2173, 2177, 2178, 3342, 3343]. The transpeptidation reactions have been studied in detail (see Sect. 4.8.1).

If the transfer reactions proceed via noncovalent enzyme-product complexes, their stability should be high enough for binding a transpeptidation acceptor and for chemical bonding between the transferable fragment and the acceptor. A feasibility of this condition can be evaluated from data on the initial transpeptidation rates [1667] and dissociation constants of the enzyme-product complexes.

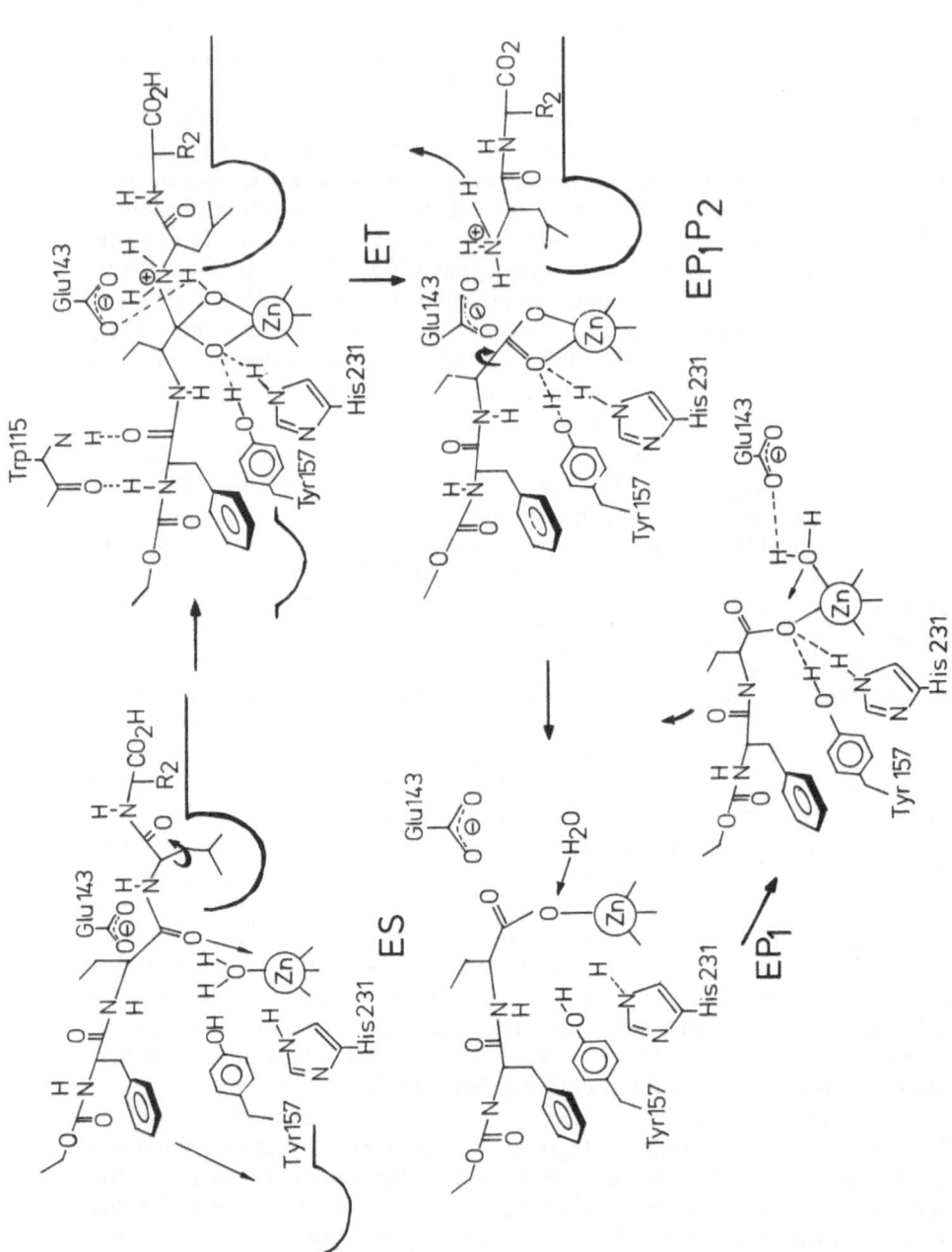

Fig. 115. Mechanism of catalysis by thermolysin [3316] including complexes EP_1P_2 and EP_1, where the formed carboxylic group of the product remains included into the coordination sphere of metal ES and ET – Michaelis complex and tetrahedral intermediate, respectively

A kinetic transpeptidation scheme can be presented as

$$E + S \underset{}{\overset{K_s}{\rightleftharpoons}} ES \xrightarrow{k_2} EP_1 \xrightarrow{k_3} E + P_1 ,$$

$$\searrow P_2$$

$$EP_1 + A \underset{k_{-4}}{\overset{k_4}{\rightleftharpoons}} EP_1 A \xrightarrow{k_5} E + S^* ,$$

where S is the substrate; S* is the transpeptidation product; P_1 and P_2 are the first (transferable) and the second hydrolysis products; A is the transpeptidation acceptor.

The ratio of the hydrolysis and transpeptidation rates is expressed by equation:

$$\frac{v_h}{v_t} = \frac{k_3 k_{-4}}{k_4 k_5} \frac{1}{[A]_0} .$$

If assumed $k_3 \approx k_{-4}$, the decomposition rate constant for the enzyme-product complex (EP_1) can be calculated:

$$k_3 = \sqrt{\frac{v_h}{v_t} k_4 k_5 [A]_0} .$$

Table 78 lists the results for several transpeptidation reactions. The calculation is performed at $k_4 = 1 \times 10^7 \, M^{-1} s^{-1}$, the rest values are derived from [1667].

Data of Table 78 show that the decomposition rate constants for the enzyme-substrate complex necessary to provide the transpeptidation rates are within $10-50 \, s^{-1}$.

Experimentally defined rate constants for the enzyme-product (or enzyme-inhibitor) complex dissociation are $1 \times 10^4 - 1 \times 10^5 \, s^{-1}$, i.e., 3 to 4 orders of magnitude higher than those for transpeptidation. Thus a usual enzyme-product complex cannot be the complex involved in the substrate fragment transfer to the acceptor.

This contradiction can be overcome. If we assume that the stepwise formation of the productive enzyme-substrate complex is not the only pathway

Table 78. Decomposition rate constants for pepsin-product complex.
Acceptor $Phe(NO_2)OH$; $[A]_0 = 0.667$ mM; pH 4.6; 37 °C

Substrate[a]	v_h/v_t	k_5, s^{-1}	k_3, s^{-1}
AcPheTyrOH	22.0	5.1×10^{-4}	8.65
AcTyrTyrOH	22.2	4.9×10^{-4}	8.5
AcPhePheApm	351	7.75×10^{-4}	42.5
GlyGlyPhePheApm	380	8.1×10^{-4}	45.3

[a] Apm – 3-(morpholinopropyl)-amide.

of the substrate conversion to the product (Sect. 6.2.3), the stepwise dissociation of the enzyme-product complex is feasible as well:

$$E + S \xrightleftharpoons{K_s} E_0S \xrightleftharpoons{K_P} E_PS \xrightarrow{k_0} E_PP_1P_2$$

$$\xrightleftharpoons{K_R} E_PP_1 \xrightleftharpoons{K_Q} E_0P_1 \xrightleftharpoons{K_i} E + P_1 .$$
$$+$$
$$P_2$$

In this case the apparent dissociation constant of the E_PP_1 complex is

$$K_i' = \frac{K_i K_Q}{1 + K_Q}, \quad \text{where } K_Q = \frac{[E_0P_1]}{[E_PP_1]}.$$

If $K_Q \gg 1$, the observed constant of inhibition by the product is $K_i' = K_i$, i.e., it characterizes only the E_0P_1 complex formation. The conversion of E_0P_1 to E_PP_1 is restricted by the low rate constant (k_{-q}) in reaction:

$$E_PP_1 \xrightleftharpoons[k_{-q}]{k_q} E_0P_1 .$$

If the conversion of E_PP_1 to E_0P_1 proceeds with the rate constant $(k_q \equiv k_3)$ of about $10-50 \, s^{-1}$, at $K_Q > 1 \, k_{-q}$ should be lower than these values.

Thus the transpeptidation occurs in the E_PP_1 complex capable of acceptor binding. The complex is readily formed from a substrate during its hydrolysis. If no substrate is available in the system, the formation of the E_PP_1 complex is restricted by the low rate of the conversion of E_0P_1 to E_PP_1. The product and acceptor thus can react yielding a new peptide, its amount being determined by the equilibrium constant of the overall process.

It is clear that the $E_PP_1P_2$ complex can decay, eliminating not only P_1 but also P_2. Depending on the relative affinity of the products, either acyl transfer or amine transfer reaction may proceed. So for pepsin and substrates with the free carboxyl group near the scissile bond the amine is transferred, for the substrates with the free amino group acyl transfer is predominantly observed [2176–2178].

According to the principle of microscopic reversibility, a new peptide bond must be formed by the mechanism which is the reverse of the hydrolysis pathway.

The stepwise decomposition of the pepsin-products complexes follows when comparing steady-state and pre-steady-state rates of hydrolysis of fluorogenic peptide substrates [2976].

7.10 Comparison of Enzymatic Hydrolysis Mechanisms

The enzymatic hydrolysis mechanisms are, in principle, the same as those in model systems. Unfortunately, we do not know the exact mechanisms, i.e., the position in space and the electron density distribution for each atom in the reaction system at each precise moment. The enzymatic systems have even some advantages as compared with the reactions in solution, since the X-ray "pictures" of certain stages can be obtained with various substrate analogs.

Figures 116–119 outline enzymatic hydrolysis and, in my opinion, fit better all the available data.

The formation of the enzyme-substrate complex is the most significant stage of enzymatic hydrolysis proceeding via several stages (at least two). The first stage includes mainly unspecific and long-range interactions (hydrophobic, electrostatic), the second fine tuning of the substrate and the enzyme active site, involving the formation of new bonds and conformational changes in the substrate and enzyme. A productive enzyme-substrate complex is a result of the process.

The presence of a definite region of binding of one side chain of the peptide substrate, which belongs usually to the amino acid residue in the vicinity of the scissile bond, is common for the majority of the amide hydrolases under study.

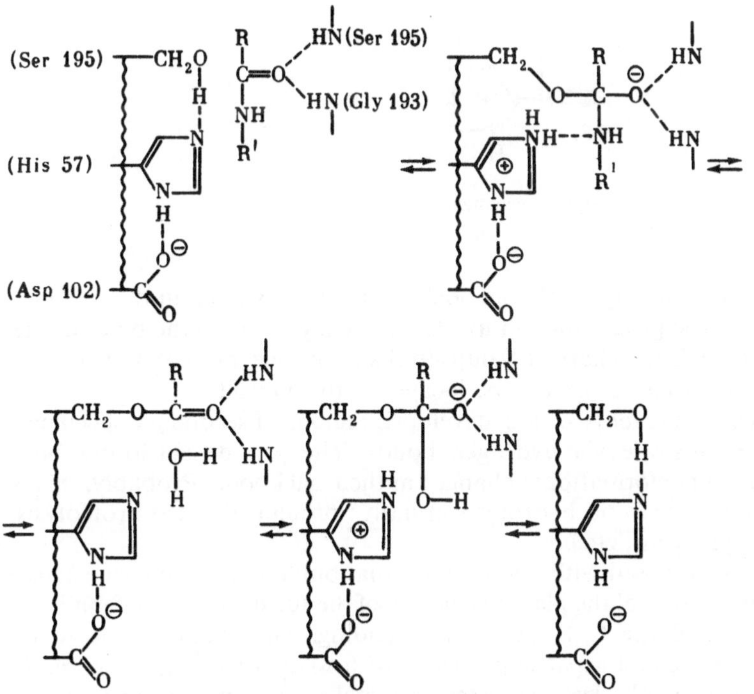

Fig. 116. Scheme for α-chymotryptic catalysis mechanism

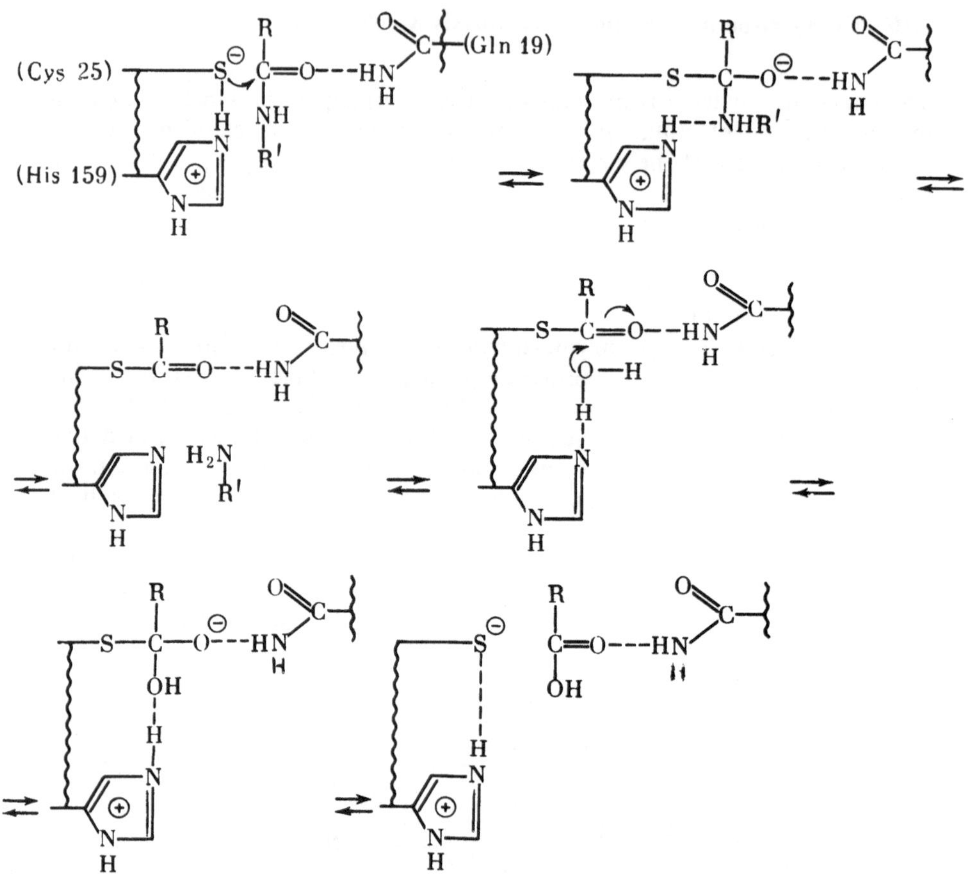

Fig. 117. Scheme for papain catalysis mechanism

These are the so-called hydrophobic pockets clearly expressed in the enzymes of the chymotryptic group, though available in enzymes with the binding site of the "cleft" type. Their charged groups (trypsin-like enzymes) or alkyl chains (elastase, etc.) limiting its size can be located in this pocket.

All proteases have more or less developed regions of secondary specificity bound to the substrate via hydrogen bonds. The peptide substrates bind without essential conformational changes in their backbone. Probably, this is of great value for the bond hydrolysis in native proteins (limited proteolysis, activation of zymogens, etc.).

Finally, it is supposed with good reason that binding in the enzyme active site causes a distortion of the planar structure of the substrate scissile bond due to rotation around the C–N bond and bending. This distortion is to be substantial in serine and aspartic proteases with weak attacking nucleophilic groups and, presumably, less expressed in metal-containing enzymes due to a strong electrophile (Zn^{2+}) that polarizes the cleavable bond.

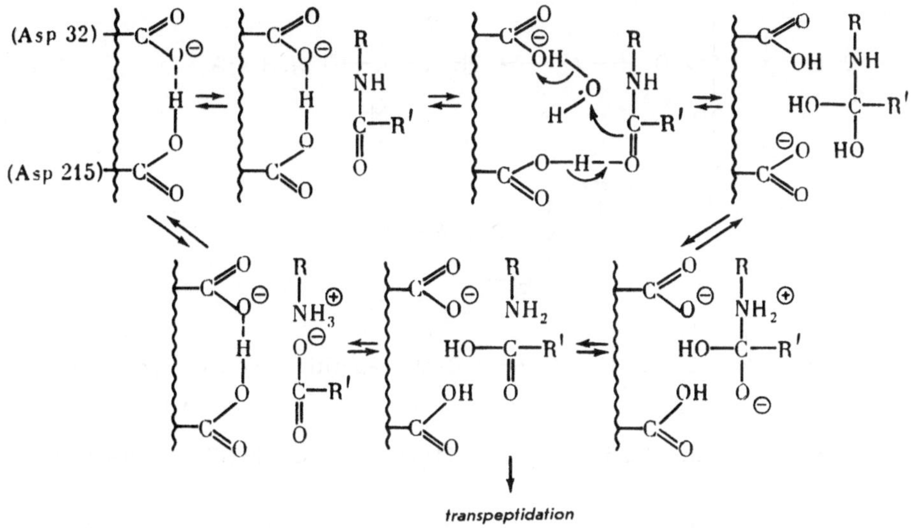

Fig. 118. Scheme for pepsin catalysis mechanism

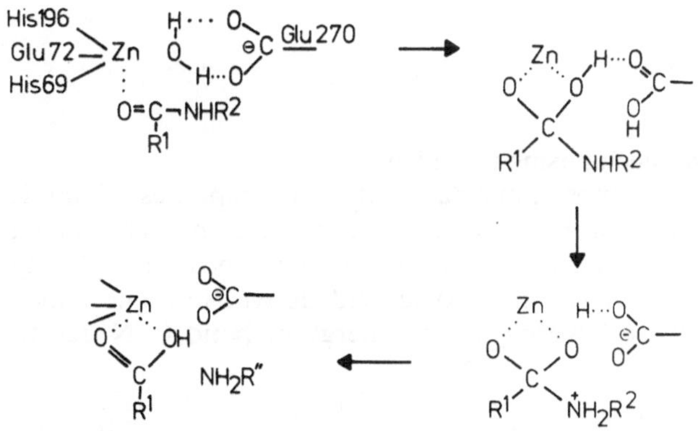

Fig. 119. Scheme for carboxypeptidase A catalysis mechanism

Amide hydrolases can be divided into two large groups according to their catalysis type – those functioning by a covalent catalysis (serine and cysteine amide hydrolases) and the others by general catalysis (aspartic and metal-containing enzymes).

Despite this difference, the catalysis mechanism and the catalytic apparatus of all amide hydrolases studied have much in common. In general, the scheme for the mechanism of the amide hydrolysis and catalytic groups typical of various amide hydrolases can be shown as follows.

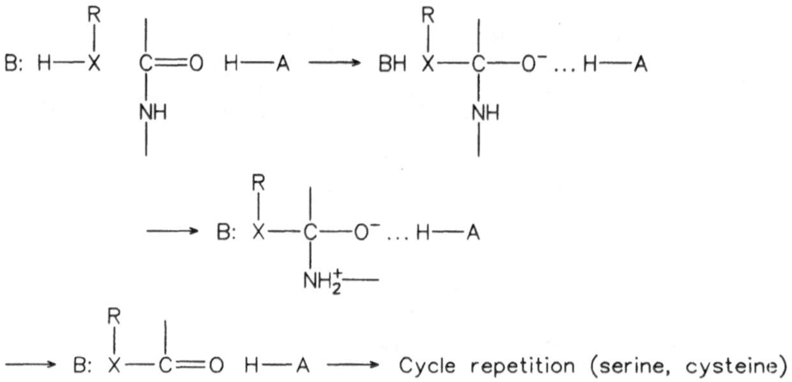

$$\text{B: H—X} \quad \underset{\underset{\text{NH}}{|}}{\overset{\overset{\text{R}}{|}}{\text{C}}}\text{=O H—A} \longrightarrow \text{BH X}—\underset{\underset{\text{NH}}{|}}{\overset{\overset{\text{R}}{|}}{\text{C}}}—\text{O}^-\dots\text{H—A}$$

$$\longrightarrow \text{B: X}—\underset{\underset{\text{NH}_2^+—}{|}}{\overset{\overset{\text{R}}{|}}{\text{C}}}—\text{O}^-\dots\text{H—A}$$

$$\longrightarrow \text{B: X}—\overset{\overset{\text{R}}{|}}{\text{C}}\text{=O H—A} \longrightarrow \text{Cycle repetition (serine, cysteine)}$$

	B	R	X	AH
Serine	>N: (His57)	—CH$_2$(Ser195)	O	NH (Gly193, Ser195)
Cysteine	>N: (His159)	—CH$_2$(Cys25)	S	NH (Gln19, Cys25)
Aspartic	$\overset{\text{O}}{\overset{\|}{—\text{C}}}$—O$^-$(Asp 32 and 215)	H	O	H$^+$(!)
Metal–containing	$\overset{\text{O}}{\overset{\|}{—\text{C}}}$—O$^-$(Glu270)	H	O	Zn^{2+} .

The similarity of these mechanisms is evident.

Let us try to analyze the most important catalytic properties of amide hydrolases from the viewpoint of the above catalysis mechanisms. The system of nucleophilic and nearby electrophilic groups provides possible splitting of the electron-labile bonds of the carboxylic acid derivatives. The amide bond hydrolysis is a stepwise lowering of the energy of bond C–N due to polarization:

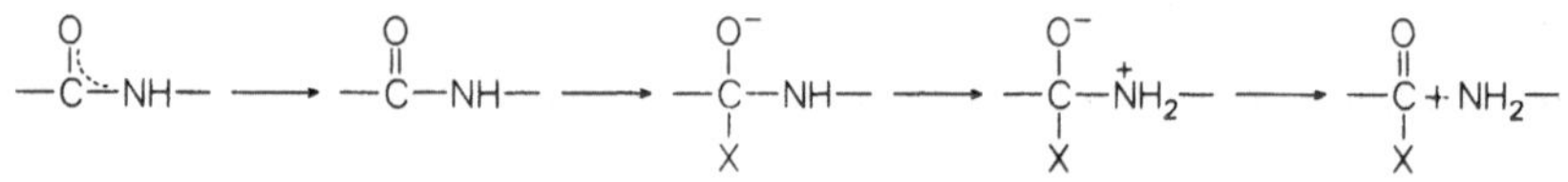

$$\overset{\overset{\text{O}}{\|}}{—\text{C}}\text{:-NH—} \longrightarrow \overset{\overset{\text{O}}{\|}}{—\text{C}}\text{—NH—} \longrightarrow \underset{\underset{\text{X}}{|}}{\overset{\overset{\text{O}^-}{|}}{—\text{C}}}\text{—NH—} \longrightarrow \underset{\underset{\text{X}}{|}}{\overset{\overset{\text{O}^-}{|}}{—\text{C}}}\overset{+}{—\text{NH}_2—} \longrightarrow \underset{\underset{\text{X}}{|}}{\overset{\overset{\text{O}}{\|}}{—\text{C}}}\text{+ NH}_2—$$

r_{C-N}(Å)	1.37	1.39	1.43	1.50	∞
ΔE_{C-N} (kcal/mol)	86	≈ 70	66	46	0

To perform such process under physiological conditions the proteins, among them amide hydrolases, contain a set of nucleophilic and electrophilic groups. In fact, they are not active enough to be attached to the saturated carbon atom. The catalysis of the saturated groups transfer is performed in the enzymes through cofactors, such as, for example, S-adenosylmethionine (see in [1405]).

At the same time, in a series of carboxylic acids derivatives the specificity (in terms k_{cat}/K_m) of various enzymes to different bonds varies (see Sect. 4.3.1.1).

The differences in the hydrolysis rates of esters and amides in serine proteases that reach 4 orders of magnitude are stipulated exclusively by the difference in the standard free energies of their hydrolysis, i.e., are explained by different states of the cleavable bond in the substrate (Fig. 120) [2009, 3344].

The energetic level of the transition state for esters and amides is virtually the same. This provides evidence in favor of the Hammond postulate about the similarity of the structures in the ground and transition states in exoergic reactions [1531] and indicates a decisive role of the ground state of the enzyme-substrate complex in the reactions catalyzed by serine proteases.

For aspartic and metal-containing amide hydrolases the ratio of the rates of the ester-amide hydrolysis tends to the unity. This witnesses [3342] the predominant role of the electrophilic catalysis (i.e., protonation of the carbonyl carbon and coordination of the metal) for these enzyme types. It is noteworthy that the ratio v_{ester}/v_{amide} is about 5 for cysteine proteases. A contribution of the electrophilic catalysis is low in this case. At least, perhaps, for aspartic and cysteine proteases the absence of pronounced distinctions in the rates of the ester and amide hydrolyses is explained, partially, if not fully, by less difference in free energies of the hydrolysis of these substrates at more acid pH values (Fig. 28). There exist amide hydrolases functioning in the neutral medium and devoid of the esterase activity. The difference in the behavior of ester and amide substrates of carboxypeptidase A can be caused by a different character of the rate-determining stage of the multistage hydrolysis [3139].

Figure 120 shows the change of AcPheGlyNH$_2$ by AcPheAlaNH$_2$, i.e., introduction of one methyl group dramatically decreases the energy level of the transition state virtually without alteration of the level of the ground-state energy. In other words, the specificity manifests itself in the maximum hydrolysis rate rather than in binding. Such a situation is typical of many proteases: it is clearly displayed in the case of pepsin (see Sect. 4.3.1.3). The secondary interactions play apparently a decisive role in the productive complex formation, the energy here being utilized to compensate for the energetically unfavorable system changes (conformational changes, destabilization of the scissile group, etc.). This contributes to understanding the unique specificity of some amide hydrolases. The representatives of each enzyme group are amide hydrolases with a wide spectrum of specificity and enzymes, which catalyze hydrolysis of one or very few one-type substrates. Kidney renin acting on angiotensinogen and weakly hydrolyzing some analogs of the N-terminal fragment of this substrate perfectly illustrates this. No other peptide effectively hydrolyzed by this enzyme has been found.

Even for a series of less specific amide hydrolases the peptide cleavage rate greatly differs, which depends on the number of amino acid residues (Table 79). Interestingly, the activity difference is caused by a change of k_{cat} but not of K_m, i.e., various enzymes have almost the same affinity to the given substrate, but convert it with the rates differing by 4 orders of magnitude.

The secondary interactions of the substrate evidently "induce" the changes differently, thus leading to a productive complex. The conformational changes

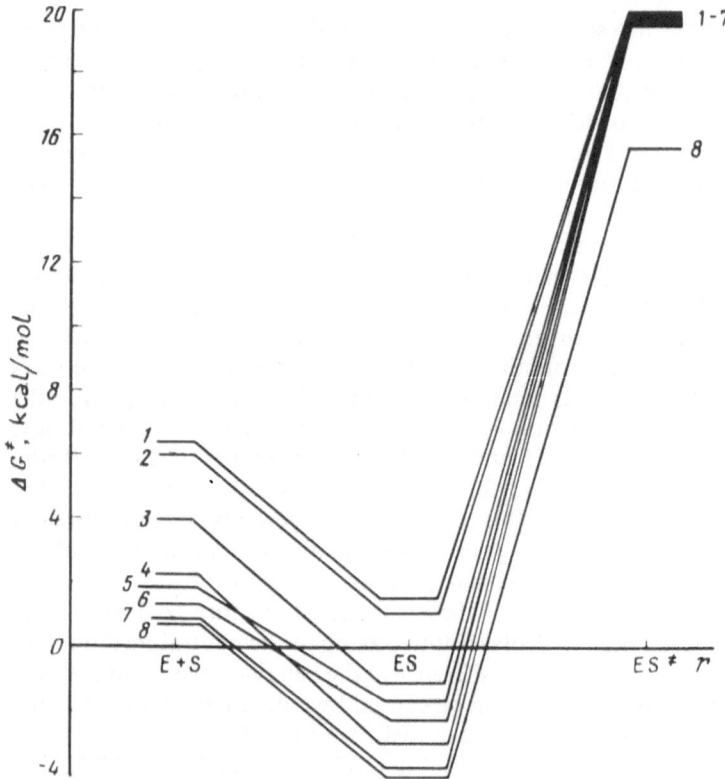

Fig. 120. Change of free energy along the reaction coordinate (r) at acylation of chymotrypsin by substrates of type AcPhe(NO$_2$)-X [3344] X: *1* OEt; *2* OMe; *3* NHC$_6$H$_4$NO$_2$-*p*; *4* NHC$_6$H$_5$; *5* NH$_2$; *6* NHNH$_2$; *7* GlyNH$_2$; *8* AlaNH$_2$

Table 79. Hydrolasis of peptide ZLeuValPhe(NO$_2$)AlaApm by aspartic proteases [1715]

Enzyme	k_{cat}, s^{-1}	$K_m \cdot 10^3$, M	k_{cat}/K_m, M^{-1} s^{-1}	Relative activity
Pepsin	43	0.3	1.43×10^{-5}	1.3×10^4
Chymosin	1.16×10^{-2}	0.25	46.4	4.2
Cathepsin D	6.6×10^{-3}	0.6	11	1

are expected to appear due to the binding of the same substrate (or the substrate-like inhibitor) being less expressed in more specific enzymes. However, comparative studies have not been carried out yet.

Amide hydrolases of the above types are distinguished by other significant properties, namely, pH optimum of the catalytic activity (see Sect. 5.2.3). These differences accord well with the distinctions of the catalytic apparatus of these enzymes. The catalytically active histidine residue, functioning as a general base and a general acid in serine, cysteine and metalloendopeptidases, limits the pH range by the ionization state of its imidazole group. In cysteine proteases at pH 5-6 the pK_a value of the imidazole group in the acylenzyme falls to 4-4.5. In a free enzyme pK_a increases up to 8, which provides for the existence of the stable ion pair $Im^+ \ldots S^-$ (see Sect. 7.1).

It is evident that only the carboxylate ions can function as a nucleophile in the acid pH range. In fact, proteases functioning in acid media belong to the group of aspartic proteases. The pH increase inducing ionization of the two carboxyl groups of the acitve site deprives these groups of the proton donor properties required for general acid catalysis.

There is, at least, one example of serine proteases functioning in weak acid medium – serine carboxypeptidases. At present, the structure of the catalytic apparatus of these enzymes is practically unknown. Perhaps the carboxyl group in these enzymes may function as a general acid-base group, but not the imidazole group, as in true serine proteases.

Some amide hydrolases tend to level the energy barriers of multistage hydrolysis reactions [2461], which can be associated with the compensating changes of the enzyme conformation during the catalysis. This peculiarity is supposed to be common for the enzymes of this group.

As to the mode of action of amide hydrolase, there are more unsolved problems than established facts. The aforementioned data are referred only to the simplest representatives of each enzyme group, but even they cannot be considered comprehensive.

Table 12 demonstrates a diversity of amide hydrolases in nature. Extrapolation of our knowledge to all of them seems unreasonable.

7.11 Some Aspects of Evolution of Chemical Mechanisms of Amide Hydrolases

Undoubtedly, the proteolysis process may have proceeded at the earliest stages of chemical evolution. At that time hydrolysis of peptide bonds could be catalyzed by relatively simple precursors of proteolytic enzymes, probably by the systems utilizing atoms of heavy metals. Today we know, at least, four types of rather similar hydrolysis mechanisms. But it is hardly possible to trace evolutionary relations between them. First, it is difficult to establish the hierarchy of mechanisms, i.e., to state that one of them is perfect. Apparently, there is no essential distinction in the efficiency of catalysis by different proteases if an "ideal" substrate is employed in each case. Secondly, only a few data are available on the types of proteolytic enzymes in various species differing in their evolutionary prehistory. Still, there are some ideas on the evolutionary interrelations of amide hydrolases. So aspartic proteases have been not found in prokaryotes yet. This enzyme type is widespread in fungi, mammals, fish and other animals, as well as in some phanerogamous (insectivorous) plants. Presumably, aspartic proteases appeared at later stages of evolution not because of their perfection, but due to their accommodation to certain conditions the primitive organisms had not faced.

Plants contain predominantly cysteine enzymes. The latter are found also in animal tissues, and are met with in some bacteria and even viruses.

Prokaryotes and eukacaryotes contain serine proteases. Insect proteases have been slightly studied, still it is known that they are, mainly, of the serine type. Metal-containing proteases are widely distributed – from microbes to brain tissues. Of the ideas how to consider these proteases, the earliest one seems attractive: the similarity of structure of the active sites of metal-containing and serine proteases, on the one hand, and of metal-containing and aspartic proteases on the other, has advanced the concept that metal-dependent proteases can be a common evolutionary precursor of these two enzyme types.

At present we do hope that further research into amide hydrolases of various types of living organisms will provide more comprehensive knowledge about the evolution of the chemical mechanisms of proteolysis. This problem will be approached below (see Sect. 8.7) in connection with the notion of all "ideal" enzyme.

7.12   Conclusion

Chemical transformation of substrates of amide hydrolases is carried out via the catalytically active enzyme groups functioning as nucleophilic, general base, and general acid catalysts. A characteristic feature of the catalysis is evidently the action of the nucleophilic group on the resonance-destabilized bond in the productive enzyme-substrate complex. The elementary step of the substrate chemical change should proceed with the rates close to those of the diffusion-controlled processes. The hydrolytic reaction is a multistep process running through tetrahedral intermediates. The rates of their formation and decomposition are determined by the substrate structure and, especially, properties of the leaving group.

According to the catalysis type, amide hydrolases are divided into two large groups – enzymes functioning by a covalent catalysis type, when the enzyme catalytic group forms a covalent intermediate with a substrate, and enzymes functioning by a general base catalysis type, when the enzyme group promotes proton release from a water molecule, acting as a nucleophilic reagent. Despite these distinctions in the mechanisms of action of amide hydrolases under study have much in common. The evolution of chemical mechanisms of amide hydrolases probably has its roots in a common precursor.

Chapter 8 Specificity and Efficiency. Concepts and Hypotheses

At present, we are confronted with many theories and hypotheses, the aim of which is to explain unique properties of enzymes – their specificity and efficiency (see reviews [9, 1402, 1613, 2410, 2450, 2846, 3345–3348]). But the theories and hypotheses often differ only in terminology, they interprete similarly the physical nature of the phenomena under study.

A comparison of the catalytic action of the enzyme on various substrates (specificity) and its correlation with the model nonenzymatic reaction (efficiency) imply a use of the first-(k_{cat}) or the second-(k_{cat}/K_m) order catalytic constants as well as the rate constant of the nonenzymatic reaction, which can be of the pseudo first or second order. Comparing it with the model system, the model reaction is supposed to run through the same intermediates as the enzymatic process, i.e., it is congruent.

Requirements for the model system envisage closeness of the pK_a values involved in grouping, besides a structural similarity of the reactive groups and intermediates, and identity of the reaction products in the systems compared [3349]. But it is not so easy to satisfy these requirements, and the rate constants ratio of the enzymatic and model reactions can change much depending on the model chosen.

The efficiency substantially depends on the choice of the substrate in the enzymatic reaction. So the value 1×10^5 (k_{cat}/K_m) for chymotrypsin often cited [2424, 2453, 3350], if amides on N-acetylamino acids are utilized as substrates, differs fivefold from that obtained with one of the best peptide substrates (see Sect. 4.4).

8.1 General Considerations

Let us consider the diagram of a change in the free energy of enzymatic and nonenzymatic reactions (Fig. 121)[1]. The major difference between them is that in the course of the enzymatic catalysis the enzyme and substrate form a complex, the free energy of which is lower in regard to that of initial reagents. *If free energy of activation of the chemical stage remains the same for the two reactions*, an increase in the rate of the catalyzed process is exactly propor-

---

[1] As standard state here and later on, the changes in free energy of 1 mol of substances at their concentration 1 M are used [3351].

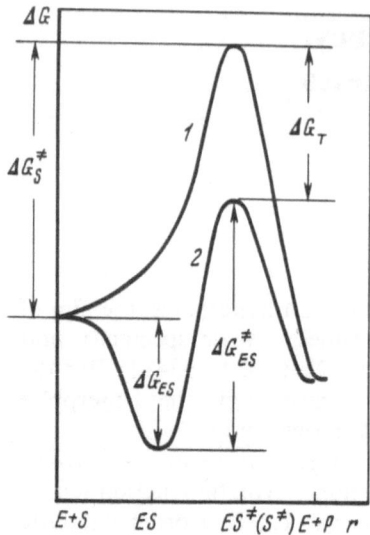

Fig. 121. Diagram of the free energy change along the reaction coordinate (*r*) for nonenzymatic (*1*) and enzymatic (*2*) processes

tional to the dissociation constant of the enzyme-substrate complex. Let us consider two ways of the substrate conversion – catalytic and noncatalytic:

$$\text{E} + \text{S} \quad \overset{\Delta G_{ES}}{\underset{\Delta G_s^*}{\rightleftarrows}} \quad \overset{\text{ES}}{\underset{\text{S}^* + \text{E}}{}} \quad \overset{\Delta G_{ES}^*}{\underset{\Delta G_T}{\rightleftarrows}} \quad \text{ES}^* \longrightarrow \text{products} \tag{1}$$

In this cycle the equation

$$\Delta G_{ES}^{\ddagger} - \Delta G_{ES} = \Delta G_S^{\ddagger} - \Delta G_T \tag{2}$$

is true.

As changes in free energies of activation of the two processes are identical $\Delta G_{ES}^{\ddagger} = \Delta G_S^{\ddagger}$), then $\Delta G_{ES} = \Delta G_T$. Thus, a change in the level of free energy of the transition state of the enzymatic reaction as regards the state of the initial reagents (ΔG_T) is exactly equal to a change in the free energy due to the complex formation.

For the simplest scheme of the enzymatic catalysis:

$$\text{E} + \text{S} \underset{}{\overset{K_s}{\rightleftharpoons}} \text{ES} \overset{k_2}{\longrightarrow} \text{products}$$

the rate is

$$v_f = \frac{k_2[\text{E}]_0[\text{S}]_0}{K_s + [\text{S}]_0}.$$

The rate of the corresponding congruent model reaction will be:

$$S \xrightarrow{\ k_n\ } products \quad v_n = k_n[S]_0 .$$

The ratio of the rates of the enzymatic and model reactions is

$$\frac{v_f}{v_n} = \frac{k_2[E]_0}{k_n(K_s + [S]_0)}$$

or taking into account that $\Delta G_{ES}^{\ddagger} = \Delta G_S^{\ddagger}$ (i.e., $k_2 = k_n$),

$$\frac{v_f}{v_n} = \frac{[E]_0}{K_s + [S]_0} .$$

When $[S]_0 \ll K_s$, $v_f/v_n = [E]_0/K_s$, and at $[S]_0 \gg K_s$ $v_f/v_n = [E]_0/[S]_0$, i.e., the reaction is not accelerated and, what is more, proceeds slower, because usually $[E]_0 < [S]_0$.

It is quite clear that the formation of the enzyme-substrate complex itself without a decrease in the energy of activation of the chemical process cannot provide the efficiency of the enzymatic catalysis.

Lowering of free activation energy of the enzymatic reaction as compared with the congruent model can be, in principle, realized in two ways: either increasing the free energy of the ground state or decreasing the energy of the transition state (Fig. 122). In both cases Eq. (2) and the values of $\Delta G_{ES}^{\ddagger}$, $\Delta G_{ES}^{\ddagger}$ and $\Delta G_S^{\ddagger}$ according to the theory of the absolute reaction rates show that the differences in free energies of activation of the two processes (ΔG_T) are defined as:

$$\Delta G_T = -RT\ln\frac{k_f}{k_n}, \quad \text{where } k_f = \frac{k_2}{K_s} .$$

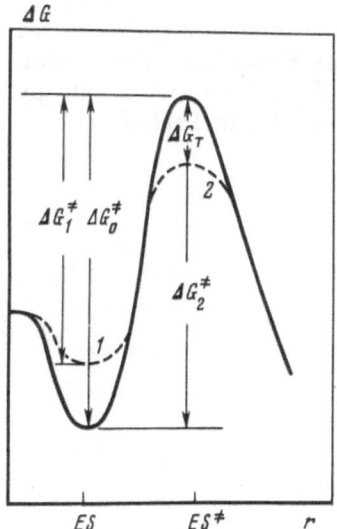

Fig. 122. Two ways of decreasing the activation energy: increase in the ground state energy (*1*) and decrease in the transition state energy (*2*)

For this reason, theories of the enzymatic catalysis are aimed at explaining the difference in k_f and k_n, some enzymologists associating this with a change in the level of the free energy of the ground state, while others with a decrease in the level of the free energy (stabilization) of the transition state. This "classification" is certainly very conventional, since it does not provide a clearly defined division, but at the same time it seems useful. Before stating the existing concepts it is necessary to consider the factors responsible for contributing to the difference between k_f and k_n.

8.2   Factors of Catalysis

The efficiency, i.e., the ratio of $k_{cat}/K_m k_n$ for amide hydrolases may reach $10^{11}-10^{12}$ (Table 50). For certain enzymes the estimation of the efficiency provides the values $10^{16}-10^{18}$ [3352, 3353]. It is of interest here, what factors of enzymatic catalysis can promote a high acceleration of the process?

8.2.1   Approximation and Orientation

Unlike the reactions in solution, the enzymatic reactions run within a narrow region of the active site, where the reactants approach each other. This results in enhancement of the local concentration of the reactive groups, which is equivalent to a decrease in $\Delta G_{ES}^{\ddagger}$ (Scheme 1).

Many researchers have evaluated differently the contribution of this factor. So taking into account the fact that the maximum concentration is 55 M for the molecules corresponding to those of water by size, Koshland [1615] has accepted this value as maximum "effective concentration" in enzymatic reactions. Such a concentration effect corresponds to a change in the free energy by 2.4 kcal/mol or in entropy by 8 e.u. Naturally, the orientation effect is not taken into account here, this can reach higher values. So studies of the reaction of diene condensation [1616, 1617] give values of the total loss of entropy equal to -45 e.u. This value is derived from those listed in Table 80.

Table 80. Changes in entropy for various types of motion

Type of motion		ΔS, e.u.
Translational motion (3 degrees of freedom)		29–36
Rotatory motion (3 degrees of freedom)[a]		5–6
Internal rotation		3–5
Vibrations with frequency (cm^{-1})	1000	0.1
	400	1.0
	100	3.4

[a] Per one bond.

The role of orientation in catalysis (see Chap. 3) is also illustrated by data on intramolecular reactions, when acceleration as compared with the corresponding intermolecular processes reaches 10^6–10^7.

Identical values of approximation and orientation effects have been [3354] theoretically derived from statistical thermodynamics. The ratios of rate constants of intra- (k_m) reactions and intermolecular (k_b) are expressed by:

$$k_m/k_b = \frac{1}{N\delta V},\qquad(3)$$

where N is the Avogadro number, δV the element of volume, which embraces the mass center of the reactive groups. The δV values may correspond to a sphere of the diameter equal to the amplitude of vibrations of the formed bond, i.e., ≈ 0.05 Å; this gives the ratio $k_m/k_b = 3 \times 10^7$ M.

The effects of approximation and orientation are essential at high orders of reactions. If two substrates are transformed into a product in the presence of three catalysts, the reaction in solution is of the fifth order. This process on the enzyme can proceed 10^{13}–10^{17} times faster [1615]. A simple formula taking into account the orientation effects has been proposed to evaluate a relative reaction rate in solution and on the enzyme [1615]:

$$\frac{v_f}{v_n} = \frac{[E]_0(55.5)^{n-1} \cdot \prod\limits^{n-1} \theta}{\prod\limits^{n} [S] \cdot \prod\limits^{n-1} \cdot f_{xy} C},\qquad(4)$$

where θ is a contact solid angle between neighboring molecules, f_{xy} is the factor of correction of the number of neighboring molecules around each reactive molecule.

A substantial contribution to the efficiency of catalysis can be made by the effect of "dynamic concentration" [3355, 3356]. The period of a molecular pairs existence in solution is about 10^{-11} s. If this complex is not converted into a product, it dissociates, and the system reaches the lowest energetic level. The period of existence of enzyme-substrates complexes according to NMR data [3357–3359] is of 10^{-4}–10^{-7} s, i.e., the probability of fluctuation leading to the activation state drastically rises. This effect can enhance the reaction rate by 6–9 orders of magnitude [3356].

Similar ideas have been developed [3360, 3361] in the "space and time hypothesis", according to which the enzyme can retain the reactive molecules at the minimum for interaction distance long enough from the viewpoint of the reaction time. It is noteworthy that in this case the high entropy expenditures necessary for retaining are underrated [3347].

The contribution of the concentration and orientation effects influences the level of free energy of the ground state (Fig. 123). The system seems to move throughout the reaction pathway, thus lowering an activation barrier. This effect may be both entropic and enthalpic.

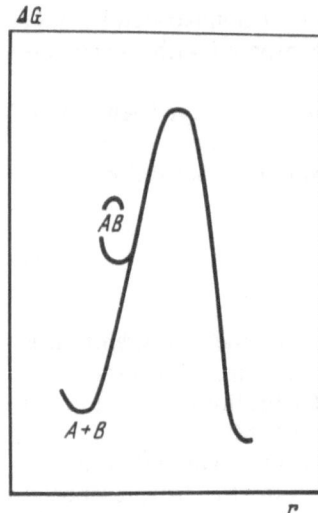

Fig. 123. Energy profile of the reaction demonstrating moving of the reactants throughout the reaction coordinate upon the complex formation between them

8.2.2 Orbital Steering

The investigation of intramolecular reactions of lactonization and thio-lactonization has yielded a consensus that correct orientation of orbitals of the reactive atoms is rather significant [3362–3365].

As shown in Fig. 124, the deviation of the reactive atoms from the ideal position (angle $\Delta\theta$) by 15° entails a lowering of the lactonization rate more

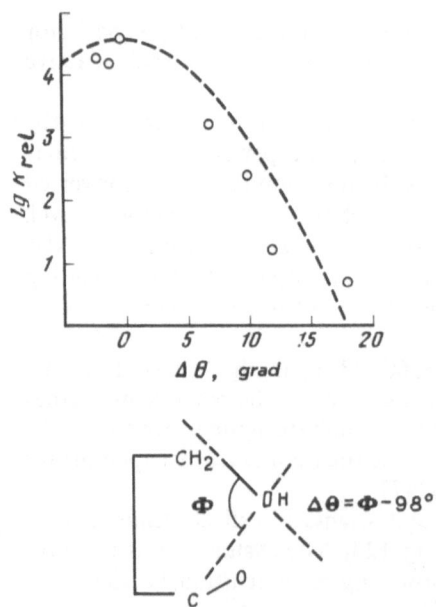

Fig. 124. Relative lactonization rate vs the degree of deviation from ideal location of reacting atoms [3363]

than 10^4 times. Hence a conclusion has been made that this is also possible at the enzymatic catalysis if the orbitals of the reactive atoms are correctly oriented in the ground state of the system.

This concept has been criticized [3366–3370] because: (1) to account for relative accelerations of 10^6–10^7 one assumes that the accuracy of orientation of the reactive atoms constitutes hundredths of a grade; (2) a chemical bond is insensitive to high angle deviations (distortion of bond C–C by $10°$ is related with energy expenditures of only 2.7 kcal/mol [1617]); (3) the differences in the rates of lactonization depend on the difference in the "rigidity" of the molecules so that the most reactive compounds show no losses caused by freezing of translational and rotatory degrees of freedom in the ground state.

8.2.3 Solvation Effects

The formation of the substrate-enzyme complex entails a change in solvation of the catalytically active substrate and enzyme groups. The change in the solvation shell influences much the nucleophilic reactivity of the charged groups [1404, 3371]. The reaction runs, as it were, in the gas phase. Thus it has been assumed [3372] that the conditions are similar to those for the gas phase in the enzyme active sites upon the substrate binding. This produces a high enhancement of the reaction rate (for example, [3373]).

In water, where the charge can be delocalized by the formation of hydrogen bonds, the anions are more stable than in the aprotic solvent. Therefore, in the aprotic medium they add a proton or form transition states more effectively. The differences in free energies of the ground states of ions in water and in nonaqueous solvents may cause the differences in the reaction rate, which reach about 17 orders of magnitude [3374]. Because enzymes changes in solvation can hardly produce high effects. Though desolvation of an anion enhances its nucleophilicity, it also increases its basicity. The ratio of nucleophilicity and basicity actually does not change [2353]. Besides, the loss of desolvation water can be compensated by reorientation of dipoles of the groups of the protein molecule close to the desolvated charge.

An unfavorable change in the free energy at desolvation and the charge appearance in the aprotic medium are partially compensated due to an increase in translational degrees of freedom, because here a few water molecules are replaced.

Thus solvation influences the level of the energy of the ground state of the system, moving it throughout the reaction to the transition state. This is true if the transition state has a lesser charge than the ground one. For a series of hydrolytic reactions the transition state has a higher charge, and the reaction in the aprotic medium runs slower than in water. Hence the significance of the charge compensation mechanisms in the transition state is quite obvious. The effect of the medium apparently makes a negative contribution to the efficiency of the catalysis by hydrolytic enzymes, but it can be compensated by "stabilization" of charges during the formation of the transition state [3352].

The solvation state of the surface of the protein globule even in the region far apart from the active site, is of essential value for the enzymes catalytic efficiency [3375–3377].

As shown, the ligand binding alters the volume of activation (ΔV^{+}) of the chemical conversion of the substrate, so that a linear correlation between the rate and the volume of activation exists [3375]. Let us suppose that the changes in the protein conformation alter the volume of the protein molecule as well as the exposure of the side chains and amide bonds of the polypeptide backbone in solution. In turn, this alters the hydrate shell of the protein, which may increase the entropy here. If the rearrangement in the hydrate shell affects the rate-determining stage, the above changes might lead to a change in the free energy of activation of the chemical conversion of the substrate. Thus the equation for the rate of the chemical reaction is [3375]

$$k = \frac{kT}{h} e^{-E^{+}/RT} e^{-P\Delta V^{*}/RT} e^{-\Delta S^{+}/R} . \tag{5}$$

So a decrease in the volume of activation at constant pressure entails an exponential increase in the rate constant.

These phenomena apparently play a crucial role in the reaction with a positive change in the entropy of activation [3377], and are also essential for understanding the nonspecific action of ions on the catalytic activity and temperature adaptation of enzymes in various organisms [3376].

As shown [3378], there exists a minimum level of hydration of the protein for a display of its catalytic activity.

8.2.4 Electrostatic Effects

These effects are closely related to those of solvation and already have been partially discussed in the previous section. Here it is important to describe the stabilization itself of the charge on the enzyme and its contribution to catalysis [2939, 3379–3382]. The reactions in aqueous solution are actually insensitive to electrostatic stabilization of the charge [3383], since in media of high dielectric permeability the energy of interaction of two oppositely charged particles is small (for proton and electron at a distance of 3.3 Å $\Delta H = -1.3$ kcal/mol). However, according to calculations for catalysis by lysozyme [2939] in the apolar medium, stabilization of the carbonium ion (reaction intermediate product) by a carboxylate ion of residue Asp52 is 9 kcal/mol, and in serine proteases the stabilization of the hydroxy anion of a tetrahedral intermediate is 30 kcal/mol [3381].

It should be noted that this effect stems from an indirect interaction of the charges as well as the interaction between a charge and permanent and induced dipoles of the enzyme [3380]. The concepts on stabilization of the transition state due to electrostatic interactions have been developed in a series of studies [3190, 3257, 3382]. However, if the information on possible nonadiabatic "tunnel" transfer of proton is reliable (see Sect. 8.2.7), the idea of the transition state charge itself seems rather vague.

Microheterogeneity in dielectric permeability is of much importance for enzymes; this makes possible interactions of the charge with water surrounding, and also preorganization of the polar medium – highly polar groups in the protein are fixed and cannot alter their orientation under the action of the external field. The charge appearing inside the protein globule, for example in the course of the substrate transformation, cannot effectively rearrange the protein "medium" due to fixation of dipoles. However, the outer medium (water around the globule) can be effectively polarized by this charge.

Calculations show that the energy of the medium rearrangement by the charge buried in the protein globule is less as compared with the charge located as a whole in the aqueous phase. This may be substantial for lowering the energy of activation (see reviews [3384, 3385]).

8.2.5 The Substrate Distortion

Any change in the bond length or valent angle in the reactive group of the substrate entails its destabilization and enables the break of the bond. In the course of the chemical reaction these parameters essentially change. If the enzyme active site has a rigid "construction", the substrate binding due to interactions with the regions remote from the reactive group may induce distortions of the reactive group itself. This results in the moving of the system to the transition state. If the active site is conformationally mobile and the substrate is "rigid", the latter can induce a distortion (probably energetically unfavorable conformational changes) in the enzyme. These concepts underlie the majority of modern theories and hypotheses of enzymatic catalysis, which will be thoroughly described later. Here we want to note that the concept of the substrate distortion has first found experimental confirmations by the studies on X-ray analysis of lysozyme and its complexes with inhibitors [3379, 3386]. Further investigations have revealed [3387] that binding of a sugar residue is not associated with the energy expenditures in the catalytic region of this enzyme (region D). All this does not indicate the absence of distortion, but can occur due to the compensation of energetic expenditures by other energetically advantageous factors.

Recall from Sect. 6.5.2 and 7.3 that the inhibitor binding by amide hydrolases also entails a distortion of a potentially hydrolyzed amide bond due to rotation around bond C–N and pyramidalization of the amide group.

The distortion mechanism of catalysis can be also illustrated by *cis-trans* isomerism of certain azo dyes, which is not actually accelerated by low-molecular substances, but strongly catalyzed by such proteins as serum albumin [3388]. According to spectral methods, the azo group is distorted in the dye-protein complex towards the transition state of *cis-trans* isomerization.

Unusual spectral characteristics of numerous metal-containing proteins and their ligand complexes show that the metal atom here is in a special altered state (as regards the distribution of the electron density) termed "entatic state" [3389, 3390]. It is supposed that this state has important advantages relative to the catalytic efficiency.

The spectral methods have defined polarization of the substrate carbonyl groups in citrate synthase [3391, 3392], yeast aldolase [3393] and triozophosphate isomerase [3394].

To distinguish between the terms "strain" and "stress", it has been proposed [1613, p. 342] to use the term "strain" only to designate the physical distortion (a change in lengths of bonds, valent angles, etc.), while the term "stress" for the states with an increased, às compared to normal, free energy. The reason for the stress can embrace solvation and electrostatic effects. It is not always easy to trace the distinction to strain or stress.

8.2.6 Polyfunctional Catalysis

The involvement of several functional groups in catalysis can make a substantial contribution to the efficiency of the process due to concentration and orientation. Simultaneous involvement of certain catalytic groups in one and the same enzyme decreases the reaction order as compared with the reaction in solution, the reaction order (n) and a change in entropy of the activation of the process being related by the ratio [1578]:

$$- T\Delta S^{+} = n.4.5 \text{ kcal/mol} .$$

Therefore, the involvement of two groups and one substrate in the enzymatic catalysis could lower the entropy of activation, as compared with a relevant trimolecular process, by more than 13 kcal/mol. In fact, such an acceleration is not observed really. One of the possible reasons is that the formation of the transition state from a higher number of interacting atoms demands a realization of synchronous vibrations; however, their probability lessens with an increase of the number of atoms in the system (see Sect. 3.4.8). The problem of the existence of the concerted general base and acid general catalysis in enzymatic reactions has not been tackled yet (see [1404, Chap. 3]).

8.2.7 Quantum-Chemical Effects

Usually, chemical reactions in solution and gas phase follow the Frank-Condon principle, i.e., the motion of atomic nuclei occurs much slower than that of electrons, and at each position of nuclei the electrons move so that the nuclei remain fixed.

For the majority of simple reactions of the electron transfer the adiabatic run is probable when at the coordinates of heavy nuclei close to the activation barrier the electron orbitals of the two reagents form the single system, and the electron at each precise moment possesses the minimum energy. But if the electron transfer occurs with difficulty, the single therm is not formed, and the electrons start tunneling. A typical quantum behavior is characteristic also of the proton [1512, 3384, 3395]. It behaves not like a classic particle (with frequency $\omega \ll kT/h \approx 1 \times 10^{-12} \text{ s}^{-1}$), but as a quantum particle (with $\omega \gg kT/h$). This means that the proton under certain conditions enables

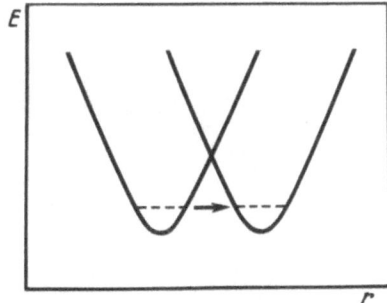

Fig. 125. Proton transfer by tunneling at equilibrium of electron terms of donor and acceptor atoms

underbarrier, tunnel transition. The vicinity of electron therms of atoms, where a proton transfer occurs, predetermines this process (Fig. 125). The substrate binding by the enzyme may result in rearrangement of the "medium" of the active site, i.e., in reorientation of dipoles so that the electron levels of atoms at the proton transfer will be aligned. The underbarrier transfer will proceed here much faster than the "classic" one. This, in turn, can enhance the efficiency of catalysis, when the proton transfer is a rate-determining stage. The studies on the temperature dependence of the kinetic isotope effect of the solvent in chymotrypsin and trypsin catalyzed reactions (see Sect. 5.5) accord with these concepts quite well.

8.3 Theories and Hypotheses. Destabilization of the Ground State

Theories and hypotheses of the enzymatic catalysis derived from the idea about "destabilization" of the ground state are widespread now and go back to the notion of catalysis introduced by Fischer. The term "destabilization" means here an increase of the free energy of the ground state stipulated by the many reasons discussed below.

Note that all the models based on destabilization of the ground state somehow or other arise from the concept that enzymatic and model reactions pass through one and the same transition state.

8.3.1 Nonproductive Binding

As early as 1892 Tamman [3396] spoke about the difference between enzymatic hydrolysis and hydrolysis under the action of acids, which covers the selectivity of the former and nonselectivity of the latter. Fischer has elaborated these ideas [3397] in his famous "lock and key" concept.

According to this idea, the enzyme has a restricted region (called now the active site) of a rigid configuration, which can bind only the substrate complementary to it by its configuration (Fig. 126). Thus the enzyme can bind and transform quite a distinct substrate or a group of rather similar substances.

E   S   ES

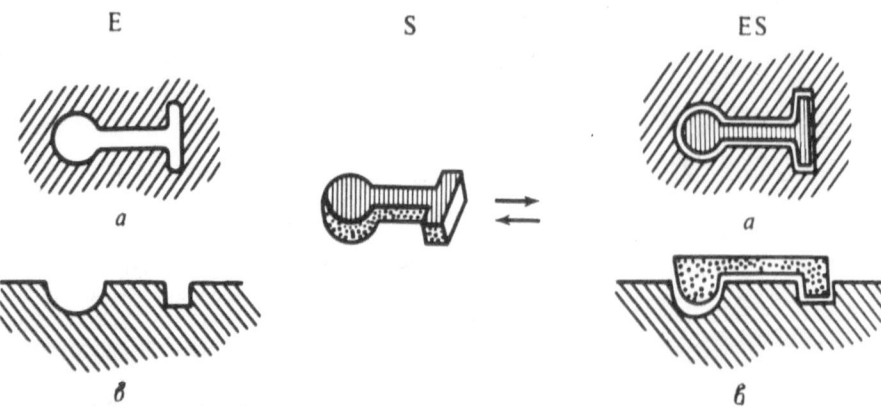

Fig. 126. Illustration to concept "key-lock"; *a* view from above; *b* view from the back side

Actually, the efficiency of the enzyme has been related to the complex formation, as discussed in Sect. 8.1.

Fischer's ideas have been developed by Bergmann and Fruton in their
multisite attachment theory [3398] and then by Ogston [3399]. The concepts
became known as the theory of three point attachment and gave an explanation of enzyme stereospecificy.

The 1950s are famous for the ideas about nonproductive binding [1717,
3400–3402]. They imply that the rate of the enzymatic reaction depends on
how "properly" the substrate binds in the active site. A "poor" substrate can
bind more effectively in the "wrong" way, thus impeding the formation of the
productive enzyme-substrate complex:

$$ES_n \; \underset{}{\overset{\bar{K}_n}{\rightleftharpoons}} \; E + S \; \underset{}{\overset{\bar{K}_p}{\rightleftharpoons}} \; ES_p \; \overset{k_{2p}}{\longrightarrow} \; \text{products} \,, \tag{6}$$

where $\bar{K}_p$ is the association constant of the productive complex (ES_p) and
$\bar{K}_n$ the association constant of the nonproductive complex (ES_n).

These have been developed by Hein and Niemann in their "tetradental"
model of specificity of chymotrypsin [1718, 3403, 3404].

According to this model, the enzyme has four tetrahedrally located loci:
ρ_1, ρ_2, ρ_3, and ρ_H being complementary as regards the substituting groups of
the substrate in *L*-configuration (Fig. 127).

The enzyme loci ρ_1, ρ_2, and ρ_3 accept the groups NHCOR', the side chain,
and the cleavable bond of the substrate, while locus ρ_H has a "negative
specifity", i.e., it cannot bind groups more bulky than the proton.

This model can embrace 16 interactions resulting in 12 complexes. Of them
productive are those that realize a proper interaction with loci ρ_3 and ρ_H.
Relative stability of complexes here depends on individual R-ρ-interactions,
while the most intrinsic position in the theory is postulating additivity of all
interactions.

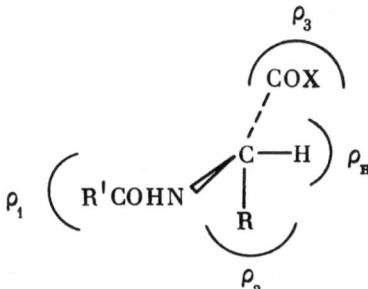

Fig. 127. "Tetradentate" model of interaction of the substrate in L-configuration with the chymotrypsin active site

The rate of the enzymatic reaction in the Hein-Niemann model is written by equation (Scheme 6):

$$v = \frac{\sum\limits_{p} \frac{k_{2p}\bar{K}_p}{\bar{K}_0}[E]_0[S]_0}{1/\bar{K}_0 + [S]_0}, \tag{7}$$

where k_{2p} is the constant of the chemical conversion of the productive complex p, $\bar{K}_p$ its association constant, $\bar{K}_0 = \sum_n \bar{K}_n + \sum_p \bar{K}_p$ is a sum of the association constants of nonproductive and productive complexes. For each complex the association constant due to additivity is equal to the product of individual (microscopic) constants of R-ρ-interactions:

$$\bar{K}_a = \prod_{ij} k_{ij}. \tag{8}$$

Microscopic constants due to symmetry of the model can be written as matrix

$$\begin{vmatrix} k_{11} & k_{12} & k_{13} & k_{1H} \\ k_{21} & k_{22} & k_{23} & k_{2H} \\ k_{31} & k_{32} & k_{33} & k_{3H} \\ k_{H1} & k_{H2} & k_{H3} & k_{HH} \end{vmatrix} = \bar{K}_{0(L)} - \bar{K}_{0(D)}, \tag{9}$$

where indices 1, 2, 3 and H are related to the interactions of R_1, R_2, R_3 and H to ρ_1, ρ_2, ρ_3 and ρ_H. In an algebraic form of this matrix positive members correspond to the association constants of the L-isomer of the substrate, while negative ones to the association constants of D-isomer. Microscopic constants (Table 81) are selected empirically, and condition (9) is fulfilled by the method for sequential approaches. It is employed to calculate the association constants of some amides of N-acylamino acids with chymotrypsin.

Formula (7) shows that the catalytic constant is

$$k_{cat} = \frac{\sum\limits_{p} k_{2p}\bar{K}_p}{\sum\limits_{n} \bar{K}_n + \sum\limits_{p} \bar{K}_p}. \tag{10}$$

In the simplest case, if one productive and one nonproductive complex are

Table 81. Microscopic constants of binding of the compounds of type $R_1R_2CHR_3$ [3403][a]

Substituent	i	k_{ij}			
		$j=3$	$j=2$	$j=1$	$j=H$
H	H	0.826	0.826	0.826	0.826
$CONH_2$	3	0.0371	0.007	0.037	0.479
$C_6H_5CH_2$	2	1.22	72.9	0.84	0.282
$p\text{-}HOC_6H_4CH_2$	2	1.08	63.3	0.649	0.301
3-$IndCH_2$	2	60.8	340.2	24.6	0.136
CH_3CONH	1	3.67	0.727	0.788	0.400
$\alpha\text{-}C_5H_5NCONH$	1	3.97	0.795	1.59	0.333
$ClCH_2CONH$	1	5.9	12.7	0.731	0.350
CF_3CONH	1	1.9	1.0	1.3	0.400
$C_2H_5OOCCONH$	1	1.38	1.37	5.28	0.300

[a] Conditions: pH 7.9; 25 °C; 0.1 N NaCl.

available, the expression for k_{cat} is

$$k_{cat} = \frac{k_{2p}\bar{K}_p}{\bar{K}_n + \bar{K}_p} = \frac{k_{2p}}{\dfrac{\bar{K}_n}{\bar{K}_p} + 1} .$$ (11)

Thus, if productive binding prevails ($\bar{K}_p \gg \bar{K}_n$), the apparent catalytic constant is equal to that of the chemical conversion of the productive complex ($k_{cat} = k_{2p}$).

Nonproductive complexes lower the magnitude by factor $\bar{K}_p/\bar{K}_n + \bar{K}_p$.

It is clear that $\bar{K}_n$ makes a contribution to the free energy of the ground state:

$$\Delta G_{cat}^{\ddagger} = (\Delta G_{2p}^{\ddagger} + \Delta G_p) - \Delta G_n .$$ (12)

According to Eq. (12), the productive binding would "destabilize" the ground state of the reaction (ΔG_p and ΔG_n are negative values) that is outlined in Fig. 128.

Prediction of the k_{cat} values in the Hein and Niemann theory seems possible if choice of the k_{2p} and definition of values of $\bar{K}_p$ and $\bar{K}_n$ are right when using the values of microscopic constants. It is supposed that k_{2p} is much higher than the rate constant of the relative nonenzymatic reaction, though the theory does not account for the reasons.

Note that this concept is employed only for a restricted number of substrates. But it does not outline the role of secondary interactions, and the magnitude of microscopic constants is established by a set of the substrates utilized at this precise moment. An increase of experimental data in number will significantly alter these values.

Besides, as all models of the nonproductive binding, the Hein–Niemann model does not account for the difference in the second-order rate constant (k_{cat}/K_m) for a diversity of the substrates.

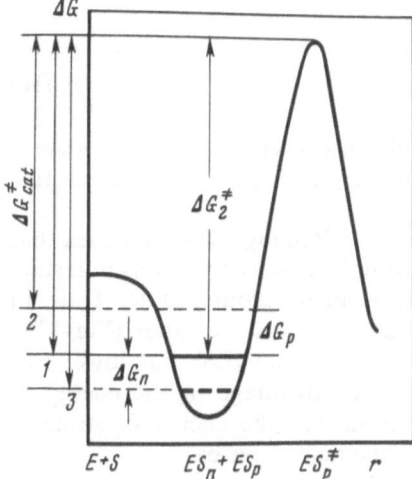

Fig. 128. Energy profile of enzymatic reaction demonstrating the role of nonproductive binding

In accordance with Eq. (7), for $p = 1$

$$\frac{k_{cat}}{K_m} = k_{2p}\bar{K}_p,$$ (13)

i.e., the second-order rate constant does not depend on the level of non-productive binding.

The ideas of nonproductive binding applied also to the distinction of the rates of hydrolysis of intermediate acyl enzymes [1328, 3405]. As known, acyl chymotrypsins formed by D-amino acids undergo hydrolysis very slowly: the more specific is the amino acid, the slower will be the cleavage. Apparently, the acyl residue in D-acylamino acylchymotrypsins is located mainly in a wrong way, but to explain these data better, the notion of strain or induced fit is introduced (see below).

8.3.2 Theory of Strain

According to Scheme 1 (see Sect. 8.1)

$$\frac{\bar{K}_{ES}K_{ES}^{\ddagger}}{\bar{K}_T K_s^{\ddagger}} = 1 \quad \text{and} \quad \frac{\bar{K}_T}{\bar{K}_{ES}} = \frac{K_{ES}^{\ddagger}}{K_s^{\ddagger}},$$ (14)

where $\bar{K}_{ES}$ and $\bar{K}_T$ are the association constants of the enzyme with the substrate and "transition state", respectively, while $K_{ES}^{\ddagger}$ and $K_s^{\ddagger}$ are the equilibrium constants for the formation of the activation complexes in the catalyzed and noncatalyzed reactions. Whereupon, according to the theory of the absolute rates of reactions

$$k = \frac{kT}{h}K^{\ddagger},$$ (15)

hence

$$\frac{\bar{K}_T}{\bar{K}_{\mathrm{ES}}} = \frac{k_f}{k_n}. \tag{16}$$

This means that catalysis ($k_f \gg k_n$) is possible only when the transition state is bound to the enzyme more effectively than the ground state of the reactive system (i.e., $\bar{K}_T \gg \bar{K}_{\mathrm{ES}}$).

All this gave Haldane [1744, p. 182] and also Pauling [2843] the idea that the active site of the enzyme is structurally complementary to a transition state of the reactants rather than to the substrate in its ground state. Thus, in binding the substrate, an enzyme strains and distorts the latter (Fig. 129). A hypothetic enzyme-substrate complex, free of strain (ES), has lower free energy than the strained complex (ES*). The advantage in the energy of activation corresponds to the energy of strain, i.e., the energy of strain is expended for destabilization, thus lowering the activation barrier.

The concept of strain has been evolved from a postulate of geometric distortion of the substrate ("rack" theory) [1375, 2449], leading to the interpretation of strain in the course of desolvation, electrostatic and entropy effects [9, 1404].

It is important that all effects in the concept of strain are observed in the ground state. Further development of this idea has resulted in the conviction that they are of more importance for the transition state than for the ground state (see Sect. 8.4).

The arguments in favor of strain or distortion in the enzyme-substrate complexes have been already discussed in Sect. 8.2.5. The concept of strain accords well with a series of experimental data, though facing certain problems. Strain evokes a change in the free energy of the ground state, i.e., it increases K_m and k_{cat} in the same degree. Thus the value of k_{cat}/K_m designating the enzyme specificity must not change. It is quite clear from Fig. 129, where does not the value of the activation barrier, as regards the initial state of the

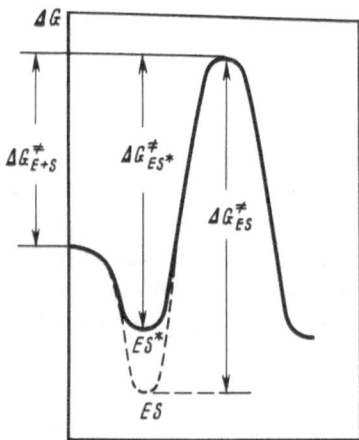

Fig. 129. Energy profile of enzymatic reaction demonstrating the role of strain or deformation in the ground state

free enzyme and substrate, depend on the strain in the ground state of the enzyme-substrate complex. This does not fit experimental data on the kinetics of hydrolysis catalyzed by many proteases. Very often a change in the length of the peptide substrate does not influence the value of K_m, but changes k_{cat} and consequently k_{cat}/K_m (see Sect. 4.3.1.3).

8.3.3 Induced Fit

If the energy of the substrate binding to the enzyme by the concept of strain is expended to distort the substrate, in the induced fit theory the energy is expended to alter the enzyme [3019, 3352, 3406]. Let us suppose that a proper substrate effectively changes the conformation of the enzyme to an energetically unfavorable but catalytically more active structure. Various substrates initiate a different "conformational response" of the enzyme, thus determining their ability for transformation.

A kinetic scheme of such a process can be written as

$$
\begin{array}{ccccc}
E_p & \underset{}{\overset{K_s}{\rightleftharpoons}} & E_pS & \xrightarrow{k_2} & \text{products}\\[4pt]
K_p \Big\updownarrow & & \Big\updownarrow K_p^* & & \\[4pt]
E_n & \underset{}{\overset{K_s'}{\rightleftharpoons}} & E_nS & &
\end{array}
\tag{17}
$$

where E_p is a catalytically active form of the enzyme, E_n its inactive form, $K_p = [E_p]/[E_n]$ and $K_p^* = [E_pS]/[E_nS]$. Thus between the two forms there exists an equilibrium shifted to the inactive form (E_n), since it is energetically preferential ($K_p \ll 1$), and the substrate alters the equilibrium in favor of the complex with the active enzyme, so that $K_p^* \gg K_p$.

The conversion rate of the substrate is defined by equation:

$$
v = \frac{\dfrac{k_2 K_p^*}{1 + K_p^*}[E_0][S_0]}{K_s \dfrac{K_p^*}{K_p(1 + K_p^*)} + [S_0]}
\tag{18}
$$

(taking into account $K_p \ll 1$).

Hence, the observed second-order rate constant (k_{cat}/K_m) is

$$
\frac{k_{cat}}{K_m} = \frac{k_2 K_p}{K_s},
\tag{19}
$$

i.e., this constant and, consequently, the enzyme specificity do not depend on the substrate structure, as the value of K_p characterizes an equilibrium between the two forms of the *free* enzyme.

According to Fig. 130, the higher is the difference in the energy of the two states of the free enzyme (when K_p is less), the more will K_m and the less k_{cat}/K_m

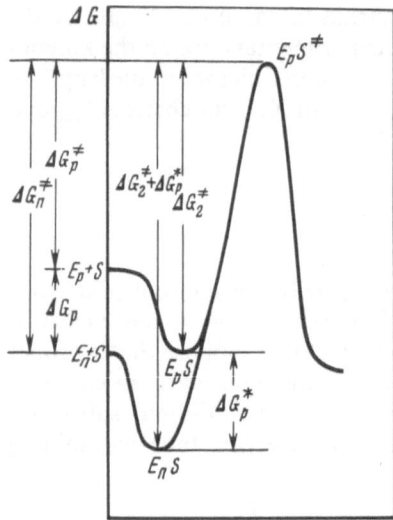

Fig. 130. Effect of induced fit on free energy of ground and transition states of the enzymatic reaction

be. If the enzyme itself is in the active conformation, the expression for K_m in the denominator of formula (18) is

$$K_m = K_s \frac{K_p^*}{1 + K_p^*},$$ (20)

and $k_{cat}/K_m = k_2/K_s$, i.e., it has the same magnitude as without an induced fit.

Though the induced fit theory generally does not shed light on the problem of the efficiency and specificity of catalysis, cases have been analyzed when the specificity can be related to the induced substrate via the conformational changes [3407]. No doubt, induction of conformational changes by a substrate in the enzyme is real. These changes can play a positive role in catalysis, thus enhancing orientation factors and changing solvation or inducing the substrate distortion. The importance of the theory was that it attracted the attention of scientists the first time to the role of conformational changes of the enzyme in catalysis.

Besides, the induced fit theory is applied to account for the allosteric process. In the Koshland-Nemethy-Filmer model [3408] the subunit protein without ligands exists in one conformation, while ligand binding induces conformational changes transmitted to the other subunit. This model accounts well for the peculiarities of the binding of oxygen by hemoglobin.

8.4   Stabilization of the Transition State

As previously established, none of the concepts considered explains a decrease in the activation barrier of enzymatic reaction (k_{cat}/K_m) relative to that of the nonenzymatic process, if it is assumed that both run through the same transition state. The above mechanisms only change the level of the ground state of the enzymatic reaction. Evidently, to explain enzyme efficiency and specificity, the transition states of enzymatic and nonenzymatic reactions should be considered as different.

Let us suppose that substrate bound to an enzyme realizes only some of the possible interactions, because the enzyme is not complementary to the substrate ground state (see Sect. 8.2) [1613]. This occurs only in the transition state, when substrate utilizes its ability of bonding to the enzyme, which has not been employed yet.

Besides, Fig. 131 shows that the free-energy level of the ground state increases by ΔG_R as compared with the level of complete complementarity, and the free-energy level of the transition state falls by the same value.

To evaluate the catalyst efficiency, let us consider Scheme 1 once again. Here values of ΔG_{ES} and $\Delta G_{ES}^{\ddagger}$ ($\Delta G_{ES'}$ and $\Delta G_{ES'}^{\ddagger}$) can be expressed as

$$\Delta G_{ES'} = \Delta G_{tot} - \Delta G_R \quad \text{and} \tag{21}$$

$$\Delta G_{ES'}^{\ddagger} = \Delta G_{ES}^{\ddagger} - \Delta G_R , \tag{22}$$

where ΔG_{tot} stands for the total energy of enzyme-substrate interaction. In accordance with Eq. (2) one can write

$$\Delta G_{ES'}^{\ddagger} - \Delta G_{ES'} = \Delta G_S^{\ddagger} - \Delta G_T . \tag{23}$$

If we place Eqs. (21) and (22) in this equatioin and take into account that $\Delta G_{ES}^{\ddagger} = \Delta G_S^{\ddagger}$, we obtain $\Delta G_T = \Delta G_{tot}$.

Thus, a decrease in the transition state energy of the enzymatic reaction as compared to the congruent model reaction is equal to the overall energy of the enzyme-substrate interaction. The evaluation of this energy for acylamino acid derivatives (see Sect. 6.6) yields values up to 25 kcal/mol. This is enough, and even more than necessary, to cover the differences in the rates of enzymatic and nonenzymatic hydrolyses.

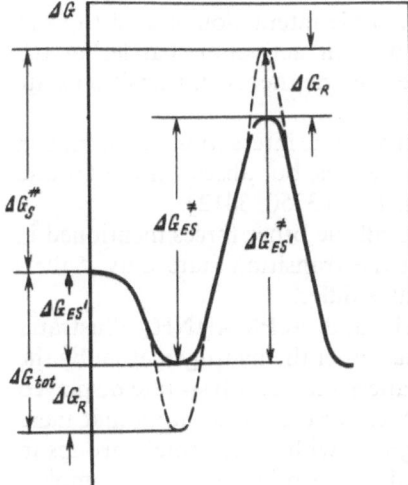

Fig. 131. Energy profile of the reaction if the enzyme is complementary to the substrate transition state

Such a mechanism, unlike those mentioned above, accounts for the alteration of the second-order rate constant in a series of substrates (specificity) and relative to the congruent model (efficiency).

The same may be derived from the concept of strain, if the ground state strain is relieved when the system reaches the transition state.

The concept of stabilization of the transition state is supported by the following arguments: (1) the evaluated overall energy of the enzyme-substrate interaction usually exceeds the observed change (in K_s) of the free energy of the complex formation; (2) there are many examples of the catalytic constant changing in a substrate series at practically constant K_s (see Sect. 4.3.4). Pepsin-catalyzed hydrolysis of peptides is typical here, when K_m (in this instance equal to K_s) changes not more than ten fold, and k_{cat} increases 10^5-fold at transition from di- to tetrapeptides; (3) site-directed mutagenesis of the groups involved in the substrate binding often affects catalysis (k_{cat}), but not binding (K_s). Tyrosyl-tRNA synthetase has been thoroughly studied in this respect [1613, 3409–3411]; (4) the so-called transition state analogs (see Sect. 5.8.7) have, as a rule, extremely low K_i. However, as mentioned above, this interpretation is not unequivocal.

As regards the catalysis by chymotrypsin and some other proteases these ideas have been formulated as an extraction-conformation model of their action (see review [1963]).

The basis of this model is that for substrates of chymotrypsin – esters of N-acylamino acids – a change in the free energy at enzyme acylation ($\Delta G_2^{\ddagger}$) and deacylation ($\Delta G_3^{\ddagger}$) corresponds to that at transfer of the substrate side chain from water to nonpolar surrounding (ΔG_{ex}^R):

$$\Delta G_2^{\ddagger} = \Delta G_3^{\ddagger} \approx \Delta G_{ex}^R . \tag{24}$$

At the same time in the bimolecular process (k_{cat}/K_m) the following equation is observed:

$$\Delta G_s^{\ddagger} \approx 2\Delta G_{ex}^R . \tag{25}$$

This indicates that the energy of hydrophobic interaction is used to form the enzyme-substrate complex, and to lower an activation barrier of the chemical state, which has been well predicted by the theory of transition state stabilization.

The effect is explained by additional hydrophobic interactions appearing in the transition state due to conformational changes, but absent in the ground state of the enzyme-substrate complex (Fig. 132) [3350, 3412].

In addition to hydrophobic interactions, all the other forces mentioned in Sect. 8.2 can contribute to stabilization of the transition state only if their efficiencies in the ground and transition states differ.

Hydrolysis of dipeptides AcPheGlyNH$_2$ and AcPheAlaNH$_2$ illustrates quite well these interactions [2011]. Substitution of the hydrogen atom by the methyl group (Gly → Ala) dramatically enhances the hydrolysis rate displayed in k_{cat}, but not in K_m. Analysis of the X-ray enzyme models shows that there are unfavorable interactions of the methyl group with some protein groups in the ground state of the chymotrypsin-alanine dipeptide substrate complex,

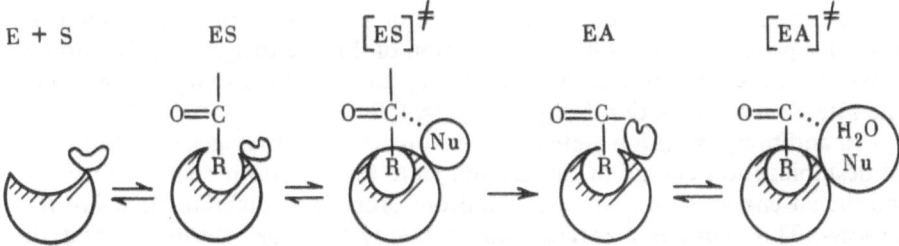

Fig. 132. Mechanism of appearance of additional hydrophobic interactions of the enzyme and substrate at the transition state formation [3350, p. 157]

absent at the tetrahedral intermediate formation. This result can be interpreted in terms of the concept on transition state stabilization as well as by the theory of strain including transition state stabilization as a result of a relief of the strain of the ground state.

Transition state stabilization in serine proteases implies the major contribution of hydrogen bonds evaluated in [3413]. Two stabilization mechanisms are proposed: one of them considers involvement of three hydrogen bonds formed by the transition state, though absent in the Michaelis complex [1330]. These are the bonds formed by the carbon oxygen atom to NH groups of Ser195 and Gly193 and the bond of the substrate acylamino group to the Ser214 carbonyl. The electrostatic field created by ion pair Asp102-His57 possibly stabilizes the tetrahedral transition state in serine proteases [3190, 3257]. The other mechanism [2011, 3414] implies the same bonding in the Michaelis complex and in the transition state. These bonds do not contribute to the enthalpy of the reaction activation, but essentially alter the entropy of the transition state. The second mechanism has been experimentally supported [3415].

The theory of transition state stabilization accords well with experimental results, though it has serious shortcomings.

The most dramatic hindrance is that transition state stabilization requires new interactions between the enzyme and substrate, which are absent in the ground state of the enzyme-substrate complex. This means that the transition state formation should be accompanied either by conformational changes in the enzyme [1963] or, at least, by reorientation of a number of enzyme groups. However, the transition state formation, i.e., the reaction elementary step is fluctuating within 10^{-12}–10^{-13} s [3416]. This time is not enough for colliding particles to exchange their energy, and the overall energy of atoms remains unchanged during the elementary step. Evidently, the conformational changes in the protein with rate constants of about 10^6–10^4 s^{-1} cannot take place *during* the elementary step.

Probably, the groups responsible for the transition state stabilization seem to "oscillate" between the two states. If the elementary step takes place at a moment favorable for stabilization of the transition state of group positions, the latter is stabilized. Though in this case noncoordinated time intervals lower, in my opinion, the probability of stabilization.

The synchronization effect on the rate of a process can be evaluated by formulae given in Sect. 3.4.8. The formation of three hydrogen bonds requires a synchronous motion of at least six heavy atoms, which sharply diminishes the rate even at activation energy $E_{\text{synch}}^{+} \approx kT$.

In addition, transition state stabilization actually envisages the utilization of destabilization energy of the ground state (or the strain energy). In other words, an energy gain and its expenditure proceed at different stages of the process. This requires a mechanism of energy "storage" during a catalytic event.

Such a mechanism has been proposed in [3417, 3418]. A protein molecule is considered as a solid, nonhomogeneous by its "density". The lowest density has been ascribed to the active site area. According to this model, the energy of the substrate sorption is stored by a protein globule as elastic deformations and "leaks" into the active site area by the nondissipative way as a large number of phonons coherent by their frequency and phase. It is still obscure whether a protein can be modeled by an elastic solid.

Finally, the model can hardly explain the constancy of K_s values in many substrates strongly differing in the number of possible contacts with the enzyme (for instance, pepsin substrates). Why is a practically constant portion of the overall energy of interaction utilized in K_s values? It seems reasonable to suggest that the larger is the overall energy of the substrate-enzyme interaction in the ground state, the lower should be the portion used at the stage of the transition state formation. This mechanism is observed for acylamino acid esters – substrates of chymotrypsin, where both K_s and k_{cat} are proportional to hydrophobicity of the side chain. However, this is not the case for peptide substrates.

8.5   Enzyme – "Machine"

The theory of the activated complex applied to reactions of biopolymers has been considered in [2450, 3419–3422]. The theory and its empirical predecessors (Arrhenius and Vant-Hoff equations) have been thought to be inapplicable to these reactions and, presumably, to those in a condensed phase. The point is that any thermal fluctuation changes the structure of the "medium" – enzyme and its "construction". As a result, the process is kinetical nonequilibrium, so it is no use to speak about complementarity and an activation barrier. The change of the latter is associated with conformation and configuration transformations of a macromolecule, and the chemical conversion itself – with an increase in the number of energy-rich molecules, i.e., with a phenomenon not related with the state of a macromolecule. The enzyme macromolecule is a kind of machine, which integrates in space mechanical and statistical components and possesses certain predetermined degrees of freedom. Such machine is capable of utilizing internal energy to perform useful work.

All these underlie *a relaxation concept* of enzymatic catalysis, which implies that conformational changes of the enzyme-substrate complex induced by the substrate attachment are of the relaxation character accompanied by the chemical conversion of the substrate molecule.

According to this hypothesis, the substrate attachment causes local, very rapid (for the time of vibrational relaxation 10^{-12}–10^{-13} s) changes in the geometry and electron structure of the active site. The substrate and its surrounding exist in a new equilibrium state, whereas the whole globule remains in the same state. Thus sterical strain appears. Relaxation of the whole macromolecule to a new state takes milliseconds, and this is just the stage of the substrate catalytic transformation.

The concept is not in contradiction to experimental data on the electron transfer processes (see, for example, [3423]), though it is of little use to explain the efficiency and specificity of hydrolytic enzymes. Particularly, the rate constants of the relaxation stages, being in line with the relaxation hypothesis, can be decreased near the equilibrium position. But data on the rates of isotope exchange at the equilibrium for pepsin- [1732] and chymotrypsin- [1731]-catalyzed reactions do not support these predictions.

The rates of the ES conversion to EP measured in the reactions catalyzed by piruvate kinase [3424] under steady-state and equilibrium conditions are the same. Apparently, the enzyme increases a population of vibration-excited states of the enzyme-substrate complexes, the potential surface of the reaction being unchanged [3424].

The driving forces of chemical conversion are still not identified. Usually, the substrate-enzyme contact is rather weak, whereas a break of the chemical bond requires a large energy expenditure. Moreover, hydrolytic reactions start with the bond formation (attack of the oxygen atom on the carbonyl carbon), but not with its break. The relaxation concept is hardly relevant to these processes.

A idea of the presence of certain "constructive" (mechanical) degrees of freedom has been contemplated also by other investigators [3417, 3425]. The mode of the enzyme action has been interpreted as a result of the conformationally disequilibrium state of the enzyme and the enzyme-substrate complex after the first event of the substrate conversion and as being preserved without energy dissipation before the next event.

8.6 Quantum-Chemical Description of Enzymatic Reactions

The quantum-chemical theory of redox reactions [3426] has been extended to the reactions with chemical bond distortion [3427] and further to enzymatic processes [3384, 3395, 3428–3432]. The essence of the theory is that the activation process is a fluctuation change in orientation of the protein macromolecule dipoles not accompanied by bond stretching. Rearrangement of the active site region leads to redistribution of electron levels of reactive atoms, while approximately equal electron levels correspond to the transition state. In this state the underbarrier (tunnel) transfer of a proton is feasible.

Thus, the probability of the elementary step depends on the conformational state of a medium and the electron state of the enzyme's active site:

$$w = \sum_{\substack{\alpha, \alpha' \\ n, n'}} \Phi_{\alpha, n} W_{\substack{\alpha, \alpha' \\ n, n'}} , \tag{26}$$

where $\Phi_{\alpha, n}$ is the probability that before the reaction the medium is characterized by conformational state α and electron state n, and $w_{\substack{\alpha, \alpha' \\ n, n'}}$ is the probability of the system transfer to states α' and n'. The overall result of electron-conformational interactions is an electron reorganization, which forces a change of the macromolecule conformation. Interpretation of these phenomena by the theory of the solid body [3433] has given birth to ideas of the existence of a quasi-particle – "conformon" (analogous to polaron) – involving an electron and the macromolecule deformation with the energy lower than that of a free electron. This is the reason for the barrier lowering.

These accord with the so-called compensation (additivity) principle [1375], according to which a conformational energy change throughout the reaction coordinate is complementary to changes of the chemical free energy, so that the total change appears to be "smoothed".

An application of the mathematical apparatus of quantum mechanics to these concepts is beyond the scope of our book; the reader can find it in the above cited-references.

8.7 "Ideal" Enzyme

The enzyme catalytic action is shown to be dependent on the enzyme affinity to the substrate at various stages of a process. Three types of enzyme-substrate interactions can be distinguished by the affinity of the reaction product to the enzyme (Fig. 133) [3348, 3431]. The first of these is *a uniform* binding, when the

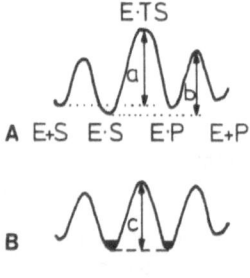

Fig. 133A–C. Three types of changes in free energy throughout the reaction at enzymatic hydrolysis [3348]; **A** uniform binding; **B** differential binding; **C** binding (catalysis) of elementary steps (see text)

enzyme does not discriminate the substrate, the transition state and the product, and binds them practically in a similar way. The second is a *differential* binding, when the enzyme binds differently the substrate and the product. The third is *a binding* (catalysis) *of elementary steps*, when the enzyme bind the transition state stronger than the substrate and the product. In the first case the optimum catalysis is observed at equal activation barriers of the transition state and dissociation of the enzyme-product complex ($a = b$). In the second, the catalytic efficiency is maximum at equal levels of the free energy of complexes ES and EP, i.e., when the equilibrium between "internal" complexes is about a unity. Finally, the "ideal" enzyme with the lowest energy barrier occurs when the transition state is stabilized. It is noteworthy that the thermodynamic analysis of these cases is true if actual concentrations in vivo are chosen as a standard state. Apparently, there exists enzymes corresponding to each binding type [3348]. β-Lactamase is an "ideal" enzyme, for example [3435]. A quantitative analysis of these cases and conditions of the congruence of free energies of "internal" stages is described in [3436].

8.8 Ground State of the Elementary Step of Enzymatic and Model Reactions

The enzymatic process is a multistage reaction proceeding through a series of sequential elementary steps. Every step changes the substrate state more or less, therefore the ground state of the very first elementary step is identical to that in a model congruent system. A comparison of activation barriers of the following steps with those of nonenzymatic reaction is possible only with a proper choice of identical or very similar ground states.

The above theories and hypotheses directly or indirectly have ignored this. So the difficulties arise when the efficiency and specificity of enzymes are explained by destabilization of the ground state of the enzyme-substrate complex, and they appear mainly because the barrier of a chemical stage in the absence of destabilization is accepted to be equal to a barrier of the chemical step of the nonenzymatic process. This is understandable, because by excluding any factor affecting the reaction rate we assume that the rates of the processes in the absence of the factor must be similar. Further more, by comparing the conversion rates of the substrate destabilized on the enzyme and those of the nonenzymatic reaction, the substrate state free of destabilizing factors is chosen as the ground state of the latter. Thus two elementary steps with different ground states are compared. *It is obvious that for evaluation of a decrease of the activation barrier of the elementary step of the chemical stage (and any other) the identical ground states of reactants must be chosen.*

This can be formalized as follows. If the substrate in state S_a is on the enzyme, one can assume that a portion of molecules in solution may be in the same state, i.e., there exists equilibrium

$$S_n \underset{}{\overset{k_n}{\rightleftharpoons}} S_a, \quad \text{where } K_n = \frac{[S_a]}{[S_n]} \ll 1 . \tag{27}$$

Forms S_n and S_a can have different reactivities. For the sake of simplicity, the reaction is expected to involve only the molecules in active form (S_a) with rate constant k_a. Then the observed rate of the *overall substrate* conversion is

$$v_n = k_a K_n [S]_0 . \tag{28}$$

Consequently, a choice of the identical ground state means that the k_a constant for a nonenzymatic process and the rate constant of the corresponding stage of a catalytic reaction are considered so as to compare their activation barriers.

8.9 Enzyme Does Not Decrease Activation Energy of the Elementary Step of the Chemical Reaction

The concept the author put forward here has been formulated jointly with M. Karpeisky and G. Yakovlev [3437]. It embraces the above ideas and develops those described in [1715, 3438]. Its major points are: (a) the enzyme does not decrease the activation energy of *an elementary step* of the substrate chemical conversion if substrate states identical for catalyzed and noncatalyzed reactions are chosen as the ground state; (b) the enzyme stabilizes a substrate state thermodynamically unfavorable but reactive in solution and forms a productive enzyme-substrate complex; (c) the productive enzyme-substrate complex is formed stepwise so that at each stage the enzyme structure is complementary to that of the substrate. Every stage proceeds under optimum conditions established when the previous stage is completed (see [3438]).

8.9.1 Description of the Model

Let us amplify the enzyme-substrate interactions (Fig. 134). As mentioned, a substrate state in solution, which is very similar in the structure and distribution of the electron density to that in a productive enzyme-substrate complex, can always be found. This state (S_a) is in an equilibrium with "inactive" form (S_n), the equilibrium being shifted to the latter ($K_n \ll 1$). The peculiarities of this state in the case of amide hydrolase-catalyzed reactions are discussed in Sect. 8.10.

At the first stage of the catalytic process the two substrate forms are capable of binding to the enzyme, yielding a nonspecific sorption complex ($E_0 S_n$ and $E_0 S_a$), the association constants of both substrate forms with the enzyme being virtually the same. This complex is formed due to a collision of a substrate molecule and the enzyme active site. The system loses here the translational degrees of freedom. However, rotation around single bonds of fragments of the substrate molecule and functional groups of the enzyme active site remains free, i.e., the substrate can be differently oriented relative to the

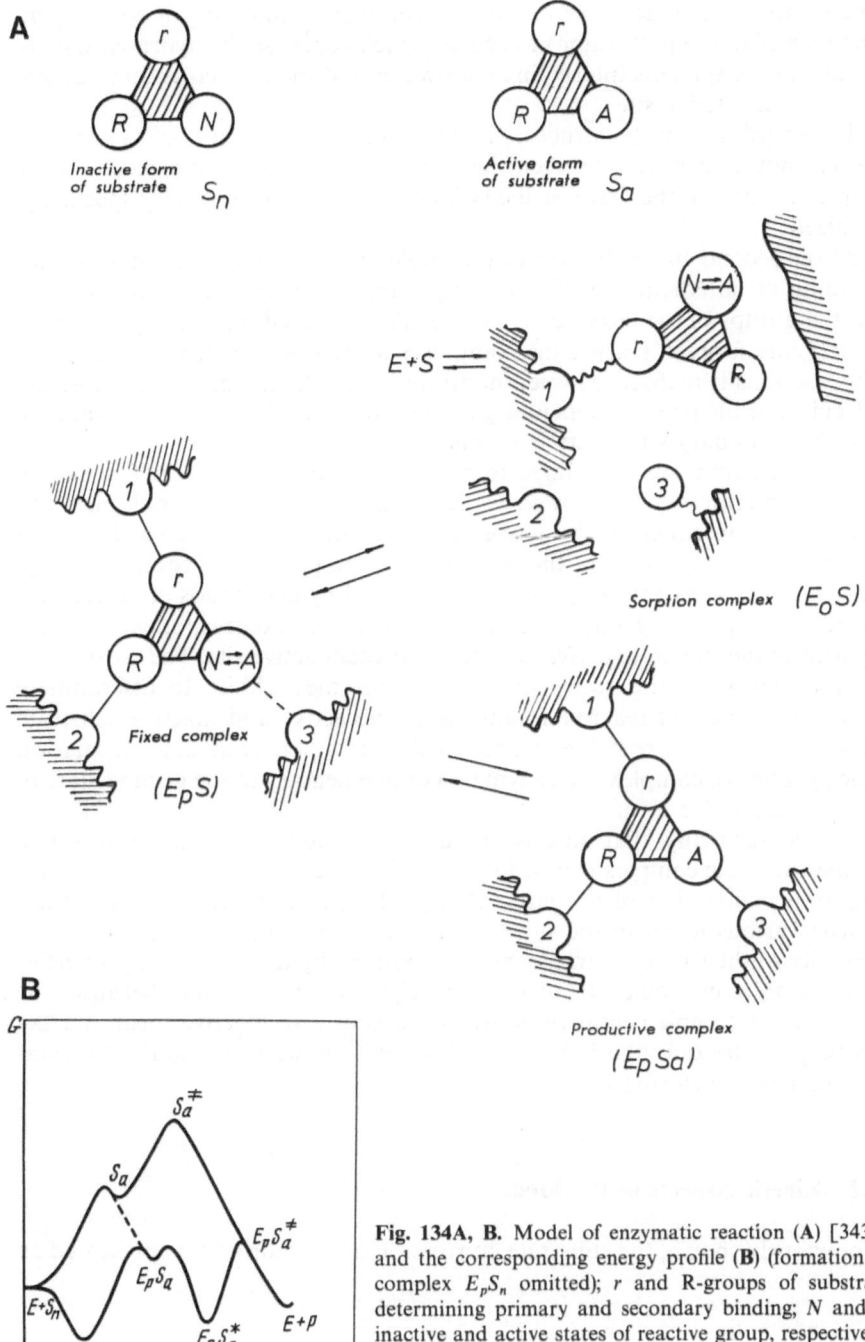

Fig. 134A, B. Model of enzymatic reaction (**A**) [3437] and the corresponding energy profile (**B**) (formation of complex $E_p S_n$ omitted); r and R-groups of substrate determining primary and secondary binding; N and A inactive and active states of reactive group, respectively; *1,2* primary and secondary binding sites of the enzyme; *3* catalytic sie of the enzyme

protein. An entropy decrease due to the complex formation can be compensated by replacement of water molecules, which compose the hydrated shell of the substrate fragments interacting with water and the molecules immobilized in the enzyme active site.

The sorption complex structure is characterized by mobility of its components, i.e., not all contacts between the substrate and the enzyme occur. At the complex formation the first specificity level of the enzyme (primary specificity) is utilized.

At the second stage, the sorption complex is fixed in a conformation thus providing an interaction of the reacting fragment of the substrate molecule with the groups of the enzyme active site. The so-called fixed complexes E_pS_n and E_pS_a are formed. During the complexes formation the maximum number of interactions is realized between nontransformed fragments of the substrate molecule and the protein, depending on a topochemical matching of reagents. Here the secondary specificity is manifested.

The fixed complex is assumed to make no contribution to the free energy caused by interaction of *the reactive fragment* of the substrate molecule with the protein. Consequently, the active substrate form has not been stabilized at this stage yet. This occurs at the next stage of the productive complex ($E_pS_a^*$) formation. The substrate active form is stabilized by new bonds of *the reactive* fragment with protein groups or due to a topochemical correspondence of this fragment of the substrate *active form* to the protein active site. The meaning of the stabilization of the active form is the enzyme's ability to discriminate between the states of reactive fragments of the active and inactive substrate forms or to form a more stable complex with the former. Thus, at the formation of the productive complex the enzyme is complementary to the ground state of the substrate active form.

It is noteworthy that this is rather conventional, and the productive enzyme-substrate complex can be formed from the complex with inactive form E_pS_n by isomerization of the latter ($E_pS_n \rightarrow E_pS_a^*$). In general, the difference between various forms of the substrate molecule regarding the structure and properties of their reactive fragments is stipulated by the structural properties of the whole molecule. Isomerization $E_pS_n \rightarrow E_pS_a^*$ occurs, because the topochemical complementarity between the substrate inactive form and the enzyme provides only the E_pS_n complex formation, whereas the $E_pS_a^*$ formation requires transformation $S_n \rightarrow S_a$.

8.9.2 Kinetic Aspects of the Model

As previously shown, the substrate conversion in solution can be presented as

$$S_n \xrightleftharpoons{k_n} S_a \xrightarrow{k_a} P \ , \quad \text{where } K_n = \frac{[S_a]}{[S_n]} \ll 1 \ .$$

Then $v = k_a K_n [S]_0$ or $v = k_n [S]_0$, where $k_n = k_a K_n$ is the observed rate constant of the nonenzymatic reaction.

According to this model, the enzymatic reaction runs by the scheme:

$$E + S_n \xrightleftharpoons[k_{-1}]{k_1} E_0 S_n \xrightleftharpoons[k_{-2}]{k_2} E_p S_n \xrightleftharpoons[k_{-3}]{k_3} E_p S_a$$

$$\xrightleftharpoons[k_{-4}]{k_4} E_n S_a^* \xrightarrow{k_5} E + P .$$

(29)

Since form S_a is stabilized only at stage k_4, the equilibrium between complexes $E_p S_n$ and $E_p S_a$ does not depend on the enzyme, and therefore k_3/k_{-3} is equal to K_n. There is, of course, a parallel pathway, which starts with the free enzyme's interactions with the substrate active form and yields complex $E_p S_a$. However, since $[S_a] \ll [S_n]$, this pathway can be neglected.

If the equilibria are established quicker than the substrate chemical conversion occurs, the rate of the enzymatic reaction can be presented by the formula:

$$v = \frac{k_5 K_p K_n K_Q [E]_0 [S]_0}{K_s + [1 + K_p(1 + K_n K_Q)] + [S]_0},$$

(30)

where $K_s = \dfrac{k_{-1}}{k_1}$, $K_p = \dfrac{k_2}{k_{-2}}$ and $K_Q = \dfrac{k_4}{k_{-4}}$.

Experimentally determined kinetic parameters are:

$$k_{cat} = \frac{k_5 K_p K_n K_Q [E]_0 [S]_0}{1 + K_p(1 + K_n K_Q)},$$

(31)

$$K'_m = \frac{K_s}{1 + K_p(1 + K_n K_Q)},$$

(32)

$$\frac{k_{cat}}{K'_m} = \frac{k_5 K_p K_n K_Q}{K_s}.$$

(33)

If the chemical stage does not determine the overall rate of the process, the expression for the reaction rate is rather complex and hardly analyzed. However, it shows that k_{cat} is the rate constant of one (the slowest) stage or the combination of rate constants of several stages, and K'_m is equal to K_s.

Efficiency. The enzymatic catalysis efficiency can be expressed as ratio k_{cat}/k_n or $k_{cat}/K'_m k_n$. Formulae (31) and (33) and $k_n = k_a K_n$ (taking into account $k_a = k_5$) give:

$$\frac{k_{cat}}{k_n} = \frac{K_p K_Q}{1 + K_p(1 + K_n K_Q)},$$

(34)

$$\frac{k_{cat}}{k_n K'_m} = \frac{K_p K_Q}{K_s}.$$

(35)

Attempts to evaluate the values of K_n and K_Q have been made in Sect. 8.10. The K_p value can be higher or lower than the unity depending on whether a contribution of secondary interactions is enough to exceed a loss associated

Table 82. Ratio of kinetic constants of two substrates

Number of case	Conditions			Value	
	K_p	$K_n K_Q$	$K_p K_n K_Q$	$k_{cat}^{(1)}/k_{cat}^{(2)}$	$K_m^{(1)} K_m^{(2)}$
1	$\gg 1$	$\gg 1$	–	1	$\dfrac{K_s^{(1)} K_p^{(2)} K_Q^{(2)}}{K_s^{(2)} K_p^{(1)} K_Q^{(1)}}$
2	$\gg 1$	$\ll 1$	–	$K_Q^{(1)}/K_Q^{(2)}$	$\dfrac{K_s^{(1)} K_p^{(2)}}{K_s^{(2)} K_p^{(1)}}$
3	$\ll 1$	–	$\gg 1$	1	Similarly (1)
4	$\ll 1$	–	$\ll 1$	$\dfrac{K_p^{(1)} K_Q^{(1)}}{K_p^{(2)} K_Q^{(2)}}$	$\dfrac{K_s^{(1)}}{K_s^{(2)}}$

with freezing of rotational degrees of freedom. At $K_p \gg 1$ and $K_n K_Q < 1$ $k_{cat}/k_n = K_Q$, and if $K_n K_Q > 1$, $k_{cat}/k_n = 1/K_n$. At the same time, if $K_p \ll 1$, k_{cat}/k_n is equal to $K_p K_Q$ or $1/K_n$, respectively.

Specificity. The specificity, i.e., the ratio of kinetic constants in a substrate series, is usually evaluated from the ratio of second-order rate constants (k_{cat}/K_m'). For a series of similar substrates the hydrolysis rate in solution is virtually independent of the substrate type, i.e., values of $k_a = k_5$ and K_n are to be the same for all the substrance in the series. Hence, the ratio of second-order rate constants for two substrates should be:

$$\frac{(k_{cat}/K_m')_1}{(k_{cat}/K_m')_2} = \frac{K_p^{(1)} K_Q^{(1)} K_s^{(2)}}{K_p^{(2)} K_Q^{(2)} K_s^{(1)}}. \tag{36}$$

The relative values of k_{cat} and K_m' depend on constants K_p and $K_n K_Q$ (Table 82).

Table 82 presents three possible cases: (a) specificity is displayed only in K_m' (1 and 3); (b) specificity is manifested both in k_{cat} and K_m' (2 and 4); in these cases the principle "better binding – better catalysis" can be fulfilled; (c) if the primary specificity regions in a number of substrates are the same or similar, $K_s^{(1)}$ is approximately equal to $K_s^{(2)}$. Here the specificity is observed only in values of catalytic constants (k_{cat}), and values of K_m' are practically the same for all the substrates (case 4).

8.9.3 Thermodynamic Aspects of the Model

Let us depict the equilibrium cycle corresponding to scheme (29):

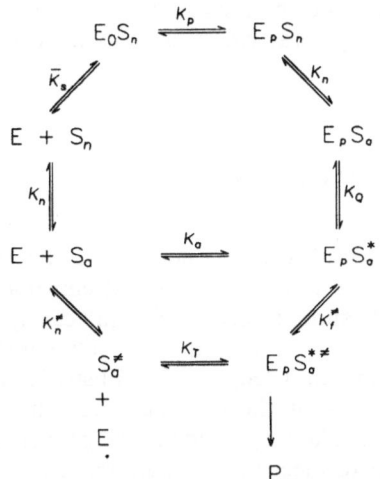

where $K_T = \dfrac{[E_p S_a^{* \dagger}]}{[E][S_a^{\dagger}]}$ characterizes the enzyme affinity to the transition state of the substrate active form and $K_a = \dfrac{[E_p S_a^*]}{[E][S_a]}$ reflects the enzyme affinity to the substrate ground state; $K_n^{\dagger}$, $K_f^{\dagger}$ are the equilibrium constants of activated complexes for nonenzymatic and enzymatic reactions, respectively; other constants have the same meaning as in scheme (29). When we consider the outer equilibrium cycle, it is clear (if $K_f^{\dagger} = K_n^{\dagger}$ according to the model) that

$$\frac{K_T}{\bar{K}_s K_p} = K_Q .$$

Thus, the transition state of nonenzymatic conversion of *the inactive substrate form* to the product (within the model this notion is hypothetical, since the substrate inactive form is not converted to the product directly) is stabilized as regards the inactive form in the ground state. If the lower internal equilibrium cycle is considered, one can see that the stabilization of the transition state of *the active form* relative to the ground state of this form does not occur, i.e., $K_a = K_T$.

So the model envisages that the enzyme stabilizes the ground state of the substrate *active form* and the "transition state" of *the inactive form*. This means that there is no strain (in thermodynamic sense) of the *active substrate form* in the enzyme-substrate complex.

An essential difference of this model from those discussed earlier is that all the factors determining the efficiency and specificity of catalysis are related to *the substrate ground state* in the productive enzyme-substrate complex. At the same time a decrease in *the apparent* activation barrier of the reaction is explained.

All effects mentioned in Sect. 8.2 can contribute to catalysis, if its manifestation in the ground state stabilizes the substrate *active form* in the complex with the enzyme.

An additional source of acceleration can be the factors displayed at the stage of the chemical conversion, such as general catalysis, polyfunctional catalysis, etc.

8.10   Approaches to the General Theory of Proteolysis

The above concept analyzed in previous sections and concrete data on the chemistry of the enzyme active may underlie a general theory of enzymatic hydrolysis of carboxylic acid derivatives. It should account for: (a) catalytic efficiency of amide hydrolases from a single viewpoint; predict the value of the catalytic acceleration of the reaction from the knowledge of the substrate and enzyme structures; (b) various manifestations of the enzyme specificity in a substrate series and in different amide hydrolases; (c) regulatory properties of enzymes. Nowadays no model is known to meet these requirements.

Let us consider the model proposed from this point of view. The question here arises: what is the substrate active form postulated by the model if applied to substrates of amide hydrolases?

The major reason for low reactivity of amides in hydrolytic reactions is resonance stabilization of the amide group. Distortion of n-π-coupling leads to a sharp increase of reactivity, which is clear from, for instance, a comparison of hydrolysis rates of amides and bicyclic β-lactams (of the penicillin type).

Analysis of some X-ray data shows that distortions of the amide group structure are quite possible in the enzyme-substrate complexes, mainly due to its pyramidalization. This results in breaking the resonance stabilization of the amide, alteration of the orders of bonds C–N and C=O, and increase of the positive charge on carbonyl carbon.

Apparently, precisely the state of the deformed (pyramidalized) amide bond is the active substrate form (S_a) which is being stabilized in the productive enzyme-substrate complex. In solution this form is thermodynamically unfavorable. Figure 2b shows that at the bending degree $\Delta_C \approx 0.2$ Å an enthalpy change is about 20 kcal/mol, which is within the frame of the overall energy change at the substrate-enzyme interaction. A probable entropy increase should be taken into account in this process owing to an increase of degrees of freedom of internal rotation, as well as the probable stabilization of a pyramidalized structure due to hydration of the charges increasing in this process. I think that evaluation of the free-energy change close to the real one at the indicated bending degree in solution gives $+15$ kcal/mol. Hence K_n [in formula (34)] is evaluated as to be about 7×10^{-10}.

Naturally, not every substrate forming the productive complex realizes the bending degree of about 0.2 Å. The same is true for the substrate that forms the complex with various enzymes. The problem is how to determine the degree of optimum distortion of resonance stability for diverse enzyme-substrate complexes.

Let us try to evaluate other parameters that define the catalysis efficiency in our model. The K_s value is readily assayed from data on steady-state kinetics, if the equality $K'_m = K_s$ is true. Within the model frames this condition is realized when $K_p < 1$ and $K_n K_Q < 1$, i.e., the equilibrium is displaced to the sorption complex but no to the fixed complex. The identity of K'_m for a series of single-type substrates is presumably associated with the fact that at the sorption complex formation the same or a similar substrate fragment interacts with the enzyme. Since the K_p value here does not exceed unity, the specificity is displayed only in maximum rates. Then a tendency to "saturation" of kinetic specificity ($k_{cat} \approx$ const) and a specificity regarding the binding should be observed. These conclusions have been experimentally supported by the k_{cat} dependences on K'_m for some amide hydrolases (Fig. 67).

It is more difficult to evaluate the K_p value. A free-energy change on this stage is determined by the efficiency of secondary interactions, which should compensate for a loss of the rotational degrees of freedom in the system. Experimental results obtained by the stopped flow or temperature jump methods (sse Sect. 6.2.3) may characterize the apparent K_p including the free energy of stabilization of the reactive fragment of the molecule. For many compounds these values are small and range within K_p equal to 5–20 (Table 68). Only for penta- and hexapeptides – papain substrates – is the value of the equilibrium constant of the second complex 120–250, which conforms to the free energy change of 2.8–3.3 kcal/mol. However, the fact that numerous inhibitors, proteins and "transition state analogs" have association constants of about 1×10^{10} M^{-1} shows that a full value of the free-energy change for the complex formation can be much higher.

Theoretical evaluation of the minimal value of K_p can be made from a proposal about the structure of the $E_0 S_n$ complex and the number of rotational degrees of freedom, which the system loses at the complex $E_p S_n$ formation. For instance, suppose that substrate AcPheAlaNH$_2$ in the $E_0 S_n$ complex interacts with chymotrypsin only by its aromatic group, then its transformation into the $E_p S_n$ complex is accompanied by freezing of the free rotation around 9th σ-bonds. This results in the entropy loss of about 45 e.u. [2453], i.e., approximately 13 kcal/mol. This value is partially compensated by the entropy increase due to solvation and must be completely compensated due to binding to the enzyme. These are very approximate values, but the calculation of K_p values is quite exact.

The value of stabilization energy of the active substrate form can be derived from formula (35), if the K_p and K_s as well as the ratio of the second-order rate constants for enzymatic and nonenzymatic reactions are known. If the latter is 10^{10} (see Table 50), and $K_s \approx 1 \times 10^{-4}$ M, then even at $K_p = 1$ the value of K_Q is 10^6, which fits the free energy change by 13.5 kcal/mol. This value seems reasonable.

What is the value of the rate constant (k_5) of the substrate chemical conversion? The fact that the amide hydrolases can transform substrates with the rates close to those of diffusion-controlled processes evidences that this constant can be comparable with the rate constants of these reactions (10^8–10^9 s^{-1}). Thus the above assumption that the equilibrium is rapidly reached in scheme (29) is hardly fulfilled; actually the rate-determining stage is one of the nonchemical stages [1716].

In this case the pK_a values of the enzyme functional groups determined from pH-dependencies of reaction rates are, in fact, kinetic pK_a (see Sect. 5.2.2). Presumably, this accounts for a frequently occurred dependence of pK_a of the free enzyme groups on the type of a substrate used for the assay.

This model shows that the nonproductive binding must proceed predominantly at the formation of the fixed enzyme complex with inactive (E_pS_n) or active (E_pS_a) substrate form. Here the corresponding value of the equilibrium constant (K_n or K_p) changes by the value of $K_i = [E_pS']/[E_pS]$, where E_pS' is the nonproductive complex, i.e., $K'_p = K_p + K_i$, while these changes similarly influence k_{cat} and K'_m rather than their ratio.

The kinetic isotope effect of solvent is usually assigned to the rate-determining proton transfer, which defines the rate of process in any case. Upon the rapid chemical conversion of the substrate this effect is not manifested. If the stages preceding the chemical conversion are associated with the proton transfer, the effect can be observed. Besides, the model postulates destabilization of the scissile bond in the productive complex that can be related to the partial proton transfer.

So the concept covering these events – a reason for the efficiency and the specificity of catalysis and the peculiarities of the substrate ground state in the enzyme-substrate complex – underlies the proposed model applicable to the amide hydrolases. These peculiarities are deformation and resonance destabilization of a cleavable bond due to its location near a nucleophile at a distance *shorter* than the sum of Van der Waals radii of reactive atoms. The energy loss here is compensated both by the interaction of the nonreactive substrate fragment with the enzyme and by the partial overlapping of binding orbitals of a nucleophile and the carbonyl carbon atom (see Sect. 7.6). This does not lead to the strain in the productive complex, as compared to the analogous ground state of the substrate in solution.

It seems to me a major advantage of the model that it does not operate with the transition state, the study of which is impossible, in principle, by physical and chemical methods, but it deals with the states that can be easily studied. Of course, at this stage the model is far from perfection yet. So the deformation of the reactive .group has been found in the complexes of few enzymes with inhibitors, and it is not evident that it reflects the state of true enzyme-sub strate complexes. There are also no data on the character of interactions at the formation of sorption and fixed complexes, though the fact of the stepwise complex formation itself is undoubted. The problems on pH-dependence of catalysis, isotope effects, etc., require further investigations.

Of much value is the elaboration of methods for quantitative prediction of the efficiency and the specificity of catalysis by the amide hydrolases, especially in the case of peptide substrates. An encouraging way here is opened by linear free-energy relationships. However, researchers investigating this way will face considerable difficulties, particularly in explaining the unique specificity of some proteases.

The elementary step of chemical conversion of the substrate has been very poorly investigated. Here quantum-chemical methods are of great value.

The problems enumerated here outline, though tentatively, a program of further studies of proteolytic reactions – one of the most essential processes of vital activity. To shed light on these questions means also to make a valid contribution to chemical enzymology as a whole.

8.11 Conclusion

A disadvantage of the majority of available theories of the enzymatic catalysis is the structural and chemical distinction between the ground state of the productive enzyme-substrate complex and that of a congruent model reaction. Accounting for this factor eliminates many difficulties in explaining enzyme efficiency and specificity. The enzyme "prepares", so to speak, the substrate for the elementary step by changing its complementarity to different states of the substrate throughout the reaction. As a result, the elementary step occurs with the low activation energy. Such ideas are of help in finding a way for the creation of a general theory of the enzymatic hydrolysis of carboxylic acid derivatives and, probably, in substantiating the reasons for the unique efficiency and specificity of enzymes.

References

1. Dean RT (1978) Cellular degradative processes. Chapman and Hall, London
2. Rechsteiner M, Rogers S, Rote K (1987) TIBS 12: 390–394
3. Reich E, Rifkin DB, Shaw E (eds) (1975) Proteases and biological control. Cold Spring Harbor Laboratory, New York
4. Neurath H, Walsh KA (1977) Proc FEBS Meet 47: 1–14
5. Lokshina LA (1979) Mol Biologia 13: 1205–1229
6. Cunningham DD, Loug GL (eds) (1987) Proteases in biological control. Liss, New York
7. Shamin AN (1971) Biocatalysis and biocatalysts. Nauka, Moscow
8. Bender ML (1960) Chem Rev 60: 53–113
9. Jencks WP (1975) Adv Enzymol 43: 219–410
10. Enzyme Nomenclature, Recommendations of the Nomenclature Committee of the IUB (1984). Academic Press, New York; a) Suppl 1 (1986) Europ J Biochem 157: 1–26; b) Suppl 2 (1989) 179: 489–533; c) Suppl 3 (1990) 187: 263–281
11. Ramachandran GN, Sasisekharan V (1968) Adv Protein Chem 23: 283–437
12. Robin MB, Bovey FA, Basch H (1970) In: Zabicky J (ed) The chemistry of amides. Interscience, London
13. Schulz GE, Schirmer RH (1979) Principles of protein structure. Springer, Berlin Heidelberg New York
14. Pauling L (1962) The nature of the chemical bond, 3rd edn. Cornell University Press, Ithaca
15. Abbreviations and symbols for the description of conformation of polypeptide chains (1970) J Mol Biol 52: 1–17
16. Ramachandran GN, Lakshminarayanan AV, Kolaskar AS (1973) Biochim Biophys Acta 303: 8–13
17. Winkler FK, Dunitz JD (1971) J Mol Biol 59: 169–182
18. Benedetti E (1977) In: Goodman M, Meinhofer J (eds) Peptides. Proc 5th Am Peptides Symp, Wiley, New York, pp 257–273
19. Wiberg KB, Laidig KE (1987) J Am Chem Soc 109: 5935–5943
20. Costain CC, Dowling JM (1960) J Chem Phys 32: 158–165
21. Hirota E, Sugisaki R, Nielsen CJ, Sorensen GO (1974) J Mol Spectr 49: 251–267
22. Ladell J, Post B (1954) Acta Crystallogr 7: 559–564
23. Kimura M, Aoki N (1953) Bull Chem Soc 26: 429–431
24. Senti F, Harker D (1940) J Am Chem Soc 62: 2008–2019
25. Lide DR (1957) J Chem Phys 27: 343–352
26. Swalen JD, Costain CC (1959) J Chem Phys 31: 1562–1574
27. Levitt M, Lifson S (1969) J Mol Biol 46: 269–279
28. Scarsdale JN, Alsenoy CV, Klimkowski VJ, Schäfer L, Momany FA (1983) J Am Chem Soc 105: 3438–3445
29. Kurland RJ, Wilson EB (1957) J Chem Phys 27: 585–590
30. Ramachandran GN (1968) Biopolymers 6: 1494–1496
31. Pedone C, Benedetti E, Immirzi A, Allegra G (1970) J Am Chem Soc 92: 3549–3552
32. Koetzle TF, Hamilton WC, Parthasarathy R (1972) Acta Crystallogr Sect B 28: 2083–2090
33. Ramachandran GN, Kolascar AS (1973) Biochim Biophys Acta 303: 385–388
34. Dunitz JD, Winkler FK (1975) Acta Crystallogr Sect B 31: 251–263
35. Kolascar AS, Lakshminarayanan AV, Sarathy KP, Sasisekharan V (1975) Biopolymers 14: 1081–1084
36. Zimmerman SS, Scheraga HA (1976) Macromolecules 9: 408–416

37. Peters D, Peters J (1978) J Mol Struct 50: 133–139
38. Carlsen NR, Radom L, Riggs NV, Rodwell WR (1979) J Am Chem Soc 101: 2233–2234
39. Benedetti E, Di Blasio B, Pavone V, Pedone C, Toniolo C, Bonora GM (1981) J Biol Chem 256: 9229–9234
40. Cieplak AS (1985) J Am Chem Soc 107: 271–273
41. Jeffrey GA, Houk KN, Paddon-Row MN, Rondau NG, Mitra J (1985) J Am Chem Soc 107: 321–326
42. Gilli G, Bertolasi V, Bellucci F, Ferretti V (1986) J Am Chem Soc 108: 2420–2424
43. Kolascar AS, Sarathy KP (1980) Biopolymers 19: 1345–1355
44. Deisenhofer J, Steigemann W (1975) Acta Crystallogr Sect B 31: 238–250
45. Drakenberg T, Forsen S (1970) J Phys Chem 74: 1–7
46. Neuman RC, Young LB (1965) J Phys Chem 69: 2570–2576
47. Rogers MT, Woodbrey JC (1962) J Phys Chem 66: 540–546
48. Wunderlich MD, Leung LK, Sandberg JA, Meyer KD, Yoder CH (1978) J Am Chem Soc 100: 1500–1503
49. Gutowsky HS, Jonas J, Siddall TH (1967) J Am Chem Soc 89: 4300–4304
50. Nagakura S (1960) Mol Phys 3: 105–112
51. Tanaka J (1958) J Chem Soc Jap 78: 1636–1643
52. Martin GJ, Gouesnard JP, Dorie J, Rabiller C, Martin ML (1977) J Am Chem Soc 99: 1381–1388
53. Yoder CH, Gardner RD (1981) J Org Chem 46: 64–66
54. Rao CNR, Rao KG, Goel A, Balasubramanian D (1971) J Chem Soc Sect A 3077–3083
55. Christensen DH, Kortzeborn RN, Bak B, Led JJ (1970) J Chem Phys 53: 3912–3922
56. Jorgensen WL, Gao J (1988) J Am Chem Soc 110: 4212–4216
57. Hughes DO (1968) Tetrahedron 24: 6423–6431
58. Antonov VK, Shkrob AM, Shemyakin MM (1965) Zhurn Obsch Khimii 35: 1380–1389
59. Mizushima S, Shimonouchi T (1961) Adv Enzymol 23: 1–27
60. Poland D, Scheraga HA (1967) Biochemistry 6: 3791–3800
61. Popov EM, Zheltova VN (1971) J Mol Struct 10: 221–230
62. Momany FA, McGuive RF, Yan JF, Scheraga HA (1970) J Phys Chem 74: 2424–2438
63. Lifson S, Hagler AT, Dauber P (1979) J Am Chem Soc 101: 5111–5121
64. Homer RB, Johnson CD (1970) In: Zabisky J (ed) The chemistry of amides. Interscience, London, pp 187–243
65. Katritzky AR, Jones RAY (1961) Chem Ind 722–727
66. Chirgadze Yu. N (1965) Infrared spectra and the structure of polypeptides and proteins. Nauka, Moscow
67. Gillespie RJ, Birchall T (1963) Can J Chem 41: 148–155
68. Fraenkel G, Franconi C (1960) J Am Chem Soc 82: 4478–4483
69. Higuchi T, Barnstein CH, Ghassemi H, Perez WE (1962) Anal Chem 34: 400–412
70. Edward JT, Chang HS, Yates K, Steward R (1960) Can J Chem 38: 1518–1525
71. Hine J, Hine M (1952) J Am Chem Soc 74: 5266–5271
72. Pimentel GC, McClellan AL (1960) The hydrogen bond. Freeman, San Francisco
73. Arnett EM, Mitchell EJ (1971) J Am Chem Soc 93: 4052–4053
74. Vinogradov SN (1979) Int J Peptide Protein Res 14: 281–289
75. Gramsted T, Fuglevik WJ (1962) Acta Chem Scand 16: 1369–1377
76. Takahashi F, Li NC (1965) J Phys Chem 69: 1622–1627
77. Klotz IM, Farnham SB (1968) Biochemistry 7: 3879–3882
78. Spencer JN, Garrett RC, Mayer FJ, Markle JE, Powell CR, Tran MT, Berger SK (1980) Can J Chem 58: 1372–1375
79. Johansson A, Kollmak PA (1972) J Am Chem Soc 94: 6196–6198
80. Umeyama H, Morokuma K (1977) Acta Crystallogr Sect B 32: 994–1001
81. Del Bene JE (1978) J Am Chem Soc 100: 1387–1394
82. Gramstad T, Fuglevik WJ (1965) Spectrochim Acta 21: 343–344
83. Joesten MD, Drago RS (1962) J Am Chem Soc 84: 2696–2699
84. Chen CYS, Swenson CA (1969) J Phys Chem 73: 1363–1370
85. Taylor R, Kennard O, Versichel W (1983) J Am Chem Soc 105: 5761–5766
86. Neuman RC, Woolfenden WR, Jonas V (1969) J Phys Chem 73: 3177–3180

87. Taylor R, Kennard O, Versichel W (1984) J Am Chem Soc 106: 244–248
88. Jorgensen WL, Swenson CJ (1985) J Am Chem Soc 107:1489–1496
89. Jorgensen WL (1989) J Am Chem Soc 111: 3770–3771
90. Levitt M, Perutz MF (1988) J Mol Biol 201: 751–754
91. Sigel H, Martin RB (1982) Chem Rev 82: 385–426
92. Freeman HC (1967) Adv Protein Chem 22: 257–424
93. Drago RS, Meek DW, Joesten MD, La Roche L (1963) Inorg Chem 2: 124–132
94. Temussi PA, Tancredi T, Quadrifoglio F (1969) J Phys Chem 73: 4227–4232
95. Armbruster AM, Pullman A (1974) FEBS Lett 49: 18–21
96. Collins TJ, Coots RJ, Furutani TT, Keech JT, Peake GT, Santarsiero BD (1986) J Am Chem Soc 108: 5333–5339
97. Hatton JV, Richards RE (1962) Mol Phys 5: 139–141
98. Worlester DL (1978) Proc Natl Acad Sci USA 75: 5475–5477
99. Pauling L (1979) Proc Natl Acad Sci USA 76: 2293–2294
100. Gra R, Suarez H, Gonzalez R (1976) Rev CENIC Cienc Fis 7: 125–134 (C.A. 1978, 89, 42180b)
101. Donohue JA, Scott RH, Menger FM (1970) J Org Chem 35: 2035–2036
102. Idoux JP, Kiefer GE, Baker GR, Pockett WE, Spence FJ, Simmons KS, Constant RB, Watlock DJ, Fuhrman SL (1980) J Org Chem 45: 441–444
103. Potapov VM, Demyanovich VM, Terentyev AP (1961) Zhurn Obsch Khimii 31: 3046–3050
104. Skulski D, Palmer GC, Calvin M (1963) Tetrahedron Lett 1773–1776
105. Flynn EH (ed) (1972) Cephalosporins and penicillins, chemistry and biology. Academic Press, New York
106. Page MI (1984) Accounts Chem Res 17: 144–151
107. Crowfoot D, Bunn CW, Rogers-Low BW, Turner-Jones A (1949) The chemistry of penicillin. Princeton Univ Press, Princeton, p 310
108. Sweet RM, Dahl LF (1970) J Am Chem Soc 92: 5489–5507
109. Rao SR, Vasudevan TK (1983) CRC Crit Rev Biochem 14: 173–206
110. Ballard SA, Malstrom DS, Smith CW (1949) The chemistry of penicillin. Princeton Univ Press, Princeton, p 973
111. Liebman JF, Greenberg A (1974) Biophys Chem 1: 222–226
112. Bailey J, North AM (1968) Trans Faraday Soc 64: 1499–1504
113. Perricaudet M, Pullman A (1973) Int J Peptide Protein Res 5: 99–107
114. O'Gorman JM, Shand W, Schomaker V (1950) J Am Chem Soc 72: 4222–4228
115. Bailey M (1949) Acta Crystallogr 2: 120–126
116. Laurence C, Guihenent G, Wojtkowiak B (1979) J Am Chem Soc 101: 4793–4801
117. Arnett EM (1967) Modern problems of physical organic chemistry. Mir, Moscow, pp 195–341
118. Tanford C (1964) J Am Chem Soc 86: 2050–2053
119. Hansch C, Leo A (1979) Substituent constants for correlation analysis in chemistry and biology, Wiley, New York
120. Wolfenden R, Anderson L, Cullis PM, Southgate CCB (1981) Biochemistry 20: 849–855
121. Fauchére JL, Pliska V (1983) Eur J Med Chem 18: 369–375
122. Parker JM, Guo D, Hodges RS (1986) Biochemistry 25: 5425–5432
123. Roseman MA (1988) J Mol Biol 200: 513–522
124. Némethy G, Scheraga HA (1977) Q Rev Biophys 10: 239–352
125. Matthews BW (1977) In: Neurath H, Hill RL, Boeder C-L (eds) The proteins, vol 3. Academic Press, San Francisco, pp 404–590
126. Rossman MG, Argos P (1981) Annu Rev Biochem 50: 497–532
127. Chothia C (1984) Annu Rev Biochem 53: 537–572
128. Baici A, Rizzo V, Skrabal P, Luisi PL (1979) J Am Chem Soc 101: 5170–5178
129. Toma F, Monnot M, Piriou F, Savrda J, Fermandjan S (1980) Biochem Biophys Res Commun 97: 751–758
130. Madison V, Kopple KD (1980) J Am Chem Soc 102: 4855–4863
131. Rizzo V, Jackle H (1983) J Am Chem Soc 105: 4195–4205
132. Popov EM, Lipkind GM, Arkhipova SF, Dashevsky GM (1968) Mol Biologia 2: 622–630
133. Platzer KEB, Momany FA, Scheraga HA (1972) Int J Peptide Protein Res 4: 187–200

134. Hagler AT, Leiserowitz L, Tuval M (1976) J Am Chem Soc 98: 4600–4612
135. Némethy G, Hodes ZI, Scheraga HA (1978) Proc Natl Acad Sci USA 75: 5760–5764
136. Hagler AT, Stern PS, Sharon R, Becker JM, Naider F (1979) J Am Chem Soc 101: 6842–6852
137. Peters D, Peters J (1980) J Mol Struct 62: 229–247
138. Weiner SJ, Singh UC, O'Donnell TJ, Kollman PA (1984) J Am Chem Soc 106: 6243–6245
139. Brooks BR, Karplus M (1983) Proc Natl Acad Sci USA 80: 6571–6575
140. Brady J, Karplus M (1985) J Am Chem Soc 107: 6103–6105
141. Peters D, Peters J (1980) J Mol Struct 64: 103–115
142. Kopple KD, Marz DH (1967) J Am Chem Soc 89: 6193–6200
143. Kreissler MA, Akhmedov NA, Arkhipova SF, Lipkind GM, Popov EM (1974) J Chem Phys 71: 913–919
144. Boussard G, Marraud M (1981) Biopolymers 20: 169–185
145. Rossky PJ, Karplus M (1979) J Am Chem Soc 101: 1913–1937
146. Ooi T, Oobatake M, Némethy G, Scheraga HA (1987) Proc Natl Acad Sci USA 84: 3086–3090
147. Newmark RA, Miller MA (1971) J Phys Chem 75: 505–508
148. Drenth J, Jansonius JN, Koekoek R, Wolthers BG (1971) Adv Prot Chem 25: 79–116
149. Quiocho FA, Lipscomb WN (1971) Adv Prot Chem 25: 1–78
150. Ramachandran GN, Mitra AK (1976) J Mol Biol 107: 85–92
151. Easthope PL, Brayer GD (1986) Int J Peptide Protein Res 27: 666–672
152. Levitt M, Chothia C (1976) Nature 261: 552–558
153. Popov EM (1989) The structure organization of proteins. Nauka, Moscow
154. Nakashima H, Nishikawa K, Ooi T (1986) J Biochem 99: 153–162
155. Lesk AM, Chothia G (1980) J Mol Biol 136: 225–270
156. Wilson IA, Haft DH, Getzolf ED, Tainer JA, Lerner RA, Brenner S (1985) Proc Natl Acad Sci USA 82: 5255–5259
157. Warme PK, Morgan RS (1978) J Mol Biol 118: 273–287
158. Ponnuswamy PK, Prabhakaran M (1980) Biochem Biophys Res Commun 97: 1582–1590
159. Rose GD (1979) J Mol Biol 134: 447–470
160. Barlow DJ, Thornton JM (1983) J Mol Biol 168: 867–885
161. Rashin AA, Honig B (1984) J Mol Biol 173: 515–521
162. Janin J, Miller S, Chothia C (1988) J Mol Biol 204: 155–164
163. Glazer AN (1976) In: Neurath H, Hill RL, Boeder CL (eds) The Proteins, vol 2. Academic Press, San Francisco, pp 2–105
164. West CM (1986) Mol Cell Biochem 72: 3–20
165. Levitt M (1983) J Mol Biol 168: 595–620
166. Levitt M (1983) J Mol Biol 168: 621–657
167. Lakowicz JR, Maliwal BP, Cherek H, Balter A (1983) Biochemistry 22: 1741–1752
168. Joly M (1965) A physico-chemical approach to the denaturation of proteins. Academic Press, London New York
169. Tanford C (1970) Adv Protein Chem 24: 1–95
170. Eisenberg D, Kautsman V (1975) Structure and function of water. Nauka, Leningrad
171. Malenkov GG (1984) Physical chemistry (yearbook). Nauka, Moscow, pp 41–76
172. Dewar MJS, Thiel W (1977) J Am Chem Soc 99: 4907–4917
173. Dewar MJS, Dieter KM (1986) J Am Chem Soc 108: 8075–8086
174. Barrett AJ, McDonald JK (1986) Biochem J 237: 935
175. Hartley BS (1960) Annu Rev Biochem 29: 45–72
176. Hagehara B (1976) In: Boyer P (ed) The enzymes, 3rd edn, vol 4. Academic Press, New York, pp 193–213
177. Glover D (ed) (1985) DNA Cloning a practical approach. IRL Press, Oxford
178. Perlmann GE, Lorand L (eds) (1970) Methods in enzymology, vol 19. Academic Press, New York
179. Lorand L (ed) (1976) Methods in enzymology, vol 45. Academic Press, New York
180. Lorand L (ed) (1981) Methods in enzymology, vol 80. Academic Press, New York
181. Barrett AJ, McDonald JK (1980) Mammalian proteases: a glossary and bibliography, vol 1. Endopeptidases. Academic Press, New York

182. McDonald JK, Barrett AJ (1986) Mammalian proteases: a glossary and bibliography, vol 2. Exopeptidases. Academic Press, New York
183. Azaryan AV (1989) Peptidhydrolases of the nervous system and their biological functions. Aiastan, Erevan
184. Estell DA, Laskowski M (1980) Biochemistry 19: 124–131
185. Lokshina LA, Dilakyan EA (1986) Mol Biologia 20: 1157–1175
186. Suzuki K (1987) TIBS 12: 103–105
187. Kostrov SV, Strongin AYa (1989) Mol Biologia 23: 255–265
188. Goldberg AL, St John AC (1976) Annu Rev Biochem 45: 747–804
189. Martino GND, Goldberg AL (1979) J Biol Chem 254: 3712–3715
190. Voellmy R, Murakami K, Goldberg AL (1979) In: Cohen GN, Holzer H (eds) Limited proteolysis in microorganisms. DHEW, Washington, DC, pp 7–16
191. Edmunds T, Goldberg AL (1986) J Cell Biochem 32: 187–191
192. Ciechanover A, Hod Y, Hershko A (1978) Biochem Biophys Res Commun 81: 1100–1105
193. Wilkinson KD, Urban MK, Haas AL (1980) J Biol Chem 255: 7529–7532
194. Hershko A, Ciechanover A, Heller H, Hass AL, Rose JA (1980) Proc Natl Acad Sci USA 77: 1783–1786
195. Etlinger JD, Goldberg AL (1977) Proc Natl Acad Sci USA 74: 54–58
196. Rivett AJ (1989) Arch Biochem Biophys 268: 1–8
197. Hase JI, Kobashi K, Nakai N, Mitsui K-I, Iwata K, Takadera T (1980) Biochim Biophys Acta 611: 205–213
198. Data not available
199. Alt J, Heymann E, Krisch K (1975) Eur J Biochem 53: 357–369
200. Holloway MR, Ticho T (1979) FEBS Lett 106: 185–188
201. Ambler RP, Auffret AD, Clarke PH (1987) FEBS Lett 215: 285–290
202. Verret CR, Roser MR, Khorana HG (1982) J Biol Chem 257: 10222–10227
203. Iwayama A, Kimura T, Adachi O, Ameyama M (1983) Agric Biol Chem 47: 2483–2486
204. Tsushida T, Takeo T (1985) Agric Biol Chem 49: 2913–2917
205. Wittwer CT, Burkhard D, Ririe K, Rasmussen R, Brown J, Wyse RW, Hansen RG (1983) J Biol Chem 258: 9733–9738
206. Takahashi N, Nishibe H (1981) Biochim Biophys Acta 657: 457–467
207. Sugiyama K, Ishihara H, Tajima S, Takahashi N (1983) Biochem Biophys Res Commun 112: 155–160
208. Tarentino AL, Quinones G, Trumble A, Changchien L-M, Duelman B, Maley F, Plummer TH (1990) J Biol Chem 265: 6961–6966
209. Fukumura T, Talbot G, Misono H, Taramura Y, Kato K, Soda K (1978) FEBS Lett 89: 298–300
210. Nickerson WJ, Durand SC (1963) Biochim Biophys Acta 77: 87–99
211. Suzuki Y, Uchida K (1985) Agric Biol Chem 49: 1573–1579
212. Jackson RL, Matsueda GR (1970) Meth Enzymol 19: 591–599
213. McDonald JK, Leibach FH, Crindeland RE, Ellis S (1968) J Biol Chem 243: 4143–4150
214. Eisenhauer DA, McDonald JK (1986) J Biol Chem 261: 8859–8865
215. Mollay C, Vilas U, Hutticher A, Kreil G (1986) Eur J Biochem 160: 31–35
216. Alvařez NG, Bordallo C, Gascoń S, Rendueles PS (1985) Biochim Biophys Acta 832: 119–125
217. Yoshimoto T, Tsuru D (1982) J Biochem 91: 1899–1906
218. Swanson AA, Albers-Jackson B, McDonald JK (1978) Biochem Biophys Res Commun 84: 1151–1159
219. Gorenstein C, Snyder SH (1980) Proc R Soc Lond B 210: 123–132
220. Lee CM, Snyder SH (1982) J Biol Chem 257: 12043–12050
221. Saison M, Verlinden J, Van Leuven F, Cassiman JJ, Van Den Berghe H (1983) Biochem J 216: 177–183
222. Kenny J, Booth AG, George SG, Ingram J, Kershaw D, Wood EJ, Young AR (1976) Biochem J 157: 169–182
223. Fukasawa KM, Fukasawa K, Hiraoka BY, Harada M (1981) Biochim Biophys Acta 657: 179–189
224. Yoshimoto T, Kita T, Ichinose M, Tsuru D (1982) J Biochem 92: 275–282

225. Yoshimoto T, Walter R (1977) Biochim Biophys Acta 485: 391–401
226. MacNair RDC, Kenny AJ (1979) Biochem J 179: 379–395
227. Kikuchi M, Fukuyama K, Epstein WL (1988) Arch Biochem Biophys 266: 369–376
228. Ogata S, Misumi Y, Ikehara Y (1989) J Biol Chem 264: 3596–3601
229. Barth A, Schulz H, Neubert K (1974) Acta Biol Med Germ 32: 157–174
230. Fukasawa KM, Fukasawa K, Hiraoka BY, Harada M (1983) Experientia 39: 1005–1007
231. McDonald JK, Hoisington AR, Eisenhauer DA (1985) Biochem Biophys Res Commun 126: 63–71
232. Bålow R-M, Ragnarsson U, Zetterqvist O (1983) J Biol Chem 258: 11622–11628
233. Bålow R-M, Tomkinson B, Ragnarsson U, Zetterqvist O (1986) J Biol Chem 261: 2409–2417
234. Bålow R-M, Eriksson I (1987) Biochem J 241: 75–80
235. Matsuda K, Misaka E (1974) J Biochem 76: 639–649
236. Steve LJ, Jappel AL (1974) Biochim Biophys Acta 341: 99–111
237. Kawamura Y, Matoba T, Doi E (1980) J Biochem 88: 1559–1561
238. Mikola L (1983) Biochim Biophys Acta 747: 241–252
239. Sørensen SB, Breddam K, Svendsen I (1986) Carlsberg Res Commun 51: 475–485
240. Sørensen SB, Svendsen I, Breddam K (1987) Carlsberg Res Commun 52: 285–295
241. Breddam K, Sørensen SB (1987) Carlsberg Res Commun 52: 275–283
242. Breddam K, Sørensen SB, Svendsen I (1987) Carlsberg Res Commun 52: 297–311
243. Sarbakanova ST, Dunaevsky YE, Belozersky MA, Rudenskaya GN (1987) Biokhimiya 52: 1263–1269
244. Wells JRE (1968) Biochim Biophys Acta 167: 388–398
245. Zuber H (1968) Hoppe-Seyler's Z Physiol Chem 349: 1337–1352
246. Zuber H (1976) Meth Enzymol 45B: 561–568
247. Funakoshi T, Shoji S, Hayata S, Kubota Y (1985) Chem Pharm Bull 33: 5428–5436
248. Shoji S, Narimatsu S, Morita T, Funakoshi T, Ueki H, Kubota Y (1985) Chem Pharm Bull 33: 4963–4972
249. Kuhn RW, Walsh KA, Neurath H (1974) Biochemistry 13: 3871–3877
250. Hayashi R (1976) Meth Enzymol 45B: 568–587
251. Svendsen I, Martin BM, Viswanatha T, Johansen JT (1982) Carlsberg Res Commun 47: 15–27
252. Hofmann T (1976) Meth Enzymol 45B: 587–599
253. Kumamoto K, Stewart TA, Johnson AR, Erdös EG (1981) J Clin Invest 67: 210–215
254. Odya CE, Erdös EG (1981) Meth Enzymol 80: 460–466
255. Yang HYT, Erdös EG, Chiang TS, Jenssen TA, Rodgers JG (1970) Biochem Pharmacol 19: 1201–1211
256. Frére J-M, Legh-Bouille M, Ghuysen JM, Nieto M, Perkins HR (1976) Meth Enzymol 45B: 610–636
257. Ambler RP, Meadway RJ (1969) Nature 222: 24–26
258. Waxman DJ, Strominger JL (1980) J Biol Chem 255: 3964–3976
259. Frére JM, Joris B (1985) CRC Crit Rev Microbiol 11: 299–335
260. Joris B, Ghuysen JM, Dive G, Renard A, Dideberg O, Charlier P, Frére J-M, Kelly JA, Boyington JC, Moews PC, Knox JR (1988) Biochem J 250: 313–324
261. Yocum RR, Blumberg PM, Stominger JL (1974) J Biol Chem 249: 4863–4871
262. Blumberg PM, Strominger JL (1972) Proc Natl Acad Sci USA 69: 3751–3755
263. Georgopapadakou N, Hammarstrom S, Strominger JL (1977) Proc Natl Acad Sci USA 74: 1009–1012
264. Schonberger OL, Tschesche H (1981) Hoppe-Seyler's Z Physiol Chem 362: 865–873
265. Tsunasawa S, Narita K (1976) Meth Enzymol 45B: 552–561
266. Kobayashi K, Smith JA (1987) J Biol Chem 262: 11435–11445
267. Nardacci NJ, Mukhopadhyay S, Campbell BJ (1975) Biochim Biophys Acta 377: 146–157
268. Simmons WH, Walter R (1980) Biochemistry 19: 39–48
269. Coan MH, Roberts RC, Travis J (1971) Biochemistry 10: 2711–2717
270. Travis J (1972) Biochem Biophys Res commun 48: 1111–1116
271. Desnuelle P (1960) In: Boyer P (ed) The enzymes, vol 4. Academic Press, New York, pp 93–118

272. Hartley BS (1964) Nature 201: 1284–1288
273. Meloun B, Kluh J, Kostka V, Morávek L, Prusík Z, Vaneček J, Keil B, Šorm F (1966) Biochim Biophys Acta 130: 543–546
274. Smillie LB, Furka A, Nagabhushan N, Stevenson KJ, Parkes CO (1968) Nature 218: 343–345
275. Charles M, Gratecos D, Rovery M, Desnuelle P (1967) Biochim Biophys Acta 140: 395–409
276. Koide A, Kataoka T, Matsuoka Y (1969) J Biochem 65: 475–477
277. Matsuoka Y, Koide A (1966) Arch Biochem Biophys 114: 422–424
278. Koide A, Matsuoka Y (1970) J Biochem 68: 1–7
279. Ryan CA, Clary JJ, Jomimatsu Y (1965) Arch Biochem Biophys 110: 175–183
280. Bell GI, Quinto C, Quiroga M, Valenzuela P, Craik CS, Rutter WJ (1984) J Biol Chem 259: 14265–14270
281. Möckel W, Barnard EA (1969) Biochim Biophys Acta 191: 370–378
282. Barnard EA, Hope WC (1969) Biochim Biophys Acta 178: 364–369
283. Tsai I-H, Liu H-C, Chuang K-L (1986) FEBS Lett 203: 257–261
284. Law JH, Dunn PE, Kramer KJ (1977) Adv Enzymol 45: 389–425
285. Jany K-D, Haug H (1983) FEBS Lett 158: 98–102
286. Jany K-D, Haug H, Pfleiderer G, Ishay J (1978) Biochemistry 17: 4675–4682
287. Folk JE, Schirmer EW (1965) J Biol Chem 240: 181–197
288. Tobita T, Folk JE (1967) Biochim Biophys Acta 147: 15–25
289. Gratecos D, Desnuelle P (1971) Biochem Biophys Res Commun 42: 857–864
290. Gibson D, Dixon GM (1969) Nature 222: 753–756
291. Travis J, Roberts RC (1969) Biochemistry 8: 2884–2889
292. Schroeder DD, Shaw E (1968) J Biol Chem 243: 2943–2949
293. Emi M, Nakamura Y, Ogawa M, Yamamoto T, Nishide T, Mori T, Matsubara K (1986) Gene 41: 305–310
294. Walsh KA, Neurath H (1964) Proc Natl Acad Sci USA 52: 884–889
295. Mikeš O, Tomášek V, Holeyšovsky V, Šorm F (1966) Biochim Biophys Acta 117: 281–284
296. Hermodson MA, Ericsson LH, Neurath H, Walsh KA (1973) Biochemistry 12: 3146–3153
297. Smith RA, Liener IE (1967) J Biol Chem 242: 4033–4039
298. Travis J, McElroy WD (1968) Biochem Biophys Res Commun 30: 730–734
299. Stevenson BJ, Hedenbuechle O, Wellauer PK (1986) Nucl Acid Res 14: 8307–8330
300. Craik CS, Choo Q-L, Swift GH, Quinto C, MacDonald RJ, Rutter WJ (1984) J Biol Chem 259: 14255–14264
301. Fletcher TS, Alnadeff M, Craik CS, Largman C (1987) Biochemistry 26: 3081–3086
302. Gendry P, Launay J-F (1988) Biochim Biophys Acta 955: 243–249
303. Lacko AG, Neurath H (1967) Biochem Biophys Res Commun 26: 272–277
304. Bradshaw RA, Neurath H, Tye RW, Walsh KA, Winter WP (1970) Nature 226: 237–239
305. Ásgeirsson B, Fox JW, Bjarnason JB (1989) Eur J Biochem 180: 85–94
306. Sawada H, Miura M, Yokosawa H, Ishii S-I (1984) Biochem Biophys Res Commun 121: 598–604
307. Sawada H, Yokosawa H, Ishii S-I (1984) J Biol Chem 259: 2900–2904
308. Camacho L, Brown JR, Kitto GB (1970) J Biol Chem 245: 3964–3972
309. Pfleiderer J, Zwilling R, Sonneborn H-H (1967) Hoppe-Seyler's Z Physiol Chem 348: 1319–1331
310. Titani K, Sasagawa T, Woodbury RG, Ericsson LH, Dörsam H, Kraemer M, Neurath H, Zwilling R (1983) Biochemistry 22: 1459–1465
311. Jany KD, Bekelar K, Pfleiderer G, Ishay J (1983) Biochem Biophys Res Commun 110: 1–7
312. Kafotos FC, Tartakoff AM, Law JH (1967) J Biol Chem 242: 1477–1487
313. Lundblad RL, Kingdon HS, Mann KG (1976) Meth Enzymol 45B: 156–176
314. Butkowski RJ, Elion J, Downing MR, Mann KG (1977) J Biol Chem 252: 4942–4957
315. Mann KG, Elion J, Butkowski RJ, Downing M, Nesheim ME (1981) Meth Enzymol 80C: 286–302
316. Elion J, Downing MR, Butkowski RJ, Mann KG (1977) In: Lundblad RL, Fenton JW, Mann KG (eds) Chemistry and biology of thrombin. Ann Arbor Scientific Publishing, Ann Arbor, pp 97–111
317. Magnusson S, Petersen TE, Sottrup-Jensen L, Claeys H (1975) In: Reich E, Rifkin DB,

Shaw E (eds) Proteases and biological control. Cold Spring Harbor Laboratory, New York, pp 123–145
318. MacGillivray RTA, Davie EW (1984) Biochemistry 23: 1626–1634
319. Irwin DM, Robertson KA, MacGillivray RTA (1988) J Mol Biol 200: 31–45
320. Titani K, Fujikawa K, Enfield DL, Ericsson LH, Walsh KA, Neurath H (1975) Proc Natl Acad Sci USA 72: 3082–3086
321. DiScipio RG, Hermondson MA, Yates SG, Davie EW (1977) Biochemistry 16: 698–706
322. Robinson DJ, Furie B, Furie BC, Bing DH (1980) J Biol Chem 255: 2014–2021
323. Hashimoto N, Morita T, Iwanaga S (1985) J Biochem 97: 1347–1355
324. Robbins KC, Summaria L (1976) Meth Enzymol 45B: 257–273
325. Petersen TE, Martzen MR, Ichinose A, Davie EW (1990) J Biol Chem 265: 6104–6111
326. Malinowski DP, Sadler JE, Davie EW (1984) Biochemistry 23: 4243–4250
327. Forsgren M, Råden B, Israelsson M, Larsson K, Hedén L-O (1987) FEBS Lett 213: 254–260
328. Brunisholz RA, Rickli EE (1981) Eur J Biochem 119: 15–22
329. Grant DAW, Magee AI, Hermon-Taylor J (1978) Eur J Biochem 88: 183–189
330. Liepnieks JJ, Light A (1979) J Biol Chem 254: 1677–1683
331. Fonseca P, Light A (1983) J Biol Chem 258: 14516–14520
332. Baratti J, Maroux S, Louvard D, Desnuelle P (1976) Biochim Biophys Acta 452: 488–496
333. Baba T, Watanabe K, Kashiwabara S.-I, Arai Y (1989) FEBS Lett 244: 296–300
334. Schleuning WD, Fritz H (1976) Meth Enzymol 45B: 330–342
335. Brown CR, Andani Z, Hartree EF (1975) Biochem J 149: 133–146
336. Baba T, Kashiwabara S-I, Watanabe K, Itoh H, Michikawa Y, Kimura K, Takada M, Fukamizu A, Arai Y (1989) J Biol Chem 264: 11920–11927
337. Whitaker DR, Jurasek L, Roy C (1966) Biochem Biophys Res Commun 24: 173–178
338. Olson MOJ, Nagabhushan N, Dzwiniel M, Smillie LB, Whitaker DR (1970) Nature 228: 438–442
339. Borovikova VP, Aksenovskaya VE, Lavrenova VE, Kislukhina OV, Kapunyants KA, Stepanov VM (1980) Biokhimiya 45: 1524–1533
340. Jacobs M, Eliasson M, Uhlen M, Flock J-I (1985) Nucl Acid Res 13: 8913–8926
341. Smith EL, DeLange RJ, Ewans WH, Landon M, Markland FS (1968) J Biol Chem 243: 2184–2191
342. Ottesen M, Svendsen I (1970) Meth Enzymol 19: 199–215
343. Markland FS, Smith EL (1971) In: Boyer P (ed) The enzymes, vol 3. Academic Press, New York, pp 561–608
344. Shoer R, Rappoport HP (1972) J Bacteriol 109: 575–583
345. Stahl ML, Ferrari E (1984) J Bacteriol 158: 411–418
346. Wells JA, Ferrari E, Henner DJ, Estell DA, Chen EY (1983) Nucl Acid Res 11: 7911–7925
347. Matsubara H, Kasper CB, Brown DM, Smith EL (1965) J Biol Chem 240: 1125–1130
348. Olaitan SA, De Lange RJ, Smith EL (1968) J Biol Chem 243: 5296–5301
349. Nedkov P, Oberthur W, Braunitzer G (1983) Hoppe-Seyler's Z Physiol Chem 364: 1537–1540
350. Rappaport HP, Riggsby WS, Holden DA (1965) J Biol Chem 240: 78–86
351. Stepanov VM, Strongin AY, Isotova LS, Abramov ZT, Lyublinskaya LA, Ermakova LM, Baratova LA, Belyanova LP (1977) Biochem Biophys Res Commun 77: 298–305
352. Koide Y, Nakamura A, Uozumi T, Beppu T (1986) J Bacteriol 167: 110–116
353. Pacaud M, Uriel J (1971) Eur J Biochem 23: 435–442
354. Pacaud M, Sibilli L, Le Bras G (1976) Eur J Biochem 69: 141–151
355. Pacaud M, Richaud C (1975) J Biol Chem 250: 7771–7779
356. Pacaud M (1976) Eur J Biochem 64: 199–204
357. Regnier P (1981) Biochem Biophys Res Commun 99: 1369–1376
358. Ichihara S, Suzuki T, Suzuki M, Mizushima S (1986) J Biol Chem 261: 9405–9411
359. Palmer SM, St John AC (1987) J Bacteriol 169: 1474–1479
360. Pacaud M (1982) J Biol Chem 257: 4333–4339
361. Sugimura E, Nishihara T (1988) J Bacteriol 170: 5625–5632
362. Goldberg AL, Sreedhara-Swamy KH, Chung C-H, Larimore FS (1981) Meth Enzymol 80: 680–702
363. Chung C-H, Goldberg AL (1983) J Bacteriol 154: 231–238

364. Svendsen I, Genov N, Idakieva K (1986) FEBS Lett 196: 228–232
365. Matsubara H, Feder J (1970) Meth Enzymol 19: 721–795
366. Turková J, Mikeš O, Hayashi K, Danno G, Polgar L (1972) Biochim Biophys Acta 257: 257–263
367. Mikeš O, Turková J, Toan NB, Šorm F (1969) Biochim Biophys Acta 178: 112–117
368. Trop M, Birk Y (1968) Biochem J 109: 475–482
369. Narahashi Y (1970) Meth Enzymol 19: 651–664
370. Henderson G, Krygsman P, Liu CJ, Davey CC, Malek LT (1987) J Bacteriol 169: 3778–3784
371. Jurasek L, Carpenter MR, Smillie LB, Gertler A, Levy S, Ericsson LH (1974) Biochim Biophys Res Commun 61: 1095–1100
372. Narahashi Y, Yoda K (1977) J Biochem 81: 587–597
373. Kaluger SV, Borovikova VP, Lavrenova GI, Stepanov VM, Shpokene AP, Gureeva MP, Uzhkurenas AP (1983) Biokhimiya 48: 1483–1490
374. Lavrenova GI, Gul'nik SV, Kaluger SV, Borovikova VP, Revina LP, Stepanov VM (1984) Biokhimiya 49: 531–539
375. Kreier VG, Rudenskaya GN, Landau NS, Pokrovskaya SS, Stepanov VM, Egorov NS (1983) Biokhimiya 48: 1365–1373
376. Hofsten BV, Van Kley H, Eaker D (1965) Biochim Biophys Acta 110: 585–598
377. Wahlby S (1968) Biochim Biophys Acta 151: 409–413
378. Ebeling W, Hennrich N, Klockow M, Metz H, Orth HD, Lang H (1974) Eur J Biochem 47: 91–97
379. Jany KD, Mayer B (1985) Biol Chem Hoppe-Seyler 366: 485–492
380. Jany KD, Lederer G, Mayer B (1986) FEBS Lett 199: 139–144
381. Gaucher GM, Stevenson KJ (1976) Meth Enzymol 45B: 415–433
382. Gusek TW, Kinsella JE (1987) Biochem J 246: 511–517
383. Taguchi H, Hamaoki M, Matsuzawa H, Ohta T (1983) J Biochem 93: 7–13
384. Matsuzawa H, Tokugawa K, Hamaoki M, Mizoguchi M, Taguchi H, Terada I, Kwon ST, Ohta T (1988) Eur J Biochem 171: 441–447
385. Kwon ST, Terada I, Matsuzawa H, Ohta T (1988) Eur J Biochem 173: 491–497
386. Kwon ST, Matsuzawa H, Ohta T (1988) J Biochem 104: 557–559
387. Sloma A, Rufo GA, Rudolph CF, Sullivan BJ, Theriault KA, Pero J (1990) J Bacteriol 172: 1470–1477
388. Boguslawski G, Shultz JL, Yehle CO (1983) Anal Biochem 132: 41–49
389. Leger RJS, Charnley AK, Cooper RM (1987) Arch Biochem Biophys 253: 221–232
390. Asahi M, Lindquist R, Fukuyama K, Apodaca G, Epstein WL, McKerrow JH (1985) Biochem J 232: 139–144
391. Cowan DA, Smolenski KA, Daniel RM, Morgan HW (1987) Biochem J 247: 121–133
392. Lockau W, Massalski M, Dirmeier A (1988) Eur J Biochem 172: 433–438
393. Vos P, Simons G, Siezen RJ, De Vos WM (1989) J Biol Chem 264: 13579–13585
394. Kleine R, Rothe U, Kettmann U, Schelle H (1981) In: Turk V, Vitale L (eds) Proteinases and their inhibitors. Proc Int Symp, Portoroz, Yugoslavia. Mladinska Knjiga-Pergamon Press, Oxford, pp 201–212
395. Stepanov VM, Chestukhina GG, Rudenskaya GN, Epremyan AS, Osterman AL, Khodova OM, Belyanova LP (1981) Biochem Biophys Res Commun 100: 1680–1687
396. Meloun B, Baudyš M, Kostka V, Hausdorf G, Frömmel C, Höhne WE (1985) FEBS Lett 183: 195–200
397. Jönsson AG (1969) Arch Biochem Biophys 129: 62–67
398. Zwilling VR (1968) Hoppe-Seyler's Z Physiol Chem 349: 326–332
399. Tong NT, Imhoff JM, Lecroisey A, Keil B (1981) Biochim Biophys Acta 658: 209–219
400. Lecroisey A, Tong NT, Keil B (1983) Eur J Biochem 134: 261–267
401. Rydén AC, Rydén L, Philipson L (1974) Eur J Biochem 44: 105–114
402. Drapeau GR (1976) Meth Enzymol 45B: 469–475
403. Drapeau GR (1978) J Biol Chem 253: 5899–6001
404. Carmona C, Gray GL (1987) Nucl Acid Res 15: 6757
405. Arvidson S, Holme T, Lindholm B (1973) Biochim Biophys Acta 302: 135–148
406. Yoshida N, Tsuruyama S, Nagata K, Hirayama K, Noda K, Makisumi S (1988) J Biochem 104: 451–456

407. Mosolova OV, Rudenskaya GN, Stepanov VM, Khodova OM, Caplina IA (1987) Biokhimiya 52: 414–422
408. Khaidarova NV, Rudenskaya GN, Revina LP, Stepanov VM, Egorov NS (1989) Biokhimiya 54: 46–53
409. Vacil'eva KI, Rudenskaya GN, Krect'yanova IN, Khodova OM, Bartoshevich YE, Stepanov VM (1985) Biokhimiya 50: 355–362
410. Rudenskaya GN, Stepanov VM, Zakharova YA, Revina LP, Khodova OM (1987) Biokhimiya 52: 1735–1755
411. Feinstein G, Janoff A (1975) Biochim Biophys Acta 403: 477–492
412. Salvesen G, Farley D, Shuman J, Przybyla A, Reilly C, Travis J (1987) Biochemistry 26: 2289–2293
413. Starkey PM, Barrett AJ (1976) Biochem J 155: 255–263
414. Marossy K, Hauck M, Elödi P (1981) Biochim Biophys Acta 662: 36–40
415. Kraeva LN, Kokryakov VN, Chesnokov IN, Yakovleva MF, Lislova SN (1988) Biokhimiya 53: 655–662
416. Radeliff R, Nemerson Y (1976) Meth Enzymol 45B: 49–56
417. Davie EW, Fujikawa K, Kurachi K, Kisiel W (1979) Adv Enzymol 48: 277–318
418. Kisiel W, Fujikawa K, Davie EW (1977) Biochemistry 16: 4189–4194
419. Byrne R, Link RP, Castellino FJ (1980) J Biol Chem 255: 5336–5341
420. Yoshitake S, Schach BG, Foster DC, Davie EW, Kurachi K (1985) Biochemistry 24: 3736–3750
421. Esmon CT (1979) J Biol Chem 254: 964–973
422. Kane WH, Majerus PW (1981) J Biol Chem 256: 1002–1007
423. Jenny RJ, Pittman DD, Toole JJ, Kriz RW, Aldape RA, Hewick RM, Kaufman RJ, Mann KG (1987) Proc Natl Acad Sci USA 84: 4846–4850
424. Fay PY, Chavin SI, Shroeder D, Young FE, Marder VJ (1982) Proc Natl Acad Sci USA 79: 7200–7204
425. Eaton DL, Hass PE, Riddle L, Mather J, Wiebe M, Gregory T, Vehar GA (1987) J Biol Chem 262: 3285–3290
426. Toole JJ, Knopf JL, Wozney JM, Sultzman LA, Buecker JL, Pittman DD, Kaufman RJ, Brown E, Shoemaker L, Orr EC, Amphlett GW, Foster WB, Coe ML, Knutson GJ, Fass DN, Hewick RM (1984) Nature 312: 342–347
427. Truett MA, Blacher R, Burke RL, Caput D, Chu C, Dina D, Hartog K, Cuo CH, Masiarz FR, Merryweather JP, Najarian R, Pachl C, Potter SJ, Puma J, Ouiroga M, Rall LB, Randolph A, Urdea MS, Valenzuela P, Dahl HH, Favalaro J, Hansen J, Nordfang O, Ezban M (1985) DNA 4: 333–349
428. Bonthron D, Orr EC, Mitsock LM, Ginsburg D, Handin RI, Orkin SH (1986) Nucl Acids Res 14: 7125–7127
429. Kisiel W (1979) J Clin Invest 64: 761–769
430. Long GL, Belagaje RM, MacGillivary RTA (1984) Proc Natl Acad Sci USA 81: 5653–5656
431. Plutzky J, Hoskins JA, Long GL, Crabtree GR (1986) Proc Natl Acad Sci USA 83: 546–550
432. Nakamura T, Morita T, Iwanaga S (1986) Eur J Biochem 154: 511–521
433. Tokunaga F, Miyata T, Nakamura T, Morita T, Kuma K-I, Miyata T, Iwanaga S (1987) Eur J Biochem 167: 405–416
434. Furie BC, Furie B (1976) Meth Enzymol 45B: 191–205
435. Witheiler J, Wilson DB (1972) J Biol Chem 247: 2217–2221
436. Kaneda M, Tominaga N (1975) J Biochem 78: 1287–1296
437. Soeda S, Ohyama M, Yamakawa N, Shimeno H, Nagamatsu A (1984) Chem Pharm Bull 32: 4061–4069
438. Soeda S, Ohyama M, Nagamatsu A (1984) Chem Pharm Bull 32: 1510–1516
439. Yamakawa N, Soeda S, Shimeno H, Nagamatsu A (1986) Chem Pharm Bull 34: 256–263
440. Moriyama A, Nakanishi M, Sasaki M (1988) J Biochem 104: 112–117
441. Knisatschek H, Bauer K (1979) J Biol Chem 254: 10936–10943
442. Rupnow JH, Taylor WL, Dixon JE (1979) Biochemistry 18: 1206–1212
443. Andrews PC, Hines CM, Dixon JE (1980) Biochemistry 19: 5494–5500
444. Yoshimoto T, Simmons WH, Kita T, Tsuru D (1981) J Biochem 90: 325–334

445. Yoshimoto T, Nishimura T, Kita T, Tsuru D (1983) J Biochem 94: 1179–1190
446. Yoshimoto T, Sattar AKMA, Hirose W, Tsuru D (1987) Biochim Biophys Acta 916: 29–37
447. Yoshimoto T, Sattar AKMA, Hirose W, Tsuru D (1988) J Biochem 104: 622–627
448. Yoshimoto T, Walter R, Tsuru D (1980) J Biol Chem 255: 4786–4792
449. Fujikawa K, Chung DW, Hendrickson LE, Davie EW (1986) Biochemistry 25: 2417–2424
450. Nolan C, Hall LS, Barlow GH (1976) Meth Enzymol 45B: 205–213
451. Orthner CL, Bhattacharya P, Strickland DK (1988) Biochemistry 27: 2558–2564
452. Kisiel W, Kondo S, Smith KJ, McMullen BA, Smith LF (1987) J Biol Chem 262: 12607–12613
453. Holleman WH, Weiss LJ (1976) J Biol Chem 251: 1663–1669
454. Itoh N, Tanaka N, Mihashi S, Yamashina I (1987) J Biol Chem 262: 3132–3135
455. Markland FS (1976) Meth Enzymol 45B: 223–236
456. Pirkle H, Markland FS, Theodor I, Baumgartner R, Bajwa SS, Kirakossian H (1981) Biochem Biophys Res Commun 99: 715–721
457. Bjarnason JB, Barish A, Direnzo GS, Campbell R, Fox JW (1983) J Biol Chem 258: 12566–12573
458. Sapru ZZ, Tu AT, Bailey GS (1983) Biochim Biophys Acta 747: 225–231
459. Pandya BV, Budzynski AZ (1984) Biochemistry 23: 460–470
460. Shu Y-Y, Moran JB, Geren CR (1983) Biochim Biophys Acta 748: 236–244
461. Shieh T-C, Tanaca S, Kihara H, Ohno M, Makisumi S (1985) J Biochem 98: 713–721
462. Shieh T-C, Kawabata S-I, Kihara H, Ohno M, Iwanaga S (1988) J Biochem 103: 596–605
463. Barlow GH (1976) Meth Enzymol 45B: 239–244
464. Schaller J, Nick H, Rickli EE, Gillessen D, Lergier W, Studer RO (1982) Eur J Biochem 125: 251–257
465. Winkler ME, Blaber M (1986) Biochemistry 25: 4041–4045
466. Belin D, Vassalli J-D, Cambèpine C, Godeau F, Nagamine Y, Reich E, Kocher HP, Duvorsin RM (1985) Eur J Biochem 148: 225–232
467. Kasai S, Arimura H, Nishida M, Suyama T (1985) J Biol Chem 260: 12382–12389
468. Degen SJF, Rajput B, Reich E (1986) J Biol Chem 261: 6972–6985
469. Eaton DL, Scott RW, Baker JB (1984) J Biol Chem 259: 6241–6247
470. Eisen AZ, Henderson KO, Jeffrey JJ, Bradshaw RA (1973) Biochemistry 12: 1814–1822
471. Grant GA, Sacchettini JC, Welgus HG (1983) Biochemistry 22: 354–358
472. Grant GA, Henderson KO, Eisen AZ, Dradshaw RA (1980) Biochemistry 19: 4653–4659
473. Hurion N, Fromentin H, Keil B (1977) Comp Biochem Physiol B 56: 259–264
474. Hurion N, Fromentin H, Keil B (1979) Arch Biochem Biophys 192: 438–445
475. Sakharov IY, Litvin FE, Artyukov AA, Kofanova NN (1988) Biokhimia 53: 1844–1849
476. Yoshinaka R, Sato M, Itoko M, Yamashita M, Ikeda S (1986) J Biochem 99: 459–467
477. Colman RW, Bagdasarian A (1976) Meth Enzymol 45B: 303–322
478. Sampaio C, Wong S-C, Shaw E (1974) Arch Biochem Biophys 165: 133–139
479. Mandle R, Kaplan AP (1977) J Biol Chem 252: 6097–6104
480. Scott CF, Liu CY, Colman RW (1979) Eur J Biochem 100: 77–83
481. Nagase H, Barrett AJ (1981) Biochem J 193: 187–192
482. Chao J, Chao L, Margolius HS (1984) Biochem Biophys Res Commun 121: 722–729
483. Takahashi S, Irie A, Miyake Y (1988) J Biochem 104: 22–29
484. Takahashi H, Nagasawa S. Suruki T (1972) J Biochem 71: 471–483
485. Kikuno Y, Takahashi H, Suruki T (1983) J Biochem 93: 235–241
486. Lundwall A (1989) Biochem Biophys Res Commun 161: 1151–1159
487. Lazure C, Ledue R, Seidah NG, Chritien M (1984) FEBS Lett 175: 1–7
488. Matsas R, Proud D, Nustad K, Bailey GS (1981) Anal Biochem 113: 264–270
489. Swift GH, Dagorn JC, Ashley PL, Cummings SW, MacDonald RJ (1982) Proc Natl Acad Sci USA 79: 7263–7267
490. Mason AJ, Evans BA, Cox DR, Shine J, Richards RI (1983) Nature 303: 300–307
491. Fahnestock M, Brundage S, Shooter EM (1986) Nucl Acids Res 14: 4823–4835
492. Ban Y, Wang MC, Watt KWK, Loor R, Chu TM (1984) Biochem Biophys Res Commun 123: 482–488
493. Watt KWK, Lee PJ, M'Timkulu T, Chan WP, Loor R (1986) Proc Natl Acad Sci USA 83: 3166–3170

494. Sziegoleit A, Linder D, Schbuter M, Ogawa M, Nishibe S, Fujimoto K (1985) Eur J Biochem 151: 595–599
495. Largman C (1983) Biochemistry 22: 3763–3770
496. Shotton DM, Hartley BS (1970) Nature 225: 802–805
497. McDonald RJ, Swift GH, Quinto C, Swain W, Pictet RL, Nikovits W. Rutter WJ (1982) Biochemistry 21: 1453–1463
498. Swift GH, Craik CS, Stary SJ, Quinto C, Lahaie RG, Rutter WJ, MacDonald RJ (1984) J Biol Chem 259: 14271–14278
499. Hartley BS, Shotton DH (1971) In: Boyer P (ed) The enzymes, 3rd edn, vol 3. Academic Press, New York, pp 323–373
500. Ardelt W (1975) Biochim Biophys Acta 393: 267–273
501. Tani T, Ohsumi J, Mita K, Takigushi Y (1988) J Biol Chem 263: 1231–1239
502. Mallory PA, Travis J (1975) Biochemistry 14: 722–729
503. Kerfelec B, Cambillau C, Puigsever A, Chapus C (1986) Eur J Biochem 157: 531–538
504. Cambillau C, Kerfelec B, Scraky M, Chapus C (1988) FEBS Lett 232: 91–95
505. Shirasu Y, Takemura K, Yoshida H, Sato Y, Iijima H, Shimoda Y, Mikayama T, Ozawa T, Ikeda N, Ishida A, Tamai Y, Matsuki S, Tanaka J-I, Ikenaga H, Ogawa M (1988) J Biochem 104: 259–264
506. Shen W-F, Fletcher TS, Largman C (1987) Biochemistry 26: 3447–3452
507. Feinstein G, Janott A (1975) Biochim Biophys Acta 403: 493–505
508. Baugh RY, Travis J (1976) Biochemistry 15: 836–841
509. Sinha S, Watorek W, Karr S, Biles J, Bode W, Travis J (1987) Proc Natl Acad Sci USA 84: 2228–2232
510. Takahashi H, Nukiwa T, Yoshimura K, Quick CD, States DJ, Holmes MD, Whang-Peng J, Knutsen T, Crystal RG (1988) J Biol Chem 263: 14739–14747
511. Aoki Y (1978) J Biol Chem 253: 2026–2032
512. Okano K, Aoki Y, Sakurai T, Kajitani M, Kanai S, Shimazu T, Shimizu H, Naruto M (1987) J Biochem 102: 13–16
513. Okano K, Aoki Y, Shimizu H, Naruto M (1990) Biochem Biophys Res Commun 167: 1326–1332
514. McKerrow JH, Pino-Heiss S, Lindquist R, Werb Z (1985) J Biol Chem 260: 3703–3707
515. Newport CR, McKerrow JH, Hedstrom R, Petitt M, McGarrigle L, Barr PJ, Agabian N (1988) J Biol Chem 263: 13179–13184
516. Fujikawa K, Walsh KA, Davie EW (1977) Biochemistry 16: 2270–2278
517. Claeys H, Collen D (1978) Eur J Biochem 87: 69–74
518. McMullen BA, Fujikawa K (1985) J Biol Chem 260: 5328–5341
519. Tripodi M, Citarella F, Guida S, Galeffi P, Fantoni A, Cortese R (1986) Nucl Acids Res 14: 3146
520. Johnson LA, Moon KE, Eisenberg M (1986) Anal Biochem 155: 358–364
521. Woodbery RG, Neurath H (1980) FEBS Lett 114: 189–196
522. Woodbery RG, Katunuma N, Kobayashi K, Titani K, Neurath H (1978) Biochemistry 17: 811–819
523. Le Trong H, Parmelee DC, Walsh KA, Neurath H, Woodbury RG (1987) Biochemistry 26: 6988–6994
524. Everitt MT, Neurath H (1979) Biochemie 61: 653–662
525. Kido H, Fukusen N, Katunuma N (1984) Anal Biochem 137: 449–453
526. Woodbury RG, Neurath H (1978) Biochemistry 17: 4298–4304
527. Kido H, Fukusen N, Katunuma N (1985) Arch Biochem Biophys 239: 436–443
528. Benfey PN, Yin FH, Leder P (1987) J Biol Chem 262: 5377–5384
529. Schwartz LB, Lewis RA, Austen KF (1981) J Biol Chem 256: 11939–11943
530. Cromlish JA, Seidah NG, Marcinkiewicz M, Hamelin J, Johnson DA, Curétien M (1987) J Biol Chem 262: 1363–1373
531. Vanderslice P, Craik CS, Nadel JA, Caughey GH (1989) Biochemistry 28: 4148–4155
532. Tanaka K, Nakamura T, Ichihara A (1986) J Biol Chem 261: 2610–2615
533. Meier M, Kwong PC, Fregeau CJ, Atkinson EA, Burrington CJ, Ehrman N, Sorensen O, Lin CC, Wilkins J, Bleackley RC (1990) Biochemistry 29: 4042–4049

534. Simon MM, Hsochutzky H, Fruth U, Simon H-G, Kramer MD (1986) EMBO J 5: 3267–3274
535. Masson D, Tschopp J (1987) Cell 49: 679–685
536. Fruth U, Eckerskorn C, Lottspeich F, Kramer MD, Prester M, Simon MM (1988) FEBS Lett 237: 45–48
537. Poe M, Bennett CD, Biddison WE, Blake JT, Norton GP, Rodkey JA, Sigal MH, Turner RV, Wu JK, Zweerink HJ (1988) J Biol Chem 263: 13215–13222
538. Lobe CG, Finlay BB, Paranchych W, Paetkau VH, Bleackley RC (1986) Science 232: 858–861
539. Gershenfeld HK, Weissman IL (1986) Science 232: 854–858
540. Jenne D, Rey C, Haefeiger J-A, Qiao B-Y, Groscurth P, Tschopp J (1988) Proc Natl Acad Sci USA 85: 4814–4818
541. Jenne DE, Masson D, Zimmer M, Haefliger J-A, Li W-H, Tschopp J (1989) Biochemistry 28: 7953–7961
542. La Bombardi VJ, Shaw E, DiStefano JF, Beck G, Brown F, Zucker S (1983) Biochem J 211: 695–700
543. Cook KS, Groves DL, Min HY, Spiegelman BM (1985) Proc Natl Acad Sci USA 82: 6480–6484
544. Min HY, Spiegelman BM (1986) Nucl Acids Res 14: 8879–8892
545. Riekkinen PJ, Ekfors TO, Hopsu VK (1966) Biochim Biophys Acta 118: 604–620
546. Levy M, Fishman L, Schenkein I (1970) Meth Enzymol 19: 672–681
547. Bothwell MA, Wilson WH, Shooter EH (1979) J Biol Chem 254: 7287–7294
548. Ikeno K, Ikeno T, Kuzuya H, Ishiguro I (1986) J Biochem 99: 1219–1226
549. Kato H, Nakanishi E, Enjyoji K-I, Hayashi I, Oh-Ishi S, Iwanaga S (1987) J Biochem 102: 1389–1404
550. Ashley PL, McDonald RJ (1985) Biochemistry 24: 4512–4520
551. Thibault G, Genest J (1981) Biochim Biophys Acta 660: 23–29
552. Shai S-Y, Woodley-Miller C, Chao J, Chao L (1989) Biochemistry 28: 5334–5343
553. Porter RR, Reid KBM (1979) Adv Protein Chem 33: 1–71
554. Arland GJ, Villiers CL, Chesne S, Colomb MG (1980) Biochim Biophys Acta 616: 116–129
555. Muller-Eberhard HJ (1988) Annu Rev Biochem 57: 321–347
556. Arland GJ, Gagnon J (1983) Biochemistry 22: 1758–1764
557. Leytus SP, Kurachi K, Sakarrassen KS, Davie EW (1986) Biochemistry 25: 4855–4863
558. Arland GJ, Reboul A, Meyer CM, Colomb MG (1977) Biochim Biophys Acta 485: 215–226
559. Porter RR, Sim RB, Reid KBM, Reboul A, Kerr MA, Tack BF, Janatova J, Thomas ML, Harrison RA, Hammer CH, Gail Crossley L, Nussenzweig V, Melton R, Johnson DMA, Gagnon J, Prohaska R (1981) Meth Enzymol 80: 3–153
560. Sim RB, Porter RR, Reid KBM, Gigli J (1977) Biochem J 163: 219–227
561. Carter PE, Dunbar B, Fothergill JE (1983) Biochem J 215: 565–571
562. Thielens NM, Villiers M-B, Reboul A, Villiers CL, Colomb MG (1982) FEBS Lett 141: 19–24
563. Sottrup-Jensen L, Stepanik TM, Kristensen T, Lonblad PB, Jones CM, Wierzbicki DH, Magnusson S, Domdey H, Wetsel RA, Lundwall A, Tack BF, Fey GH (1985) Proc Natl Acad Sci USA 82: 9–13
564. Bentley DR (1986) Biochem J 239: 339–345
565. Goldberger G, Bruns GA, Rits M, Edge MD, Kwiatkowski DJ (1987) J Biol Chem 262: 10065–10071
566. Davis AE (1980) Proc Natl Acad Sci USA 77: 4938–4942
567. Johnson DMA, Gagnon J, Reid KBM (1980) Biochem J 187: 863–874
568. Campbell D, Porter RR (1983) Proc Natl Acad Sci USA 80: 4464–4468
569. Niemann MA, Bhown AS, Bennett JC, Volanakis JE (1984) Biochemistry 23: 2482–2486
570. Johnson DMA, Gagnon J, Reid KBM (1984) FEBS Lett 166: 347–351
571. Mole JE, Niemann MA (1980) J Biol Chem 255: 8472–8476
572. De Bruijn MHL, Fey GH (1985) Proc Natl Acad Sci USA 82: 708–712
573. Hammer CH, Wirtz GH, Renfer L, Gresham HD, Tack BF (1981) J Biol Chem 256: 3995–4006

574. Lambris JD, Muller-Eberhard HJ (1984) J Biol Chem 259: 12685–12690
575. Moehle CM, Tizard R, Lemmon SK, Smart J, Jones EW (1987) Mol Cell Biol 7: 4390–4399
576. Sanada Y, Fujishiro K, Tanaka H, Katunuma N (1979) Biochem Biophys Res Commun 86: 815–821
577. Fujishiro K, Sanada Y, Tanaka H, Katunuma N (1980) J Biochem 87: 1321–1326
578. Kominami E, Hoffschulte H, Holzer H (1981) Biochim Biophys Acta 661: 124–135
579. Farley PC, Shepherd MG, Sullivan PA (1986) Biochem J 236: 177–184
580. Semeekens SP, Steiner DF (1990) J Biol Chem 265: 2997–3000
581. Wagner JC, Escher C, Wolf DH (1987) FEBS Lett 218: 31–34
582. Mizuno K, Nakamura T, Ohshima T, Tanaka S, Matsuo H (1989) Biochem Biophys Res Commun 159: 305–311
583. Fuller RS, Brake A, Thorner J (1989) Proc Natl Acad Sci USA 86: 1434–1438
584. Mizuno K, Nakamura T, Ohshima T, Tanaka S, Matsuo H (1988) Biochem Biophys Res Commun 156: 246–254
585. Lecroisey A, Boulard C, Keil B (1979) Europ J Biochem 101: 385–393
586. Lecroisey A, Gilles AM, De Wolf A, Keil B (1987) J Biol Chem 262: 7546–7551
587. Masaki T, Tanabe M, Nakamura K, Soejima M (1981) Biochim Biophys Acta 660: 44–50
588. Morihara K, Oka T, Tsuzuki H, Tochino Y, Kanaya T (1980) Biochem Biophys Res Commun 92: 396–402
589. Tsunasawa S, Masaki T, Hirose M, Soejima H, Sakiyama F (1989) J Biol Chem 264: 3832–3839
590. Elliott BW, Cohen C (1986) J Biol Chem 261: 11259–11265
591. Allen RJ, Scott GK (1983) Int J Biochem 15: 151–154
592. Levyant MT, Bilinkina VS, Trudolyubova MG, Orekhovich VN (1975) Biokhimiya 40: 1322–1324
593. Langer J, Ansorge S, Bohley P, Welfle H, Bielka H (1977) Acta Biol Med Germ 36: 1729–1733
594. Langer J, Korant BD (1981) In: 4th Symp Intracellular protein catabolism, 1981. Halle, 129 pp
595. Fodor EJB, Ako H, Walsh KA (1975) Biochemistry 14: 4923–4927
596. Carroll EJ (1976) Meth Enzymol 45B: 343–354
597. Urch UA, Hedrick JL (1981) Arch Biochem Biophys 206: 424–431
598. Lepage T, Gache C (1989) J Biol Chem 264: 4787–4793
599. Saklatvala J, Bond JS, Barrett AJ (1981) Biochem J 193: 251–259
600. Charette MF, Henderson GW, Markovitz A (1981) Proc Natl Acad Sci USA 78: 4728–4732
601. Chung CH, Goldberg AL (1981) Proc Natl Acad Sci USA 78: 4931–4935
602. Larimore FS, Waxman L, Goldberg AL (1982) J Biol Chem 257: 4187–4195
603. Amerik AY, Chistyakova AG, Ostroumova NI, Rotanova TV, Gurevich AI, Antonov VK (1988) Bioorgan Khimiya 14: 408–411
604. Chin DT, Goff SA, Webster T, Smith T, Goldberg AL (1988) J Biol Chem 263: 11718–11728
605. Hwang BJ, Park WJ, Chung CH, Goldberg AL (1987) Proc Natl Acad Sci USA 84: 5550–5554
606. Hwang BJ, Woo KM, Goldberg AL, Chung CH (1988) J Biol Chem 263: 8727–8734
607. Katayama Y, Gottesman S, Pumphrey J, Rudikoff S, Clark WP, Maurizi MR (1988) J Biol Chem 263: 15226–15236
608. Swamy KHS, Chung CH, Goldberg AL (1983) Arch Biochem Biophys 224: 543–554
609. Dahlman D, Kopp F, Kuehn L, Niedel B, Pfeifer G, Hegerl R, Baumeister W (1989) FEBS Lett 251: 125–131
610. Ishiura S, Sugita H (1986) J Biochem 100: 753–763
611. Tanaka K, Yoshimura T, Kumatori A, Ichihara A, Ikai A, Nishigai H, Kameyama K, Takagi T (1988) J Biol Chem 263: 16209–16217
612. Tanaka K, Ichihara A (1988) FEBS Lett 236: 159–162
613. Saitoh Y, Yokosawa H, Takahashi K, Ishii S-I (1989) J Biochem 105: 254–260
614. Fujiwara T, Tanaka K, Kumatori A, Shin S, Yoshimura T, Ichihara A, Tokunaga F, Aruga R, Iwanaga S, Kakizuka A, Nakanishi S (1989) Biochemistry 28: 7332–7340
615. Kumatori A, Tanaka K, Tamura T, Fujiwara T, Ichihara A, Tokunaga F, Onikura A, Iwanaga S (1990) FEBS Lett 264: 279–282

616. Tanaka K, Fujiwara T, Kumatori A, Shin S, Yoshimura T, Ichihara A, Tokunaga F, Aruga R, Iwanaga S, Kakizuka A, Nakanishi S (1990) Biochemistry 29: 3777–3785
617. Gottesman S, Squires C, Pichersky E, Carrington M, Hobbs M, Mattick JS, Dalrymple B, Kuramitsu H, Shiroza T, Foster T, Clark WP, Ross B, Squires CL, Maurizi MR (1990) Proc Natl Acad Sci USA 87: 3513–3517
618. Watabe S, Kimura T (1985) J Biol Chem 260: 5511–5517
619. Desautels M, Goldberg AL (1982) J Biol Chem 257: 11673–11679
620. Fried VA, Smith HT, Hildebrandt E, Weiner K (1987) Proc Natl Acad Sci USA 84: 3685–3689
621. Poe M, Wu JK, Florance JR, Rodkey JA, Bennett CD, Hoogsteen K (1983) J Biol Chem 258: 2209–2216
622. Pirkle H,Theodor I, Miyada D, Simmons G (1986) J Biol Chem 261: 8830–8835
623. Lynn KR, Clevette-Radford NA (1985) Phytochemistry 24: 2843–2845
624. Lynn KR, Clevette-Radford NA (1983) Biochim Biophys Acta 746: 154–159
625. Aducci P, Ascenzi P, Pierini M, Ballio A (1986) Plant Physiol 81: 812–818
626. Cromlish JA, Seidah NG, Chretien M (1986) J Biol Chem 261: 10850–10858
627. Cromlish JA, Seidah NG, Chretien M (1986) J Biol Chem 261: 10859–10870
628. Leytus SP, Loeb KR, Hayen FS, Kurachi K, Davie EW (1988) Biochemistry 27: 1067–1074
629. Zisfein JB, Graham RM, Dreskin SW, Wildey GM, Fischman AJ, Homey CJ (1987) Biochemistry 26: 8690–8697
630. Beynon RJ, Kay J (1978) Biochemistry 17: 291–298
631. Beinfeld MC, Bourdais J, Kuks P, Morel A, Cohen P (1989) J Biol Chem 264: 4460–4465
632. Satoh M, Yokosawa H, Ishii S-I (1988) J Biochem 103: 493–498
633. Silberring J, Nyberg F (1989) J Biol Chem 264: 11082–11086
634. Charriaut-Marlangue C, Barel M, Frade R (1986) Biochem Biophys Res Commun 140: 1113–1120
635. Kohle D, Kauss H (1984) Biochim Biophys Acta 799: 59–67
636. Hagiwara H, Miyazaki K, Matuo Y, Yamashita J, Horio T (1981) Biochim Biophys Acta 660: 73–82
637. Tsurugi K, Ogata K (1982) J Biochem 92: 1369–1381
638. Wilson WH, Shooter EM (1979) J Biol Chem 254: 6002–6009
639. Thomas KA, Baglan NC, Bradshaw RA (1981) J Biol Chem 256: 9156–9166
640. Taylor JM, Mitchell WM, Cohen S (1974) J Biol Chem 249: 2188–2194
641. Lundren S, Ronne H, Rask L, Peterson PA (1984) J Biol Chem 259: 7780–7784
642. Legrand Y, Pignaud G, Caen JP (1977) FEBS Lett 76: 294–298
643. Hatcher VB, Lazarus GS, Levine N, Burk PG, Yost FJ (1977) Biochim Biophys Acta 483: 160–171
644. Orlowski M, Flick MR, Rand J, Lesser M (1987) Arch Biochem Biophys 254: 156–169
645. Zwizinski C, Wickner W (1980) J Biol Chem 255: 7973–7977
646. Wolfe PB, Wickner W, Goodman JM (1983) J Biol Chem 258: 12073–12080
647. Greenburg G, Shelness GS, Blobel G (1989) J Biol Chem 264: 15762–15765
648. Innis MA, Tokunaga M, Williams ME, Loranger JM, Chang S-Y, Chang S, Wu HC (1984) Proc Natl Acad Sci USA 81: 3708–3712
649. Yu F, Yamada H, Daishima K, Mizushima S (1984) FEBS Lett 173: 264–268
650. McEntee K, Hesse JE, Epstein W (1976) Proc Natl Acad Sci USA 73: 3979–3983
651. Roberts JW, Roberts CW, Craig NL (1978) Proc Natl Acad Sci USA 75: 4714–4718
652. Horii T, Ogawa T, Ogawa H (1980) Proc Natl Acad Sci USA 77: 313–317
653. Chatterjee PK, Flint SJ (1987) Proc Natl Acad Sci USA 84: 714–718
654. Kräusslich.H-G, Wimmer E (1988) Annu Rev Biochem 57: 701–754
655. Gorbalenya AE, Koonin EV, Blinov VM, Donchenko AP (1988) FEBS Lett 236: 287–290
656. Bazan JF, Fletterick RJ (1989) FEBS Lett 249: 5–7
657. Boege U, Wengler G, Wengler Gerd, Wittman-Liebold B (1981) Virology 113: 293–303
658. Wriston JC, Yellin TO (1973) Adv Enzymol 39: 185–248
659. Jennings MP, Beacham IR (1990) J Bacteriol 172: 1491–1498
660. Maita T, Matsuda G (1980) Hoppe-Seyler's Z Physiol Chem 361: 105–117
661. Tosa T, Sano R, Yamamoto K, Nakamura M, Chilafa T (1973) Biochemistry 12: 1075–1079
662. Yonei M, Mitsui Y, Iitaka Y (1977) J Mol Biol 110: 179–186

663. Davidson L, Burcom M, Ahn S, Chang L-C, Kitto B (1977) Biochim Biophys Acta 480: 282–294
664. Dunlop PC, Meyer GM, Ban D, Roon RJ (1978) J Biol Chem 253: 1297–1304
665. Dunlop PC, Meyer GM, Roon RJ (1980) J Biol Chem 255: 1542–1546
666. Cole M (1964) Nature 203: 519–520
667. Kutzbach C, Rauenbusch E (1974) Hoppe-Seyler's Z Physiol Chem 355: 45–53
668. Chiang C, Bennett RE (1967) J Bacteriol 93: 302–308
669. Furth AJ (1975) Biochim Biophys Acta 377: 431–443
670. Fisher JF, Knowles JR (1978) Annu Rep Med Chem 13: 239–248
671. Knott-Hunziker V, Redhead K, Petursson S, Waley SG (1980) FEBS Lett 121: 8–10
672. Duez C, Frére J-M, Klein D, Noël M, Ghuysen JM, Delcame L, Dierickx L (1981) Biochem J 193: 75–82
673. Mezes PSF, Yang YQ, Hussain M, Lampen JO (1983) FEBS Lett 161: 195–200
674. Dale JW, Godwin D, Mossakowska D, Stephenson P, Wall S (1985) FEBS Lett 191: 39–44
675. Joris B, De Meester F, Galleni M, Masson S, Dusart J, Frére J-M, Van Beeumen J, Bush K, Sykes R (1986) Biochem J 239: 581–586
676. Jaurin B, Grundstrom T (1981) Proc Natl Acad Sci USA 78: 4897–4901
677. Yoshimoto T, Saeki T, Tsuru D (1983) J Biochem 93: 469–477
678. Yoshimoto T, Tsuru D (1985) J Biochem 97: 1477–1485
679. Ninomiya K, Kawatani K, Tanaka S, Kawata S, Makisumi S (1982) J Biochem 92: 413–421
680. Mantle D, Lauffart B, McDermott JR, Kidd AM, Pennington RJT (1985) Eur J Biochem 147: 307–312
681. Mäkinen KK, Mäkinen P-L (1978) Biochem J 175: 1051–1067
682. Kawata S, Takayama S, Ninomiya K, Makisumi S (1980) J Biochem 88: 1025–1032
683. Hiraoka BY, Fukasawa K, Harada M (1983) J Biochem 94: 1201–1208
684. Sebti SM, Mignano JE, Jani JP, Srimatkandada S, Lazo JS (1989) Biochemistry 28: 6544–6548
685. Chulkova TM, Orekhovich VN (1982) Biokhimiya 47: 1091–1095
686. Morgan JG, Donlon J (1985) Eur J Biochem 146: 429–435
687. Endo F, Tanoue A, Nakai H, Hata A, Indo Y, Titani K, Matsuda I (1989) J Biol Chem 264: 4476–4481
688. Davis NC, Smith EL (1957) J Biol Chem 224: 261–275
689. Yoshimoto T, Matsubara F, Kawano E, Tsuru D (1983) J Biochem 94: 1889–1896
690. Browne P, O'Cuinn G (1983) J Biol Chem 258: 6147–6154
691. Haley EE (1970) Meth Enzymol 19: 737–741
692. Haley EE (1970) Meth Enzymol 19: 730–737
693. Haley EE (1968) J Biol Chem 243: 5748–5752
694. Bouma JMW, Scheper A, Duursma A, Gruber H (1976) Biochim Biophys Acta 444: 853–862
695. McDonald JK, Callahan PX, Ellis S (1972) Meth Enzymol 25: 272–281
696. Lynn KR, Labow RS (1985) Can J Biochem Cell Biol 62: 1301–1308
697. Hui K-S (1988) J Biol Chem 263: 6613–6618
698. Dunn NW, McQuillan HT (1971) Biochim Biophys Acta 235: 149–158
699. Tamura T, Imal Y, Strominger JL (1976) J Biol Chem 251: 414–423
700. Ninjoor V, Taylor SL, Tappel AA (1974) Biochim Biophys Acta 370: 308–321
701. Lones M, Chatterjee R, Singh H, Kalnitsky G (1983) Arch Biochem Biophys 221: 64–78
702 Lipperheide C, Otto K (1986) Biochim Biophys Acta 880: 171–178
703. Szewczuk A, Kwaitkowska J (1970) Eur J Biochem 15: 92–96
704. Bauer K, Nowak P, Kleinkauf H (1981) Eur J Biochem 118: 173–176
705. Doolittle RF, Armentrout RW (1968) Biochemistry 7: 516–521
706. Doolittle RF (1970) Meth Enzymol 19: 555–569
707. Tsuru D, Fujiwara K, Kado K (1978) J Biochem 84: 467–476
708. Fujiwara K, Kitogawa T, Tsuru D (1981) Biochim Biophys Acta 658: 10–16
709. Barrett AJ (1973) Biochem J 131: 809–822
710. Ritonja A, Popovic T, Turk V, Wiedenmann K, Machleidt W (1985) FEBS Lett 181: 169–172

711. Chan SJ, San Segundo B, McCormick MB, Steiner DF (1986) Proc Natl Acad Sci USA 83: 7721–7725
712. Fong D, Calhoun DH, Hsieh W-T, Lee B, Wells RD (1986) Proc Natl Acad Sci USA 83: 2909–2913
713. Suhar A, Marks N (1979) Eur J Biochem 101: 23–30
714. MacGregor RR, Hamilton JW, Shofstall RE, Cohn DV (1979) J Biol Chem 254: 4423–4427
715. Towatari T, Kawabata Y, Katunuma N (1979) Eur J Biochem 102: 279–289
716. Olstein AD, Liener IE (1983) J Biol Chem 258: 11049–11056
717. Takahashi T, Dehdarani AH, Schmidt PG, Tang J (1984) J Biol Chem 259: 9874–9882
718. Takahashi T, Yonezawa S, Dehdarani AH, Tang J (1986) J Biol Chem 261: 9368–9374
719. Mort JS, Leduc MS, Recklies AD (1983) Biochim Biophys Acta 755: 369–375
720. Takio K, Towatari T, Katunuma N, Titani K (1980) Biochem Biophys Res Commun 97: 340–346
721. Takio K, Towatari T, Katunuma N, Teller DC, Titani K (1983) Proc Natl Acad Sci USA 80: 3666–3670
722. Gay NJ, Walker JE (1985) Biochem J 225: 707–712
723. Meloun B, Baudys M, Pohl J, Pavlik M, Kostka V (1988) J Biol Chem 263: 9087–9093
724. Wada K, Tanabe T (1988) J Biochem 104: 472–476
725. Blumberg S, Schechter I, Berger A (1970) Eur J Biochem 15: 97–102
726. Husain SS, Lowe G (1969) Biochem J 114: 279–288
727. Liener JE, Friedenson B (1970) Meth Enzymol 19: 261–273
728. Husain SS, Lowe G (1970) Biochem J 117: 333–340
729. Metrione RM, Johnston RB, Seng R (1967) Arch Biochem Biophys 122: 137–143
730. Murachi T (1976) Meth Enzymol 45B: 475–485
731. Ota S, Muta E, Katahira Y, Okamoto Y (1985) J Biochem 98: 219–228
732. Ritonja A, Rowan AD, Buttle DJ, Rawlings ND, Turk V, Barrett AJ (1989) FEBS Lett 247: 419–424
733. Pal G, Sinha NK (1980) Arch Biochem Biophys 202: 321–329
734. Rowan AD, Buttle DJ, Barrett AJ (1988) Arch Biochem Biophys 267: 262–270
735. Watson DC, Yaguchi M, Lynn KR (1990) Biochem J 266: 75–81
736. Glazer AN, Smith EL (1971) In: Boyer P (ed) The enzyme, 3rd edn, vol 3. Academic Press, New York, pp 501–546
737. Baines BS, Brockleh;urst K (1974) Biochem J 177: 541–548
738. Polgár L (1981) Biochim Biophys Acta 658: 262–269
739. Khan IU, Polgár L (1983) Biochim Biophys Acta 760: 350–356
740. Lynn KR (1979) Biochim Biophys Acta 569: 193–201
741. Dubois T, Kleinschmidt T, Schnek AG, Looze Y, Braunitzer G (1988) Biol Chem Hoppe-Seylers 369: 741–754
742. Buttle DJ, Kembhavi AA, Sharp SL, Shute RE, Rich DH, Barrett AJ (1989) Biochem J 261: 469–476
743. Gilles A-M, Imhoff J-M, Keil B (1979) J Biol Chem 254: 1462–1468
744. Gilles A-M, Lecroisey A, Keil B (1984) Eur J Biochem 145: 469–476
745. Liu T-Y, Neumann NP, Elliott SD, Moor S, Stein WH (1963) J Biol Chem 238: 251–256
746. Liu T-I, Elliott SD (1971) In: Boyer P (ed) The enzyme, 3rd edn, vol 3. Academic Press, New York, pp 609–647
747. Tai JY, Kortt AA, Liu T-Y, Elliott SD (1976) J Biol Chem 251: 1955–1959
748. Silink M, Reddel R, Bethel M, Rowe PB (1975) J Biol Chem 250: 5982–5994
749. Saini PK, Rosenberg IH (1974) J Biol Chem 249: 5131–5134
750. Elsenhans B, Ahmad O, Rosenberg IH (1984) J Biol Chem 259: 6364–6368
751. McDowall MA (1970) Eur J Biochem 14: 214–221
752. Carne A, Moore CH (1978) Biochem J 173: 73–83
753. Mason RW, Green GDJ, Barrett AJ (1985) Biochem J 226: 233–242
754. Ritonja A, Popovic T, Kotnik M, Machleidt W, Turk V (1988) FEBS Lett 228: 341–345
755. Kirschke H, Langner J, Wiederanders B, Ansorge S, Bohley P (1977) Eur J Biochem 74: 293–301
756. Okitani A, Matsukura U, Kato H, Fujimaki M (1980) J Biochem 87: 1133–1143

757. Mason RW, Taylor MAJ, Etherington DJ (1982) FEBS Lett 146: 33–36
758. Bando Y, Kominami E, Katunuma N (1986) J Biochem 100: 35–42
759. Wada K, Takai T, Tanabe T (1987) Eur J Biochem 167: 13–18
760. Dufour E, Obled A, Valin C, Béchet D, Ribadeau-Dumas B, Huet JC (1987) Biochemistry 26: 5689–5695
761. Towatari T, Katunuma N (1988) FEBS Lett 236: 57–61
762. Tanaka K, Ikegaki N, Ichihara A (1984) J Biol Chem 259: 5937–5944
763. Singh H, Kalnitsky G (1978) J Biol Chem 253: 4319–4326
764. Schwartz WN, Barrett AJ (1980) Biochem J 191: 487–497
765. Ishidoh K, Kominami E, Katunuma N, Suzuki K (1989) FEBS Lett 253: 103–107
766. Whellock MJ (1982) J Biol Chem 257: 12471–12474
767. Yoshimura N, Kikuchi T, Sasaki T, Kitahara A, Hatanaka M, Murachi T (1983) J Biol Chem 258: 8883–8889
768. Malik MN, Fenko MD, Igbal K, Wisniewski HM (1983) J Biol Chem 258: 8955–8962
769. Pontremoli S, Melloni E, Salamino F, Sparatore B, Michetti M, Horecker BL (1984) Proc Natl Acad Sci USA 81: 53–56
770. Kitahara A, Sasaki T, Kikuchi T, Yumoto N, Yoshimura N, Hatanaka M, Murachi T (1984) J Biochem 95: 1759–1766
771. Sorimashi H, Imajoh-Ohmi S, Emori Y, Kawasaki H, Ohno S, Minami Y, Suzuki K (1989) J Biol Chem 264: 20106–20111
772. Ohno S, Emori Y, Imajoh S, Kawasaki H, Kisaragi M, Suzuki K (1984) Nature 312: 566–570
773. Aoki K, Imajoh S. Ohno S, Emori Y, Koike M, Kosaki G, Suzuki K (1986) FEBS Lett 205: 313–317
774. Malik MN, Feuko MD, Sheikh AM, Kascsak RJ, Tonna-De Masi MS, Wisniewski HM (1987) Biochim Biophys Acta 916: 135–144
775. Mizutani S, Sumi S, Suzuki O, Narita O, Tomoda Y (1984) Biochim Biophys Acta 786: 113–117
776. Alekseenko LI, Zolotov NN, Orekhovich VH (1976) Bioorgan Khimiya 2: 942–949
777. Mizuno K, Miyata A, Kangawa K, Matsuo H (1982) Biochem Biophys Res Commun 108: 1235–1242
778. Devi L, Goldstein A (1984) Proc Natl Acad Sci USA 81: 1892–1896
779. Achstetter T, Ehmann C, Osaki A, Wolf DH (1984) J Biol Chem 259: 13344–13348
780. Achstetter T, Ehmann C, Wolf DH (1985) J Biol Chem 260: 4585–4590
781. Portnoy DA, Erickson AH, Kochan J, Ravetch JV, Unkeless JC (1986) J Biol Chem 261: 14697–14703
782. Evans P, Etherington DJ (1978) Europ J Biochem 83: 87–97
783. Evans P (1979) FEBS Lett 99: 55–58
784. Kirschke H, Ločnikar P, Turk V (1984) FEBS Lett 174: 123–127
785. Kirschke H, Schmidt I, Wiederanders B (1986) Biochem J 240: 455–459
786. Gohda E, Pitot HC (1981) J Biol Chem 256: 2567–2572
787. Gohda E, Pitot HC (1981) Biochim Biophys Acta 659: 114–122
788. Liao JCR, Lenney JF (1984) Biochem Biophys Res Commun 124: 909–916
789. Rogers JC, Dean D, Heck GR (1985) Proc Natl Acad Sci USA 82: 6512–6516
790. Kageyama T, Takahashi SY, Takahashi K (1981) J Biochem 90: 665–671
791. Hayashi H (1975) Int Rev Cytol 40: 101–151
792. Shroyer LA, Varandani PT (1985) Arch Biochem Biophys 236: 205–219
793. McKenzie RA, Burghen GA (1984) Arch Biochem Biophys 229: 604–611
794. Falanga A, Gordon SG (1985) Biochemistry 24: 5558–5567
795. Darby NJ, Smyth DG (1988) Biochem Biophys Res Commun 153: 1193–1200
796. Molla A, Yamamoto T, Maeda H (1988) J Biochem 104: 616–621
797. Gustafson GL, Thon LA (1979) J Biol Chem 254: 12471–12478
798. Williams JG, North MJ, Mahbubani H (1985) EMBO J 4: 999–1006
799. Busconi L, Folco EJ, Martone C, Trucco RE, Sánger JJ (1984) FEBS Lett 176: 211–214
800. Nordwig A (1971) Adv Enzymol 34: 155–205
801. Aswanikumar S, Radhakrishnan AN (1975) Biochim Biophys Acta 384: 194–202
802. Morales TT, Wolssuer JF (1977) J Biol Chem 252: 4855–4860

803. Tisljar U, Barrett AJ (1989) Arch Biochem Biophys 274: 138–144
804. Tisljar U, De Camargo ACM, Da Costa CA, Barrett AJ (1989) Biochem Biophys Res Commun 162: 1460–1464
805. Wilk S, Pearce S, Orlowski M (1979) Life Sci 24: 457–464
806. Okitani A, Nishimura T, Kato H (1981) Eur J Biochem 115: 269–274
807. Shinagawa T, Do YS, Baxter JD (1990) Proc Natl Acad Sci USA 87: 1927–1931
808. Ismail F, Gevers W (1983) Biochim Biophys Acta 742: 399–408
809. Rivett AJ (1985) J Biol Chem 260: 12600–12606
810. McGuire MJ, De Martino GN (1986) Biochim Biophys Acta 873: 279–289
811. Zolfaghari R, Baker CRF, Amirgholami A, Canizaro PC, Behal FJ (1987) Arch Biochem Biophys 258: 42–50
812. McGuire MJ, McCullough ML, Croall DE, De Martino GN (1989) Biochim Biophys Acta 995: 181–186
813. McGuire MJ, Croall DE, De Martino GN (1988) Arch Biochem Biophys 262: 273–285
814. Waxman L, Fagan JM, Tanaka K, Goldberg AL (1985) J Biol Chem 260: 11994–12000
815. Bazan JF, Fletterick RJ (1988) Proc Natl Acad Sci USA 85: 7872–7876
816. Gorbalenya AE, Donchenko AP, Blinov VM, Koonin EV (1989) FEBS Lett 243: 103–114
817. Gorbalenya AE, Svitkin YV (1983) Biokhimiya 48: 442–453
818. Konig H, Rosenwirth B (1988) J Virol 62: 1243–1250
819. Hardy WR, Strauss JH (1989) J Virol 63: 4653–4664
820. Kordel W, Schneider F (1975) Hoppe-Seyler's Z Physiol Chem 356: 915–920
821. Kordel W, Schneider F (1976) Biochim Biophys Acta 445: 446–457
822. Reitz MS, Rodwell VW (1970) J Biol Chem 245: 3091–3096
823. Broom MF, Sherriff RM, Tate WP, Collings J, Chadwick VS (1989) Biochem J 257: 51–56
824. Imai T (1973) J Biochem 73: 139–153
825. Kikuchi M, Sakaguchi K (1976) Meth Enzymol 45B: 485–492
826. Fukuda H, Iwade S, Kimura A (1982) J Biochem 91: 1731–1738
827. Matsumoto J, Nagai S (1972) J Biochem 72: 269–279
828. Schmid PC, Zuzarte-Augustin ML, Schmid HHO (1985) J Biol Chem 260: 14145–14149
829. Osada H, Isono K (1986) Biochem J 233: 459–463
830. Tang J (1970) Meth Enzymol 19: 406–421
831. Roberts NB, Taylor WH (1978) Biochem J 169: 607–615
832. Sogawa K, Fujii-Kuriyama Y, Mizukami Y, Ichihara Y (1983) J Biol Chem 258: 5306–5311
833. Athauda SBP, Tanji M, Kageyama T, Takahashi K (1989) J Biochem 106: 920–927
834. Ryle AP (1970) Meth Enzymol 19: 316–336
835. Furihata C, Saito D, Fujiki H, Kanai Y, Matsushima T, Sugimura T (1980) Eur J Biochem 105: 43–50
836. Stepanov VM, Lavrenova GI, Rudenskaya GI, Gonchar MI, Gonchar MI, Lobareva LS, Kotlova EK, Strongin AY, Baratova LI, (1976) Biokhimiya 41: 1285–1290
837. Harboe MK, Andersen PM, Foltmann B, Kay J, Kassell B (1974) J Biol Chem 249: 4487–4494
838. Sepulveda P, Marciniszyn J, Liu D, Tang J (1975) J Biol Chem 250: 5082–5088
839. Kageyama T, Moriyama A, Takahashi K (1983) J Biochem 94: 1557–1567
840. Kageyama T, Takahashi K (1986) J Biol Chem 261: 4395–4405
841. Ishihara T, Ichihara Y, Hayano T, Katsura I, Sogawa K, Fujii-Kuriyama Y, Takahashi K (1989) J Biol Chem 264: 10193–10199
842. Merrett TR, Bar-Eli E, Van Vunakis H (1969) Biochemistry 8: 3696–3702
843. Bohak Z (1969) J Biol Chem 244: 4638–4648
844. Bohak Z (1970) Meth Enzymol 19: 347–363
845. Baudyš M, Kostka V (1983) Eur J Biochem 136: 89–99
846. Taggart RT, Cass LG, Mohandas TK, Derby P, Barr PJ, Pals G, Bell GI (1989) J Biol Chem 264: 375–379
847. Ryle AP, Hamilton MP (1966) Biochem J 101: 176–183
848. Ichihara Y, Sogawa K, Morohashi K-I, Fujii-Kuriyama Y, Takahashi K (1986) Eur J Biochem 161: 7–12
849. Minamiura N, Ito K, Kobayashi M, Kobayashi O, Yamamoto T (1984) J Biochem 96: 1061–1069

850. Foltmann B (1970) Meth Enzymol 19: 421–436
851. Foltmann B, Pedersen VB, Kauffman D, Wybrandt G (1979) J Biol Chem 254: 8447–8456
852. Danley DE, Geoghegan KF (1988) J Biol Chem 263: 9785–9789
853. Jensen T, Axelsen NH, Foltmann B (1982) Biochim Biophys Acta 705: 249–256
854. Barrett AJ (1977) Adv Exp Biol Med 95: 291–299
855. Faust PL, Kornfeld S, Chirgwin JM (1985) Proc Natl Acad Sci USA 82: 4910–4914
856. Huang JS, Huang SS, Tang J (1979) J Biol Chem 254: 11405–11417
857. Yamamoto K, Katsuda N, Himeno M, Kato K (1979) Eur J Biochem 95: 459–467
858. Barkhudaryan NA, Akopyan TN, Galoyan AA (1980) Biokhimiya 45: 1293–1297
859. Oikawa T, Iwaguchi T (1984) Chem Pharm Bull 32: 5059–5063
860. Figueiredo AFS, Takii Y, Tsuji H, Kato K, Inagami T (1983) Biochemistry 22: 5476–5481
861. Takahashi T, Tang J (1983) J Biol Chem 258: 6435–6443
862. Shewale JG, Tang J (1984) Proc Natl Acad Sci USA 81: 3703–3707
863. Muto N, Tani S (1985) Biochim Biophys Acta 843: 114–122
864. Srivastava PN, Ninjoor V (1982) Biochem Biophys Res Commun 109: 63–69
865. Watabe S, Taguchi S, Ikeda T, Takada M, Yago N (1982) J Biol Chem 92: 45–55
866. Fukushima K, Gnoh GH, Shinano S (1971) Agric Biol Chem 35: 1495–1502
867. Kageyama T, Takahashi K (1980) J Biochem 87: 725–735
868. Muto N, Arai KM, Tani S (1983) Biochim Biophys Acta 745: 61–69
869. Muto N, Yamamoto M, Tani S (1987) J Biochem 101: 1069–1075
870. Antonov VK, Zilberman MI, Vorotyntseva TI (1985) In: Kostka V (ed) Aspartic proteinases and their inhibitors. De Gruyter, Berlin, pp 129–135
871. Roberts NB, Taylor WH (1978) Biochem J 169: 617–624
872. Yamamoto K, Katsuda N, Kato K (1978) Eur J Biochem 92: 499–508
873. Yamamoto K, Ueno E, Uemura H, Kato Y (1987) Biochem Biophys Res Commun 148: 267–272
874. Yonezawa S, Tanaka T, Miyauchi T (1987) Arch Biochem Biophys 256: 499–508
875. Azuma T, Pals G, Mohandas TK, Couvreur JH, Taggart RT (1989) J Biol Chem 264: 16748–16753
876. Ruenwongsa P, Chulavatnatol M (1977) Adv Exp Biol Med 95: 329–341
877. Afting E-G, Becker M-L (1979) Hoppe-Seyler's Z Phys Chem 360: 222
878. Lobareva LS, Stepanov VM (1978) Uspekhi Biol Khimii 19: 83–105
879. Davidson R, Gertler A, Hofmann T (1975) Biochem J 147: 45–53
880. Tanaka N, Takeuchi M, Ichishima E (1977) Biochim Biophys Acta 485: 406–416
881. Kovaleva GG, Shimanskaya MP, Stepanov VM (1972) Biochem Biophys Res Commun 49: 1075–1081
882. Kovaleva GG, Neuitsova ER, Syrova IA, Revina LP, Timokhina GA, Ostoslavskaya VI, Kotlova EK, Levin ED, Baratova LA, Belyanova LP, Stepanov VM (1979) In: FEBS Spec Meet on Enzymes, 1979, Dubrovnic-Cavtat, S 3, p 53
883. Kumagai I, Yamasaki M, Ui N (1981) Biochim Biophys Acta 659: 334–343
884. Södek J, Hofmann T (1970) Can J Biochem 48: 1014–1016
885. Cunningham A, Wang H-M, Jones SR, Kurosky A, Rao L, Harris CI, Rhee SH, Hofmann T (1976) Can J Biochem 54: 902–914
886. Houmard J, Raymond H-N (1979) Biochemie 61: 979–982
887. Graham JES, Södek J, Hofmann T (1973) Can J Biochem 51: 789–796
888. Delaney R, Wong RNS, Meng G-Z, Wu N-h, Tang J (1987) J Biol Chem 262: 1461–1467
889. Takahashi K (1987) J Biol Chem 262: 1468–1478
890. Takahashi K (1988) J Biochem 103: 162–167
891. Horiuchi H, Yanai K, Okuzaki T, Takagi M, Yano K (1988) J Bacteriol 170: 272–278
892. Holzer H, Bunning P, Neussdoerffer F (1977) Adv Exp Biol Med 95: 271–289
893. Neussdoerffer F, Tortora P, Holzer H (1980) J Biol Chem 255: 12087–12093
894. Magni G, Natalini P, Santarelli I, Vita A (1982) Arch Biochem Biophys 213: 426–433
895. Dreyer T, Halkier B, Svendsen I, Ottensen M (1986) Carlsberg Res Commun 51: 27–41
896. Rucher R (1981) Biochim Biophys Acta 659: 99–113
897. Hagemeyer K, Fawwal I, Whitaker JR (1968) J Dairy Sci 51: 1916–1922
898. Barkholt V (1987) Eur J Biochem 167: 327–338
899. Ottesen M, Rickert W (1970) Meth Enzymol 19: 459–460

References   421

900. Beck A (1981) Neth Milk Dairy 35: 275–280
901. Tonouchi N, Shoun H, Uozumi T, Beppu T (1986) Nucl Acids Res 14: 7557–7568
902. Baudyš M, Foundling S, Pavlik M, Blundell T, Kostka V (1988) FEBS Lett 235: 271–274
903. Carias J-R, Raingeaud J, Mazaud C, Vachon G, Lucas N, Cenatiempo Y, Julien A (1990) FEBS Lett 262: 97–100
904. Gaida AV, Osterman AL, Rudenskaya GN, Stepanov VM (1981) Biokhimiya 46: 181–189
905. Lin X-l, Tang J (1990) J Biol Chem 265: 1490–1495
906. Dickie N, Liener IE (1962) Biochim Biophys Acta 64: 41–51
907. Levy MR, Chou SC (1974) Biochim Biophys Acta 334: 423–430
908. Rudenskaya GN, Gaida AV, Stepanov VM (1980) Biokhimiya 45: 561–568
909. Oda K, Murao S (1974) Agric Biol Chem 38: 2435–2444
910. Morihara K, Tsuzuki H, Murao S, Oda K (1979) J Biochem 85: 661–668
911. Maita T, Nagata S, Matsuda G, Maruta S, Oda K, Murao S, Tsuru D (1984) J Biochem 95: 465–475
912. McKay VL, Welch SK, Insley MT, Manney TR, Holly J, Saari GC, Parker HL (1988) Proc Natl Acad Sci USA 85: 55–59
913. Smith GD, Murray MA, Nichol LW, Trikojus VM (1969) Biochim Biophys Acta 171: 288–298
914. Loh YP, Parish DC, Tuteja R (1985) J Biol Chem 260: 7194–7205
915. Parish DC, Tuteja R, Altstein M, Gainer H, Loh YP (1986) J Biol Chem 261: 14392–14397
916. Berlet HH (1986) FEBS Lett 194: 297–300
917. Amagase S (1972) J Biochem 72: 73–81
918. Shinano S, Fukushima K (1971) Agric Biol Chem 35: 1488–1494
919. Garg GK, Virupaksha TK (1970) Eur J Biochem 17: 4–12
920. Belozersky MA, Dunaevsky YaE, Rudenskaya GN, Stepanov VM (1984) Biokhimiya 49: 479–485
921. Sarbakanova ST, Belozersky MA, Dunaevsky YaE, Zairov SZ (1988) Biokhimiya 53: 768–775
922. Rodrigo I, Vera P, Conejero V (1989) Eur J Biochem 184: 663–669
923. Inagami T, Murakami K (1977) J Biol Chem 252: 2978–2983
924. Yokosawa H, Inagami T, Hass E (1978) Biochim Biophys Acta 83: 306–312
925. Dzau VJ, Slater EE, Haber E (1979) Biochemistry 18: 5224–5237
926. Iwao H, Minami T, Ikemoto F, Takaori K, Nakamura N, Yamamoto K (1982) Biochem Biophys Res Commun 106: 933–939
927. Hirose S, Ohsawa T, Inagami T, Murakami K (1982) J Biol Chem 257: 6316–6321
928. Takahashi S, Ohsawa T, Miura R, Miyake Y (1983) J Biochem 93: 265–274
929. Misono KS, Chang J-J, Inagami T (1982) Proc Natl Acad Sci USA 79: 4858–4862
930. Miyazaki H, Fukamizu A, Hirose S, Hayashi T, Hori H, Ohkubo H, Nakanishi S, Murakami K (1984) Proc Natl Acad Sci USA 81: 5999–6003
931. Shinagawa T, Do Y-S, Tam H, Hsueh WA (1986) Biochem Biophys Res Commun 139: 446–454
932. Holm I, Ollo R, Panthier J-J, Rougeon F (1984) EMBO J 3: 557–562
933. Pontremoli S, Salamino F, Sparatiore B, Melloni E, Morelli A, Benatti U, De Flora A (1979) Biochem J 181: 559–568
934. Yamamoto K, Marchesi VT (1984) Biochim Biophys Acta 790: 208–218
935. Yamamoto K, Takeda M, Yamamoto H, Taysuni M, Kato Y (1985) J Biochem 97: 821–830
936. Tarasova NI, Szecsi PB, Foltmann B (1986) Biochim Biophys Acta 880: 96–100
937. Takeda M, Ueno E, Kato Y, Yamamoto K (1986) J Biochem 100: 1269–1277
938. Gardell SJ, Tate SS (1979) J Biol Chem 254: 4942–4945
939. Pearl LH, Taylor WR (1987) Nature 329: 351–354
940. Seelmeier S, Schmidt H, Turk V, Helm K (1988) Proc Natl Acad Sci USA 85: 6612–6616
941. Graves MC, Lim JJ, Heimer EP, Kramer RA (1988) Proc Natl Acad Sci USA 85: 2449–2453
942. Darke PL, Leu C-T, Davis LJ, Heimbach JC, Dieche RE, Hill WS, Dixon RAF, Sigal IS (1989) J Biol Chem 264: 2307–2312
943. Nam SH, Hatanaka M (1986) Biochem Biophys Res Commun 139: 129–135
944. Shimotohno K, Takahashi Y, Shimizu N, Gojobori T, Golde DW, Chen ISY, Miwa M, Sugimura T (1985) Proc Natl Acad Sci USA 82: 3101–3105

945. Yoshinaka Y, Katoh I, Copeland TD, Oroszlan S (1985) Proc Natl Acad Sci USA 82: 1618-1622
946. Yoshinaka Y, Katoh I, Copeland TD, Smytherss GW, Oroszlan S (1986) J Virology 57: 826-832
947. Sedláček J, Strop P, Kaprálek F, Pečenka V, Kostka V, Travniček M, Riman J (1988) FEBS Lett 237: 187-190
948. Roberts J, Holcenberg JS, Dolowy WC (1972) J Biol Chem 247: 84-90
949. Tanaka S, Robinson EA, Appella E, Miller M, Ammon HL, Roberts J, Weber IT, Wlodawer A (1988) J Biol Chem 263: 8583-8590
950. Himmelhoch SR (1970) Meth Enzymol 19: 508-514
951. Delange RJ, Smith EL (1971) In: Boyer P (ed) The enzymes, 3rd edn, vol 3. Academic Press, New York, pp 81-118
952. Ledeme N, Hennon G, Vincent-Figuet O, Ploguet R (1981) Biochim Biophys Acta 660: 262-270
953. Dounce AL, Allen PZ (1987) Arch Biochem Biophys 257: 13-16
954. Hanson H, Frohne M (1976) Meth Enzymol 45B: 504-521
955. Oettgen HC, Taylor A (1985) Anal Biochem 146: 238-245
956. Cuypers HT, Loon-Klaassen LAH, Vree Egberts WTM, De Jong WW, Bloemendal H (1982) J Biol Chem 257: 7077-7085
957. Metz G, Röhm K-H (1976) Biochim Biophys Acta 429: 933-949
958. Chang Y-H, Smith JA (1989) J Biol Chem 264: 6979-6983
959. Machuga EJ, Ives DH (1984) Biochim Biophys Acta 789: 26-36
960. Kim YS, Brophy EJ (1976) J Biol Chem 251: 3199-3205
961. Gray GM, Santiago NA (1977) J Biol Chem 252: 4922-4928
962. Benajiba A, Maroux S (1980) Eur J Biochem 107: 381-388
963. Feracci H, Maroux S (1980) Biochim Biophys Acta 599: 448-463
964. Svensson B, Sjöström H, Norén O (1982) Eur J Biochem 126: 481-488
965. Olsen J, Cowell GN, Konigshofer E, Danielson EM, Moller J, Laustsen L, Hansen OC, Welinder KG, Endberg J, Hunziker W, Spiess M, Sjöström H, Norén D (1988) FEBS Lett 238: 307-314
966. Behal FJ, Story MN (1969) Arch Biochem Biophys 131: 74-82
967. Desnuelle P (1979) Eur J Biochem 101: 1-11
968. Nakanishi M, Moriyama A, Narita Y, Sasaki M (1989) J Biochem 106: 818-825
969. Watt VM, Yip CC (1989) J Biol Chem 264: 5480-5487
970. Wagner GW, Tavianini MA, Herrmann KM, Dixon JE (1981) Biochemistry 20: 3884-3890
971. Gros C, Giros B, Schwartz J-C (1985) Biochemistry 24: 2179-2185
972. Mantle D, Hardy ME, Lauffart B, McDermott JR, Smith AI, Pennington RJT (1983) Biochem J 211: 567-573
973. Jacobs-Sturm A, Dahlmann B, Reinauer H (1982) Biochim Biophys Acta 715: 34-41
974. Verlinden J, Van Leuven F, Cassiman JJ, Berghe H (1981) FEBS Lett 123: 287-290
975. Yman L (1970) Acta Pharm Suec 7: 75-80
976. Lalu K, Lampelo S, Vanha-Perttula T (1986) Biochim Biophys Acta 873: 190-197
977. Doumeng C, Maroux S (1979) Biochem J 177: 801-808
978. Sach L, Marks N (1982) Biochim Biophys Acta 706: 229-238
979. Nagatsu I, Nagatsu T, Yamamoto T, Glenner GG, Mehl JW (1970) Biochim Biophys Acta 198: 255-270
980. Lalu K, Lampelo S, Nummelin-Kortelainen M, Vanha-Perttula T (1984) Biochim Biophys Acta 789: 324-333
981. Danielsen EM, Norén O, Sjöström H, Ingram J, Kenny AJ (1980) BIochem J 189: 591-603
982. Tobe H, Kojima F, Aoyagi T, Umezawa H (1980) Biochim Biophys Acta 613: 459-468
983. Gorvel JP, Banajiba A, Maroux S (1980) Biochim Biophys Acta 615: 271-274
984. Maze M, Gray GM (1980) Biochemistry 19: 2351-2358
985. Feracci H, Banajiba A, Gorvel JP, Doumeng C, Maroux S (1981) Biochim Biophys Acta 658: 148-157
986. Chulkova TM, Orekhovich VN (1978) Biokhimiya 43: 964-969
987. Dehm P, Nordwig A (1970) Eur J Biochem 17: 364-371

988. Lasch J, Koelsch R, Steinmetzer T, Neumann U, Demuth HU (1988) FEBS Lett 227: 171–174
989. Hooper NM, Hryszko J, Turner AJ (1990) Biochem J 267: 509–515
990. Yaron A, Berger A (1970) Meth Enzymol 19: 521–534
991. Yoshimoto T, Tone H, Honda T, Osatami K, Kobayashi R, Tsuru D (1989) J Biochem 105: 412–416
992. Prescott JM, Wilkes SH (1976) Meth Enzymol 45B: 530–543
993. Ruffin P, Van Brussel E, Biguet J, Breserhe G (1979) Biochemie 61: 495–500
994. Vaganova TI, Ivanova NM, Klepikova FS, Lyublinskaya LA, Yusupova MP, Stepanov VM (1984) Biokhimiya 49: 1899–1907
995. Yusupova MP, Vaganova TI, Lyublinskaya LA, Stepanov VM (1985) Biokhimiya 50: 270–278
996. Spungin A, Blumberg S (1989) Eur J Biochem 183: 471–477
997. Vosbeck KD, Chow K-F, Awad WM (1973) J Biol Chem 248: 6029–6034
998. Miller C, Strauch KL, Kukral AM, Miller JL, Wingfield PT, Mazzei GJ, Werlen RC, Graber P, Movva NR (1987) Proc Natl Acad Sci USA 84: 2718–2722
999. Wingfield P, Graber P, Turcatti G, Movva NR, Pelletier M, Craig S, Rose K, Miller CG (1989) Eur J Biochem 180: 23–32
1000. Wagner FW, Ray LE, Ajabnoor MA, Ziemba PL, Hall RL (1979) Arch Biochem Biophys 197: 63–72
1001. Lazdunski C, Busuttil J, Lazdunski A (1975) Eur J Biochem 60: 363–369
1002. McCaman MT, Villarejo MR (1982) Arch Biochem Biophys 213: 384–394
1003. Bally M, Foglino M, Bruschi M, Murgier M, Lazdunski A (1986) Eur J Biochem 155: 565–569
1004. McCaman MT, Gabe JD (1986) Mol Gen Genet 204: 148–152
1005. Coletti-Previero MA, Mattras H, Descomps B, Previero A (1980) Biochim Biophys Acta 657: 122–127
1006. Traficante LJ, Rotrosen J, Siekierski J, Tracer H, Gerhon S (1980) Life Sci 26: 1697–1706
1007. Nyberg F, Thörnwall M, Hetta J (1990) Biochim Biophys Res Commun 167: 1256–1262
1008. Hersh LB (1981) Biochemistry 20: 2345–2350
1009. Hui K-S, Wang Y-Y, Lajtha A (1983) Biochemistry 22: 1062–1067
1010. Kolesanova EF, Rotanova TV, Amerik AY, Ginodman LM, Antonov VK (1989) Bioorgan Khimiya 12: 340–347
1011. Roncari G, Stoll E, Zuber H (1976) Meth Enzymol 45B: 522–530
1012. Chapuis R, Zuber H (1970) Meth Enzymol 19: 552–555
1013. Rabier D, Desmazland MJ (1973) Biochimie 55: 389–404
1014. Lee CCT, Markel JR (1981) Biochim Biophys Acta 661: 39–44
1015. Markel JR, Lee CCT, Freund TS (1981) Biochim Biophys Acta 661: 32–38
1016. Kessler E, Yaron A (1976) Meth Enzymol 45B: 544–552
1017. Vogt VM (1970) J Biol Chem 245: 4760–4769
1018. Little GH, Starnes WL, Behal FJ (1976) Meth Enzymol 45B: 495–503
1019. Freitas JO, Guimarães JA, Borges DR, Prado JL (1979) Int J Biochem 10: 81–89
1020. Kohno H, Kanda S, Kanno T (1986) J Biol Chem 261: 10744–10748
1021. Kawata S, Imamura T, Ninomiya K, Makisumi S (1982) J Biochem 92: 1093–1101
1022. Sharma KK, Ortwerth BJ (1986) J Biol Chem 261: 4295–4301
1023. Achstetter T, Ehmann C, Wolf DH (1982) Biochem Biophys Res Commun 109: 341–347
1024. Lenney JF (1976) Biochim Biophys Acta 429: 214–219
1025. Kunze N, Kleinkauf H, Bauer K (1986) Eur J Biochem 160: 605–613
1026. Kumon A, Matsuoka Y, Kakimoto Y, Nakajima T, Sano I (1970) Biochim Biophys Acta 200: 466–474
1027. Lenney JF, Baslow MH, Sugiyama GH (1978) Comp Biochem Physiol B 61: 253–259
1028. Semenza G (1957) Biochim Biophys Acta 24: 401–413
1029. Pratt AG, Crawford EJ, Friedkin M (1968) J Biol Chem 243: 6367–6372
1030. Mayer H, Nordwig A (1973) Hoppe-Seyler's Z Physiol Chem 354: 371–379
1031. Akrawi AF, Bailey GS (1976) Biochim Biophys Acta 422: 170–178
1032. Adachi H, Kubota I, Okamura N, Iwata H, Tsujimoto M, Nakazato H, Nishihara T, Noguchi T (1989) J Biochem 105: 957–961

1033. Harper C, René A, Campbell BJ (1971) Biochim Biophys Acta 242: 446–458
1034. Hirota T, Nishikawa Y, Tanaka M, Igarashi T, Kitagawa H (1986) Eur J Biochem 160: 521–525
1035. Patterson EK (1976) Meth Enzymol 45B: 386–393
1036. Patterson EK (1976) Meth Enzymol 45B: 377–386
1037. Ito Y, Sugiura M, Sawaki S (1983) J Biochem 94: 871–877
1038. Brown JL (1973) J Biol Chem 248: 409–416
1039. Lenney JF, Kan S-C, Siu K, Sugiyama GH (1977) Arch Biochem Biophys 184: 257–266
1040. De Lapp NW, Dieckman DK (1988) J Invest Dermatol 90: 490–499
1041. Gee NS, Kenny AJ (1987) Biochem J 246: 97–102
1042. Takeuchi K, Shimizu T, Ohoshi N, Seyama Y, Takaku F, Yotsumoto H (1989) J Biochem 106: 442–445
1043. Das M, Hartley JL, Soffers RL (1977) J Biol Chem 252: 1316–1319
1044. Harris RB, Ohlsson JT, Wilson JB (1981) Anal Biochem 111: 227–234
1045. Takeda Y, Hiwada K, Kokubu T (1981) J Biochem 90: 1309–1319
1046. Lanzillo JJ, Steves J, Dasarathy Y, Yotsumoto H, Fanburg BL (1985) J Biol Chem 260: 14938–14944
1047. Eliseeva YE, Pavlikhina LV, Orekhovich·VN (1974) Dokl Acad Nauk SSSR 217: 953–957
1048. Soffer RL (1976) Annu Rev Biochem 45: 73–94
1049. Nishimura K, Hiwada K, Ueda E, Kokubu T (1976) Biochim Biophys Acta 445: 158–160
1050. Rohrbach MS, Williams EB, Rolstad RA (1981) J Biol Chem 256: 225–230
1051. Kumar RS, Kusari J, Roy SN, Softer RL, Sen GC (1989) J Biol Chem 264: 16754–16758
1052. Sharma M, Singh US (1988) J Biochem 104: 57–61
1053. Yaron A (1976) Meth Enzymol 45B: 599–610
1054. Peterson LM, Sokolovsky M, Vallee BL (1976) Biochemistry 15: 2501–2508
1055. Bazzone TJ, Sokolovsky M, Cueni LB, Vallee BL (1979) Biochemistry 18: 4362–4366
1056. Reynolds DS, Stevens RL, Curley DS (1989) J Biol Chem 264: 20094–20099
1057. Bradshaw RA, Ericsson LH, Walsh KA, Neurath H (1969) Proc Natl Acad Sci USA 63: 1389–1394
1058. Bodwell JE, Meyer WL (1981) Biochemistry 20: 2767–2777
1059. Marinkovic DV, Marinkovic JN, Erdös EG, Robinson CJG (1977) Biochem J 163: 253–260
1060. Folk JE (1971) In: Boyer P (ed) The enzymes, 3rd edn, vol 3. Academic Press, New York, pp 57–79
1061. Prahl JW, Neurath H (1966) Biochemistry 5: 4137–4145
1062. Titani K, Ericsson LH, Walsh KA, Neurath H (1975) Proc Natl Acad Sci USA 72: 1666–1670
1063. Trapeznikova SS, Paskhina TS (1968) Biokhimiya 33: 1012–1022
1064. Plummer TH, Hurwitz MY (1978) J Biol Chem 253: 3907–3912
1065. Tan F, Weerasinghe DK, Skidgel RA, Tamei H, Kaul RK, Roninson IB, Schilling JW, Erdös EG (1990) J Biol Chem 265: 13–19
1066. Gebhard W, Schube M, Eulitz M (1989) Eur J Biochem 178: 603–607
1067. Skidgel RA, Davis RM, Erdös EG (1984) Anal Biochem 140: 520–531
1068. Félix F, Brouillet N (1966) Biochim Biophys Acta 122: 127–144
1069. Levy CC, Goldman P (1968) J Biol Chem 243: 3507–3511
1070. Levy CC, Goldman P (1969) J Biol Chem 244: 4467–4472
1071. Young FE (1966) J Biol Chem 241: 3462–3467
1072. Garnier M, Vacheron M-J, Gumand M, Michel G (1985) Eur J Biochem 148: 539–543
1073. Dideberg O, Joris B, Frére JM, Ghuysen J-M, Weber G, Robaye R, Delbrouck JM, Roelandts J (1980) FEBS Lett 117: 215–218
1074. Joris B, Beeumen J, Casagrande F, Gerday C, Frére JM, Ghuyusen J-H (1983) Eur J Biochem 130: 53–69
1075. Wolf DH, Weiser U (1977) Eur J Biochem 73: 553–556
1076. Manser E, Fernandez D, Loo L, Goh PY, Montries G, Hall C, Lim L (1990) Biochem J 267: 517–525
1077. Fricker LD, Evans CJ, Esch FS, Herbert E (1986) Nature 323: 461–464
1078. Rodriguez C, Brayton KA, Brownstein M, Dixon JE (1989) J Biol Chem 264: 5988–5995
1079. Davidson HW, Hutton JC (1987) Biochem J 245: 575–582

1080. Hook VYH, Affolter H-U (1988) FEBS Lett 238: 338–342
1081. Tan F, Chan SJ, Steiner DF, Schilling JW, Skidgel RA (1989) J Biol Chem 264: 13165–13170
1082. Osterman AL, Stepanov VM, Rudenskaya GN, Khodova OM, Caplina IA, Yakovleva MB, Loginova LG (1984) Biokhimiya 49: 292–301
1083. Narahashi Y (1990) J Biochem 107: 879–886
1084. Rudenskaya GN, Kreiyer VG, Landau NS, Tarasova NI, Timokhina EA, Egorov NS, Stepanov VM (1987) Biokhimiya 52: 2002–2008
1085. Hedeager-Sorensen S, Kenny AJ (1985) Biochem J 229: 251–257
1086. Zwilling R, Jakob F, Bauer H, Neurath H, Enfield DL (1979) Eur J Biochem 94: 223–229
1087. O'Connor B, O'Cuinn G (1985) Eur J Biochem 150: 47–52
1088. Pfleiderer G, Sumyk G (1961) Biochim Biophys Acta 51: 482–493
1089. Bjarnason JB, Tu AT (1978) Biochemistry 17: 3395–3404
1090. Shannon JD, Baramova EN, Bjarnason JB, Fox JW (1989) J Biol Chem 264: 11575–11583
1091. Civello DJ, Duong HL, Geren CR (1983) Biochemistry 22: 749–755
1092. Kurecki T, Laskowski M, Kress LF (1978) J Biol Chem 253: 8340–8345
1093. Takeya H, Arakawa M, Miyata T, Iwanaga S, Omori-Satoh T (1989) J Biochem 106: 151–157
1094. Rothe W, Pfleiderer G, Zwilling R (1970) Hoppe-Seyler's Z Physiol Chem 351: 629–634
1095. Seifter S, Harper E (1971) In: Boyer P (ed) The enzymes, 3rd edn, vol 3. Academic Press, New York, pp 649–697
1096. Bond MD, Van Wart HE (1984) Biochemistry 23: 3085–3091
1097. Wilkes SH, Prescott JM (1976) Meth Enzymol 45B: 404–415
1098. Vaganova TI, Ivanova NM, Stepanov VM (1988) Biokhimiya 53: 1344–1351
1099. Paberit NY, Pank MS, Liiders MA, Vanatalu KP (1984) Biokhimiya 49: 275–283
1100. Yasunobu KT, McConn J (1970) Meth Enzymol 19: 569–575
1101. Yang MY, Ferrari E, Henner DJ (1984) J Bacteriol 160: 15–21
1102. Matsubara H (1970) Meth Enzymol 19: 642–651
1103. Titani K, Hermodson MA, Ericsson LH, Walsh KA, Neurath H (1972) Nature New Biol 238: 35–38
1104. Takagi M, Imanaka T, Aiba S (1985) J Bacteriol 163: 824–831
1105. Dalhammar G, Steiner H (1984) Eur J Biochem 139: 247–252
1106. Vasantha N, Thompson LD, Rhodes G, Banner C, Nagle J, Filpula D (1984) J Bacteriol 159: 811–819
1107. Stoeva S, Kleinschmidt T, Mesrob B, Braunitzer G (1990) Biochemistry 29: 527–534
1108. Cheng Y-SE, Zipser D (1979) J Biol Chem 254: 4698–4706
1109. Finch PW, Wilson RE, Brown K, Hickson ID, Emmerson PT (1986) Nucl Acids Res 14: 7695–7703
1110. Roseman JE, Levine RL (1987) J Biol Chem 262: 2101–2110
1111. Nakajima M, Mizusawa K, Yoshida F (1974) Eur J Biochem 44: 87–96
1112. Gul'nik SV, Lavrenova GI, Stepanov VM (1987) Biokhimiya 52: 1387–1396
1113. Gripon JC, Hermier J (1974) Biochimie 56: 1323–1332
1114. Morihara K, Tsuzuki H, Oka T, Inoue H, Ebata M (1965) J Biol Chem 240: 3295–3304
1115. Fukushima J, Yamamoto S, Morihara K, Atsumi Y, Takeuchi H, Kawamoto S, Okuda K (1989) J Bacteriol 171: 1698–1704
1116. Morihara K (1963) Biochim Biophys Acta 73: 113–124
1117. Miyata K (1970) Agric Biol Chem 34: 1457–1462
1118. Lee IS, Wakabayashi S, Miyata K, Tomoda K, Yoneda M, Kangawa K, Minamini N, Matsuo H, Matsubata H (1984) J Biochem 96: 1409–1418
1119. Nakahama K, Yoshimura K, Marumoto R, Kikuchi M, Lee IS, Hase T, Matsubata H (1986) Nucl Acids Res 14: 5843–5855
1120. Arvidson S (1973) Biochim Biophys Acta 302: 149–157
1121. Yan T-R, Azuma N, Kaminogawa S, Yamauchi K (1987) Eur J Biochem 163: 259–265
1122. Mäkinen P-L, Clewell DB, An F, Mäkinen KK (1989) J Biol Chem 264: 3325–3334
1123. Chandrasekaran S, Dhar SC (1987) Arch Biochem Biophys 257: 402–408
1124. Cowan DA, Daniel RM (1982) Biochim Biophys Acta 705: 293–305
1125. Day WC, Toncic P, Stratman SL, Leeman V, Harmon SR (1968) Biochim Biophys Acta 167: 597–606

1126. Blow AMJ, Van Heyningen R, Barrett AJ (1975) Biochem J 145: 591–599
1127. Prescott JM, Wagner FW (1976) Meth Enzymol 45B: 397–404
1128. Woolley DC, Tucker JS, Green G, Evanson JM (1976) Biochem J 153: 119–126
1129. Harris ED, Cartwright E (1977) In: Barrett AJ (ed) Proteinases of mammalian cells and tissues. North Holland, Amsterdam, pp 249–283
1130. Goldberg GI, Wilhelm SM, Kronberger A, Bauer EA, Grant GA, Eisen AZ (1986) J Biol Chem 261: 6600–6605
1131. Jeffrey JJ, Gross J (1970) Biochemistry 9: 268–273
1132. Iijima K-I, Kishi J-I, Hayakawa T (1981) J Biochem 89: 1101–1106
1133. Lecroisey A, Keil-Dlouha V, Woods DR, Perrin D, Keil B (1975) FEBS Lett 59: 167–172
1134. Keil-Dlouha V (1976) Biochim Biophys Acta 429: 239–251
1135. Trocheris I, Herry P, Keil-Dlouha V, Keil B (1980) Biochim Biophys Acta 615: 436–448
1136. Rippon JW (1968) J Bacteriol 95: 43–46
1137. Landsperger WJ, Peters EH, Dresden MH (1981) Biochim Biophys Acta 661: 213–220
1138. Seltzer JL, Adams SA, Grant GA, Eisen AZ (1981) J Biol Chem 256: 4662–4668
1139. Mikuni-Takagaki Y, Gross J (1984) J Biol Chem 259: 6739–6747
1140. Nethery A, O'Grady RL (1989) Biochim Biophys Acta 994: 149–160
1141. Galloway WA, Murphy G, Sandy JD, Gavrilovic J, Cawston TE, Reynolds JJ (1983) Biochem J 209: 741–752
1142. Chin JR, Murphy G, Werb Z (1985) J Biol Chem 260: 12367–12376
1143. Whitham SE, Murphy G, Angel P, Rahmsdorf H-J, Smith BJ, Lyons A, Harris TJR, Reynolds JJ, Herrlich P, Docherty AJP (1986) Biochem J 240: 913–916
1144. Okada Y, Nadase H, Harris ED (1986) J Biol Chem 261: 14245–14255
1145. Saus J, Quinones S, Otani Y, Nagase H, Harris ED, Kurkinen M (1988) J Biol Chem 263: 6742–6745
1146. Wilhelm SM, Collier IE, Kronberger A, Eisen AZ, Marmer BL, Grant GA, Bauer EA, Goldberg GI (1987) Proc Natl Acad Sci USA 84: 6725–6729
1147. Murphy G, McAlpine CG, Poll CT, Reynolds JJ (1985) Biochim Biophys Acta 831: 49–58
1148. Yu RJ, Harmon SR, Blank F (1968) J Bacteriol 96: 1435–1436
1149. Mumford R, Pierzchala PA, Strauss AW, Zimmerman M (1981) Proc Natl Acad Sci USA 78: 6623–6627
1150. Fulcher IS, Chaplin MF, Kenney AJ (1983) Biochem J 215: 317–323
1151. Fulcher IS, Kenney AJ (1983) Biochem J 211: 743–753
1152. Gee NS, Matsas R, Kenney AJ (1983) Biochem J 214: 377–386
1153. Malfroy B, Schwartz J-C (1984) J Biol Chem 259: 14365–14370
1154. Devault A, Lazure C, Nault C, Le Moual H, Seidah NG, Chrétien M, Kahn P, Powell J, Mallet J, Beaumont A, Roques BP, Crine P, Boileau G (1987) EMBO J 6: 1317–1322
1155. Malfroy B, Schofield PR, Kuang W-J, Seeburg PH, Mason AJ, Henzel WJ (1987) Biochem Biophys Res Commun 144: 59–66
1156. Malfroy B, Kuang W-J, Seeburg PH, Mason AJ, Schofield PR (1988) FEBS Lett 229: 206–210
1157. Beynon RJ, Shannon JD, Bond JS (1981) Biochem J 199: 591–598
1158. Butler PE, McKay MJ, Bond JS (1987) Biochem J 241: 229–235
1159. Orlowski M, Wilk S (1981) Biochemistry 20: 4942–4950
1160. Relton JM, Gee NS, Matsas R, Turner AJ, Kenny AJ (1983) Biochem J 215: 519–523
1161. Lee C-M, Sandberg BEB, Hanley MR, Iversen LL (1981) Eur J Biochem 114: 315–327
1162. Endo S, Yokosawa H, Ishii S-I (1988) J Biochem 104: 999–1006
1163. Miura S, Amaya Y, Mori M (1986) Biochem Biophys Res Commun 134: 1151–1159
1164. Kumamoto T, Ito A, Omura T (1986) J Biochem 100: 247–254
1165. Sterchi EE, Naim HY, Lentze MJ, Hauri H-P, Fransen JAM (1988) Arch Biochem Biophys 265: 105–118
1166. Kirschner RY, Goldberg AL (1983) J Biol Chem 258: 967–976
1167. Barrett D, Edwards BF (1976) Meth Enzymol 45B: 354–373
1168. Yasumasu S, Iuchi I, Yamagami K (1989) J Biochem 105: 204–211
1169. Yasumasu S, Iuchi I, Yamagami K (1989) J Biochem 105: 212–218
1170. Hotez PJ, Le Traug N, McKerrow JH, Cerami A (1985) J Biol Chem 260: 7343–7348
1171. Stöcker W, Wolz RL, Zwilling R, Strydom DJ, Auld DS (1988) Biochemistry 27: 5026–5032

1172. Titani K, Torff H-J, Hormel S, Kumar S, Walsh KA, Rögl J, Neurath H, Zwilling R (1987) Biochemistry 26: 222–226
1173. Taniguchi Y, Akasaka K, Shiroya T, Shimada H (1986) J Biochem 99: 931–938
1174. Kornfeld SJ, Plant AG (1981) Rev Infect Dis 3: 521–533
1175. Halila R, Peltonen L (1984) Biochemistry 23: 1251–1256
1176. Hojima Y, McKenzie J, Rest M, Prockop DJ (1989) J Biol Chem 264: 11336–11345
1177. Njieha FK, Morikawa T, Tuderman L, Prockop DJ (1982) Biochemistry 21: 757–764
1178. Ishida M, Ogawa M, Kosaki G, Mega T, Ikenaka T (1983) J Biochem 94: 17–24
1179. Katayama K, Kuwada M (1984) Biochim Biophys Acta 787: 138–145
1180. Hiroi Y, Natori Y (1987) J Biochem 102: 985–992
1181. Gluschankof P, Morel A, Benoit R, Cohen P (1985) Biochem Biophys Res Commun 128: 1051–1057
1182. Gluschankof P, Gomez S, Morel A, Cohen P (1987) J Biol Chem 262: 9615–9620
1183. Brakch N, Boussetta H, Rholam M, Cohen P (1989) J Biol Chem 264: 15912–15916
1184. Checler F, Vincent J-P, Kitabgi P (1986) J Biol Chem 261: 11274–11281
1185. Kenny AJ, Ingram J (1987) Biochem J 245: 515–524
1186. Vodkin LO, Scandalios JG (1980) Biochemistry 19: 4660–4667
1187. Robinson C, Ellis RJ (1984) Eur J Biochem 142: 337–342
1188. Voskoboinikova NE, Dunaevsky YaE, Belozersky MA, Mishanov-Golikov AV, Vol'fson AD (1989) Biokhimiya 54: 1965–1973
1189. Kohno M, Yamashina I (1972) Biochim Biophys Acta 258: 600–617
1190. Schär HP, Holzmann W, Ramos Tombo GM, Ghisalba O (1986) Eur J Biochem 158: 469–475
1191. Miyake M, Innami T, Kakimoto Y (1983) Biochim Biophys Acta 760: 206–214
1192. Kikuchi M, Koshiyama I, Fukushima D (1983) Biochim Biophys Acta 744: 180–188
1193. Kuwabara S (1970) Biochem J 118: 457–465
1194. Davies RB, Abraham EP (1974) Biochem J 143: 115–127
1195. Baldwin GS, Edwards G, Kiener PA, Tully MJ, Waley SG, Abraham EP (1980) Biochem J 191: 111–116
1196. Shvyadas V, Galaev IYu, Kozlova EV (1984) Biokhimiya 49: 1486–1491
1197. Dayhoff MO (1972) Atlas of protein sequence and structure, vol 5. Natl Biomed Res Found, Washington, DC
1198. Kreil G (1981) Annu Rev Biochem 50: 317–348
1199. Craik CS, Sprang S, Fletterick R, Rutter WJ (1982) Nature 299: 180–182
1200. Rogers J (1985) Nature 315: 458–459
1201. Neurath H, Walsh KA, Winter WP (1967) Science 158: 1638–1640
1202. Cornish-Bowden A (1980) Biochem J 191: 349–354
1203. De Haën C, Neurath H, Teller DC (1975) J Mol Biol 92: 225–259
1204. IUPAC-IUB Commission on biochemical nomenclature (1968) J Biol Chem 243: 3557–3559
1205. Sharma SK, Hopkins TR (1979) Biochemistry 18: 1008–1013
1206. Maroux S, Desnuelle P (1969) Biochim Biophys Acta 181: 59–72
1207. Berezin IV, Kazanskaya NF, Larionova NI (1970) Biokhimiya 35: 983–988
1208. Bender ML, Begué-Cantón ML, Blakeley RL, Brubacher LJ, Feder J, Gunter CR, Kézdy FJ, Killheffer JV, Marshall TH, Miller CG, Roeske RW, Stoops JK (1966) J Am Chem Soc 88: 5890–5913
1209. Metrione RM (1986) TIBS 11: 117–118
1210. Brenner S (1988) Nature 334: 528–530
1211. Foltman B, Pedersen VB (1977) Adv Exp Med Biol 95: 3–22
1212. Tang J (1979) Mol Cell Biochem 26: 93–109
1213. Bargetzi JP, Thompson EOP, Sampath Kumar KSV, Walsh KA, Neurath H (1964) J Biol Chem 239: 3767–3774
1214. Wintersberger E, Cox DJ, Neurath H (1965) Biochemistry 4: 1526–1532
1215. Hooper NM, Turner AJ (1988) FEBS Lett 229: 340–344
1216. Clement GE, Rooney J, Zakheim D, Eastman J (1970) J Am Chem Soc 92: 186–189
1217. Williams SP, Bridger WA, James MNG (1986) Biochemistry 25: 6655–6659
1218. Vermeer C (1990) Biochem J 266: 625–636

1219. Blundell TL, Jonson LN (1976) Protein crystallography. Academic Press, New York
1220. Matthews BW, Sigler PB, Henderson R, Blow DM (1967) Nature 214: 652–656
1221. Birktoft JJ, Blow DM (1972) J Mol Biol 68: 187–240
1222. Blow DM (1974) Isr J Chem 12: 483–494
1223. Raghavan NV, Tulinsky A (1979) Acta Crystallogr 35B: 1776–1785
1224. Blevins RA, Tulinsky A (1985) J Biol Chem 260: 4264–4275
1225. Blevins RA, Tulinsky A (1985) J Biol Chem 260: 8865–8872
1226. Tsukada H, Blow DM (1985) J Mol Biol 184: 703–711
1227. Cohen GH, Silverton EW, Davies DR (1981) J Mol Biol 148: 449–479
1228. Dixon MM, Matthews BW (1989) Biochemistry 28: 7033–7038
1229. Birktoft JJ, Kraut J, Freer ST (1976) Biochemistry 15: 4481–4485
1230. Wang D, Bode W, Huber R (1985) J Mol Biol 185: 595–624
1231. Stroud RM, Kay LM, Dickerson RE (1974) J Mol Biol 83: 185–208
1232. Fehlhammer H, Bode W (1975) J Mol Biol 98: 683–692
1233. Bode W, Schwager P (1975) J Mol Biol 98: 693–717
1234. Chambers JL, Stroud RM (1977) Acta Crystallogr 33B: 1824–1837
1235. Huber R, Bode W (1978) Acc Chem Res 11: 114–122
1236. Chambers JL, Stroud RM (1979) Acta Crystallogr 35B: 1861–1875
1237. Kossiakoff AA, Chambers JL, Kay LM, Stroud RM (1977) Biochemistry 16: 654–664
1238. Fehlhammer H, Bode W, Huber R (1977) J Mol Biol 111: 415–438
1239. Bode W (1979) J Mol Biol 127: 357–374
1240. Sawyer L, Shottou DM, Wendell CPL, Muirhead H, Watson HC, Diamond R, Ladner RC (1978) J Mol Biol 118: 137–208
1241. Meyer EF, Cole G, Radhakrishnan R, Epp O (1988) Acta Crystallogr 44B: 26–38
1242. Bode W, Meyer E, Powers JC (1989) Biochemistry 28: 1951–1963
1243. Bode W, Chen Z, Bartels K, Kutzbach C, Schmidt-Kastner G (1983) J Mol Biol 164: 237–282
1244. Fujinaga M, James MNG (1987) J Mol Biol 195: 373–396
1245. Remington SJ, Woodbury RG, Reynolds RA (1988) Biochemistry 27: 8097–8105
1246. Tulinsky A, Park CH, Skrzypczak-Jankun E (1988) J Mol Biol 202: 885–901
1247. Read RJ, James MNG (1988) J Mol Biol 200: 523–551
1248. Bott R, Ultsch M, Kossiakoff A, Graycar T, Katz B, Power S (1988) J Biol Chem 263: 7895–7906
1249. Matthews DA, Alden RA, Birktoft JJ, Freer ST, Kraut J (1977) J Biol Chem 252: 8875–8883
1250. Drenth J, Hol WGJ, Jansonius JN, Koekoek R (1972) Eur J Biochem 26: 177–181
1251. Teplyakov AV, Kuranova IP, Harutyunyan EN, Frömmel C, Höhne WE (1989) FEBS Lett 244: 208–212
1252. Brayer GD, Delbaere LTJ, James MNG (1978) J Mol Biol 124: 261–283
1253. Sielecki AR, Hendrickson WA, Broughton CG, Delbaere LTJ, Brayer GD, James MNG (1979) J Mol Biol 134: 781–804
1254. Moult J, Sussman F, James MNG (1985) J Mol Biol 182: 555–566
1255. James MNG, Delbaere LTJ, Brayer GD (1978) Can J Biochem 56: 396–402
1256. Greenblatt HH, Ryan CA, James MNG (1989) J Mol Biol 205: 201–228
1257. Brayer GD, Delbaere LTJ, James MNG (1979) J Mol Biol 131: 743–775
1258. Fujinaga M, Delbaere LTJ, Brayer GD, James MNG (1985) J Mol Biol 184: 479–502
1259. Betzel C, Pal GP, Saenger W (1988) Acta Crystallogr 44B: 163–172
1260. Kelly JA, Knox JR, Moews PC, Hite GJ, Bartoloue JB, Zhao H, Joris B, Frére J-M (1985) J Biol Chem 260: 6449–6458
1261. Moews PC, Knox JR, Dideberg O, Charlier P, Frére J-H (1990) Proteins 7: 156–171
1262. Kamphuis IG, Kalk KH, Swarte MBA, Drenth J (1984) J Mol Biol 179: 233–256
1263. Kamphuis IG, Drenth J, Baker EN (1985) J Mol Biol 182: 317–329
1264. Baker EN (1977) J Mol Biol 115: 263–277
1265. Baker EN (1980) J Mol Biol 141: 441–484
1266. Heinemann U, Pal GP, Hildenfeld R, Saenger W (1982) J Mol Biol 161: 591–606
1267. Andreeva NS, Gustchina AE, Fedorov AA, Shutzkever NE, Volnova TV (1977) Adv Exp Med Biol 95: 23–31

1268. Andreeva NS, Fedorov AA, Gustchina AE, Riskulov RR, Shutzkever NE, Safro MG (1978) Mol Biologia 12: 922–935
1269. Andreeva NS, Zhdanov AS, Fedorov AA, Gustchina AE, Shutzkever NE, Nikitina LV, Goryunov AI (1984) Mol Biologia 18: 313–322
1270. Sielecki AR, Fedorov AA, Boodhoo A, Andreeva NS, James MNG (1990) J Mol Biol 214: 148–170
1271. James MNG, Sielecki AR (1986) Nature 319: 33–38
1272. Hartsuck JA, Remington SJ (1988) The 18th Linderstrom-Lang Conf on Aspartic proteinases, 4–8 July 1988, Elsinor, p 28
1273. Safro MG, Andreeva NS, Zhdanov AS (1985) Mol Biologia 19: 400–411
1274. Gilliland GL, Winborne EL, Nachman J, Wlodawer A (1988) The 18th Linderstrom-Lang Conf on Aspartic proteinases, 4–8 July 1988, Elsinor, p 29
1275. Sielecki AR, Hayakawa K, Fujinaga M, Murphy MEP, Fraser M, Muir AK, Carilli CT, Lewicki JA, Baxter JD, James MNG (1989) Science 243: 1346–1351
1276. Hsu IN, Delbaere LTJ, James MNG, Hofmann T (1977) Nature 266: 140–145
1277. James MNG, Sielecki AR (1983) J Mol Biol 163: 299–361
1278. Subramanian E, Swan IDA, Liu M, Davies DR, Jenkins JA, Tickle IJ, Blundell TL (1977) Proc Natl Acad Sci USA 74: 556–559
1279. Wong C-H, Lee TJ, Lu T-H, Yang I-H (1979) Biochemistry 18: 1638–1640
1280. Blundell T, Jenkins JA, Sewell BT, Pearl LH, Cooper JB, Tickle IJ, Veerapandian B, Wood SP (1990) J Mol Biol 211: 919–941
1281. Subramanian E, Liu M, Swan IDA, Davies DR (1977) Adv Exp Biol Med 95: 33–41
1282. Suguna K, Bott RR, Padlan EA, Subramanian E, Sheriff S, Cohen GH, Davies DR (1987) J Mol Biol 196: 877–900
1283. Fraser ME, Hayakawa K, Yoshinaka Y, Katoh I, James MNG (1989) In: Viral proteinases as targets for chemotherapy. Cold Spring Harbor Laboratory, New York (in press)
1284. Miller M, Jaskólski M, Mohana Rao JK, Leis J, Wlodawer A (1989) Nature 337: 576–579
1285. Navia MA, Fitzgerald PMD, McKeever BM, Leu CT, Heimbach JC, Herber WK, Sigal IS, Drake PL, Springer JP (1989) Nature 337: 615–620
1286. Wlodawer A, Miller M, Jaskólski M, Sathyanarayana BK, Baldwin E, Weber IT, Selk LM, Clawson L, Schneider J, Kent SBH (1989) Science 245: 616–621
1287. Rees DC, Lewis M, Lipscomb WN (1983) J Mol Biol 168: 367–387
1288. Rees DC, Lipscomb WN (1983) Proc Natl Acad Sci USA 80: 7151–7154
1289. Hardman KD, Lipscomb WN (1984) J Am Chem Soc 106: 463–464
1290. Schmid MF, Herriott JR (1976) J Mol Biol 103: 175–190
1291. Matthews BW, Jansonius JN, Colman PM, Schoehborn BP, Dupourque D (1972) Nat New Biol 238: 37–41
1292. Kester WR, Matthews BW (1977) J Biol Chem 252: 7704–7710
1293. Holmes MA, Matthews BW (1982) J Mol Biol 160: 623–639
1294. Pauptit RA, Karlsson R, Picot D, Jenkins JA, Niklaus-Reimer A-S, Jansonius JN (1988) J Mol Biol 199: 525–537
1295. Polyakov KM, Obmolova GV, Kuranova IP, Starokonvetov BV, Vainstein BK, Mosolova OV, Stepanov VM, Rudenskaya GN (1989) Mol Biologia 23: 266–272
1296. Dideberg O, Charlier P, Dive G, Joris B, Frére J-M, Ghuysen JM (1982) Nature 299: 469–470
1297. Delbaere LTJ, Hutcheon WLB, James MNG, Thiessen WE (1975) Nature 257: 758–763
1298. Kopp F, Steiner R, Dahlmann B, Kuehn L, Reinauer H (1986) Biochim Biophys Acta 872: 253–260
1299. Baumeister W, Dahlmann B, Hegerl R, Kopp F, Kuehn L, Pfeifer G (1988) FEBS Lett 241: 239–245
1300. Andreeva NS, Gustchina AE (1979) Biochem Biophys Res Commun 87: 32–42
1301. Blundell TL, Sewell BT, McLachlan AD (1979) Biochim Biophys Acta 580: 24–31
1302. Tang J, James MNG, Hsu IN, Jenkins JA, Blundell TL (1978) Nature 271: 618–621
1303. Andreeva NS (1985) Molekular Biol 19: 218–224
1304. Meek TD, Dayton BD, Metcalf BW, Dreyer GB, Strickler JE, Gorniak JG, Rosenberg M, Moore ML, Magaard VM, Debouck C (1989) Proc Natl Acad Sci USA 86: 1841–1845

1305. Kiselev NA, Stel'mashchuk VY, Tsuprun VL, Ludewig M, Hanson H (1977) J Mol Biol 115: 33–43
1306. Taylor A, Volz KW, Lipscomb WN, Takemoto LJ (1984) J Biol Chem 259: 14757–14761
1307. Leatherbarrow RJ, Fersht AR (1987) In: Page MI, Williams A (eds) Enzyme mechanism. R Soc Chem, London, pp 78–96
1308. Jansen EF, Nutting M-DF, Jang R, Balls AK (1949) J Biol Chem 179: 189–199
1309. Schaffer NK, Harshman S, Engle RR (1955) J Biol Chem 214: 799–806
1310. Oosterbaun RA, Kunst P, Rotterdam J, Cohen JA (1958) Biochim Biophys Acta 27: 556–563
1311. Naughton MA, Sanger F, Hartley BS, Shaw DC (1960) Biochem J 77: 149–163
1312. Kramer KJ, Felsted RL, Law JH (1973) J Biol Chem 248: 3021–3028
1313. Gross J, Harper H, Harris ED, McCroskery PA, Highberger JH, Corbett C (1974) Biochem Biophys Res Commun 61: 605–612
1314. Lecroisey A, Keil B (1983) Biochem Biophys Res Commun 112: 907–910
1315. Kaneda M, Ohmine H, Yonezawa H, Tominaga N (1984) J Biochem 95: 825–829
1316. Tomkinson B, Wernstedt C, Hellman U, Zetterqvist Ö (1987) Proc Natl Acad Sci USA 84: 7508–7512
1317. Isackson PJ, Ullrich A, Bradshaw RA (1984) Biochemistry 23: 5997–6002
1318. Duez C, Joris B, Frére J-M, Ghuyusen JM, Van Beeumen J (1981) Biochem J 193: 83–86
1319. Knott-Hunziker V, Petursson S, Jayatilake GS (1982) Biochem J 201: 621–627
1320. Yocum RR, Waxman DJ, Rasmussen JR, Srominger JL (1979) Proc Natl Acad Sci USA 76: 2730–2734
1321. Heckler TG, Day RA (1983) Biochim Biophys Acta 745: 292–300
1322. Balls AK, Aldrich FL (1955) Proc Natl Acad Sci USA 41: 190–194
1323. Schoellmann G, Shaw E (1963) Biochemistry 2: 253–258
1324. Nakagawa Y, Bender ML (1970) Biochemistry 9: 259–263
1325. Henderson R (1971) Biochem J 124: 13–18
1326. Antonov VK, Vorotintseva TI (1972) Dokl Akad Nauk SSSR 202: 1441–1442
1327. Blow DM, Birktoft JJ, Hartley BS (1969) Nature 221: 337–340
1328. Henderson R (1970) J Mol Biol 54: 341–354
1329. Steitz TA, Henderson R, Blow DM (1969) J Mol Biol 46: 337–348
1330. Robertus JD, Kraut J, Alden RA, Birktoft JJ (1972) Biochemistry 11: 4293–4303
1331. Kimmel JR, Smith EL (1954) J Biol Chem 207: 515–531
1332. Balls AK (1939) Nature 144: 513–515
1333. Lowe G, Williams A (1965) Biochem J 96: 194–204
1334. Gilles A-M, DeWolf A, Keil B (1983) Eur J Biochem 130: 473–479
1335. Suzuki K, Hayashi H, Hayashi T, Iwai K (1983) FEBS Lett 152: 67–70
1336. Gilles A-M, Keil B (1984) FEBS Lett 173: 58–62
1337. Clark PI, Lowe G (1978) Eur J Biochem 84: 293–299
1338. Nakayama S-I, Hoshino M, Takahashi K, Watanabe T, Yoshida M (1981) Biochem Biophys Res Commun 98: 471–475
1339. Nakayama S-I, Watanabe T, Takahashi K, Hoshino M, Yoshida M (1987) J Biochem 102: 531–535
1340. Delpierre GR, Fruton JS (1965) Proc Natl Acad Sci USA 54: 1161–1167
1341. Delpierre GR, Fruton JS (1966) Proc Natl Acad Sci USA 56: 1817–1822
1342. Kozlov LV, Ginodman LM, Orekhovich VN (1967) Biokhimiya 32: 1011–1019
1343. Lundblad RL, Stein WH (1969) J Biol Chem 244: 154–160
1344. Hartsuck JA, Tang J (1972) J Biol Chem 247: 2575–2580
1345. James MNG, Hsu I-N, Delbaere LTJ (1977) Nature 267: 808–813
1346. Tsuru D, Shimada S, Maruta S, Yoshimoto T, Oda K, Murao S, Miyata T, Iwanga S (1986) J Biochem 99: 1537–1539
1347. Pechik IV, Gustchina AE, Andreeva NS, Fedorov AA (1989) FEBS Lett 247: 118–122
1348. Argos P, Garavito RM, Eventoff W, Rossman MG, Bränden CI (1978) J Mol Biol 126: 141–158
1349. Vallee BL, Auld DS (1990) Biochemistry 29: 5648–5659
1350. Soubrier F, Alhene-Gelas F, Hubert C, Allegrini J, John M, Tregar G, Corvol P (1988) Proc Natl Acad Sci USA 85: 9386–9390

1351. Harris RB, Wilson IB (1985) J Biol Chem 260: 2208–2211
1352. Wilkes SH, Bayliss ME, Prescott JM (1988) J Biol Chem 263: 1821–1825
1353. Devault A, Nault C, Zollinger M, Fournie-Zaluski M-C, Roques BP, Crine P, Boileau G (1988) J Biol Chem 263: 4033–4040
1354. Frohne M, Kettmann V (1976) Acta Biol Med Germ 35: 353–357
1355. Cuypers HT, Loon-Klaassen LAH, Vree Egberts WTM, De Jong WW, Bloemendal H (1982) J Biol Chem 257: 7086–7091
1356. Imamura T, Kawata S, Ninomiya K, Makisumi S (1983) J Biochem 94: 267–273
1357. Garavito RM, Rossman MG, Argos P, Eventoff W (1977) Biochemistry 16: 5065–5071
1358. Andreeva NS, pers. comm.
1359. Kramer G, Argos P (1981) Biochim Biophys Acta 669: 93–97
1360. Ruply JA (1969) In: Timasheff SN, Fasman GD (eds) Structure and stability of biological macromolecules. Dekker, New York
1361. Rossi GL, Bernhardt SA (1970) J Mol Biol 49: 85–91
1362. Corey RB, Battfay O, Brueckner DA, Mark FG (1965) Biochim Biophys Acta 94: 535–545
1363. Matthews BW, Cohen GN, Silverton EW, Braxton H, Davies DR (1968) J Mol Biol 36: 179–195
1364. Segal DM, Cohen GH, Davies DR, Powers JC, Wilcox PE (1972) Cold Spring Harbor Syinp Quant Biol 36: 85–90
1365. Huang T-H, Bachovchin WW, Griffin RG, Dobson CM (1984) Biochemistry 23: 5933–5937
1366. Spilburg CA, Bethune JL, Vallee BL (1974) Proc Natl Acad Sci USA 71: 3922–3926
1367. Spilburg CA, Bethune JL, Vallee BL (1977) Biochemistry 16: 1142–1150
1368. Johansen JT, Vallee BL (1973) Proc Natl Acad Sci USA 70: 2006–2010
1369. Riordan JF, Muszynska G (1974) Biochem Biophys Res Commun 57: 447–451
1370. Johansen JT, Klyosov AA, Vallee BL (1976) Biochemistry 15: 296–303
1371. Lipscomb WN (1973) Proc Natl Acad Sci USA 70: 3797–3801
1372. Alter GM, Leussing DL, Neurath H, Vallee BL (1977) Biochemistry 16: 3663–3668
1373. Anderson WF, Fletterick RJ, Steitz TA (1974) J Mol Biol 86: 261–269
1374. Hajdu J, Acharya KR, Stuart DI, Barford D, Johnson LN (1988) TIBS 13: 104–109
1375. Lumry R, Biltonen R (1969) In: Timasheff SN, Fasman GD (eds) Structure and stability of biological macromolecules. Dekker, New York, pp 7–173
1376. Likhtenstein GI, Avilova TV (1973) Uspekhi Sovr Biol 75: 26–45
1377. Gurd FRN, Rothgeb TM (1979) Adv Prot Chem 33: 73–165
1378. Kaivariyainen AI (1980) Dynamical behaviour of proteins in water medium and its functions. Nauka, Leningrad
1379. Abaturov LV (1983) Mol Biologia 17: 683–704
1380. Abaturov LV, Varshavsky YM (1978) Mol Biologia 12: 36–46
1381. Harrison LW, Auld DS, Vallee BL (1975) Proc Natl Acad Sci USA 72: 4356–4360
1382. Rebeiro AA, King R, Restivo C, Jardetzky O (1980) J Am Chem Soc 102: 4040–4051
1383. Artymiuk PJ, Blake CCF, Grace DEP, Oatley SJ, Phillips DC, Sternberg MJE (1979) Nature 280: 563–568
1384. McCammon JA, Wolynes PG, Karplus M (1979) Biochemistry 18: 927–942
1385. Scheule RK, Van Wart HE, Vallee BL, Scheraga HA (1980) Biochemistry 19: 759–766
1386. Popov EM (1975) Mol Biologia 9: 578–593
1387. Karplus M (1981) CRC Crit Rev Biochem 9: 283–320
1388. Von Gunsteren WF, Karplus M (1982) Biochemistry 21: 2259–2274
1389. Piszkiewicz D, Bruice TC (1968) Biochemistry 7: 3037–3047
1390. Oliver CN, Stadtman ER (1983) Proc Natl Acad Sci USA 80: 2156–2160
1391. Chatonnet A, Masson P (1985) FEBS Lett 182: 493–498
1392. Boopathy R, Balasubramanian AS (1987) Eur J Biochem 162: 191–197
1393. Tramontano A, Janda KD, Lerner RA (1987) Science 234: 1566–1570
1394. Pollack SJ, Jacobs JW, Schultz PG (1987) Science 234: 1570–1573
1395. Jacobs J, Schultz PG, Sugasawara R, Powell M (1987) J Am Chem Soc 109: 2174–2176
1396. Janda KD, Lerner RA, Tramontano A (1988) J Am Chem Soc 110: 4835–4837
1397. Benkovic SJ, Napper AD, Lerner RA (1988) Proc Natl Acad Sci USA 85: 5355–5358
1398. Tramontano A, Ammann AA, Lerner RA (1988) J Am Chem Soc 110: 2282–2286
1399. Pollack SJ, Schultz PG (1989) J Am Chem Soc 111: 1929–1931

1400. Lerner RA, Tramontano A (1987) TIBS 12: 427–430
1401. Bruice TC, Benkovic SJ (1966) Bioorganic mechanisms. WA Benjamin Inc, New York
1402. Bender ML (1971) Mechanisms of homogeneous catalysis from proton to proteins. Wiley, New York
1403. O'Connor C (1970) Q Rev 24: 553–564
1404. Jencks WP (1969) Catalysis in chemistry and enzymology. McGraw Hill, New York
1405. Hegazi MF, Quinn DM, Schowen RL (1978) In: Gandour RD, Schowen RL (eds) Transition states of biochemical processes. Plenum Press, New York, pp 355–428
1406. Dugas H, Penney C (1981) Bioorganic chemistry: a chemical approach to enzyme action. Springer, Berlin Heidelberg New York
1407. Fersht AR, Requena Y (1971) J Am Chem Soc 93: 3499–3504
1408. Morawetz H, Otaki PS (1963) J Am Chem Soc 85: 463–468
1409. Jencks WP, Schaffhausen B, Tornheim K, White H (1971) J Am Chem Soc 93: 3917–3922
1410. D'yachenko ED, Kozlov LV, Antonov VK (1977) Bioorgan Khimiya 3: 99–104
1411. D'yachenko ED, Kozlov LV, Antonov VK (1971) Biokhimiya 36: 981–984
1412. Fastrez J, Fersht AR (1973) Biochemistry 12: 2025–2034
1413. Kozlov LV, Ginodman LM, Orekhovich VN, Valueva TA (1966) Biokhimiya 31: 315–321
1414. Carpenter FH (1960) J Am Chem Soc 82: 1111–1122
1415. Kozlov LV, Ginodman LM (1965) Biokhimiya 30: 1051–1054
1416. Jencks WP, Cordes S, Carriuolo J (1960) J Biol Chem 235: 3608–3614
1417. Fastrez J (1977) J Am Chem Soc 99: 7004–7013
1418. Sander EG, Jencks WP (1968) J Am Chem Soc 90: 6154–6162
1419. Dobry A, Fruton JS, Sturtevant JM (1952) J Biol Chem 195: 149–154
1420. Sturtevant JM (1955) J Am Chem Soc 77: 1495–1498
1421. Forrest WW, Gutfreund H, Sturtevant JM (1956) J Am Chem Soc 78: 1349–1352
1422. Rawitscher M, Wadsö I, Sturtevant JM (1961) J Am Chem Soc 83: 3180–3184
1423. Bell RP (1973) The proton in chemistry, 2nd edn. Chapman & Hall, London
1424. Martinek K, Semenov AN, Berezin IV (1981) Biochim Biophys Acta 658: 76–89
1425. Kahne D, Still WC (1988) J Am Chem Soc 110: 7529–7534
1426. Yamana T, Mizukami Y, Tsuji A, Yasuda Y, Masuda K (1972) Chem Pharm Bull 20: 881–891
1427. Sluyterman LAE (1968) Biochim Biophys Acta 151: 178–187
1428. Bender ML, Turnquest BW (1955) J Am Chem Soc 77: 4271–4275
1429. Hine J, King RS-M, Midden WR, Sinha A (1981) J Org Chem 46: 3186–3189
1430. Jaffé HH (1953) Chem Rev 53: 191–212
1431. Taft RU (1956) In: Newman MS (ed) Steric effects in organic chemistry. Wiley, New York
1432. Palm VA (1967) Principles of quantitative theory of organic reactions. Khimiya, Leningrad
1433. Shorter J (1970) Q Rev 24: 433–453
1434. Ritchie CD, Sager WF (1964) Progr Phys Org Chem 2: 323–355
1435. De Los De Tar F (1980) J Org Chem 45: 5166–5174
1436. Dmitrieva MG, Khurgin YI (1965) Izves Acad Nauk SSSR, Ser Khim 7: 1174–1180
1437. Idoux JP, Scandrett JM, Sikorski JA (1977) J Am Chem Soc 99: 4577–4583
1438. Charton M, Charton BI (1979) J Org Chem 44: 2284–2287
1439. Koppel IA (1965) Organic Reactivity (Tartu State Univ) 2: 26–32
1440. Hancock CK, Meyers EA, Yager BJ (1961) J Am Chem Soc 83: 4211–4214
1441. Newman MS (1950) J Am Chem Soc 72: 4783–4786
1442. Robinson JR, Matheson LE (1969) J Org Chem 34: 3630–3633
1443. Antonov VK, Rumsh LD (1970) FEBS Lett 9: 67–71
1444. Menger FM, Smith JN (1970) Tetrahedron Lett 4163–4168
1445. Stauffer CE (1972) J Am Chem Soc 94: 7887–7891
1446. Holley AD, Holley RW (1950) J Am Chem Soc 72: 2771–2772
1447. Frére J-M, Kelly JA, Klèin D, Ghuysen JM, Claes P, Vanderhaenhe H (1982) Biochem J 203: 223–234
1448. Rao SN, O'Ferrall RAM (1990) J Am Chem Soc 112: 2729–2735
1449. Pracejus H (1959) Chem Ber 92: 988–998
1450. Somayaji V, Brown RS (1986) J Org Chem 51: 2676–2686
1451. Somayaji V, Skorey KI, Brown RS, Ball RG (1986) J Org Chem 51: 4886–4872

1452. Skorey KI, Somayaji V, Brown RS (1989) J Am Chem Soc 111: 1445–1452
1453. Karve DD, Kelkar BW (1946) Proc Indian Acad Sci 24A: 254–258
1454. Price FP, Hammett LP (1941) J Am Chem Soc 63: 2387–2393
1455. Ikeda K, Kunugi S (1980) J Biochem 88: 977–986
1456. Knudson SK, Idoux JP (1979) J Org Chem 44: 520–523
1457. Weiner SJ, Singh UC, Kollman PA (1985) J Am Chem Soc 107: 2219–2229
1458. Fukuda EK, McIver RT (1979) J Am Chem Soc 101: 2498–2499
1459. Kim JK, Caserio MC (1981) J Am Chem Soc 103: 2124–2127
1460. Scatchard G (1932) Chem Rev 10: 229–245
1461. Amis ES (1966) Solvent effects on reaction rates and mechanisms. Academic Press, New York
1462. Tommila E, Murto ML (1963) Acta Chim Scand 17: 1947–1956
1463. Richards JH (1970) In: Boyer P (ed) The enzymes, vol 2. Academic Press, New York, pp 321–333
1464. Melander L, Saunders WH (1980) Reaction rates of isotopic molecules. Wiley, New York
1465. Bruice TC (1976) Annu Rev Biochem 45: 331–373
1466. Miller AR (1978) J Am Chem Soc 100: 1984–1992
1467. Murdoch JR (1983) J Am Chem Soc 105: 2667–2672
1468. Maggiora GM, Christoffersen RE (1978) In: Gandour RD, Schowen RL (eds) Transition states of biochemical processes. Plenum Press, New York, pp 119–163
1469. Bürgi HB, Dunitz JD, Sheffer E (1973) J Am Chem Soc 95: 5065–5067
1470. Bürgi HB, Dunitz JD, Lehn JM, Wipff G (1974) Tetrahedron 30: 1563–1572
1471. Bürgi HB, Lehn JM, Wipff G (1974) J Am Chem Soc 96: 1956–1957
1472. Lehn JM, Wipff G (1978) Helv Chim Acta 61: 1274–1286
1473. Lehn JM, Wipff G (1980) J Am Chem Soc 102: 1347–1354
1474. Stone AJ, Erskine RW (1980) J Am Chem Soc 102: 7185–7192
1475. Williams IH, Maggiora GM, Schowen RL (1980) J Am Chem Soc 102: 7831–7839
1476. Deslongchamps P (1977) Heterocycles 7: 1271–1317
1477. Alexandrov SL, Antonov VK (1984) Mol Biologia 18: 1576–1582
1478. Pollack RM, Bender ML (1970) J Am Chem Soc 92: 7190–7194
1479. McClelland RA, Santry LJ (1983) Acc Chem Res 16: 394–399
1480. Bender ML, Ginger RD, Kemp KC (1954) J Am Chem Soc 76: 3350–3351
1481. Bender ML (1953) J Am Chem Soc 75: 5986–5990
1482. Rogers GA, Bruice TC (1974) J Am Chem Soc 96: 2481–2488
1483. Ott H, Frey AJ, Hofmann A (1963) Tetrahedron 19: 1675–1689
1484. Shemyakin MM, Antonov VK, Shkrob AM, Shchelokov VI, Agadzhanyan ZE (1965) Tetrahedron 21: 3537–3572
1485. Bender ML, Ginger RD, Unik JP (1958) J Am Chem Soc 80: 1044–1048
1486. Bender ML, Heck H d'A (1967) J Am Chem Soc 89: 1211–1220
1487. Deslongchamps P, Barlet R, Taillefer RJ (1980) Can J Chem 58: 2167–2172
1488. Ślebocka-Tilk H, Brown RS, Olekczyk J (1987) J Am Chem Soc 109: 4620–4622
1489. Eigen M (1963) Angew Chem 75: 489–508
1490. Gutman M, Nachliel E, Gershon E (1985) Biochemistry 24: 2937–2941
1491. Scheiner S (1985) Acc Chem Res 18: 174–180
1492. Hillenbrand EA, Scheiner S (1986) J Am Chem Soc 108: 7178–7186
1493. Cybulski SM, Scheiner S (1989) J Am Chem Soc 111: 23–31
1494. Sirkin YK, Moiseev II (1958) Uspekhi Khimii 27: 717–732
1495. Williams IH, Spangler D, Femec DA, Maggiora GM, Schowen RL (1980) J Am Chem Soc 102: 6619–6621
1496. Venkatasubban KS, Davis KR, Hogg JL (1978) J Am Chem Soc 100: 6125–6128
1497. Huskey W, Warren CT, Hogg JL (1981) J Org Chem 46: 59–63
1498. Williams IH, Spangler D, Femec DA, Maggiora GM, Schowen RL (1983) J Am Chem Soc 105: 31–40
1499. Kirby AJ, McDonald RS (1974) J Chem Soc Perkin II: 1495–1504
1500. Lam CH, Kluger RH, Csizmadia IG (1977) Tetrahedron Lett: 1365–1368
1501. Komiyama M, Bender ML (1979) Proc Natl Acad Sci USA 76: 557–560
1502. Ballinger P, Long FA (1960) J Am Chem Soc 82: 795–798

1503. Klinman JP (1978) Adv Enzymol 46: 416–494
1504. Kresge AJ (1964) Pure Appl Chem 8: 243–258
1505. Hogg JL, Phillips MK, Jergens DE (1977) J Org Chem 42: 2459–2461
1506. Komiyama M, Bender ML (1979) Bioorg Chem 8: 141–145
1507. Tseng L, Stewart JA (1971) J Am Chem Soc 93: 1273–1275
1508. Bergman N-A, Chiang Y, Kresge AJ (1978) J Am Chem Soc 100: 5954–5956
1509. Murdoch JR (1980) J Am Chem Soc 102: 71–78
1510. Paneth P (1985) J Am Chem Soc 107: 7070–7071
1511. Kresge AJ (1980) J Am Chem Soc 102: 7797–7798
1512. Dogonadze RR, Kuzentsov AM, Levich VG (1967) Elektrokhimiya 3: 739–745
1513. McClelland RA (1978) J Am Chem Soc 100: 1844–1849
1514. Jensen JL, Jencks WP (1979) J Am Chem Soc 101: 1476–1488
1515. Sawyer CB, Kirsch JF (1975) J Am Chem Soc 97: 1963–1964
1516. O'Leary MH, Marlier JF (1979) J Am Chem Soc 101: 3300–3306
1517. Marlier JF, O'Leary MH (1981) J Org Chem 46: 2175–2177
1518. Deslongchamps P, Atlani P, Frehel D, Malaval A (1972) Can J Chem 50: 3405–3408
1519. Deslongchamps P (1975) Tetrahedron 31: 2463–2490
1520. Lehn JM, Wipff G (1974) J Am Chem Soc 96: 4048–4050
1521. Evans CM, Glenn R, Kirby AJ (1982) J Am Chem Soc 104: 4706–4707
1522. Perrin CL, Nuñez O (1986) J Am Chem Soc 108: 5997–6003
1523. Jencks WP (1972) J Am Chem Soc 94: 4731–4733
1524. Jencks WP (1976) Acc Chem Res 9: 425–432
1525. Guthrie JP (1980) J Am Chem Soc 102: 5286–5293
1526. Bronsted JN, Pedersen K (1924) Z Physik Chem 108: 185–190
1527. Bruice TC, Fife TH, Bruno JJ, Brandon NE (1962) Biochemistry 1: 7–11
1528. Jencks WP, Gilchrist M (1962) J Am Chem Soc 84: 2910–2913
1529. Biechler SS, Taft RW (1957) J Am Chem Soc 79: 4927–4935
1530. Schowen RL, Zuorich GW (1966) J Am Chem Soc 88: 1223–1225
1531. Hammond GS (1955) J Am Chem Soc 77: 334–338
1532. Blackburn GM, Jencks WP (1968) J Am Chem Soc 90: 2638–2645
1533. Fersht AR (1971) J Am Chem Soc 93: 3504–3515
1534. Bender ML, Thomas RJ (1961) J Am Chem Soc 83: 4183–4189
1535. Lienhard GE, Jencks WP (1965) J Am Chem Soc 87: 3855–3862
1536. Kershner LD, Schowen RL (1971) J Am Chem Soc 93: 2014–2024
1537. Jencks WP, Carriuolo J (1960) J Am Chem Soc 82: 675–681
1538. Bruice TC, Fedor LR (1964) J Am Chem Soc 86: 4886–4897
1539. Bruice TC, Willis RG (1965) J Am Chem Soc 87: 531–536
1540. Jencks WP, Gilchrist M (1966) J Am Chem Soc 88: 104–108
1541. Bruice TC, Lapinski R (1958) J Am Chem Soc 80: 2265–2267
1542. Hupe DJ, Jencks WP (1977) J Am Chem Soc 99: 451–464
1543. Whitaker JR (1962) J Am Chem Soc 84: 1900–1904
1544. Howard AE, Kollman PA (1988) J Am Chem Soc 110: 7195–7200
1545. Edward JT, Wong SC (1979) J Am Chem Soc 101: 1807–1809
1546. Bender ML, Neveu MC (1958) J Am Chem Soc 80: 5388–5391
1547. Hajdu J, Smith GM (1980) J Am Chem Soc 102: 3960–3962
1548. Butler AR, Gold V (1962) J Chem Soc: 1334–1339
1549. Bell RP (1941) Acid-base catalysis. Clarendon Press, Oxford, p 94
1550. Gandour RD (1981) Bioorg Chem 10: 169–176
1551. Allinger NL, Chang SHM (1977) Tetrahedron 33: 1561–1567
1552. Wyness KG (1958) J Chem Soc: 2934–2938
1553. Kroll H (1952) J Am Chem Soc 74: 2036–2039
1554. Bender ML, Turnquest BW (1957) J Am Chem Soc 79: 1889–1893
1555. Westheimer FH, Shookhoff MW (1940) J Am Chem Soc 62: 269–275
1556. Aksnes G, Prue JE (1959) J Chem Soc: 103–107
1557. Sayre LM (1986) J Am Chem Soc 108: 1632–1635
1558. Chaffee E, Dasgupta TP, Harris GM (1973) J Am Chem Soc 95: 4169–4173
1559. Buckingham DA, Engelhardt LM (1975) J Am Chem Soc 97: 5915–5917

1560. Martin RB (1967) J Am Chem Soc 89: 2501–2502
1561. Futrell JH, Tiernan TO (1968) Science 162: 415–422
1562. Nakon R, Recham PR, Angelici RJ (1974) J Am Chem Soc 96: 2117–2120
1563. Groves JT, Chambers RR (1984) J Am Chem Soc 106: 630–638
1564. Fife TH, Przystas TJ (1986) J Am Chem Soc 108: 4631–4636
1565. Buckingham DA, Keene FR, Sargeson AM (1974) J Am Chem Soc 96: 4981–4983
1566. Groves JT, Dias RM (1979) J Am Chem Soc 101: 1033–1035
1567. Fife TH, Przystas TJ, Squillacote VL (1979) J Am Chem Soc 101: 3017–3026
1568. Fife TH, Przystas TJ (1980) J Am Chem Soc 102: 7297–7300
1569. Chin J, Banaszczyk M (1989) J Am Chem Soc 111: 2724–2726
1570. Meriwether L, Westheimer FH (1956) J Am Chem Soc 78: 5119–5123
1571. Jencks WP, Carriuollo J (1959) J Biol Chem 234: 1280–1285
1572. Bruice TC, Schmir GL (1957) J Am Chem Soc 79: 1663–1667
1573. Bender ML, Turnquest BW (1957) J Am Chem Soc 79: 1656–1662
1574. Schneider F, Wenck H (1967) Hoppe-Seyler's Z Physiol Chem 348: 1221–1224
1575. Langenbeck W, Mahrwald R (1957) Chem Ber 90: 2423–2431
1576. Bruice TC, Schmir GL (1958) J Am Chem Soc 80: 148–156
1577. Bruice TC, Herz JL (1964) J Am Chem Soc 86: 4109–4116
1578. BruiceTC, Benkovic SJ (1964) J Am Chem Soc 86: 418–426
1579. Cordes EH, Jencks WP (1962) Biochemistry 1: 773–778
1580. Kirsch JF, Jencks WP (1964) J Am Chem Soc 86: 837–846
1581. Komiyama M, Bender ML (1978) J Am Chem Soc 100: 5977–5978
1582. Komiyama M, Bender ML (1978) Bioorg Chem 7: 133–139
1583. Kirby AJ, Fersht AR (1971) Progr Bioorg Chem 1: 1–81
1584. Kunitake T (1977) Bioorg Chem 1: 153–172
1585. Kirby AJ (1987) In: Page MI, Williams A (eds) Enzyme mechanisms. R Soc Chem, London, pp 67–77
1586. Bender ML, Kèzdy FK, Zerner B (1963) J Am Chem Soc 85: 3017–3024
1587. Bruice TC, Tanner DW (1965) J Org Chem 30: 1668–1669
1588. Belke CJ, Su SCK, Shafer JA (1971) J Am Chem Soc 93: 4552–4560
1589. Schneider F (1967) Hoppe-Seyler's Z Physiol Chem 348: 1034–1042
1590. Garrett ER (1960) J Am Chem Soc 82: 711–718
1591. Fersht AR, Kirby AJ (1967) J Am Chem Soc 89: 4857–4863
1592. Fersht AR, Kirby AJ (1967) J Am Chem Soc 89: 4853–4857
1593. Hurst GH, Bender ML (1971) J Am Chem Soc 93: 704–711
1594. Burrows HD, Topping RM (1970) Chem Commun: 1389–1390
1595. Kirby AJ, Lancaster PW (1972) J Chem Soc Perkin Trans II: 1206–1214
1596. Kluger R, Lam C-H (1978) J Am Chem Soc 100: 2191–2197
1597. Kluger RK, Chin J, Choy WW (1979) J Am Chem Soc 101: 6976–6980
1598. Somayaji V, Keillor J, Brown RS (1988) J Am Chem Soc 110: 2625–2629
1599. Menger FM, Ladika M (1988) J Am Chem Soc 110: 6794–6796
1600. Skorey KI, Somayaji V, Brown RS (1988) J Am Chem Soc 110: 5205–5206
1601. Zimmerman SC, Cramer KD (1988) J Am Chem Soc 110: 5906–5908
1602. Hulf JB, Askew B, Duff RJ, Rebek J (1988) J Am Chem Soc 110: 5908–5909
1603. Fife TH, Przystas TJ (1982) J Am Chem Soc 104: 2251–2257
1604. Schepartz A, Breslow R (1987) J Am Chem Soc 109: 1814–1826
1605. Groves JT, Baron LA (1989) J Am Chem Soc 111: 5442–5448
1606. Overberger CG, Shen C-M (1971) J Am Chem Soc 93: 6992–6998
1607. Mallick JM, DiSousa VT, Yamagichi M, Lee J, Chalabi P, Gadwood RC (1984) J Am Chem Soc 106: 7252–7255
1608. Fife TH, Bambery RJ, De Mark BR (1978) J Am Chem Soc 100: 5500–5507
1609. Killian FL, Bender ML (1969) Tetrahedron Lett: 1255–1258
1610. Komiyama M, Roesel TR, Bender ML (1977) Proc Natl Acad Sci USA 74: 23–25
1611. Bender ML, Chloupek F, Neveu M (1958) J Am Chem Soc 80: 5384–5387
1612. Bruice TC, Sturtevant JM (1959) J Am Chem Soc 81: 2860–2870
1613. Fersht A (1985) Enzyme structure and mechanism, 2nd edn. Freeman, New York
1614. Fersht AR, Kirby AJ (1968) J Am Chem Soc 90: 5833–5844

1615. Koshland DE (1962) J Theor Biol 2: 75–86
1616. Page MI, Jencks WP (1971) Proc Natl Acad Sci USA 68: 1678–1683
1617. Page MI (1973) Chem Soc Rev 2: 295–323
1618. Milstein S, Cohen LA (1972) J Am Chem Soc 94: 9158–9165
1619. Caswell M, Schmir GL (1980) J Am Chem Soc 102: 4815–4821
1620. King MM, Cohen LA (1983) J Am Chem Soc 105: 2752–2760
1621. Dorigo AE, Houk KN (1987) J Am Chem Soc 109: 3698–3708
1622. Menger FM, Glass LE (1980) J Am Chem Soc 102: 5404–5406
1623. Bruice TC, Benkovic SJ (1963) J Am Chem Soc 85: 1–8
1624. Alexandrov IV (1976) Teor Ekcperim Khimiya 12: 299–303
1625. Likhtenstein GI (1979) Multinuclear oxidation-reduction metalloenzymes. Nauka, Moscow, p 237
1626. Dewar MJS (1984) J Am Chem Soc 106: 209–219
1627. Knowles JR, Parsons CA (1967) Chem Comm: 755–757
1628. Berezin IV, Martinek K, Yatsimirsky AK (1973) Uspekhi Khimii 42: 1729–1756
1629. Martinek K, Osipov AP, Yatsimirsky AK, Berezin IV (1975) Tetrahedron 31: 709–718
1630. Fendler JH, Fendler EJ (1975) Catalysis in micellar and macromolecular systems. Academic Press, New York
1631. Martinek K (1989) Biochem Int 18: 871–893
1632. Okahata Y, Ando R, Kunitake T (1977) J Am Chem Soc 99: 3067–3075
1633. Moss RA, Lee Y-S, Alwis KW (1980) J Am Chem Soc 102: 6646–6648
1634. Ihara Y, Nango M, Kuroki N (1980) J Org Chem 45: 5009–5011
1635. Murakami Y, Nakono A, Yoshimatsu A, Fukuya K (1981) J Am Chem Soc 103: 728–730
1636. Ihara Y, Nango M, Kimura Y, Kuroki N (1983) J Am Chem Soc 105: 1252–1255
1637. Menger FM, Whitesell LC (1985) J Am Chem Soc 107: 707–708
1638. O'Connor CJ, Tan A-L (1980) Aust J Chem 33: 747–755
1639. Letsinger RL, Savereidl TJ (1962) J Am Chem Soc 84: 3122–3127
1640. Kirsh YE, Pluzhnov SK, Shomina TS, Kabanov VA, Kargin VA (1970) Visokomolekular Soedineniya 12: 186–204
1641. Klotz IM (1978) Adv Chem Phys 39: 109–128
1642. Klotz IM (1987) In: Page MI, Williams A (eds) Enzyme mechanisms. R Soc Chem, London, pp 14–34
1643. Kunitake T, Okahata Y (1976) Adv Polym Sci 20: 217–245
1644. Belikov VM, Latov VK, Fastovskaya MI (1980) J Mol Catal 8: 443–453
1645. Latov VK, Fastovskaya MI, Belikov VM (1987) Visokomolekular Soedineniya 29A: 2571–2576
1646. Stoddart JF (1987) In: Page MI, Williams A (eds) Enzyme mechanisms. R Soc Chem, London, pp 35–55
1647. Cram DJ, Katz HE (1983) J Am Chem Soc 105: 135–137
1648. Bender ML (1987) In: Page MI, Williams A (eds) Enzyme mechanisms. R Soc Chem, London, pp 56–66
1649. Czarniecki MF, Breslow R (1978) J Am Chem Soc 100: 7771–7772
1650. D'Souza VT, Hanabuss K, O'Leary T, Gadwood RC, Bender ML (1985) Biochem Biophys Res Commun 129: 727–732
1651. D'Souza VT, Lu XL, Ginder RD, Bender ML (1987) Proc Natl Acad Sci USA 84: 673–674
1652. Bülow L, Mosbach K (1987) FEBS Lett 210: 147–152
1653. Atassi MZ (1985) Biochem J 226: 477–485
1654. Menger FM, Ladika M (1987) J Am Chem Soc 109: 3145–3146
1655. Segel IH (1975) Enzyme kinetics. Wiley & Sons, New York
1656. Wong JT-F (1975) Kinetics of enzyme mechanisms. Academic Press, London
1657. Fromm HJ (1975) Initial rate enzyme kinetics. Springer, Berlin Heidelberg New York
1658. Cornish-Bowden A (1976) Principles of enzyme kinetics. Butterworths, London
1659. Berezin IV, Klyosov AA (1976) Practical course of chemical and enzymic kinetics. Moscow Univ, Moscow
1660. Cleland WW (1977) Adv Enzymol 45: 273–388

1661. Dixon M, Webb EC, Thorne CJR, Tipton KF (1979) Enzymes, 3rd edn. Longman, London, pp 138–163
1662. Michaelis L, Menten ML (1913) Biochem Z 49: 333–341
1663. Briggs GE, Haldane JBS (1925) Biochem J 19: 338–343
1664. Symbolism and terminology in enzyme kinetics. Recommendation 1981 (1982) Eur J Biochem 128: 281–291
1665. Symbolism and terminology in enzyme kinetics (1983) Biochem J 213: 561–571
1666. Zerner B, Bender ML (1964) J Am Chem Soc 86: 3669–3674
1667. Antonov VK, Rumsh LD, Tikhodeeva AG (1974) FEBS Lett 46: 29–33
1668. King EL, Altman C (1956) J Phys Chem 60: 1365–1372
1669. Fromm HJ (1970) Biochem Biophys Res Commun 40: 692–699
1670. Chou KC (1980) Eur J Biochem 113: 195–198
1671. Popova SV, Goldstein BN (1982) Mol Biologia 16: 1271–1278
1672. Lineweaver H, Burk D (1934) J Am Chem Soc 56: 658–670
1673. Hofstee BHJ (1959) Nature 184: 1296–1298
1674. Nimmo IA, Atkins GL (1974) Biochem J 141: 913–914
1675. Cornish-Bowden A (1975) Biochem J 149: 305–312
1676. Mannervik B, Jakobson I, Warholm M (1979) Biochim Biophys Acta 567: 43–48
1677. Matyska L, Kovář J (1985) Biochem J 231: 171–177
1678. Darvey IG, Williams JF (1964) Biochim Biophys Acta 85: 1–10
1679. Boeker EA (1985) Biochem J 226: 29–36
1680. Klyosov AA, Berezin IV (1972) Biokhimiya 37: 170–183
1681. Dugglely RG (1986) Biochem J 235: 613–615
1682. Evans SA, Olson ST, Shore JD (1982) J Biol Chem 257: 3014–3017
1683. Hartley BS, Kilby BA (1954) Biochem J 56: 288–297
1684. Ascenzi P, Menegatti E, Guarneri M, Amiconi G (1987) Biochim Biophys Acta 915: 421–425
1685. Livingston DC, Brocklehurst JR, Cannon JF, Leytus SP, Wehrly JA, Peltz SW, Peltz GA, Mangel WF (1981) Biochemistry 20: 4298–4306
1686. Leytus SP, Patterson WL, Mangel WF (1983) Biochem J 215: 253–260
1687. Mehlich A, Beckman J, Wenzel HR, Tschesche H (1988) Biochim Biophys Acta 957: 420–429
1688. Roodyn DB (1970) Automated enzyme assays. North-Holland, Amsterdam
1689. Roffman S, Sanocka U, Troll W (1970) Anal Biochem 36: 11–17
1690. Latt SA, Auld DS, Vallee BL (1972) Anal Biochem 50: 56–62
1691. Sachdev GP, Brownstein AD, Fruton JS (1973) J Biol Chem 248: 6292–6299
1692. Spencer PW, Titus JS, Spencer RD (1975) Anal Biochem 64: 556–566
1693. Zimmerman M, Ashe B, Yurewicz EC, Patel G (1977) Anal Biochem 78: 47–51
1694. Furihata C, Senma T, Saito D, Matsushima T, Sugimura T (1978) Anal Biochem 84: 479–485
1695. Zimmerman M, Quigley JP, Ashe B, Dorn C, Goldfarb R, Troll W (1978) Proc Natl Acad Sci USA 75: 750–753
1696. Castillo MJ, Nakajima K, Zimmerman M, Powers JC (1979) Anal Biochem 99: 53–64
1697. Green GDJ, Shaw E (1979) Anal Biochem 93: 223–226
1698. Garesse R, Castell JV, Vallejo CG, Marco R (1979) Eur J Biochem 99: 253–259
1699. Branchini BR, Salituro FG, Hermes JD, Post NJ (1980) Biochem Biophys Res Commun 97: 334–339
1700. Varani J, Johnson K, Kaplan J (1980) Anal Biochem 107: 377–384
1701. Andrews AT (1982) FEBS Lett 141: 207–210
1702. Takamiya S, Ohshima T, Tanizawa K, Soda K (1983) Anal Biochem 130: 266–270
1703. Pohl J, Baudyš M, Kostka V (1983) Anal Biochem 133: 104–109
1704. Florentin D, Sassi A, Roques BP (1984) Anal Biochem 141: 62–69
1705. Kuromizu K, Shimokawa Y, Abe O, Izumiya N (1985) Anal Biochem 151: 534–539
1706. Green JD (1986) Anal Biochem 152: 83–88
1707. Hwang SY, Kingsbury WG, Hall NM, Jakas DR, Dunn GL, Gilvarg C (1986) Anal Biochem 154: 552–558

1708. Kokotos G, Tzougraki C (1986) Int J Pept Prot Res 28: 186–191
1709. Bratovanova EK, Petkov DD (1987) Anal Biochem 162: 213–218
1710. Beattie RE, Guthrie DJS, Elmore DT, Williams CH, Walker B (1987) Biochem J 242: 281–283
1711. Kawabata S-I, Miura T, Morita T, Kato H, Fujikawa K, Iwanaga S, Takada K, Kimura T, Sakakibara S (1988) Eur J Biochem 172: 17–25
1712. Stack MS, Gray RD (1989) J Biol Chem 264: 4277–4281
1713. Rossier J, Barrès E, Hutton JC, Bicknell RJ (1989) Anal Biochem 178: 27–31
1714. Janska H, Light A (1989) Anal Biochem 176: 132–136
1715. Antonov VK (1980) Bioorgan Khimiya 6: 805–839
1716. Cleland WW (1975) Acc Chem Res 8: 145–151
1717. Bernhard SA (1959) J Cell Comp Physiol (suppl 1) 54: 256–260
1718. Hein GE, Niemann C (1962) J Am Chem Soc 84: 4495–4503
1719. Berezin IV, Kazanskaya NF, Klyosov AA (1971) Biokhimiya 36: 227–235
1720. Keleti T (1986) FEBS Lett 208: 109–112
1721. Gutfreund H, Hammond BR (1959) Biochem J 73: 526–530
1722. Bender ML, Clement GE, Gunter CR, Kézdy FJ (1964) J Am Chem Soc 86: 3697–3703
1723. Berezin IV, Kazanskaya NF, Klyosov AA (1971) FEBS Lett 15: 121–124
1724. Dorovska-Taran V, Raykova D (1980) Biochim Biophys Acta 615: 509–513
1725. Martinek K, Yatsimirsky AK, Berezin IV (1971) Mol Biologia 5: 96–109
1726. Martinek K, Klyosov AA, Kazanskaya NF, Berezin IV (1974) Int J Chem Kinet 6: 801–811
1727. Boyer PD, Silverstein E (1963) Acta Chem Scand 17: 195–212
1728. Cleland WW (1975) Biochemistry 14: 3220–3224
1729. Lyakisheva AG, Ginodman LM, Antonov VK (1973) Mol Biologia 7: 810–816
1730. Gurova AG, Ginodman LM, Antonov VK (1977) Mol Biologia 11: 1155–1159
1731. Antonov VK, Ginodman LM, Gurova AG (1977) Mol Biologia 11: 1160–1166
1732. Ginodman LM, Lutsenko NG (1972) Biokhimiya 37: 101–111
1733. Eigen M, Hammes GG (1963) Adv Enzymol 25: 1–38
1734. Chance B, Eisenhardt RH, Gibson QH, Lonberg-Holm KK (eds) (1964) Rapid mixing and sampling techniques in biochemistry. Academic Press, New York
1735. Hammes GG, Schimmel PR (1970) In: Boyer P (ed) The enzymes, vol 2. Academic Press, New York, pp 67–114
1736. Hammes G (ed) (1974) Investigation of rates and mechanisms of reaction, 3rd edn. Wiley & Sons, New York
1737. Kurganov BI (1978) Allosteric enzymes. Nauka, Moscow
1738. Klyosov AA, Vallee BL (1977) Bioorgan Khimiya 3: 806–815
1739. Hill R (1925) Proc R Soc Lond 100B: 419–425
1740. Simonova GV, Livanova NV, Kurganov BI (1969) Mol Biologia 3: 768–784
1741. Bardsley WG, Wright AJ (1983) J Mol Biol 165: 163–182
1742. Hill TL, Levitzki A (1980) Proc Natl Acad Sci USA 77: 5741–5745
1743. Kurganov BI (1974) Mol Biologia 8: 525–535
1744. Haldane JBS (1930) Enzymes. Longmans, London
1745. Schechter J, Berger A (1967) Biochem Biophys Res Commun 27: 157–162
1746. Boyer PD (ed) (1971) The enzymes, vol 3. Academic Press, London
1747. Spickett RGW (1971) Chem Ind: 83–94
1748. Mosolov VV (1971) Proteolytic enzymes. Nauka, Moscow
1749. Clement GE (1973) In: Kaiser ET, Kezdy FJ (eds) Progress in bioorganic chemistry, vol 2. Wiley, New York, pp 177–238
1750. Fruton JS (1976) Adv Enzymol 44: 1–35
1751. Lasch J (1979) Ophthalmic Res 11: 372–376
1752. Harper E (1980) Annu Rev Biochem 49: 1063–1078
1753. Dahlqvist U, Wählby S (1975) Biochim Biophys Acta 391: 410–414
1754. Crestfield AM, Moore S, Stein WH (1963) J Biol Chem 238: 622–627
1755. Naughton MA, Sanger F (1961) Biochem J 78: 156–163
1756. Johansen JT, Ottesen M (1968) C R Trav Lab Carlsberg 36: 265–273
1757. Kärgel H-J, Debtmer R, Etzold G, Kirschke H, Bohley P, Languer J (1980) FEBS Lett 114: 257–260

1758. Ryle AP, Leclerc J, Falla F (1968) Biochem J 110: 4p
1759. Keilová H, Keil B (1968) Coll Czech Chem Commun 33: 131–140
1760. Reichelt D, Jacobson E, Haschen RJ (1974) Biochim Biophys Acta 341: 15–26
1761. Zerner B, Bond RPM, Bender ML (1964) J Am Chem Soc 86: 3674–3679
1762. Seely JH, Benoiton NL (1970) Can J Biochem 48: 1122–1131
1763. Thompson RC, Blout ER (1970) Proc Natl Acad Sci USA 67: 1734–1740
1764. Thompson RC, Blout ER (1973) Biochemistry 12: 57–65
1765. Gertler A, Weiss Y, Burstein Y (1977) Biochemistry 16: 2709–2715
1766. Baker EN, Boland MJ, Calder PC, Hardman MJ (1980) Biochim Biophys Acta 616: 30–34
1767. Inouye K, Fruton JS (1967) J Am Chem Soc 89: 187–188
1768. Auld DS, Holmquist B (1974) Biochemistry 13: 4355–4361
1769. Keung W-M, Holmquist B, Riordan JF (1980) Biochem Biophys Res Commun 96: 506–513
1770. McFarlane ND, Brammar WJ, Clarke PH (1965) Biochem J 95: 24c
1771. Woods MJ, Findlater JD, Orsi BA (1979) Biochim Biophys Acta 567: 225–237
1772. Margolin AĹ, Svedas VK, Berezin IV (1980) Biochim Biophys Acta 616: 283–289
1773. Suda H, Yamamoto K, Aoyagi T, Umezawa H (1980) Biochim Biophys Acta 616: 60–67
1774. Fukuda M, Kunugi S (1987) J Biochem 101: 233–240
1775. Rumsh LD, Volkova LI, Antonov VK (1970) FEBS Lett 9: 64–66
1776. Yasutake A, Miyazaki K, Aoyagi H, Kato T, Izumiya N (1979) FEBS Lett 100: 241–243
1777. Yasutake A, Miyazaki K, Aoyagi H, Kato T, Izumiya N (1980) Int J Pept Protein Res 16: 61–65
1778. Ohno M, Izumiya N (1965) Bull Chem Soc Jpn 38: 1831–1840
1779. Kenner GW, Laird AH (1965) Chem Commun: 305–306
1780. Keilova H, Blaha K, Keil B (1968) Eur J Biochem 4: 442–447
1781. Tarada S, Kato T, Izumia N (1975) Eur J Biochem 52: 273–282
1782. Antonov VK, Vanyukova NA (1970) Biokhimiya 35: 202–204
1783. Powers JC, Carroll DL (1975) Biochem Biophys Res Commun 67: 639–644
1784. Gupton BF, Carroll DL, Tuhy PM, Kam C-M, Powers JC (1984) J Biol Chem 259: 4279–4287
1785. King SW, Lum VR, Fife TH (1987) Biochemistry 26: 2294–2300
1786. Lowe G, Williams A (1965) Biochem J 96: 189–193
1787. Frankfater A, Kézdy FJ (1971) J Am Chem Soc 93: 4039–4043
1788. Vencill CF, Rasnick D, Crumley KV, Nishino N, Powers JC (1985) Biochemistry 24: 3149–3157
1789. Beattie RE, Elmore DT, Williams CH, Guthrie DJS (1987) Biochemistry 25: 285–288
1790. Polgár L (1972) Acta Biochim Biophys Acad Sci Hung 7: 29–34
1791. McRae BJ, Kurachi K, Heimark RL, Fujikawa K, Davie EW, Powers JC (1981) Biochemistry 20: 7196–7206
1792. Tanaka T, McRae BJ, Cho K, Cook R, Fraki JE, Johnson DA, Powers JC (1983) J Biol Chem 258: 13552–13557
1793. Kam C-M, McRae BJ, Harper JW, Niemann MA, Volanakis JE, Powers JC (1987) J Biol Chem 262: 3444–3451
1794. Lands WEM, Niemann C (1959) J Am Chem Soc 81: 2204–2208
1795. Barrett AJ (1980) In: Mildner P, Ries B (eds) Enzyme regulation and mechanism of action. Pergamon, Oxford, pp 307–315
1796. Towatari T, Katunuma N (1983) J Biochem 93: 1119–1128
1797. Marks N, Berg MJ, Benuck M (1986) Arch Biochem Biophys 249: 489–499
1798. Takahashi T, Dahdarani AH, Tang J (1988) J Biol Chem 263: 10952–10957
1799. Skidgel RA, Erdös EG (1985) Proc Natl Acad Sci USA 82: 1025–1029
1800. Kunugi S, Fukuda M, Hayashi R (1985) Eur J Biochem 153: 37–40
1801. Johnson BA, Aswad DW (1990) Biochemistry 29: 4373–4380
1802. Alecio MR, Dann ML, Lowe G (1974) Biochem J 141: 495–501
1803. Fujiwara K, Kobayashi R, Tsuru D (1979) Biochim Biophys Acta 570: 140–148
1804. Nomura K (1986) FEBS Lett 209: 235–237
1805. Heins J, Welker P, Schönlein C, Born I, Hartrodt B, Neubert K, Tsuru D, Barth A (1988) Biochim Biophys Acta 954: 161–169

1806. Kuroda K, Akanuma H, Sukenaga Y, Sugihara H, Yamasaki M (1980) J Biochem 87: 1690–1701
1807. Grant DAW, Harmon-Taylor J (1979) Biochim Biophys Acta 567: 207–215
1808. Christensen U, Ipsen H-H (1979) Biochim Biophys Acta 569: 177–183
1809. Lottenberg R, Hall JA, Blinder M, Binder EP, Jackson GM (1983) Biochim Biophys Acta 742: 539–557
1810. Castillo MJ, Kurachi K, Nishino N, Ohkubo I, Powers JC (1983) Biochemistry 22: 1021–1029
1811. Stone SR, Hofsteenge J (1985) Biochem J 230: 497–502
1812. Lottenberg R, Hall JA, Pautler E, Zupan A, Christensen U, Jackson CM (1986) Biochim Biophys Acta 874: 326–336
1813. Bezeaud A, Guillin M-C (1988) J Biol Chem 263: 3576–3581
1814. McRae BJ, Lin T-Y, Powers JC (1981) J Biol Chem 256: 12362–12366
1815. Steiner DF, Docherty K, Carroll R (1984) J Cell Biochem 24: 121–130
1816. Schwartz TW (1986) FEBS Lett 200: 1–10
1817. Sugimura K (1988) Biochem Biophys Res Commun 153: 753–759
1818. Bayle-Lacoste M, Tinguy-Moreaud ED, Geoffre S, Neuzil E (1987) Int J Pept Protein Res 29: 392–405
1819. Lowbridge J, Fruton JS (1974) J Biol Chem 249: 6754–6761
1820. Mattis JA, Fruton JS (1976) Biochemistry 15: 2191–2194
1821. Sasaki T, Kikuchi T, Yumoto N, Yoshimura N, Murachi T (1984) J Biol Chem 259: 12489–12494
1822. Zucker S, Buttle DJ, Nicklin MJH, Barrett AJ (1985) Biochim Biophys Acta 828: 196–204
1823. Allen MP, Yamada AH, Carpenter FH (1983) Biochemistry 22: 3778–3783
1824. Bunting J, Kabir SH (1977) J Am Chem Soc 99: 2775–2780
1825. Baumann WK, Bizzozero SA, Dutler H (1970) FEBS Lett 8: 257–261
1826. Bizzozero SA, Baumann WK, Dutler H (1982) Eur J Biochem 122: 251–258
1827. Harper JW, Cook RR, Roberts J, McLaughlin BJ, Powers JC (1984) Biochemistry 22: 2995–3002
1828. Yoshida N, Everitt MT, Neurath H, Woodbury RG, Powers JC (1980) Biochemistry 19: 5799–5804
1829. Powers JC, Tanaka T, Harper JW, Minematsu Y, Barker L, Lincoln D, Crumley KV, Kraki JE, Schechter NM, Lazarus GG, Nakajima K, Nakashino K, Neurath H, Woodbury RG (1985) Biochemistry 24: 2048–2058
1830. Visser S, Von Rooijen PJ, Schattenkerk C, Kerling KET (1977) Biochim Biophys Acta 481: 171–176
1831. Morihara K (1981) In: Turk V, Vitale L (eds) Proteinases and their inhibitors. Int Symp Portorož, Yugoslavia, 1981, Mladinska Knjiga, Ljubljana; Pergamon Press, Oxford, pp 213–222
1832. Martin P (1984) Biochim Biophys Acta 791: 28–36
1833. Dunn BM, Jimenez M, Parten BF, Valler MJ, Rolph CE, Kay J (1986) Biochem J 237: 899–906
1834. Johnson EK, Lamkin RM, Jordan KJ, Gordon JA, Lindley BR (1987) Int J Pept Protein Res 30: 170–176
1835. Waxman L, Goldberg AL (1985) J Biol Chem 260: 12022–12028
1836. Woo KM, Chung WJ, Ha DB, Goldberg AL, Chung CH (1989) J Biol Chem 264: 2088–2091
1837. Driscoll J, Goldberg AL (1989) Proc Natl Acad Sci USA 86: 787–791
1838. Kotler M, Danho W, Katz RA, Leis J, Skalka AM (1989) J Biol Chem 264: 3428–3435
1839. Kräusslich H-G, Ihgraham RH, Skoog MT, Wimmer E, Pallai PV, Carter CA (1989) Proc Natl Acad Sci USA 86: 807–811
1840. Atlas D, Berger A (1972) Biochemistry 11: 4719–4722
1841. Bauer C-A, Thompson RC, Blout ER (1976) Biochemistry 15: 1296–1299
1842. Thomson A, Kapadia SB (1979) Eur J Biochem 102: 111–116
1843. Del Mar EG, Largman C, Brodrick JW, Fassett M, Gcokas MC (1980) Biochemistry 19: 468–472
1844. Stein RL, Strimpler AM, Hori H, Powers JC (1987) Biochemistry 26: 1301–1305

1845. Bieth JG, Dirrig S, Jung M-L, Bondier C, Papamichael E, Sakarellos C (1989) Biochim Biophys Acta 994: 64–74
1846. Morihara K, Oka T, Tsuzuki H (1970) Arch Biochem Biophys 138: 515–525
1847. Brömme D, Peters K, Fink S, Fittkau S (1986) Arch Biochem Biophys 244: 439–446
1848. Weingarten H, Feder J (1986) Biochem Biophys Res Commun 139: 1184–1187
1849. Mookhtiar KA, Steinbrink DR, Van Wart HE (1985) Biochemistry 24: 6527–6533
1850. Malfroy B, Schwartz J-C (1982) Biochem Biophys Res Commun 106: 276–285
1851. Buttle DJ, Ritonja A, Pearl LH, Turk V, Barrett AJ (1990) FEBS Lett 260: 195–197
1852. Houmard J, Drapeau GR (1972) Proc Natl Acad Sci USA 69: 3506–3509
1853. Revina LP, Khaidarova NV, Rudenskaya GN, Grebetshikov NI, Baratova LA, Stepanov VM (1989) Biokhimiya 54: 846–850
1854. Heijne G (1985) J Mol Biol 184: 99–105
1855. Bizzozero SA, Baumann WK, Dutler H (1975) Eur J Biochem 58: 167–176
1856. Fisher J, Belasco JG, Khosla S, Knowles JR (1980) Biochemistry 19: 2895–2901
1857. Matagne A, Misselyn-Bauduin A-M, Joris B, Erpicum T, Granier B, Frére J-M (1990) Biochem J 265: 131–146
1858. Baumann WK, Bizzozero SA, Dutler H (1973) Eur J Biochem 39: 381–391
1859. Bauer C-A, Thompson RC, Blout ER (1976) Biochemistry 15: 1291–1295
1860. Bauer C-A (1980) Eur J Biochem 105: 565–570
1861. Hill CR, Tomalin G (1983) Arch Biochem Biophys 221: 324–328
1862. Lobo AP, Wos JD, Yu SM, Lawson WB (1976) Arch Biochem Biophys 177: 235–244
1863. Bizzozero SA, Dutler H (1987) Arch Biochem Biophys 256: 662–676
1864. Fiedler F (1987) Eur J Biochem 163: 303–312
1865. Lonsdale-Eccles JD, Hogg DH, Elmore DT (1980) Biochim Biophys Acta 612: 395–400
1866. Kibirev VK, Romanova VI, Serebryanii SB (1983) Biokhimiya 48: 937–943
1867. Marsh HC, Meinwald YC, Thannhauser TW, Scheraga HA (1983) Biochemistry 22: 4170–4174
1868. Cho K, Tanaka T, Cook RR, Kisiel W, Fujikawa K, Kurochi K, Powers JC (1984) Biochemistry 23: 644–650
1869. Pesquero JL, Boschcov P, Oliveira MCF, Paiva ACM (1982) Biochem Biophys Res Commun 108: 1441–1446
1870. Levinson PR, Tomalin G (1982) Biochem J 203: 299–302
1871. Atlas D, Levit S, Schechter J, Berger A (1970) FEBS Lett 11: 281–283
1872. Renaud A, Lestienne P, Hughes DL, Bieth JG, Dimicoli J-L (1983) J Biol Chem 258: 8312–8316
1873. Stein R (1985) J Am Chem Soc 107: 5767–5775
1874. Schellenberger V, Schellenberger U, Mitin Y, Jakubke H-D (1989) Eur J Biochem 179: 161–163
1875. Kaplan H, Whitaker DR (1967) J Am Chem Soc 89: 3352–3353
1876. Morihara K, Tsuzuki H (1971) Arch Biochem Biophys 146: 291–296
1877. Bratovanova EK, Petkov DD (1987) Biochem J 248: 957–960
1878. Medzihradszky K, Voynick IM, Medzihradszky-Schweiger H, Fruton JS (1970) Biochemistry 9: 1154–1162
1879. Raymond MN, Garnier J, Bricas E, Cilianu S, Blasnic M, Chaix A, Lefrancier P (1972) Biochimie 54: 145–154
1880. Sampath-Kumar PS, Fruton JS (1974) Proc Natl Acad Sci USA 71: 1070–1072
1881. Fruton JS (1974) Isr J Chem 12: 505–513
1882. Auffret CA, Ryle AP (1979) Biochem J 179: 239–246
1883. Dunn BM, Kammermann B, McCurry KR (1984) Anal Biochem 138: 68–73
1884. Dunn BM, Valler MJ, Rolph CE, Foundling SI, Jimenez M, Kay J (1987) Biochim Biophys Acta 913: 122–130
1885. Stammers DK, Dann JG, Harris CJ, Smith DR (1987) Arch Biochem Biophys 258: 413–420
1886. Burton J, Quinn T (1988) Biochim Biophys Acta 952: 8–12
1887. Kotler M, Katz RA, Danho W, Leis J, Skalka AM (1988) Proc Natl Acad Sci USA 85: 4185–4189
1888. Morihara K, Tsuzuki H (1970) Eur J Biochem 15: 374–380

1889. Hersh LB, Morihara K (1986) J Biol Chem 261: 6433–6437
1890. Kania R, Santiago NA, Gray GM (1977) J Biol Chem 252: 4929–4934
1891. Ötvös L, Moravcsik E, Mády G (1971) Biochem Biophys Res Commun 44: 1056–1064
1892. Matsuda K (1976) J Biochem 80: 659–669
1893. Walter R, Yoshimoto T (1978) Biochemistry 17: 4139–4144
1894. Cheung H-S, Wang F-L, Ondetti MA, Sabo EF, Cushman DW (1980) J Biol Chem 255: 401–407
1895. De Camargo ACM, Da Fonseca MJV, Caldo H, De Morias Carvalho K (1982) J Biol Chem 257: 9265–9267
1896. Fischer G, Heins J, Barth A (1983) Biochim Biophys Acta 742: 452–462
1897. Deschodt-Lanckman M, Strosedrg AD (1983) FEBS Lett 152: 109–113
1898. Matsas R, Turner AJ, Kenney AJ (1984) FEBS Lett 175: 124–128
1899. Connelly JC, Skidgel RA, Schulz WW, Johnson AR, Erdös EG (1985) Proc Natl Acad Sci USA 82: 8737–8741
1900. Haroda M, Fukasawa K, Hiraoka BY, Mogi M, Barth A, Neubert K (1985) Biochim Biophys Acta 830: 341–344
1901. Maziak L, Lajoic G, Belleau B (1986) J Am Chem Soc 108: 182–183
1902. Gluschankof P, Gomez S, Lepage A, Créminou C, Nyberg F, Terenius L, Cohen P (1988) FEBS Lett 234: 149–152
1903. Morihara K, Oka T (1973) Arch Biochem Biophys 156: 764–771
1904. Morihara K, Oka T (1973) FEBS Lett 33: 54–56
1905. Gertler A, Hofmann T (1970) Can J Biochem 48: 384–386
1906. Fruton JS (1975) In: Reich E, Rifkin DB, Shaw E (eds) Proteases and biological control. Cold Spring Harbor Laboratory, pp 33–50
1907. Zinchenko AA, Rumsh LD, Antonov VK (1977) Bioorgan Khimiya 3: 1663–1670
1908. Hill CR, Tomalin G (1981) Biochim Biophys Acta 660: 65–72
1909. Mihalyi E (1978) Application of proteolytic Enzymes to protein structure studies, 2nd edn, vols 1 and 2. CRC Press, West Palm Beach
1910. Hershko A, Ciechanover A (1982) Annu Rev Biochem 51: 335–364
1911. Bennett WS, Huber R (1984) CRC Crit Rev Biochem 15: 291–384
1912. Mayer RJ, Doherty F (1986) FEBS Lett 198: 181–193
1913. Bek E, Berry R (1990) Biochemistry 29: 178–183
1914. Linderström-Lang KV (1952) Proteins and enzymes. Lane medical lectures, vol 6. Stanford Univ Press, Stanford, pp 53–72
1915. Vorob'ev MM, Paskolova SV, Vitt SV, Belikov VM (1986) Die Nahrung 30: 995–1001
1916. Imoto T, Yamada H, Ueda T (1986) J Mol Biol 190: 647–649
1917. Wright HT (1977) Eur J Biochem 73: 567–578
1918. Bicknell R, Schaeffer A, Auld DS, Riordan JF, Monwanni R, Bertini I (1985) Biochem Biophys Res Commun 133: 787–793
1919. Linderstrom-Lang KV, Offesen M (1949) CR Trav Lab Carlsberg 26: 403–412
1920. Anderson LE, Walsh KA, Neurath H (1977) Biochemistry 16: 3354–3360
1921. Light A, Savithri HS, Liepnieks JJ (1980) Anal Biochem 106: 199–206
1922. Harris RB (1989) Arch Biochem Biophys 275: 315–333
1923. Mentlein R (1988) FEBS Lett 234: 251–256
1924. Revina LP, Vakhitova EA, Baratova LA, Belyanova LP, Stepanov VM (1975) Bioorgan Khimiya 1: 958–964
1925. Rholam M, Nicolas P, Cohen P (1986) FEBS Lett 207: 1–6
1926. Bewly TA, Li CH (1985) Biochemistry 24: 6568–6571
1927. Ogasahara K, Tsunasawa S, Soda Y, Yutani K, Sugino Y (1985) Eur J Biochem 150: 17–21
1928. Fontana A, Fassina G, Vita C, Dalzoppo D, Zamani M, Zambonin M (1986) Biochemistry 25: 1848–1851
1929. Rodionov MA, Galaktionov SG, Akhrem AA (1987) FEBS Lett 223: 402–404
1930. Novotný J, Bruccoleri RE (1987) FEBS Lett 211: 185–189
1931. Visser S, Rooijen PJ, Slangen CJ (1980) Eur J Biochem 108: 415–421
1932. Carles C, Ribadeau-Dumas B (1984) Biochemistry 23: 6839–6843
1933. Carles C, Martin P (1985) Arch Biochem Biophys 242: 411–416
1934. Carles C, Dumas BR (1985) FEBS Lett 185: 282–286

1935. Drohse HB, Foltmann B (1989) Biochim Biophys Acta 995: 221–224
1936. Visser S, Slangen CJ, Rooijen PJ (1987) Biochem J 244: 553–558
1937. Safro MG, Andreeva NS, Blundell TL (1987) Mol Biologia 21: 1582–1589
1938. Higgins DL, Lewis SD, Shafer JA (1983) J Biol Chem 258: 9276–9282
1939. Estell DA, Wilson KA, Laskowski M (1980) Biochemistry 19: 131–137
1940. Abita JP, Delaage M, Lazdunski M, Savrda J (1969) Eur J Biochem 8: 314–324
1941. Maier M, Austen KF, Spragg J (1983) Proc Natl Acad Sci USA 80: 3928–3932
1942. Maroux S, Baratti J, Desnuelle P (1971) J Biol Chem 246: 5031–5039
1943. Perlman D, Halvorson HO (1983) J Mol Biol 167: 391–409
1944. Zhu H-Y, Dalbey RE (1989) J Biol Chem 264: 11833–11838
1945. Zhirnov OP (1988) Mol Biologia 22: 581–600
1946. Hershko A, Leshinsky E, Ganoth D, Heller H (1984) Proc Natl Acad Sci USA 81: 1619–1623
1947. Evans AC, Wilkinson KD (1985) Biochemistry 24: 2915–2923
1948. Hershko A, Heller H, Eytan E, Kaklij G, Rose IA (1984) Proc Natl Acad Sci USA 81: 7021–7025
1949. Matthews W, Tanaka K, Driscoll J, Ichihara A, Golgberg AL (1989) Proc Natl Acad Sci USA 86: 2597–2601
1950. Schulze IT, Colowick SP (1969) J Biol Chem 244: 2306–2310
1951. Graves DJ, Mann AS, Philip G, Oliveira RJ (1968) J Biol Chem 243: 6090–6098
1952. Miles EW, Higgins W (1978) In: Cohen GN, Holzer H (eds) Limited proteolysis in microorganisms. NIH Publishing, Washington, p 225
1953. Tang J (1963) Nature 199: 1094–1095
1954. Hill RL (1965) Adv Protein Chem 20: 37–107
1955. Zinchenko AA, Rumsh LD, Antonov VK (1976) Bioorgan Khimiya 2: 803–810
1956. Powers JC, Harley D, Myers DV (1977) Adv Exp Med Biol 95: 141–158
1957. Free SM, Wilson JM (1964) J Med Chem 7: 395–399
1958. Pozsgay M, Gáspár R, Bajusz S, Elödi P (1978) Acta Biochim Biophys Acad Sci Hung 13: 184–187
1959. Pozsgay M, Gáspár R, Bajusz S, Elödi P (1979) Eur J Biochem 95: 115–119
1960. Marossy K, Szabó GC, Pozsgay M, Elödi P (1980) Biochem Biophys Res Commun 96: 762–769
1961. Pozsgay M, Szbó GC, Bajusz S, Simonsson R, Gáspár R, Elödi P (1981) Eur J Biochem 115: 491–495
1962. Knówles JR (1965) J Theor Biol 9: 213–228
1963. Berezin IV, Klyosov AA, Martinek K (1979) In: Chemistry reviews (USSR), Sect B, vol 1. Harwood Acad Publ, Chur, pp 206–277
1964. Klyosov AA, Berezin IV (1980) The enzymic catalysis. Moscow State Univ Press, Moscow
1965. Dorovska VN, Varfolomeyev SD, Kazanskaya NF, Klyosov AA (1972) FEBS Lett 23: 122–124
1966. Klyosov AA, Vallee BL, Berezin IV (1977) Dokl Acad Nauk SSSR 234: 964–967
1967. Milstein JBM, Fife TH (1969) Biochemistry 8: 623–627
1968. Kogan RL, Fife TH (1985) Biochemistry 24: 2610–2614
1969. Fife TH, Milstein JB (1967) Biochemistry 6: 2901–2907
1970. Berezin IV, Kazanskaya NF, Klyosov AA, Martinek K (1971) FEBS Lett 15: 125–128
1971. Hansch C (1972) J Org Chem 37: 92–95
1972. Dupaix A, Bechet J-J, Roucous C (1973) Biochemistry 12: 2559–2566
1973. Hansch C, Grieco C, Silipo C, Vittoria A (1977) J Med Chem 20: 1420–1435
1974. Hansch C, Smith RN, Rockoff A, Calef DF, Jow PYC, Fukunaga J (1977) Arch Biochem Biophys 183: 383–392
1975. Hansch C, Coats E (1970) J Pharm Sci 59: 731–748
1976. Martinek K, Dorovska VN, Varfolomeev SD (1972) Biokhimiya 37: 1245–1250
1977. Bender ML, Killheffer JV, Kézdy FJ (1964) J Am Chem Soc 86: 5330–5331
1978. Seydoux F, Yon J (1967) Eur J Biochem 3: 42–56
1979. Klyosov AA (1977) Bioorgan Khimiya 3: 1100–1110
1980. Klyosov AA, Fedoseév VN, Kirret OG (1977) Biokhimiya 42: 1939–1951
1981. Carotti A, Smith RN, Wong S, Hansch C, Blaney JM, Langridge R (1984) Arch Biochem Biophys 229: 112–125

1982. Pank M, Kirret O, Paberit N, Aaviksaar A (1982) FEBS Lett 142: 297–300
1983. Marshall TH, Akgün A (1971) J Biol Chem 246: 6019–6023
1984. Bender ML, Brubacher LJ (1964) J Am Chem Soc 86: 5333–5334
1985. Bernhard SA, Malhotra OP (1974) Isr J Chem 12: 471–481
1986. Smolarsky M (1978) Biochemistry 17: 4606–4615
1987. Bender ML, Kézdy FJ, Gunter CR (1964) J Am Chem Soc 86: 3714–3721
1988. Cohen SG, Weinstein SY (1964) J Am Chem Soc 86: 5326–5330
1989. Caplow M, Harper C (1972) J Am Chem Soc 94: 6508–6512
1990. Rumsh LD, Tikhodeeva AG, Antonov VK (1974) Biokhimiya 39: 899–902
1991. Cohen SG, Torem B, Vaidya V, Ehret A (1976) J Biol Chem 251: 4722–4728
1992. Martinek K, Dorovska VN, Varfolomeyev SD, Berezin IV (1972) Biochim Biophys Acta 271: 80–86
1993. Cohen SG, Milovanović A (1968) J Am Chem Soc 90: 3495–3502
1994. Cohen SG, Lo LW (1970) J Biol Chem 245: 5718–5727
1995. Cohen SG, Neuwirth Z, Weinstein SY (1966) J Am Chem Soc 88: 5306–5315
1996. Inagami T, York SS, Patchornik A (1965) J Am Chem Soc 87: 126–127
1997. Parker L, Wang JH (1968) J Biol Chem 243: 3729–3734
1998. Inagami T, Patchornik A, York SS (1969) J Biochem 65: 809–819
1999. Caplow M (1969) J Am Chem Soc 91: 3639–3645
2000. Fastrez J, Fersht AR (1973) Biochemistry 12: 1067–1074
2001. Philipp M, Pollack RM, Bender ML (1973) Proc Natl Acad Sci USA 70: 517–520
2002. Petkov D, Christova E, Stoineva J (1978) Biochim Biophys Acta 527: 131–141
2003. Whitaker JR, Bender ML (1965) J Am Chem Soc 87: 2728–2737
2004. Brubacher LJ, Bender ML (1966) J Am Chem Soc 88: 5871–5880
2005. Fink AL, Bender ML (1969) Biochemistry 8: 5109–5118
2006. Klyosov AA, Andreev VM, Berezin IV (1974) Biokhimiya 39: 1222–1230
2007. Grieco C, Hansch C, Silipo C, Smith RN, Vittoria A, Yamada K (1979) Arch Biochem Biophys 194: 542–551
2008. Kozlov LV, Antonov VK, D'jachenko ED (1975) In: 7th Meet FEBS, Varna 1975, p 110
2009. Kozlov LV, D'jachenko ED, Antonov VK (1977) Bioorgan Khimiya 3: 105–110
2010. Kozlov LV (1979) Biokhimiya 44: 166–171
2011. Fersht AR, Blow DM, Fastrez J (1973) Biochemistry 12: 2035–2041
2012. Kozlov LV (1980) Biokhimiya 45: 1442–1447
2013. Kozlov LV (1979) Biokhimiya 44: 172–176
2014. Yamamoto T, Izumiya N (1966) Arch Biochem Biophys 114: 459–464
2015. Moffitt MJ, Means GE (1978) Biochem Biophys Res Commun 83: 1415–1421
2016. Bergman M, Fruton J.S. (1937) J Biol Chem 117: 193–198
2017. Béchet J-J, Dupaix A, Roucous C (1973) Biochemistry 12: 2566–2572
2018. Yoshimoto M, Hansch C (1976) J Org Chem 41: 2269–2273
2019. Polgar L, Fejes J (1979) Eur J Biochem 102: 531–536
2020. Hein GE, McGriff RB, Niemann C (1960) J Am Chem Soc 82: 1830–1831
2021. Hein GE, Niemann C (1962) J Am Chem Soc 84: 4487–4494
2022. Cohen SG, Milovanović A, Schultz RM, Weinstein SY (1969) J Biol Chem 244: 2664–2674
2023. Silver MS, Stoddard M, Sone T, Matta MS (1970) J Am Chem Soc 92: 3151–3160
2024. Schwartz HM, Wu W-S, Marr PW, Jones JB (1978) J Am Chem Soc 100: 5199–5203
2025. Pattabiraman TN, Lawson WB (1972) J Biol Chem 247: 3029–3038
2026. Belleau B, Chevalier R (1968) J Am Chem Soc 90: 6864–6866
2027. Abdulaev ND, Bystrov VF, Rumsh LD, Antonov VK (1969) Tetrahedron Lett: 5287–5290
2028. Dugas H (1969) Can J Biochem 47: 985–987
2029. Antonov VK (1972) Dokl Acad Nauk SSSR 205: 839–842
2030. Kazanskaya NF, Slobodyanskaya EM, Tsetlin VI, Shepel EN, Ivanov VT, Ovchinnikov YA (1970) Biokhimiya 35: 1147–1152
2031. Kowlessur D, Thomas EW, Topham CM, Templeton W, Brocklehurst K (1990) Biochem J 266: 653–660
2032. Sakurai T, Margolin AL, Russell AJ, Klibanov AM (1988) J Am Chem Soc 110: 7236–7237
2033. Lin L-N, Brandts JF (1979) Biochemistry 18: 5037–5042
2034. Lin L-N, Brandts JF (1983) Biochemistry 22: 4480–4485

2035. Fisher G, Bang H, Berger G, Schellenberger A (1984) Biochim Biophys Acta 791: 87–97
2036. Lin L-N, Brandts JF (1985) Biochemistry 24: 6533–6538
2037. Berezin IV, Varfolomeev SD, Martinek K (1970) Dokl Acad Nauk SSSR 193: 932–935
2038. Nikol'skaya II, Petrova VV, Khludova MS, Mkhitarov RA, Oreshin MM, Kazanskaya NF (1979) In: Berezin IV, Martinek K (eds) The progress in bioorganic catalysis. Moscow State Univ Press, Moscow, pp 275–285
2039. Aisina RB, Andrianova II, Kazanskaya NF, Lukasheva EV (1979) Biokhimiya 44: 2227–2233
2040. Antonov VK (1981) In: Turk V, Vitale L (eds) Proteinases and their inhibitors. Int Symp Portoroż Yugoslavia, 1980. Mladinska Knjiga, Ljubljana; Pergamon Press, Oxford, pp 141–150
2041. Berezin IV, Varfolomeev SD (1979) Biokinetics. Nauka, Moscow
2042. Aaviksaar A, Paris J, Palm V (1971) Organic Reactivity (Tartu State Univ) 8: 817–832
2043. Kosover E (1964) Molecular biochemistry. Mir, Moscow
2044. Kunugi S, Hirohara H, Nishimura E, Ise N (1978) Arch Biochem Biophys 189: 298–308
2045. Karasaki Y, Ohno M (1979) J Biochem 86: 563–567
2046. McKay TJ, Phelan AW, Plummer TH (1974) Arch Biochem Biophys 197: 487–492
2047. Philipp M, Bender ML (1973) Nature 244: 44–46
2048. Hardy LW, Kirsch JF (1984) Biochemistry 23: 1275–1282
2049. Brouwer AC, Kirsch JF (1982) Biochemistry 21: 1302–1307
2050. Ogura Y (1955) Arch Biochem Biophys 57: 288–295
2051. Veselova MN, Chukhrai ES, Poltorak OM (1969) Vestnic Moscow Univ, Ser Khim 10: N6 8–12
2052. Chang GL, Davis LC, Reeck GR (1990) Anal Biochem 186: 269–272
2053. Morihara K, Oka T (1977) Arch Biochem Biophys 178: 188–194
2054. Dorer FE, Lentz KE, Kahn JR, Levine M, Skeges LT (1978) J Biol Chem 23: 3140–3142
2055. Nakajima K, Powers JC, Ashe BM, Zimmerman M (1979) J Biol Chem 254: 4027–4032
2056. Ryle AP, Auffret CA (1979) Biochem J 179: 247–249
2057. Martin P, Raymond M-N, Bricas E, Ribadeau Dumas B (1980) Biochim Biophys Acta 612: 410–420
2058. McRae B, Nakajima K, Travis J, Powers JC (1980) Biochemistry 19: 3973–3978
2059. Ascenzi P, Menegatti E, Guarneri M, Bortolotti F, Antonini E (1982) Biochemistry 21: 2483–2490
2060. Robinson NC, Neurath H, Walsh KA (1973) Biochemistry 12: 420–426
2061. Lonsdale-Eccles JD, Neurath H, Walsh KA (1978) Biochemistry 17: 2805–2809
2062. Lonsdale-Eccles JD, Kerr HA, Walsh KA, Neurath H (1979) FEBS Lett 100: 157–160
2063. Plese PC, Behnké WD (1977) Biochim Biophys Acta 483: 172–180
2064. Kozlov LV, Krilova YI, Antonov VK (1979) Bioorgan Khimiya 5: 1083–1090
2065. Pedersen VB, Christensen KA, Foltmann B (1979) Eur J Biochem 94: 573–580
2066. Valenzuela P, Bender ML (1971) J Am Chem Soc 93: 3783–3784
2067. Hofsteenge J, Braun PJ, Stone SR (1988) Biochemistry 27: 2144–2151
2068. Sereiskaya AA, Osadchuk TV, Kornelyuk AI, Pechnik IV, Serebryanii SB (1989) Biokhimiya 54: 542–548
2069. Pratt RE, Ouellette AJ, Dzau VJ (1983) Proc Natl Acad Sci USA 80: 6809–6813
2070. Light A, Fonseca P (1984) J Biol Chem 259: 13195–13198
2071. Tam L-T, Engelbrecht S, Talent JM, Gracy RW, Erdös EG (1985) Biochem Biophys Res Commun 133: 1187–1192
2072. Hathaway DR, Werth DK, Haeberle JR (1982) J Biol Chem 257: 9072–9077
2073. Wu H-L, Shi G-Y, Bender ML (1987) Proc Natl Acad Sci USA 84: 8292–8295
2074. Wu H-L, Shi G-Y, Wohl RC, Bender ML (1987) Proc Natl Acad Sci USA 84: 8793–8795
2075. Vita C, Dalzoppo D, Fontana A (1985) Biochemistry 24: 1798–1806
2076. Cohen LA (1970) In: Boyer PD (ed) The enzymes, 3rd edn, vol 1. Academic Press, New York, pp 147–211
2077. Sigman DS, Mooser G (1975) Annu Rev Biochem 44: 889–931
2078. Glazer AN, De Lange RJ, Sigman DS (1975) Chemical modification of proteins. Selected method and analytical procedures. North-Holland, Amsterdam
2079. Inada Y, Yoshimoto T, Matsushima A, Saito Y (1986) Trends Biotechnol 4: 68–73

2080. Feeney RE (1987) Int J Pept Protein Res 29: 145–161
2081. Mosolov VV, Valueva TV, Sokolova GI, Antonov VK (1975) Dokl Acad Nauk SSSR 221: 484–487
2082. Fujiwara K, Tsuru D (1978) Biochim Biophys Acta 522: 195–204
2083. Polgár L, Sajgo M (1981) Biochim Biophys Acta 667: 351–354
2084. Philipp M, Isai IH, Bender ML (1979) Biochemistry 18: 3769–3773
2085. Wu Z-P, Hilvert D (1989) J Am Chem Soc 111: 4513–4514
2086. Gorecki M, Wilchek M, Blumberg S (1978) Biochim Biophys Acta 535: 90–99
2087. Cueni L, Riordan JF (1978) Biochemistry 17: 1834–1841
2088. Blumberg S (1979) Biochemistry 18: 2815–2820
2089. Urdea MS, Legg JI (1979) J Biol Chem 254: 11868–11874
2090. Avraham RB, Shalitin Y (1985) FEBS Lett 180: 239–242
2091. Remy MH, Bourdillon C, Thomas D (1985) Biochim Biophys Acta 829: 69–75
2092. Mozhaev VV, Šikšnis VA, Melik-Nubarov NS, Galkantaite NZ, Denis GJ, Butkus EP, Zaslavsky BY, Mestechkina NH (1988) Eur J Biochem 173: 147–154
2093. Auld DS, Vallee BL (1970) Biochemistry 9: 602–609
2094. Ajabnoor MA, Ray LE, Wagner FW (1980) Arch Biochem Biophys 202: 540–545
2095. Prescott JM, Wagner FW, Holmquist B, Vallee BL (1983) Biochem Biophys Res Commun 114: 646–652
2096. Margolis HC, Nakagawa Y, Douglas KT, Kaiser ET (1978) J Biol Chem 253: 7891–7897
2097. Beeley JG (1977) Biochem Biophys Res Commun 76: 1051–1055
2098. Geisow MJ (1978) FEBS Lett 87: 111–114
2099. Gráf L, Hollós M (1980) Biochem Biophys Res Commun 93: 1089–1093
2100. Loh YP, Gainer H (1978) FEBS Lett 96: 269–272
2101. Hartley B (1974) Sympos Soc Gen Microbiol 24: 151–165
2102. Betz JL, Brown PR, Smyth MJ, Clarke PH (1974) Nature 247: 261–264
2103. Gregoriou M, Brown PR (1979) Eur J Biochem 96: 101–108
2104. Zoller MJ, Smith M (1983) Meth Enzymol 100: 468–475
2105. Carter P (1986) Biochem J 237: 1–7
2106. Itakura K, Rossi JJ, Wallace RB (1984) Annu Rev Biochem 53: 323–356
2107. Fersht AR, Shi J-P, Wilkinson AJ, Blow DM, Carter P, Waye MHY, Winter GP (1984) Angew Chem Int Ed Engl 23: 467–473
2108. Smith M (1985) Annu Rev Genet 19: 423–462
2109. Estell DA, Graycar TP, Miller JV, Powers DB, Burnier JP, Ng PG, Wells JA (1986) Science 233: 659–663
2110. Carter P, Wells JA (1988) Nature 332: 564–568
2111. Estell DA, Graycar TP, Wells JA (1985) J Biol Chem 260: 6518–6521
2112. Wells JA, Powers DB (1986) J Biol Chem 261: 6564–6570
2113. Katz BA, Kossiakoff A (1986) J Biol Chem 261: 15480–15485
2114. Wong C-H, Chen S-T, Hennen WJ, Bibbs JA, Wang Y-F, Liu JL-C, Pantoliano MW, Whitlow M, Bryan PN (1990) J Am Chem Soc 112: 945–953
2115. Matchinson C, Wells JA (1989) Biochemistry 28: 4807–4815
2116. Sternberg MJE, Hayes FRF, Russell AJ, Thomas PG, Fersht AR (1987) Nature 330: 86–88
2117. Hilvert D, Gardell SJ, Rutter WJ, Kaiser ET (1986) J Am Chem Soc 108: 5298–5304
2118. Wells JA, Powers DB, Bott RR, Graycar TP, Estell DA (1987) Proc Natl Acad Sci USA 84: 1219–1223
2119. Sprang S, Standing T, Fletterick RJ, Stroud RM, Finer-Moore J, Xuong N-H, Hamlin R, Rutter WJ, Craik CS (1987) Science 237: 905–913
2120. Wells JA, Cunningham BC, Graycar TP, Estell DA (1987) Proc Natl Acad Sci USA 84: 5167–5171
2121. Pantoliano MW, Whitlow M, Wood JF, Rollence ML, Finzel BC, Gillilaud GL, Poulos TL, Bryan PN (1988) Biochemistry 27: 8311–8317
2122. Bryan P, Pantoliano MW, Quill SG, Hsiao H-Y, Poulos T (1986) Proc Natl Acad Sci USA 83: 3743–3745
2123. Craik CS, Gardell S, Roczniak S, Fletterick R, Rutter WJ (1985) Biochemistry 24: 3373
2124. Craik CS, Roczniak S, Sprang S, Fletterick R, Rutter W (1987) J Cell Biochem 33: 199–211
2125. Warshel A, Sussman F (1986) Proc Natl Acad Sci USA 83: 3806–3810

2126. Graf L, Craik CS, Patthy A, Roczniak S, Fletterick RJ, Rutter WJ (1987) Biochemistry 26: 2616–2623
2127. McCaman MT, Cummings DB (1986) J Biol Chem 261: 15345–15348
2128. Yamauchi T, Nagahama M, Hori N, Murakami K (1988) FEBS Lett 230: 205–208
2129. Devault A, Sales V, Nault C, Beaumont A, Roques B, Crine P, Boileau G (1988) FEBS Lett 231: 54–58
2130. Ivanoff LA, Towatari T, Ray J, Korant BD, Petteway SR (1986) Proc Natl Acad Sci USA 83: 5392–5396
2131. Mous J, Heimer EP, Le Grice SFJ (1988) J Virol 62: 1433–1436
2132. Dalbadie-McFarland G, Cohen LW, Riggs AD, Morin C, Itakura K, Richards JH (1982) Proc Natl Acad Sci USA 79: 6409–6413
2133. Sigal IS, Harwood BG, Arentzen R (1982) Proc Natl Acad Sci USA 79: 7157–7160
2134. Sigal IS, De Grado WF, Thomas BJ, Petteway SR (1984) J Biol Chem 259: 5327–5332
2135. Gardell SJ, Craik CS, Hilvert D, Urdea MS, Rutter WJ (1985) Nature 317: 551–555
2136. Bergmann M, Fraenkel-Conrat H (1937) J Biol Chem 119: 707–712
2137. Fruton JS (1982) Adv Enzymol 53: 239–305
2138. Morihara K, Oka T (1977) Biochem J 163: 531–542
2139. Petkov DD, Stoineva I (1984) Biochem Biophys Res Commun 118: 317–323
2140. Gololobov MY, Shvyadas VK (1988) Biokhimiya 53: 1174–1180
2141. Cheng E, Miranda MTM, Tominaga M (1988) Int J Pept protein Res 31: 116–125
2142. Shuster Y, Mitin YV, Yakubke G-D (1989) Bioorgan Khimiya 15: 345–347
2143. Wayne SI, Fruton JS (1983) Proc Natl Acad Sci USA 80: 3241–3244
2144. Reichmann L, Kasche V (1986) Biochim Biophys Acta 872: 269–276
2145. Nakanishi K, Matsuno R (1986) Eur J Biochem 161: 533–540
2146. Breddam K, Widmer F, Johansen JT (1980) Carlsberg Res Commun 45: 237–247
2147. Kapitannikov YV, Popov AA, Shimbarevich EV, Rumsh LD, Antonov VK (1988) Bioorgan Khimiya 14: 797–802
2148. Kapune A, Kasche V (1978) Biochem Biophys Res Commun 80: 955–962
2149. Konopinska D, Muzalewski F (1983) Mol Cell Biochem 51: 165–175
2150. Nakatsuka T, Sasaki T, Kaiser ET (1987) J Am Chem Soc 109: 3808–3810
2151. West JB, Scholten J, Stolowich NJ, Hogg JL, Scott AI, Wong C-H (1988) J Am Chem Soc 110: 3709–3710
2152. Riechmann L, Kasche V (1984) Biochem Biophys Res Commun 120: 686–691
2153. Gololobov MY, Borisov IL, Shvyadas VK (1987) Biokhimiya 52: 584–591
2154. Homandberg GA, Mattis JA, Laskowski M (1978) Biochemistry 17: 5220–5227
2155. Martinek K, Semenov AN (1981) Biochim Biophys Acta 658: 90–101
2156. Bednarski MD, Chenault HK, Simon ES, Whitesides GH (1987) J Am Chem Soc 109: 1283–1285
2157. Reslow M, Adlercreutz P, Mattiasson B (1988) Eur J Biochem 177: 313–318
2158. Gaertner H, Puigserver A (1989) Eur J Biochem 181: 207–213
2159. Riva S, Chopineau J, Kieboom APG, Klibanov AM (1988) J Am Chem Soc 110: 584–589
2160. Kitaguchi H, Klibanov AM (1989) J Am Chem Soc 111: 9272–9273
2161. Kullmann W (1980) J Biol Chem 255: 8234–8238
2162. Isowa Y, Ohmori M, Sato M, Mori K (1977) Bull Chem Soc Jpn 50: 2766–2772
2163. Isowa Y, Ichikawa T, Ohmori M (1978) Bull Chem Soc Jpn 51: 271–276
2164. Sealock RW, Laskowski M (1969) Biochemistry 8: 3703–3710
2165. Homandberg GA, Laskowski M (1979) Biochemistry 18: 586–592
2166. Morihara K, Ueno Y, Sakina K (1986) Biochem J 240: 803–810
2167. Fruton JS, Johnston RB, Fried M (1951) J Biol Chem 190: 39–53
2168. Neumann H, Levin Y, Berger A, Katchalski E (1959) Biochem J 73: 33–41
2169. Takahashi M, Wang TT, Hofmann T (1974) Biochem Biophys Res commun 57: 39–46
2170. Dowmont YP, Fruton JS (1952) J Biol Chem 197: 271–283
2171. Blau K, Waley SG (1954) Biochem J 57: 538–541
2172. Yagisawa S, Watanabe S, Takaoka T, Azuma H (1990) Biochem J 266: 771–775
2173. Morihara K, Tsuzuki H, Oka T (1978) Biochem Biophys Res Commun 84: 95–101
2174. Fruton JS, Fujii S, Knappenberger MH (1961) Proc Natl Acad Sci USA 47: 759–761
2175. Kitson TM, Knowles JR (1971) Biochem J 122: 249–256

2176. Silver MS, James SLT (1980) J Biol Chem 255: 555–560
2177. Newmark AK, Knowles JR (1975) J Am Chem Soc 97: 3557–3559
2178. Wang TT, Hofmann T (1977) Can J Biochem 55: 286–294
2179. Hanson H, Lasch J (1967) Hoppe-Seyler's Z Physiol Chem 348: 1525–1531
2180. Nishino T, Kozarich JW, Strominger JL (1977) J Biol Chem 252: 2934–2939
2181. Perkins HR, Frére J-M, Ghuysen JM (1981) FEBS Lett 123: 75–78
2182. Malt'sev NI, Ginodman LM, Orekhovich VN, Valueva TA, Akimova LM (1966) Biokhimiya 31: 983–987
2183. Sharon N, Grisaro V, Neumann H (1962) Arch Biochem Biophys 97: 219–221
2184. Grisaro V, Sharon N (1964) Biochim Biophys Acta 89: 152–155
2185. Kozlov LV, Ginodman LM, Orekhovich VN (1965) Dokl Acad Nauk SSSR 161: 1455–1457
2186. Kozlov LV (1974) Biokhimiya 39: 512–515
2187. Jackson WT, Schlamowitz M, Shaw A (1966) Biochemistry 5: 4105–4110
2188. Breslow R, Wernic DL (1977) Proc Natl Acad Sci USA 74: 1303–1307
2189. Antonov VK, Ginodman LM, Rumsh LD, Kapitannikov YV, Barshevskaya TN, Yavashev LP, Gurova AG, Volkova LI (1980) Bioorgan Khimiya 6: 436–446
2190. Sugimoto T, Kaiser ET (1978) J Am Chem Soc 100: 7750–7751
2191. Sugimoto T, Kaiser ET (1979) J Am Chem Soc 101: 3946–3951
2192. Spratt TE, Sugimoto T, Kaiser ET (1983) J Am Chem Soc 105: 3679–3683
2193. Nashed NT, Kaiser ET (1981) J Am Chem Soc 103: 3611–3612
2194. Nashed NT, Kaiser ET (1986) J Am Chem Soc 108: 2710–2715
2195. Spratt TE, Kaiser ET (1984) J Am Chem Soc 106: 6440–6442
2196. Doherty DG (1955) J Am Chem Soc 77: 4887–4892
2197. Neurath H, Walsh KA (1976) Proc Natl Acad Sci USA 73: 3825–3832
2198. Walsh KA (1975) In: Reich E, Rifkin DB, Shaw E (eds) Proteases and biological control. Proc Cold Spring Harbor Conf on cell proliferation, vol 2. Cold Spring Harbor Laboratory, New York, pp 1–11
2199. Holzer H (1980) Annu Rev Biochem 49: 63–91
2200. Gierasch LM (1989) Biochemistry 28: 923–930
2201. Tomita N, Izumoto Y, Horii A, Doi S, Yokouchi H, Ogawa M, Mori T, Matsubara K (1989) Biochem Biophys Res Commun 158: 569–575
2202. Ikemura H, Takagi H, Inouye M (1987) J Biol Chem 262: 7859–7864
2203. Matheja J, Degens ET (1971) Adv Enzymol 34: 1–39
2204. Neurath H (1957) Adv Prot Chem 12: 319–386
2205. Charles M, Rovery M, Guidoni A, Desnuelle P (1963) Biochem Biophys Acta 69: 115–129
2206. Jacobsen CF (1947) C R Trav Lab Carlsberg 25: 325–334
2207. Sharma SK, Hopkins TR (1978) J Biol Chem 253: 3055–3061
2208. Parkes CO, Smillie LB (1966) Biochim Biophys Acta 113: 629–631
2209. Prahl JW, Neurath H (1966) Biochemistry 5: 2131–2146
2210. Rovery M, Bianchetta J (1972) Biochim Biophys Acta 268: 212–220
2211. Degen SJF, Davie EW (1987) Biochemistry 26: 6165–6177
2212. Suttie JW, Jackson CM (1977) Physiol Rev 57: 1–70
2213. Ferlund P, Stenflo J, Roepstorff P, Thomsen J (1975) J Biol Chem 250: 6125–6133
2214. Stenn KS, Blout ER (1972) Biochemistry 11: 4502–4515
2215. Krishnaswami S, Church WR, Nesheim ME, Mann KG (1987) J Biol Chem 262: 3291–3299
2216. Geiger R (1981) In: Turk VC, Vitale LJ (eds) Proteinases and their inhibitors. Structure, function and applied aspects. Proc Int Symp Portoroz, Yugoslavia 1980. Mladinska Knjiga, Ljubljana; Pergamon Press, Oxford, pp 353–376
2217. Movat HZ (1978) Rev Physiol Biochem Pharmacol 84: 143–202
2218. Austen KF, Fearon DT (1979) Adv Exp Med Biol 120B: 3–17
2219. Morgan PH, Nair JG (1975) Biochem Biophys Res Cummun 66: 1037–1041
2220. Kasahara Y, Odai H, Takahashi K, Nagasawa S, Koyama J (1985) J Biol Chem 97: 365–372
2221. Tankersley DL, Finlayson JS (1984) Biochemistry 23: 273–279
2222. Baba T, Michikawa Y, Kawakura K, Arai Y (1989) FEBS Lett 244: 132–136
2223. Terada I, Kwon S-T, Miyata Y, Matsuzawa H, Ohta T (1990) J Biol Chem 265: 6576–6581
2224. Salvesen G, Enghild JJ (1990) Biochemistry 29: 5304–5308

2225. Eisenberg M, Johnson L, Moon KE (1984) Biochem Biophys Res Commun 125: 279–285
2226. Müller M, Müller H (1981) J Biol Chem 256: 11962–11965
2227. Ikemura H, Inouye M (1988) J Biol Chem 263: 12959–12963
2228. Silen JL, Franc D, Fujishige A, Boul R, Agard DA (1989) J Bacteriol 171: 1320–1325
2229. Young M, Jo Koroly M (1980) Biochemistry 19: 5316–5321
2230. Lo S-S, Fraser BA, Liu T-Y (1984) J Biol Chem 259: 11041–11045
2231. Sluyterman LA (1967) Biochim Biophys Acta 139: 430–438
2232. Klein IB, Kirsch JF (1969) Biochem Biophys Res Commun 34: 575–581
2233. De Martino GN, Huff CA, Croall DE (1986) J Biol Chem 261: 12047–12052
2234. Herriott RM (1938) J Gen Physiol 21: 501–540
2235. Hayano T, Sogawa K, Ichihara Y, Fujii-Kuriyama Y, Takahashi K (1988) J Biol Chem 263: 1382–1385
2236. Lin X-l, Wong RNS, Tang J (1989) J Biol Chem 264: 4482–4489
2237. Auer HE, Glick DM (1984) Biochemistry 23: 2735–2739
2238. Glick DM, Auer HE, Rich DH, Kawai M, Kamath A (1986) Biochemistry 25: 1858–1864
2239. Glick DM, Shalitin Y, Hilt CR (1989) Biochemistry 28: 2626–2630
2240. Hartsuck JA, Marciniszyn J, Huang JS, Tang J (1977) Adv Exp Med Biol 95: 85–102
2241. Sanny CG, Hartsuck JA, Tang J (1975) J Biol Chem 250: 2635–2639
2242. Kay J, Dykes CW (1977) Adv Exp Med Biol 95: 103–127
2243. Kageyama T, Takahashi K (1983) J Biochem 93: 743–754
2244. Kageyama T, Takahashi K (1987) Eur J Biochem 165: 483–490
2245. Kostka V, Keilová H, Baudyš M (1981) In: Turk VC, Vitale Lj (eds) Proteinases and their inhibitors. Structure, function and applied aspects. Proc Int Symp Portoroz, Yugoslavia 1980. Mladinska Knjiga, Ljubljana; Pergamon Press, Oxford, pp 125–140
2246. Pichová I, Pohl J, Štrop P, Kostka V (1985) In: Kostka V (ed) Aspartic proteinases and their inhibitors. Gruyter, Berlin, pp 301–308
2247. Kageyama T, Ichinose M, Miki K, Athauda SB, Tanji M, Takahashi K (1989) J Biochem 105: 15–22
2248. Marciniszyn JP, Kassell B (1971) J Biol Chem 246: 6560–6565
2249. Kumar PM, Ward PH, Kassell B (1977) Adv Exp Med Biol 95: 211–221
2250. Herriott RM (1941) J Gen Physiol 24: 325–338
2251. Kumar PM, Kassell B (1977) Biochemistry 16: 3846–3849
2252. Dunn BM, Deyrup C, Moesching WG, Gilbert W, Nolan RJ, Trach ML (1978) J Biol Chem 253: 7269–7275
2253. Petrova EN, Revina LP, Timokhina EA, Lavrenova TI, Vaganova T, Stepanov VM (1988) Biokhimiya 53: 1389–1396
2254. Barrett AJ (1977) In: Barrett AJ (ed) Proteinases in mammalian cells and tissues. North-Holland, Amsterdam, pp 209–233
2255. Turk V, Puizdar V, Lah T, Gubenšek F, Kregar I (1981) In: Turk VC, Vitale Lj (eds) Proteinases and their inhibitors. Structure, function and applied aspects. Proc Int Symp Portoroz, Yugoslavia 1980. Mladinska Knjiga, Ljubljana; Pergamon Press, Oxford, pp 117–124
2256. Puizdar V, Turk V (1981) FEBS Lett 132: 299–304
2257. Samarel AM, Worobec SW, Ferguson AG, Lesch M (1984) J Biol Chem 259: 4702–4705
2258. Gieselmann V, Hasilik A, Figura K (1985) J Biol Chem 260: 3215–3220
2259. Yonezawa S, Takahashi T, Wang X-J, Wong RNS, Hartsuck JA, Tang J (1988) J Biol Chem 263: 16504–16511
2260. Imai T, Miyazaki H, Hirose S, Hori H, Hayashi T, Kageyama R, Ohkubo H, Nakanishi S, Murakami K (1983) Proc Natl Acad Sci USA 80: 7405–7409
2261. Inagami T, Murakami K, Misono K, Workman RJ, Cohen S, Suketa Y (1977) Adv Exp Med Biol 95: 225–247
2262. Takii Y, Inagami T (1982) Biochem Biophys Res Commun 104: 133–140
2263. Kim S-J, Hirose S, Miyazaki H, Ulno N, Higashimori K, Morinaga S, Kimura T, Sakakibara S, Murakami K (1985) Biochem Biophys Res Commun 126: 641–645
2264. Fritz LC, Arfsten AE, Dzau VJ, Alteas SA, Baxter JD, Fiddes JC, Shine J, Cofer CL, Kushner P, Poute PA (1986) Proc Natl Acad Sci USA 83: 4114–4118
2265. Leckie BJ, McConnell A, Jordan J (1977) Adv Exp Med Biol 95: 249–269

2266. Higashimori K, Mizuno K, Nakajo S, Boehm FH, Marcotte PA, Egan DA, Holleman WH, Heusser C, Poisner AM, Inagami T (1989) J Biol Chem 264: 14662–14667
2267. Derkx FHM, Schalekamp MPA, Schalekamp MADH (1987) J Biol Chem 262: 2472–2477
2268. Anson ML (1937) J Gen Physiol 20: 781–793
2269. Keller PJ, Cohen E, Neurath H (1958) J Biol Chem 230: 905–915
2270. Brown JR, Greenshields RN, Yamasaki M, Neurath H (1963) Biochemistry 2: 867–876
2271. Freisheim JH, Walsh KA, Neurath H (1967) Biochemistry 6: 3010–3019
2272. Martínez MC, Nieuwenhuysen P, Clauwaert J, Cuchillo CM (1983) Biochem J 215: 23–27
2273. Vendrell J, Cuchillo CM, Avilés FX (1990) J Biol Chem 265: 6949–6953
2274. Freisheim JH, Walsh KA, Neurath H (1967) Biochemistry 6: 3020–3028
2275. Quinto C, Quiroga M, Swain WF, Nikovits WC, Standring DN, Pictet RL, Valenzuela P, Rutter WJ (1982) Proc Natl Acad Sci USA 79: 31–35
2276. Chapus C, Kerfelec B, Foglizzo E, Bonicel J (1987) Eur J Biochem 166: 379–385
2277. Chang C-N, Nielsen JBK, Izui K, Blobel G, Lampen JO (1982) J Biol Chem 257: 4340–4344
2278. Hayashi S, Chang S-Y, Chang S, Wu HC (1984) J Biol Chem 259: 10448–10454
2279. Zwilling R, Pfleiderer G, Zonnenborn HH, Kraft V, Stucky I (1970) J Compar Biochem Physiol 28: 1275–1281
2280. James MNG (1976) In: Ribbons DW, Brew K (eds) Proteolysis and physiological regulation. Academic Press, New York, pp 125–135
2281. Freer ST, Kraut J, Robertus JD, Wright HT, Xuong NH (1970) Biochemistry 9: 1997–2009
2282. Wright HT (1973) J Mol Biol 79: 13–23
2283. Wright HT (1973) J Mol Biol 79: 1–11
2284. Morgan PH, Robinson NC, Walsh KA, Neurath H (1972) Proc Natl Acad Sci USA 69: 3312–3316
2285. Gertler A, Walsh KA, Neurath H (1974) Biochemistry 13: 1302–1310
2286. Chauvet J, Asher R (1974) Int J Pept Protein Res 6: 37–41
2287. Bode W, Huber R (1976) FEBS Lett 68: 231–236
2288. Nagendra K, Reddy N, Markus G (1972) J Biol Chem 247: 1683–1691
2289. Carraway KL, Spoerl P, Koshland DE (1969) J Mol Biol 42: 133–137
2290. Fersht AR (1973) FEBS Lett 29: 283–285
2291. Hollis DP, McDonald G, Biltonen RL (1967) Proc Natl Acad Sci USA 58: 758–765
2292. Kosman DJ (1972) J Mol Biol 67: 247–264
2293. Surovaya AN, Slobodyanskaya EM, Kozlov LV, Antonov VK (1972) Mol Biologia 6: 106–110
2294. Bazzone TJ, Vallee BL (1976) Biochemistry 15: 868–875
2295. Dixon M (1953) Biochem J 55: 161–171
2296. Alberty RA, Massey V (1954) Biochim Biophys Acta 13: 347–353
2297. Renard M, Fersht AR (1973) Biochemistry 12: 4713–4718
2298. Fersht AR, Renard M (1974) Biochemistry 13: 1416–1426
2299. Knowels JR (1976) Crit Rev Biochem 4: 165–205
2300. Edsall JT, Wyman J (1958) Biophysical chemistry. Academic Press, New York, p 510
2301. Shiao DDF, Lumry R, Rajender S (1972) Eur J Biochem 29: 377–385
2302. Fersht AR, Requena Y (1971) J Am Chem Soc 93: 7079–7087
2303. Petterson G (1987) Eur J Biochem 166: 163–165
2304. Bender ML, Gibian MJ, Whelan DJ (1966) Proc Natl Acad Sci USA 56: 833–839
2305. Hess GP, McConn J, Ku E, McConkey G (1970) Philos Trans R Soc Lond 257B: 89–97
2306. Fersht AR (1972) J Mol Biol 64: 497–509
2307. Garel J-R (1975) FEBS Lett 50: 339–345
2308. Heremans L, Heremans K (1989) Biochim Biophys Acta 999: 192–197
2309. Murachi T, Yamazaki M (1970) Biochemistry 9: 1935–1938
2310. Chang WT, Douglas KT (1980) Biochem J 187: 843–849
2311. Bender ML, Clement GE, Kézdy FJ, Heck Hd'A (1964) J Am Chem Soc 86: 3680–3690
2312. Johnson PE, Stewart JA (1972) Arch Biochem Biophys 149: 295–306
2313. Antonini E, Ascenzi P (1981) J Biol Chem 256: 12449–12455
2314. Ascenzi P, Menagatti E, Guarneri M, Bolognesi M, Amiconi G (1984) Biochim Biophys Acta 789: 99–103

2315. Stewart JA, Anderson JK, Tseng JK, Hallada RM (1971) Biochem Biophys Res Commun 42: 1220–1227
2316. Kasserra HP, Laidler KJ (1969) Can J Chem 47: 4021–4029
2317. Stein RL, Strimpler AM (1987) J Am Chem Soc 109: 6538–6540
2318. Hayashi R, Bai Y, Hata T (1975) J Biochem 77: 69–79
2319. Schack P, Kaarsholm NC (1984) Biochemistry 23: 631–635
2320. Kortt AA, Liu TY (1973) Biochemistry 12: 338–345
2321. Mole JE, Horton HR (1973) Biochemistry 12: 816–822
2322. Cole PW, Murakami K, Inagami T (1971) Biochemistry 10: 4246–4252
2323. Tikhodeeva AG, Zinchenko AA, Rumsh LD, Antonov VK (1974) Dokl Acad Nauk SSSR 214: 355–358
2324. Denburg JL, Nelson R, Silver MS (1968) J Am Chem Soc 90: 479–486
2325. Hollands TR, Fruton JS (1968) Biochemistry 7: 2045–2053
2326. Clement GE, Cashell GS (1972) Biochem Biophys Res Commun 47: 328–332
2327. Auld DS, Vallee BL (1971) Biochemistry 10: 2892–2897
2328. Bunting JW, Kabir SH (1978) Biochim Biophys Acta 527: 98–107
2329. Feder J, Schuck JM (1970) Biotechnol Bioeng 12: 313–319
2330. Müller-Frohne M, Lasch J, Damerau W, Fittkau S (1979) Wiss Z Univ Halle 28(6): 81–97
2331. Baker JO, Prescott JM (1983) Biochemistry 22: 5322–5331
2332. Bicknell R, Knott-Hunziker V, Waley SG (1983) Biochem J 213: 61–66
2333. Hollands TR, Voynick IM, Fruton JS (1969) Biochemistry 8: 575–585
2334. Hofmann T, Hodges RS, James MNG (1984) Biochemistry 23: 635–643
2335. Zeffren E, Kaiser ET (1967) J Am Chem Soc 89: 4204–4208
2336. Clement GE, Snyder SL, Price H, Cartmell R (1968) J Am Chem Soc 90: 5603–5610
2337. Jackson WT, Schlamowitz M, Shaw A, Trujillo R (1969) Arch Biochem Biophys 131: 374–385
2338. Knowels JR, Sharp H, Greenwell P (1969) Biochem J 113: 343–351
2339. Tahin QS, Rocha TL, Paiva ACM (1969) Enzymologia 37: 153–156
2340. Ascenzi P, Amiconi G, Bolognesi M, Guarneri M, Menegatti E, Antonini E (1984) Biochim Biophys Acta 785: 75–80
2341. Ascenzi P, Aducci P, Torroni A, Amiconi G, Ballio A, Menegatti E, Guarneri M (1987) Biochim Biophys Acta 912: 203–210
2342. Halász P, Polgar L (1977) Eur J Biochem 79: 491–494
2343. Broklenhurst K, Malthouse JPG (1980) Biochem J 191: 707–718
2344. Creighton DJ, Gassouroun MS, Heapes JM (1980) FEBS Lett 110: 319–322
2345. Shaked Z, Szajewski RP, Whitesides GM (1980) Biochemistry 19: 4156–4166
2346. Nicholson EM, Shafer JA (1980) Arch Biochem Biophys 200: 560–566
2347. Johnson FA, Lewis SD, Shafer JA (1981) Biochemistry 20: 44–48
2348. Markley JL (1975) Acc Chem Res 8: 70–80
2349. Robillard G, Shulman RG (1974) J Mol Biol 86: 519–540
2350. Bachovchin WW, Roberts JD (1978) J Am Chem Soc 100: 8041–8047
2351. Sakiyama F, Kawata Y (1983) J Biochem 94: 1661–1669
2352. Malthouse JPG, Primrose WV, Mackenzie NE, Scott AI (1985) Biochemistry 24: 3478–3487
2353. Susi H, Zell T, Timasheff SN (1959) Arch Biochem Biophys 85: 437–443
2354. Timasheff SN, Rupley JA (1972) Arch Biochem Biophys 150: 318–323
2355. Koeppe RE, Stroud RM (1976) Biochemistry 15: 3450–3458
2356. Timasheff SN (1970) In: Boyer PD (ed) The enzymes, vol 2. Academic Press, New York, pp 371–443
2357. Gorbunoff MJ (1971) Biochemistry 10: 250–257
2358. Cowgill RW (1965) Biochim Biophys Acta 94: 81–88
2359. Cruickshank WH, Kaplan H (1972) Biochem J 130: 1125–1131
2360. Kitson TM, Knowles JR (1971) FEBS Lett 16: 337–338
2361. Migliorini M, Creighton DJ (1986) Eur J Biochem 156: 189–192
2362. Willenbrock F, Brocklehurst K (1986) Biochem J 238: 103–107
2363. Polgár L, Csoma C (1987) J Biol Chem 262: 14448–14453

2364. Desnuelle P, Gabeloteau C (1957) Arch Biochem Biophys 69: 475–479
2365. Nomoto M, Narahashi Y, Murakami M (1960) J Biochem 48: 453–463
2366. Chao L-P, Liener IE (1965) Biochim Biophys Acta 96: 508–516
2367. Delaage M, Lazdunski M (1967) Biochem Biophys Res Commun 28: 390–394
2368. Voordouw G, Milo C, Roche RS (1976) Biochemistry 15: 3716–3724
2369. De Jersey J, Lahue RS, Martin RB (1980) Arch Biochem Biophys 205: 536–542
2370. Clitte SGR, Grant DAW (1981) Biochem J 193: 655–658
2371. Briedigket L, Frömmel C (1989) FEBS Lett 253: 83–87
2372. Martin RB, Richardson FS (1979) Q Rev Biophys 12: 181–209
2373. Zolin VF, Koreneva LG (1980) The rare earths labels in chemistry and biology. Nauka, Moscow
2374. Snyder AP, Sudnick DR, Arkle VK, Horrocks W (1981) Biochemistry 20: 3334–3339
2375. Birnbaum ER, Abbott F, Gomez JE, Darnall DW (1977) Arch Biochem Biophys 179: 469–476
2376. Bush WA, Stromer MH, Goll DE, Suzuki A (1972) J Cell Biol 52: 367–381
2377. Waxman L (1979) Fed Proc 38: 479
2378. Szpacenko A, Otsuka Y, Goll DE, Stromer MN (1980) Fed Proc 39: 2044
2379. Szpacenko A, Kay J, Goll DE, Otsuka Y (1981) In: Turk VC, Vitale Lj (eds) Proteinases and their inhibitors. Structure, function and applied aspects. Proc Int Symp Portoroz, Yugoslavia 1980. Mladinska Knjiga Ljubljana; Pergamon Press, Oxford, pp 151–161
2380. Fabiato A, Fabiato F (1979) Annu Rev Physiol 41: 473–484
2381. Suzuki K, Tsuji S (1982) FEBS Lett 140: 16–18
2382. Tufty RM, Kretsinger RH (1975) Science 187: 167–169
2383. Suzuki K, Tsuji S, Kubota S, Kimura Y, Imahori K (1981) J Biochem 90: 275–278
2384. Suzuki K, Ishiura S (1983) J Biochem 93: 1463–1471
2385. Okitani A, Goll DE, Stromer MH, Robson RH (1976) Fed Proc 35: 1746
2386. Otsuka Y, Goll DE (1980) Fed Proc 39: 2044
2387. Pontremoli S, Melloni E, Michetti M, Salamino F, Sparatore B, Horecker BL (1988) Proc Natl Acad Sci USA 85: 1740–1743
2388. Melloni E, Sparatore B, Salamino F, Michetti M, Ponremoli S (1982) Biochem Biophys Res Commun 107: 1053–1059
2389. Latt SA, Holmquist B, Vallee BL (1969) Biochem Biophys Res Commun 37: 333–339
2390. Feder J, Garrett LR, Wildi BS (1971) Biochemistry 10: 4552–4556
2391. Shimizu T, Hatano M (1982) Biochem Biophys Res Commun 104: 1356–1362
2392. Voordouw G, Gaucher GH, Roche RS (1974) Can J Biochem 52: 981–990
2393. Matthews BW, Colman PM, Jansonius JN, Titani K, Walsh KA, Neurath H (1972) Nat New Biol 238: 41–43
2394. Colman PM, Weaver LH, Matthews BW (1972) Biochem Biophys Res Commun 46: 1999–2005
2395. Ludewig M, Lasch J, Kettmann U, Frohne M, Hanson H (1971) Enzymologia 41: 59–63
2396. Lasch J, Kudernatsch W, Hanson H (1973) Eur J Biochem 34: 53–57
2397. Krebs HA (1930) Biochem Z 220: 289–298
2398. Berezin IV, Will H, Martinek K (1967) Dokl Akad Nauk SSSR 175: 230–233
2399. Savin YV, Martinek K, Berezin IV (1971) Dokl Akad Nauk SSSR 197: 717–720
2400. Martinek K, Will H, Strel'tsova ZA, Berezin IV (1969) Mol Biologia 3: 554–565
2401. Dorer FE, Kahn JR, Lentz KE, Levine M, Skeggs LT (1974) Circ Res 34: 824–827
2402. Söderling E (1982) Arch Biochem Biophys 216: 105–115
2403. McDonald JK, Reilly TJ, Zeitman BB, Ellis S (1966) Biochem Biophys Res Commun 24: 771–775
2404. Shapiro R, Holmquist B, Riordan JF (1983) Biochemistry 22: 3850–3857
2405. Oshima G, Gecse A, Erdös EG (1974) Biochim Biophys Acta 350: 26–37
2406. Shapiro R, Riordan JF (1983) Biochemistry 22: 5315–5321
2407. Watabe S, Tarada A, Ikeda T, Kouyama H, Taguchi S, Yago N (1979) Biochem Biophys Res Commun 89: 1161–1167
2408. Chin JF, Boeker EA (1979) Arch Biochem Biophys 196: 493–500
2409. Svenneby G (1971) J Neurochem 18: 2201–2208
2410. Von Hippel PH, Schleich T (1969) Acc Chem Res 2: 251–257

2411. Martin RB, Niemann C (1957) J Am Chem Soc 79: 4814
2412. Martin RB, Niemann C (1958) J Am Chem Soc 80: 1481–1486
2413. Miles JL, Robinson DA, Canady WY (1963) J Biol Chem 238: 2932–2937
2414. Valenzuela P, Bender ML (1969) Proc Natl Acad Sci USA 63: 1214–1221
2415. Kahana L, Shalitin Y (1974) Isr J Chem 12: 573–589
2416. Royer G, Cuppett CC, Williams E, Resnick H, Canady WJ (1969) Arch Biochem Biophys 134: 253–255
2417. Zolton RP, Mertz ET (1971) Can J Biochem 49: 529–537
2418. Williams AC, Auld DS (1986) Biochemistry 25: 94–100
2419. Bicknell R, Schäffer A, Bertini I, Luchinat C, Vallee BL, Auld DS (1988) Biochemistry 27: 1050–1057
2420. Hirose J, Ando S, Kidani Y (1987) Biochemistry 26: 6561–6565
2421. Douzou P (1977) Cryobiochemistry. An introduction. Academic Press, London
2422. Harned HS, Owen BB (1958) The physical chemistry of electrolytic solutions. Reinhold, New York, pp 755–756
2423. Inagami T, Sturtevant JM (1960) Biochim Biophys Acta 38: 64–79
2424. Laidler KJ (1958) The chemical kinetics of enzyme action. Oxford Univ Press, London
2425. Webb JL (1963) Enzyme and metabolic inhibitors. General principles of inhibition. Academic Press, New York
2426. Castañeda-Agulló M, Del Castillo LM (1959) J Gen Physiol 42: 617–623
2427. Mares-Guia M, Figueiredo AFS (1972) Biochemistry 11: 2091–2098
2428. Fink AL (1974) Biochemistry 13: 277–280
2429. Barbas CF, Matos JR, West JB, Wong C-H (1988) J Am Chem Soc 110: 5162–5166
2430. Clement GE, Bender ML (1963) Biochemistry 2: 836–843
2431. Maurel P (1978) J Biol Chem 253: 1677–1683
2432. Zaks A, Klibanov AM (1986) J Am Chem Soc 108: 2767–2768
2433. Zaks A, Klibanov AM (1988) J Biol Chem 263: 3194–3201
2434. Zaks A, Klibanov AM (1988) J Biol Chem 263: 8017–8021
2435. Russell AJ, Klibanov AM (1988) J Biol Chem 263: 11624–11626
2436. Schowen RL (1977) In: Cleland WW, O'Leary MH, Northrop DB (eds) Isotope effects on enzyme-catalyzed reactions. Univ Park Press, Baltimore, pp 64–99
2437. Mason R, Ghiron CA (1961) Biochim Biophys Acta 51: 377–378
2438. Kresege AJ (1973) J Am Chem Soc 95: 3065–3067
2439. Pollock E, Hogg JL, Schowen RL (1973) J Am Chem Soc 95: 968–969
2440. Quinn DM, Elrod JP, Ardis R, Friesen P, Schowen RL (1980) J Am Chem Soc 102: 5358–5365
2441. Elrod JP, Hogg JL, Quinn DM, Venkatasubban KS, Schowen RL (1980) J Am Chem Soc 102: 3917–3922
2442. Northrop DB (1981) J Am Chem Soc 103: 1208–1212
2443. Khoshtariya DE (1978) Bioorgan Khimiya 4: 1673–1677
2444. Clement GE, Snyder SL (1966) J Am Chem Soc 88: 5338–5339
2445. Hollands R, Fruton JS (1969) Proc Natl Acad Sci USA 62: 1116–1120
2446. Kaiser BL, Kaiser ET (1969) Proc Natl Acad Sci USA 64: 36–41
2447. Quinn DM, Venkatasubban KS, Kise M, Schowen RL (1980) J Am Chem Soc 102: 5365–5369
2448. Khoshtariya DE, Topolev VV, Krishtalik LI (1978) Bioorgan Khimiya 4: 1341–1451
2449. Lumry R (1959) In: Boyer P, Lardy H, Myrbäck K (eds) The enzymes, vol 1. Academic Press, New York
2450. Blumenfeld LA (1977) Problems of biological physics. Nauka, Moscow
2451. Laidler KJ, Banting PS (1973) The chemical kinetics of enzyme action, 2nd edn. Oxford Univ Press, Oxford, p 220
2452. Adams PA, Swart ER (1977) Biochem J 161: 83–92
2453. Dorovska-Taran V, Momtcheva R, Gulubova N, Martinek K (1982) Biochim Biophys Acta 702: 37–53
2454. Rajender S, Lumry R, Han M (1971) J Phys Chem 75: 1375–1386
2455. Slighton JL, Bolen DW (1980) J Am Chem Soc 102: 6318–6324
2456. Kunugi S, Hirohara H, Ise N (1979) J Am Chem Soc 101: 3640–3646

2457. De Souza Otero A, Rogana E, Mares-Guia M (1980) Arch Biochem Biophys 204: 109–116
2458. Baines NJ, Baird JB, Elmore DT (1964) Biochem J 90: 470–476
2459. Marshall TH, Chen V (1973) J Am Chem Soc 95: 5400–5405
2460. Fukuda M, Kunugi S (1984) Eur J Biochem 142: 565–570
2461. Rajender S, Han M, Lumry R (1970) J Am Chem Soc 92: 1378–1385
2462. Baggott JE, Klapper MH (1976) Biochemistry 15: 1473–1481
2463. Londesborough J (1980) Eur J Biochem 105: 211–215
2464. Wang C-LA, Calvo KS, Klapper MH (1981) Biochemistry 20: 1401–1408
2465. Kozlov LV, Mescheryakova EA, Zavada LL, Efremov ES, Rashkovetskii LG (1979) Biokhimiya 44: 338–349
2466. Sturtevant JM (1977) Proc Natl Acad Sci USA 74: 2236–2240
2467. Shiao DDF, Sturtevant JM (1969) Biochemistry 8: 4910–4918
2468. Shiao DDF (1970) Biochemistry 9: 1083–1090
2469. Schultz RM, Konovessi-Panayotatos A. Peters JR (1977) Biochemistry 16: 2194–2202
2470. Mares-Guia M, Figueiredo AFS (1970) Biochemistry 9: 3223–3227
2471. Liu CY, Kaplan KL, Markowitz AH, Nossel HL (1980) J Biol Chem 255: 7627–7630
2472. Kunugi S, Fukuda M, Ise N (1982) Biochem Biophys Acta 704: 107–113
2473. Makimoto S, Taniguchi Y (1987) Biochim Biophys Acta 914: 304–307
2474. Likhtenshtein GI, Sukhorukov BI (1964) Zhurnal Fizicheskoi Khimii 38: 747–751
2475. Likhtenshtein GI (1966) Biofizika 11: 24–28
2476. Lumry R, Rajender S (1970) Biochemistry 9: 1125–1227
2477. Northrop JN (1932) Ergebn Enzymforsch 1: 302–316
2478. Antonov VK, Vorotintseva TI, Kogan GA (1970) Mol Biologia 4: 240–245
2479. Brandts DF (1969) In: Timasheff SN, Fasman GD (eds) Structure and stability of biological macromolecules. Marcel Dekker, New York, Chap 2
2480. Sumner JB, Somers GF (1953) Chemistry and methods of enzymology, 3rd edn. Academic Press, New York
2481. Symposium on growth of microorganisms at extremes of temperature. (1968) J Appl Bacteriol 31: 1–120
2482. Proc Int Symp on Enzymes and proteins from thermophilic microorganisms (1976), Zürich, 1975. Birkhäuser Verlag, Basel
2483. Bilai TI (1979) Thermostable enzymes of fungi. Naukova dumka, Kiev
2484. Schreier E, Fittkau S, Höhne WE (1984) Int J Pept Protein Res 23: 134–141
2485. Thomas DJ, Richards AD, Jupp RA, Uenò E, Yamamoto K, Samloff IM, Dunn BM, Kay J (1989) FEBS Lett 243: 145–148
2486. McConn JD, Tsuru D, Yasunobu KT (1964) J Biol Chem 239: 3706–3715
2487. Whitaker DR (1970) Meth Enzymol 19: 599–613
2488. Topfler H, Reische K (1970) Folia Haematol 101: 91–98
2489. Kundu AK, Manna S (1975) Appl Microbiol 30: 507–513
2490. Hubert FM, Barnwell PA, Baltz RH (1971) Mycopathol Mycol Appl 44: 149–157
2491. Bilai TI, Zakharchenko VA, Zhernova II (1971) In: Metabolites of soil fungi. Naukova dumca, Kiev, pp 177–182
2492. Somkuti GA, Babel FI (1968) J Bacteriol 95: 1407–1414
2493. Craveri R, Manachini PL, Aragozzini F (1974) Mycopathol Mycol Appl 54: 193–204
2494. Hashimoto H, Iwaasa T, Yokotsuka T (1973) Appl Microbiol 25: 578–583
2495. Karavaeva NI, Mukhiddinova GN (1975) Prikladnaya Biokhimiya 11: 704–710
2496. Perutz MF (1978) Science 201: 1187–1191
2497. Argos P, Rossmann MG, Grau UM, Zuber H, Frank G, Tratschin JD (1979) Biochemistry 18: 5698–5703
2498. Dill KA, Alonso DOV, Hutchinson K (1989) Biochemistry 28: 5439–5449
2499. Menéndez-Arias L, Argos P (1989) J Mol Biol 206: 397–406
2500. Melik-Nubarov NS, Shikshnis VA, Slepnyov VI, Stchoygolev AA, Mozhaev VV (1990) Mol Biologia 24: 346–357
2501. Imanaka T, Shibazaki M, Takagi M (1986) Nature 324: 695–697
2502. Berezin IV, Martinek K, Antonov VK (eds) (1976) Immobilized enzymes, vol 2. Moscow Univ Press, Moscow, pp 7–17

2503. Martinek K, Klibanov AM, Chernishova AV, Berezin IV (1975) Dokl Acad Nauk SSSR 223: 233–236
2504. Fink AL (1976) J Theor Biol 61: 419–445
2505. Fink AL (1977) Acc Chem Res 10: 233–239
2506. Douzou P (1977) Adv Enzymol 45: 157–272
2507. Bielky BHJ, Freed S (1964) Biochim Biophys Acta 89: 314–323
2508. Galdes A, Auld DS, Vallee BL (1983) Biochemistry 22: 1888–1893
2509. Geoghegan KF, Galdes A, Martinelli RA, Holmquist B, Auld DS, Vallee BL (1983) Biochemistry 22: 2255–2262
2510. Cartwright SJ, Waley SG (1987) Biochemistry 26: 5329–5337
2511. Douzou P, Balny C (1977) Proc Natl Acad Sci USA 74: 2297–2300
2512. Fink AL (1973) Biochemistry 12: 1736–1742
2513. Alexandrov SL, Antonov VK (1974) Biokhimiya 39: 96–101
2514. Fink AL (1976) Biochemistry 15: 1580–1586
2515. Fink AL, Feldman R, Zehnder J (1979) Biochem J 181: 733–736
2516. Douzou P, Keh E, Balny C (1979) Proc Natl Acad Sci USA 76: 681–684
2517. Malthouse JPG, Scott AI (1983) Biochem J 215: 555–563
2518. Compton P, Fink AL (1984) Biochemistry 23: 2989–2994
2519. Compton PD, Coll RJ, Fink AL (1986) J Biol Chem 261: 1248–1252
2520. Fink AL, Groyn C (1979) Biochemistry 13: 1190–1194
2521. Angelides KJ, Fink AL (1978) Biochemistry 17: 2659–2668
2522. Angelides KJ, Fink AL (1979) Biochemistry 18: 2355–2363
2523. Makinen MW, Yamamura K, Kaiser ET (1976) Proc Natl Acad Sci USA 73: 3882–3886
2524. Sander ME, Witzel H (1985) Biochem Biophys Res Commun 132: 681–687
2525. Dunn BM, Fink AL (1984) Biochemistry 23: 5241–5247
2526. Hofmann Th, Fink AL (1984) Biochemistry 23: 5247–5256
2527. Lin SH, Von Wart HE (1982) Biochemistry 21: 5528–5533
2528. Von Wart HE, Lin SH (1983) Proc Natl Acad Sci USA 80: 7506–7509
2529. Lin W-Y, Lin SH, Morris RJ, Von Wart HE (1988) Biochemistry 27: 5068–5074
2530. Virden R, Tan AK, Fink AL (1990) Biochemistry 29: 145–153
2531. Compton P, Fink AL (1980) Biochem Biophys Res Commun 93: 427–431
2532. Douzou P, Hoa GHB, Petsko GA (1974) J Mol Biol 96: 367–380
2533. Petsko GA (1975) J Mol Biol 96: 381–392
2534. Douzou P, Petsko GA (1984) Adv Protein Chem 36: 245–361
2535. Mackenzie NE, Malthouse JPG, Scott AI (1984) Biochem J 219: 437–444
2536. Martinek K, Levashov AV, Klyachko NL, Berezin IV (1977) Dokl Akad Nauk SSSR 236: 920–923
2537. Peng Q, Luisi PL (1990) Eur J Biochem 188: 471–480
2538. Walde P, Peng Q, Fodnavis NW, Battistel E, Luisi PL (1988) Eur J Biochem 173: 401–409
2539. Berezin IV, Martinek K (1971) Mol Biologia 5: 347–350
2540. Vrzhestch PV (1988) Biokhimiya 53: 1704–1711
2541. Straus OH, Goldstein A (1943) J Gen Physiol 26: 559–568
2542. Sculley MJ, Morrison JF (1986) Biochim Biophys Acta 874: 44–53
2543. Krupyanko VI (1990) Vector method of description of enzymatic reactions. Nauka, Moscow
2544. Berezin IV, Martinek K (1967) Theor Eksp Khimiya 3: 458–462
2545. Shaw E (1970) Physiol Rev 50: 243–296
2546. Kitz R, Wilson IB (1962) J Biol Chem 237: 3245–3249
2547. Leytus SP, Toledo DL, Mangel WF (1984) Biochim Biophys Acta 788: 74–86
2548. Gray PJ, Duggleby RG (1989) Biochem J 257: 419–424
2549. Ray WJ, Koshland DE (1961) J Biol Chem 236: 1973–1979
2550. Tsou Chen-Lu (1962) Sci Sin 11: 1535–1558
2551. Horiike K, Tojo H, Yamano T, Nozaki M (1984) J Biochem 95: 605–609
2552. Walace RA, Kurtz AN, Niemann C (1963) Biochemistry 2: 824–836
2553. Martinek K, Levashov AV, Berezin IV B. (1970) Mol Biologia 4: 517–528
2554. Martinek K, Levashov AV, Berezin IV B. (1970) Mol Biologia 4: 339–347

2555. Hymes AJ, Robinson DA, Canady WJ (1965) J Biol Chem 240: 134–138
2556. Berezin IV, Levashov AV, Martinek K (1970) FEBS Lett 7: 20–22
2557. Berezin IV, Levashov AV, Martinek K (1970) Eur J Biochem 16: 472–476
2558. Geratz JD, Stevens FM, Polakoski KL, Parrish RF, Tidwell RR (1979) Arch Biochem Biophys 197: 551–559
2559. Nakayama T, Okutome T, Matsui R, Kurumi M, Sakurai Y, Aoyama T, Fujii S (1984) Chem Pharm Bull 32: 3968–3980
2560. Glazer AN (1965) Proc Natl Acad Sci USA 54: 171–176
2561. Bernhard SA, Lee BF, Toshjian ZH (1966) J Mol Biol 18: 405–420
2562. Glaser AN (1967) J Biol Chem 242: 3326–3331
2563. Antonov VK, D'yakov VA (1975) Bioorgan Khimiya 1:1324–1331
2564. Varfolomeev SD, Martinek K, Berezin IV (1972) Mol Biologia 6:148–152
2565. Varfolomeev SD, Martinek K, Berezin IV (1973) Mol Biologia 7: 115–123
2566. Glaser AN (1967) J Biol Chem 242: 4528–4533
2567. Kalos J, Avatis K (1966) Biochemistry 5: 1979–1983
2568. Samokish VA, Anufrieva EV, Volkenshtein MV (1971) Mol Biologia 5: 711–717
2569. Tang J (1965) J Biol Chem 240: 3810–3815
2570. Hill RL, Smith EL (1957) J Biol Chem 224: 209–223
2571. Feder J, Garrett LR, Kochavi D (1971) Biochim Biophys Acta 235: 370–377
2572. Chao J, Tanaka S, Margolius HS (1983) J Biol Chem 258: 6461–6465
2573. Seol JH, Park SC, Ha DB, Chung CH, Tanaka K, Ichihara A (1979) FEBS Lett 247: 197–200
2574. Antonov VK (1977) Adv Exp Med Biol 95: 179–198
2575. Bender ML, Wedler FC (1967) J Am Chem Soc 89: 3052–3054
2576. McConn J, Ku E, Odele C, Czerlinski G, Hess GP (1968) Science 161: 274–276
2577. Valenzuela P, Bender ML (1970) Biochemistry 9: 2440–2446
2578. Martin CJ, Frazier AR (1963) J Biol Chem 238: 3268–3273
2579. Friedberg F, Long JE, Brecher AS (1969) Proc Soc Exp Biol Med 130: 1046–1047
2580. Nichol LW, Jackson WJH, Winzor DJ (1972) Biochemistry 11: 585–591
2581. Simpson RT, Riordan JF, Vallee BL (1963) Biochemistry 2: 616–622
2582. McClure WO, Neurath H, Walsh KA (1964) Biochemistry 3: 1897–1901
2583. Bunting JW, Chu SS-T (1978) Biochem Biophys Acta 524: 142–155
2584. Bender ML, Whitaker JR, Menger F (1965) Proc Natl Acad Sci USA 53: 711–716
2585. Cleland WW (1963) Biochim Biophys Acta 67: 104–137
2586. Kitson TM, Knowles JR (1971) Biochem J 122: 241–247
2587. Kozlov LV, Zavada LL B (1975) Mol Biologia 9: 735–741
2588. Silver MS, Stoddard M (1975) Biochemistry 14: 614–620
2589. Byers LD, Wolfenden R (1973) Biochemistry 12: 2070–2078
2590. McKay JT, Plummer TH (1978) Biochemistry 17: 401–405
2591. Plummer TH, Ryan TJ (1981) Biochem Biophys Res Commun 98: 448–454
2592. Tamura Y, Hirado M, Okamura K, Minato Y, Fujii S (1977) Biochim Biophys Acta 484: 417–422
2593. Pfuetzner RA, Chan WW-C (1988) J Biol Chem 263: 4056–4058
2594. DiGregorio MD, Pickering DS, Chan WW-C (1988) Biochemistry 27: 3613–3617
2595. Inouye K, Fruton JS (1967) Biochemistry 6: 1765–1777
2596. Feder J, Brougham LR, Wildi BS (1974) Biochemistry 13: 1186–1189
2597. D'yachenko RD, Volkova LI, Kozlov LV, Antonov VK (1974) Mol Biologia 8: 879–885
2598. Burton J, Poulsen K, Haber E (1975) Biochemistry 14: 3892–3898
2599. Klyosov AA, Tseitlin VI (1977) Bioorgan Khimiya 3: 1523–1538
2600. Llorens C, Gacel G, Swerts J-P, Perlrisot R, Fournie-Zaluski M-C, Schartz J-C, Roques BP (1980) Biochem Biophys Res Commun 96: 1710–1716
2601. Lin T-Y, Williams HR (1979) J Biol Chem 254: 11875–11883
2602. Okada Y, Tsuda Y, Teno N, Nagamatsu Y, Okamoto U, Nishi N (1985) Chem Pharm Bull 33: 5301–5309
2603. Murphy J, Rowlett R, Smith SB, Hoeferlin J (1980) Arch Biochem Biophys 202: 405–413
2604. Inouye K, Fruton JS (1968) Biochemistry 7: 1611–1615

2605. Lestienne P, Dimicoli JL, Wermuth CG, Bieth JG (1981) Biochim Biophys Acta 658: 413–416
2606. Dunlap RP, Stone PJ, Abeles RH (1987) Biochem Biophys Res Commun 145: 509–513
2607. Imperiali B, Abeles RH (1987) Biochemistry 26: 4474–4477
2608. Stein RL, Strimpler AM, Edwards PD, Lewis JJ, Mauger RC, Schwartz JA, Stein MM, Trainor DA, Wildonger RA, Zottola MA (1987) Biochemistry 26: 2682–2689
2609. Sham HL, Stein H, Rempel CA, Cohen J, Plattner JJ (1987) FEBS Lett 220: 299–301
2610. Liang T-C, Abeles RH (1987) Arch Biochem Biophys 252: 626–634
2611. Plattner JJ, Greer J, Fung AKL, Stein H, Kleinert HD, Shaw HL, Smital JR, Perun TJ (1986) Biochem Biophys Res Commun 139: 982–990
2612. Cooper J, Foundling S, Hemmings A, Blundell T, Jones DM, Hallett A, Szelke M (1987) Eur J Biochem 169: 215–221
2613. Dann JG, Stammers DK, Harris CJ, Arrowsmith RJ, Davies DE, Hardy GW, Morton JA (1986) Biochem Biophys Res Commun 134: 71–77
2614. Epps DE, Schostarez H, Argoudelis CV, Poorman R, Hinzmann J, Sawyer TK, Mandel F (1989) Anal Biochem 181: 172–181
2615. Natarajan S, Gordon EM, Sabo EF, Goldfrey JD, Weller HN, Pluščec J, Rom MB, Gushman DW (1984) Biochem Biophys Res Commun 124: 141–147
2616. Grobelny D, Goli UB, Galardy RE (1985) Biochemistry 24: 7612–7617
2617. Wallace DA, Bates SRE, Walker B, Kay G, White J, Guthrie DJS, Blumsom NL, Elmore DT (1986) Biochem J 239: 797–799
2618. Bartlett PA, Kezer WB (1984) J Am Chem Soc 106: 4282–4283
2619. Nishino N, Powers JC (1979) Biochemistry 18: 4340–4347
2620. Holmquist B, Vallee BL (1979) Proc Natl Acad Sci USA 76: 6216–6220
2621. Ondetti MA, Condon ME, Reid J, Sabo EF, Cheung HS, Cushman DW (1979) Biochemistry 18: 1427–1430
2622. Galardy RE (1980) Biochem Biophys Res Commun 97: 94–99
2623. Jacobsen NE, Bartlett PA (1981) J Am Chem Soc 103: 654–657
2624. Weller HN, Gordon EM, Rom MB, Pluščec J (1984) Biochem Biophys Res Commun 125: 82–89
2625. Ocain TD, Rich DH (1987) Biochem Biophys Res Commun 145: 1038–1042
2626. Bartlett PA, Marlowe CK (1987) Science 235: 569–571
2627. Benchetrit T, Fournié-Zaluski MC, Roques BP (1987) Biochem Biophys Res Commun 147: 1034–1040
2628. Christianson DW, Lipscomb WN (1988) J Am Chem Soc 110: 5560–5565
2629. Roderick SL, Fournié-Zaluski MC, Roques BP, Matthews BW (1989) Biochemistry 28: 1493–1497
2630. Lejczak B, Kafarski P, Zygmunt J (1989) Biochemistry 28: 3549–3555
2631. King GF, Crossley MJ, Kuchel PW (1989) Eur J Biochem 180: 377–384
2632. Okuda M, Arakawa K (1985) J Biochem 98: 621–628
2633. Vallee BL, Riordan JF, Johansen JT, Livingstone DM (1971) Cold Spring Harbor Symp Quant Biol 36: 517–531
2634. Nesheim ME, Prendergast FG, Mann KG (1979) Biochemistry 18: 996–1003
2635. Lobb RR, Auld DS (1980) Biochemistry 19: 5297–5302
2636. Baker BR (1967) Design of active-site-directed irreversible enzyme inhibitors. The organic chemistry of the enzymic active site. Wiley, New York
2637. Shaw E, Green GDJ (1981) Meth Enzymol 80C: 820–826
2638. Kattner C, Shaw E (1981) Meth Enzymol 80C: 826–842
2639. Hartley BS, Kilby BA (1950) Nature 166: 784–786
2640. Boone BJ, Becker EL, Canham DH (1964) Biochim Biophys Acta 85: 441–445
2641. Mounter LA, Shipley BA, Mounter M-E (1963) J Biol Chem 238: 1979–1983
2642. Kovach IM, Larson M, Schowen RL (1986) J Am Chem Soc 108: 5490–5495
2643. Gorenstein DG, Shah D, Chen R, Kallick D (1989) Biochemistry 28: 2050–2058
2644. Sikk PF, Aaviksaar AA, Godovikov NN, Morozova NA, Palm VA (1970) Organic Reactivity (Tartu State Univ) 7: 986–995
2645. Silipo C, Hansch C, Grieco C, Vittoria A (1979) Arch Biochem Biophys 194: 552–557

2646. Cohen W, Erlanger BF (1960) J Am Chem Soc 82: 3928–3934
2647. Cohen W, Lache M, Erlanger BF (1962) Biochemistry 1: 686–693
2648. Fahrney DE, Gold AM (1963) J Am Chem Soc 85: 997–1000
2649. Gold AM (1965) Biochemistry 4: 897–901
2650. Laura R, Robison DJ, Bing DH (1980) Biochemistry 19: 4859–4864
2651. Erlander BF, Cooper AG, Cohen W (1966) Biochemistry 5: 190–196
2652. Shaw E (1970) In: Boyer P (ed) The enzymes, 3rd edn, vol 1. Academic Press, New York, pp 91–146
2653. Yoshimoto T, Orlowski RC, Walter R (1977) Biochemistry 16: 2942–2948
2654. Powers JC, Gupton BF, Harley AD, Nishino N, Whitley RJ (1977) Biochim Biophys Acta 485: 156–166
2655. Collen D, Lijnen HR, De Cock F, Durieux JP, Loffet A (1980) Biochim Biophys Acta 615: 158–166
2656. Renaud A, Dimicoli JL, Lestienne P, Bieth J (1978) J Am Chem Soc 100: 1005–1007
2657. Kettner Ch, Shaw E (1978) Biochemistry 17: 4778–4784
2658. Pelton JT, Johnston RB, Balk JL, Schmidt CJ, Roche EB (1980) Biochem Biophys Res Commun 97: 1391–1398
2659. Johnston LA, Moon KE, Eisenberg M (1988) Biochim Biophys Acta 953: 269–279
2660. Ong EB, Shaw E, Schoellmann G (1965) J Biol Chem 240: 694–698
2661. Stevenson KJ, Smillie LB (1965) J Mol Biol 12: 937–941
2662. Blair TT, Marini MA, Martin CJ (1972) FEBS Lett 20: 41–43
2663. Penny GS, Dyckes DF (1980) Biochemistry 19: 2888–2894
2664. Shaw E, Ruscica J (1971) Arch Biochem Biophys 145: 484–489
2665. McMurray JS, Dyckes DF (1986) Biochemistry 25: 2298–2301
2666. Finucane MD, Hudson EA, Malthouse JPG (1989) Biochem J 258: 853–859
2667. Stein MJ, Liener IE (1967) Biochem Biophys Res Commun 26: 376–379
2668. Rasnick D (1985) Anal Biochem 149: 461–465
2669. Rauber P, Angliker H, Walker B, Shaw E (1986) Biochem J 239: 633–640
2670. Smith RA, Coop LJ, Donnelly SL, Spencer RW, Krautz A (1988) Biochemistry 27: 6568–6573
2671. Lawson WB, Rao GJS (1980) Biochemistry 19:2133–2139
2672. Groutas WC, Brubaker MJ, Zandler ME, Stanga MA, Huang TL, Castrisos JC, Crowley JP (1985) Biochem Biophys Res Commun 128: 90–93
2673. Segal D, Shalitin Y, Wingert H, Kitamura T, Stang PJ (1989) FEBS Lett 247: 217–220
2674. Schenkein DP, Pratt RF (1980) J Biol Chem 255: 45–48
2675. Leary R, Larsen D, Watanabe H, Shaw E (1977) Biochemistry 16: 5857–5861
2676. Fujiwara K, Matsumoto E, Kitagawa T, Tsuru D (1982) Biochim Biophys Acta 702: 149–154
2677. Zumbrunn A, Stone S, Shaw E (1988) Biochem J 250: 621–623
2678. Brocklehurst K, Malthouse JPG (1978) Biochem J 175: 761–764
2679. Watanabe H, Green GD, Shaw E (1979) Biochem Biophys Res Commun 89: 1354–1360
2680. Kirshke H, Wikstrom P, Shaw E (1988) FEBS Lett 228: 128–130
2681. Husain SS, Lowe G (1968) Biochem J 110: 53–57
2682. Smith RA, Copp LJ, Coles PJ, Pauls HW, Robinson VJ, Spencer RW, Heard SB, Krantz A (1988) J Am Chem Soc 110: 4429–4431
2683. Smith RA, Coles PJ, Spencer RW, Copp LJ, Jones CS, Krantz A (1988) Biochem Biophys Res Commun 155: 1201–1206
2684. Shaw E (1988) J Biol Chem 263: 2768–2772
2685. Rajagopalan TG, Stein WH, Moore S (1966) J Biol Chem 241: 4295–4297
2686. Ong EB, Perlmann GE (1967) Nature 215: 1492–1494
2687. Stepanov VM, Lobareva LS, Mal'tsev NI (1968) Biochim Biophys Acta 151: 719–721
2688. Husain SS, Ferguson JB, Fruton JS (1971) Proc Natl Acad Sci USA 68: 2765–2768
2689. Trost BM (1967) J Am Chem Soc 89: 138–142
2690. Nakayama S-J, Nagashima Y, Hoshino M, Moriyama A, Takahashi K, Watanabe T, Yoshida M (1983) J Biochem 93: 1297–1304
2691. Tang J (1971) J Biol Chem 246: 4510–4517
2692. Ross WCJ (1960) J Chem Soc: 2257–2261

2693. Chen KCS, Tang J (1972) J Biol Chem 247: 2566–2574
2694. Rasnick D, Powers JC (1978) Biochemistry 17: 4363–4369
2695. Holmes MA, Tronrud DE, Matthews BW (1983) Biochemistry 22: 236–240
2696. Fittkau S, Förster U, Pascual C, Schunck W-H (1974) Eur J Biochem 44: 523–528
2697. Valenty VB, Wos JD, Lobo AP, Lawson WB (1979) Biochem Biophys Res Commun 88: 1375–1381
2698. Zimmerman M, Morman H, Mulvey D, Jones H, Frankshun R, Ashe BM (1980) J Biol Chem 255: 9848–9851
2699. Hedstrom L, Moorman AR, Dobbs J, Abeles RH (1984) Biochemistry 23: 1753–1759
2700. Ringe D, Mottonen JH, Gelb MH, Abeles RH (1986) Biochemistry 25: 5633–5638
2701. Kam C-M, Fujikawa K, Powers JC (1988) Biochemistry 27: 2547–2557
2702. Baek D-J, Reed PE, Daniels SB, Katzenellenbogen JA (1990) Biochemistry 29: 4305–4311
2703. Groutas WC, Badger RC, Ocain TD, Felker D, Frankson J, Theodoravis M (1980) Biochem Biophys Res Commun 95: 1890–1894
2704. Naruto S, Motoc I, Marshall GR, Daniels SB, Sofia MJ, Katzenellenbogen JA (1985) J Am Chem Soc 107: 5262–5270
2705. White EH, Roswell DF, Politzer LR, Branchini BR (1975) J Am Chem Soc 97: 2290–2291
2706. Mobashery S, Ghosh SS, Tamura SY, Kaiser ET (1990) Proc Natl Acad Sci USA 87: 578–582
2707. Charnas RL, Knowles JR (1981) Biochemistry 20: 2732–2737
2708. Cohen SA, Pratt RF (1980) Biochemistry 19: 3996–4003
2709. Umezawa H (1972) Enzyme inhibitors of microbiological origin. Tokyo Univ Press, Tokyo
2710. Umezawa H (1976) Meth Enzymol 45B: 678–695
2711. Laskowski M, Sealock RW (1970) In: Boyer P (ed) The enzymes, vol 3. Academic Press, New York, pp 375–473
2712. Proc Symp Enzyme inhibitors of the kallikrein and renin systems, Dallas, April 4, 1979. Fed Proc 38: 2751–2791
2713. Mosolov VV (1981) Progress in biological chemistry, vol 22. Nauka, Moscow, pp 100–115
2714. Barrett AJ, Salvesen G (eds) (1986) Proteinase inhibitors. Elsevier, New York
2715. Barrett AJ, Dingle JT (1972) Biochem J 127: 439–441
2716. McKoron MM, Workman RJ, Gregerman RI (1974) J Biol Chem 249: 7770–7774
2717. Suda H, Aoyagi T, Takeuchi T, Umezawa H (1976) Arch Biochem Biophys 177: 196–200
2718. Aoyagi T, Tobe H, Kojima F, Hamada M, Takeuchi T, Umezawa H (1978) J Antibiot 31: 636–638
2719. Hanada K, Tamai M, Yamagishi M, Ohmura S, Sawada J, Tanaka I (1978) Agric Biol Chem 42: 523–528
2720. Hashida S, Towatari T, Kominami E, Katunuma N (1980) J Biochem 88: 1805–1811
2721. Aoyagi T (1986) Proc Jpn Soc Invest Dermatol 10: 77–82
2722. Stein RL, Strimpler AM (1987) Biochemistry 26: 2611–2615
2723. Kunimoto S, Aoyagi T, Morishima H, Takeuchi T, Umezawa H (1972) J Antibiot 25: 251–255
2724. Rich DH, Sun E, Singh J (1977) Biochem Biophys Res Commun 74: 762–767
2725. Marciniszyn J, Hartsuck JA, Tang J (1977) Adv Exp Med Biol 95: 199–200
2726. Rich DH, Bernatowicz MS, Agarwal NS, Kawai M, Salituro FG, Schmidt PG (1985) Biochemistry 24: 3165–3173
2727. Jupp RA, Dunn BM, Jacobs JW, Vlasuk G, Arcuri KE, Veber DF, Perlow DS, Payne LS, Boger J, Laszlo S, Chakravarty PK, Broeke J, Hangauer DG, Ondeyka D, Greenlee WJ, Kay J (1990) Biochem J 265: 871–878
2728. Holden HM, Tronrud DE, Monzingo AF, Weaver LH, Matthews BW (1987) Biochemistry 26: 8542–8553
2729. Kay J (1985) In: Kostka V (ed) Aspartic proteinases and their inhibitors. De Gruyter, Berlin, pp 1–17
2730. Katoh I, Yasunaga T, Ikawa Y, Yoshinaka Y (1987) Nature 329: 654–656
2731. Richards AD, Roberts R, Dunn BM, Graves MC, Kay J (1989) FEBS Lett 247: 113–117
2732. Barrett AJ (1981) In: Turk VC, Vitale Lj (eds) Proteinases and their inhibitors. Structure, function and applied aspects. Proc Int Symp Portoroz, Yugoslavia 1980. Mladinska Knjiga, Ljubljana; Pergamon Press, Oxford, pp 103–107

2733. Katunuma N, Towatari T, Kominami E, Hashida S (1981) In: Turk VC, Vitale Lj (eds) Proteinases and their inhibitors. Structure, function and applied aspects. Proc Int Symp Portoroz, Yugoslavia 1980. Mladinska Knjiga, Ljubljana; Pergamon Press, Oxford, pp 83–92

2734. Barrett AJ, Kembhavi AA, Brown MA, Kirschke H, Knight CG, Tamai M, Hanada K (1982) Biochem J 201: 189–198

2735. Matsumoto K, Yamamoto D, Ohishi H, Tomoo K, Ishida T, Inoue M, Sadatome T, Kitamura K, Mizuno H (1989) FEBS Lett 245: 177–180

2736. Wilkes SH, Prescott JM (1985) J Biol Chem 260: 13154–13162

2737. Harbeson SL, Rich DH (1988) Biochemistry 27: 7301–7310

2738. Kaise H, Shinohara M, Miyazaki W, Izawa T, Nakano Y, Sugawara M, Sugiura K, Sasaki K (1979) J Chem Soc Chem Commun: 726–727

2739. McMurry JE, Erion MD (1985) J Am Chem Soc 107: 2712–2720

2740. Iio K, Yamasaki M (1976) Biochim Biophys Acta 429: 912–924

2741. Sutherland JHR, Greenbaum LM (1983) Biochem Biophys Res Commun 110: 332–338

2742. Harpel PC (1976) Meth Enzymol 45B: 751–760

2743. Damus PS, Rosenberg RD (1976) Meth Enzymol 45B: 653–669

2744. Schiessler H, Fink E, Fritz H (1976) Meth Enzymol 45B: 847–859

2745. Greene LJ, Pubols MH, Bartelt DC (1976) Meth Enzymol 45B: 813–825

2746. Čechová O (1976) Meth Enzymol 45B: 806–813

2747. Bagdy D, Barabas E, Gráf L, Petersen T, Magnusson S (1976) Meth Enzymol 45B: 669–678

2748. Iwanaga S, Takahashi H, Suzuki T (1976) Meth Enzymol 45B: 874–881

2749. Laskowski M, Kato I (1980) Annu Rev Biochem 49: 593–626

2750. Barrett AJ (1981) Meth Enzymol 80C: 737–754

2751. Travis J, Morii M (1981) Meth Enzymol 80C: 765–771

2752. Barrett AJ (1981) Meth Enzymol 80C: 771–778

2753. Hass GM, Ryan CA (1981) Meth Enzymol 80C: 778–792

2754. Čechová D, Jonáková V (1981) Meth Enzymol 80C: 792–804

2755. Seemüller U, Fritz H, Eulitz M (1981) Meth Enzymol 80C: 804–816

2756. Wunderer G, Machleidt W, Fritz H (1981) Meth Enzymol 80C: 816–820

2757. Travis J, Salvesen GS (1983) Annu Rev Biochem 52: 655–709

2758. Carrell R, Travis J (1985) TIBS 10: 20–24

2759. Sottrup-Jensen L (1989) J Biol Chem 264: 11539–11542

2760. Sato S, Murao S (1973) Agric Biol Chem 35: 1067–1074

2761. Heimburger N, Haupt H, Schwick H (1971) In: Fritz H. et al. (eds) Proc Int Res Conf on Proteinase inhibitors. De Gruyter, Berlin, pp 1–21

2762. Matheson NR, Gibson HL, Hallewell RA, Barr PJ, Travis J (1986) J Biol Chem 261: 10404–10409

2763. Long GL, Chandra T, Woo SLC, Davie EW, Kurachi K (1984) Biochemistry 23: 4828–4837

2764. Loeberman H, Tokuoka R, Deisenhofer J, Huber R (1984) J Mol Biol 177: 531–556

2765. Huber R. Carrell RW (1989) Biochemistry 28: 8951–8966

2766. Rubin H, Wang Z, Nickbarg EB, McLarney S, Naidoo N, Schoenberger OL, Johnson JL, Cooperman BS (1990) J Biol Chem 265: 1199–1207

2767. Holmes WE, Nelles L, Lijnen HR, Collen D (1987) J Biol Chem 262: 1659–1664

2768. Holmes WE, Lijnen HR, Collen D (1987) Biochemistry 26: 5133–5140

2769. Suzuki K, Deyashiki Y, Nishioka J, Kurachi K, Akira M, Yamamoto S, Hashimoto S (1987) J Biol Chem 262: 611–616

2770. Bock SC, Skriver K, Nielsen E, Thogersen H-C, Wiman B, Donaldson VH, Eddy RL, Marrinan J, Radziejewska E, Huber R, Shows TB, Magnusson S (1986) Biochemistry 25: 4292–4301

2771. Hsiung L-m, Barelay AN, Brandon MR, Sim E, Porter RR (1982) Biochem J 203: 293–298

2772. Dahlbäck B, Podack ER (1985) Biochemistry 24: 2368–2374

2773. Kanost MR, Prasad SV, Wells MA (1989) J Biol Chem 264: 965–972

2774. Horii A. Kobayashi T, Tomita N, Yamamoto T, Fukushige S, Murotsu T, Ogawa M, Mori T, Matsubara K (1987) Biochem Biophys Res Commun 149: 635–641

2775. Fioretti E, Iacopino G, Angeletti M, Barra D, Bossa F, Ascoli F (1985) J Biol Chem 260: 11451–11455

2776. Fioretti E, Angeletti M, Citro G, Barra D, Ascoli F (1987) J Biol Chem 262: 3586–3589
2777. Tschesche H, Wittig B, Decker G, Müller-Esterl W, Fritz H (1982) Eur J Biochem 126: 99–104
2778. Weber E, Paramokos E, Bode W, Huber R, Kato I, Laskowski M (1981) J Mol Biol 149: 109–123
2779. Ardelt W, Laskowski M (1985) Biochemistry 24: 5313–5320
2780. Vanchugova LV, Valueva TA, Romashkin VI, Rozenfeld MA, Valuev LI, Mosolov VV, Plate NA (1988) Biokhimiya 53: 1455–1461
2781. Mao SJT, Yates MT, Blankenship DT, Cardin AD, Krstenansky JL, Lovenberg W, Jackson RL (1987) Anal Biochem 161: 514–518
2782. Mao SJT, Yates MT, Owen TJ, Krstenansky JL (1988) Biochemistry 27: 8170–8173
2783. Dodt J, Köhler S, Baici A (1988) FEBS Lett 229: 87–90
2784. Braun PJ, Dennis S, Hofsteenge J, Stone SR (1988) Biochemistry 27: 6517–6522
2785. Dunwiddie C, Thornberry NA, Bull HG, Sardana M, Friedman PA, Jacobs JW, Simpson E (1989) J Biol Chem 264: 16694–16699
2786. Nutt E, Gasic T, Rodkey J, Gasic GJ, Jacobs JW, Friedman PA, Simpson E (1988) J Biol Chem 263: 10162–10167
2787. Ritonja A, Turk V, Gubenšek F (1983) Eur J Biochem 133: 427–432
2788. McGrogan M, Kennedy J, Li MP, Hsu C, Scott RW, Simonsen CC, Baker JB (1988) Biotechnology 6: 172–177
2789. Birk Y (1976) Meth Enzymol 45B: 695–728
2790. McPhalen CA, James MNG (1987) Biochemistry 26: 261–269
2791. McWherter CA, Walkenhorst WF, Campbell EJ, Glover GI (1989) Biochemistry 28: 5708–5714
2792. Christensen U, Ishida S, Ishii S-I, Mitsui Y, Iitaka Y, McClarin J, Langridge R (1985) J Biochem 98: 1263–1274
2793. Franke AE, Danley DE, Koczmarek FS, Hawrylik SJ, Gerard RD, Lee SE, Geoghegan KF (1990) Biochim Biophys Acta 1037: 16–23
2794. Müller-Esterl W, Fritz H, Kellermann J, Lottspeich F, Machleidt W, Turk V (1985) FEBS Lett 191: 221–226
2795. Turk V, Brzin J, Conger M, Ritonja A, Eropkin M, Borchart V, Machleidt W (1983) Hoppe-Seyler's Z Physiol Chem 364: 1487–1496
2796. Abrahamson M, Ritonja A, Brown MA, Grubb A, Machleidt W, Barrett AJ (1987) J Eiol Chem 262: 9688–9694
2797. Bode W, Engh R, Musil D, Thiele V, Huber R, Karshikov A, Brzin J, Kos J, Turk V (1988) EMBO J 7: 2593–2599
2798. Lenney JF, Tolan JR, Sugai WJ, Lee AE (1979) Eur J Biochem 101: 153–161
2799. Kominami E, Wakamatsu N, Katunuma N (1982) J Biol Chem 257: 14648–14652
2800. Evans HJ, Barrett AJ (1987) Biochem J 246: 795–797
2801. Isemura S, Saitoh E, Sanada K (1987) J Biochem 102: 693–704
2802. Hiwasa T, Yokoyama S, Ha J-M, Noguchi S, Sakiyama S (1987) FEBS Lett 211: 23–26
2803. Ritonja A, Kopitar M, Jerala R, Turk V (1989) FEBS Lett 255: 211–214
2804. Laber B, Krieglstein K, Henshen A, Kos J, Turk V, Huber R, Bode W (1989) FEBS Lett 248: 162–168
2805. Machleidt W, Borchart V, Fritz H, Brzin J, Ritonja A, Turk V (1983) Hoppe-Seyler's Z Physiol Chem 364: 1481–1486
2806. Lenney JF, Liao JR, Sugg SL, Gopalakrishnan V, Wong HCH, Ouye KH, Chan PWH (1982) Biochem Biophys Res Commun 108: 1581–1587
2807. Rohlich ST, Levy H, Rifkin DB (1985) Biol Chem Hoppe-Seyler's 366: 147–155
2808. Ritonja A, Machleidt W, Barrett AJ (1985) Biochem Biophys Res Commun 131: 1187–1192
2809. Brzin J, Popovic T, Turk V, Borchart U, Machleidt W (1984) Biochem Biophys Res Commun 118: 103–109
2810. Salvesen G, Parkes C, Abrahamson M, Grubb A, Barrett AJ (1986) Biochem J 234: 429–434
2811. Higashiyama S, Ohkubo I, Ishiguro H, Kunimatsu M, Sawaki K, Sasaki M (1986) Biochemistry 25: 1669–1675
2812. Kellermann J, Haupt H, Auerswald EA, Müller-Esterl W (1989) J Biol Chem 264: 14121–14128

2813. Björk J, Ylinenjärvi K (1990) Biochemistry 29: 1770–1776
2814. Ievleva EV, Zimachova AV, Mosolov VV (1987) Biokhimiya 52: 1786–1791
2815. Emori Y, Kawasaki H, Imajoh S, Minami Y, Suzuki K (1988) J Biol Chem 263: 2364–2370
2816. Kawasaki H, Emori Y, Imajoh-Ohmi S, Minami Y, Suzuki K (1989) J Biochem 106: 274–281
2817. Heinrikson RL, Kézdy FJ (1976) Meth Enzymol 45B: 740–751
2818. Zimachova AV, Ievleva EV, Mosolov VV (1984) Biokhimiya 49: 1153–1158
2819. Zimachova AV, Ievleva EV, Mosolov VV (1988) Biokhimiya 53: 740–745
2820. Valler MJ, Kay J, Aoyagi T, Dunn BM (1985) J Enzyme Inhibition 1: 77–82
2821. Homandberg GA, Litwiller RD, Peanasky RJ (1989) Arch Biochem Biophys 270: 153–161
2822. Ueno N, Miyazaki H, Hirose S, Murakami K (1981) J Biol Chem 256: 12023–12027
2823. Birk Y (1976) Meth Enzymol 45B: 728–739
2824. Mareš M, Meloun B, Pavlik M, Kostka V, Baudyš M (1989) FEBS Lett 251: 94–98
2825. Hass GM, Hermdson MA (1981) Biochemistry 20: 2256–2260
2826. Murai H, Hara S, Ikenaka T, Oda K, Murao S (1985) J Biochem 97: 173–180
2827. Cawston TE, Murphy G, Mercer E, Galloway WA, Hazleman BL, Reynolds JJ (1983) Biochem J 211: 313–318
2828. Boone TC, Johnson MJ, De Clerck YA, Langley KE (1990) Proc Natl Acad Sci USA 87: 2800–2804
2829. Virca GD, Travis J (1984) J Biol Chem 259: 8870–8874
2830. Tschesche H, Dietl T (1976) Meth Enzymol 45B: 772–792
2831. Beckmann J, Mehlich A, Schröder W, Wenzel HR, Tschesche H (1988) Eur J Biochem 176: 675–682
2832. Richarz R, Tschesche H, Wüthrich K (1980) Biochemistry 19: 5711–5715
2833. Nikawa T, Towatari T, Ike Y, Katunuma N (1989) FEBS Lett 255: 309–324
2834. Salvesen GS, Barrett AJ (1980) Biochem J 187: 695–701
2835. Barrett AJ, Starkey PM (1973) Biochem J 133: 709–724
2836. Christensen U, Sottrup-Jensen L (1984) Biochemistry 23: 6619–6626
2837. Feldman SR, Gonias SL, Pizzo SV (1985) Proc Natl Acad Sci USA 82: 5700–5704
2838. Favandon V, Tourbez M, Pochon F, Mareix R, Tourbez H (1987) Eur J Biochem 165: 31–37
2839. Sottrup-Jensen L, Sand O, Kristensen L, Fey GH (1989) J Biol Chem 264: 15781–15789
2840. Hass GM, Ryan CA (1980) Biochem Biophys Res Commun 97: 1481–1486
2841. Hass GM, Ako H, Grahn DT, Neurath H (1976) Biochemistry 15: 93–100
2842. Homandberg GA, Minor ST, Peanasky RJ (1980) Biochim Biophys Acta 612: 384–394
2843. Pauling L (1946) Chem Eng News 24: 1375–1379
2844. Wolfenden R (1969) Nature 223: 704–705
2845. Lienhard GE (1973) Science 180: 149–154
2846. Wolfenden R (1976) Annu Rev Biophys Bioeng 5: 271–306
2847. Antonov VK, Ivanina TV, Berezin IV, Martinek K (1968) Dokl Akad Nauk SSSR 183: 1435–1438
2848. Antonov VK, Ivanina TV, Ivanova AG, Berezin IV, Levashov AV, Martinek K (1972) FEBS Lett 20: 37–40
2849. Lindquist RN, Nguyen AC (1977) J Am Chem Soc 99: 6435–6437
2850. Matteson DS, Sadhu KM, Lienhard GE (1981) J Am Chem Soc 103: 5241–5242
2851. Philipp M, Maripuri S (1981) FEBS Lett 133: 36–38
2852. Kettner CA, Shenvi AB (1984) J Biol Chem 259: 15106–15114
2853. Kato Y, Kido H, Fukusen N, Katunuma N (1988) J Biochem 103: 820–822
2854. Kiener PA, Waley SG (1978) Biochem J 169: 197–204
2855. Beesley T, Gascoyne N, Knott-Hunziker V, Petursson S, Waley SG, Jaurin B, Grundström Th (1983) Biochem J 209: 229–233
2856. Shenvi AB (1986) Biochemistry 25: 1286–1291
2857. Rotanova TV, Klaus R, Ivanova AG, Ginodman LM, Antonov VK (1976) Bioorgan Khimiya 2: 837–845
2858. Yarv YY, Speek MA, Langel YP, Rotanova TV (1978) Bioorgan Khimiya 4: 1364–1371
2859. Imperiali B, Abeles RH (1986) Biochemistry 25: 3760–3767
2860. Breaux EJ, Bender ML (1975) FEBS Lett 56: 81–84
2861. Kettner CA, Bone R, Agard DA, Bachovchin WW (1988) Biochemistry 27: 7682–7688

2862. Rotanova TV, Ivanova AG, Antonov VK, Rakodjiova A, Blagoev BB (1976) Int J Pept Protein Res 8: 225–231

2863. Rotanova TV, Vasil'eva NV, Ginodman LM, Antonov VK (1978) Bioorgan Khimiya 4: 694–697

2864. Nakatani H, Morita T, Hiromi K (1978) Biochim Biophys Acta 525: 423–428

2865. Shvyadas V-YK, Klyosov AA, Berezin IV, Rotanova TV, Vasil'eva NV, Ginodman LM, Antonov VK (1976) Proc 3rd All-Union Symp Structure and function of enzyme active site, 1976, Pustchino, p 25

2866. Iomptova VM, Blagoev BB, Rotanova TV, Antonov VK (1982) Bioorgan Khimiya 8: 1659–1665

2867. Tanizawa K. Kanaoka Y, Wos JD, Lawson WB (1985) Biol Chem Hoppe-Seyler's 366: 871–878

2868. Westerik JO'C, Wolfenden R (1974) J Biol Chem 249: 6351–6353

2869. Lewis CA, Wolfenden R (1977) Biochemistry 16: 4890–4895

2870. Tsujinaka T, Kajiwara Y, Kambayashi J, Sakon M, Higuchi N, Tanaka T, Mori T (1988) Biochem Biophys Res Commun 153: 1201–1208

2871. Galardy RE, Kartylewicz ZP (1984) Biochemistry 23: 2083–2087

2872. Gorenstein DG, Shah DO (1982) Biochemistry 21: 4679–4686

2873. Tarnus C, Jung MJ, Rémy J-M, Baltzer S, Schirlin DG (1989) FEBS Lett 249: 47–50

2874. Tulinsky A, Mavridis I, Mann RF (1978) J Biol Chem 253: 1074–1078

2875. Matthews BW, Alden RA, Birktoft JJ, Freer ST, Kraut J (1975) J Biol Chem 250: 7120–7126

2876. Bone R, Shenvi AB, Kettner CA, Agard DA (1987) Biochemistry 26: 7609–7614

2877. Takahashi LH, Radhakrishnan R, Rosenfield RE, Meyer EF (1989) Biochemistry 28: 7610–7617

2878. Adebodun F, Jordan F (1988) J Am Chem Soc 110: 309–310

2879. Bachovchin WW, Wong WYL, Farr-Jones S, Shenvi AB, Kettner CA (1988) Biochemistry 27: 7689–7697

2880. Nakatani H, Hiromi K (1978) Biochim Biophys Acta 524: 413–417

2881. Andersson L, Isley TC, Wolfenden R (1982) Biochemistry 21: 4177–4180

2882. Fehrentz J-A, Heitz A, Castro B, Cazanbon C, Nisato D (1984) FEBS Lett 167: 273–276

2883. Grobelny D, Galardy RE (1986) Biochemistry 25: 1072–1078

2884. Frick L, Wolfenden R (1985) Biochim Biophys Acta 829: 311–318

2885. Mackenzie NE, Grant SK, Scott AI, Malthouse JPG (1986) Biochemistry 25: 2293–2298

2886. Mattis JA, Henes J, Fruton JS (1977) J Biol Chem 252: 6776–6782

2887. Takahashi LH, Radhakrishnan R, Rosenfield RE, Mayer EF, Trainor DA (1989) J Am Chem Soc 111: 3368–3374

2888. Liang T-Ch, Abeles RH (1987) Biochemistry 26: 7603–7608

2889. Globelny D, Teater C, Galardy RE (1989) Biochem Biophys Res Commun 159: 426–431

2890. Mookhtiar KA, Grobelny D, Galardy RE, Von Wart HE (1988) Biochemistry 27: 4299–4304

2891. Bartlett PA, Marlow CK (1983) Biochemistry 22: 4618–4624

2892. Yamauchi K, Ohtsuki S, Kinoshita M (1985) Biochim Biophys Acta 827: 275–282

2893. Tronrud DE, Monzingo AF, Matthews BW (1986) Eur J Biochem 157: 261–268

2894. Nakatani H, Hiromi K, Satoi S, Oda K, Murao S, Ichishima E (1975) Biochim Biophys Acta 391: 415–421

2895. Marciniszyn J, Hartsuck JA, Tang J (1976) J Biol Chem 251: 7088–7094

2896. Steiner SA, Castellino FJ (1985) Biochemistry 24: 609–617

2897. Steiner SA, Castellino FJ (1985) Biochemistry 24: 1136–1141

2898. Lestienne P, Bieth JG (1980) J Biol Chem 255: 9289–9294

2899. Schechter J, Zazepizki E (1971) Eur J Biochem 18: 469–473

2900. Wang TT, Dorrington KJ, Hofmann Th (1974) Biochem Biophys Res Commun 57: 865–869

2901. Zinchenko AA, Rumsh LD, Antonov VK (1978) Bioorgan Khimiya 4: 1122–1128

2902. Zinchenko AA, Rumsh LD, Ginodman LM, Antonov VK (1981) Bioorgan Khimiya 7: 1195–1200

2903. Silver MS, James SLT (1981) Biochemistry 20: 3183–3189

2904. Silver MS, James SLT (1981) Biochemistry 20: 3177–3182

2905. Inagami T, Hatano H (1969) J Biol Chem 244: 1176–1182

2906. Seydoux F, Counly G, Yon J (1971) Biochemistry 10: 2284–2290
2907. Vajda T, Szabó T (1978) Eur J Biochem 85: 121–124
2908. Huang JS, Tang J (1977) Fed Proc 36: 892
2909. Dionyssion-Asterion A, Rakiteis ET (1979) Biochem J 177: 355–356
2910. Watabe S, Yago N (1983) Biochem Biophys Res Commun 110: 934–939
2911. Pillai S, Botti R, Zull JE (1983) J Biol Chem 258: 9724–9728
2912. Pillai S, Zull JE (1985) J Biol Chem 260: 8384–8389
2913. Chung CH, Goldberg AL (1982) Proc Natl Acad Sci USA 79: 795–799
2914. Erlanger BF, Wasserman NH, Cooper AG (1973) Biochem Biophys Res Commun 52: 208–215
2915. Goto S, Hess GP (1979) J Biochem 85: 961–965
2916. Akanuma H, Yamasaki M (1979) J Biochem 85: 775–783
2917. Christensen U (1978) Biochim Biophys Acta 526: 194–201
2918. Neet KE, Ainslic GR (1976) TIBS 1: 145
2919. Kurganov BI (1982) Mol Biologia 16: 424–433
2920. Kvamme E, Tveit B, Svenneby G (1970) J Biol Chem 245: 1871–1877
2921. Kvamme E, Svenneby G, Tveit B (1968) In: Kvamme E, Pihl A (eds) Regulation of enzyme activity and allosteric interactions. Univ forlaget, Oslo, p 89
2922. Olsen BR, Svenneby G, Kvamme E, Tveit B, Eskeland T (1970) J Mol Biol 52: 239–245
2923. Na K-J, Lee H-L (1983) Arch Biochem Biophys 227: 580–586
2924. Röhm K-H (1985) Arch Biochem Biophys 239: 216–225
2925. Goldberg AL, Waxman L (1985) J Biol Chem 260: 12029–12034
2926. Gorter J, Gruber M (1970) Biochim Biophys Acta 198: 546–555
2927. Lin T-Y, Fletcher DS (1980) J Biol Chem 255: 7756–7762
2928. Camiolo SM, Markus G, Evers JL, Hobika GH (1980) Thromb Res 17: 697–706
2929. Martin CJ (1964) Biochemistry 3: 1635–1642
2930. Papaioannou S. Liener IE (1968) J Chromatogr 32: 746–756
2931. Ingram VM (1951) Nature 167: 83–84
2932. Blumenfeld OO, Léonis J, Perlmann GE (1960) J Biol Chem 235: 379–382
2933. Foltmann B (1962) C R Trav Lab Carlsberg 32: 425–437
2934. Sluyterman LAA, Wijdenes J (1980) Eur J Biochem 113: 189–193
2935. Glaser AN (1969) Proc Natl Acad Sci USA 64: 235–240
2936. Maroux S, Rovery M, Desnuelle P (1967) Biochim Biophys Acta 140: 377–380
2937. London F (1937) Uspekhi Fis Nauk 18: 421–435
2938. Brant DA, Flory PJ (1965) J Am Chem Soc 87: 2791–2800
2939. Warshel A, Levitt M (1976) J Mol Biol 103: 227–249
2940. Levitt M (1974) J Mol Biol 82: 393–420
2941. Némethy G (1967) Angew Chem 79: 260–271
2942. Leo A, Hansch C, Elkins D (1971) Chem Rev 71: 525–616
2943. Hermann RB (1972) J Phys Chem 76: 2754–2759
2944. Chothia C (1974) Nature 248: 338–339
2945. Abraham MN (1980) J Am Chem Soc 102: 5910–5912
2946. Cordes EH, Gitler C (1973) In: Kaiser ET, Kézdy FJ (eds) Progress in bioorganic chemistry, vol 2. Wiley, New York, pp 1–53
2947. Fersht AR (1987) TIBS 12: 301–304
2948. Schowen RL (1973) J Pharm Sci 56: 931–943
2949. Tamres M, Searles S, Leighly EM, Mohrman DW (1954) J Am Chem Soc 76: 3983–3987
2950. Rubin J, Panson GS (1965) J Phys Chem 69: 3089–3091
2951. Tabushi I, Kiyosuke Y, Yamamura K (1981) J Am Chem Soc 103: 5255–5257
2952. Gráf L, Jancsó A, Szilagyi L, Hegyi G, Pintér K, Naray-Szabo G, Hepp J, Medzihradszy K, Rutter WJ (1988) Proc Natl Acad Sci USA 85: 4961–4965
2953. Deranleau DA (1969) J Am Chem Soc 91: 4044–4049
2954. Deranleau DA, Schwyzer R (1969) Biochemistry 9: 126–134
2955. Burley SK, Petsko GA (1986) J Am Chem Soc 108: 7995–8001
2956. Némethy G, Scheraga NA (1981) Biochem Biophys Res Commun 98: 482–487
2957. Reid KSC, Lindley PF, Thornton JM (1985) FEBS Lett 190: 209–213
2958. Smoluchowski MV (1917) Z Phys Chem 92: 129–152

2959. Waite TR (1957) Phys Rev 107: 463–470
2960. Bronsted JN (1925) Z Phys Chem 115: 337–351
2961. Alberty RA, Hammes GG (1958) J Phys Chem 62: 154–159
2962. Scatchard G (1943) In: Cohu EJ, Edsall JT (eds) Proteins, amino acids and peptides. Reinold, New York
2963. Hammes GG, Alberty RA (1959) J Phys Chem 63: 274–279
2964. Hippel PH, Berg OG (1989) J Biol Chem 264: 675–678
2965. Chou K-C, Li T-T, Zhou G-Q (1980) Biochim Biophys Acta 657: 304–308
2966. Moelwyn-Hughes EA (1961) Physical chemistry, Pergamon Press, London
2967. Hirohara H, Philipp M, Bender ML (1977) Biochemistry 16: 1573–1580
2968. Smallcombe SH, Ault B, Richards JH (1972) J Am Chem Soc 94: 4585–4590
2969. Havsteen BH (1967) J Biol Chem 242: 769–771
2970. Quast U, Engel J, Heumann H, Krause G, Steffen E (1974) Biochemistry 13: 2512–2520
2971. Koren R, Hammes GG (1976) Biochemistry 15: 1165–1171
2972. Johannin G, Yon J (1966) Biochem Biophys Res Commun 25: 320–325
2973. Haynes R, Feeney RE (1968) Biochemistry 7: 2879–2875
2974. Vincent J-P, Peron-Renner M, Pudles J, Lazdunski M (1974) Biochemistry 13: 4205–4211
2975. Uehara Y, Tonomura B, Hiromi K (1980) Arch Biochem Biophys 202: 250–258
2976. Sachdev GP, Fruton JS (1975) Proc Natl Acad Sci USA 72: 3424–3427
2977. Kitagishi K, Nakatani H, Hiromi K (1980) J Biochem 87: 573–579
2978. Kam C-M, Nishino N, Powers JC (1979) Biochemistry 18: 3032–3038
2979. Hammes GG (1974) Pure Appl Chem 40: 525–543
2980. Hammes GG, Knoche W (1966) J Chem Phys 45: 4041–4048
2981. Grunwald E, Ralph EK (1967) J Am Chem Soc 89: 4405–4411
2982. Hess GP (1971) In: Boyer PD (ed) The enzymes, vol 3. Academic Press, New York, pp 213–248
2983. Rodier FJL, Ilgenfritz G (1982) Eur J Biochem 128: 451–454
2984. Maehler R, Whitaker JR (1982) Biochemistry 21: 4621–4633
2985. Goto S, Hess GP (1979) J Biochem 86: 619–625
2986. Kuramochi H, Nakata H, Ishii S-I (1979) J Biochem 86: 1403–1410
2987. Zhou J-M, Liu C, Tsou C-L (1989) Biochemistry 28: 1070–1076
2988. Nilsson T, Wiman B (1983) Eur J Biochem 129: 663–667
2989. Harrison LW, Vallee BL (1978) Biochemistry 17: 4359–4363
2990. Kitagishi K, Hiromi K (1986) J Biochem 99: 191–197
2991. Kitagishi K, Hiromi K (1983) J Biochem 93: 55–59
2992. Yon JM (1976) Biochimie 58: 61–69
2993. Lobb RR, Auld DS (1977) Fed Proc 36: 877
2994. Dunn BM, Pham C, Raney L, Abayasekara D, Gillespie W, Hsu A (1981) Biochemistry 20: 7206–7211
2995. Dunn BM (1982) Arch Biochem Biophys 214: 763–771
2996. Geoghegan KF, Galdes A, Hanson G, Holmquist B, Auld DS, Vallee BL (1986) Biochemistry 25: 4669–4674
2997. Nakatani H, Hiromi K, Kitagishi K (1988) Arch Biochem Biophys 263: 311–314
2998. Rosenberry T (1975) Adv Enzymol 43: 103–218
2999. Knorre DG (1975) FEBS Lett 58: 50–52
3000. Antonov IV, Dudkin SM, Karpeisky MYa, Platonov AL, Protasevich II, Yakovlev GI (1978) Bioorgan Khimiya 4: 388–396
3001. Schuber F, Travo P, Pascal M (1979) Bioorg Chem 8: 83–89
3002. Scheider W (1979) Proc Natl Acad Sci USA 76: 2283–2287
3003. Garland F, Cheung HC (1979) Biochemistry 18: 5281–5289
3004. Raju EV, Humphreys RE, Fruton JS (1972) Biochemistry 11: 3533–3536
3005. Pohl J, Dunn BM (1988) Biochemistry 27: 4827–4834
3006. Thompson RC, Bauer C-A (1979) Biochemistry 18: 1552–1558
3007. Grobelny D, Goli UB, Galardy RE (1989) Biochemistry 28: 4948–4951
3008. Ross PD, Subramanian S (1981) Biochemistry 20: 3096–3102
3009. Menegatti E, Guarneri M, Bolognesi M, Ascenzi P, Amiconi G (1984) J Mol Biol 176: 425–430

3010. Takahashi K, Fukada H (1985) Biochemistry 24: 297–300
3011. Platzer KEB, Momany FA, Scheraga HA (1972) Int J Pept Protein Res 4: 201–219
3012. Náray-Szabó G (1984) J Am Chem Soc 106: 4584–4589
3013. Goodford PJ (1985) J Med Chem 28: 849–857
3014. Wong CF, McCammon JA (1986) J Am Chem Soc 108: 3830–3832
3015. Wipff G, Dearing A, Weiner PK, Blaney JM, Kollman PA (1983) J Am Chem Soc 105: 997–1005
3016. Schultz RM, Varma-Nelson P, Peters JR, Treadway WJ (1979) J Biol Chem 254: 12411–12418
3017. Yung BYK, Trowbridge CG (1980) J Biol Chem 255: 9724–9730
3018. Tellam R, Jersey J, Winzor DJ (1979) Biochemistry 18: 5316–5321
3019. Koshland DE (1960) Adv Enzymol 22: 45–98
3020. Bode W, Schwager P, Huber R (1978) J Mol Biol 118: 99–112
3021. Huber R (1979) In: Holzer H, Tschesche H (eds) Biological function of proteinases. 30th Colloq Mosbach. Springer, Berlin Heidelberg New York, pp 1–16
3022. Griffith OW, Meister A (1982) J Biol Chem 257: 4392–4397
3023. Lesk AM, Chothia C (1984) J Mol Biol 174: 175–191
3024. Popov EM (1977) Mol Biologia 11: 5–41
3025. Ulmer DD, Vallee BL (1965) Adv Enzymol 27: 37–104
3026. Perkins HR, Ghuysen JM, Frére JM, Nieto M (1975) In: Sund H, Blauer G (eds) Protein-ligand interactions. de Gruyter, Berlin, pp 372–384
3027. Nakatani H, Kitagishi K, Hiromi K (1976) Biochim Biophys Acta 452: 521–524
3028. Marini JL, Caplow M (1971) J Am Chem Soc 93: 5560–5566
3029. Böning W, Havsteen BH (1982) J Biol Chem 257: 6836–6843
3030. Burr ME, Koshland DE (1964) Proc Natl Acad Sci USA 52: 1017–1020
3031. Likhtenshtein GI (1974) Method of spin labels in molecular biology. Nauka, Moscow
3032. Brand L, Seliskar CJ, Turner DC (1971) Probes of structure and function of macromolecules and membranes. Academic Press, New York, pp 17–40
3033. Klyosov AA, Vallee BL (1977) Bioorgan Khimiya 3: 958–963
3034. Scheule RK, Han SL, Van Wart HE, Vallee BL, Scheraga HA (1981) Biochemistry 20: 1778–1784
3035. Van Wart HE, Vallee BL, Scheule RH, Scheraga HA (1981) TIBS 6: 316–318
3036. Dimicoli JL, Lam-Tanh H, Toma F, Fermandjian S (1984) Biochemistry 23: 3173–3180
3037. Kitson DH, Hagler AT (1988) Biochemistry 27: 7176–7180
3038. Stoddart BL, Bruhnke J, Porter N, Ringe D, Petsko GA (1990) Biochemistry 29: 4871–4879
3039. Tulinsky A, Blevins RA (1987) J Biol Chem 262: 7737–7743
3040. Grütter MG, Fendrich G, Huber R, Bode W (1988) EMBO J 7: 345–351
3041. Blow DH, Wright CS, Kukla D, Rühlmann A, Steigemann W, Huber R (1972) J Mol Biol 69: 137–144
3042. Fujinaga M, Sielecki AK, Read RJ, Ardelt W, Laskowski M, James MNG (1987) J Mol Biol 195: 397–418
3043. Rühlmann A, Kukla D, Schwager P, Bartels K, Huber R (1973) J Mol Biol 77: 417–436
3044. Huber R, Kukla D, Bode W, Schwager P, Bartels K, Deisenhofer J, Steigemann W (1974) J Mol Biol 89: 73–101
3045. Marquart M, Walter J, Deisnhofer J, Bode W, Huber R (1983) Acta Crystallogr B 39: 480–490
3046. Sweet RM, Wright HT, Janin J, Chothia CH, Blow DM (1974) Biochemistry 13: 4212–4218
3047. Hughes DL, Sieker LC, Bieth J, Dimicoli JL (1982) J Mol Biol 162: 645–658
3048. Meyer EF, Radhakrishnan R, Cole GM, Presta LG (1986) J Mol Biol 189: 533–539
3049. Radhakrishnan R, Presta LG, Meyer EF, Wildonger R (1987) J Mol Biol 198: 417–424
3050. Meyer EF, Clore GM, Gronenborn AM, Hansen HAS (1988) Biochemistry 27: 725–730
3051. Navia NA, McKeever BM, Springer JP, Lin T-Y, Williams HR, Fluder EM, Dorn CP, Hoogsteen K (1989) Proc Natl Acad Sci USA 86: 7–11
3052. Bode W, Wei A-Z, Huber R, Meyer E, Travis J, Neumann S (1986) EMBO J 5: 2453–2458
3053. Bode W, Mayr I, Baumann U, Huber R, Stone SR, Hofsteenge J (1989) EMBO J 8: 3467–3475
3054. Chen Z, Bode W (1983) J Mol Biol 164: 283–311

3055. Bone R, Frank D, Kettner CA, Agard DA (1989) Biochemistry 28: 7600–7609
3056. Bode W, Papamokos E, Musil D (1987) Eur J Biochem 166: 673–692
3057. McPhalen CA, Svendsen I, Jonassen I, James MNG (1985) Proc Natl Acad Sci USA 82: 7242–7246
3058. Hirono S, Akagawa H, Mitsui Y, Iitaka Y (1984) J Mol Biol 178: 389–413
3059. Wright CS (1972) J Mol Biol 67: 151–163
3060. Robertus JD, Alden RA, Birktoft JJ, Kraut J, Powers JC, Wilcox PE (1972) Biochemistry 11: 2439–2449
3061. Poulos TL, Alden RA, Freer ST, Biktoft JJ, Kraut J (1976) J Biol Chem 251: 1097–1103
3062. Delbaere LTJ, Brayer GD (1985) J Mol Biol 183: 89–103
3063. James MNG, Sielecki AR, Brayer GD, Delbaere LTJ, Bauer C-A (1980) J Mol Biol 144: 43–88
3064. Brayer GD, Delbaere LTJ, James MNG, Bauer C-A, Thompson RC (1979) Proc Natl Acad Sci USA 76: 96–100
3065. James MNG, Brayer GD, Delbaere LTJ, Sielecki A, Gerteer A (1980) J Mol Biol 139: 423–438
3066. Read RJ, Fujinaga M, Sielecki AR, James MNG (1983) Biochemistry 22: 4420–4433
3067. Gros P, Betzel C, Dauter Z, Wilson KS, Hol WGJ (1989) J Mol Biol 210: 347–367
3068. Betzel C, Pal GP, Struck M, Jany K-D, Saenger W (1986) FEBS Lett 197: 105–110
3069. Varughese KI, Ahmed FR, Carey PR, Hasnain S, Huber CP, Storer AC (1989) Biochemistry 28: 1330–1332
3070. Andreeva NS, Zdanov AS, Gustchina AE, Fedorov AA (1984) J Biol Chem 259: 11353–11365
3071. James MNG, Sielecki A, Salituro F, Rich DH, Hofmann T (1982) Proc Natl Acad Sci USA 79: 6137–6141
3072. James MNG, Sielecki A, Hofman T (1985) In: Kostka V (ed) Aspartic proteinases and their inhibitors. de Gruyter, Berlin, pp 163–177
3073. Sali A, Veerapandian B, Cooper JB, Foundling SI, Hoover DJ, Blundell TL (1989) EMBO J 8: 2179–2188
3074. Cooper JB, Foundling SI, Blundell TL, Boger J, Jupp RA, Kay J (1989) Biochemistry 28: 8596–8603
3075. Bott R, Subramanian E, Davies D (1982) Biochemistry 21: 6956–6962
3076. Suguna K, Padlan EA, Smith CM, Carlson WD, Davis DR (1987) Proc Natl Acad Sci USA 84: 7009–7013
3077. Christianson DW, Lipscomb WN (1987) J Am Chem Soc 109: 5536–5538
3078. Christianson DW, Lipscomb WN (1986) J Am Chem Soc 108: 4998–5003
3079. Christianson DW, Lipscomb WN (1986) Proc Natl Acad Sci USA 83: 7568–7572
3080. Shoham G, Christianson DW, Oren DA (1988) Proc Natl Acad Sci USA 85: 684–688
3081. Kim H, Lipscomb WN (1990) Biochemistry 29: 5546–5555
3082. Rees DS, Lipscomb WN (1982) J Mol Biol 160: 475–498
3083. Kester WR, Matthews BW (1977) Biochemistry 16: 2506–2516
3084. Bolognesi MC, Matthews BW (1979) J Biol Chem 254: 634–639
3085. Weaver LH, Kester WR, Matthews BW (1977) J Mol Biol 114: 119–132
3086. Holden HM, Matthews BW (1988) J Biol Chem 263: 3256–3260
3087. Monzingo AF, Matthews BW (1984) Biochemistry 23: 5724–5729
3088. Kelly JA, Knox JR, Zhao H, Frére J-M, Ghuysen J-H (1989) J Mol Biol 209: 281–295
3089. Morgan RS, Miller SL, McAdon JM (1979) J Mol Biol 127: 31–39
3090. Mizusaki K, Sugahara Y, Tsunematsu H, Makisumi S (1986) J Biochem 100: 21–25
3091. Gerig JT, Rimerman RA (1972) J Am Chem Soc 94: 7565–7569
3092. Carey PR, Ozaki Y, Storer AC (1983) Biochem Biophys Res Commun 117: 725–731
3093. Storer AC, Lee H, Caray PR (1983) Biochemistry 22: 4789–4796
3094. Godzhaev NN, Aliev PE, Popov EM (1986) Mol Biologia 20: 102–119
3095. Bolognesi M, Gatti G, Menegatti E, Guarneri M, Marquart M, Paramokos E. Huber R (1982) J Mol Biol 162: 839–868
3096. Bode W, Walter J, Huber R, Wenzel HR, Tschesche H (1984) Eur J Biochem 144: 185–190
3097. Hunkapiller MW, Forgac MD, Yu EH, Richards JH (1979) Biochem Biophys Res Commun 87: 25–31

3098. Baillargeon MW, Laskowski M, Neves DE, Porubcan MA, Santini RE, Markley JL (1980) Biochemistry 19: 5703–5710
3099. Shah DO, Gorenstein DG (1983) Biochemistry 22: 6096–6101
3100. Argade PV, Gerke GK, Weber YP, Peticolas WL (1984) Biochemistry 23: 299–304
3101. Weber JA, Turpin P-Y, Bernhard SA, Peticolas WL (1986) Biochemistry 25: 1912–1917
3102. Huber CP, Ozaki Y, Pliura DH, Storer AC, Carey PR (1982) Biochemistry 21: 3109–3115
3103. Pearl LH (1987) FEBS Lett 214: 8–12
3104. Hofmann T, Allen B, Bendiner M, Blum M, Cunningham A (1988) Biochemistry 27: 1140–1146
3105. Harrison RK, Stein RL (1990) Biochemistry 29: 1684–1689
3106. Gerig JT, Halley BA (1984) Arch Biochem Biophys 232: 467–476
3107. McPhalen CA, James MNG (1988) Biochemistry 27: 6582–6598
3108. Shotton DM, Watson HC (1970) Philos Trans R Soc Lond Ser B 257: 111–125
3109. Papamokos E, Weber E, Bode W, Huber R, Empie MW, Kato I, Laskowski M (1982) J Mol Biol 158: 515–537
3110. Bauer C-A (1978) Biochemistry 17: 375–380
3111. Bode W, Schwager P, Huber R (1976) In: Ribbons DW, Brew K (eds) Proteolysis and physiological regulation: Miami Winter Symposium, vol 11. Academic Press, New York, pp 43–76
3112. Treadway WJ, Schultz RM (1976) Biochemistry 15: 4171–4174
3113. Landis BH, Berliner LJ (1980) J Am Chem Soc 102: 5350–5354
3114. Landis BH, Berliner LJ (1980) J Am Chem Soc 102: 5354–5358
3115. Bittner EW, Gerig JT (1970) J Am Chem Soc 92: 5001–5003
3116. Gammon KL, Smallcombe SH, Richards JH (1972) J Am Chem Soc 94: 4573–4580
3117. Smallcombe SH, Gammon KL, Richards JH (1972) J Am Chem Soc 94: 4581–4584
3118. Gerig JT, Loehr DT, Luk KFS, Roe DC (1979) J Am Chem Soc 101: 7482–7487
3119. Vandlen RL, Tulinsky A (1973) Biochemistry 12: 4193–4200
3120. Aliev RE, Godzhaev NN, Popov EM (1986) Mol Biologia 20: 346–356
3121. Popov EM, Godzhaev NN, Aliev RE (1986) Mol Biologia 20: 357–368
3122. Drenth J, Kalk KH, Swen HM (1976) Biochemistry 15: 3731–3738
3123. Lowe G, Yuthavong Y (1971) Biochem J 124: 107–115
3124. Wolthers BG, Drenth J, Jansonius JN, Koekoek R, Swen HM (1970) In: Desnuelle P, Neurath H, Ottesen M (eds) Int Symp on Structure-function of proteolytic enzymes. Munksgaard, Copenhagen, pp 272–288
3125. Evans BLB, Knopp JA, Horton HR (1981) Arch Biochem Biophys 206: 362–371
3126. Pickersgill RW, Goodenough PW, Summer IS, Collins ME (1988) Biochem J 254: 235–238
3127. Brocklenhurst K, O'Driscoll M, Kowlessur D, Phillips IR, Templton W, Thomas EW, Topham CH, Wharton CW (1989) J Biochem 257: 309–312
3128. Andreeva NS, Gustchina AE, Fedorov AA (1981) Dokl Akad Nauk SSSR 259: 1261–1264
3129. James MNG (1980) Can J Biochem 58: 251–271
3130. Blundell TL, Cooper J, Foundling SI, Jones DM, Atrash B, Szelke M (1987) Biochemistry 26: 5585–5590
3131. James MNG, Sielecki AR (1985) Biochemistry 24: 3701–3713
3132. Schmidt PG, Bernatowicz MS, Rich DH (1982) Biochemistry 21: 1830–1835
3133. Thomas KA, Smith GM, Thomas TB, Feldmann RJ (1982) Proc Natl Acad Sci USA 79: 4843–4847
3134. Lipscomb WN (1974) Tetrahedron 30: 1725–1730
3135. Nakagawa S, Umeyama H (1978) J Am Chem Soc 100: 7716–7725
3136. Mock WL (1975) Bioorg Chem 4: 270–278
3137. Mock WL (1976) Bioorg Chem 5: 403–414
3138. Lipscomb WN (1970) Acc Chem Res 3: 81–89
3139. Lipscomb WN (1980) Proc Natl Acad Sci USA 77: 3875–3878
3140. Johansen JT, Vallee BL (1975) Biochemistry 14: 649–659
3141. Klyosov AA, Vallee BL (1977) Bioorgan Khimiya 3: 964–980
3142. Rees DC, Lipscomb WN (1981) Proc Natl Acad Sci USA 78: 5455–5459
3143. Lipkind GM, Paslen VV (1980) Mol Biologia 14: 1142–1150
3144. Bender ML (1973) CRC Crit Rev Biochem 1: 149–199

3145. Stroud RM, Krieger M, Koeppe RE, Kossiakoff AA, Chambers JL (1975) In: Reich E, Rifkin DB, Shaw E (eds) Proteases and biological control. Cold Spring Harbor Laboratory, pp 13–32

3146. Kraut J (1977) Annu Rev Biochem 46: 331–358

3147. Shefer E (1978) In: Gandour RD, Schowen RL (eds) Transition states of biochemical processes. Plenum Press, New York, pp 339–352

3148. Polgár L, Hálasz P (1982) Biochem J 207: 1–10

3149. Page MI, Williams A (eds) (1987) Enzyme mechanisms. Soc Chem, London

3150. Zannis VI, Kirsch JF (1978) Biochemistry 17: 2669–2674

3151. Hardman MJ, Valenzuela P, Bender ML (1978) J Biol Chem 246: 5907–5913

3152. Lawson WB (1980) Biochemistry 19: 2140–2144

3153. Brocklenhurst K, Baines BS, Mushiri MS (1980) Biochem J 189: 189–192

3154. Mock WL, Tsay J-T (1986) Biochemistry 25: 2920–2927

3155. Chan VWF, Jorgensen AM, Borders CL (1988) Biochem Biophys Res Commun 151: 709–718

3156. Mock WL, Tsay J-T (1988) J Biol Chem 263: 8635–8641

3157. Robillard G, Schulman RG (1972) J Mol Biol 71: 507–511

3158. Markley JL, Porubcan MA (1976) J Mol Biol 102: 487–509

3159. Lewis SD, Johnson FA, Shafer JA (1976) Biochemistry 15: 5009–5017

3160. Porubcan MA, Westler WM, Ibañes IB, Markley JL (1979) Biochemistry 18: 4108–4116

3161. Primrose WV, Scott AI, Mackenzie NE, Malthouse JPG (1985) Biochem J 231: 677–682

3162. Jordan F, Polgár L, Tous G (1985) Biochemistry 24: 7711–7717

3163. Warshel A (1981) Biochemistry 20: 3167–3177

3164. Gilson MK, Honig BH (1988) Proteins 3: 32–52

3165. Goldblum A (1988) Biochemistry 27: 1653–1658

3166. Epstein J, Michel HO, Mosher WA (1968) J Theor Biol 19: 320–325

3167. Willenbrock F, Brocklehurst K (1985) Biochem J 227: 521–528

3168. Adams EQ (1916) J Am Chem Soc 38: 1503–1510

3169. Lewis SD, Johnson FA, Shafer JA (1981) Biochemistry 20: 48–51

3170. Cornish-Bowden AJ, Knowles JR (1969) Biochem J 113: 353–361

3171. Johansen JT, Vallee BL (1971) Proc Natl Acad Sci USA 68: 2532–2535

3172. Codding PW, Delbaere LTJ, Hayakawa K, Hutcheon WLB, James NMG, Jurášek L (1974) Can J Biochem 52: 208–220

3173. Byers LD, Koshland DE (1978) Bioorg Chem 7: 15–33

3174. Antonov VK, Vorotyntseva TI (1972) FEBS Lett 23: 361–363

3175. Fersht AR, Sperling J (1973) J Mol Biol 74: 137–149

3176. Hunkapiller MW, Smallcombe SH, Whitaker DR, Richards JH (1973) Biochemistry 12: 4732–4743

3177. Stein RL, Elrood JP, Schowen RL (1983) J Am Chem Soc 105: 2446–2452

3178. Schowen RL (1984) CRC Crit Rev Biochem 17: 1–44

3179. Scholten JD, Hogg JL, Raushel FM (1988) J Am Chem Soc 110: 8246–8247

3180. Bachovchin WW, Kaiser R, Richards JH, Roberts JD (1981) Proc Natl Acad Sci USA 78: 7323–7326

3181. Westler WM, Markley JL, Bachovchin WW (1982) FEBS Lett 138: 233–235

3182. Kossiakoff AA, Spencer SA (1981) Biochemistry 20: 6462–6474

3183. Bachovchin WW (1985) Proc Natl Acad Sci USA 82: 7948–7951

3184. Bachovchin WW (1986) Biochemistry 25: 7751–7759

3185. Craik CS, Roczniak S, Largman C, Rutter WJ (1987) Science 237: 909–913

3186. Khurgin YI, Burshtein KI (1974) Dokl Akad Nauk SSSR 217: 965–969

3187. Scheiner S, Kleier DA, Lipscomb WN (1975) Proc Natl Acad Sci USA 72: 2606–2610

3188. Umeyama H, Nakagawa S (1980) Chem Pharm Bull 28: 2292–2300

2189. Umeyama H, Nakagawa S, Kudo T (1981) J Mol Biol 150: 409–421

3190. Nakagawa S, Umeyama H (1984) J Mol Biol 179: 103–123

3191. Umeyama H, Hirono S, Nakagawa S (1984) Proc Natl Acad Sci USA 81: 6266–6270

3192. Hamilton SE, Zerner B (1981) J Am Chem Soc 103: 1827–1831

3193. Roberts JD, Flanagan CYC, Birdseye TR (1982) J Am Chem Soc 104: 3945–3949

3194. Pearl L, Blandell T (1984) FEBS Lett 174: 96–101

3195. Rees DC, Lewis M, Honzatko RB, Lipscomb WN, Hardman VD (1981) Proc Natl Acad Sci USA 78: 3408–3412
3196. Radom L, Pople JA, Mock WL (1972) Tetrahedron Lett: 479–482
3197. Voityuk AA, Vasil'ev VV (1987) Mol Biologia 21: 807–813
3198. Epand RM, Wilson IB (1963) J Biol Chem 238: 1718–1723
3199. Williams A (1970) Biochemistry 9: 3383–3390
3200. Kézdy FJ, Bender ML (1964) J Am Chem Soc 86: 937–938
3201. Kézdy FJ, Bender ML (1964) J Am Chem Soc 86: 938–940
3202. Kézdy FJ, Clement GE, Bender ML (1964) J Am Chem Soc 86: 3690–3696
3203. Himoe A, Brandt KG, Desa RJ, Hess GP (1969) J Biol Chem 244: 3483–3493
3204. Breaux JE, Bender ML (1976) Biochem Biophys Res Commun 70: 235–240
3205. Swedberg SA, Pesek JJ, Fink AL (1990) Anal Biochem 186: 153–158
3206. Makinen MW, Kuo LC, Dymowski JJ, Jaffer S (1979) J Biol Chem 254: 356–366
3207. Lachman H, Mauser H, Schneider F (1972) Hoppe-Seyler's Z Physiol Chem 353: 730–731
3208. Malthouse JPG, Gamcsik MP, Boyd ASF, Mackenzie NE, Scott AI (1982) J Am Chem Soc 104: 6811–6813
3209. Belasco JG, Knowles JR (1980) Fed Proc 39: 1667
3210. Anderson EG, Pratt RF (1981) J Biol Chem 256: 11401–11404
3211. Cartwright SJ, Fink AL (1982) FEBS Lett 137: 186–188
3212. Epand RM, Wilson IB (1965) J Biol Chem 240: 1104–1107
3213. Inagami T, Sturtevant JM (1964) Biochem Biophys Res Commun 14: 69–74
3214. Epand RM (1969) Biochem Biophys Res Commun 37: 313–318
3215. Rasmussen JR, Strominger JL (1978) Proc Natl Acad Sci USA 75: 84–88
3216. Bajkowski AS, Frankfater A (1983) J Biol Chem 258: 1645–1649
3217. Fitzpatrick PF (1989) Biochem Biophys Acta 995: 201–203
3218. Akhtar M, Al-Janabi JM (1969) Chem Commun: 1002–1003
3219. Kitson TM, Knowles JR (1970) Chem Commun: 361–362
3220. Takahashi M, Hofmann T (1975) Biochem J 147: 549–563
3221. Nakagawa Y, Ko Sun L-H, Kaiser ET (1976) J Am Chem Soc 98: 1616–1617
3222. Bresler SE, Krutyakov VM, Vlasov GP (1971) Eur J Biochem 18: 131–139
3223. Berezin IV, Kazanskaya NF, Klyosov AA, Svedas VK (1973) Eur J Biochem 38: 529–535
3224. Stauffer CE, Zeffren E (1970) J Biol Chem 245: 3282–3284
3225. Bender ML, Kemp KC (1957) J Am Chem Soc 79: 111–116
3226. Bender ML, Kemp KC (1957) J Am Chem Soc 79: 116–120
3227. Antonov VK, Yavashev LP, Volkova LI, Sadovskaya VL, Ginodman LM (1979) Bioorgan Khimiya 5: 1427–1429
3228. Antonov VK, Ginodman LM, Rumsh LD, Kapitannikov YV, Barshevskaya TN, Yavashev LP, Gurova AG, Volkova LI (1981) Eur J Biochem 117: 195–200
3229. Antonov VK (1985) In: Kostka V (ed) Aspartic proteinases and their inhibitors. de Gruyter, Berlin, pp 203–220
3230. Popov AA, Kapitannikov YV, Rumsh LD, Antonov VK (1986) Bioorg Khimiya 12: 467–469
3231. Antonov VK, Ginodman LM, Kapitannikov YV, Barshevskaya TN, Gurova AG, Rumsh LD (1978) FEBS Lett 88: 87–90
3232. Antonov VK, Ginodman LM, Rumsh LD (1984) Bioorg Khimiya 10: 1044–1058
3233. Kozlov LV, Ginodman LM, Orekhovich VN (1967) Dokl Akad Nauk SSSR 172: 1207–1209
3234. Williams DC, Whitaker JR (1968) Biochemistry 7: 2562–2569
3235. Ginodman LM, Kpitannikov YV, Rumsh LD, Antonov VK (1988) Bioorg Khimiya 14: 1641–1649
3236. Nau H, Riordan JF (1975) Biochemistry 14: 5285–5294
3237. Röhm KH, Van Effen RL (1986) Eur J Biochem 160: 327–332
3238. Röhm KH, Van Effen RL (1986) Arch Biochem Biophys 244: 128–136
3239. Paul G, Heusden HV, Von Der Bosch H (1979) Biochem Biophys Res Commun 90: 1000–1006
3240. Hofmann T, Dunn BM, Fink AL (1984) Biochemistry 23: 5253–5256
3241. Hoffman SJ, Chu SST, Lee HH, Kaiser ET, Carey PR (1983) J Am Chem Soc 105: 6971–6973

3242. Auld DS, Geoghegan KG, Galdes A, Vallee BL (1986) Biochemistry 25: 5156–5159
3243. Galdes A, Auld DS, Vallee BL (1986) Biochemistry 25: 646–651
3244. Auld DS (1987) In: Page MI, Williams A (eds) Enzyme mechanisms. R Soc Chem, London, pp 240–258
3245. Makinen MW, Fukuyama JM, Kuo LC (1982) J Am Chem Soc 104: 2667–2669
3246. Kuo LC, Fukuyama JM, Makinen MW (1983) J Mol Biol 163: 63–105
3247. Suh J, Cho W, Chung S (1985) J Am Chem Soc 107: 4530–4535
3248. Suh J, Hong S-B, Chung S (1986) J Biol Chem 261: 7112–7114
3249. Femfert U, Pfleiderer G (1970) FEBS Lett 8: 65–67
3250. May SW, Kaiser ET (1972) Biochemistry 11: 592–600
3251. Kaiser ET, Nakagawa Y (1977) Adv Exp Med Biol 95: 159–178
3252. Alexandrov SL, Antonov VK, Mel'nikov PN (1984) Mol Biologia 18: 1569–1575
3253. Scheiner S, Lipscomb WN (1976) Proc Natl Acad Sci USA 73: 432–436
3254. Fastrez J (1983) Eur J Biochem 135: 339–341
3255. Weiner SJ, Seibel GL, Kollman PA (1986) Proc Natl Acad Sci USA 83: 649–653
3256. Warshel A, Russel S (1986) J Am Chem Soc 108: 6569–6579
3257. Warshel A, Náray-Szabó G, Sussman F, Hwang J-K (1989) Biochemistry 28: 3629–3637
3258. Brocklenhurst K (1987) In: Page MI, Williams A (eds) Enzyme mechanisms. R Soc Chem, London, pp 140–158
3259. Asbóth B, Polgár L (1983) Biochemistry 22: 117–122
3260. Asbóth B, Stokum E, Khan IV, Polgár L (1985) Biochemistry 24: 606–609
3261. Holmes MA, Matthews BW (1981) Biochemistry 20: 6912–6920
3262. Kuo LC, Makinen MW (1982) J Biol Chem 257: 24–27
3263. Van Wart HE, Vallee BL (1978) Biochemistry 17: 3385–3394
3264. Christianson DW, Alexander RS (1989) J Am Chem Soc 111: 6412–6419
3265. Alexandrov SL, Antonov VK (1990) Bioorg Khimiya 16: 464–475
3266. Antonov VK (1985) Proc 16th FEBS Congr 1984, Part A. VNU Science Press, pp 31–40
3267. Voituyk AA (1987) Mol Biologia 21: 882–887
3268. Mao B, Pear MR, McCammon JA, Northrup SH (1981) J Mol Biol 151: 199–202
3269. Tsai I-H, Bender ML (1979) Biochemistry 18: 3764–3768
3270. Brocklenhurst K, Malthouse JPG (1981) Biochem J 193: 819–823
3271. Likhtenshtein GI (1977) Kinetika i Cataliz 18: 878–882
3272. Likhtenshtein GI (1977) Kinetika i Cataliz 18: 1255–1260
3273. Schowen RL (1972) Progr Phys Org Chem 9: 275–332
3274. Minor SS, Schowen RL (1973) J Am Chem Soc 95: 2279–2281
3275. Choi M-U, Thornton ER (1974) J Am Chem Soc 96: 1428–1436
3276. Scheiner S, Lipscomb WN (1977) J Am Chem Soc 99: 3466–3472
3277. Kapitannikov YV, Yavashev LP, Ginodman LM, Antonov VK (1983) Bioorg Khimiya 9: 228–231
3278. Wang JH, Parker L (1967) Proc Natl Acad Sci USA 58: 2451–2454
3279. Wang JH (1968) Science 161: 328–334
3280. Wang JH (1970) Proc Natl Acad Sci USA 66: 874–881
3281. Hubbard CD, Shoupe TS (1977) J Biol Chem 252: 1633–1638
3282. Fersht AR (1972) J Am Chem Soc 94: 293–295
3283. Lucas EC, Caplow M, Bush KJ (1973) J Am Chem Soc 95: 2670–2673
3284. O'Leary MH, Kluetz MD (1970) J Am Chem Soc 92: 6089–6090
3285. O'Leary MH, Kleutz MD (1972) J Am Chem Soc 94: 3585–3589
3286. Wang C-L, Trout CM, Calvo KC, Klapper MH, Wong LK (1980) J Am Chem Soc 102: 1221–1223
3287. Mishra AK, Klapper MH (1986) Biochemistry 25: 7328–7336
3288. Stein RL (1988) J Am Chem Soc 110: 7907–7908
3289. O'Leary MH, Urberg M, Young AP (1974) Biochemistry 13: 2077–2081
3290. Inward PW, Jencks WP (1965) J Biol Chem 240: 1986–1996
3291. Zeeberg B, Caswell M, Caplow M (1973) J Am Chem Soc 95: 2734–2735
3292. Zeeberg B Caplow M (1973) J Biol Chem 248: 5887–5891
3293. Hess GP, Seybert D, Lewis A, Spoonhower J, Cookingham R (1975) Science 189: 384–386
3294. Kennedy WP, Schultz RM (1979) Biochemistry 18: 349–356

3295. Chen R, Gorenstein DG, Kennedy WP, Lowe G, Nure D, Schultz RM (1979) Biochemistry 18: 921–926

3296. Schmidt PG, Holladay MW, Salituro FG, Rich DH (1985) Biochem Biophys Res Commun 129: 597–602

3297. Christianson DW, David PR, Lipscomb WN (1987) Proc Natl Acad Sci USA 84: 1512–1515

3298. Brady K, Liang T-C, Abeles RH (1989) Biochemistry 28: 9066–9070

3299. Hunkapiller MW, Forgac MD, Richards JH (1976) Biochemistry 15: 5581–5588

3300. Balny C, Bieth JG (1977) FEBS Lett 80: 182–186

3301. Fink AL, Meehan P (1979) Proc Natl Acad Sci USA 76: 1566–1569

3302. Angelides KJ, Fink AL (1979) Biochemistry 18: 2363–2369

3303. Markley JL, Travers F, Balny C (1981) Eur J Biochem 120: 477–485

3304. Lehrmann G, Quinn D, Cordes EH (1980) J Am Chem Soc 102: 2491–2492

3305. Malthouse JPG, Mackenzie NE, Boyd ASF, Scott AI (1983) J Am Chem Soc 105: 1685–1686

3306. De Los De Tar F (1981) J Am Chem Soc 103: 107–110

3307. Wells JA, Cunningham BC, Graycar TP, Estelle DA (1986) Philos Trans R Soc Lond Ser A 317: 415–419

3308. Sumi H, Ulstrup J (1988) Biochim Biophys Acta 955: 26–42

3309. Bringar WS, Chao TL (1975) Biochem Biophys Res Commun 63: 78–83

3310. Polgár L (1979) Eur J Biochem 98: 369–374

3311. Alexandrov SL, Antonov VK (1987) Mol Biologia 21: 147–158

3312. Fastrez J, Houyet N (1977) Eur J Biochem 81: 515–522

3313. Petkov DD (1978) Biochim Biophys Acta 523: 538–541

3314. Grunwald E, Puar MS (1967) J Phys Chem 71: 1842–1845

3315. Polgár L (1987) FEBS Lett 219: 1–4

3316. Hangauer DG, Monzingo AF, Matthews BW (1984) Biochemistry 23: 5730–5741

3317. Bünning P, Riordan JF (1985) J Inorg Biochem 24: 183–198

3318. Hinkle PM, Kirsch JF (1971) Biochemistry 10: 3700–3707

3319. Nakamura T (1979) J Biochem 86: 751–755

3320. Hubbard CD, Kirsch JF (1970) Biochemistry 11: 2483–2490

3321. Riddles PW, De Jersy J, Zerner B (1978) Proc Natl Acad Sci USA 75: 172–174

3322. Adams PA (1980) Biochem J 191: 653–655

3323. Bender ML, Brubacher LJ (1966) J Am Chem Soc 88: 5880–5889

3324. Mullen GP, Dunlap RB, Odom JD (1986) Biochemistry 25: 5625–5632

3325. Tonge PJ, Carey PR (1989) Biochemistry 28: 6701–6709

3326. Phelpes DJ, Schneider H, Carey PR (1981) Biochemistry 20: 3447–3454

3327. Charney E, Bernhard SA (1967) J Am Chem Soc 89: 2726–2733

3328. Brown WE (1975) Biochemistry 14: 5079–5084

3329. Berezin IV, Martinek K (1975) Bioorg Khimiya 1: 520–531

3330. Vishnu, Caplow M (1969) J Am Chem Soc 91: 6754–6758

3331. Stein RL, Strimpler AM (1987) J Am Chem Soc 109: 4387–4390

3332. Volini M, Wang S-F (1978) Arch Biochem Biophys 187: 163–169

3333. Tanizawa K, Kasaba Y, Kanaoka J (1980) J Biochem 87: 417–427

3334. Nakayama H, Tanizawa K, Kanaoka J (1980) J Am Chem Soc 102: 3214–3218

3335. Tonge PJ, Lee H, Sans Cartier LR, Ruzsicska BP, Carey PR (1989) J Am Chem Soc 111: 1496–1497

3336. Carey PR, Angus RH, Lee H-H, Storer AC (1984) J Biol Chem 259: 14357–14360

3337. Paberit M, Peips M, Aaviksaar A (1984) Biochim Biophys Acta 789: 257–265

3338. Cairi M, Gerig JT (1983) J Am Chem Soc 105: 4793–4800

3339. Cairi M, Gerig JT (1984) J Am Chem Soc 106: 3640–3643

3340. Gerig JT, Hammond SJ (1984) J Am Chem Soc 106: 8244–8251

3341. Smolarsky M (1980) Biochemistry 19: 478–484

3342. Knowles JR (1970) Philos Trans R Soc Lond Ser B 257: 135–146

3343. Silver MS, Stoddard M (1972) Biochemistry 11: 191–200

3344. Antonov VK (1978) Symp Biol Hung 21: 63–79

3345. Bosshard HR (1976) Experientia 32: 949–963

3346. Page MI (1984) In: Page MI (ed) The chemistry of enzyme action. Elsevier, Amsterdam, pp 1–54
3347. Page MI (1987) In: Page MI, Williams A (eds) Enzyme mechanisms. R Soc Chem, London, pp 1–13
3348. Knowles JR (1987) Science 236: 1252–1258
3349. Byers LD (1977) J Am Chem Soc 99: 4146–4149
3350. Berezin IV, Martinek K (1977) Bases of physical chemistry of enzymatic catalysis. Visshaya shkola, Moscow, p 6
3351. Page MI (1980) Int J Biochem 11: 331–335
3352. Koshland DE, Neet KE (1968) Annu Rev Biochem 37: 359–410
3353. Koshland DE (1956) J Cell Comp Physiol (Suppl 1) 47: 217–232
3354. Delisi C, Crothers DM (1973) Biopolymers 12: 1689–1704
3355. Westheimer FH (1962) Adv Enzymol 24: 441–482
3356. Reuben J (1971) Proc Natl Acad USA 68: 563–565
3357. Marshall AG (1968) Biochemistry 7: 2450–2453
3358. Gerig JT (1968) J Am Chem Soc 90: 2681–2686
3359. Gerig JT, Reinheimer JD (1970) J Am Chem Soc 92: 3146–3150
3360. Menger FM, Venkataram UV (1985) J Am Chem Soc 107: 4706–4709
3361. Menger FM (1985) Acc Chem Res 18: 128–134
3362. Dafforn A, Koshland DE (1971) Proc Natl Acad Sci USA 68: 2463–2467
3363. Storm DR, Koshland DE (1972) J Am Chem Soc 94: 5805–5814
3364. Storm DR, Koshland DE (1972) J Am Chem Soc 94: 5815–5825
3365. Hoare DG (1972) Nature 236: 437–440
3366. Bruice TC, Brown A, Harris DO (1971) Proc Natl Acad Sci USA 68: 658–661
3367. Bruice TC (1972) Nature 237: 335–336
3368. Capon B (1971) J Chem Soc B 6: 1207–1209
3369. Port GNJ, Richards WG (1971) Nature 231: 312–313
3370. Jencks WP, Page MI (1974) Biochem Biophys Res Commun 57: 887–892
3371. Parker AJ (1967) Adv Phys Org Chem 5: 173–235
3372. Dewar MJS, Storch DM (1985) Proc Natl Acad Sci USA 82: 2225–2229
3373. Madura JD, Jorgensen WL (1986) J Am Chem Soc 108: 2517–2527
3374. Alexander R, Parker AJ (1967) J Am Chem Soc 89: 5549–5551
3375. Low PS, Somero GN (1975) Proc Natl Acad Sci USA 72: 3305–3309
3376. Low PS (1978) Symp Biol Hung 21: 217–247
3377. Loftfield RB, Eigner EA, Pastuszyn A, Lövgren TNE, Jakubowski H (1980) Proc Natl Acad Sci USA 77: 3374–3378
3378. Khurgin YI, Roslyakov VYa, Klyachko-Gurvich AL, Brueva TR (1972) Biokhimiya 37: 485–492
3379. Dunn BM, Bruice TC (1973) Adv Enzymol 37: 1–59
3380. Warshel A (1978) Proc Natl Acad Sci USA 75: 5250–5254
3381. Meot-Ner M (1988) J Am Chem Soc 110: 3075–3080
3382. Warshel A, Sussman F, Hwang J-K (1988) J Mol Biol 201: 139–159
3383. Johannin G, Kellershohn N (1972) Biochem Biophys Res Commun 49: 321–327
3384. Krishtalik LI (1986) Charge transfer reactions in electrochemical and chemical processes. Consultant Bureau, New York
3385. Krishtalik LI (1988) In: Dogonadze RR, Kalman E, Kerayshev AA, Ulstrup J (eds) Chemical physics of solvation, vol 3. Elsevier, Amsterdam
3386. Phillips DC (1967) Proc Natl Acad Sci USA 57: 484–495
3387. Schindler M, Assaf Y, Sharon N, Chipman DM (1977) Biochemistry 16: 423–431
3388. Lovrien R, Linn T (1967) Biochemistry 6: 2281–2293
3389. Williams RJP (1966) In: Peeters H (ed) Protides. Biol Fluids Proc Colloq Bruges, vol 14. Elsevier, Amsterdam, pp 25–30
3390. Vallee BL, Williams RJP (1968) Proc Natl Acad Sci USA 59: 498–505
3391. Kurz LC, Ackerman JJH, Drysdale GR (1985) Biochemistry 24: 452–457
3392. Kurz LC, Drysdale GR (1987) Biochemistry 26: 2623–2627
3393. Belasco JG, Knowles JR (1983) Biochemistry 22: 122–129

3394. Belasco JG, Knowles JR (1980) Biochemistry 19: 472–477
3395. Vol'kenshtein MV, Dogonadze RR, Madumarov AK, Urushadze ZD, Kharkts YI (1972) Mol Biologia 6: 431–439
3396. Tamman G (1892) Zhurn Russ Fiz. -Khim Obshtsestva Ser Khim 24: 698–718
3397. Fischer E, Abderhalden E (1902) Z Physiol Chem 36: 276
3398. Bergmann M, Fruton JS (1941) Adv Enzymol 1: 63–98
3399. Ogston AG (1948) Nature 162: 963–965
3400. Bernhard SA, Gutfreund H (1958) Proc Int Symp Enzyme Chem, Tokyo, Kyoto, p 124
3401. Spencer T, Sturtevant JM (1959) J Am Chem Soc 81: 1874–1882
3402. Hein GE, Niemann C (1961) Proc Natl Acad Sci USA 47: 1341–1350
3403. Niemann C (1964) Science 143: 1287–1296
3404. Hamilton CL, Niemann C, Hammond GS (1966) Proc Natl Acad Sci USA 55: 664–669
3405. Ingles DW, Knowles JR (1967) Biochem J 104: 369–377
3406. Koshland DE (1958) Proc Natl Acad Sci USA 44: 98–102
3407. Hershlag D (1988) Bioorg Chem 16: 62–96
3408. Koshland DE, Némethy G, Filmer D (1966) Biochemistry 5: 365–385
3409. Fersht AR, Leatherbarrow RJ, Wells TNC (1986) TIBS 11: 321–325
3410. Borgford TJ, Gray TE, Brand NJ, Fersht AR (1987) Biochemistry 26: 7246–7250
3411. Fersht AR (1987) Biochemistry 26: 8031–8041
3412. Berezin IV, Martinek K (1974) Structure and function of enzymes active sites. Nauka, Moscow, pp 5–21
3413. Voityuk AA (1987) Mol Biologia 21: 1671–1676
3414. Blow DM (1976) Acc Chem Res 9: 145–152
3415. Campbell P, Nashed NT (1978) Bioorg Chem 7: 69–76
3416. Emmanuel NM, Knorre DT (1969) Course of chemical kinetics. Visshaya shkola, Moscow, p 61
3417. Chernavsky DS, Khurgin YI, Shnol SE (1967) Mol Biologia 1: 419–424
3418. Fröhlich M (1970) Nature 228: 1093–1097
3419. Blumenfeld LA (1971) Biofizika 16: 724–727
3420. Blumenfeld LA (1972) Biofizika 17: 954–956
3421. Sidorenko NP (1972) Biofizika 17: 907–908
3422. Averbukh IS, Blumenfeld LA, Kovarsky VA, Perelman NF (1986) Biochim Biophys Acta 873: 290–296
3423. Belovolova LV, Blumenfeld LA, Burbaev DS, Vanin AF (1975) Mol Biologia 9: 934–940
3424. Stackhouse J, Nambiar KP, Burlaum JJ, Stauffer DM, Brenner SA (1985) J Am Chem Soc 107: 2757–2763
3425. Shnol SE (1967) Vibration processes in biological and chemical systems. Nauka, Moscow, p 22
3426. Marcus RA (1964) Annu Rev Phys Chem 15: 155–180
3427. Dogonadze RR, Kuznet'sov AM (1969) Itogi nauki Electrokhimiya VINITI. Moscow
3428. Volkenshtein MV, Dogonadze RR, Madumarov AK, Kharkats YI (1971) Dokl Akad Nauk SSSR 199: 124–127
3429. Levich VG, Madumarov AK, Kharkats YI (1972) Dokl Akad Nauk SSSR 203: 1351–1353
3430. Krishtalik LI (1972) Dokl Akad Nauk SSSR 205: 469–472
3431. Krishtalik LI (1979) Mol Biologia 13: 577–581
3432. Krishtalik LI (1981) Mol Biologia 15: 290–297
3433. Volkenshtein MV (1972) J Theor Biol 34: 193–195
3434. Albery WL, Knowles JR (1976) Biochemistry 15: 5631–5640
3435. Christensen H, Martin MT, Waley SG (1990) Biochem J 266: 853–861
3436. Tian G (1987) Biochem J 248: 619–620
3437. Karpeisky MY, Yakovlev GI, Antonov VK (1980) Bioorg Khimiya 6: 645–654
3438. Ivanov VI, Karpeisky MY (1969) Adv Enzymol 32: 21–53

Subject Index